Ultrashort Pulse Lasers and Ultrafast Phenomena

This book describes the basic physical principles and structure of the special technique used to obtain such ultrashort pulse lasers. It introduces the basics of non-linear optics that provides their generation and measurement and provides application examples of ultrafast spectroscopy to solid-state materials. The chapters explains the basic physical principles of ultrashort pulse laser construction and describes the detailed structure of the world's shortest visible laser and DUV lasers. Finally, it also provides several examples of the applications of ultrafast spectroscopy to solid-state materials.

Ultrashort Pulse Lasers and Ultrafast Phenomena

Takayoshi Kobayashi

CRC Press
Taylor & Francis Group
Boca Raton London New York

CRC Press is an imprint of the
Taylor & Francis Group, an **informa** business

Designed cover image: Takayoshi Kobayashi

First edition published 2023
by CRC Press
6000 Broken Sound Parkway NW, Suite 300, Boca Raton, FL 33487-2742

and by CRC Press
4 Park Square, Milton Park, Abingdon, Oxon, OX14 4RN

CRC Press is an imprint of Taylor & Francis Group, LLC

© 2023 Taylor & Francis Group, LLC

ISBN: 978-0-367-18471-1 (hbk)
ISBN: 978-1-032-41172-9 (pbk)
ISBN: 978-0-429-19657-7 (ebk)

DOI: 10.1201/9780429196577

Typeset in Times
by codeMantra

Contents

SECTION 1   Generation of Ultrashort Pulses in Deep Ultraviolet to Near Infrared
SECTION 1.1   Ultrashort Visible Near-Infrared Pulses

SECTION 1.2   *Ultrashort Ultraviolet, Deep-Ultraviolet, and Infrared Pulses*

SECTION 2   *Generation of Ultrashort Pulses in Terahertz*

SECTION 3   CEP (Octave-Span)

SECTION 4   Simple NLO Processes with a Few Colors

SECTION 5   Multi-Color Involved NLO Processes

SECTION 6 *Broadband Ultrashort Pulse Generation*

SECTION 7 *NLO Materials*

SECTION 8   NLO Processes in Time-Resolved Spectroscopy

SECTION 9 Low Dimensional (D) Materials
SECTION 9.1 0D

SECTION 9.2 1D CNT

SECTION 9.3   1D Oligomers and Polymers

SECTION 9.4 2D Topological Materials

SECTION 10   Conductors and Superconductors
SECTION 10.1   Super Conductors

SECTION 10.2 THz, MIR Spectroscopy of Materials

SECTION 11   Chemical Reactions and Material Processing
SECTION 11.1   Chemical Reactions

SECTION 11.2   Material Processing

SECTION 12 *Photobiological Reactions*

Preface

The contents of this book are composed of products of research conducted for more than 20 years with many collaborators in The Department of Physics of the University of Tokyo, The University of Electro-Communications, National Chiao-Tung University, Tokyo University of Science. They contain almost every research in the field of ultrashort pulse lasers and ultrafast spectroscopy in various systems such solid-state materials, solution, liquids, organic and inorganic materials and biologically important systems. I hope this book is useful for those young people who are interested in getting involved in this field.

The spectral ranges are ultraviolet, visible, near-infrared and mind infrared including the generation of pulses, pulse characterization and methods of spectroscopic measurements and scientific meanings of the studied ultrafast time spectroscopies. The contents of this book were partially published in the journals published by the following publishing companies and associations. The author would like to express thanks to them all for their support at the time of their publications. They include the following:

Applied Physics, J. Chemical Physics, and AIP Advances; from American Institute of Physics

Opt. Letters, Opt. Express, and Applied Optics; from The Optical Society of America (presently Optica)

J. Phys. Chem. A, B, C, Inorganic Chemistry, Nano Letters; from American Chemical Society,

Phys. Rev. B, New J. Physics, Superconductor Science and Technology; from IOP Publishing,

J. Luminescence and Chem. Phys. Lett.; Elsevier

Molecules, Photonics; MDPI (Multidisciplinary Digital Publishing Institute)

Phys. Chem. Chem. Phys.; Royal Society of London

Pure and Applied Chemistry IUPAC; IUPAC.

J. Nanomaterials; Hindawi Publishing Company

Scientific Reports: Nature.

Biophysical Journal; Biophysical Society

Special thanks to my wife Toshino for her continuing support to my long research life.

Author

Takayoshi Kobayashi was born in Niigata Prefecture in 1944. He graduated from the University of Tokyo and got the bachelor, master, and doctor degrees for m eh same university. He joined the Institute of Physical and Chemical Research (Riken). Between 1977 and 1979 he had been a temporary member of the Technical Staff in Bell Laboratories. In 1980, he joined the Department of Physics, University of Tokyo, as an associate professor and promoted to a full professor in 1994. In 2006 March, he was appointed as an ICORP (International Cooperative Research Project) Research Director of Ultrashort Pulse Laser Project. In 2006 March, he retired from the university and in April of the same year he moved to the Department of Applied Physics and Chemistry in the University of Electro-Communications, which is another national university in Tokyo. From April in 2006, he started to be a chair professor of National Chiao-Tung University and the director of the Advanced Ultrafast Laser Center.

AWARDS:

1995:   Scientific Achievement Award from the Chemical Society of Japan
2000:   Fellow of the Optical Society of America
2003:   Scientific Achievement Award from the Spectroscopy Society of Japan
2005:   Scientific Achievement Award from the International Vibrational Spectroscopy
2005:   Scientific Achievement Award from the Matsuo Foundation
2005:   Chair Professor, National Chiao-Tung University in Taiwan.
2006:   Professor Emeritus of the University of Tokyo
2006:   Outstanding Science Scholar Award from Foundaton for the Advancement of Outstanding Scholarship in Taiwan.
2010:   Outstanding Contribution to Science and Technology Award from the Shimadzu Science and Technology Foundation
2010:   Japan Minister of Education, Culture, Science Award for Outstanding Contribution to Science and Technology
2010:   Outstanding Visiting Professor of Chinese Academy of Science in Beijing,
2011:   Outstanding Visiting Professor of Chinese Academy of Science, Shanghai Institute of Optics and Fine Mechanics in Shanghai
2011:   Humboldt Award, Outstanding Research in the Development of Ultrashort Pulse Laser and Ultrafast Processes in Molecules
2012   The Chemical Society of Japan Award
2013   The Chemical Society Japan Fellow
2015:   Senior Member of the Optical Society of America (at present Optica)
2015:   Fellow of the Laser Society of Japan

Section 1

Generation of Ultrashort Pulses in Deep Ultraviolet to Near Infrared

Section 1.1

Ultrashort Visible Near-Infrared Pulses

1.1.1 Noncollinearly Phase-Matched Femtosecond Optical Parametric Amplification with a $2000\,\mathrm{cm^{-1}}$ Bandwidth

Recent advances in the generation of an ultrashort optical pulse from a Ti:sapphire laser and amplifier are remarkable. The broad gain bandwidth as broad as $3000\,\mathrm{cm^{-1}}$ of the Ti-doped crystal, followed by a white-light continuum generation, enables the generation of shorter pulses than 5-fs duration [1,2]. On the other hand, the shortening of the output pulse from an optical parametric amplifier (OPA) faces the following problem of spectral acceptance of the phase-matching condition, which is mainly limited by the group-velocity mismatching (GVM) between the signal and idler pulses in the nonlinear optical crystal [3]. The bandwidths of the OPAs reported up to now are $\sim300\,\mathrm{cm^{-1}}$, which limit the shortest pulse width to about 30 fs [4,5]. The bandwidth broadening is most important to obtain sub-10 fs pulses outside the 800 nm region. Several methods were proposed for the spectral broadening in an OPA [6,7], but they are rather complicated. A noncollinear phase-matching geometry is quite attractive to modify the phase-matching condition as well as the amount of the GVM in a straightforward and simple way [8–12]. Di Trapani et al. demonstrated a noncollinear OPA with the GV matching between the pump pulse and signal or idler pulse in a β-BaB$_2$O$_4$ (BBO) crystal to obtain a higher conversion efficiency with an interaction length longer than in the case of a collinear OPA [8,9]. In a synchronously pumped optical parametric oscillator (OPO), there was reported a visible 13 fs pulse generation by utilizing the broadband phase-matching condition of a BBO in a noncollinear geometry [10,11]. Just recently, a nanosecond OPO also generated a signal with a broader spectrum than 100 nm [12]. An extreme pulse shortening is expected from a femtosecond noncollinear OPA which enables the broadband amplification.

In this subsection, we demonstrate a scheme of Ti:sapphire-based optical parametric amplification of the white-light continuum in a noncollinear phase-matching geometry in a BBO crystal. By utilizing an extraordinary noncollinear phase-matching property, equivalent to the matching of the GVs between the signal and idler pulses, a major fraction of the continuum is amplified with as broad as $2000\,\mathrm{cm^{-1}}$ bandwidth with mJ-level pulse energy. It is compressed currently to the shortest pulse duration reaching 14 fs. A simple and novel tuning scheme between 550 and 690 nm with a duration of shorter than 20 fs is attained by shifting the delay line of the pump with respect to the white-light continuum. The bandwidth is found to be limited only by the chirp of the continuum. The spatial chirp and the pulse-front tilting also limit the pulse width and a sub-10 fs source tunable in a visible region is strongly expected by compensating these factors.

Figure 1.1.1.1a shows the phase-matching curves of the noncollinear phase-matched OPA in a BBO crystal pumped at 395 nm in a type-I ($e \rightarrow o+o$) configuration. The wavelength of the signal satisfying phase-matching condition is plotted against a polar angle θ with respect to the z-axis of the BBO crystal. Figure 1.1.1.1b illustrates the arrangement of each beam in the crystal, where the signal beam is injected into the crystal with a fixed noncollinear angle α with respect to the pump direction while the idler is generated with the variable angle b to satisfy the phase-matching condition.

DOI: 10.1201/9780429196577-3

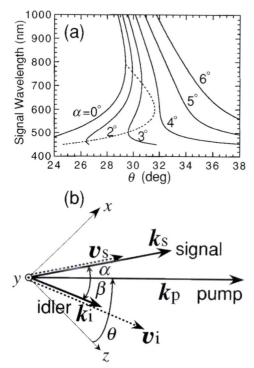

FIGURE 1.1.1.1 (a) Theoretical phase-matching curves of type-I BBO OPA pumped at 395 nm with different noncollinear angles α. The dashed line indicates the GV matching points between the signal and idler pulses. (b) Geometry of the noncollinear phase matching. The wave vectors of the pump ($\boldsymbol{k}_p$), signal ($\boldsymbol{k}_s$), and idler ($\boldsymbol{k}_i$) are shown in the BBO crystal. The group velocities of the signal ($\boldsymbol{v}_s$) and idler ($\boldsymbol{v}_i$) are also shown by dashed lines. α and bare internal angles.

The curve strongly depends on a, and there exists an $\alpha(\lambda_s)$ ($<4°$) where the inflection occurs at the signal wavelength $\lambda_s \sim$ (indicated by a dashed line in Figure 1.1.1.1a). Here a broadband amplification is expected, which is attributed to the GV matching between the signal and idler [11]. The gain bandwidth of an optical parametric conversion is inversely proportional to the GVM between the signal and idler [3]. In the case of the noncollinear phase-matched OPA, the phase mismatch contributing to the gain reduction is given by the component along the pump direction, $\Delta k = k_p - k_s \cos \alpha - k_i \cos \beta$, in terms of the wave vector k_j with the index $j = p$, s, or i corresponding to the pump, signal, and idler, respectively. By a minor calculation, the effective GVM that reduces the bandwidth is approximately given as $1/v_s - 1/v_i \cos(\alpha + \beta)$ where v_j is the GV of one of the three beams. When the signal GV is equal to the component of the idler GV projected to the signal direction, the bandwidth is only limited by the GV dispersion mismatching [3]. In the case of a collinear geometry ($\alpha = \beta = 0$), the GV matching only occurs at the degeneracy in the spectral range with a normal dispersion under a type-I interaction ($v_s < v_i$). By utilizing the noncollinear geometry, on the other hand, the GV matching can be realized at any ls by choosing $\alpha(\lambda_s)$.

Figure 1.1.1.1a shows clearly that GV matching is satisfied around $\alpha \sim 4°$ with the broadest bandwidth. At $\theta \sim 31.5°$ signals of a spectral region over ~200 nm can interact simultaneously. The GV matching realizes this enormously broadband phase matching while the collinear geometry suffers the GVM of 110 fs/mm at 600 nm. The synchronously pumped OPO with a bandwidth exceeding 40 nm (>1000 cm^{-1}) was reported based on this configuration. While the strict restriction for the synchronization in the femtosecond OPO with the pump laser as well as the reflectivity of the cavity mirrors limits the bandwidth [11], an OPA with only single or a few-stage amplification has the potential to realize a broader and shorter pulse generation straightforwardly. The schematic of the noncollinearly phase-matched broadband OPA is shown in Figure 1.1.1.2.

A 1 kHz Ti:sapphire regenerative amplifier (Clark-MXR, CPA-1000) produces 300 mJ pulses of 120 fs duration at 790 nm. A small fraction of the pulse energy of about 1 mJ is converted to a single filament white-light continuum in a 2-mm-thick sapphire plate of which function in the OPA is a signal. The signal beam is collimated and passed through a notch filter with a peak reflectance (~90%) at 800 nm to reduce the fundamental pulse energy and suppress an unwanted subharmonic amplification since the difference of the phase-matching angle is only ~0.4°. The second harmonics (SH, 100 mJ at 395 nm) generated in a 1-mm-thick BBO crystal is separated from the fundamental and utilized as a pump source. The pulse width of the SH is estimated to be ~150 fs by the cross-correlation measurement with the fundamental pulse. After passing through a synchronizing delay line, the pump beam is telescoped to obtain a peak intensity ~300 GW/cm^2 and passes through a 1-mm-thick type-I BBO crystal (cut at $\theta=30°$). Even though higher amplification is desired, the thickness of the BBO crystal is limited by the GVM between the pump and signal pulses (~115 fs/mm at 600 nm), which is reduced to a half of the value in the case of a collinear configuration but still remains under the relatively small noncollinear angle [8,9]. The small diameter (0.5 mm) of the pump reduces the effect of the pulse-front tilting inevitable in the noncollinear configuration [8,9]. A noncollinearly phase-matched parametric fluorescence is strongly emitted in a conical plane with a spectral dispersion. The dispersion is minimized around u 531.5°, which is the signal-idler GV matching point [13,14]. The fluorescence covering a large portion of the visible region is emitted conically with the cone angle of; 3.7° in the crystal. The continuum propagates along the intersection of the xz plane of the BBO and the conical surface, and a broad spectral range of the continuum is noncollinearly amplified to 2–3 µJ pulse energy.

The center wavelength of the amplified signal depends sensitively on the position of the delay line of the pump pulse. By scanning the delay with the range of 50 µm, the signal is stably tuned between 550 and 690 nm without any significant pulse energy reduction, indicating the bandwidth is not limited by the gain bandwidth but by the chirp of the continuum. Figure 1.1.1.3a shows an extremely broad spectrum centered around 625 nm with a 66 nm full width at half maximum

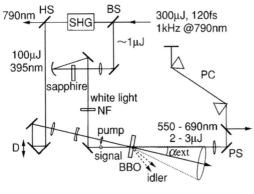

FIGURE 1.1.1.2 Schematic of the noncollinearly phase-matched OPA. SHG, second harmonic generator; BS, beam sampler; HS, fundamental and second harmonic separator; NF, notch filter centered at 800 nm; D, variable optical delay line; PS, periscope for rotating the polarization of the signal; PC, prism compressor. The conelike parametric fluorescence with the minimized dispersion as described in the text. It is illustrated with the external cone angle α_{ext}.

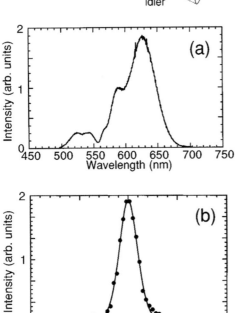

FIGURE 1.1.1.3 (a) Spectrum of the signal centered at 625 nm. The FWHM is 66 nm corresponding to the bandwidth of: ~1700 cm^{-1}. (b) Intensity autocorrelation trace (dots) after pulse compression. The sech^2 fit (solid line) with a pulse width of 14 fs (FWHM).

(FWHM) equivalent to ~1700 cm^{-1}. The small peak structure around 530 nm is due to the imperfect flatness of the phase-matching curve. In this case, the idler is also broadly generated and fan-shaped with a spanning angle of about 7° in the xz plane, which can be recognized by the fanning of the SH of the idler. The wave vector of the idler is spatially dispersed to maintain phase-matching condition to the broad signal spectrum [12,14].

Figure 1.1.1.3b shows the intensity autocorrelation trace after the pulse compression by using a BK7-prism pair with a 37 cm slant length. A fit assuming a sech^2-pulse envelope function yields a pulse width of 14 fs with relatively large wings on both sides. The time-bandwidth product of; 0.7 indicates a large residual chirp remaining in the signal pulse. The autocorrelator composed of a 100-mm-thick BBO with the GVM of 40 fs at 600 nm and dispersive media such as a lens and a beam splitter may induce overestimation of the pulse width. The time resolution of 3.3 fs/step of the pulse stage used for the delay line also disturbs the accurate measurement of the pulse width. We are now preparing a low-dispersion autocorrelator and a grating pair to compensate the higher-order dispersion.

The measured spectral width and pulse duration over the pump delay tuning range are shown in Figure 1.1.1.4. The spectral bandwidths of the pulse are varying between 700 and 2000 cm^{-1}. While the pulse duration is maintaining sub-20 fs duration. The time-bandwidth products are 0.6–1.1, except for 0.4 at 550 nm. In the long spectral region, the spectrum distributes to ~800 nm with a non-negligible spike, which degrades the products to be larger than 1. The noncollinear geometry removes the degeneracy and a phase-free stable subharmonic amplification is possible even under the type-I interaction. The bandwidth may be further increased by a chirp control of the continuum, using a prism pair.

The large time-bandwidth product may be mainly caused by a spatial chirp. It is caused by the sensitive α dependence of the phase matching causes an additional spectral broadening, accompanied by an inevitable spatial chirp of the signal beam. The ~20 nm shift of the center wavelength (measured at 600 nm by scanning a slit located perpendicular to the beam direction on the xz plane (Figure 1.1.1.1b) over the cross section after collimation) indicates the limitation of the compression. The pulse-front tilting (estimated to be ~6° just after the BBO [9]) may also increase the measured pulse width. The low conversion efficiency (~5%), mainly attributed to the residual parametric fluorescence even in the amplification regime, can be improved by the additional amplification up to ~8 mJ pulse energy, even with the slightly longer duration of ~18 fs, which may be caused by the further tilting [14]. The compensation of both the spatial chirp and the tilt angle [15] is to be a subsequent work to compress the pulse width.

In summary, a noncollinearly phase-matched OPA is demonstrated to exhibit the GV matching between the signal and idler and broadband amplification up to 2000 cm^{-1}. The simple novel configuration realizes over a 100 nm tuning range by scanning the delay line with conserving less than 20 fs duration. The tuning range with the broad gain bandwidth may be extended by changing the noncollinear angle to satisfy the GV matching. Compensating the higher-order dispersion, spatial

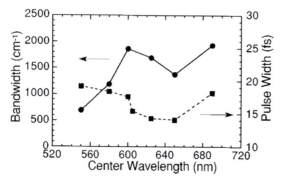

FIGURE 1.1.1.4 Wavelength dependence of the bandwidth (full circle) and pulse width ~full square, sech^2 fit).

chirp, and pulse-front tilting is quite essential and in progress for a sub-10 fs multi-μJ source tunable in a visible region.

The work presented in this subsection was conducted by the following people [16]: A. Shirakawa and T. Kobayashi.

Note added in proof. Around the time when the content in this subsection was first submitted, there was also reported a sub-20 fs OPA with a similar configuration [17]. While we utilized a simple delay line tuning for convenience, a tuning by changing the noncollinear angle was demonstrated in this chapter.

REFERENCES

1. A. Baltuška, Z. Wei, M. S. Pshenichnikov, and D. A. Wiersma, *Opt. Lett.* **22**, 102 (1997).
2. M. Nisoli, S. De Silvestri, O. Svelto, R. Szipöcs, F. Krausz, Ch. Spielmann, S. Sartania, and F. Krausz, *Opt. Lett.* **22**, 522 (1997).
3. R. Danielius, A. Piskarskas, A. Stabinis, G. P. Banfi, P. Di Trapani, and R. Righini, *J. Opt. Soc. Am. B* **10**, 2222 (1993).
4. V. V. Yakovlev, B. Kohler, and K. R. Wilson, *Opt. Lett.* **19**, 2000 (1994).
5. M. K. Reed, M. S. Armas, M. K. Steiner-Shepard, and D. K. Negus, *Opt. Lett.* **20**, 605 (1995).
6. T. S. Sosnowski, P. B. Stephens, and T. B. Norris, *Opt. Lett.* **21**, 140 (1996).
7. S. Takeuchi and T. Kobayashi, *J. Appl. Phys.* **75**, 2757 (1994).
8. P. Di Trapani, A. Andreoni, P. Foggi, C. Solcia, R. Danielius, and A. Piskarskas, *Opt. Commun.* **119**, 327 (1995).
9. P. Di Trapani, A. Andreoni, C. Solcia, P. Foggi, R. Danielius, A. Dubietis, and A. Piskarskas, *J. Opt. Soc. Am. B* **12**, 2237 (1995).
10. G. M. Gale, M. Cavallari, T. J. Driscoll, and F. Hache, *Opt. Lett.* **20**, 1562 (1995).
11. F. Hache, M. Cavallari, and G. M. Gale, *Ultrafast Phenomena X*, edited by P. F. Barbara, J. G. Fujimoto, W. H. Knox, and W. Zinth, Springer, Berlin, p. 33 (1996).
12. J. Wang, M. H. Dunn, and C. F. Rae, *Opt. Lett.* **22**, 763 (1997).
13. V. Krylov, A. Kalintsev, A. Rebane, D. Erni, and U. P. Wild, *Opt. Lett.* **20**, 151 (1995).
14. A. Shirakawa, S. Morita, K. Misawa, and T. Kobayashi, CLEO/Pacific Rim '97, Chiba, July 1997, Paper No. TuF2.
15. O. E. Martinez, *Opt. Commun.* **59**, 229 (1986).
16. A. Shirakawa and T. Kobayashi, *Appl. Phys.* **72**, 147 (1998).
17. T. Wilhelm, J. Piel, and E. Riedle, *Opt. Lett.* **22**, 1494 (1997).

1.1.2 Simultaneous Compression and Amplification of a Laser Pulse in a Glass Plate

1.1.2.1 INTRODUCTION

Over the past decade, several pulse-compression techniques based on phase modulation induced by nonlinear optical effects have been developed extensively. One of the important methods to obtain compressed pulse is using self-phase modulation (SPM) in a gas-filled hollow fiber [1,2], a filament in gas cell [3,4], a fiber [5], or a bulk medium [6,7] to broaden the spectrum and then compensating the spectral phase dispersion. This method is usually used to generate intense few-cycle pulses. Another technique for compressing pulses that have been developed in recent years is molecular phase modulation (MPM) [8,9]. In this method, a short, intense pump pulse excites impulsively coherent vibration in a Raman-active molecular gas, and then a weak, delayed pulse is phase modulated by the instantaneous refractive index change associated with the molecular vibration. A third pulse compression method is cross-phase modulation (XPM). It was extensively studied more than 20 years ago in a long fiber and in thick glass using picosecond pulses [10–13]. Recently, it has been theoretically demonstrated that XPM can compress a pulse to nearly the single-cycle regime from ultraviolet (UV) to mid-infrared (MIR) in a bulk medium [14]. Prior to the SPM effect, in the XPM process, the spectral, temporal, and spatial properties of the weak seed pulse can be controlled by manipulating both the intense pump pulse profile and the delay time between the pump pulse and the seed pulse [10,12,13]. Compared to MPM, it is much simpler and flexible to broaden the spectrum using XPM in a bulk medium. In addition, the pump and seed beams are spatially well separated, and the spectral shape is not limited by the time period of the molecular vibration [8,9].

In this chapter, we proposed and demonstrated, for the first time, a method for simultaneous compression and amplification of a weak femtosecond pulse in a bulk medium using XPM in conjunction with the four-wave optical parametric amplifier (FWOPA) [10,15–17] that is pumped by an intense femtosecond pulse. Furthermore, the spectrum of the weak pulse can be tuned by varying the delay between the pump and seed pulses. This method promises to be useful for the generation and optimization of ultrashort pulses at different wavelengths for use in pump-probe experiments that require tunable short pulses over a wide spectral range.

1.1.2.2 PRINCIPLE

The principle of this method is schematically illustrated in Figure 1.1.2.1a. An intense pump beam and a weak seed beam are focused onto a glass plate with a crossing angle α. Usually, the wavelength of the pump pulse is fixed; it should be different from that of the seed pulse to prevent interference between the two pulses. When the pump and seed pulses are synchronous in time and overlapping in space in a transparent bulk medium, the seed pulse spectrum will be broadened due to the XPM effect in the medium induced by the intense pump pulse. Furthermore, the weak seed pulse will be simultaneously amplified when the crossing angle α satisfies the phase-matching condition of four-wave mixing (FWM) [10,15–17].

DOI: 10.1201/9780429196577-4

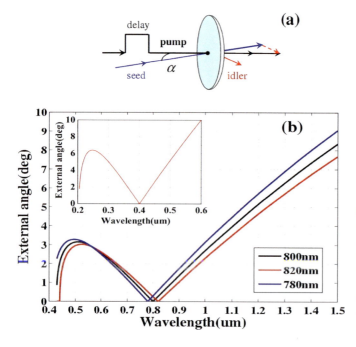

FIGURE 1.1.2.1 (a) Schematic of the experimental setup; α is the crossing angle. (b) Phase-matching curves showing the crossing angle α as a function of the seed wavelength for fused silica (or CaF_2, inset) when the pump pulse was fixed at a typical wavelength of 780, 800, and 820 nm (or 400 nm, inset).

A diagram of the vectors of FWM is also presented in Figure 1.1.2.1a. According to the phase-matching condition, the crossing angle α_{in} in the medium can be represented by

$$\cos \alpha_{in} = \left[\left(2k_p \right)^2 + k_s^2 - k_i^2 \right] / 4k_p k$$

in terms of wavenumber k with the subscripts of p, s, i indicating the pump, seed, and idler beams, respectively. Figure 1.1.2.1b and its inset show the plots of the phase-matching curves of the crossing angle in air α as a function of the seed wavelength for fused silica and CaF_2, respectively. The pump pulse is fixed at a typical wavelength of 800 and 400 nm for fused silica and CaF_2, respectively. As indicated in Figure 1.1.2.1b (or inset of Figure 1.1.2.1b), there is a broad phase-matching bandwidth around 500 nm (or 250 nm) when the crossing angle α is about 3.1° (or 6.4°).

1.1.2.3   EXPERIMENTAL SETUP

The experiment was performed using a 1-kHz Ti:sapphire regenerative amplifier laser system (Micra+Legend-USP, Coherent) with a pulse duration of 40 fs and an average power of 2.5 W. Figure 1.1.2.2 shows experimental setup. The laser pulse was split into four beams. One of the beams (Beam3) was focused by a 1-m focal length lens onto a 0.5-mm-thick fused silica plate (G2) after passing through a variable neutral-density filter and a motor-driven delay stage with a resolution of 10-nm/step. Another beam (Beam4) was used to measure the pulse duration by the cross-correlation frequency-resolved optical gating (XFROG) technique by mixing it with the two input beams or the spectrally broadened pulse in a 10-µm-thick BBO crystal. The other two beams (Beam1 and Beam2) were used to generate cascaded FWM sidebands in a similar manner as in our previous studies [18–21] in which nearly transform limited pulses were obtained when the input pulses were appropriately chirped. These cascaded FWM sidebands were used as incident seed beams and were focused onto the glass plate by a concave mirror with a focal length of 600 cm. The diameter of the pump (seed) beam on the glass plate was 300 µm (153 µm) as measured using a CCD camera (BeamStar FX33, Ophir Optronics). The pulse duration of the pump (seed) pulse was about 75±3 fs (22.6±0.5 fs) prior

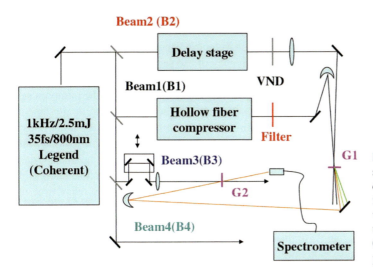

FIGURE 1.1.2.2 Experimental setup. VND, variable neutral-density filter; Filter, bandpass filter at 700 nm center wavelength with 40 nm bandwidth; G1, 1 mm-thick CaF$_2$ glass plate; G2, 0.5-mm-thick fused silica glass plate.

to the glass. The group velocity delay between pump pulse (800 nm) and seed (510/620 nm) pulse was 36/14 fs in a 0.5 mm fused silica glass plate. The pump beam was focused on a larger diameter, and its pulse duration was adjusted to be longer than that of the seed pulse to reduce the spatial chirp in XPM and ensure a good temporal overlap with the seed pulse through the glass plate.

1.1.2.4   EXPERIMENTAL RESULTS AND DISCUSSION

Initially, the third-order anti-Stokes (AS3) sideband [18–21] was used as the seed beam because its center wavelength is close to the broad phase-matching spectral range of around 500 nm. The incident pulse energies of the AS3 and pump beams were 300 nJ and 140 μJ, respectively. The spectral profile and spectral intensity of the output seed beam as a function of the delay time t_{ps} when the crossing angle α was 1.80°±0.05° are shown in Figure 1.1.2.3a. A negative delay time t_{ps} indicates that the seed pulse precedes the pump pulse. The full width at half maximum (FWHM) spectral bandwidth of the seed pulse was smoothly broadened from 18 to 50 nm (i.e., by a factor of about 2.8) at a delay time of approximately 0 fs due to XPM. This broadened spectrum can support transform-limited pulse duration of 9.6 fs. In this case, the spectral broadening of a Gaussian pulse

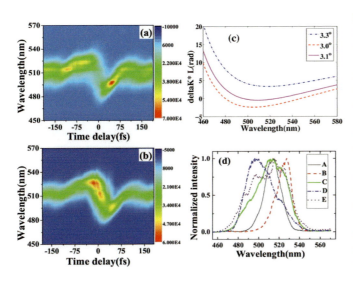

FIGURE 1.1.2.3 Dependence of the spectral profile and intensity of the output seed beam (AS1) on the delay time t_{ps} for crossing angles α of (a) 1.80°±0.05° and (b) 3.30°±0.05°. (c) The phase mismatching curves at three different external crossing angles (3.3°, 3.1°, and 3.0°) in a 0.5 mm-thick fused silica glass when the seed pulse was centered at 510 nm and pumped by 800 nm pulse. (d) A: incident spectrum of seed pulse. B, C, and D show spectra of output pulse at delay times of −30, 0, and 30 fs when α was 3.30°±0.05°, respectively. (e) Spectra of output pulse at delay times of 0 fs when α was 1.80°±0.05°.

due to XPM can be simply expressed as $\Delta\omega \approx \omega_s / c n_2 \left(|E_s|^2 + 2|E_p|^2 \right) l / T_0$ [10,13] because the intensity of the pump pulse is nearly undepleted and the group-velocity dispersion in the bulk medium is negligibly small. In the expression, ω_p is the angle frequency of the seed pulse, n_2 is the nonlinear refractive index of the medium, E_s and E_p are electric field amplitudes of seed pulse and pump pulse, respectively, l is the thickness of bulk medium, T_0 is the FWHM pulse duration of the pump pulse. Using the above expression, the spectral broadening is calculated to be about 40 nm in agreement with the experimental result. In this case, the seed beam was not amplified due to the large phase mismatch. When the crossing angle α was $3.30° \pm 0.05°$, the FWM phasematching condition was satisfied (Figure 1.1.2.1b). In this case, the spectrum of the seed pulse was smoothly broadened and its output power was simultaneously amplified, as shown in Figure 1.1.2.3b. The pulse energy of the seed pulse was amplified from 300 to 940 nJ (i.e., by a factor of about 3.1) at a delay time of approximately 0 fs. The spectrally asymmetric amplification in Figure 1.1.2.3b is due to the fact phase matching is satisfied on the longer wavelength side for broader spectral range, as shown in Figure 1.1.2.3c. The phase-matching angle in the experiment ($3.30° \pm 0.05°$) was slightly larger than the calculated value for the best phase matching as shown in Figure 1.1.2.3c and Figure 1.1.2.1b.

This is because of the nonlinear phase shift $\varphi_{NL} = \Delta kl = 2\omega_p n_2 I_p l / ck$ induced by the nonlinear index [10]. In the expression, ω_p and I_p are the angle frequency and intensity of pump pulse, respectively. This induced additional phase amounts to about π phase in a 0.5 mm fused silica glass under the experimental condition. The spectra of the output seed pulse at delay times of -30, 0, and 30 fs are shown in Figure 1.1.2.3d. The seed pulse spectrum was clearly red-shifted (blue-shifted) when the two incident pulses were overlapping in negative (positive) delay time. The peak wavelength of the seed pulse can be shifted by about 20 nm on both sides, indicating that the spectrum of the amplified output pulse can be tuned by adjusting the delay time t_{ps}. This spectral shift can be easily explained as follows. The frequency shift induced by XPM can be given by $\delta\omega(t) = -\partial\varphi_{NL}/\partial t \propto -\partial |E_p(t)|^2 / \partial t$, where $E_p(t)$ is the electric field of the pump pulse. It can be concluded from the above expression that $\delta\omega(t) < 0$ at the leading edge of the pump pulse and $\delta\omega(t) > 0$ at the trailing edge of the pump pulse. As for the negative delay time, the leading edge of the pump pulse will overlap with the seed pulse. Therefore, the induced frequency change is negative ($\delta\omega(t) < 0$) and the spectrum of the seed pulse is red-shifted; vice versa for a positive delay time.

This phenomenon was observed also when the first-order anti-Stokes (AS1, 620 nm) sideband [18–21] was used as the seed beam. When the crossing angle α was around $2.80° \pm 0.05°$, the incident seed pulse was spectrally broadened and amplified simultaneously. In this case, the incident pulse energy of AS1 was 400 nJ, and the pump energy was 140 μJ. The spectral profile and intensity of the amplified output beam as a function of the delay time t_{ps} are shown in Figure 1.1.2.4a. Cascaded FWM signals were simultaneously generated in this case, as shown in the photograph in the inset of Figure 1.1.2.4a. The maximum output pulse energies of the seed beam and the first-order cascaded signal (around 500 nm) were 1.1 μJ and 250 nJ, respectively. The output energy of the seed pulse as a function of the pump intensity at a delay time of 0 fs is shown in Figure 1.1.2.4b. Much higher output energies are expected to be obtained when cylindrical lens for focusing is used [15–17]. The small thickness of the glass plate ensured that the phase-matching spectral bandwidth was broad. Therefore, there was still broadband amplification around 620 nm. In the process, the pump pulse has a slightly steeper trailing edge, which was measured by using SHG-FROG, as shown in the inset of Figure 1.1.2.4b. Furthermore, the self-steepening effect will introduce a sharper trailing edge of the pump pulse during its propagation [10]. As a result, the rapid decrease of trailing edge induced a broader blue spectral shift in the seed pulse, as shown in Figure 1.1.2.3a and b and Figure 1.1.2.4a. The steep trailing edge of the pump pulse also induced a rapid decrease of the spectral shift in the positive delay time, as is also shown in Figure 1.1.2.3a and b and Figure 1.1.2.4a.

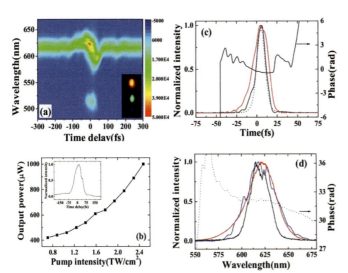

FIGURE 1.1.2.4 (a) The spectral profile and intensity of the output seed beam (AS3) as a function of the delay time t_{ps} when the crossing angle α was $2.80° \pm 0.05°$. (b) The dependence of the output energy of the output seed pulse on the pump intensity. The inset curve is the temporal profile of pump pulse. (c) The retrieved temporal profiles of the incident seed pulse (red solid line) and the compressed output seed pulse (black solid line), and the transform limited pulse of the broadened spectrum (blue dotted line). (d) The retrieved spectrum (blue solid line) and the spectral phase (blue dotted line) of the output seed pulse. The measured spectra of the incident seed pulse (black solid line) and output seed pulse (red solid line).

A quasi-linear chirp can be imposed across the weak seed pulse when the pump pulse is much wider compared with it [10]. In the same way as SPM-based compressors, the phase induced by XPM can also be compensated by using a chirped mirror pair. Furthermore, XPM-based compressor can have more flexibility than SPM ones because the phase can be tuned by the pump pulse. After passing through a pair of chirped mirrors for four bounces ($-40\,\text{fs}^2$/bounce), the pulse duration of the weak output pulse was measured by XFROG. The spectral and temporal profile and the phase were retrieved using commercial software (FROG 3.0, Femtosoft Technologies) with a 512×512 grid. The retrieval error was smaller than 0.006. The retrieved temporal profiles of both the incident seed pulse and the compressed output seed pulse, and the transform-limited pulse for the broadened seed pulse are presented in Figure 1.1.2.4c. The $22.6 \pm 0.5\,\text{fs}$ incident pulse was compressed to $12.6 \pm 0.5\,\text{fs}$, which is well close to the calculated transform-limited pulse duration of $10.5\,\text{fs}$. The retrieved spectrum and spectral phase of the output seed pulse are shown in Figure 1.1.2.4d.

We also guided three beams, AS1, AS2, and AS3, generated by cascaded FWM [18–21] into the glass at the same time. The schematic of the experimental setup was shown in the inset of Figure 1.1.2.5. The beam diameters of AS1, AS2, and AS3 on the 0.5-mm thick fused silica glass were all about 190 µm. The crossing angles between the pump beam and three anti-Stokes sidebands, AS1, AS2, and AS3 were about 2.6°, 2.9°, and 3.2°, respectively. The pump pulse energy was 160 µJ with about 350 µm diameter on the glass. When these seed and pump beams synchronized in time on the glass, the spectra of the three seed beams, AS1, AS2, and AS3 were broadened simultaneously, as shown in Figure 1.1.2.5. The pulse energy of AS1, AS2, and AS3 were also amplified from 133, 40, and 12 to 163, 50, and 15 nJ, respectively. This experiment proves that it is possible to simultaneously amplify and compress several weak pulses in a bulk medium at the same time. Then, it can be used for the cascaded FWM process to optimize the spectrum of the obtained sidebands, which was expected to obtain single cycle pulse after combining the optimized sidebands [22].

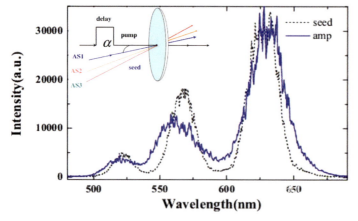

FIGURE 1.1.2.5 The spectral profile and intensity of AS1, AS2, and AS3 without pump (black dash line). The blue solid line is the curve of the spectral profile and intensity of AS1, AS2, and AS3 with pump. The inset was schematic of the experimental setup; α is the crossing angle between pump and AS1.

1.1.2.5   CONCLUSION

In summary, a novel method of simultaneously amplifying and compressing a weak pulse was demonstrated using XPM in conjunction with FWOPA in a bulk medium. It can also be used to broaden the spectra of several weak pulses with different wavelengths at the same time.

This method described here is simple and can be used to optimize ultrashort pulses (e.g., noncollinear optical parametric amplifier system and cascaded four-wave mixing system) from UV to MIR in a broad region, which is very important for applications in photochemistry, photophysics, and photobiology. An intense tunable single-cycle pulse is expected to be obtained using this method [23].

The research presented in this section is reproduced and adapted with permission from Ref. [23] ©The Optical Society. *Opt. Exp.* **18**(3), 2495–2502 (2010) and it was conducted by collaborative work of the following people: A. Baltuska, T. Fuji, and T. Kobayashi.

REFERENCES

1. M. Nisoli, S. De Silvestri, and O. Svelto, "Generation of high energy 10 fs pulses by a new pulse compression technique," *Appl. Phys. Lett.* **68**(20), 2793–2795 (1996).
2. M. Giguère, B. E. Schmidt, A. D. Shiner, M. A. Houle, H. C. Bandulet, G. Tempea, D. M. Villeneuve, J. C. Kieffer, and F. Légaré, "Pulse compression of submillijoule few-optical-cycle infrared laser pulses using chirped mirrors," *Opt. Lett.* **34**(12), 1894–1896 (2009).
3. C. P. Hauri, W. Kornelis, F. W. Helbing, A. Heinrich, A. Couairon, A. Mysyrowicz, J. Biegert, and U. Keller, "Generation of intense, carrier-envelope phase-locked few-cycle laser pulses through filamentation," *Appl. Phys. B* **79**(6), 673–677 (2004).
4. X. W. Chen, X. F. Li, J. Liu, P. F. Wei, X. C. Ge, R. X. Li, and Z. Z. Xu, "Generation of 5 fs, 0.7 mJ pulses at 1 kHz through cascade filamentation," *Opt. Lett.* **32**(16), 2402–2404 (2007).
5. W. J. Tomlinson, R. H. Stolen, and C. V. Shank, "Compression of optical pulses chirped by self-phase modulation in fibers," *J. Opt. Soc. Am. B* **1**(2), 139–149 (1984).
6. C. Rolland and P. B. Corkum, "Compression of high-power optical pulses," *J. Opt. Soc. Am. B* **5**(3), 641–647 (1988).
7. E. Mével, O. Tcherbakoff, F. Salin, and E. Constant, "Extracavity compression technique for high-energy femtosecond pulses," *J. Opt. Soc. Am. B* **20**(1), 105–108 (2003).
8. N. Zhavoronkov and G. Korn, "Generation of single intense short optical pulses by ultrafast molecular phase modulation," *Phys. Rev. Lett.* **88**(20), 203901 (2002).
9. R. A. Bartels, T. C. Weinacht, N. Wagner, M. Baertschy, C. H. Greene, M. M. Murnane, and H. C. Kapteyn, "Phase modulation of ultrashort light pulses using molecular rotational wave packets," *Phys. Rev. Lett.* **88**(1), 013903 (2001).
10. G. P. Agrawal, *Nonlinear Fiber Optics*, Academic Press, California (2007).

11. R. R. Alfano, Q. X. Li, T. Jimbo, J. T. Manassah, and P. P. Ho, "Induced spectral broadening of a weak picosecond pulse in glass produced by an intense picosecond pulse," *Opt. Lett.* **11**(10), 626–628 (1986).
12. P. L. Baldeck, R. R. Alfano, and G. P. Agrawal, "Induced-frequency shift of copropagating ultrafast optical pulses," *Appl. Phys. Lett.* **52**(23), 1939–1941 (1988).
13. R. R. Alfano, P. L. Baldeck, P. P. Ho, and G. P. Agrawal, "Cross-phase modulation and induced focusing due to optical nonlinearities in optical fibers and bulk materials," *J. Opt. Soc. Am. B* **6**(4), 824–829 (1989).
14. M. Spanner, M. Y. Ivanov, V. Kalosha, J. Hermann, D. A. Wiersma, and M. Pshenichnikov, "Tunable optimal compression of ultrabroadband pulses by cross-phase modulation," *Opt. Lett.* **28**(9), 749–751 (2003).
15. H. Valtna, G. Tamošauskas, A. Dubietis, and A. Piskarskas, "High-energy broadband four-wave optical parametric amplification in bulk fused silica," *Opt. Lett.* **33**(9), 971–973 (2008).
16. A. Dubietis, G. Tamošauskas, P. Polesana, G. Valiulis, H. Valtna, D. Faccio, P. Di Trapani, and A. Piskarskas, "Highly efficient four-wave parametric amplification in transparent bulk Kerr medium," *Opt. Express* **15**(18), 11126–11132 (2007).
17. J. Darginavičius, G. Tamošauskas, G. Valiulis, and A. Dubietis, "Broadband four-wave optical parametric amplification in bulk isotropic media in the ultraviolet," *Opt. Commun.* **282**(14), 2995–2999 (2009).
18. J. Liu and T. Kobayashi, "Wavelength-tunable, multicolored femtosecond-laser pulse generation in fused-silica glass," *Opt. Lett.* **34**(7), 1066–1068 (2009).
19. J. Liu and T. Kobayashi, "Generation of uJ-level multicolored femtosecond laser pulses using cascaded fourwave mixing," *Opt. Express* **17**(7), 4984–4990 (2009).
20. J. Liu and T. Kobayashi, "Generation of sub-20-fs multicolor laser pulses using cascaded four-wave mixing with chirped incident pulses," *Opt. Lett.* **34**(16), 2402–2404 (2009).
21. J. Liu and T. Kobayashi, "Cascaded four-wave mixing in transparent bulk media," *Opt. Commun.* **283**(6), 1114–1124 (2009).
22. J. L. Silva, R. Weigand, and H. M. Crespo, "Octave-spanning spectra and pulse synthesis by nondegenerate cascaded four-wave mixing," *Opt. Lett.* **34**(16), 2489–2491 (2009).
23. J. Liu, Y. Kida, T. Teramoto, and T. Kobayashi, "Simultaneous compression and amplification of a laser pulse in a glass plate," *Opt. Exp.* **18**(3), 2495–2502 (2010).

1.1.3 Pulse-Front-Matched Optical Parametric Amplification for Sub-10-fs Pulse Generation Tunable in the Visible and Near Infrared

1.1.3.1 INTRODUCTION

Progress in extremely short pulse generation is remarkable [1] and has been stimulated in particular by reports of sub-5-fs pulse generation in 1997 [2,3], which exceeded the 6-fs record [4]. Although this traditional method of continuum compression lacks tuning capability, noncollinearly phase-matched optical parametric conversion [5–9] seems to be a leading candidate for a tunable several-femtosecond light source, which is strongly desired for various spectroscopic applications. Recent reports of femtosecond noncollinear optical parametric oscillators [5] (NOPA's) and optical parametric amplifiers [6–8] based on B-BaB$_2$O$_4$ (BBO) crystals indicated successful generation of sub-20-fs pulses tunable in wide visible ranges. However, it was difficult to generate transform-limited (TL) pulses, and only a large time–bandwidth product ($\Delta t \Delta \nu \sim 1$) could be obtained, even though sub-10-fs durations are expected from the broad spectra [6,8]; this is due mainly to pulse-front tilting, which is inevitable in noncollinear interaction [6,9]. In this Letter generation of sub-10-fs pulses from a NOPA that is tunable in both the visible and the near IR (NIR) is presented for the first time to our knowledge. The pulse-front-matched geometry eliminates the spatial chirp of the signal pulse, resulting in tunable TL sub-9-fs pulse generation with the shortest width of 6.1 fs. This NOPA is quite useful for spectroscopy with the highest time resolution and tunability [10].

1.1.3.2 EXPERIMENTAL

A schematic of the sub-10-fs NOPA system is shown in Figure 1.1.3.1. In the system, a small intensity fraction of the output f is split from a Ti:sapphire regenerative amplifier (Clark-MXR CPA-1000; 400 mJ, 130 fs, 1 kHz at 790 nm). The split beam is then converted to a single filament of white-light continuum in a 2-mm-thick sapphire plate and used as a signal of the OPA. The output passes through a thin (175-μm) cutoff filter ($T \sim 0\%$ at >750 nm), which prevents undesired amplification of the fundamental spikes [6] and leads to generate a beam with smooth spectra by adjusting the incident angle. The second harmonic (SH; 150 fs, 130 mJ) is generated in a 1-mm-thick LiB$_3$O$_5$ (LBO) crystal, which pumps a 1-mm-thick BBO crystal (type I, $\theta = 30°$). The signal beam is aligned in such a way to propagate on the surface of dispersion-managed fluorescence cone [6] and amplified with a noncollinear internal angle α of 3.7°. This configuration matches the group velocities of the signal and the idler and permits an amplification bandwidth as broad as over 2000 cm^{-1} [5–8].

DOI: 10.1201/9780429196577-5

1.1.3.3   RESULTS AND DISCUSSION

The pump beam passes through an antireflection coated 45° fused-silica (FS) prism with an incident angle of 49° so that the pulse front is tilted 2.3° (see Figure 1.1.3.1) [11,12]. A telescope with a longitudinal magnification factor of 2.8 images the front with a tilt angle of 6.4° at the BBO crystal. The corresponding internal tilt angle of 3.7° is equal to a, ensuring a maximum spatial overlap between the pump and the signal fronts. The diameter of each beam at the BBO crystal is, 1.5 mm; however, owing to mode mismatch and nonuniform intensity across the pump beam, the signal is amplified only in the limited area up to 3-mJ pulse energy. In the conventional nontilted-pump (NTP) geometry [6–9] the pulse-front tilting of a signal with an estimated beam diameter of, 0.3 mm in the BBO crystal can cause signal-pulse broadening to 100 fs, accompanied by spatial chirp [6,9,11–13]. In Figure 1.1.3.2 the chirp of the amplified signal is compared with that of a NTP geometry reported in Ref. 6. The large chirp of the signal in the NTP case can be reasonably explained by the tilted signal generation, where the relation of $\tan \gamma = \lambda \times d\varepsilon/d\lambda$ between the predicted external tilt angle ($\gamma = 6.3°$ for the signal) and exit angle ε is used [12]. The tilt-matched interaction dramatically reduces the spatial chirp, indicating nontilted signal amplification. Pulse shortening from ~14 to <10 fs was observed by the use of the same prism compressor as in Ref. 6, as expected. Both the pump and the signal beams are reflected back to the crystal by concave mirrors located at the confocal positions, and pulse fronts are imaged and matched again in the second-stage amplification, with an insignificant spectral change. The final pulse energy is 5 mJ with peak–peak fluctuation of 5%–10%. We can tune the signal from 550 to 700 nm by scanning the delay line of the seed that is due to the seed chirp [6,8].

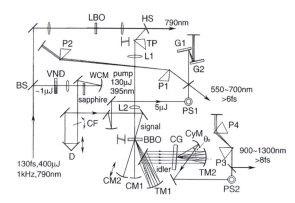

FIGURE 1.1.3.1 Schematic of the sub-10-fs NOPA. BS, beam sampler; HS, harmonic separator; TP, FS prism for pulse-front tilting; L1, L2, f-200 and 71-mm lenses for tilt imaging; CM1, CM2, concave mirrors (r-100 mm); VND, variable neutral-density filter; WCM, concave mirror ($r = 120$ mm); CF, cutoff filter; D, delay line; TM1, concave mirror ($r = 250$ mm); TM2, cylindrical mirror ($r = 115$ mm); CG, grating (600 lines/mm); CyM, cylindrical mirror; PS1, PS2, periscopes, P1, P2, 45° FS prisms; P3, P4, 60.2° SF10 prisms (both pairs are used with the minimum deviation); G1, G2, gratings (150 lines/mm; 550-nm blaze; incident angle, 10°). The pulse fronts of the pump at TP and the BBO crystal are shown by the thick gray lines on the right sides of TP and BBO.

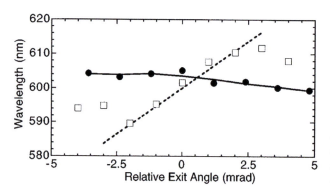

FIGURE 1.1.3.2 Comparison of the spatial chirp of the amplified signal centered at 600 nm in the horizontal direction for nontilted (open squares) and tilted (filled circles) pump geometries. A center-of-mass wavelength is shown for each case. The dashed line indicates the expected spatial chirp of the tilted signal in the NTP geometry. The solid curve is a guide for the eye.

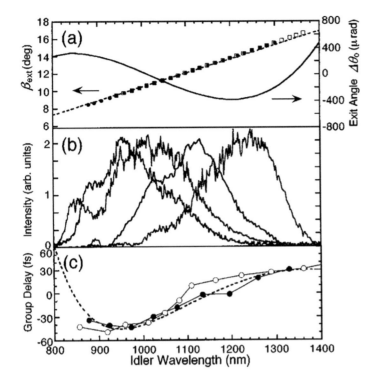

FIGURE 1.1.3.3 Properties of the idler. (a) Spectral angular dispersion of the calculated (dashed curve) and the measured (filled circles at λ_S-680 nm and open squares at λ_S-600 nm) external noncollinear angles (β_{ext}) with respect to the pump. The distribution of the exit angle (θ_0; see Figure 1.1.3.1) after the pulse passes through the telescope–grating compensator is also shown (solid curve). At zero, $\theta_0 = 32.6°$. (b) Spectra of the dispersion-compensated idler. (c) Measured (filled circles) and calculated (open circles) GD. The former is fitted by the GD of a 60.2° SF10 prism pair with the signs inverted (dashed curve).

The idler is generated with a large spectral angular dispersion so that it is phase-matched with the broad spectrum of the signal (Figure 1.1.3.3a). This is a characteristic property of the group-velocity-matched NOPA between the signal and idler [5–7,14]; i.e., it is an achromatic down conversion analogous to achromatic phase matching in SH generation [15] (SHG) or upconversion [16]. A collimated idler beam is obtained by the design of an angular-dispersion compensator composed of a reflective-type telescope and a grating [15]. They are aligned so that the 1100-nm component is back reflected with a small vertical tilt angle, which minimizes astigmatism. A spherical mirror recollimates the idler in both the vertical and the horizontal directions, and a cylindrical mirror focuses the idler horizontally upon the grating. Ray-trace analysis shows that the angular distribution of the first-order diffraction at the exit is well suppressed within 600 mrad across the tuning range from 800 to 1400 nm in the case of −7° incidence (Figure 1.1.3.3a). The deviation is due mainly to the nonlinear angular dispersion of the grating, which can be further optimized by a fine optical design in which nonspherical mirrors are used. An additional cylindrical mirror after the grating collimates the wave-diffracted idler horizontally. The pulse energy is currently limited to, 0.5 mJ owing to the low diffraction efficiency of the grating blazed at 600 nm. The visible non-phase-matched SH that is coaxially generated from the BBO crystal can be used for total alignment [6]. The resulting broad spectra of ~200-nm bandwidths tunable from 900 to 1300 nm are shown in Figure 1.1.3.3b.

Each pulse compressor is designed to cancel the group delay (GD) across the whole tuning range. The GD is measured by upconversion with 130-fs gate pulses with a spectral peak at 790 nm[2]. The signal passes through a grating–prism compressor [2,4] with ~15-mm separation of the grating pair and 550-mm separation and ~3.3-mm interprism length (IPL) of the 45° FS prism pair at 600 nm. The current beam intensity throughput of ~10% will be improved by the replacement of the grating pair with chirped mirrors [17]. The idler pulse is negatively chirped (Figure 1.1.3.3c), compensating for the positive seed chirp through parametric interaction under the chirp-free pump

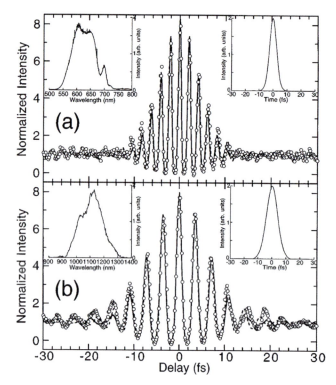

FIGURE 1.1.3.4 FRAC traces of (a) signal (at 630 nm) and (b) idler (at 1100 nm). Measured (open circles), calculated TL FRAC (solid curves), and sech^2-fit (dashed curves) traces are shown. The spectra and the calculated TL intensity profiles are shown in the insets at the upper left and the upper right, respectively. The sech^2-fit (TL) pulse widths are (a) 6.2 fs (6.0 fs) and (b) 8.4 fs (9.4 fs). Each bandwidth is 98 and 203 nm FWHM.

condition. The measured and the calculated GD by inversion of the seed chirp agree well within the limited time resolution. A 60.2° SF10 prism pair with a positive GD dispersion toward 1400 nm can compensate for the chirp fairly well over a 500-nm width with a prism spacing of 70 mm and an IPL of 4 mm at 1100 nm.

The pulse shape is measured by a dispersion-free fringe-resolved autocorrelator (FRAC). Each arm is perfectly balanced by two 0.5-mm-thick broadband Cr beam splitters, and the fine delay is on-line calibrated; 10- and 100-μm-thick BBO crystals are used for SHG of the signal and the idler, respectively. A typical FRAC trace of the signal at 630 nm is shown in Figure 1.1.3.4a. While the sech^2-shape-assumed pulse width is 6.2 fs, fitting by Fourier transformation of the spectrum with the parameters of GD dispersion and third-order dispersion [3] gives a 7.3-fs width. Both widths are fairly close to a TL width of 6.0 fs, which indicates the importance of tilt control. A FRAC trace of the idler at 1100 nm (Figure 1.1.3.4b) also shows a nearly TL pulse shape, where the sech^2-fit pulse width of 8.4 fs is also reasonably matched with the TL width of 9.4 fs within the uncertainty. The sech^2-fit pulse widths at 550, 680, and 1250 nm are 7.1, 6.1, and 9.0 fs, respectively, all of which are also nearly TL within 10% deviation. The longer width of the idler than that of the signal is due mainly to spectral narrowing caused by the grating and partly to the residual angular dispersion (see Figure 1.1.3.3a). The tilt mismatch of the idler in the BBO seems to be less effective because of the intrinsic large angular dispersion. Optimization of the system will also provide sub-7-fs pulses in the NIR.

1.1.3.4   CONCLUSION

Further plans were in progress for the chirped mirrors to be introduced not only for a high-throughput compressor but also for sub-5-fs TL pulse generation by suppression of the seed chirp instead of the 6-fs tunability reported above [18].

The research presented in this section was conducted by collaborative work of the following people: A. Shirakawa, I. Sakane, and T. Kobayashi [18].

The research presented in this section is reproduced and adapted with permission from [Ref. 18] ©The Optical Society. *Opt. Lett.* 23 1292–1294 (1998) was conducted by collaborative work of the following people: A. Shirakawa, I. Sakane, and T. Kobayashi [18].

REFERENCES

1. I. D. Jung, F. X. Ka¨rtner, N. Matuschek, D. H. Sutter, F. Morier-Genoud, G. Zhang, U. Keller, V. Scheuer, M. Tilsch, and T. Tschudi, *Opt. Lett.* **22**, 1009 (1997); L. Xu, Ch. Spielmann, F. Krausz, and R. Szipocs, *Opt. Lett.* **21**, 1259 (1996); J. Zhou, G. Taft, C. P. Huang, M. M. Murnane, H. C. Kapteyn, and I. P. Christov, *Opt. Lett.* **19**, 1149 (1994); D. Steinbach, W. Hu¨gel, and M. Wegner, *J. Opt. Soc. Am. B* **15**, 1231 (1998).
2. A. Baltusˇka, Z. Wei, M. S. Pshenichnikov, D. A. Wiersma, and R. Szipocs, *Appl. Phys. B* **65**, 175 (1997).
3. M. Nisoli, S. Stagira, S. De Silvestri, O. Svelto, S. Sartania, Z. Cheng, M. Lenzner, Ch. Spielmann, and F. Krausz, *Appl. Phys. B* **65**, 189 (1997).
4. R. L. Fork, C. H. Brito-Cruz, P. C. Becker, and C. V. Shank, *Opt. Lett.* **12**, 483 (1987).
5. G. M. Gale, M. Cavallari, T. J. Driscoll, and F. Hache, *Opt. Lett.* **20**, 1562 (1995); G. M. Gale, M. Cavallari, and F. Hache, *J. Opt. Soc. Am. B* **15**, 702 (1998), and references therein.
6. A. Shirakawa and T. Kobayashi, *Appl. Phys. Lett.* **72**, 147 (1998); *IEICE Trans. Electron.* **E81-C**, 246 (1998).
7. T. Wilhelm, J. Piel, and E. Riedle, *Opt. Lett.* **22**, 1494 (1997).
8. G. Cerullo, M. Nisoli, and S. De Silvestri, *Appl. Phys. Lett.* **71**, 3616 (1997).
9. P. Di Trapani, A. Andreoni, C. Solcia, P. Foggi, R. Danielius, A. Dubietis, and A. Piskarskas, *J. Opt. Soc. Am. B* **12**, 2237 (1995).
10. Q. Wang, R. W. Shoenlein, L. A. Peteanu, R. A. Mathies, and C. V. Shank, *Science* **266**, 422 (1994).
11. R. Danielius, A. Piskarskas, P. Di Trapani, A. Andreoni, C. Solcia, and P. Foggi, *IEEE J. Quantum Electron.* **34**, 459 (1998), and references therein.
12. Z. Bor and B. Ra´cz, *Opt. Commun.* **54**, 165 (1985).
13. O. E. Martinez, *Opt. Commun.* **59**, 229 (1986).
14. J. Wang, M. H. Dunn, and C. F. Rae, *Opt. Lett.* **22**, 763 (1997).
15. O. E. Martinez, *IEEE J. Quantum Electron.* **25**, 2464 (1989); R. A. Cheville, M. T. Reiten, and N. J. Halas, *Opt. Lett.* **17**, 1343 (1992); B. A. Richman, S. E. Bisson, R. Trebino, E. Sidick, and A. Jacobson, *Opt. Lett.* **23**, 497 (1998), and references therein.
16. Th. Hofman, K. Mossavi, F. K. Tittel, and G. Szabo, *Opt. Lett.* **17**, 1691 (1992).
17. R. Szipocs and A. K¨ oh´ azi-Kis, *Appl. Phys. B* **65**, 115 (1997), and references therein.
18. A. Shirakawa, I. Sakane, and T. Kobayashi, *Opt. Lett.* **23**, 1292–1294 (1998).

1.1.4 Visible 4 fs Pulse from Dispersion Control Optical Parametric Amplifier

1.1.4.1 INTRODUCTION

Discovery of the unique non-collinear phase-matching conditions in the β–barium borate (BBO) crystal [1] has opened an unprecedented opportunity for ultra-broadband parametric generation and amplification. Exploring this avenue, several groups have attained bandwidths approaching 200 THz and reported pulse durations as short as 4.7 fs [2–4].

To date sub-5-fs pulse compression in the NOPA has relied on a combination of broadband dielectric chirped mirrors and a prism sequence [3] or solely on custom-made ultrabroad-chirped mirrors [4]. Implementation of adaptive control over the spectral phase would be of a great advantage, as has been shown previously in an automated compression of tunable NOPA pulses down to 16 fs by a liquid crystal mask [5]. Recently developed inexpensive micro-machined flexible mirrors facilitated another attractive technique for fine group delay tuning [6]. Using this novel method, Plachta et al. [7] achieved versatile compression of the visible NOPA output to the spectrum-limited pulse duration of 7 fs.

In this subsection, we report modifications in the layout of a double-pass NOPA, which allow the generation of a visiblenear-IR spectrum that covers as broad as 300 nm at its FWHM. The schematic of the experimental layout is shown in Figure 1.1.4.1. The system is pumped by 120-fs 150-µJ pulses from a CPA-1000 regenerative amplifier (Clark-MXR) seeded by a Femtolite fiber oscillator

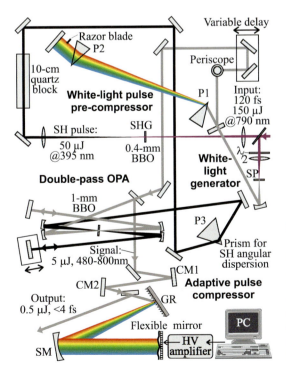

FIGURE 1.1.4.1 Schematic of experimental setup. $\lambda/2$, 800 nm wave-plate; SP, 2-mm sapphire plate; P1, 2, 45° quartz prisms; P3, 69° quartz prism; CM1, CM2, ultrabroadband chirped mirrors (Hamamatsu Photonics); GR, 300 lines/mm ruled diffraction grating (Jobin Yvon); SM, spherical mirror, $R = -400$ mm; SHG, 0.4-mm $\theta = 29°$ type I BBO (EKSMA); NOPA crystal, 1-mm $\theta = 31.5°$ type I BBO (Casix); Spherical mirrors around the NOPA crystal are $R = -200$ nm. HV, high voltage.

DOI: 10.1201/9780429196577-6

(IMRA). Compared with the NOPA design developed previously by A. Shirakawa et al. [3], in our group, we have implemented several important improvements that enabled further extension of the spectral bandwidth allowing even shorter pulse duration and furnished additional means of spectral shaping to have a better function as a probe beam in pump/probe spectroscopy experiment.

1.1.4.2 CONFIGURATION OF THE SYSTEM

We have implemented three important improvements that enabled the generation of ultrashort pulses with better quality suited to applications including pump/probe spectroscopy experiment.

1. First, a pair of 45° fused silica prisms has been introduced to pre-compress the seed pulse, the white light continuum generated in a 2-mm sapphire plate. An adjustable razor blade behind the inner prism is used to cut off the intense spectral components that are close to the fundamental 790-nm light. On the other hand, the insertion depth of the inner prism adjusts the cut-off wavelength on the blue side of the spectrum. The use of non-Brewster prisms with a more acute apex angle helps reducing higher-order phase distortion [8]. Nevertheless, the cubic dispersion of the prism material dominates the phase properties of the prism pre-compressor. Consequently, the spectral components at 500 nm and 750 nm are roughly synchronized while 600-nm light is retarded with respect to them.

2. Second, because of the abundant intensity of the second harmonic (SH) light, we have chosen to stretch the pump pulse in time by down-chirping it in a 100-mm block of fused silica. As a result, the peak intensity of the pump pulse is decreased below the damage threshold of the BBO crystal that is used in the NOPA. This allowed switching to a confocal configuration that greatly improved the output mode pattern. At the same time, the lengthened temporal envelope of the pump pulse dramatically lowered the requirement for an accurate pre-compression of the seed light.

3. Thirdly, the angular dispersion of the pump is utilized in such a way that various frequency components of the SH spectrum intersect with the seed beam in the NOPA BBO crystal at slightly different angles, which helps broadening the effective phase-matching bandwidth. Conversely, we find this effect to be more significant than the concerns for the tilted pump geometry and pulse-front matching, emphasized previously [3]. While tilted-pulse pumping is essential for increasing interaction length and pulse overlap in longer crystals [9], it plays a secondary role in the case of a 1-mm BBO crystal used in this work. Figure 1.1.4.2 illustrates the idea behind the enhancement of the phase-matching bandwidth. Since adjacent pump wavelengths correspond to slightly off-set phase-matching curves, the resulting bandwidth that can be simultaneously amplified in each direction of the seed light is extended by the use of a broadband SH pump. This becomes apparent from the comparison of a quasi-monochromatic pump (Figure 1.1.4.2, dashed curve) and a broadband pump (Figure 1.1.4.2, light-shaded contour). To maximize the usable SH bandwidth, we have employed a relatively thin, 0.4-mm, BBO SHG crystal that ensures a 30% frequency-doubling efficiency and provides a 6-nm-wide pump spectrum, shown in the inset in Figure 1.1.4.2. The FWHM of the angular phase-matching, depicted by shaded contours in Figure 1.1.4.2, was computed according to Ref. [10] and represents the case of low efficiency of parametric frequency conversion [11]. As can be seen from the breadth of the dark-shaded contour in Figure 1.1.4.2, an additional broadening can be obtained by manipulating the incidence angle for various pump frequencies. The calculation in Figure 1.1.4.2 describes the actual conditions of our experiment where a Brewster-angled fused silica prism is used to introduce angular dispersion in the pump beam. The SH light is subsequently focused onto the BBO crystal by an $R = -200$ mm mirror placed at an 80-cm distance from the prism (Figure 1.1.4.1). Analogous ideas about phase-matching extension have been previously implemented in achromatic frequency doubling [12,13] and a multi-pass OPA [14].

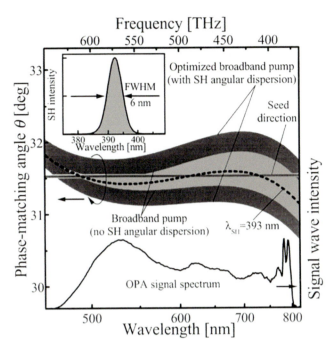

Frequency [THz]

FIGURE 1.1.4.2 Broadband non-collinear phase matching in a 1-mm type I BBO crystal, $\theta = 31.5°$. Dashed curve shows the exact phase-matching solution for quasi-monochromatic pump at $\lambda_{SH} = 393$ nm. Light-shaded contour represents FWHM of angular phase matching corresponding to the pump by the entire available SH bandwidth. Dark-shaded contour displays FWHM of angular phase matching in the case of dispersed SH beam (see text for details). Inset depicts experimentally measured SH spectrum used to pump the NOPA. Solid curve shows uncompressed spectrum of the amplified signal wave obtained under optimal phase-matching conditions. Straight horizontal line gives seed direction (with respect to the optical axis of BBO).

The improvements listed above have led to the generation of a smooth parametrically amplified spectrum (Figure 1.1.4.2, solid curve), which corresponds to a 3.5-fs FWHM pulse duration, if ideally compressed.

Amplitude-phase characterization of both chirped and compressed pules was carried out by SHG FROG in a very thin BBO wedge. The crystal thickness at the point used in the measurement was estimated to be about 5 μm. To reverse the effect of spectral filtering a post-experiment correction [15] has been applied to all measured FROG traces. In 2002, crystal angle dithering has been suggested to solve the problem of the insufficient phase-matching bandwidth in an SHG FROG measurement [16]. Nevertheless, we avoided using this method in our work since it is very difficult to ensure that the axis, around which the crystal is being tilted, coincides precisely with the beam intersection. This concern becomes vital in the measurement of the compressed-long change) NOPA pulses (having duration shorter than 2-μm spatial distance, in which crystal dithering can easily result in the unwanted time-delay jittering of the FROG/autocorrelation trace.

The pulse compressor consists of a pair of chirped mirrors CM1, 2 (Figure 1.1.4.1) and a grating dispersion line with a flexible mirror [6,7] (OKO Technologies) positioned in the focal plane. The combination of the grating dispersion line and the chirped mirrors has been designed through a dispersive ray-tracing analysis aiming to match the spectral phase of the chirped parametrically amplified pulse, which was obtained from the FROG measurement. The total throughput of the pulse shaper is smaller than 12% due to the low diffraction efficiency of the grating, which limits the energy of the compressed pulses to ~500 nJ.

The FWHM pulse duration corresponding to the optimal grating–spherical mirror separation and the "switched off" state of the flexible mirror (no bias voltage applied to the actuators) is 5.3 fs. The corresponding measured and recovered SHG FROG traces are displayed in Figure 1.1.4.3(a) and (c), while the spectral phase and temporal intensity profile are shown in Figure 1.1.4.3e and f, respectively.

Provided the actual deflection of the membrane is well-calibrated as a function of applied voltage, the task of attaining the ultimate pulse compression becomes straightforward [6] as the phase setting opposite to the one depicted in Figure 1.1.4.3e (dash-dotted curve) corresponds to an ideal pulse compression. In practice, however, the perfect calibration of the membrane deflection is very cumbersome. Therefore, we have relied on a feedback with iterative optimization based on SHG

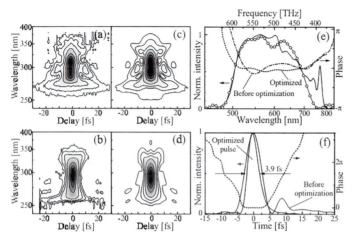

FIGURE 1.1.4.3 Overview of pulse shaping results. (a) and (b) Depict measured SHG FROG traces before and after adaptive phase correction, respectively. Corresponding retrieved traces are displayed in (c) and (d). Contour lines in (a–d) are drawn at values 0.02, 0.05, 0.1, 0.2, 0.4, 0.6, and 0.8 of the FROG peak intensity. (e) Shows fundamental spectrum (shaded contour) measured at the crystal location in the FROG apparatus and the spectrum recovered by the FROG retrieval algorithm (open circles). Dash-dotted curve represents spectral phase prior to the shaping, while dashed curve shows the optimized phase. (f) Initial (solid curve) and optimized (shaded contour) temporal intensity profiles. Dashed curve depicts temporal phase of the optimized pulse.

FROG [6,17]. To maximize the SH signal at every wavelength, our algorithm employs Brent's search method [18], looped over 13 independent actuator settings. The automated iterative phase adjustment is typically complete within several minutes. An example of a FROG measurement following the described optimization is presented in Figure 1.1.4.3(b) and (d).

1.1.4.3 ANALYSIS AND DISCUSSION

The retrieved spectral phase shows almost negligible deviations from a flat phase throughout most of the bandwidth. However, it must be noted, that the 5-μm thickness of the BBO crystal used for pulse diagnostic was not sufficient to cover the whole bandwidth at one time. Abrupt spectral variations of SH conversion efficiency, regardless of any FROG data corrections, lead to a total loss of information from both the high and low-frequency wings of the spectrum, as is evident from the comparison of the measured and retrieved spectra in Figure 1.1.4.3e. The truncated SH bandwidth introduces system error on the FROG trace [19], causes the inversion algorithm to distort the real pulse shape, and generally results in poor convergence. The FROG error was calculated to be 0.0068 on a 128×128 matrix for the trace in Figure 1.1.4.3b. Considering this experimental uncertainty, it can be safely concluded that the FWHM duration of the pulse obtained in our system is 4 fs within a 10% error margin. The accuracy is estimated from numerical simulations [19] and the dispersion of FROG inversion results of the experimental data. Further reduction of the pulse duration could be expected after a modification of the dispersion line.

1.1.4.4 CONCLUSION

Thus, it demonstrated achromatic phase matching of a 300-nm-wide parametrically amplified spectrum [20]. The use of adaptive optics allowed compressing the resulting pulses to a 4-fs duration thereby providing a tool for ultrafast nonlinear spectroscopy with the shortest time resolution available to date. The research presented in this section is reproduced and adapted with permission from [Ref. 20] ©The Optical Society. *Opt. Lett.* **27**, 306 (2002) was conducted by collaborative work of the following people: A. Baltuska, T. Fuji, and T. Kobayashi.

REFERENCES

1. G. M. Gale, M. Cavallari, T. J. Driscoll, and F. Hache, *Opt. Lett.* **20**, 1562 (1995).
2. G. Cerullo, M. Nisoli, S. Stagira, and S. De Silvestri, *Opt. Lett.* **23**, 1283 (1998).
3. A. Shirakawa, I. Sakane, M. Takasaka, T. Kobayashi, *Appl. Phys. Lett.* **74**, 2268 (1999).
4. M. Zavelani-Rossi, G. Cerullo, S. De Silvestri, L. Gallmann, N. Matuschek, G. Steinmeyer, U. Keller, G. Angelow, V. Scheuer, and T. Tschudi, *Opt. Lett.* **26**, 1155 (2001).
5. D. Zeidler, T. Hornung, D. Proch, and M. Motzkus, *Appl. Phys. B* **70S**, S125 (2000).
6. E. Zeek, K. Maginnis, S. Backus, U. Russek, M. Murnane, G. R. Mourou, H. Kapteyn, and G. Vdovin, *Opt. Lett.* **24**, 493 (1999).
7. M. Armstrong, P. Plachta, E. Ponomarev, and R. J. D. Miller, *Opt. Lett.* **26**, 1152 (2001).
8. A. Baltuška, Z. Wei, M. S. Pshenichnikov, D. A. Wiersma, R. Szipöcs, *Appl. Phys. B* **65**, 175 (1997).
9. P. Di Trapani, A. Andreoni, C. Solcia, P. Foggi, R. Danielius, A. Dubietis, and A. Piskarskas, *J. Opt. Soc. Am. B* **12**, 2237 (1995).
10. V. G. Dmitriev, G. G. Gurzadyan, and D. N. Nikogosyan, *Handbook of Nonlinear Optical Crystals*, 3rd ed., Springer-Verlag, Berlin, p. 96 (1999).
11. The effective phase-matching width can be additionally broadened in the regime of strong pump depletion.
12. B. A. Richman, S. E. Bisson, R. Trebino, M. G. Mitchell, E. Sidick, and A. Jacobson, *Opt. Lett.* **22**, 1223 (1997).
13. B. A. Richman, S. E. Bisson, R. Trebino, E. Sidick, and A. Jacobson, *Opt. Lett.* **23**, 497 (1998).
14. T. S. Sosnowski, P. B. Stephens, and T. B. Norris, *Opt. Lett.* **21**, 140 (1996).
15. T. Taft, A. Rundquist, M. Murnane, I. Christov, H. Kapteyn, K. DeLong, D. Fittinghoff, M. Krumbügel, J. Sweetser, and R. Trebino, *IEEE J. Select. Topics Quantum Electron.* **2**, 575 (1996).
16. P. O'Shea, M. Kimmel, X. Gu, and R. Trebino. *Opt. Express.* **7**, 342 (2000).
17. T. Brixner, M. Strehle, and G. Gerber, *Appl. Phys. B* **68**, 281 (1999).
18. W. H. Press, S. A. Teukolsky, W. T. Vetterling, and B. P. Flannery, *Numerical Recipes in C*, 2nd ed., Cambridge University Press, New York, p. 402 (1996).
19. A. Baltuška, M. S. Pshenichnikov, D. A. Wiersma, *IEEE J. Quantum Electron.* **35**, 459 (1999).
20. A. Baltuska, T. Fuji, and T. Kobayashi, *Opt. Lett.* **27**, 306 (2002).

1.1.5 Ultrafast Laser System Based on Noncollinear Optical Parametric Amplification for Laser Spectroscopy

1.1.5.1 INTRODUCTION

Ultrafast optical science is rapidly evolving in multidisciplinary fields: it has the capability of performing pump/probe-type experiments; namely excite(pump) samples with femtosecond light pulses and detect(probe) the subsequent evolution on ultrashort time scales, which opened up new fields of research in physics, chemistry, and biology [1–5], particularly for the study of dynamics and reactions in various organic molecules [6–8]. In this kind of research, the dynamics process is considered to be from the femtosecond to the picosecond time scale. Therefore, to investigate these processes, laser pulses with a shorter duration than the time range of ultrafast dynamics phenomena in the interesting samples are desirable. Most organic materials have absorptions in the visible ranges [9], and sub-15 fs ultrafast visible pulses generated from a tunable noncollinear optical parametric amplification (NOPA) have been widely used to study this kind of real-time spectroscopy for lots of molecules [10–13].

The available tunability can be used to probe the transient species by photo-inducing optical transitions occurring at various wavelengths [14,15]. The wider the probe spectrum, the more spectral signatures of the transient species involved in the photo process can be monitored simultaneously, which significantly facilitates the assignment of the underlying reaction mechanism. It is of great importance to generate visible-pump/near-infrared (NIR)-probe system for studying the photo-induced dynamics in many photoelectronic materials because they have strong absorptions in the visible range and the generated electron absorption is in the NIR spectral region [16–18]. The alternative approach to cover a much wider range in the NIR is to use a supercontinuum generation. This well-known phenomenon can be induced by focusing an ultrafast laser pulse under the proper conditions into a wide variety of optically transparent nonlinear media, like gases, liquids, photonic crystal fibers, and solids [19]. For example, a femtosecond white-light continuum is available at wavelengths shorter than 1.6 μm and is generated by a 10-mm path length cell containing CCl_4 or some other liquids or some transparent solid materials. A coherent supercontinuum generated from a photonic crystal fiber is applied to an NIR coherent anti-Stokes Raman spectroscopy (NIR-CARS) microscopy, and a clear CARS image of polystyrene microsphere has been obtained using the CH_2-stretching band [20]. Even though white light generation has been used in time-resolved studies, most of the generated supercontinua obtained from bulk materials are focused on the visible range.

In this subsection, we describe the generation of broad, tunable, sub-6 fs visible pulses by NOPA in a 1 mm-thick type I BBO crystal, and a white-light continuum for the NIR spectral region by a 12 mm-thick sapphire rod is also obtained. These both have very broad and smooth spectrums, high temporal and spatial coherence, and very high pulse-to-pulse energy stabilities for spectroscopies. Two pump-probe experiments are completed using these two visible-pump/NIR-probe and visible pump/visible-probe spectroscopies. The results prove that the obtained ultrafast visible pulse and

DOI: 10.1201/9780429196577-7

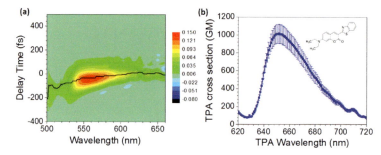

NIR white light continuum are very useful for many kinds of ultrafast time-resolved spectroscopy experiments. They also illustrate that the setup could not only be used to study the visible-pump/NIR-probe spectroscopy but also visible-pump/visible-probe spectroscopy. The NIR-probe beam that is generated by the unconverted 800 nm laser pulse of the NOPA system can fully utilize the energy light in the experiment.

1.1.5.2 EXPERIMENTAL

A schematic of the experimental pump-probe setup is illustrated in Figure 1.1.5.1. About 800 µJ after a beam splitter using a commercial Ti: sapphire regenerative amplifier (0.9 mJ, 50 fs, 5 kHz at 800 nm) [21,22] passes through a 45° fused silica prism with an incident angle of 49°, a telescope collimates the spectral lateral walk off and images the titled fronts on the focal plane, ensuring the maximum spatial overlap between the pump and the signal fronts. Then, the pulse is frequency doubled in a 200 µm-thick beta barium borate (BBO, type I, $\theta = 29.2°$) crystal. A small fraction of the fundamental energy is converted to generate a continuum of white light seed in a sapphire crystal plate (1 mm thick, used for the visible spectral region). The longer wavelength part in the seed beam is filtered out through a short-pass filter ($T > 90\%$ at 500–750 nm, $R > 99\%$ at 800–850 nm) to prevent any undesired amplification. A pair of chirped mirrors ($-25 \, \text{fs}^2$/bounce, Layertec) is used to pre-compress and temporally shape the white light, which will have a better overlap in time and in the lateral space with the pump pulse to be amplified in the NOPA. To get better beam profiles and higher pulse energies, a two amplification-stage NOPA system (1 mm-thick BBO crystal, type I) is employed with a suitable tilt angle of 6.4° between the seed and the pump. The pulses can be continuously tuned in the spectral region from 510 to 750 nm for the most interesting excitation wavelengths by making a suitable choice for the angle of noncollinearity between the pump and seed, and also by the phase-matching angle of the BBO [23].

1.1.5.3 RESULTS AND DISCUSSION

The chirped output visible pulses with energies of several microjoules are compressed with a pair of chirped mirrors and Brewster prisms. The most convenient way of compression of the chirped visible pulse is to use the fused silica as the prism material in the compressor. It is because above 550 nm, the length of the compressor can be substantially shortened with a kind of spectralite flint (SF10) instead of the fused silica prisms. A very helpful feature shown in Figure 1.1.5.1 is the use of two mirrors acting as a retro reflector inside the prism compressor. This folding allows for a quick and easy adaptation of the length of the compressor to a change of the NOPA center wavelength. In this way, the tunability of the NOPA can really be exploited with a minute change in the alignment of the whole setup. Figure 1.1.5.2 After compression, pulse durations of sub-6 fs can be routinely achieved throughout the visible region. The spectral profile of the output pulse after the prisms at different central wavelengths is shown in Figure 1.1.5.2a, and the insert is the compressed pulse duration according to the spectrum represented by the pink line. The spectral width of the

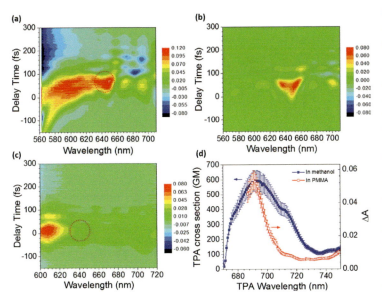

FIGURE 1.1.5.2 (a) Spectra of the different visible compressed wavelength pulses. The green line represents the spectra obtained by pumping the zinc chloride aggregate, and the red line represents the spectra for the PTB7-Th polymer in the experiments. The insert is the shortest compressed pulse duration according to the pink line (the black line is the duration measured, while the red line and blue fit are the Lorentzian fit and Gaussian fit, respectively). (b) Spectra of NIR. Insert is the stationary absorption spectrum of the PTB7-Th polymer (red line).

NOPA pulses can be adjusted depending on the specific needs of the spectroscopic experiment. For example, the green line in Figure 1.1.5.2a is the output spectrum which is adjusted according to the absorption spectrum of the zinc chlorin aggregate.

The main mechanism of ultrafast continuum generation is strong self-phase modulation enhanced by the self-steepening of the pulse [24]. In the present study, the Ti:sapphire fundamental is used for the generation of a visible seed continuum by focusing it into a thicker sapphire plate to generate the NOPA beam. We observed that the use of a longer sapphire length ($l=10$–20 mm) yields a broader continuum, from the visible into the NIR.

Therefore, to extend the spectra into the NIR to probe the exciton dynamics, we use the white light generated by focusing the unconverted 800 nm beam from the NOPA into a sapphire rod (12 mm thick, used for the infrared spectral region) with a lens F_2 ($f=500$ mm), as shown in Figure 1.1.5.1. The unconverted 800 nm beam collected after the second-harmonic generation is used as a pump beam for the visible NOPA in the NIR probe experiment. The specific needs of the spectroscopic experiment should be met by the appropriate choice and alignment of the length of the sapphire rod and filter stage. After the sapphire rod, the white light went through a high-pass filter ($T>90\%$ above 900 nm, $R>99\%$ below 900 nm), which can obtain the right spectrum to probe the excited signal and prevent the undesired amplification of the sample. The output spectrum is measured by using lens F_4 ($f=50$ mm) to image the output surface of the medium onto the entrance slit of the spectrograph. Briefly, the NIR part of the broad and smooth supercontinuum (950–1050 nm) is obtained and shown in Figure 1.1.5.2b. To measure the pulse at various stages of propagation in the sapphire rod, focusing lens F_2 is moved to a translation stage, such that the desired stage of propagation coincides with the output surface of the medium. Together with the NOPA, the setup allows us to study examples of a typical visible pump/NIR-probe experiment. Since both the NOPA and NIR beams come from the same laser source, they can synchronize with each other easily. The maximum pulse energy of the NIR-probe beam can be as high as 1 μJ. In this experiment, we do not need the full energy of the NIR beam; we just need to use a 120 nJ energy (150 fs) pulse to probe the sample, since the pump pulse of the visible beam is about 80 nJ at this moment. After passing through the sample at an external angle of ~6° with respect to the pump, the probe is dispersed in a polychromator and guided to a 128-channel lock-in amplifier connected with photodetector [25]. A chopper wheel in the pump beam blocks every second excitation pulse so that changes in the optical density of the sample can be measured.

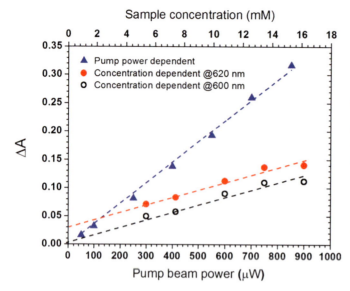

FIGURE 1.1.5.3 (a)Two-dimensional pattern of different absorbances (ΔA), which are dependent on the delay time (t) from −200 to 2800 fs in the whole probe spectral region (670–760 nm); (right) time trace of different absorbance rates. (b) Two-dimensional pattern of different absorbances (ΔA) dependent on the time delay from −0.4 to 100 ps in the whole probe spectral region (960–1040 nm); (right) time trace of different absorbances.

As an example, the obtained NOPA pulses are divided into two beams and used for a visible-pump and visible probe experiment. A sample of zinc chloride aggregate [26] is used in this experiment. The pulse energies of the pump and probe are about 30 and 3 nJ, respectively, with spectral range extending from 518 to 786 nm. The time trace of the normalized transmittance changes $\Delta T/T$ is obtained as a function of the pump-probe time delay from −200 to 2800 fs with every 1 fs step. The experiment is performed at physiologically relevant temperatures (295 ± 1 K). Figure 1.1.5.3a shows that the two-dimensional different absorptions $\Delta A = -\log(1 + \Delta T/T)$ is dependent on the pump-probe time delay in the spectral range between 670 and 780 nm, and the right side shows an example of a ΔA trace. We can observe that the fast relaxation process from the higher multi-exciton state to the one-exciton state is found to take place in 100 ± 5 fs, which shows that only the sub-6 fs ultrafast pulse can accurately detect such a short relaxation time. Detailed analyses have been made in another chapter [27].

A sample of organic photovoltaic material PTB7-Th polymer [28,29] is used in the visible-pump and NIR-probe experiment. According to the peak absorption of the PTB7-Th polymer, which is shown to be located around 700 nm in Figure 1.1.5.2b, the center wavelength of the pump beam is shifted to a little longer wavelength, as shown in the red line in Figure 1.1.5.2a, which proves there is a very good overlap between the laser spectrum and the absorption spectrum of the sample. The pulse energies of the pump beam and probe beam are 80 and 120 nJ, respectively. The time trace of the normalized transmittance changes ($\Delta A(\lambda, t) = -\log(1 + \Delta T(t)/T(t<0))$) is obtained as a function of the pump-probe time delay from −0.4 to 100 ps with every 20 fs step. The experiment is performed at physiologically relevant temperatures (295.1 K). Figure 1.1.5.3b shows the different absorption spectra $\Delta A(\lambda, t)$ of the exciton depending on the pump-probe time delay (t) in the spectral range between 960 and 1040 nm. On the right is an example of a ΔA trace, and the corresponding decay time of the PTB7-Th polymer is determined to be about 5 ps.

1.1.5.4 CONCLUSION

In conclusion, stable sub-6 fs pulses at an optimal central wavelength are obtained using a tunable NOPA source compressed with a pair of chirped mirrors and Brewster prisms. Meanwhile, a white light continuum in the NIR range from 900 to 1100 nm is successfully generated by focusing the unconverted 800 nm beam of the NOPA onto a sapphire rod. The visible-pump/visible probe and visible-pump/NIR-probe experiments using a zinc chloride aggregate and organic photovoltaic material PTB7-Th polymer as samples are studied with these short pulses, respectively. The results show that the induced absorptions and oscillating features in the time traces of the absorbance change with different periods can be clearly observed, which proves that the ultrashort laser pulses extremely are useful for many kinds of spectroscopy experiments. It was proposed that the dynamics observed using these ultrashort pulses in many materials can potentially be extended to study even more complicated processes, such as novel artificial synthetic materials, energy conversion systems in a natural photosynthesis system, and vision process in photopigments [30], biomimetic systems, and artificial materials. The cooperation research work presented in this subsection was performed by the following people: D. Han, Y. Li, J. Du, K. Wang, Y. Li, T. Miyatake, H. Tamiaki, T. Kobayashi, and Y. Leng [30].

REFERENCES

1. G. Cerullo and S. De Silvestri, *Rev. Sci. Instrum.* **74**, 1 (2003).
2. J. Du, Z. Li, B. Xue, T. Kobayashi, D. Han, Y. Zhao, and Y. Leng, *Opt. Express* **23**, 17653 (2015).
3. Y. Li, J. Hou, Z. Jiang, and L. Huang, *Chin. Opt. Lett.* **12**, 031901 (2014).
4. T. Kobayashi, J. Du, W. Feng, and K. Yoshino, *Phys. Rev. Lett.* **101**, 037402 (2008).
5. Z. Guo, D. Lee, R. D. Schaller, X. Zuo, B. Lee, T. Luo, H. Gao, and L. Huang, *J. Am. Chem. Soc.* **136**, 10024 (2014).
6. J. Du, Z. Wang, W. Feng, K. Yoshino, and T. Kobayashi, *Phys. Rev. B* **77**, 195205 (2008).
7. S. M. Falke, C. A. Rozzi, D. Brida, M. Maiuri, M. Amato, E. Sommer, A. De Sio, A. Rubio, G. Cerullo, E. Molinari, and C. Lienau, *Science* **344**, 1001 (2014).
8. T. Kobayashi, Z. Nie, B. Xue, H. Kataura, Y. Sakakibara, and Y. Miyata, *J. Phys. Chem. C* **118**, 3285 (2014).
9. J. Liu, Y. Kida, T. Teramoto, and T. Kobayashi, *Opt. Express* **18**, 4664 (2010).
10. H. Kano, T. Saito, and T. Kobayashi, *J. Phys. Chem. B* **105**, 413 (2001).
11. O. A. Sytina, I. H. M. van Stokkum, D. J. Heyes, C. N. Hunter, R. van Grondelle, and M. L. Groot, *J. Phys. Chem. B* **114**, 4335 (2010).
12. J. Du, K. Nakata, Y. Jiang, E. Tokunaga, and T. Kobayashi, *Opt. Express* **19**, 22480 (2011).
13. J. Du, T. Teramoto, K. Nakata, E. Tokunaga, and T. Kobayashi, *Biophys. J.* **101**, 995 (2011).
14. V. de Waele, M. Beutter, U. Schmidhammer, E. Riedle, and J. Daub, *Chem. Phys. Lett.* **390**, 328 (2004).
15. U. Megerle, I. Pugliesi, C. Schriever, C. F. Sailer, and E. Riedle, *Appl. Phys. B* **96**, 215 (2009).
16. U. Schmidhammer, P. Jeunesse, G. Stresing, and M. Mostafavi, *Appl. Spectrosc.* **68**, 1137 (2014).
17. C. Yi and K. L. Knappenberger, *Nanoscale* **7**, 5884 (2015).
18. J. Du and T. Kobayashi, *Chin. Opt. Lett.* **9**, S010601 (2011).
19. R. R. Alfano, *The Supercontinuum Laser Source*, Springer (2005).
20. V. Nagarajan, E. Johnson, P. Schellenberg, W. Parson, and R. Windeler, *Rev. Sci. Instrum.* **73**, 4145 (2002).
21. A. Shirakawa, I. Sakane, and T. Kobayashi, *Opt. Lett.* **23**, 1292 (1998).
22. A. Baltuska, T. Fuji, and T. Kobayashi, *Opt. Lett.* **27**, 306 (2002).
23. E. Riedle, M. Beutter, S. Lochbrunner, J. Piel, S. Schenkl, S. Spörlein, and W. Zinth, *Appl. Phys. B* **71**, 457 (2000).
24. G. Yang and Y. R. Shen, *Opt. Lett.* **9**, 510 (1984).
25. T. Teramoto, J. Du, Z. Wang, J. Liu, E. Tokunaga, and T. Kobayashi, *J. Am. Chem. Soc. B* **28**, 1043 (2011).
26. T. Miyatake and H. Tamiaki, *Coord. Chem. Rev.* **254**, 2593 (2010).
27. D. Han, J. Du, T. Kobayashi, T. Miyatake, H. Tamiaki, Y. Li, and Y. Leng, *J. Phys. Chem. B* **119**, 12265 (2015).
28. Y. Lin, Z. Zhang, H. Bai, J. Wang, Y. Yao, Y. Li, D. Zhu, and X. Zhan, *Energy Environ. Sci.* **8**, 1 (2015).
29. P. Cheng, Y. Li, and X. Zhan, *Energy Environ. Sci.* **7**, 2005 (2014).
30. D. Han, Y. Li, J. Du, K. Wang, Y. Li, T. Miyatake, H. Tamiaki, T. Kobayashi, and Y. Leng, *Chinese Opt. Lett.* **13**, 1271401(1-4) (2015).

1.1.6 Development of Ultrashort Pulse Lasers for Ultrafast Spectroscopy

1.1.6.1 INTRODUCTION

Curiosity is a driving force of science development. Humans have been motivated to investigate microscopic structures of materials even from the ancient Greek era. The elementary composites of the material are called "atom" of which meaning is "not possible (a) to divide (tom)". This motivation of research to divide materials into most fundamental "inseparable" elements was a driving force. This motivation and movement are forming the frontier in science even now. Together with the interest in the elementary particles in the ultimate size, the molecular systems are of interest from the viewpoint of material properties that are sensed by human eyes and other sensing organs such as ears, noses, toungs, and skins in everyday life. Among these sensing, vision is special. It provides highest fraction of information from the surrounding which is reaching nearly 70%. In the vison electronic excitation in receptor molecules is involved. Surrounding information from the furthest distance can be sensed in vision, even though the energy density on the earth is relatively low. Among animals, dogs are more sensitive to smell than light.

The optical properties of material are determined by the static properties and dynamics of electrons in the outmost orbital in an atom. To understand these, it is necessary to investigate the motion of the outmost electrons whose fastest processes are in the femtosecond range of which the Fourier transform is in the near ultraviolet (UV), visible, and near infrared (NIR) ranges. The femtosecond dynamics study using ultrashort optical pulses can provide the most direct information of the electronic states.

In the last two decades, the development of the mode-locked femtosecond laser has enabled the observation of ultrafast dynamics to elucidate the ultrafast processes in molecules and various condensed-phase materials, opening a new research field of femtoscience [1]. To disclose the processes in nanomaterials and molecular systems where elementary processes are taking place with ultrahigh speed, it is necessary to time-resolve the processes with high temporal resolution. The primary processes in electronic relaxations in molecules and solid-state materials like metals and semiconductors are in the range of femtosecond. Nuclear motion associated with molecular vibration and lattice vibration in the optical phonon branch are in the femtosecond and subpicosecond-picosecond regime, respectively. The most direct method to study such processes is to use pump/probe-type spectroscopy using pump and probe pulses with femtosecond duration. Efficient photochemical reactions competing with electronic relaxation can also take place in the subpicosecond range. Hence, to observe the processes directly, it is required to utilize a short enough pulse. To fulfil this requirement, we have been developing various techniques to generate ultrashort pulses in the NIR-visible and UV-DUV (deep UV) spectral ranges as described in this compact review article. An intense FT (Fourier transform) limited pulse in these wavelengths triggers the excitation of electrons in the outmost molecular orbital in most of the organic aromatic molecules and interband transition in semiconductors. Another FT limited pulse probes the change induced by the pump. A chirped pulse instead of an FT limited pulse can be used by using a multi-channel broad spectral range, to be discussed later in the present paper, utilizing simultaneous resolution of time and spectrum.

DOI: 10.1201/9780429196577-8

This paper is organized as follows. The explanation of importance of the ultrashort pulse is described in Section 1.1.6.1. Section 1.1.6.2 briefly introduces the development of the ultrashort pulse laser as a tool for studying ultrafast processes. In Section 1.1.6.3, electronic relaxation and vibrational dynamics which are the principle information obtained by ultrafast spectroscopy, are described. In Section 1.1.6.4, the principles and advantages of broad-band time-resolved spectroscopy that can provide both electronic relaxation and vibrational dynamics are explained. In the following Sections 1.1.6.5 and 1.1.6.6, ultrashort visible pulse laser and ultrashort ultraviolet (UV) laser, respectively, both based on parametric processes, are discussed. Finally, concluding remarks are made in Section 1.1.6.7.

1.1.6.2   LIGHT SOURCES FOR STUDYING ULTRAFAST PROCESSES

To study the mechanism of fast chemical reactions, flash photolysis was developed by R. G. W. Norrish and G. Porter, who were awarded the Nobel Prize in Chemistry in 1967 [2]. By flash photolysis, the time-resolution is limited by the flash duration, which is in the order of a few nanoseconds at best. Development of the mode-locking method has enabled to generate an ultrashort laser pulse much shorter than 1 ns. Among various mode-locked lasers operated by several mechanisms, a self-mode locked Ti:sapphire laser can generate an ultrashort pulse stably and has been used most widely. It generates visible-NIR pulse with a wavelength range of around 800 nm. The Ti:sapphire laser system can generate an ultrashort visible-NIR pulse with sub-10 fs width or its second harmonics (SH), which can have sub-20 fs duration. Ultrafast dynamics of the various chemical reactions and carrier dynamics in a condensed matter like semiconductors have been studied using the ultrashort visible-NIR pulse or its SH. Elucidation of ultrafast dynamics in various systems accelerates development in various applications like photosensors [3–5], ultrafast optical switches [6–8], and ultrafast optical memories [9–11]. Many interesting ultrafast dynamics in biological systems including photosynthesis [12–14] and vision processes [15–17], were studied by ultrafast spectroscopy using visible or UV pulse lasers. In these experiments, sometimes the intensities of the generated pulses are not stable due to the various nonlinearities in the generation process. Due to instability-induced intensity noise, the results do not have enough signal-to-noise ratio (S/N), which is desired to be improved to discuss the detailed mechanism of the ultrafast processes of interest. Therefore, it is quite important to develop an ultrashort visible pulse laser whose intensity is high and stable enough to be used as a light source for time-resolved spectroscopy of visible and UV photo-induced dynamics.

To extend the available ultrashort laser wavelength and to obtain even shorter pulse, several nonlinear optical (NLO) processes of the second order and the third order are frequently utilized. Examples of the second-order NLO processes are the second-harmonic generation (SHG) and optical parametric amplification (OPA). Examples of the third-order process are third harmonic generation and parametric four-wave mixing. For the time-resolved spectroscopic measurement, various NLO processes such as optical gating by optical Kerr effect are utilized.

1.1.6.3   ELECTRONIC RELAXATION AND VIBRATIONAL DYNAMICS

Thanks to the high time resolution of ultrafast spectroscopy, it is possible to time-resolve not only ultrafast electronic relaxation among different electronic states including some chemical reactions but also vibration dynamics in real time. The meaning of real-time here is that the time-course of vibration amplitude can be visualized in the time domain. One of the typical molecular vibration modes, i.e., the C–C double bond stretching, has an oscillation period of about 20 fs. Therefore, the transient absorption (TA) signal measured by the ultrashort pulsed laser can reflect both the electronic relaxation and vibrational dynamics, temporally resolving the nuclear motion in the molecular vibrations. A pulse duration as short as sub-10 fs is short enough to impulsively excite almost all molecular vibration modes in organic molecules except some OH and CH stretching modes with vibration periods shorter than 10 fs. Even though these high-frequency stretching modes cannot be fully resolved, they are in the range of detection with reduced amplitudes by the ratio factor of about (pulse duration time)/(vibration period). Compared with other methods, such as time-resolved vibration spectroscopy in the frequency domain including time-resolved Raman scattering and IR absorption spectroscopies, TA spectroscopy by ultrashort pulse has the advantage that it is possible to obtain vibrational phases and to study the electronic dynamics and vibrational dynamics simultaneously in the same experimental condition. Taking advantage of this method, it is possible to elucidate ultrafast dynamics in various photo-induced dynamic processes. Both sub-5 fs visible-near infrared (VIS-NIR) pulse and sub-10 fs deep UV (DUV) pulse are developed as discussed later and it has become possible to study ultrafast dynamics in various materials.

1.1.6.4   PRINCIPLES AND ADVANTAGES OF BROAD-BAND ULTRAFAST SPECTROSCOPY

The ultrafast signal change in the femtosecond region can be measured by the pump-probe method using the ultrashort femtosecond pulse. Spectroscopic data are discussed in terms of ultrafast TA signal in the extended spectral range, which contains contributions from various mechanisms depending on the probed wavelength. Quantitative analysis of TA is made by difference absorbance change defined by. Here, $\Delta A(\omega) = \log [I_n(\omega)/I_w(\omega)]$ Here $I_n(\omega)$ and $I_w(\omega)$ are the transmitted probe light intensity without and with excitation, respectively, at the probe photon angular frequency ω. $\Delta A(\omega)$ can be positive or negative depending on the phenomenon triggered by the pump. Positive values correspond to photoexcitation-induced absorption (IA) due to transient absorption from the excited state to a higher state or due to some (photo) chemical species created by the pump pulse. Positive values are induced by the stimulated emission (SE) from the excited state to the ground state or by the ground-state bleaching (GSB) due to depletion of the ground state, which are mixed together in the measured spectral range. They are difficult to be identified from the measurement at a single probe wavelength. The probe wavelength dependence can be studied in principle by repeating the pump-probe measurement at each probe wavelength. However, it takes a long time to cover the spectral region of interest, and the conditions of sample and pump and probe pulses for each wavelength measurement may not be maintained during the long time measurement.

Therefore, it is desirable to obtain broadband spectral information at one time because of the long measurement time and systematic errors introduced by the fluctuation of the ultrashort pulse in the case of one-by-one wavelength measurement by single point detection using a photomultiplier or photodiode. We have developed two detection schemes: the multi-channel lock-in detector [18] and the diode array detection system [19,20].

The broadband probe spectra of femtosecond pulse and white light continuum are of great advantage for spectroscopy. The broadband spectroscopic data provide invaluable information of the species appearing in the time-resolved data. For example, they enable discrimination of the congested spectrum and separate the signal into IA, SE, and GSB from the wavelength-dependent dynamics of the ultrafast TA signal [18–20].

To study the electronic relaxation and vibrational dynamics simultaneously, the transient absorption (TA) signal should be measured with a fine delay step (to resolve molecular vibration dynamics) and broad delay region (to observe electronic relaxation). Typically, the measurement of broad-band time-resolved spectrum takes about 1 h to a few 10 min. Then, in some cases, especially in biological samples, photo-damage effects may be accumulated in the samples, even on a single-scan time scale. Also, it has been hard to perform the TA spectroscopy using ultrashort pulse that cannot maintain stability enough to guarantee the reliability of electronic relaxation dynamics information. This is because the gradual intensity change of the laser may distort the slow electronic dynamics, which can take place in the delay-stage scanning time similar to the time of laser intensity change, even if all controllable environmental conditions are carefully adjusted for about 1 h to maintain the stability of the ultrashort pulse. This is because of the unavoidable changes in the intensity of pump and probe pulses obtained by many nonlinear processes. Then, the time-trace data thus obtained may suffer from artifacts in the electronic relaxation dynamics. To overcome this difficulty, we have developed a fast-scan femtosecond time-resolved broadband spectroscopy system, which can complete the full-trace scan measurement more than 100 times faster than the previous method. Intensity changes of pump and probe pulses during such a short scan are more than an order of magnitude reduced. Therefore, scan trace can be averaged without the artefacts. The details of this system are described elsewhere [19,20].

The result of TA spectroscopy shows the probe wavelength dependence reflecting the contributions from the ground state depletion, the excited states, and intermediates. In the case of single wavelength measurement, their lifetimes are hard to be evaluated at the wavelength where the fractional signal intensity of the corresponding species may be too small due to the overlap of other contributions. Thus, it is necessary to analyze the TA spectroscopy data obtained simultaneously at all probe wavelengths. Global fitting analysis—for example, [21]—is one of the solutions. However, the global fitting analysis cannot be used when the lifetime itself is dependent on the probe wavelength. We have introduced another method of two-dimensional correlation spectroscopy, which is described elsewhere [22].

Various nonlinear optical (NLO) phenomena are involved not only in the generation and characterization of ultrashort pulse but also in the ultrafast spectroscopy measurement. In the following section, how NLO processes are utilized is described in some detail.

1.1.6.5 ULTRASHORT VISIBLE PULSE GENERATION BASED ON NON-LINEAR OPTICAL PARAMETRIC AMPLIFIER (NOPA)

Our group and other groups developed a broadband ultrashort pulse with a spectral range covering 520–720 nm achieving sub-10 fs pulse duration by the optical parametric amplifier (OPA) process in noncollinear configuration [23–29]. The scheme of the noncollinear optical parametric amplifier (NOPA) is illustrated in Figure 1.1.6.1.

The linear polarization of the NIR pulse was rotated by an achromatic half-wave plate, and then the NIR pulse was separated into two copies by a polarization beam splitter (PBS). Adjusting the rotation angle of the half-wave plate, we set the power ratio between the two NIR pulses as 10:1. The higher intensity pulse of about 680 mW from the PBS was focused into a β-BaB$_2$O$_4$ (BBO) crystal to

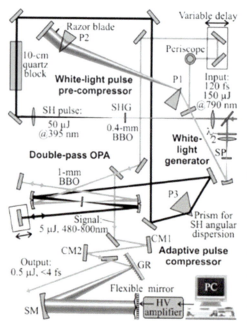

FIGURE 1.1.6.1 Diagram of NOPA system. SP, a plate of sapphire for broad-band continuum generation to be used as a signal beam with spectrum >450 nm; 0.4-mm BBO, nonlinear crystal to generate second harmonic used for signal; 1-mm BBO, nonlinear crystal for noncollinear parametric amplification; 10-cm Quartz block, for pulse stretching; TP, a prism for pulse-front tilting; cm, chirped mirror; P1, P2, prism pair for pulse compression; P3, prism for introducing angular dispersion to the SH pulse; the elements of thin rectangle shape without a label are plane >99% reflection mirrors.

generate second harmonic UV pulse [23]. The UV pulse was used as a pump in the NOPA process. The lower intensity pulse from the PBS was focused into a 2 mm-thick sapphire plate to generate spectrally broadband pulse by self-phase modulation (SPM), which is the third-order NLO process. This 10:1 ratio is to obtain as high an NOPA pulse energy as possible to be used in the pump–probe experiment, which requires many optical components such as chirped mirrors and prism pairs (see Figure 1.1.6.1). The shortest edge can reach up to ~450 nm close to the UV region, and the longest wavelength is beyond 800 nm extending to >1.2 μm. Thus, the broadband SPM spectrum covers nearly all of the visible spectral region a part of NIR. However, the intensity of the visible spectral component is much lower than that of the original spectral component around 800 nm. For the application of the time-resolved spectroscopy, the broadband visible SPM light pulse should be amplified. For this purpose, this broadband pulse is used as a seed of signal in the OPA process, which is the second-order NLO process. A long wavelength cut-off filter (CF) blocking >750 nm is set after the sapphire plate to eliminate the intense 800 nm spike in the SPM spectrum. The other beam propagates through a 0.4 mm-thick BBO crystal (29.2° z-cut) to yield a second harmonic UV pulse for the pump of NOPA, satisfying the type-I ($o+o \rightarrow e$) phase match condition. BBO crystal is an excellent NLO crystal widely used in the field of ultrashort pulse laser research [30]. To achieve efficient and stable OPA, a 10-cm quartz block is used to stretch the pump pulse duration to 200 fs to be comparable to the white-light continuum seed (signal) pulse. This condition makes the interaction efficiency between the pump and signal robust against the jitter between them. A prism (TP) with 45° apex is set after the quartz block for pre-tilting the pulse front of the pump beam to satisfy the pule-front matching condition with the signal beam co-propagating with the pump beam in the crystal. Then, the pulse front pre-tilted seed is focused into a 1 mm-thick BBO crystal (31.5° z-cut) together with the UV pump pulse, which satisfies good spatial overlap with white-light continuum signal beam during the co-propagation. The optical delay line in white-light continuum path is adjusted to attain temporal overlap. To achieve higher signal pulse energy, the remaining pump and signal pulse energies are reflected after the first OPA processing and focused again in the BBO crystal at a slightly lower position. The temporal overlapping of pump and signal can be controlled by the optical delay line in the pump beam path.

The amplified signal just after the BBO crystal is about 5 μJ. The time duration of the amplified broadband visible pulse was compressed using a set of chirped mirrors and a deformable mirror, to be shorter than 10 fs in the sample. A pair of ultra-broadband chirped mirrors (CM1 and CM2) with the combination of the prism pair (P1 and P2), air and the beam splitter was designed to compensate GDD and TOD in the NOPA system. The chirped mirrors have high reflectivity ($R > 99\%$). The best noncollinear angle was obtained by adjusting the fluorescence ring width to be thinnest. Ultrashort visible pulse with 4.7 fs duration was generated by pulse-front-matched noncollinear optical parametric amplification [23]. The pulse width was as short as 3.9 fs with the spectrum 520–770 nm [24]. The optimized output pulse from the NOPA was characterized by the SHG FROG (second harmonic frequency-resolved optical gating) method [31], the results are shown in Figure 1.1.6.2.

The NOPA is tunable in wavelength at the expense of bandwidth, pulse duration, and/or energy. For example in real experimental conditions, the spectrum of non-collinear NOPA can be adjusted in such a way that the wavelength range extends to longer region, 556–753 nm, to achieve higher absorbance of some samples. In this case, the pulse duration is 7.0 fs. By using the NOPA system developed, the time resolution of the measurement to observe ultrafast electronic dynamics has reached as high as sub-10 fs. Moreover, the measured signal also contains periodically modulating component, which reflects wave packet motion moving on the potential energy surface in the period of molecular vibration. Thus, ultrafast electronic dynamics and vibrational dynamics can be studied simultaneously in the common experimental condition [32–37].

Further, we have developed a shorter pulse covering 430–850 nm, corresponding to nearly one-octave bandwidth. Also, the carrier envelope phase (CEP) of the pulse is stabilized, and it makes the pulse useful for studying the CEP effect on electronic and vibrational dynamics [38].

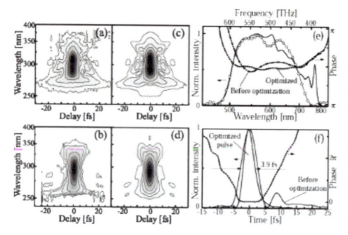

FIGURE 1.1.6.2 Characterization of output from NOPA (noncollinear optical parametric amplifier). (a,b) Measured SHG FROG (second harmonic frequency-resolved optical gating traces before and after adaptive phase correction using deformable mirror, respectively. Corresponding retrieved traces are displayed in (c,d). Contour lines in (a–d) are at the values of 0.02, 0.05, 0.1, 0.2, 0.4, 0.6, and 0.8 of the FROG peak intensity. (e) Shaded area, fundamental spectrum measured at the crystal location in the FROG apparatus; open circles, spectrum recovered by the FROG retrieval algorithm; dashed-dotted curve, spectral phase before shaping; dash-dotted curve, the optimized phase, Initial (solid curve) and optimized (shades area) temporal intensity profiles; dashed curve, temporal phase of the optimized pulse.

1.1.6.6   ULTRASHORT DEEP ULTRAVIOLET LASER

1.1.6.6.1   DUV PULSE GENERATION

Generally, the DUV laser pulse based on the Ti:sapphire laser is generated by the SFG or four-wave mixing (FWM) method through nonlinear materials. Both cases are usually two-step processes. The former is the combination of SHG and the following SFG (SH plus fundamental). The latter case is the combination of SHG and parametric FWM (SH and the following interaction of two SH photons and fundamental photon) and is going to be discussed in detail in the following. Previously, in several systems for sub-10 fs DUV pulse generation was reported in a group developing high-power lasers aiming attosecond pulse generation [39,40]. Also, an intense sub-10 fs DUV pulse is generated [41], but it has not yet been well applied to time-resolved spectroscopy. Such a situation may be because of the following several difficulties. First, in such high-power systems, the intense amplified spontaneous emission can be easily amplified to generate satellite pulses resulting in the multi-pulse structure. Second, when the DUV pulse is transmitted in the air, the effect of group velocity dispersion (GVD) is much more severe than the visible case, which leads to substantial pulse broadening, resulting in a distorted non-clean shape. These make the ultrashort pulse compensation more complicated and difficult. For the spectroscopy, a clean FT limited pulse is required for application. In the following, we describe the novel method developed to obtain such a clean nearly FT limited pulse.

If the pump and idler pulses in the FWM processes are both linearly chirped by appropriate control, the output of the FWM can be designed to be an FT-limited pulse. It is performed by a method called chirped pulse FWM (CPFWM). Energy conservation in the FWM process can be expressed by $\omega_{pump} + \omega_{pump} - \omega_{idler} = \omega_{signal}$, where ω_{pump}, ω_{idler}, and ω_{signal} are the pump, idler, and signal angular frequencies, respectively. In case when the pump and idler pulses are linear chirped, it is possible to obtain nearly-FT limited pulse as follows. At first, let us assume that the instantaneous angular frequencies of the pump and idler pulses are represented by $\omega_{pump}(t) = 2\omega_0 + \beta_{pump}t$ and $\omega_{idler}(t) = \omega_0 + \beta_{idler}t$, respectively. Here, ω_0 is the center angular frequency of the idler and β_{pump} and β_{idler} are the chirp rates of the pump and idler pulses, respectively. Then the energy conservation condition gives the DUV signal pulse frequency, $\omega_{signal}(t) = 3\omega_0 + (2\beta_{pump} - \beta_{idler})$ t in case their time origin is common. The frequency chirps of the idler and pump pulses cancel out each other when the frequency chirps of the two input pulses have the same signs. When $2\beta_{pump} - \beta_{idler} = 0$ is satisfied, no frequency chirp is induced in the amplified signal pulse. In the present experiment, frequency chirps were induced in the input pulses by spectrum broadening in the hollow fiber. The pump pulses are frequency chirped by a pair of gratings.

1.1.6.6.2 SUB-10 FS DUV LASER PULSE OBTAINED BY BROAD-BAND CPFWM

To generate a high-quality DUV pulse suitable for ultrafast spectroscopy, a broad-band chirped-pulse four-wave mixing method (BBCPFWM) was demonstrated [42,43]. The scheme of the system is depicted in Figure 1.1.6.3. The basic pulse source is a Ti: Sapphire laser (Spitfire Ace, Spectra Physics, Santa Clara, CA, USA.). The energy, duration, and repetition of the output pulse beam are 2.5 mJ, 35 fs, and 1 kHz, respectively. It is split into two beams. They have pulse energies of 900 and 300 μJ, the sum of which is smaller than 2.5 mJ due to beam pointing stabilizer mirror (PS), lens, beam splitter (BS), and half-wave-plate (λ/2) as shown in Figure 1.1.6.3. The 900 μJ pulse is introduced to a BBO crystal to generate a second harmonic (SH) near UV pulse. The UV pulse is negatively chirped after passing through two phase-delay controlled chirped mirrors (PM). The other fundamental pulse of 300 μJ is focused into hollow core fiber 2 filled with krypton gas (630 Torr) to obtain broadband supercontinuum in NIR. Then, the positive chirped NIR pulse and NUV pulse are both coupled into another hollow core fiber filled with argon gas (61 Torr), where they are spatially and temporally overlapping. In this second hollow fiber, a broadband and negative chirped DUV pulse is generated by the BBCPFWM process. Just after the second hollow fiber, a clean and well-compressed DUV pulse could be obtained at a certain distance assisted by the positive GVD in the air just after this second hollow fiber. After the collimation of the output beam, the NIR and NUV pulses are carefully eliminated by a set of four dichroic mirrors (DM) step by step.

The pulse characterization of the generated DUV pulse is performed using the frequency-resolved optical gating (FROG) method [31]. The results of FROG measurement are shown in Figure 1.1.6.4. Figure 1.1.6.4a and b depicts the measured and retrieved FROG traces, respectively. Figure 1.1.6.4c shows the spectrum measured with the spectrometer, the spectrum, and the spectral phase retrieved from the FROG traces. Figure 1.1.6.4d presents the temporal intensity profile of the transform-limited pulse corresponding to the measured spectrum and the retrieved temporal spectral intensity profile and spectral phase.

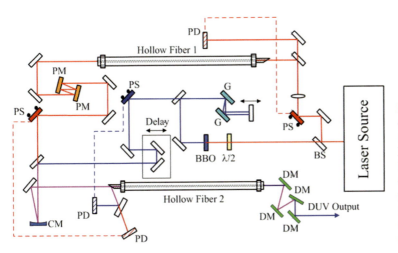

FIGURE 1.1.6.3 DUV (deep UV) system setup scheme. BS, beam splitter; PS, beam pointing stabilizer; PD, beam pointing detector; G, grating; PM, phase-delay controlled chirped mirror; CM, concave mirror, DM, dichroic mirror.

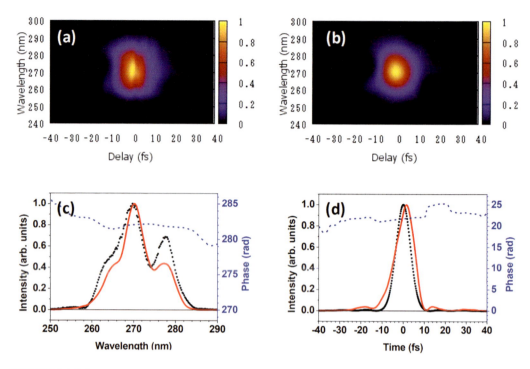

FIGURE 1.1.6.4   (a) Measured and (b) retrieved FROG traces. (c) The spectrum measured with the spectrometer (broken line), the spectrum (solid line), and the spectral phase (dotted line) retrieved from the FROG traces are shown. (d) The temporal intensity profile of the transform-limited pulse corresponding to the measured spectrum is shown by the broken line, and the retrieved temporal intensity profile and phase are indicated by the solid and dotted lines, respectively.

1.1.6.6.3 DUV Pulse Stability Optimization

For the spectroscopy application based on the pump-probe experiment, the long-time stability of the intensity of laser spectrum is of vital importance to obtain reliable data. However, the stability of the DUV laser pulse energy and spectrum after the hollow fiber is often not good enough to be used in a real-time spectroscopy experiment [38]. This is because the fluctuations of the laser beam pointing before the hollow fiber induce substantial variations in the pulse duration, spectrum, and energy of the output pulse. Especially in the present two-step scheme as shown in Figure 1.1.6.3, since two hollow fibers are introduced, the stability situation becomes even more serious. Therefore, it is highly desirable to stabilize the coupling of beam into the hollow fibers. The instability is induced by the change in the propagation direction and point of incidence of the beam before the hollow fibers which can be controlled by the reflection mirror of the beam to be coupled. The stabilization is attained by the feedback systems of the beam intensity in front of the entrances of the fibers to the mirror pairs in front of the fibers [42,43]. In the left part of Figure 1.1.6.5, the stability of DUV output power intensity is measured for 2000 s without the feedback system. The RMS noise is over 10%, which is not acceptable for the pump probe experiment. To maintain a stable DUV pulse output, three beam pointing stabilizers (PS) were used for each input beam before it was focused into the hollow fiber. Pointing detectors (PD) sense the change of intensity and feedback to PS to adjust the azimuthal and elevation angles of PS mirrors. As shown in the right part of Figure 1.1.6.5, after the stabilization, the DUV laser output is improved to 1.04% RMS, being measured for the same time length as without the stabilizer.

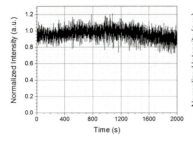 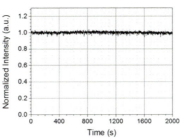

FIGURE 1.1.6.5 DUV laser long-time intensity stability performance before (left) and after (right) introducing the stabilization system.

1.1.6.7 CONCLUSION

In the present paper, we described two types of femtosecond laser developed by using two types of optical parametric processes. One is in the visible range based on optical parametric amplification (OPA), and the other is in the DUV range based on optical parametric four-wave mixing.

The width of the visible pulse is as short as 4.7 fs with a spectrum of 520–720 nm. To satisfy the broad phase-matching condition, noncollinear PA (NOPA) was used. Even further pulse shortening to 3.9 fs duration was attained by compensating group-delay dispersion (GDD) and third-order dispersion (TOD) in the NOPA system using a pair of ultra-broadband chirped mirrors (CM) with the combination of the prism pair. In real experimental conditions, the spectrum of non-collinear NOPA can be adjusted in such a way that the wavelength range extends to longer region, 556–753 nm, to achieve higher absorbance of some samples. In this case, the pulse duration is 7.0 fs.

An ultrashort DUV pulse laser is developed using the broad-band chirped-pulse four-wave mixing method (BBCPFWM). The laser pulse suffered from instability induced by the fluctuation of coupling efficiency to the two hollow core fibers containing krypton and argon gas in one of the two fibers each. The former is for the introduction of chirping to obtain a broad spectrum, and the latter is for the BBCPFWM processes.

The problem of the instability of the DUV pulse output introduced by the unstable coupling efficiency to the fibers due to optical components was solved by the coupling stabilizer to the level of 1.04% RMS. The energy and duration of the DUV pulse are 300-nJ and 9.6 fs, respectively. The research results described in this subsection is obtained by the members of Takayoshi Kobayashi [44]:

REFERENCES

1. J. C. Polanyi and A. H. Zewail, "Direct observation of the transition state," *Acc. Chem. Res.* **28**, 119–132 (1995), doi:10.1021/ar00051a005.
2. *The Nobel Prize in Chemistry* (1967). Available online: https://www.nobelprize.org/nobel_prizes/chemistry/laureates/1967/ (accessed on 18 July 2018).
3. A. Boyer, M. Déry, P. Selles, C. Arbour, and F. Boucher, "Colour discrimination by forward and reverse photocurrents in bacteriorhodopsin-based photosensor," *Biosens. Bioelectron.* **10**, 415–422 (1995), doi:10.1016/0956-5663(95)96888-6.
4. K. J. Hellingwerf, J. Hendriks, and T. Gensch, "On the configurational and conformational changes in photoactive yellow protein that leads to signal generation in ectothiorhodospira halophile," *J. Biol. Phys.* **28**, 395–412 (2002), doi:10.1023/A:1020360505111.
5. Y. Imamoto and M. Kataoka, "Structure and photoreaction of photoactive yellow protein, a structural prototype of the PAS domain superfamily," *Photochem. Photobiol.* **83**, 40–49 (2007), doi:10.1562/2006-02-28-IR-827.
6. G. Eichmann, Y. Li, and R. R. Alfano, "Optical binary coded ternary arithmetic and logic," *Appl. Opt.* **25**, 3113 (1986), doi:10.1364/AO.25.003113.
7. D. Hulin, A. Mysyrowicz, A. Antonetti, A. Migus, W. T. Masselink, H. Morkoc, H. M. Gibbs, and N. Peyghambarian, "Ultrafast all-optical gate with subpicosecond ON and OFF response time," *Appl. Phys. Lett.* **49**, 749 (1986), doi:10.1063/1.97535.

8. V. De Waele, U. Schmidhammer, T. Mrozek, J. Daub, and E. Riedle, "Ultrafast bidirectional dihydroazulene/vinylheptafulvene (DHA/VHF) molecular switches: Photochemical ring closure of vinylheptafulvene proven by a two-pulse experiment," *J. Am. Chem. Soc.* **124**, 2438–2439 (2002), doi:10.1021/ja017132s.

9. L. Kuhnert, "A new optical photochemical memory device in a light-sensitive chemical active medium," *Nature* **319**, 393–394 (1986), doi:10.1038/319393a0.

10. T. Ikeda, S. Horiuchi, D. B. Karanjit, S. Kurihara, and S. Tazuke, "Photochemical image storage in polymer liquid crystals," *Chem. Lett.* **17**, 1679–1682 (1988), doi:10.1246/cl.1988.1679.

11. N. E. Korolev, I. Y. Mokienko, A. E. Poletimov, and A. S. Shcheulin, "Optical storage material based on doped fluoride crystals," *Phys. Status Solidi* **127**, 327–333 (1991), doi:10.1002/pssa.2211270205.

12. F. Pellegrino, "Ultrafast energy transfer processes in photosynthetic systems probed by picosecond fluorescence spectroscopy," *Opt. Eng.* **22**, 225508 (1983), doi:10.1117/12.7973190.

13. V. Z. Paschenko, A. A. Kononenko, S. P. Protasov, A. B. Rubin, L. B., Rubin, and N. Y. Uspenskaya, "Probing the fluorescence emission kinetics of the photosynthetic apparatus of Rhodopseudomonas sphaeroides, strain 1760–1, on a picosecond pulse fluorometer," *BBA-Bioenergy* **461**, 403–412 (1977), doi:10.1016/0005-2728(77)90229-8.

14. M. R. Wasielewski, P. A. Liddell, D. Barrett, T. A. Moore, and D. Gust, "Ultrafast carotenoid to pheophorbide energy transfer in a biomimetic model for antenna function in photosynthesis," *Nature* **322**, 570–572 (1986), doi:10.1038/322570a0.

15. A. Yabushita, T. Kobayashi, and M. Tsuda, "Time-resolved spectroscopy of ultrafast photoisomerization of octopus rhodopsin under photoexcitation," *J. Phys. Chem. B* **116** (2012), doi:10.1021/jp209356s.

16. K. Suzuki, T. Kobayashi, H. Ohtani, H. Yesaka, S. Nagakura, Y. Shichida, and T. Yoshizawa, "Observation of the picosecond time-resolved spectrum of squid bathorhodopsin at room temperature," *Photochem. Photobiol.* **32** (1980) doi:10.1111/j.1751-1097.1980.tb04060.x.

17. K. Peters, M. L. Applebury, and P. M. Rentzepis, "Primary photochemical event in vision: Proton translocation," *Proc. Natl. Acad. Sci. USA* **74**, 3119–3123 (1977).

18. N. Ishii, E. Tokunaga, S. Adachi, T. Kimura, H. Matsuda, and T. Kobayashi, "Optical frequency and vibrational time-resolved two-dimensional spectroscopy by real-time impulsive resonant coherent Raman scattering in polydiacetylene," *Phys. Rev. A* **70**, 023811 (2004).

19. A. Yabushita, C. Kao, Yu Lee, and T. Kobayashi, "Development and demonstration of table-top synchronized fast-scan femtosecond time-resolved spectroscopy system by single-shot scan photo detector array," *Jpn. J. Appl. Phys.* **54**, 072401 (2015).

20. A. Yabushita, Y. Lee, and T. Kobayashi, "Development of a multiplex fast-scan system for ultrafast time-resolved spectroscopy," *Rev. Sci. Instrum.* **81**, 063110 (2010).

21. *Global Fitting with Parameter Sharing.* Available online: https://www.originlab.com/doc/Tutorials/Fitting-Global (accessed on 18 July 2018).

22. C. C. Hung, A. Yabushita, T. Kobayashi, P. F. Chen, and K. S. Liang, "Ultrafast relaxation dynamics of nitric oxide synthase studied by visible broadband transient absorption spectroscopy," *Chem. Phys. Lett.* **683**, 619–624 (2017).

23. A. Shirakawa, I. Sakane, M. Takasaka, and T. Kobayashi, "Sub-5-fs visible pulse generation by pulse-front-matched noncollinear optical parametric amplification," *Appl. Phys. Lett.* **74**, 2268–2270 (1999).

24. A. Baltuška, T, Kobayashi, "Adaptive shaping of two-cycle visible pulses using a flexible mirror," *Appl. Phys. B Lasers Opt.* **75** (2002) doi:10.1007/s00340-002-1008-3.

25. S. Adachi, P. Kumbhakar, and T. Kobayashi, "Quasi-monocyclic near-infrared pulses with a stabilized carrier-envelope phase characterized by noncollinear cross-correlation frequency-resolved optical gating," *Opt. Lett.* **29** (2004), doi:10.1364/OL.29.001150.

26. T. Wilhelm, J. Piel, and E. Riedle, "Sub-20-fs pulses tunable across the visible from a blue-pumped single-pass noncollinear parametric converter," *Opt. Lett.* **22**, 1494–1496 (1997), doi:10.1364/OL.22.001494.

27. G. Cerullo, M. Nisoli, S. Stagira, and S. De Silvestri, "Sub-8-fs pulses from an ultrabroadband optical parametric amplifier in the visible," *Opt. Lett.* **23**, 1283–1285 (1998), doi:10.1364/OL.23.001283.

28. T. Kobayashi and A. Shirakawa, "Tunable visible and near-infrared pulse generator in a 5 fs regime," *Appl. Phys. B Lasers Opt.* **70** (2000), doi:10.1007/s003400000325.

29. J. Du, W. Yuan, X. Xing, T. Miyatake, H. Tamiaki, T. Kobayashi, and Y. Leng, ". *Chem. Phys. Lett. Zewail* **683**, 154–159 (2017).

30. C. Chen, Y. X. Fan, R. C. Eckardt, and R. L. Byer, "Recent developments in barium borate," *Int. Soc. Opt. Photonics* **12** (1987) doi:10.1117/12.939612.

31. R. Trebino, *Frequency-Resolved Optical Gating: The Measurement of Ultrashort Laser Pulses*, Springer, Berlin, Germany (2002). ISBN 1-4020-7066-7.

32. T. Kobayashi, "Development of ultrashort pulse lasers and their applications to ultrafast spectroscopy in the visible and Nir ranges," in *Advances in Multi-Photon Processes and Spectroscopy*, edited by S. H. Lin, World Scientific, Singapore, pp. 155–211 (2008).

33. Y. T. Wang, M. N. Chen, C. T. Lin, J. J. Fang, C J. Chang, C. W. Luo, A. Yabushita, K. H. Wu, and T. Kobayashi, "Use of ultrafast time-resolved spectroscopy to demonstrate the effect of annealing on the performance of P3HT: PCBM solar cells," *ACS Appl. Mater. Interf.* **7**, 4457–4462 (2015).

34. D. Han, J. Du, T. Kobayashi, T. Miyatake, H. Tamiaki, Y. Li, and Y. Leng, "Excitonic relaxation and coherent vibrational dynamics in zinc chlorin aggregates for artificial photosynthetic systems," *J. Phys. Chem. B* **119**, 12265–12273 (2015).

35. S. Hashimoto, A. Yabushita, T, Kobayashi, and I. Iwakura, "Real-time measurements of ultrafast electronic dynamics in the disproportionation of [TCNQ] 22–using a visible sub-10 fs pulse laser," *Chem. Phys. Lett.* **650**, 47–51 (2016).

36. C. C. Hung, X. R. Chen, Y. K. Ko, T. Kobayashi, C. S. Yang, A. Yabushita, and A. Schiff, "Base Proton acceptor assists photoisomerization of retinal chromophores in bacteriorhodopsin," *Biophys. J.* **112**, 2503–2519 (2017).

37. T. Kobayashi, "Ultrafast spectroscopy of coherent phonon in carbon nanotubes using sub-5-fs visible pulses," *Irago Conf.* (2016), doi:10.1063/1.4941200.

38. K. Okamura and T. Kobayashi, "Octave-spanning carrier-envelope phase stabilized visible pulse with sub-3-fs pulse duration," *Opt. Lett.* **36**, 226–228 (2011).

39. *Ferenc Krausz.* Available online: http://www.attoworld.de.

40. C. G. Durfee, S. Backus, H. C. Kapteyn, and M. M. Murnane, "Intense 8-fs pulse generation in the deep ultraviolet," *Opt. Lett.* **24**, 697 (1999), doi:10.1364/OL.24.000697.

41. A. Ermolov, "Ultrashort UV pulses via nonlinear light matter interaction in gas-filled hollow-core fibres: generation and characterization," Friedrich-Alexander-Universitaet Erlangen-Nuernberg (2017). Available online: https://opus4.kobv.de/opus4-fau/frontdoor/index/index/docId/8530 (accessed on 18 July 2018).

42. Y. Kida, J. Liu, T. Teramoto, and T. Kobayashi, "Sub-10 fs deep-ultraviolet pulses generated by chirped-pulse four-wave mixing," *Opt. Lett.* **35** (2010) doi:10.1364/OL.35.001807.

43. T. Kobayashi and Y. Kida, "Ultrafast spectroscopy with sub-10 fs deep-ultraviolet pulses," *Phys. Chem. Chem. Phys.* **14** (2012), doi:10.1039/c2cp23649d.

44. T. Kobayashi, "Development of ultrashort pulse lasers for ultrafast spectroscopy," *MDPI Photonics*, **5**, 19 (2018). © 2018 by the authors. "Submitted for possible open access publication under the terms and conditions of the Creative Commons Attribution (CC BY) license," (http://creativecommons.org/licenses/by/4.0/).

Section 1.2

Ultrashort Ultraviolet, Deep-Ultraviolet, and Infrared Pulses

1.2.1 Generation of Stable Sub-10 fs Pulses at 400 nm in a Hollow Fiber for UV Pump-Probe Experiment

1.2.1.1 INTRODUCTION

Ultrafast time-resolved spectroscopy is a powerful technique for the investigation of electronic and vibrational dynamics in molecules, which are key elements in various fields in physics, chemistry, biology, and materials science research. It is because it can provide important information in photophysical and photochemical processes occurring in molecules [1,2]. In this kind of research, the dynamic process is in femtosecond to picosecond time scale. To investigate these processes, the laser pulse in the experiment is required to have shorter duration than the time range of phenomena of interest. Sub-10 fs visible pulse generated from non-collinear optical parametric amplifier had been used for study the real-time vibration in lots of molecules, which have absorption in the visible range [3–9]. Ultrashort pulse in the UV spectral region is needed to study the real-time resolved spectroscopy for many samples, which have strong absorption in the UV. By using achromatic frequency doubling and sum-frequency mixing [10–12], sub-10 fs was obtained in the UV region. However, the spatial mode and stability of the output pulse were usually not good due to several nonlinear processes. Another method was using hollow fiber to compress the pulse duration in the spectral region from UV to infrared [13–15]. In 1999, sub-10 fs pulses at 400 nm were generated through guiding second harmonic (SH) pulses from Ti:sapphire chirped parameter amplified laser system into a hollow fiber to broaden the laser spectrum and then dispersion compensated by using chirped mirrors and prism pair [15]. However, the stability of laser pulse energy and spectrum after the hollow fiber is often not good enough to be used in real-time spectroscopy experiment. It is because that the fluctuations of laser beam pointing before the hollow fiber induce substantial variations in the pulse duration, spectrum, and energy of the output pulse [16]. Damage of the entrance of the hollow fiber may be induced by this fluctuation. In the pump-probe experiment, the long-time stability of the intensity of laser spectrum is of vital importance to obtain reliable data for the pump-probe experiment. Then, highly stable sub-10 fs pulses at 400 nm are required for this kind of research.

In this chapter, a beam-pointing stabilizer was used before the hollow-fiber compression system to improve the beam-pointing stability. As a result, the stability of the output power was improved from 0.72% RMS to about 0.39% RMS for about 2 h, which was about two times better than that without the pointing stabilization system. The laser after the hollow fiber was compressed to about 9.1 fs. A pump-probe experiment of perylene dissolved in cyclohexane was studied using this sub-10 fs pulse. The result proved that the obtained sub-10 fs pulse was very useful for many kinds of spectroscopy experiments.

1.2.1.2 EXPERIMENTAL SETUP

The schematic of the experimental setup is shown in Figure 1.2.1.1. About 900 μJ laser pulse after Ti:sapphire laser system was frequency doubled in a 200-μm-thick beta barium borate (BBO, type I, $\theta = 29.2°$) crystal and a pair of chirped mirrors (−25 fs^2/bounce, Layertec) was used to precompensate the chirp of the SH induced by the dispersion of the glass windows. Then, about 90 μJ SH

DOI: 10.1201/9780429196577-10

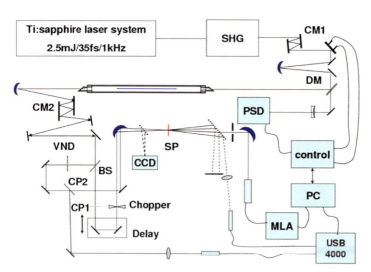

FIGURE 1.2.1.1 Schematic of the 400 nm hollow-fiber compressor and the pump–probe experimental setup. SHG, second-harmonic generation; CM1, CM2, chirped mirror; DM, dichroic mirror; VND, 0.1-mm-thick variable neutral-density filter; BS, 0.5-mm-thick beam splitter; CP1, compensate plate for VND; CP2, compensate plate for BS; SP, sample for pump–probe spectroscopy or BBO crystal for pulse characterization; MLA, 128-channel multi-lockin amplifier; PSD, position sensing detector; USB 4000, spectrometer; CCD, charge coupled device (CCD) camera.

laser pulse at 400 nm was focused into a hollow-fiber compression system with an aluminum-coated concave mirror with a focal length of 750 mm. The beam diameter at the entrance of the fiber was about 93 μm. In the experiment, the outer and inner diameters of the fused-silica hollow fiber were 3 mm and 140 μm, respectively. The hollow-fiber length was about 60 cm. The output pulse energy was about 45 μJ with around 50% transmission efficiency. After being spectrally broadened through the hollow fiber, the laser pulse was dispersion compensated through two pairs of chirped mirrors (one pair from Layertec: $-25 \, fs^2$/bounce, one pair from Femtolasers: $-20 \, fs^2$/bounce). The compressed pulse was guided into a pump-probe experiment setup. This setup was used also for pulse duration measurement through the self-diffraction frequency resolved optical gating (SD-FROG) method. A 0.5-mm-thick 50%-reflection fused-silica beam splitter (BS) was used to split the beam into pump beam and probe beam. A 0.1-mm-thick variable neutral-density filter (VND) was located in the probe beam to reduce the pulse energy. Two dispersion-compensation plates (CP) were used to cancel the group-velocity dispersion of the VND and BS. The power stability was monitored during the pump-probe experiment by detecting the intensity of the reflected light from CP2. A Charge Coupled Device (CCD) camera (Thorlab, BC106-VIS) was used to monitor the beam diameter and the overlapping between pump and probe beams. Time-resolved spectra were measured with polychromator coupled to the multi-channel detector.

1.2.1.3   EXPERIMENTAL RESULTS AND DISCUSSION

From Figure 1.2.1.1, a part of 800 nm beam was collimated to a position-sensing detector (PSD) (Thorlab, PDQ80A) photodiode sensor. The beam-pointing position was detected by a quad detector (Thorlab, TQD001), which was used to feed-back to control a mirror mount with two piezoelectric actuators that were driven by two piezoelectric controllers (Thorlab, TPZ001). The feedback system can run at 200-Hz rate, which makes it possible to control the beam guiding into the hollow fiber very well for long experimental time [16]. The beam-pointing stability at the focal point was measured using a CCD camera (BeamStar FX 33, Ophir Optronics) for about an hour, as shown in Figure 1.2.1.2. When the beam-pointing stabilization system did not work, the focal beam before the entrance of the hollow fiber wandered for about 30 μm in the X-direction and about 22 μm in the Y-direction, as shown in Figure 1.2.1.2a. The standard deviation in the X-direction and Y-direction were 4.7 and 3.1 μm, respectively.

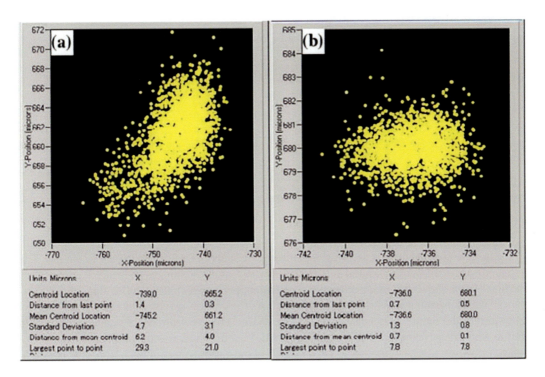

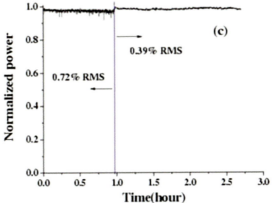

FIGURE 1.2.1.2 Beam-pointing position detected using a CCD camera at the focal point (a) Without the beam-pointing stabilizer. (b) With the beam-pointing stabilizer. (c) Output power stability after the hollow fiber without (left line) and with (right line) the beam-pointing stabilizer.

When the beam-pointing stabilizer was turned on, the focal beam wandering was reduced to about 8 μm both in the X-direction and in the Y-direction, as shown in Figure 1.2.1.2b. The standard deviation in the X-direction and Y-direction were reduced to 1.3 and 0.8 μm, respectively. It can be seen that the beam-pointing stability was improved for more than three times with the beam-pointing stabilization system. The output power stability was also measured using a power meter. Figure 1.2.1.2c shows that the output pulse power stability was improved from 0.72% to 0.39% RMS by about two times with the beam-pointing stabilizer for about 2 h. Furthermore, when the beam pointing stabilizer was turned on, there was no sudden decrease in a much longer time in Figure 1.2.1.2c. However, there were several peaks in the figure when the beam-pointing stabilization system did not work.

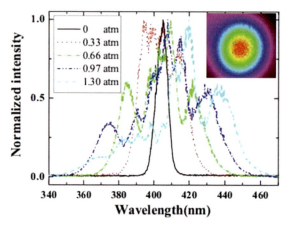

FIGURE 1.2.1.3 The spectral profile of the output pulse after the hollow fiber at different argon gas pressures. 0 atm, black solid line; 0.33 atm, red dotted line; 0.66 atm, green dash-dot line; 0.97 atm, blue dash-dot-dot line; 1.30 atm, cyan dash line. The inset pattern was the beam profile after the hollow fiber.

  The hollow-fiber chamber was vacuum-pumped at first and then filled with pure argon gas. The output-pulse spectra were measured at 0, 0.33, 0.66, 0.97, and 1.30 atm, as shown in Figure 1.2.1.3. The black solid line shows the spectrum of the incident pulse, which has a full width at half maximum (FWHM) spectral bandwidth of about 8 nm. As the gas pressure was increased, the spectral bandwidth of the output laser pulse increased gradually. The spectrum was extended from 350 to 460 nm when the gas pressure was increased to about 1.3 atm. These spectra can support transform-limited pulse durations of about 11, 7, 5.1, and 4.4 fs at the gas pressures of 0.33, 0.66, 0.97, and 1.30 atm, respectively. When the gas pressure was increased to about 1.3 atm, a filament appeared clearly near the entrance of the fiber. This filament introduced instability of the output power. Moreover, the coating of the chirped mirror and beam splitter can only support spectral range from 360 to 440 nm. Then, the argon gas pressure was fixed at around 0.90 atm in the experiment. The inset pattern in Figure 1.2.1.3 was the beam profile after the hollow fiber which showed a Gaussian mode.

  The pulse energy was reduced to about 4 μJ due to the chirped mirrors and several aluminum-coated mirrors. Two 50%-reflection beam splitters and one 0.1-mm-thick VND filter was used before the pump-probe setup to tune the pulse energy. The pulse energy of the pump pulse before the sample could be tuned from 15 to 100 nJ by the VND filter. In the measurement, a 70-μm-thick BBO crystal was used as the nonlinear medium to generate the self-diffraction signal. The crossing angle between the pump beam and the probe beam was about 1.2°. The small thickness of the crystal and small crossing angle minimized the measurement error of the pulse duration [17]. The pulse energies of the pump and probe beams were tuned to be nearly equal by tuning a VND filter in the probe beam when the pulse duration was measured. The laser spectrum before the sample and the maximum self-diffraction-signal spectrum are shown in Figure 1.2.1.4a. Both the spectra can support about 5.7 fs transform-limited pulse duration. The retrieved spectrum and the spectral phase are also shown in Figure 1.2.1.4a. The spectrum was a little narrower than the maximum SD signal spectrum and the measured spectrum due to the limited spectral bandwidth and quality of the chirped mirrors. Figure 1.2.1.4b shows the retrieved temporal intensity profile and the temporal phase of the compressed pulse with a 512×512 grid and a 0.006 retrieval error. The inset pattern showed the measured SD-FROG trace. The obtained pulse duration was 9.1 fs, as shown in Figure 1.2.1.4b. It is expected to obtain much shorter pulse duration by using a prism-pair compressor combined with a deformable-mirror compressor in the future.

  The obtained pulse was used for UV-pump/UV-probe experiment. A sample of cyclohexane solution of perylene in a 1-mm-thick cell was used in the UV pump-probe experiment. The experimental method was nearly the same as that of our previous visible-pump/visible-probe experiments [3–7]. The system is a combined system of a polychromator and a multichannel lock-in amplifier (MLA) [18]. The reference and probe pulses were dispersed by the polychromator (600 grooves/mm, 300 nm blazed) and guided by a 128 channel bundle fiber to the 128 photo-detectors before

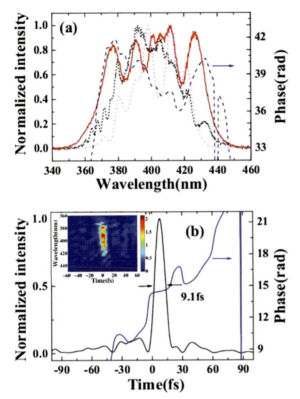

FIGURE 1.2.1.4 (a) Spectra of the compressed 400 nm pulse (red solid) before the sample, the retrieved spectrum (magenta dotted), and that of the maximum SD signal (black dashed) in the measurement. The blue dashed line is the retrieved spectral phase. (b) The retrieved temporal intensity profile and temporal phase (blue solid line) of the compressed pulse. The inset is the measured SD-FROG trace.

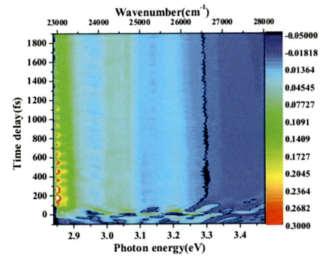

FIGURE 1.2.1.5 Two-dimensional pattern of the difference absorbance (ΔA) of probe dependent on the time delay between the pump and probe pulse from −100 to 1900 fs in the whole probe spectral region (2.84–3.48 eV).

the MLA. The spectral resolution of the system was about 0.8 nm. In the experiment, the beam diameters and pulse energies of the pump and probe beam were 115 μm and 69 nJ and 95 μm and 12 nJ, respectively. Time trace of normalized transmittance changes ($\Delta T/T$) was obtained as a function of the pump-probe delay time from −100 to 1900 fs with every 0.2-fs step. The experiment was performed at room temperature (295 ± 1 K).

Figure 1.2.1.5 shows the two-dimensional difference absorbance ($\Delta A = -\log (1 + \Delta T/T)$) vs. the pump-probe delay time in the spectral range between 356 and 436 nm. The black line was the position where dA is equal to 0. The pattern shows clearly that the absorbance oscillation depends on the delay time in the whole spectral region. Figure 1.2.1.6a shows several examples of dA traces at several probe photon energies (3.40, 3.28, 3.16, 3.05, 2.94, 2.89, and 2.86 eV). The absorbance change was negative for 3.40 eV and positive for other six probe photon energies due to bleaching and induced absorption, respectively. The vibration decay was clearly shown at 2.89 and 2.86 eV probe photon energies. In all seven probe photon energies, the oscillating features in the time traces of the absorbance change with different periods were clearly observed.

Figure 1.2.1.6b shows the Fourier transform (FT) amplitude spectra of Figure 1.2.1.6a with a broad frequency region from about 100 cm^{-1} to nearly 3000 cm^{-1}. The high FT amplitude indicated that the vibronic coupling of this mode is strong. All the seven probe photon energies show a well-known mode at 352 cm^{-1} [19–21]. It also shows that the FT amplitude is increased with the decrease of probe photon energy in the spectral region, and it reaches maximum at 2.86 eV probe photon energy. Other lower frequency modes related to mode beating were observed at around 205 and 106 cm^{-1}. This mode beating was also observed in Ref. [20]. Several higher frequency modes around 1387, 1302, and 1600 cm^{-1} are clearly found in Figure 1.2.1.6b, which could not be

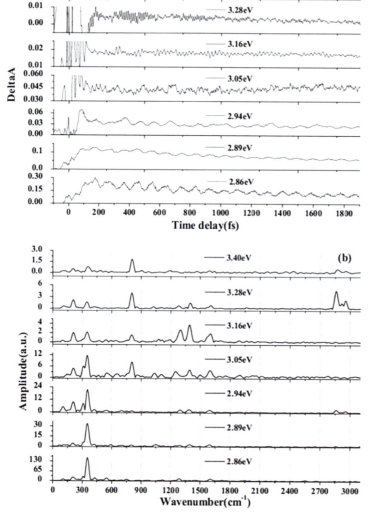

FIGURE 1.2.1.6 (a) The traces of difference absorbance (dA) plotted against the pump–probe delay time at several different probe photon energies (3.40, 3.28, 3.16, 3.05, 2.94, 2.89, and 2.86 eV). (b) the Fourier-transform amplitude spectra of the traces in (a).

observed in a previous paper because a laser pulse much longer than our research (about 50 fs) was used in the paper [20,22]. Moreover, the C–H stretching of vibration mode at around 2860, 2916, and 2955 cm^{-1} was detected at 3.28 eV probe photon energy, as shown in Figure 1.2.1.6b. This is the highest frequency of molecular vibration observed in real-time at the time of this paper being published [22]. To detect 3000 cm^{-1} mode in real-time, it is needed to use a pulse with the duration shorter than 10-fs.

1.2.1.4 CONCLUSION

In summary, stable 9.1-fs pulses at 400-nm center wavelength were obtained using a hollow fiber compression with a beam-pointing stabilizing system [22]. The beam-pointing stability was improved by about three times and the output-power stability was improved by around two times with the beam-pointing stabilizer. A UV pump-probe experiment using perylene dissolved in cyclohexane as a sample was studied by using this sub-10 fs pulse. Thanks to the stability highest molecular vibration frequency of about 3000 cm^{-1} excited by UV pulse was real-time observed. The result proved that it was useful for many kinds of spectroscopy experiments.

The contents of the work presented in this subsection is reproduced and adapted with permission from [Ref. 22]©The Optical Society. *Opt. Lett.* **27**, 306 (2002) and it was conducted by the following people in collaboration: J. Liu, Y. Kida, T. Teramoto, and T. Kobayashi [22].

REFERENCES

1. A. H. Zewail, "Femtochemistry: Atomic-scale dynamics of the chemical bond," *J. Phys. Chem. A* **104**(24), 5660–5694 (2000).
2. *Ultrafast Phenomena, IV*, edited by T. Kobayashi, T. Okada, K. A. Nelson, and S. De Silvestri, Springer, New York, Vol. **79** (2004).
3. T. Kobayashi, T. Saito, and H. Ohtani, "Real-time spectroscopy of transition states in bacteriorhodopsin during retinal isomerization," *Nature* **414**(6863), 531–534 (2001).
4. S. Adachi, V. M. Kobryanskii, and T. Kobayashi, "Excitation of a breather mode of bound soliton pairs in trans polyacetylene by sub-5-fs optical pulses," *Phys. Rev. Lett.* **89**(2), 027401 (2002).
5. T. Kobayashi, I. Iwakura, and A. Yabushita, "Excitonic and vibrational nonlinear processes in a polydiacetylene studied by a few-cycle pulse laser," *N. J. Phys.* **10**(6), 065016 (2008).
6. T. Kobayashi, Z. Wang, and I. Iwakura, "The relation between the symmetry of vibrational modes and the potential curve displacement associated with electronic transition studied by using real-time vibration spectroscopy," *N. J. Phys.* **10**(6), 065009 (2008).
7. I. Iwakura, A. Yabushita, and T. Kobayashi, "Transition states and nonlinear excitations in chloroform observed with a sub-5 fs pulse laser," *J. Am. Chem. Soc.* **131**(2), 688–696 (2009).
8. G. Cerullo, D. Polli, G. Lanzani, S. De Silvestri, H. Hashimoto, and R. J. Cogdell, "Photosynthetic light harvesting by carotenoids: Detection of an intermediate excited state," *Science* **298**(5602), 2395–2398 (2002).
9. D. Polli, M. R. Antognazza, D. Brida, G. Lanzani, G. Cerullo, and S. De Silvestri, "Broadband pump-probe spectroscopy with sub-10-fs resolution for probing ultrafast internal conversion and coherent phonons in carotenoids," *Chem. Phys.* **350**(1–3), 45–55 (2008).
10. I. Kozma, P. Baum, S. Lochbrunner, and E. Riedle, "Widely tunable sub-30 fs ultraviolet pulses by chirped sum frequency mixing," *Opt. Express* **11**(23), 3110–3115 (2003).
11. P. Baum, S. Lochbrunner, and E. Riedle, "Tunable sub-10-fs ultraviolet pulses generated by achromatic frequency doubling," *Opt. Lett.* **29**(14), 1686–1688 (2004).
12. B. Zhao, Y. Jiang, K. Sueda, N. Miyanaga, and T. Kobayashi, "Sub-15 fs ultraviolet pulses generated by achromatic phase-matching sum-frequency mixing," *Opt. Express* **17**(20), 17711–17714 (2009).
13. M. Nisoli, S. De Silvestri, and O. Svelto, "Generation of high energy 10 fs pulses by a new pulse compression technique," *Appl. Phys. Lett.* **68**(20), 2793–2795 (1996).
14. M. Giguère, B. E. Schmidt, A. D. Shiner, M. A. Houle, H. C. Bandulet, G. Tempea, D. M. Villeneuve, J. C. Kieffer, and F. Légaré, "Pulse compression of submillijoule few-optical-cycle infrared laser pulses using chirped mirrors," *Opt. Lett.* **34**(12), 1894–1896 (2009).

15. O. Dühr, E. T. J. Nibbering, G. Korn, G. Tempea, and F. Krausz, "Generation of intense 8-fs pulses at 400 nm," *Opt. Lett.* **24**(1), 34–36 (1999).
16. T. Kanai, A. Suda, S. Bohman, M. Kaku, S. Yamaguchi, and K. Midorikawa, "Pointing stabilization of a high repetition-rate high-power femtosecond laser for intense few-cycle pulse generation," *Appl. Phys. Lett.* **92**(6), 061106 (2008).
17. R. Trebino, *Frequency-Resolved Optical Grating: The Measurement of Ultrashort Laser Pulses*, Kluwer Academic Publishers, pp. 237–250 (2000).
18. T. Teramoto, E. Tokunaga and T. Kobayashi, "Two dimensional detection system for broadband spectroscopy by using multi-channel lock-in amplifiers," (in prepare).
19. B. Brüggemann, P. Persson, H.-D. Meyer, and V. May, "Frequency dispersed transient absorption spectra of dissolved perylene: A case study using the density matrix version of the MCTDH method," *Chem. Phys.* **347**(13), 152–165 (2008).
20. A. L. Dobryakov, and N. P. Ernsting, "Lineshapes for resonant impulsive stimulated Raman scattering with chirped pump and supercontinuum probe pulses," *J. Chem. Phys.* **129**(18), 184504 (2008).
21. Y. H. Meyer, and P. Plaza, "Ultrafast excited singlet state absorption/gain spectroscopy of perylene in solution," *Chem. Phys.* **200**(1–2), 235–243 (1995).
22. J. Liu, Y. Kida, T. Teramoto, and T. Kobayashi, "Generation of stable sub-10 fs pulses at 400 nm in a hollow fiber for UV pump-probe experiment," *Opt. Express* **18**(5), 4664–4672 (2010).

1.2.2 Sub-10 fs Deep-Ultraviolet Pulses Generated by Chirped-Pulse Four-Wave Mixing

1.2.2.1 INTRODUCTION

Few-cycle pulse laser systems in the visible range have been applied to ultrafast spectroscopy to clarify ultrafast mechanisms in nanosystems and molecules in which all elementary processes are considered to occur in the femtosecond regime [1–3]. There is high demand for ultrafast spectroscopy with few-cycle laser pulses in the ultraviolet (UV) to deep-UV (DUV) regions, especially for investigating ultrafast dynamics in biologically relevant molecules [4–6]. Several approaches have been proposed and demonstrated for generating ultrashort DUV pulses [7–10], including the generation of a 7 fs pulse with low energy of 120 nJ in the atmospheric environment [7] and of an intense 3:7 fs pulse in a vacuum chamber [8]. However, generation of ultrashort DUV pulses shorter than 7 fs has not been demonstrated in the atmospheric environment. It is difficult to generate such short DUV pulses because of pulse broadening due to the group-velocity dispersion (GVD) in optical components such as mirrors, beam splitters, and even in the air. Compensation cannot be performed easily using conventional devices, such as grating compressors [9] or prism compressors because they induce third- and higher-order dispersions. A deformable mirror has been used to compensate for third and higher-order dispersions [7], but the energy loss and undesirable spatial chirp induced by the deformable mirror prevent the generation of intense DUV pulses shorter than 7 fs. Wojtkiewicz et al. proposed using chirped-pulse four-wave mixing (CFWM) [10] to generate negatively chirped DUV pulses that are compressible by a normal GVD in a transparent medium. A recent numerical study has suggested that sub-10 fs DUV pulses can be generated by using temporally broadened sub-10 fs near-IR (NIR) pulses for CFWM; however, the compressed temporal profile produced by this technique does not have a single pulse structure [11].

In this Letter, a method for generating sub-10 fs DUV pulses is reported that uses a self-phase-modulated idler pulse for CFWM (SMI-CFWM). In this technique, CFWM is induced by a negatively chirped near-UV (NUV) pulse and a positively chirped NIR pulse whose spectral width has been preliminarily broadened by self-phase modulation (SPM). By temporally overlapping the positively chirped part of the modulated NIR pulse and the NUV pulse in a gas-filled hollow fiber, a DUV pulse negatively chirped with a smooth spectrum is generated by SMI-CFWM, which can be compressed into a nearly transform-limited (TL) single pulse by material dispersion. Numerical calculations have been performed to quantitatively explain the experimental results given in this Letter; these numerical results, which will be reported elsewhere [12], are in reasonable agreement with the experimental results. We were able to generate DUV pulses shorter than 10 fs by SMI-CFWM, which is, to the best of our knowledge, the first experimental demonstration of the shortest DUV pulse generation in the atmospheric environment without using any additional pulse compressors.

DOI: 10.1201/9780429196577-11

1.2.2.2 EXPERIMENTAL

For the experimental demonstration of SMI-CFWM, a horizontally polarized NIR pulse with a pulse duration of 35 fs, a repetition rate of 1 kHz, pulse energy of 1:2 mJ, and a center wavelength of 800 nm was generated by a Ti:sapphire chirped-pulse amplifier (CPA; Legend EliteUSP, Coherent), as shown in Figure 1.2.2.1. The nearly TL pulse was split into two pulses. One pulse had pulse energy of 300 μJ and was temporally broadened by a 16-mm-thick fused-silica block and focused into a Kr-filled hollow fiber (core diameter: 250 μm; length, 600 mm; pressure, 630 Torr). The spectrally broadened and positively chirped NIR pulse emerging from the fiber was used as the idler for SMI-CFWM. The polarization of the other NIR pulse with an energy of 900 μJ was changed from horizontal to vertical with a periscope and was transmitted through a 200-μm-thick beta barium borate (BBO) crystal for generating a horizontally polarized NUV pulse (400 nm). The NUV pulse was then negatively chirped by a prism compressor after the beam diameter was changed by a plano–convex lens and a plano–concave lens. The NIR (150 μJ, about 150 fs) and NUV (140 μJ, 120 fs) pulses were focused into an Ar-filled hollow fiber (core diameter, 140 μm; length, 570 mm; pressure, 61 Torr). A horizontally polarized DUV pulse was generated by CFWM by temporally overlapping the two pulses in the fiber. After being collimated by a concave mirror, the DUV pulse emerging from the fiber was sent to a dispersion-free self-diffraction frequency-resolved optical gating (SD-FROG) system [9] with a 100-μm-thick sapphire plate for generating an SD signal.

1.2.2.3 RESULTS AND DISCUSSION

The pulse energy of the DUV pulse was measured to be 300 nJ after separating the DUV pulse from the output NIR and NUV pulses using a Brewster-cut fused-silica prism. The spectrum of the DUV pulse extended from 258 to 290 nm (Figure 1.2.2.2c) and supported a TL pulse duration of 8 fs (7:8 fs 0:1 fs), which is estimated from the inverse Fourier transform of the spectrum (solid line in Figure 1.2.2.2d). The spectral shape of the DUV pulse differs greatly from that of the NIR pulse injected into the hollow fiber (HF2 in Figure 1.2.2.1). As shown in Figure 1.2.2.2c, SMI-CFWM produces a DUV pulse with a smooth spectral profile and a wide spectral width, making it ideal for spectroscopic applications [1–6].

Self-compression in the air has been considered to compensate for the negative group-delay dispersion (GDD) and, thus, compress the DUV pulse to shorter than 10 fs. In the present experiment, we demonstrated self-compression in the following manner. The positive GVD in the output window of the hollow-fiber chamber (0.87-mm-thick MgF_2, 79 fs^2) and in the air was utilized to compensate for the negative GDD of the DUV pulse generated in the hollow fiber. To optimize the

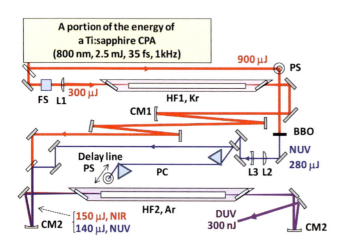

FIGURE 1.2.2.1 Experimental setup, which is composed of two hollow-fiber chambers filled with Kr (HF1) and Ar (HF2) gases, a fused-silica plate (FS), plano–convex lenses (L1, $f = 1000$; L2, $f = 800$ mm), a plano-concave lens (L3, $f = -1000$ mm), concave mirrors (CM1, $f = 1500$ mm; CM2, $f = 750$ mm), periscopes (PS), and a double-pass prism compressor (PC).

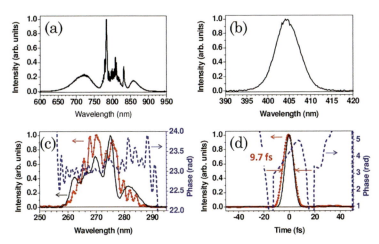

FIGURE 1.2.2.2 Spectra of the input: (a) NIR, (b) NUV, and output DUV [solid curve in (c)] pulses measured with spectrometers. The spectrum (solid curve with filled circles) and spectral phase (broken curve) measured by the FROG are shown in (c), while the corresponding temporal profile (solid curve with filled circles) and phase (broken curve) are shown in (d). The inverse Fourier transform of the solid curve in (c) is shown by the solid curve in (d).

path length in air, the spectral change of the SD signal with respect to the time delay of a replica in the FROG system was measured for path lengths in the air of 1.6, 1.9, and 2:2 m. Spectral changes of the SD signal that indicate negative and positive chirps of the DUV pulse were clearly detected at lengths of 1.6 and 2:2 m, respectively. Based on these three observations, the optimum length was determined to be 1:9 m, which is the length between the output window of the hollow fiber chamber and the sapphire plate in the FROG system. The positive GDD arising from the output window and the air was calculated to be 265 fs² [13].

Figure 1.2.2.2c and d show the spectrum and the temporal profile of the DUV pulse characterized by the SD-FROG measurement, respectively. The temporal profile of the pulse contains a single pulse with a duration of 9:70:3 fs (the average and standard deviation of three consecutive measurements). All the FROG errors in the three measurements were smaller than 0.009. The spectral phase of the retrieved spectrum shown in Figure 1.2.2.2c is almost flat, except for the small-amplitude modulation over the spectral range 258 to 290 nm. The corresponding temporal phase retrieved from the FROG trace does not contain appreciable quadratic- or higher-order phase distortions. The measured pulse duration of 9:7 fs is not far from the TL pulse duration of 7:8 fs; these two values differ by only 24%. This difference might be due to higher order dispersion than GDD; that is, the complicated structure in the retrieved spectral phase (Figure 1.2.2.2c).

The scheme for generating sub-10 fs DUV pulses by SMI-CFWM is similar to that used in a recent theoretical study [11]. The difference between SMI-CFWM and the CFWM in [11] is that a 35 fs NIR pulse with a spectrum broadened by SPM [14] is used as the idler in SMI-CFWM, whereas a temporally broadened sub-10 fs NIR pulse with a $sech^2$ spectral profile or a Gaussian profile is used as the idler in [11]. Moreover, SMI-CFWM generates a single DUV pulse with an excellent temporal profile, while the technique reported in [11] generates a sub-10 fs pulse with a strong tail. The pulse durations of the NIR and NUV pulses were optimized here by performing a simulation using the nonlinear Schrödinger equation to generate a single pulse. The experimental demonstration of the generation of a sub-10 fs DUV pulse by SMI-CFWM suggests that the generation of sub-10 fs pulses by the technique reported in [11] is feasible if the required light sources are available because the technique in [11] is simpler than SMI-CFWM.

A shorter pulse can be obtained by using NIR and NUV pulses with spectral widths broader than those used here, leading to the possibility of sub-7 fs pulse generation. The pulse energy of the DUV pulse generated in this study was limited by the pulse energies of the NIR and

NUV pulses in the hollow fiber: The pulse energy can be increased by using a NUV pulse with a higher pulse energy than that used in the current experiment. This may generate a sub-7 fs DUV pulse with an energy of several microjoules. Because SMI-CFWM is based on CFWM [10], it is possible to use input pulses with energies of several millijoules. NIR pulses with pulse energies of several millijoules, generated by a commercially available Ti:sapphire CPA or optical parametric CPA with a high pulse energy and broad spectral width, are ideal for SMI-CFWM enabling sub-10 fs DUV pulses with energies of several tens of microjoules to be generated, although the pulse energy and the pulse duration of the DUV pulse generated in the present study are respectively lower and longer than those reported in [9].

1.2.2.4 CONCLUSION

In conclusion, a negatively chirped DUV pulse with a spectral width supporting a 7:8 fs TL pulse has been generated by SMI-CFWM. The DUV pulse was compressed to 9:7 fs by propagation in air, and the pulse energy of the DUV pulse was 300 nJ. This technique is capable of generating ultrashort DUV pulses without using pulse compressors. In a future study, we will demonstrate the generation of DUV pulses shorter than 7 fs with pulse energies of several microjoules using a NIR pulse with a broader spectral width and a NUV pulse with a higher pulse energy than this study; these are small extensions of this research [12].

The work presented in this subsection is reproduced and adapted with permission from [Ref. 12] ©The Optical Society. *Opt. Lett.* **35**, 1807 (2010) and it was conducted by collaborative work of the following people: Y. Kida, J. Liu, T. Teramoto, and T. Kobayashi.

REFERENCES

1. T. Kobayashi, T. Saito, and H. Ohtani, *Nature* **414**, 531 (2001)
2. L. Lüer, C. Gadermaier, J. Crochet, T. Hertel, D. Brida, and G. Lanzani, *Phys. Rev. Lett.* **102**, 127401 (2009).
3. A. Yabushita and T. Kobayashi, *Biophys. J.* **96**, 1447 (2009).
4. C. E. Crespo-Hernández, B. Cohen, and B. Kohler, *Nature* **436**, 1141 (2005).
5. H. Kang, C. Jouvet, C. Dedonder-Lardeux, S. Martrenchard, G. Grégoire, C. Desfrançois, J.-P. Schermann, M. Barat, and J. A. Fayeton, *Phys. Chem. Chem. Phys.* **7**, 394 (2005).
6. S. Schenkl, F. van Mourik, N. Friedman, M. Sheves, R. Schlesinger, S. Haacke, and M. Chergui, *Proc. Natl. Acad. Sci. USA* **103**, 4101 (2006).
7. P. Baum, S. Lochbrunner, and E. Riedle, *Opt. Lett.* **29**, 1686 (2004).
8. U. Graf, M. Fiess, M. Schultze, R. Kienberger, F. Krausz, and E. Goulielmakis, *Opt. Express* **16**, 18956 (2008).
9. C. G. Durfee, S. Backus, H. C. Kapteyn, and M. M. Murnane, *Opt. Lett.* **24**, 697 (1999).
10. J. Wojtkiewicz, K. Hudek, and C. G. Durfee, in *Conference on Lasers and Electro-Optics* (2005), Paper CMK5.
11. I. Babushkin and J. Herrmann, *Opt. Express* **16**, 17774 (2008).
12. E. R. Peck and K. Reeder, *J. Opt. Soc. Am.* **62**, 958 (1972).
13. M. Nisoli, S. Stagira, S. De Silvestri, O. Svelto, S. Sartania, Z. Cheng, M. Lenzner, C. Spielmann, and F. Krausz, *Appl. Phys. B* **65**, 189 (1997).
14. Y. Kida, J. Liu, T. Teramoto, and T. Kobayashi, *Opt. Lett.* **35**, 1807 (2010).

1.2.3 Generation and Optimization of Femtosecond Pulses by Four-Wave Mixing Process

1.2.3.1 INTRODUCTION

Ultrashort laser pulses are powerful tools for laser spectroscopic techniques that are widely used in all fields of science (including chemistry, physics, and biology) and that provide microscopic insights into bulk materials, molecules, and chemical and biochemical reactions [1–4]. Advances in ultrashort-laser-pulse technology have enabled the generation of sub-10-fs pulses with a central wavelength around 800 nm and with nanojoule pulse energies by using Ti:sapphire lasers [5]. Such pulses can be compressed to shorter than 5 fs by expanding the spectral width in fibers followed by dispersion compensation [6]. By using gas-filled hollow-core fibers or filament compressors, sub-10-fs pulses with high pulse energies can be produced around wavelengths of 800 and 400 nm with kilohertz repetition rates [7–10]. These ultrashort pulses permit the detection of real-time electronic, phonon, and vibrational dynamics in various molecular systems and polymers and inorganic bulk materials including semiconductors and insulators with an extremely high temporal resolution. On the other hand, wavelength-tunable femtosecond pulses are required in various studies of ultrafast phenomena. Over the past decades of late 90s and early 00s, wavelength-tunable femtosecond lasers with wavelengths ranging from ultraviolet (UV) to mid-IR have been rapidly developed by using three-wave mixing in various nonlinear crystals [11–19]. In particular, wavelength-tunable few-cycle pulses have been generated at visible wavelengths using noncollinear optical parametric amplifier (NOPA) based on beta barium borate crystals. These pulses have been widely used in pump-probe experiments [20–24]. Femtosecond midIR pulses can be generated by difference frequency generation in nonlinear crystals [17,18,25], and they have been widely used in 2D-IR spectroscopy [26–28]. By achromatic broadband frequency doubling these NOPA pulses, wavelength tunable sub-10-fs pulses can be generated in the spectral region of 275–335 nm [29].

Four-wave mixing (FWM) has recently been investigated in various optically transparent media as a new method for the generation of tunable ultrashort pulses with ultrabroad spread spectral range. Tunable visible ultrashort pulses have been generated by FWM through filament generation in a laser-wave guiding long cell filled with argon gas [30]. In addition, femtosecond pulses in the deep UV (DUV) and mid-IR have been generated by FWM through filamentation in a gas cell [31–34]. Pulses at various UV wavelengths have been generated by cascaded FWM in hollow fibers filled with noble gases [35,36]. Sub-10-fs DUV pulses have also been generated by FWM and third-harmonic generation (THG) in gaseous media [37,38].

It was found that ultrabroad spectra and wavelength-tunable ultrashort pulses could be generated in bulk media by FWM, if the two pump beams have a finite crossing angle in the medium [39–61]. Wavelength-tunable mid-IR pulses could be obtained in the range of 2.4–12 μm by FWM in CaF_2 and BaF_2 plates [39,40]. An idler pulse with ~30-fs duration around 300 nm was obtained by four-wave optical parametric chirped-pulse amplification (OPCPA) in a fused silica plate [41]. This is one of the other third-order nonlinear optical processes. In the visible region, spatially separated cascaded FWM multicolored sidebands have been generated in BK7 glass [42–44], fused

DOI: 10.1201/9780429196577-12

silica [45–48], and a sapphire plate [49,50]. Up to as many as 15 sidebands can be obtained, and the spectrum of the generated sidebands can extend over more than 1.5 octaves from UV to the near-infrared (NIR) [44,45]. Multicolored sidebands have also been observed in many nonlinear crystals [51–61]. This phenomenon has been explained in terms of different-frequency resonant FWM, and processes which are known such as cascaded stimulated Raman scattering or coherent anti-Stokes Raman scattering are involved. By combining these sidebands into a single beam, isolated 25- and 13-fs pulses were obtained in $LiNbO_3$ and $KTaO_3$ crystals, respectively [53,54]. It is expected that these multicolored sidebands with broadband spectrum can be used to generate near-single-cycle pulses [43,44]. These multicolored femtosecond pulses can be conveniently used in multicolored pump-probe experiments. The generated multicolored sidebands contain very similar wavelengths as the emission wavelengths of various fluorescent proteins, such as green fluorescent protein (GFP), cyan fluorescent protein (CFP), and red fluorescent protein (RFP) [62] and some semiconductor quantum dots [63,64]. Therefore they are considered to be useful by being used in simultaneous multicolored imaging of biological samples by nonlinear optical microscopy [65–67]. In addition to these FWM processes, by inducing another intense pump pulse, a weak seed pulse can be amplified by noncollinear four-wave optical parametric amplification (FWOPA) in a transparent bulk Kerr medium in the UV and NIR spectral regions [68–74].

This subsection presents the recent research that we have done in which we used FWM to generate and optimize femtosecond laser pulses. It is organized as follows. First, the mechanism of cascaded FWM is presented. Second, we discuss the generation of wavelength-tunable multicolored 15-fs pulses by nondegenerate cascaded FWM in a bulk medium. Self-diffraction (SD) (also known as degenerate cascaded FWM) is described as a powerful method for cleaning femtosecond laser pulses. Then, we describe using FWM in a gas cell or a hollow fiber to generate ultrashort pulses in the UV region. Finally, FWOPA is described and used to simultaneously amplify and compress generated FWM signals. Finally, conclusions and future prospects are given.

1.2.3.2   CASCADED FWM IN BULK MEDIA

UV pulses were generated by cascaded FWM in a gas-filled hollow fiber with a collinear configuration in 2001 [36]. Bulk media have large dispersions compared with gaseous media. To generate cascaded FWM signals, there should be a small crossing angle between the two incident beams in the medium so as to satisfy the phase-matching condition.

1.2.3.2.1   PRINCIPLE OF CASCADED FWM

Cascaded FWM processes are schematically depicted in Figure 1.2.3.1a. The two input beams have wave vectors k_1 and k_2 that have frequencies ω_1 and ω_2 ($\omega_1 > \omega_2$), respectively. Cascaded FWM is deconstructed step by step in Figure 1.2.3.1b–e. In the first step, two $k_1^{(1)}$ photons interact with a $k_2^{(1)}$ photon to generate a first-order anti-Stokes photon k_{AS1}. A subsequent FWM process among the generated k_{AS1} photon, one $k_2^{(1)}$ photon, and one $k_1^{(1)}$ photon generates a second-order anti-Stokes photon k_{AS2}. Thus, all the processes are third-order nonlinear FWM processes. Higher-order signals are obtained from the generated lower-order signals; hence, the process is called *cascaded* FWM. The mth-order anti-Stokes sideband has the following phase-matching condition for the different mth-order components: $\boldsymbol{k}_{ASm} = \boldsymbol{k}_{AS(m-1)} + \boldsymbol{k}^{(m)} - \boldsymbol{k}_1^{(m)} \sim (m+1)\boldsymbol{k}_1^{(1)} - m\boldsymbol{k}_2^{(1)} \sim (m+1)\omega_1^{(1)} - m\omega_2^{(1)}$. In each FWM step, $\boldsymbol{k}_1^{(m)}$ and $\boldsymbol{k}_2^{(m)}$ photons have the same direction but different frequencies ($\omega_1^{(m)}$ and $\omega_2^{(m)}$) and wave vector magnitudes $\left|k_1^{(m)}\right|$ and $\left|k_2^{(m)}\right|$. On the Stokes side, the mth-order Stokes sideband will have the following phase-matching condition: $k_{s^m} = k_{S(m-1)} + k_2^{(-m)} - k_1^{(-m)} \approx (m+1)k_2^{(-1)}$

$$mk_1^{(-1)}, \ \omega_{Sm} \approx (m+1)\omega_2^{(-1)} - m\omega_1^{(-1)}$$

It will be a degenerate cascaded FWM process, if $\omega_1 = \omega_2$.

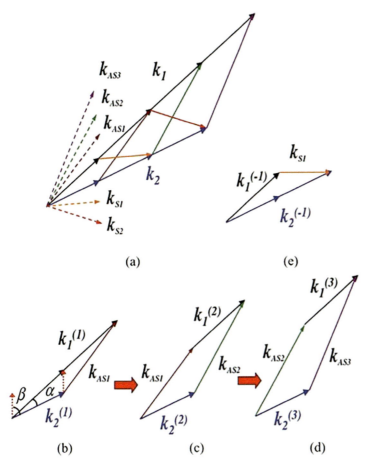

FIGURE 1.2.3.1 (a) Phase-matching geometry for cascaded FWM. Phase-matching geometries for generating (b) AS1, (c) AS2, (d) AS3, and (e) S1. k_1 and k_2 are the two input beams. The angle α is the crossing angle between the two input beams in the medium [48].

Using the phase-matching condition, the dependences of the output central wavelength and the exit angles of the cascaded FWM signals on the incident crossing angle were calculated. The results showed that the central wavelength of the generated cascaded FWM signals shift to shorter wavelengths as the incident crossing angle increases [48]. In addition to the phase-matching condition, it is also necessary to take into account the group velocity delay between the incident pulses and the generated sidebands. The calculation results reveal that using a material with a smaller dispersion and thickness will give a broader output spectrum [48]. However, a thin nonlinear medium has low energy-conversion efficiency. Therefore, for the selection of medium thickness, we must think about the tradeoff between the spectral bandwidth and the efficiency. It is determined based on the requirements for laser pulse used in the experiment to be performed.

A numerical simulation of this process revealed the main characteristics of highly nondegenerate cascaded FWM of noncollinear femtosecond pulses in the spatial, spectral, and temporal domains [44].

1.2.3.2.2 GENERATION OF WAVELENGTH-TUNABLE SELF-COMPRESSED MULTICOLORED PULSES BY NONDEGENERATE CASCADED FWM

According to the phase-matching condition, to generate wavelength-tunable multicolored pulses in a bulk medium by cascaded FWM, the two incident pulses should have different central wavelengths. In the present case, in addition to the Ti:sapphire laser pulse at 800 nm (beam 1), a pulse with a wavelength around 700 nm was generated by filtering the broadband spectrum generated in a hollow-fiber compressor (beam 2). The experimental setup is described in detail in [45–48].

Negatively chirped or nearly transform-limited output pulses can be produced by FWM when one pump beam is negatively chirped, and the other is positively chirped, both of which are properly chirp controlled [47,48]. This principle can be easily explained. Both chirped input pulses can be written as $E_j(t) \propto \exp\{i[\omega_{j0}t + \varphi_j(t)]\}$, with $j = 1,2$, where beam 1 is negatively chirped ($\partial^2\varphi_1(t)/\partial t^2 < 0$), and beam 2 is positively chirped ($\partial^2\varphi_2(t)/\partial t^2 > 0$).

The mth-order anti-Stokes signal can be expressed as

$$E_{ASm}(t) \propto \exp\{i[((m+1)\omega_{10} - m\omega_{20})t + (m+1)\varphi_1(t) - m\varphi_2(t))]\}$$

Given $\partial^2\varphi_1(t)/\partial t^2 < 0$ and $\partial^2\varphi_2(t)/\partial t^2 > 0$, we obtain

$$\partial^2\varphi_{ASm}(t)/\partial t^2 = (m+1)\partial^2\varphi_1(t)/\partial t^2 - m\partial^2\varphi_2(t)/\partial t^2 < 0$$

This implies that the mth-order anti-Stokes signal is also negatively chirped. Nearly transform-limited pulses will be produced when the negative chirp of the anti-Stokes sidebands just compensates for the dispersion of the transparent bulk medium and the phase change in the medium. This phase-transfer method has also been used for three-wave mixing [75,76].

The photograph in Figure 1.2.3.2a shows the generated multicolored sidebands viewed on a white sheet of paper placed about 30 cm behind the glass plate. These sidebands are well separated in space. The incident pulse of beam 1 was positively chirped from 35 to 75 fs by passing it through a bulk medium. The pulse of beam 2 was negatively chirped to 45 fs by chirped mirrors. The diameter of beam 1 (beam 2) on the surface of the 1-mm-thick fused silica plate was 250 μm (300 μm) in the vertical direction and 300 μm (600 μm) in the horizontal direction. Beams 1 and 2 had input pulse energies of 24 and 15 μJ.

The spectra of the sidebands from the first-order anti-Stokes (AS1) through to the fifth-order anti-Stokes (AS5) signals extend from 450 to 700 nm when the incident crossing angle was 1.78° (see Figure 1.2.3.2b). The spectra of AS2 and AS3 can be simultaneously tuned by varying the incident crossing angle in the range of 1.78°–2.78° (see Figure 1.2.3.2c). The spectrum of AS3 at 1.78° is located between the spectra of AS2 when the crossing angle is between 2.23° and 2.78°, demonstrating that the sideband spectra are continuously tunable with no gap in between. The central wavelength of the generated sideband is different in different bulk media, even at the same incident crossing angle [48].

We can generate 15-fs AS1 and 16-fs AS2 pulses by simply adding a glass plate to compensate the negative chirp. The pulses were measured using a cross-correlation frequency-resolved optical gating (XFROG) and were retrieved using commercial software (Femtosoft Technologies).

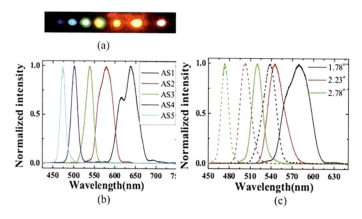

FIGURE 1.2.3.2 (a) Photograph of sidebands on a sheet of white paper placed 30 cm after the glass plate when the crossing angle between the two input beams is 1.78°. The first, second, and third spots from the right-hand side are beam 2, beam 1, and AS1, respectively. (b) Spectra of the sidebands AS1 to AS5 (from longer to shorter wavelength) when the crossing angle between the two input beams is 1.78°. (c) Spectra of AS2 (solid lines) and AS3 (dashed lines) for crossing angles of 1.78°, 2.23°, and 2.78° (from longer to shorter wavelength for AS2 (solid curves) and AS3 (dashed curves)). [47].

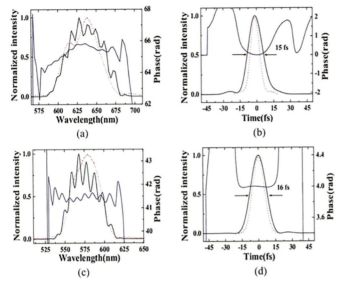

FIGURE 1.2.3.3 (a) Recovered spectrum (solid black curve), spectral phase (solid blue curve), and measured spectrum (dotted red curve) of AS1. (b) Recovered intensity profile and phase of AS1. The dotted red curve is the transform-limited (9 fs) pulse profile of AS1. (c) Recovered spectrum (solid black curve extending ~600nm to ~675nm), spectral phase (solid blue curve extending from 600nm to 685nm), and measured spectrum (dotted red curve) of AS2. (d) Recovered pulse profile (curve with 16-fs FWHM mark) and phase (nearly constant phase value around 4.0 radian) of AS2. The dotted red curve is the transform-limited (12 fs) pulse profile of AS1 [47].

A 1-mm-thick CaF2 plate is used to compensate the dispersion of AS1. The retrieved spectral phase of AS1 still exhibits a clear negative chirp. Figure 1.2.3.3a–d shows the recovered intensity profiles, spectra, and phases of AS1 and AS2. The spectral phase indicates that both pulses have a small negative chirp. A shorter pulse is expected to be obtained when the negative chirp is completely compensated. AS1 and AS2 have output pulse energies of 0.65 and 0.15 μJ, respectively. Due to self-focusing and FWM, the spatial modes of the sidebands have perfect Gaussian profiles and good beam qualities, as reported earlier [45–48]. Angular dispersion occurs in the generated cascaded FWM beams due to phase matching in the noncollinear parametric process [48].

1.2.3.2.3   PULSE CLEANING BY DEGENERATE CASCADED FWM

When the two incident pulses have the same central wavelength, the generated cascaded FWM signals will have the same central wavelength as the incident pulses. This process, which is also known as SD, has been extensively used to measure pulse duration using SD-FROG [77]. It has recently been used to smooth laser spectra and clean laser pulses after passing through a hollow-fiber compressor [10,78].

Pulses with high temporal contrast are important for generating plasmas, since they suppress the generation of undesirable preplasma [79,80]. Recently, a third-order nonlinear process, cross-polarized wave generation, has been studied extensively, and it has been used to enhance the temporal contrast of femtosecond laser pulses [81]. Like the cross-polarized wave generation process, SD is a third-order nonlinear process that can also be used to improve the temporal contrast of femtosecond pulses. SD has the advantage that the generated SD signals are spatially separated from the incident beams so that polarization discrimination is not necessary. In a nonresonant electronic Kerr medium, SD occurs over a femtosecond time scale because of inertia-free interaction [82]. Consequently, it can be used to clean even picosecond pulses. The intensity of the first-order SD signal (SD1) can be described by the following equations in both frequency and time domains [77]:

$$I_{sd1}(\omega_{sd1}) \propto \left| \begin{array}{c} \iint d\omega_1 d\omega_{-1} \chi^{(3)} \tilde{E}_1^*(z,\omega_1) \tilde{E}_{-1}(z,\omega_{-1}) \\ \tilde{E}_1\left(z, \omega_{sd1} - \omega_{-1} + \omega_1\right) \\ \sin c\left(\Delta k_z(\omega_{sd1},\omega_1,\omega_{-1})\frac{L}{2}\right) \end{array} \right|^2$$

$$I_{sd1} \propto I_1^2(t)I_{-1}(t-\tau)$$

where ω_{sd1} is the angular frequency of the SD1 signal and ω_1 and ω_{-1} are those of the two incident beams, Δk_z is the phase mismatch, and L is the path length in the medium. As can be seen, the spectral intensity of the SD1 signal is an integral of the spectral intensity of the two incident pulses. This implies that the SD1 signal intensity for each wavelength component is an average contribution over the whole spectral range of the incident pulses. Therefore, the SD1 signal spectrum is automatically smoothed. The pulse duration of the SD1 signal will be at most √3 shorter than that of the incident pulse [83].

A proof-of-principle experiment was performed with a Ti:sapphire laser. Two incident beams were focused into a 0.5 mm-thick fused silica plate with a crossing angle of about 1.5°. The silica plate was located about 20 mm behind the focal point. Both beams had $1/e^2$ diameters of about 360 μm on the glass plate. The transmission pulse energies of the two incident beams after the glass plate, beam_1 and beam_−1, were 40 and 51 μJ, respectively. The SD1 signals generated in addition to beam_1 and beam_−1 had pulse energies of about 5 and 6 μJ. The energy-conversion efficiency from the input laser beams to the two SD1 signals was about 12%. Due to the low pulse energy and low power of the generated SD signals, we performed a second-order autocorrelation (SAC) measurement to measure the temporal contrast. This measurement requires a much lower incident pulse energy and is more sensitive to low-energy noise than other measurement techniques that use third-order nonlinear processes because it involves only one second-order nonlinear process [77,84]. The pulse durations of the input pulse and the SD signal were measured using a second harmonic generation FROG (SHG-FROG).

Figure 1.2.3.4 shows that the pulse is cleaned, even in 1 ps, and that extraneous components are removed, while the main pulse remains. For the incident pulse, the SAC peak intensity around ±0.7 ps is about 1.2×10^{-2} of the main pulse. The SAC of the SD1 signal has a small peak at the same delay that is less than 1.2×10^{-6} of the main pulse, which is four orders of magnitude smaller and is less than the cube of 1.2×10^{-2} (i.e., 1.7×10^{-6}). The pulse is self-compressed in this process; in addition, self-focusing also improves the temporal contrast. In nonresonant electronic Kerr media, self-focusing is instantaneous and it has a power threshold due to competition with diffraction. The intensity-dependent self-focusing effect increases the main pulse intensity, whereas amplified spontaneous emission and noise peaks are not enhanced. Consequently, the intensity enhanced main pulse has a much improved temporal contrast. When the glass plate is located after the focal point, the generated SD1 signal has a smaller divergence angle than the scattered light due to self-focusing; this reduces the noise of the scattered light.

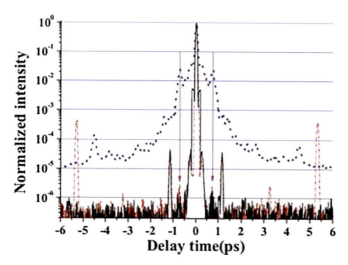

FIGURE 1.2.3.4 SAC intensity of the incident pulses (dotted blue curve) and SAC intensity of the SD1 signal when the incident beams were incident at the Brewster angle (solid black curve) and perpendicular (dash-dotted red curve) to the glass plate for delay times from −6 to 6 ps and with a 5-fs/step resolution [78].

Due to the convolution effect, the spectrum of the SD1 signal is smoother and broader than the input laser spectrum (see Figure 1.2.3.5a)

The pulse duration of the SD1 signal for a zero delay time was shortened from 75 to 54 fs relative to the input pulse (see Figure 1.2.3.5b). The retrieved temporal and spectral phases of the SD1 signal were found to be smoothed with some positive chirp. In the medium, self-phase modulation (SPM) and cross-phase modulation (XPM) accompany SD. The peak wavelength of the SD1 signal was shifted about ±10 nm at a delay time of ±33 fs (the positive sign indicates that beam_1 is ahead of beam_−1) due to XPM and the small-frequency chirps of the incident pulses (see Figure 1.2.3.5c) [74]. The retrieved spectral phase also shows that the reduction or enhancement of the chirp rates depends on the sign of the delay time for the same SD1 signal (see Figure 1.2.3.5c). Using suitable delay times and chirps of the incident pulse will induce self-compression of the SD1 signal to a nearly transform-limited pulse. As shown in Figure 1.2.3.5b, the pulse duration was shortened to 39 fs, which is close to its transform-limited pulse duration of 33 fs.

As in cascaded FWM experiments [45–48] and pulse compression experiments in bulk media [85], the spatial profile and beam quality of the SD1 signal were improved in this SD process compared with the input laser beam due to spatial filtering induced by the self-focusing in the medium which is another third-order nonlinear optical phenomenon. The 2-D beam profile of the SD1 signal is improved from an asymmetric incident beam to a nearly symmetric Gaussian beam (see inset of Figure 1.2.3.5d). M^2 of the SD1 beam was also improved from 1.6 of the input laser beam to 1.3. M^2 is the value representing the laser beam quality factor. This value of laser beam for single-mode gaussian beam it is 1.0 and provide minimal spot size at the focus plane and beam with M^2 larger than unity spot size is larger. In this case mode pattern is considered to be decomposed into multiple spatial modes which are composing orthogonal sets of mode functions in two-dimension space.

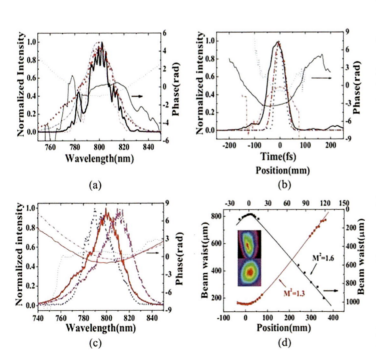

FIGURE 1.2.3.5 (a) Measured spectrum (thick curve) and retrieved spectral phase (thin curve) of the incident pulse (solid black curve) and the SD1 signal (dotted red curve). The thin dash-dotted blue and magenta curves show the retrieved spectra of the SD1 and the incident pulse, respectively. (b) Retrieved temporal profile (thick curve) and temporal phase (thin curve) of the incident pulse (solid black curve), and the SD1 signals at delay times of 0 fs (dash-dotted red curve) and + 33 fs (dotted blue curve). (c) Measured spectra (thick curve) and retrieved spectral phase (thin curve) of the SD1 signals for delay times of 0 fs (solid red curve), −33 fs (dash-dotted magenta curve), and + 33 fs (dotted blue curve). (d) M^2 and 2-D beam profiles of the incident beam (black squares; upper pattern) and the SD1 signal (red circles; lower pattern) [78].

1.2.3.3   UV PULSE GENERATION BY FWM IN HOLLOW FIBER

This subsection discusses in terms of degenerate FWM in a gas to generate ultrashort DUV pulses. Unlike bulk media, gases cannot be optically damaged making them suitable media for frequency conversion of intense pulses. The gas may be suffered from breakdown by ionization to form plasma composed of ions of gas material. This kind of phenomena takes place at much higher intensity level of focused laser.

Here we discuss the case of lower level than such processes as break down case but high enough for the nonlinear optical phenomena of frequency conversion take place. Since a gas medium has a much lower density than a bulk medium, a long interaction length or input pulses with high peak intensities are necessary to achieve a high frequency-conversion efficiency. There are two ways to realize a long interaction length: filamentation or using a hollow waveguide. The former method forms a filament by spatially and temporally overlapping two input laser pulses (i.e., the pump and idler pulses) [30,31,34]. The laser beams propagate with a constant beam size in the filament so that a long interaction length is realized. An energy-conversion efficiency from the pump to the signal of 4% and generation of a DUV pulse with an energy of 20 µJ have been demonstrated [31]. In this approach, input pulses with high peak intensities are necessary to form a filament. The signal-generation stability depends on the stability of the filament. The latter method uses a hollow waveguide to achieve a long interaction length, which is determined by the waveguide length [37] and is independent of the input pulse parameters. Unlike free-space propagation, the waveguide has a negative group-velocity dispersion (GVD) that is used to cancel the positive GVD induced by the gas medium. This allows the phase-matching condition to be satisfied for a high energy-conversion efficiency [36,37]. FWM in a hollow waveguide is thus applicable for energy conversion of laser pulses with low peak intensities as well as laser pulses with high peak intensities. Another advantage of this method is the high beam quality of the signal pulse generated in the waveguide.

FWM in a gas has been utilized to generate ultrashort DUV pulses. FWM in a filament has been used to generate 12-fs DUV pulses [31] and FWM in a hollow waveguide can produce 8-fs DUV pulses [37] after passing through a dispersion compensator which is composed of a grating compressor. Wavelength-tunable UV laser pulses can also be generated by combining FWM with NOPA [86]. Since chirped mirrors are not available in the DUV region because of poor reflectivity, dispersion compensation in this wavelength region must be performed using a grating compressor or a prism compressor. Since both compressors produce large third-order dispersion (sometimes too large for fine tuning), it may be necessary to use two of the following three elements, a grating compressor, a prism compressor, and a deformable mirror, to compensate third or higher order dispersion to generate a pulse shorter than 8 fs by fine-tuning. However, this requires a complex optical setup with some adjustable stage or mover to obtain the best chirp compensation condition, which leads to a large energy loss.

Wojtkiewicz et al. proposed a method for achieving chirped pulse FWM [87,88]. In this scheme, chirped input pulses are used to generate a negatively chirped signal pulse. The chirped signal pulse can be compressed by propagation through a transparent medium so that no external pulse compressor is required. Precise dispersion compensation is possible by selecting an appropriate transparent medium, such as magnesium fluoride that has no appreciable high-order GVD in the DUV region [87–90]. Since the input pulses are chirped and have lower peak intensities than the corresponding transform limited pulses, a hollow fiber is suitable for chirped-pulse FWM.

Because chirped input pulses are used, SPM and XPM, which broaden the signal spectrum [37], hardly occur so that the signal pulse bandwidth is mainly determined by the input pulse bandwidths. To generate a sub-10-fs DUV pulse by chirped-pulse FWM, it is necessary to use a broadband idler or a pump pulse that supports a sub-10-fs transform-limited pulse duration [90].

Our group has recently generated a sub-10-fs DUV pulse by using a self-phase-modulated pulse as the input idler for chirped-pulse FWM [91]. Using a broadband idler supporting a sub-10-fs transform-limited pulse duration leads to the generation of a DUV pulse with a pulse duration shorter than 10 fs. The following section describes the principle of broadband chirped-pulse FWM using a self-phase-modulated idler pulse. After presenting a scheme for this process, the properties of the experimentally generated sub-10-fs DUV pulses are discussed.

1.2.3.3.1 Chirped-Pulse FWM in a Gas-Filled Hollow Waveguide

In degenerate FWM with energy conservation $\omega_{\mathrm{sig}} = 2\omega_{\mathrm{pump}} - \omega_{\mathrm{idler}}$, using a negatively chirped pump pulse and a positively chirped idler pulse produces a negatively chirped signal pulse [87,88]. Figure 1.2.3.6 shows a schematic diagram of this process (it is similar to that shown in [89]. In addition, the figure shows that the negative frequency chirp in the signal is due to the positive frequency chirp in the idler. The idler frequency increases with time, whereas the signal frequency decreases with time. A negative frequency chirp in the pump pulse leads to a negative frequency chirp in the signal pulse, which can be explained in a similar manner as the relation between the frequency chirps of the idler and the signal.

Signal pulse generation by FWM in a gas-filled hollow fiber generated by input pulses propagating along the z-axis is expressed by [92]

$$\frac{\partial \varepsilon_s}{\partial z} = iD_s \varepsilon_s + i\left(\frac{\omega_s}{c}\right) n_2 T_s \left\{ \left[|\varepsilon_s|^2 + 2|\varepsilon_p|^2 + 2|\varepsilon_i|^2 \right] \varepsilon_S + \varepsilon_p^2 \varepsilon_i^* \exp\left(-i\Delta\beta z\right) \right\} \tag{1.2.3.1}$$

$$D_S = \frac{-\alpha_S}{2} - i\left(\frac{\beta_S^{(2)}}{2}\right)\left(\frac{\partial^2}{\partial t^2}\right) + i\left(\frac{\beta_S^{(3)}}{6}\right)\left(\frac{\partial^3}{\partial t^3}\right) + \ldots \tag{1.2.3.2}$$

where β_s is the propagation constant of the signal pulse inside the hollow fiber [93], the asterisks indicate complex conjugates, ω_s is the angular frequency of the electric field, and n_2 is the nonlinear refractive index of the core medium. The phase mismatch is given by $\Delta\beta = \beta_i + \beta_s - 2\beta_p$ and the higher order dispersion is $\beta_S^{(n)} = \partial^n \beta / \partial \omega^n \big|_{\omega = \omega_k}$ α_s is the loss constant of the gas-filled hollow waveguide at the signal frequency [93]. $T_s = \left\{ 1 + (i/\omega_s)(\partial/\partial t) \right\}$ contains the effect of self-steeping, A_{eff} is the effective core area, and c is the speed of light in vacuum [94,95]. The complex electric field amplitudes of the signal, idler, and pump pulses are, respectively, ε_s, ε_i, and ε_p in Eq. (1.2.3.1), which is expressed in a frame of reference propagating with the signal group velocity.

For chirped-pulse FWM with input pulses having low peak intensities and a low-density gas medium, the terms related to SPM and XPM in Eq. (1.2.3.1) may be dropped. By assuming an input pump energy that is sufficiently low to neglect the pump depletion, a negligibly small GVD due to the low gas density to satisfy phase matching [37], and a negligibly low propagation loss, the propagation of the signal pulse generated by FWM is expressed as

$$\varepsilon_S(t,z) = \left(\frac{zn_2\omega_S}{cA_{\mathrm{eff}}}\right) A(t)_p^2 \, A(t)_i \, \exp\left[i\left(2\phi_p(t) - \phi_i(t) + \frac{\pi}{2} \right) \right] \tag{1.2.3.3}$$

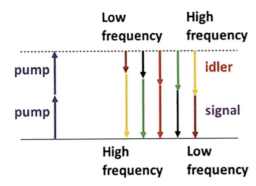

FIGURE 1.2.3.6 Energy diagram for chirped-pulse FWM.

In deriving Eq. (1.2.3.3), $\varepsilon_{p,i}(t) = A_{p,i}(t)\exp[i\phi_{p,i}(t)]$ was substituted into the coupled Eq. (1.2.3.1), where $A_{p,i}(t)$ and $\phi_{p,i}(t)$ are, respectively, the real amplitudes and phases of the pump and idler. Differentiating the phase term in Eq. (1.2.3.3) gives the time evolution of the instantaneous signal frequency, which is expressed as $2d\phi_p(t)/dt - d\phi_i(t)/dt$. This explains the relation between the frequency chirp in the input pulses and the chirp in the signal, as discussed earlier.

1.2.3.3.2 BROADBAND CHIRPED-PULSE FWM

To generate a broadband signal pulse by chirped-pulse FWM, a broadband pump pulse or a broadband idler pulse is necessary. For this purpose, a self-phase-modulated pulse is used as an input pulse for chirped-pulse FWM. By using SPM in a hollow waveguide filled with a noble gas [96], it is possible to expand the spectral width of a 30-fs pulse generated by a commercial Ti:sapphire chirped-pulse amplifier to a spectral width that supports 5-fs transform-limited pulses. However, a self-phase modulated pulse contains nonlinear phase distortion in its temporal profile, which is related to the spectral intensity and phase distribution of the pulse. Specifically, SPM induces phase modulation within the temporal profile of a laser pulse that varies proportionally with the temporal intensity distribution of the laser pulse [94]. The pulse contains a positively chirped central region that is suitable for chirped-pulse FWM, but it also contains a negatively chirped time range in both leading and trailing edges. Such nonlinear frequency modulation produces dips in the spectrum. When a self-phase-modulated pulse is used as the input for chirped-pulse FWM, the nonlinear phase distortion in the SPM idler may be transferred to the signal pulse, resulting in a nonlinear temporal phase and a nonuniform signal spectrum due to complex interference among the different time domains. However, this does not matter, if the pump pulse is much shorter (i.e., less than about 50%) than the idler pulse. Since a signal pulse is generated only in the temporal region in which the two input pulses overlap, by employing a short pump pulse, it is possible to generate a signal pulse that results from only the interaction between the central region (which is free of nonlinear temporal chirp) of the self-phase-modulated idler and the pump pulse. In this case, the idler pulse behaves as if it is a linearly chirped broadband pulse. When the amplitudes of the input idler and pump pulses are given by Gaussian functions and the pump pulse is at least two times shorter than the idler pulse, the signal becomes a linearly (negatively) chirped Gaussian pulse (see Figure 1.2.3.7) and a smooth spectral shape is obtained. A nearly Fourier transform limited (FT-limited) pulse duration is available for the signal after compensating the negative frequency chirp, provided that a positive group-delay dispersion (GDD) is added to the signal without adding substantial high-order GVD. Such dispersion compensation is realized using a transparent medium (e.g., magnesium fluoride) whose absorption wavelength is far from the signal wavelength.

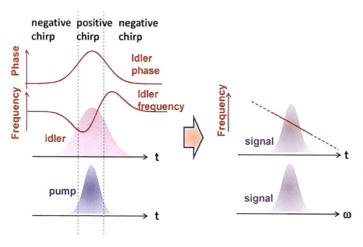

FIGURE 1.2.3.7 Frequency sweep of a SPM idler with respect to time for a pump pulse that is shorter than the idler. The shorter pump produces a linearly chirped signal and a smooth spectral profile.

1.2.3.3.3 PRACTICAL ISSUES IN BROADBAND CHIRPED-PULSE FWM

Figure 1.2.3.8 shows an example of an experimental setup for demonstrating the afore-described broadband chirped-pulse FWM. A femtosecond NIR pulse from a Ti:sapphire chirped-pulse amplifier is split into two pulses. One is used for generating a broadband idler and the other is used for generating a pump pulse by frequency doubling the NIR pulse. Before the idler is used as the input for FWM, its spectral width is expanded by SPM in a gas-filled hollow waveguide. The near-UV (NUV) pulse generated through frequency doubling is stretched by a prism pair (prism compressor) that induces a negative frequency chirp.

The NUV pump pulse duration is lengthened in this process, but it remains shorter than the broadband idler pulse duration. Finally, the SPM idler and the negatively chirped pump pulses are spatially and temporally synthesized in a gas-filled hollow waveguide to generate a signal pulse by FWM. Filamentation in a gas cell [30,31] could also be used when the input pulse energies are above about 1 mJ.

As discussed earlier in this subsection, the pump pulse should be shorter than the idler to generate a signal pulse with a smooth spectral shape. Two procedures can be considered to realize this condition: pulse broadening (t-broadening) in a transparent medium before (Procedure 1) or spectral broadening (ω-broadening) by SPM after (Procedure 2).

Procedure 1: t-broadening first: The self-phase modulated idler has a broad spectral width, and consequently the temporal profile of the self-phase modulated idler is substantially distorted by high-order dispersion in a transparent medium. Moreover, the phase of the self-phase modulated pulse is nonlinear in terms of its temporal evolution. The positive chirp in the central region is enhanced by the positive GVD in the transparent medium, while the negative chirps in the leading and trailing edges are compensated and produce intense spikes due to the GVD. This affects the temporal profile and, thus, the spectrum and spectral phase of the signal generated by FWM are distorted. Furthermore, the temporal width of the positively chirped region of the self-phase modulated idler is not effectively expanded in this case.

Procedure 2: ω-broadening first: On the other hand, pulse broadening of the idler before spectral broadening is useful. In this case, the positive frequency chirp induced by the transparent medium and the positive frequency chirp induced by SPM both contribute to generate a positive chirp. This leads to a wide positively chirped region within the idler pulse duration because of the extended pulse duration of the idler prior to SPM. The positive chirp due to the transparent medium, which is produced before SPM, partially cancels the negative chirp at the leading and trailing edges of the idler pulse induced by SPM. Since the pulse duration of the idler is already broadened by GVD in the transparent medium, it is not necessary to significantly increase the pulse duration after spectral broadening. Distortion

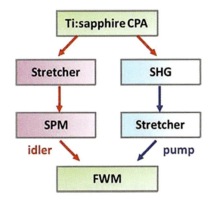

FIGURE 1.2.3.8 Scheme for broadband chirped-pulse FWM. SHG, second harmonic generation; CPA, chirped-pulse amplifier.

in the temporal profile of the broadband idler can be minimized in this case, making it possible to generate a signal with smooth temporal and spectral profiles. It also minimizes the high-order spectral phase in the signal, resulting in the generation of a single ultrashort pulse after dispersion compensation.

A prism compressor can be used for providing negative-chirping to a NUV pump pulse with a narrow spectral width whose transform limited pulse duration is longer than about 30 fs. A prism compressor generally induces a large negative third-order dispersion as well as a negative group-delay dispersion. This distorts the temporal profile and the phase of a pulse when the pulse has a wide bandwidth that can support a pulse duration shorter than 10 fs. However, this does not matter for narrowband pulses since the nonlinear spectral phase distortion due to the third-order spectral phase distortion lies outside the narrow bandwidth. Chirped mirrors can also be used to negatively chirp a NUV pulse. Although chirped mirrors for NUV pulses have oscillations in their dispersion curves, their effect may not be significant for a narrowband pulse that has a transform-limited pulse duration longer than 30 fs.

By spectral broadening (ω broadening) of the idler pulse rather than that of the NUV pump pulse and by prechirping the idler pulse prior to spectral broadening (t-broadening first), it is possible to generate a nearly linearly chirped signal pulse. This results in minimal distortion of the phases and temporal profiles of the input pulses, and, thus, provides smooth temporal profile and phase to the signal pulses. The experimental scheme discussed above is used in [91].

1.2.3.3.4   SUB-10-FS DUV PULSES GENERATED BY BROADBAND CHIRPED-PULSE FWM

The energy-conversion efficiency from the pump to the signal is determined by the input pump and idler intensities. Since the responsive efficiency is nonlinear with respect to the pump intensity, the conversion efficiency of the partial fractional intensity of the pump to the signal is much more sensitive to the pump intensity than the case of the conversion efficiency of the pump to the idler. Because of this nonlinearity when chirped input pulses are used, chirped-pulse FWM has a lower energy conversion efficiency than FWM that uses transform-limited input pulses [31,37]. Therefore, chirped-pulse FWM is not useful for generating high-energy DUV pulses when only a low-energy pump pulse is available. In the case of a chirped pump pulse with a pulse energy of about 100 μJ, the pulse energy of a signal pulse generated by broadband FWM is limited to 300 nJ [91]. Increasing the pulse energy (and thus the intensity) of the pump pulse will significantly increase the signal pulse energy. Using a 1-mJ pump pulse, it is expected to be possible to generate a signal pulse with an energy of several tens of microjoules [88,90,92].

It is possible to generate a broadband signal pulse with a spectrum extending from 260 to 290 nm using FWM with a broadband chirped pulse [91]. In general, the signal spectrum differs from both spectra of the idler and pump pulse except partial degenerate FWM case. The difference in the spectrum in some cases is introduced by the effects of pulse durations. For example, in an experiment, the idler spectrum contained two deep dips, whereas the signal spectrum had only one deep dip (see Figure 1.2.3.9) [91]. This difference in spectral shapes is related to the different pump and idler pulse durations, as discussed above. Contrary to the discussion, the spectral shape of the signal is not a single peaked (Gaussian shape) pulse. This is due to the pump pulse not being sufficiently shorter than the idler pulse. A signal pulse with a Gaussian-shaped spectrum can be generated, if a larger chirp than that in the experiment is added to the idler prior to spectral broadening by SPM [92].

A broadband negatively chirped DUV pulse is generated in broadband chirped-pulse FWM. The pulse can be easily compressed by a normal GVD induced by a transparent medium. Unlike the case when prism and grating compressors are utilized, third-order dispersion is not introduced in this case, and the pulse energy loss is small. For example, in an experiment, a DUV pulse generated by broadband chirped-pulse FWM had a negative GDD of 265 fs^2 [91]. Some of the GDD was partially compensated by the positive GDD induced by the magnesium fluoride output window of the

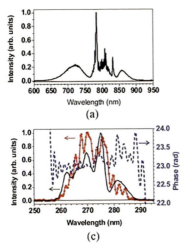

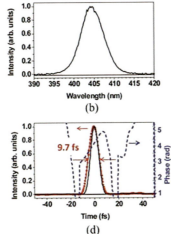

(a)

(b)

(c)

(d)

FIGURE 1.2.3.9 Spectra of the input (a) NIR and (b) NUV, and output DUV [solid line in (c)] pulses measured with spectrometers. The spectrum (solid line with filled circles) and spectral phase (broken line) measured by the SD-FROG are shown in (c), while the corresponding temporal profile (solid line with filled circles) and phase (broken line) are shown in (d). The inverse Fourier transform of the solid line in (c) is indicated by the solid line in (d) [91].

hollow-fiber chamber, which generated FWM. The residual negative GDD in the DUV pulse was compensated by propagation in the air [91]. In such a case of air compensation, experiment using the pulse output (e.g., femtosecond time-resolved pump-probe spectroscopy) must be well designed in such a way that the optimally shortened pulse duration is satisfied at the sample position. SD-FROG measurements are useful for optimizing the path length in air. In a FROG, the edge of an aluminum mirror is used for beam splitting, which results in negligible pulse broadening [37]. The spectral change of the SD signal with respect to the time delay of a replica is sensitive to the GDD of a pulse to be measured [97]. In the experiment, only three SD-FROG measurements were used to optimize the path length and, thus, the positive GDD induced by the air [91]. The DUV pulse duration was compressed to shorter than 10 fs. The same pulse compression can also be realized by varying the path length of the pulse in a pair of wedges made of a set of MgF_2 or CaF_2 thin plates.

The DUV pulse duration can be precisely compressed in broadband chirped-pulse FWM. The pulse compression procedure is relatively simple, as discussed above. The transform limited pulse duration estimated from an experimentally measured spectrum is 8 fs, while the corresponding measured pulse duration after dispersion compensation is 9.7 fs [91]. These pulse durations differ by only 24%.

The compressed signal pulse has a smooth temporal profile (see Figure 1.2.3.9c). It contains almost no satellite pulses; the energy of the satellite pulses is smaller than 5% of the total pulse energy. In broadband chirped-pulse FWM, a linearly chirped DUV pulse is obtained by using a pump pulse that is shorter than the idler pulse. The linearly chirped DUV pulse has a smooth temporal profile and, hence, has almost no high-order spectral phase distortion. An ultrashort DUV pulse with a single pulse structure can be generated by compensating the negative GDD in the DUV pulse without inducing a substantial high-order dispersion by using a transparent medium.

1.2.3.4 FOUR-WAVE OPTICAL PARAMETRIC AMPLIFICATION (FWOPA) IN BULK MEDIA

We have demonstrated using FWM to generate and clean femtosecond laser pulses in both gases and bulk media. Another FWM process, FWOPA, has recently been used to amplify ultrashort pulses in a glass plate at different wavelengths [69–74]. This method is particularly useful in the UV region, due to the lack of suitable nonlinear crystals in this region. Here, we demonstrate that a weak laser pulse can be simultaneously amplified and compressed by using another intense laser pulse [74].

The principle of this method is schematically illustrated in Figure 1.2.3.10a. An intense pump beam and a weak seed beam are focused onto a glass plate with a crossing angle α. When the pump and seed pulses are synchronous in time and overlap in space in a transparent bulk medium, the seed pulse spectrum will be broadened due to XPM in the medium induced by the intense pump pulse.

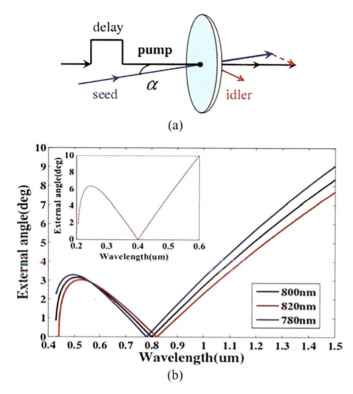

(a)

(b)

FIGURE 1.2.3.10 (a) Schematic showing experimental setup. α is the crossing angle. (b) Phase-matching curves for the crossing angle α as a function of the seed wavelength for fused silica (CaF$_2$, inset) when the pump pulse was fixed at wavelengths of 780, 800, and 820 nm (curves from the bottom to top are correspondingly for 820 nm, 800 nm, and 780 nm at the wavelength at 1um. (400 nm, inset) [74].

Furthermore, the weak seed pulse will be simultaneously amplified when the crossing angle α satisfies the phase-matching condition for FWM [68–74].

According to the phase-matching condition, the crossing angle α_{in} in the medium is given by $\cos\alpha_{in}[(2k_p)^2 + k_s^2 - k_i^2]/4k_pk_s$ terms of the wavenumber k, where the subscripts p, s, and i indicate the pump, seed, and idler beams, respectively. Figure 1.2.3.10b and its inset show phase-matching curves for the crossing angle in air α as a function of the seed wavelength for fused silica and CaF$_2$, respectively. The pump pulse has typical wavelengths of 800 and 400 nm for fused silica and CaF$_2$, respectively. There is a broad phase-matching bandwidth around 500 nm (250 nm) when the crossing angle α is about 3.1° (6.4°), as shown in Figure 1.2.3.10b.

The intense pump pulse at 800 nm simultaneously spectrally broadens and amplifies the incident seed pulse AS1 at 620 nm when the crossing angle α is around 2.80°±0.05°. Figure 1.2.3.11a shows the spectral profile and intensity of the amplified pulse as a function of the delay time t_{ps}. The 400-nJ incident pulse was amplified to 1.1 μJ with a 140 μJ pump pulse. Cascaded FWM signals were simultaneously generated around 500 nm with 250 nJ, as shown in the photograph in the inset of Figure 1.2.3.11a. Figure 1.2.3.11b shows the output energy of the seed pulse as a function of the pump intensity at a delay time of 0 fs. Much higher output energies are expected to be obtained when a cylindrical lens is used for focusing [69–71]. The thin glass plate ensured that the phase-matching spectral bandwidth was broad. Consequently, broadband amplification could still take place in the spectral range around 620 nm.

A quasi-linear chirp can be introduced across the weak seed pulse when the pump pulse is much wider than it [98]. The phase induced by XPM can also be compensated by using a pair of chirped mirrors. Furthermore, XPM-based compressors are more flexible than SPM ones because the phase can be tuned by the pump pulse independently from the idler pulse. After passing through the chirped mirrors for four bounces (−40 fs²/bounce), the pulse was compressed from 22.6 to 12.6 fs

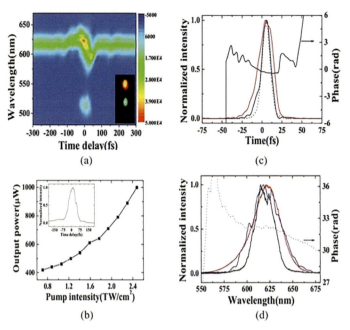

FIGURE 1.2.3.11 (a) Spectral profile and intensity of the output seed beam (AS 3) a function of the delay time t_{ps} when the crossing angle α was $2.80 \pm 0.05°$. (b) Dependence of the output energy of the output seed pulse on the pump intensity. The curve in the inset is the temporal profile of the pump pulse. (c) Retrieved temporal profiles of the incident seed pulse (solid red line) and the compressed output seed pulse (solid black line), and the transform-limited pulse of the broadened spectrum (dotted blue line). (d) Retrieved spectrum (solid blue line) and spectral phase (dotted blue line with noisy structure) of the output seed pulse. Measured spectra of the incident seed pulse (solid black line) and output seed pulse (solid red line with most smooth pulse shape) [74].

(see Figure 1.2.3.11c), which is close to the calculated transform-limited pulse width of 10.5 fs with only 20% longer duration. Figure 1.2.3.11d shows the incident laser spectrum, the broadened and retrieved spectrum, and the spectral phase of the compressed pulse. Several pulses with different wavelengths can be simultaneously guided into this system; the system will then simultaneously amplify these pulses and broaden their spectra [74].

1.2.3.5 CONCLUSION AND PROSPECTS

FWM has been used to generate and optimize ultrashort laser pulses. Self-compressed 15-fs multicolored pulses are simultaneously generated with a very broad spectral range extending from UV to NIR by cascaded FWM in a bulk medium. The wavelengths of the generated multicolored pulses can be tuned by varying the incident-crossing angle. Studies on degenerate and nondegenerate cascaded FWM reveal that the generated cascaded signals have not only broader and smoother spectra, but also shorter and cleaner pulses, and furthermore they have improved beam profiles and spatial quality. These outstanding characteristics make cascaded FWM signals useful for multicolor time-resolved spectroscopy and multicolor linear and/or nonlinear optical microscopy. The high temporal-contrast feature of cascaded FWM signals makes them suitable for seed pulses in background-free petawatt lasers.

Broadband DUV pulses are generated using self-phase modulated pulses as the idler in chirped-pulse FWM. The DUV pulse is negatively chirped and can be easily compressed by a transparent medium even to sub 10 fs. The temporal profile of the compressed pulse is smooth and useful for ultrafast spectroscopic applications in the DUV region.

The FWOPA method can be used to simultaneously amplify and compress laser pulses in a bulk medium. This method is expected to amplify DUV pulses pumped by 400-nm pulses and compress DUV pulses to below 5 fs in the near future. The contents in this subsection are product of the collaboration of the following people: T. Kobayashi, J. Liu, and Y. Kida [99].

REFERENCES

1. A. H. Zewail, "Femtochemistry: Atomic-scale dynamics of the chemical bond," *J. Phys. Chem. A* **104**, 5660–5694 (May 2000).

2. J. Shah, *Ultrafast Spectroscopy of Semiconductors and Semiconductor Nanostructures*, 1st ed., Springer, Berlin, Germany (1999).

3. T. Kobayashi, T. Okada, K. A. Nelson, and S. De Silvestri, *Ultrafast Phenomena XIV*, 1st ed., Springer, New York (2004).

4. T. Kobayashi, A. Shirakawa, and T. Fuji, "Sub-5-fs transform-limited visible pulse source and its application to real-time spectroscopy," *IEEE J. Sel. Topics Quantum Electron* **7**(4), 525–538 (Jul. 2001).

5. L. Xu, Ch. Spielmann, F. Krausz, and R. Szipocs, "Ultrabroadband ring¨ oscillator for sub-10-fs pulse generation," *Opt. Lett.* **21**, 1259–1261 (Aug. 1996).

6. A. Baltuska, Z. Wei, M. S. Pshenichnikov, and D. A. Wiersma, "Optical pulse compression to 5 fs at a 1-MHz repetition rate," *Opt. Lett.* **22**, 102–104 (Jan. 1997).

7. M. Nisoli, S. De Silvestri, O. Svelto, R. Szipocs, K. Ferencz, C. Spielmann, S. Sartania, and F. Krausz, "Compression of high-energy laser pulses below 5 fs," *Opt. Lett.* **22**, 522–524 (Apr. 1997).

8. S. Bohman, A. Suda, T. Kanai, S. Yamaguchi, and K. Midorikawa, "Generation of 5.0 fs, 5.0 mJ pulses at 1 kHz using hollow-fiber pulse compression," *Opt. Lett.* **35**, 1887–1889 (Jun. 2010).

9. C. P. Hauri, W. Kornelis, F. W. Helbing, A. Heinrich, A. Couairon, A. Mysyrowicz, J. Biegert, and U. Keller, "Generation of intense, carrier envelope phase-locked few-cycle laser pulses through filamentation," *Appl. Phys. B* **79**, 673–677 (Sep. 2004).

10. J. Liu, K. Okamura, Y. Kida, T. Teramoto, and T. Kobayashi, "Clean sub-8-fs pulses at 400 nm generated by a hollow fiber compressor for ultraviolet ultrafast pump-probe spectroscopy," *Opt. Express* **18**, 20645–20650 (Sep. 2010).

11. A. Shirakawa, I. Sakane, and T. Kobayashi, "Pulse-front-matched optical parametric amplification for sub-10-fs pulse generation tunable in the visible and near infrared," *Opt. Lett.* **23**, 1292–1294 (Aug. 1998).

12. G. Cerullo, M. Nisoli, S. Stagira, and S. De Silvestri, "Sub-8-fs pulses from an ultrabroadband optical parametric amplifier in the visible," *Opt. Lett.* **23**, 1283–1285 (Aug. 1998).

13. A. Shirakawa, I. Sakane, M. Takasaka, and T. Kobayashi, "Sub-5-fsvisible pulse generation by pulse-front-matched noncollinear optical parametric amplification," *Appl. Phys. Lett.* **74**, 2268–2270 (Apr. 1999).

14. A. Baltuska, T. Fuji, and T. Kobayashi, "Visible pulse compression to 4 fs by optical parametric amplification and programmable dispersion control," *Opt. Lett.* **27**, 306–308 (Mar. 2002).

15. G. Cerullo and S. De Silvestri, "Ultrafast optical parametric amplifiers," *Rev. Sci. Instrum.* **74**, 1–18 (Jan. 2003).

16. D. Brida, G. Cirmi, C. Manzoni, S. Bonora, P. Villoresi, S. De Silvestri, and G. Cerullo, "Sub-two-cycle light pulses at 1.6 µm from an optical parametric amplifier," *Opt. Lett.* **33**, 741–743 (Apr. 2008).

17. R. A. Kaindl, M. Wurm, K. Reimann, P. Hamm, A. M. Weiner, and M. Woerner, "Generation, shaping, and characterization of intense femtosecond pulses tunable from 3 to 20 µm," *J. Opt. Soc. Amer. B* **17**, 2086–2094 (Dec. 2000).

18. C. Heese, C. R. Phillips, L. Gallmann, M. M. Fejer, and U. Keller, "Ultrabroadband, highly flexible amplifier for ultrashort midinfrared laser pulses based on aperiodically poled $Mg:LiNbO_3$," *Opt. Lett.* **35**, 2340–2342 (Jul. 2010).

19. I. Kozma, P. Baum, S. Lochbrunner, and E. Riedle, "Widely tunable sub 30 fs ultraviolet pulses by chirped sum frequency mixing," *Opt. Express* **11**, 3110–3115 (Dec. 2003).

20. T. Kobayashi, T. Saito, and H. Ohtani, "Real-time spectroscopy of transition states in bacteriorhodopsin during retinal isomerization," *Nature* **414**, 531–534 (Nov. 2001).

21. G. Cerullo, D. Polli, G. Lanzani, S. De Silvestri, H. Hashimoto, and R. J. Cogdell, "Photosynthetic light harvesting by carotenoids: Detection of an intermediate excited states," *Science* **298**, 2395–2398 (Dec. 2002).

22. S. Adachi, V. M. Kobryanskii, and T. Kobayashi, "Excitation of a breather mode of bound soliton pairs in trans-polyacetylene by sub-5-fs optical pulses," *Phys. Rev. Lett.* **89**, 027401-1–027401-4 (Jul. 2002).

23. P. Kukura, D. W. McCamant, S. Yoon, D. B. Wandschneider, and R. A. Mathies, "Structural observation of the primary isomerization in vision with femtosecond-stimulated Raman," *Science* **310**, 1006–1009 (Nov. 2005).

24. D. Polli, P. Altoe, O. Weingart, K. M. Spillane, C. Manzoni, D. Brida, G. Tomasello, G. Orlandi, P. Kukura, R. A. Mathies, M. Garavelli, and G. Cerullo, "Conical intersection dynamics of the primary photoisomerization event in vision," *Nature* **467**, 440–443 (Sep. 2010).
25. J. A. Gruetzmacher and N. F. Scherer, "Few-cycle mid-infrared pulse generation, characterization, and coherent propagation in optically dense media," *Rev. Sci. Instrum.* **73**, 2227–2236 (Jun. 2002).
26. Y. S. Kim and R. M. Hochstrasser, "Applications of 2D IR spectroscopy to peptides, proteins, and hydrogen-bond dynamics," *J. Phys. Chem. B* **113**, 8231–8251 (Dec. 2009).
27. R. M. Hochstrasser, "Two-dimensional spectroscopy at infrared and optical frequencies," *Proc. Nat. Acad. Sci. USA* **104**, 14190–14196 (Jul. 2007).
28. J. M. Anna, M. J. Nee, C. R. Baiz, R. McCanne, and K. J. Kubarych, "Measuring absorptive two-dimensional infrared spectra using chirped pulse upconversion detection," *J. Opt. Soc. Amer. B* **27**, 382–393 (Mar. 2010).
29. P. Baum, S. Lochbrunner, and E. Riedle, "Tunable sub-10-fs ultraviolet pulses generated by achromatic frequency doubling," *Opt. Lett.* **29**, 1686–1688 (Jul. 2004).
30. F. Theberge, N. Ak´ozbek, W. Liu, A. Becker, and S. L. Chin, "Tunable¨ ultrashort laser pulses generated through filamentation in gases," *Phys. Rev. Lett.* **97**, 023904-1–023904-4 (Jul. 2006).
31. T. Fuji, T. Horio, and T. Suzuki, "Generation of 12 fs deep-ultraviolet pulses by four-wave mixing through filamentation in neon gas," *Opt. Lett.* **32**, 2481–2483 (Sep. 2007).
32. T. Fuji and T. Suzuki, "Generation of sub-two-cycle mid-infrared pulses by four-wave mixing through filamentation in air," *Opt. Lett.* **32**, 3330–3332 (Nov. 2007).
33. P. Zuo, T. Fuji, and T. Suzuki, "Spectral phase transfer to ultrashort UV pulses through four-wave mixing," *Opt. Express* **18**, 16183–16192 (Jul. 2010).
34. M. Beutler, M. Ghotbi, F. Noack, and I. V. Hertel, "Generation of sub-50 fs vacuum ultraviolet pulses by four-wave mixing in argon," *Opt. Lett.* **35**, 1491–1493 (May 2010).
35. G. C. Durfee III, S. Backus, M. M. Murnane, and C. H. Kapteyn, "Ultrabroadband phase-matched optical parametric generation in the ultraviolet by use of guided waves," *Opt. Lett.* **22**, 1565–1567 (Oct. 1997).
36. L. Misoguti, S. Backus, C. G. Durfee, R. Bartels, M. M. Murnane, and H. C. Kapteyn, "Generation of broadband VUV light using third-order cascaded processes," *Phys. Rev. Lett.* **87**, 013601-1–013601-4 (Jul. 2001).
37. C. G. Durfee, S. Backus, H. C. Kapteyn, and M. M. Murnane, "Intense 8-fs pulse generation in the deep ultraviolet," *Opt. Lett.* **24**, 697–699 (May 1999).
38. U. Graf, M. Fieß, M. Schultze, R. Kienberger, F. Krausz, and E. Goulielmakis, "Intense few-cycle light pulses in the deep ultraviolet," *Opt. Express* **16**, 18956–18963 (Nov. 2008).
39. H. Okamoto and M. Tatsumi, "Generation of ultrashort light pulses in the mid-infrared (3000–800 cm^{-1}) by four-wave mixing," *Opt. Commun.* **121**, 63–68 (Nov. 1995).
40. H. K. Nienhuys, P. C. M. Planken, R. A. van Santen, and H. J. Bakker, "Generation of mid-infrared pulses by $\chi^{(3)}$ difference frequency generation in CaF_2 BaF_2," *Opt. Lett.* **26**, 1350–1352 (Sep. 2001).
41. J. Darginavicius, G. Tamoauskas, A. Piskarskas, and A. Dubietis, "Generation of 30-fs ultraviolet pulses by four-wave optical parametric chirped pulse amplification," *Opt. Express* **18**, 16096–16101 (Jul. 2010).
42. H. Crespo, J. T. Mendonc¸a, and A. Dos Santos, "Cascaded highly nondegenerate four-wave-mixing phenomenon in transparent isotropic condensed media," *Opt. Lett.* **25**, 829–831 (Jun. 2000).
43. R. Weigand, J. T. Mendonca, and H. Crespo, "Cascaded nondegenerate four-wave mixing technique for high-power single-cycle pulse synthesis in the visible and ultraviolet ranges," *Phys. Rev. A* **79**, 063838-1–063838-5 (Jun. 2009).
44. J. L. Silva, R. Weigand, and H. Crespo, "Octave-spanning spectra and pulse synthesis by non-degenerate cascaded four-wave mixing," *Opt. Lett.* **34**, 2489–2491 (Aug. 2009).
45. J. Liu and T. Kobayashi, "Wavelength-tunable multicolored femtosecond laser pulse generation in fused silica glass," *Opt. Lett.* **34**, 1066–1068 (Apr. 2009).
46. J. Liu and T. Kobayashi, "Generation of uJ-level multicolored femtosecond laser pulses using cascaded four-wave mixing," *Opt. Express* **17**, 4984–4990 (Mar. 2009).
47. J. Liu and T. Kobayashi, "Generation of sub-20-fs multicolor laser pulses using cascaded four-wave mixing with chirped incident pulses," *Opt. Lett.* **34**, 2402–2404 (Aug. 2009).
48. J. Liu and T. Kobayashi, "Cascaded four-wave mixing in transparent bulk media," *Opt. Commun.* **283**, 1114–1123 (2010).
49. J. Liu and T. Kobayashi, "Cascaded four-wave mixing and multicolored arrays generation in a sapphire plate by using two crossing beams of femtosecond laser," *Opt. Express* **16**, 22119–22125 (Dec. 2008).
50. J. Liu, T. Kobayashi, and Z. G. Wang, "Generation of broadband two dimensional multicolored arrays in a sapphire plate," *Opt. Express* **17**, 9226–9234 (May 2009).

51. M. Zhi and A. V. Sokolov, "Broadband coherent light generation in a Raman-active crystal driven by two-color femtosecond laser pulses," *Opt. Lett.* **32**, 2251–2253 (Aug. 2007).

52. M. Zhi and A. V. Sokolov, "Broadband generation in a Raman crystal driven by a pair of time-delayed linearly chirped pulses," *New J. Phys.* **10**, 025032-1–025032-12 (Feb. 2008).

53. E. Matsubara, T. Sekikawa, and M. Yamashita, "Generation of ultrashort optical pulses using multiple coherent anti-Stokes Raman scattering in a crystal at room temperature," *Appl. Phys. Lett.* **92**, 071104-1–071104-3 (Feb. 2008).

54. E. Matsubara, Y. Kawamoto, T. Sekikawa, and M. Yamashita, "Generation of ultrashort optical pulses in the 10 fs regime using multicolor Raman sidebands in $KTaO_3$," *Opt. Lett.* **34**, 1837–1839 (Jun. 2009).

55. H. Matsuki, K. Inoue, and E. Hanamura, "Multiple coherent antiStokes Raman scattering due to phonon grating in $KNbO_3$ induced by crossed beams of two-color femtosecond pulses," *Phys. Rev. B* **75**, 024102-1–024102-4 (Jan. 2007).

56. K. Inoue, J. Kato, E. Hanamura, H. Matsuki, and E. Matsubara, "Broadband coherent radiation based on peculiar multiple Raman scattering by laser-induced phonon grating in TiO_2," *Phys. Rev. B* **76**, 041101(R)–041101-4 (Jul. 2007).

57. E. Matsubara, K. Inoue, and E. Hanamura, "Violation of Raman selection rules induced by two femtosecond laser pulses in $KTaO_3$," *Phys. Rev. B* **72**, 134101-1–134101-5 (Oct. 2005).

58. J. Takahashi, E. Matsubara, T. Arima, and E. Hanamura, "Coherent multistep anti-Stokes and stimulated Raman scattering associated with third harmonics in $YFeO_3$ crystals," *Phys. Rev. B* **68**, 155102-1–155102-5 (Oct. 2003).

59. J. Takahashi, M. Keisuke, and Y. Toshirou, "Raman lasing and cascaded coherent anti-Stokes Raman scattering of a two-phonon Raman band," *Opt. Lett.* **31**, 1501–1503 (May 2006).

60. M. Zhi, X. Wang, and A. V. Sokolov, "Broadband coherent light generation in diamond driven by femtosecond pulses," *Opt. Express* **16**, 12139–12147 (Aug. 2008).

61. J. Liu, J. Zhang, and T. Kobayashi, "Broadband coherent anti-Stokes Raman scattering light generation in BBO crystal by using two crossing femtosecond laser pulses," *Opt. Lett.* **33**, 1494–1496 (Jul. 2008).

62. G. Patterson, R. Day, and D. Piston, "Fluorescent protein spectra," *J. Cell Sci.* **114**, 837–838 (May 2001).

63. M. Bruchez Jr., M. Moronne, P. Gin, S. Weiss, and A. P. Alivisatos, "Semiconductor nanocrystals as fluorescent biological labels," *Science* **281**, 2013–2016 (Sep. 1998).

64. X. Michalet, F. F. Pinaud, L. A. Bentolila, J. M. Tsay, S. Doose, J. J. Li, G. Sundaresan, A. M. Wu, S. S. Gambhir, and S. Weiss, "Quantum dots for live cells, in vivo imaging, and diagnostics," *Science* **307**, 538–544 (Jan. 2005).

65. S. Shrestha, B. E. Applegate, J. Park, X. Xiao, P. Pande, and J. A. Jo, "High-speed multispectral fluorescence lifetime imaging implementation for in vivo applications," *Opt. Lett.* **35**, 2558–2560 (Aug. 2010).

66. H. Kobayashi, M. Ogawa, R. Alford, P. L. Choyke, and Y. Urano, "New strategies for fluorescent probe design in medical diagnostic imaging," *Chem. Rev.* **110**, 2620–2640 (May 2009).

67. C. W. Freudiger, W. Min, B. G. Saar, S. Lu, G. R. Holtom, C. He, J. C. Tsai, J. X. Kang, and X. S. Xie, "Label-free biomedical imaging with high sensitivity by stimulated Raman scattering microscopy," *Science* **322**, 1857–1861 (Dec. 2008).

68. A. Penzkofer and H. J. Lehmeier, "Theoretical investigation of noncollinear phase-matched parametric four-photon amplification of ultrashort light pulses in isotropic media," *Opt. Quantum Electron.* **25**, 815–844 (Nov. 1993).

69. H. Valtna, G. Tamosauskas, A. Dubietis, and A. Piskarskas, "High-energy broadband four-wave optical parametric amplification in bulk fused silica," *Opt. Lett.* **33**, 971–973 (May 2008).

70. A. Dubietis, G. Tamosauskas, P. Polesana, G. Valiulis, H. Valtna, D. Fac-cio, P. Di Trapani, and A. Piskarskas, "Highly efficient four-wave parametric amplification in transparent bulk Kerr medium," *Opt. Express* **15**, 11126–11132 (Sep. 2007).

71. J. Darginavicius, G. Tamosauskas, G. Valiulis, and A. Dubietis, "Broadband four-wave optical parametric amplification in bulk isotropic media in the ultraviolet," *Opt. Commun.* **282**, 2995–2999 (Jul. 2009).

72. A. Dubietis, J. Darginavicius, G. Tamosauskas, G. Valiulis, and A. Piskarskas, "Generation and amplification of ultrashort UV pulses via parametric four-wave interactions in transparent solid-state media," *Lithuanian J. Phys.* **49**, 421–431 (Apr. 2009).

73. D. Faccio, A. Grun, K. P. Bates, O. Chalus, and J. Biegert, "Optical amplification in the near-infrared in gas-filled hollow-core fibers," *Opt. Lett.* **34**, 2918–2920 (Oct. 2009).

74. J. Liu, Y. Kida, T. Teramoto, and T. Kobayashi, "Simultaneous compression and amplification of a laser pulse in a glass plate," *Opt. Express* **18**, 4665–4672 (Feb. 2010).

75. S. Shimizu, Y. Nabekawa, M. Obara, and K. Midorikawa, "Spectral phase transfer for indirect phase control of sub-20-fs deep UV pulses," *Opt. Express* **13**, 6345–6353 (Aug. 2005).

76. M. T. Seidel, S. Yan, and H.-S. Tan, "Mid-infrared polarization pulse shaping by parametric transfer," *Opt. Lett.* **35**, 478–480 (Feb. 2010).

77. R. Trebino, *Frequency-Resolved Optical Grating: The Measurement of Ultrashort Laser Pulses*, Kluwer, Norwell, MA (2000).

78. J. Liu, K. Okamura, Y. Kida, and T. Kobayashi, "Temporal contrast enhancement of femtosecond pulses by a self-diffraction process in a bulk Kerr medium," *Opt. Express* **18**, 22245–22254 (Oct. 2010).

79. G. A. Mourou, T. Tajima, and S. V. Bulanov, "Optics in the relativistic regime," *Rev. Mod. Phys.* **78**, 309–371 (Apr./Jun. 2006).

80. S. V. Bulanov, T. Esirkepov, D. Habs, F. Pegoraro, and T. Tajima, "Relativistic laser-matter interaction and relativistic laboratory astrophysics," *Eur. Phys. J. D* **55**, 483–507 (Nov. 2009).

81. A. Jullien, O. Albert, F. Burgy, G. Hamoniaux, J.-P. Rousseau, J.-P. Chambaret, F. A. Rochereau, G. Che´riaux, J. Etchepare, N. Minkovski, and S. M. Saltiel, "10^{-10} temporal contrast for femtosecond ultraintense lasers by cross-polarized wave generation," *Opt. Lett.* **30**, 920–922 (Apr. 2005).

82. T. Schneider, D. Wolfframm, R. Mitzner, and J. Reif, "Ultrafast optical switching by instantaneous laser-induced grating formation and self-diffraction in barium fluoride," *Appl. Phys. B* **68**, 749–751 (Nov. 1999).

83. A. Jullien, L. Canova, O. Albert, D. Boschetto, L. Antonucci, Y. H. Cha, J. P. Rousseau, P. Chaudet, G. Cheriaux, J. Etchepare, S. Kourtev, N. Minkovski, and S. M. Saltiel, "Spectral broadening and pulse duration reduction during cross-polarized wave generation: Influence of the quadratic spectral phase," *Appl. Phys. B* **87**, 595–601 (Jun. 2007).

84. K. W. DeLong, R. Trebino, J. Hunter, and W. E. White, "Frequency resolved optical gating with the use of second-harmonic generation," *J. Opt. Soc. Amer. B* **11**, 2206–2215 (Nov. 1994).

85. X. W. Chen, Y. X. Leng, J. Liu, Y. Zhu, R. X. Li, and Z. Z. Xu, "Pulse self-compression in normally dispersive bulk media," *Opt. Commun.* **259**, 331–335 (Mar. 2006).

86. A. E. Jailaubekov and S. E. Bradforth, "Tunable 30-femtosecond pulses across the deep ultraviolet," *Appl. Phys. Lett.* **87**, 021107-1–021107-3 (Jul. 2005).

87. J. Wojtkiewicz and C. G. Durfee, "Hollow-fiber OP-CPA for energetic ultrafast ultraviolet pulse generation," in *Proc. Conf. Lasers and Electro Optics*, California, pp. 423–424 (2002).

88. J. Wojtkiewicz, K. Hudek, and C. G. Durfee, "Chirped-pulse frequency conversion of ultrafast pulses to the deep-UV," in *Proc. Conf. Lasers and Electro-Optics*, Maryland, pp. 186–188 (2005).

89. P. Tzankov, O. Steinkellner, J. Zheng, A. Husakou, J. Herrmann, W. Freyer, V. Petrov, and F. Noack, "Generation and compression of femtosecond pulses in the vacuum ultraviolet by chirped-pulse four-wave difference frequency mixing," in *Proc. Conf. Lasers and Electro-Optics*, California, pp. 1–2 (2006).

90. I. Babushkin and J. Herrmann, "High energy sub-10 fs pulse generation in vacuum ultraviolet using chirped four wave mixing in hollow waveguides," *Opt. Express* **16**, 17774–17779 (Oct. 2008).

91. Y. Kida, J. Liu, T. Teramoto, and T. Kobayashi, "Sub-10 fs deep-ultraviolet pulses generated by chirped-pulse four-wave mixing," *Opt. Lett.* **35**, 1807–1809 (Jun. 2010).

92. Y. Kida and T. Kobayashi, "Generation of sub-10-fs ultraviolet Gaussian pulses," *J. Opt. Soc. Amer. B* **28**, 139–148 (Jan. 2011).

93. E. J. Marcateli and R. Schmeltzer, "Hollow metallic and dielectric waveguides for long distance optical transmission and lasers," *Bell Syst. Tech. J.* **43**, 1783–1809 (Jul. 1964).

94. J.-C. Diels and W. Rudolph, *Ultrashort Laser Pulse Phenomena*, 2nd ed., Academic Press, San Diego (2006).

95. G. P. Agrawal, *Nonlinear Fiber Optics*, 4th ed., Academic Press, San Diego (2006).

96. M. Nisoli, S. D. Silvestri, and O. Svelto, "Generation of high energy 10 fs pulses by a new pulse compression technique," *Appl. Phys. B* **68**, 2793–2795 (May 1996).

97. D. J. Kane and R. Trebino, "Characterization of arbitrary femtosecond pulses using frequency-resolved optical gating," *IEEE J. Quantum Electron.* **29**(2), 571–579 (Feb. 1993).

98. M. Spanner, M. Y. Ivanov, V. Kalosha, J. Hermann, D. A. Wiersma, and M. Pshenichnikov, "Tunable optimal compression of ultrabroadband pulses by cross-phase modulation," *Opt. Lett.* **28**, 749–751 (May 2003).

99. T. Kobayashi, J. Liu, and Y. Kida, "Generation and optimization of femtosecond pulses by four-wave mixing process," *IEEE J. of Selected topics in Quantum Electron.* **18**(1), 54–65 (Jan./Feb. 2012).

Section 2

Generation of Ultrashort Pulses in Terahertz

2.1 Sellmeier Dispersion for Phase-Matched Terahertz Generation in Nonlinear Optical Crystal
An Example of ZnGeP$_2$

2.1.1 INTRODUCTION

In recent years, among researchers, significant interest has been seen in the development of systems for the generation and detection of coherent tunable terahertz THz radiation. THz radiation has several potential applications in, e.g., space communications, time-domain far-infrared spectroscopy, THz ranging, and THz imaging [1]. Generation of such radiation is possible by various techniques, namely, by optical rectification in semiconductors with femtosecond laser radiation sources [2,3], from photoconductive antennas [4], and also by nonlinear optical NLO frequency-conversion techniques, such as optical parametric oscillation and difference-frequency mixing DFM in various NLO crystals [5–9]. Zinc germanium diphosphide ZnGeP$_2$, ZGP is one of the most promising crystals for NLO applications [5–10] because of its high second-order nonlinearity d_{36}=75 pmV, low absorption in the infrared, and wide THz spectral transmission [7,8]. Since this crystal in addition has a high thermal conductivity relative to other infrared crystals, it is attractive for use in high-average-power nonlinear optics [5–10]. The generation of THz radiation in this crystal was reported by Boyd et al. [7] and also by Apollonov et al. [8] using CO$_2$ lasers as input pump radiation sources. Shi and Ding [9] reported the generation of continuously tunable coherent THz radiation in this crystal by phasematched DFM between the 1.064-m Nd:YAG laser radiation and the tunable idler radiations of a Nd: YAG-pumped optical parametric oscillator. They used an annealed sample of ZGP crystal, which has a smaller absorption coefficient of 1.52 cm^{-1} than the 5.63 cm^{-1} for an unannealed sample at 1.064 m. However, no difference was found in the absorption property of the two crystal samples at other wavelengths, including the THz range covering 66.5–300 m. It was noted by Shi and Ding [9] that the absolute value for the refractive index of the crystal sample at 1.064 m is not affected by the annealing process. The change in absorption in the crystal only at 1.064 m caused by the annealing process is due to the reduction of the impurity density of the crystal without modification of the phonon modes [9].

However, it is observed that the experimental phase-matching data of Ref. [9] differ substantially, by 5°–18°, from those calculated from the Sellmeier dispersion relations of Bhar et al. [10] which provide the best fit. Shi and Ding [9] noted this discrepancy and speculated only qualitatively that the dispersion of the crystal had been modified by the annealing process only at 1.064 m, and so the experimental phase-matching data are different from the results calculated from the Sellmeier dispersion relations of Bhar et al. [10] which were formulated with the measured refractive indices of an unannealed crystal sample. Shi and Ding [9] also speculated that the dispersion of the crystal in the THz range has a negligible effect on the determination of the phasematching angle. However, it is shown here that the dispersion of the crystal in the THz region has a significant role in determining the phase-matching angle of the considered NLO process. We also find that the refractive indices of the crystal in the relevant THz spectral region cannot be obtained appropriately with the Sellmeier

DOI: 10.1201/9780429196577-14

dispersions of Bhar et al. [10] which were formulated with mid-infrared refractive indices. An earlier, separate dispersion relation for determining the ordinary indices of the ZGP crystal in the THz spectral region was formulated by Boyd et al. [7] to calculate the phase-matching angle for the generation of THz radiation by DFM between different CO_2 laser wavelengths. Apollonov et al. [8] also previously reported some refractive indices of ZGP crystal at THz wavelengths. It is observed that with these [7,8] refractive-index dispersion data for the THz region the theoretical fit to the experimental phase-matching data [9] is improved to some extent but cannot be explained properly over the entire THz wavelength region. However, an appropriate Sellmeier dispersion is formulated here to determine the refractive index of an ordinary polarized ray in the ZGP crystal in the THz region. With the Sellmeier dispersion formulated here, the experimental phase-matching data of Shi and Ding [9] as well as those of Boyd et al. [7] are explained satisfactorily over the entire THz range.

2.1.2 DERIVATION OF THE SELLMEIER DISPERSION

The Sellmeier dispersion relation model [11] for the refractive index of NLO crystals is very useful for the calculation of NLO properties. Here, the Sellmeier dispersion for the ordinary refractive index of ZGP crystal in the THz spectral range is formulated. The ordinary refractive indices n_o of ZGP crystal in the THz wavelength region were measured earlier by Boyd et al. [7] 66.7 m and Apollonov et al. [8] 100 m. The former fitted a Sellmeier dispersion with their measured indices, while the latter did not. Based on the measured data of Boyd et al. [7] and Apollonov et al. [8] in the shorter and longer wavelengths, respectively, we have formulated a new Sellmeier dispersion relation, Eq. (2.1.1), for n_o of ZGP crystal, for the determination of the phase-matching properties for the generation of THz radiation. The formulated Sellmeier dispersion relation is n_0^2 10.93904 0.60675^{22} 1600, (Eq. 2.1.1) where 60 m is in micrometers.

In Figure 2.1.1 the dependence of n_o on the THz wavelength is shown, and curves 1, 2, 3, and 4 correspond to Refs. [7,8,10] and our Eq. (2.1.1), respectively. The variation of n_o obtained from the Sellmeier dispersion for n_o in Ref. [10] has also been shown in Figure 2.1.1 for comparison with the measured data of Boyd et al. [7] and Apollonov et al. [8]. From Figure 2.1.1 it is clearly observed that the values of n_o that are obtained from Ref. [10] are different from the measured data [7,8]. The Sellmeier dispersion of Ref. [10] was obtained earlier, with the measured refractive indices covering only the 0.64–12-m wavelength range, so it is not surprising that this Sellmeier dispersion [10] does not reproduce the refractive indices of the crystal properly in the THz region. In the following sections, it will be shown that our formulated Sellmeier dispersion relation, given by Eq. (2.1.1), provides a satisfactory explanation of the experimental phase-matching data [7,9] for the generation of THz radiation in this crystal by DFM, whereas the other data [7,8] for the THz refractive-index dispersion fail.

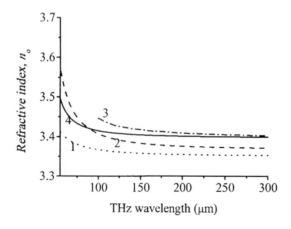

FIGURE 2.1.1 Dispersion of the ordinary refractive index n_o of the ZnGeP$_2$ crystal in the THz spectral range. Curves 1, 2, 3, and 4 are obtained from the dispersion data of Refs. [7], 8, 10, and Eq. (2.1.1) of the present paper, respectively.

2.1.3   GENERATION OF TERAHERTZ RADIATION WITH A ND:YAG LASER

By collinear phase-matched DFM between the ordinary- (*o*) polarized 1.064 μm (λ_1) Nd:YAG laser radiation and the extraordinary (*e*) polarized tunable ($\lambda_2 > 1.064$ μm) idler radiation of an 1.064-μm pumped optical parametric oscillator, the generation of continuously tunable THz(λ_3) radiation is possible in ZGP crystal. Two different DFM configurations, *oe-o* and *oe-e,* were used [9]; in the first case, the polarization of the generated THz radiation is *o*, whereas for the latter one it is *e*. The experimental and computed dependence of the external phase-matching angle (θ_{ext}) with λ_3 for *oe-o* and *oe-e* DFM configurations are shown in Figures 2.1.2 and 2.1.3, respectively. In Figures 2.1.2 and 2.1.3 the refractive indices at the input pump wavelengths have been calculated by use of the Bhar et al. [10] dispersion equations, whereas the ordinary refractive indices at the generated THz wavelengths have been taken from Refs. [7,8,10] for the computed curves 1, 2, and 3, respectively. Curves 4 of Figures 2.1.2 and 2.1.3 are obtained with our Eq. (2.1.1). In the case of the *oe-e* configuration, the phase-matching angle depends on both the ordinary and extraordinary indices at the generated THz wavelength, unlike for the *oe-o* configuration, for which the phase-matching angle depends only on the ordinary index at the generated THz wavelength. Therefore, to obtain curves 2, 3, and 4 in Figure 2.1.3, the extraordinary index has been calculated as in Ref. [12] and with the corresponding ordinary index from Refs. [7,8] and Eq. (2.1.1), respectively.

However, for curve 1 both the ordinary and the extraordinary indices for the input pump as well as for the generated THz wavelengths have been calculated from Ref. [10]. The vertical and horizontal error bars in Figures 2.1.2 and 2.1.3 show the probable errors in the experimental data [9].

In Figure 2.1.2 it can be seen that the experimental phase-matching data [9] differ from those curves 1 calculated with the Bhar et al. [10] Sellmeier dispersion over the entire wavelength region. This Sellmeier dispersion [10] is not able to reproduce the experimental phase-matching data of Ref. [9] because, as was shown in Section 2.1.2, the Sellmeier dispersion of Ref. [10] could not reproduce the refractive indices of the crystal in the relevant THz spectral region. Curve 2 in Figure 2.1.2 shows that the experimental data also cannot be explained with the Boyd et al. [7] dispersion, except for small improvements at wavelengths shorter than ~100 m.

On the other hand, curve 3 in Figure 2.1.2 shows that the fit to the experimental data is improved with the Apollonov et al. [8] refractive index data only in the longer THz wavelength region. Moreover, no refractive-index data are available in Ref. [8] for 100-m wavelengths. However, in Figure 2.1.2 it is observed that the calculated phase-matching data curve 4 with our formulated Sellmeier dispersion Eq. (2.1.1) excellently matches the experimental data of *oe-o* DFM [9] over the entire wavelength region. In the case of the *oe-e* DFM, for which the phasematching curves are shown in Figure 2.1.3, it is observed that curve 4, obtained with the Sellmeier dispersion Eq. (2.1.1),

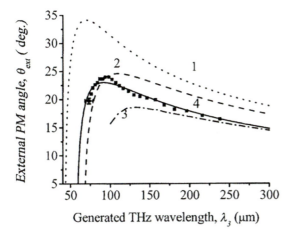

FIGURE 2.1.2 External phase-matching angle θ_{ext} versus λ_3 for *oe-o* DFM. THz ordinary refractive indices are obtained from Refs. [7,8,10] and our Eq. (2.1.1) for curves 1, 2, 3, and 4, respectively. Ref. [10] has been used for the calculation of the refractive indices at the input pump wavelengths. The horizontal and vertical error bars show the uncertainties in the experimental data squares [9].

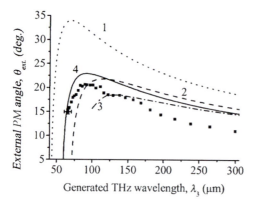

FIGURE 2.1.3 External phase-matching angle θ_{ext} versus λ_3 for *oe-e* DFM. For curves 2, 3, and 4, the ordinary refractive index n_o at a generated THz wavelength λ_3 has been obtained from Refs. [7,8] and Eq. (2.1.1), respectively, and the corresponding extraordinary THz index n_e has been calculated as in Ref. [12]. For curve 1, both n_o and n_e for each λ_3 have been calculated with the Sellmeier dispersion of Ref. [10], while those for the input pump wavelengths have been calculated from Ref. [10] for all curves. The horizontal and vertical error bars show the uncertainties in the experimental data (squares) [9].

fits the experimental phase-matching data well, particularly, in the shorter-wavelength region, whereas results curves 1, 2, and 3 calculated with other dispersion data [7,8,10] deviate from the experiment over a broad spectral region. In the case of *oe-e* DFM in the longer-wavelength region, there are still some discrepancies between the experimental data and our calculated results curve 4; however, these might be within the uncertainty in the experimental data. For example, for the generation of 157-m radiation by *oe-e* DFM the deviation in the external phase-matching angle is 2.2°, which corresponds to an internal angle deviation of only 0.6°.

2.1.4 GENERATION OF TERAHERTZ RADIATION WITH CO_2 LASERS

The generation of THz radiation in ZGP crystal is also possible by DFM between different wavelengths of two CO_2 laser sources [7,8]. The experimental data, together with the computed phase-matching characteristics of ZGP crystal for the generation of THz radiation from CO_2 laser radiations, are shown in Figure 2.1.4. Figure 2.1.4 corresponds to the *eo-o* DFM configuration; i.e., in this case, the input radiations are orthogonally *e* and *o* polarized CO_2 laser lines, and the generated THz radiation is *o* polarized.

To yield the theoretical curves in Figure 2.1.4, the refractive indices at the input pump wavelengths have been calculated with the Bhar et al. [10] dispersion equations, whereas the refractive indices at the generated THz wavelengths have been taken from Refs. [7,8,10] and our Eq. (2.1.1) for curves 1, 2, 3, and 4, respectively. In Figure 2.1.4 the wavelength of the *e*-polarized radiation $\lambda_1 = 9.588\,\mu m$, and that of the *o*-polarized radiation λ_2 is varied to obtain the generation of phase-matched wavelength-tunable *o*-polarized THz radiation. It may be noted that in the present case the input radiation sources are CO_2 lasers that are line tunable, and so the wavelength of the generated THz radiation is not continuously tunable, unlike the cases presented above in Section 2.1.3.

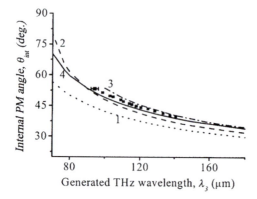

FIGURE 2.1.4 Internal phase-matching angle$_{int}$ vs. λ_3 for *eo-o* DFM between CO_2 laser radiations. Computed curves 1, 2, 3, and 4 are based on THz ordinary index data from Refs. [7,8,10] and our Eq. (2.1.1), respectively, while Ref. [10] has been used for the calculation of the refractive indices at the input wavelengths. The error bar shows the uncertainty in the experimental data squares [7].

The phase-matching configuration considered in Figure 2.1.4 is the same as that of forward-wave phase-matching in Ref. [7]. Unlike in Figures 2.1.2 and 2.1.3, the internal phase-matching angle$_{int}$ has been plotted in Figure 2.1.4 to show similarity with Ref. [7]. Figure 2.1.4 shows that our Sellmeier dispersion Eq. (2.1.1) provides the best to the experimental data [7]. Deviations of the experimental data from the calculated results with the available THz Sellmeier dispersion of Boyd et al. [7] are within $2.5°–2.9°$. On the other hand, the deviation of the experimental data from the calculated results with our formulated Sellmeier dispersion Eq. (2.1.1) is within $1.0°–1.5°$, the same as the reported uncertainty in obtaining the original phasematching data, because of the refractive-index uncertainty [7]. It may also be noted that refractive-index dispersion data of Apollonov et al. [8] also provide a better fit to the experimental phase-matching data, particularly in the longer-wavelength region; however, the deviations at shorter wavelengths are larger, and there is no refractive-index data available for wavelengths shorter than 100 mμm.

2.1.5 DISCUSSION

Now we again consider Figures 2.1.2 and 2.1.3 for some detailed discussion. As was mentioned above, the measured data of Shi and Ding [9] deviate as widely as $5°–18°$ from their data calculated with the Bhar et al. [10] Sellmeier dispersion relations. In the experiment [9] an annealed sample of ZGP crystal had been used to reduce 1.064 m absorption in the crystal. Shi and Ding [9] speculated that through the annealing process the dispersion property and the phase-matching angle of the crystal at 1.064 m had been modified. They also presumed that the dispersion of the crystal in the THz domain had a negligible effect on the phase-matching angle. Here we have found that the origin of the phase-matching discrepancy in Ref. [9] is the use of the inappropriate Sellmeier dispersion relations for the calculation [9] of the refractive-index dispersion of the crystal in the THz range. As was mentioned above, the dispersion equations of Bhar et al. [10] were constructed with measured refractive indices up to only 12 μm in the long-wavelength range, and hence it is not surprising that these dispersion relations cannot reproduce the refractive index dispersion of the crystal in the THz region as shown in Section 2.1.2 with $\lambda \gg 12$ μm. It is clearly observed from curves 2 and 3 of Figures 2.1.2 and 2.1.3 that, when separate dispersion data from Refs. [7,8] are used for the THz region, the calculated phase-matching data reverse their trend of variation in the shorter wavelength region and also that the absolute values of the phase-matching angles are reduced substantially, coming closer to the experimental data [9]. However, it is observed from Figures 2.1.2 and 2.1.3 that the experimental data of Shi and Ding [9] over the whole THz region considered can be explained neither by the THz dispersion data of Apollonov et al. [8] nor by that of Boyd et al. [7]. On the other hand, the calculated results shown as curves 4 in Figures 2.1.2 and 2.1.3 with our formulated Sellmeier dispersion Eq. (2.1.1) satisfactorily explain the measured phase-matching data throughout the entire wavelength region.

The importance of the THz dispersion for determining the phase-matching angle can also be seen in Figure 2.1.4. It is observed in Figure 2.1.4 that the calculated phasematching data curve 1 for Bhar et al. [10] dispersion differ as widely as ~$5°–10°$ from the measured data [7]; the deviations are almost similar to the DFM cases with Shi and Ding [9]. However, it has been seen in Section 2.1.4 that our calculation results curve 4 in Figure 2.1.4 with the Sellmeier dispersion Eq. (2.1.1) satisfactorily explains the experimental phase-matching data of Boyd et al. [7].

2.1.6 CONCLUSION

In conclusion, we have presented a Sellmeier dispersion of ZGP crystal, which is used to obtain the refractive-index dispersion of the crystal in the THz spectral range. We have also presented the phasematching characteristics of the crystal for the generation of widely tunable THz radiation by phase-matched DFM with Nd:YAG and CO_2 lasers as fundamental radiation sources. The computed results, with our formulated Sellmeier dispersion Eq. (2.1.1) and other available THz refractive index dispersion data [7,8], have been compared with the available experimental data [7,9]. Our formulated Sellmeier dispersion Eq. (2.1.1) explains well the phase-matching data of Boyd et al. [7] and the results computed with Eq. (2.1.1) also provide a satisfactory explanation of phase-matching discrepancies as large as ~5°–18° reported recently in Ref. [9]. Here, we note that the refractive index and birefringence of the material under consideration, particularly near the band-edge region, are subject to variations from sample to sample owing to differences in stoichiometry and impurity concentrations [6,11]. However, the Sellmeier dispersion of Bhar et al. [10] for near-infrared and Eq. (2.1.1) in this subsection for the THz range presented here provides a good reproduction of the currently available data for nonlinear experiments for THz generation in the crystal samples used [7,9,13]. The presented Sellmeier dispersion would be useful for determining the NLO properties of this material for different NLO applications in the THz spectral region. The contents described in this subsection was collaborative work conducted by thefollowing people: P. Kumbhakar, T. Kobayashi, and G. C. Bhar [13].

REFERENCES

1. B. Ferguson and X.-C. Zhang, "Materials for terahertz science and technology," *Nat. Mater.* 1, 26–33 (2002).
2. A. Bonavalet, M. Joffre, J.-L. Martin, and A. Migus, "Generation of ultrabroadband femtosecond pulses in the mid-infrared by optical rectification of 15 fs light pulses at 100 MHz repetition rate," *Appl. Phys. Lett.* 67, 2907–2909 (1995).
3. A. Nahata, A. S. Weling, and T. F. Heinz, "A wideband coherent terahertz spectroscopy system using optical rectification and electro-optic sampling," *Appl. Phys. Lett.* 69, 2321–2323 (1996).
4. M. S. Tani, M. Herrmann, and K. Sakai, "Generation and detection of terahertz pulsed radiation with photoconductive antennas and its application to imaging," *Meas. Sci. Technol.* 13, 1739–1745 (2002).
5. K. L. Vodopyanov, F. Ganikhanov, J. P. Maffetone, I. Zwieback, and W. Ruderman, "ZnGeP$_2$ optical parametric oscillator with 3.8–12.4 m," *Opt. Lett.* 25, 841–843 (2000).
6. G. D. Boyd, E. Buehler, and F. G. Storz, "Linear and nonlinear optical properties of ZnGeP$_2$ and CdSe," *Appl. Phys. Lett.* 18, 301–304 (1971).
7. G. D. Boyd, T. J. Bridges, C. K. N. Patel, and E. Buehler, "Phase-matched submillimeter wave generation by difference frequency mixing in ZnGeP$_2$," *Appl. Phys. Lett.* 21, 553–555 (1972).
8. V. V. Apollonov, A. I. Gribenyukov, V. V. Korotkova, A. G. Suzdal'tsev, and Yu. A. Shakir, "Subtraction of the CO$_2$ laser radiation frequencies in a ZnGeP$_2$ crystal," *Sov. J. Quantum Electron.* 26, 469–470 (1996).
9. W. Shi and Y. J. Ding, "Continuously tunable and coherent terahertz radiation by means of phase-matched difference frequency generation in zinc germanium phosphide," *Appl. Phys. Lett.* 83, 848–850 (2003).
10. G. C. Bhar, L. K. Samanta, D. K. Ghosh, and S. Das, "Tunable parametric crystal oscillator," *Sov. J. Quantum Electron.* 17, 860–861 (1987).
11. G. C. Bhar, "Refractive index interpolation in phasematching," *Appl. Opt.* 15, 305–307 (1976).
12. The extraordinary index n_e $n_o n$; n 0.0397, which is the long-infrared birefringence, from Ref. 6.
13. P. Kumbhakar, T. Kobayashi, and G. C. Bhar, "Sellmeier dispersion for phase-matched terahertz generation in ZnGeP$_2$," *Appl. Opt.* 43, 3324–3328 (2004).

2.2 Saturation of the Free Carrier Absorption in ZnTe Crystals

2.2.1 INTRODUCTION

Terahertz (THz) science has seen significant developments over the past several decades and has a large number of applications in a variety of fields such as plasma physics, astronomy, medical imaging, biology and communication [1]. It is certainly important for the development of bright and broadband THz sources. Of various techniques used, optical rectification in crystals (typically ZnTe crystals) is mostly used for the generation of THz waves. However, the saturation of THz output power generated from ZnTe crystals limits the development of intense THz sources [2]. There have been many studies on the saturation mechanism of THz output power [3–6]. Some researchers focus on the study of large-area THz emitters, which avoid the creation of too many carriers and reduce the THz output power [7,8]. Recently, it has been proved that the saturated THz conversion efficiency of ZnTe crystals mainly depends on free carrier absorption, rather than the pumping power, which is attenuated by two-photon absorption [5,6]. These studies used ultrafast laser amplifiers with high pulse energy. The red shift in the THz spectra of ZnTe crystals and other nonlinear effects due to high pumping fluences (below the damage threshold) has not been explained because of the lack of wide range and finely tuned pumping fluences.

In this study, a high-power Ti:sapphire laser with fine-tuning of the fluences was used as a pumping source for THz generation. To the authors' knowledge, this is the first time that these issues have been studied over a wide range of pumping fluences. To determine the relationship between free carriers and THz generation in ZnTe crystals, the dependence of the THz temporal waveforms (spectra) and the photoluminescence (PL) radiated from ZnTe crystals on the pumping fluences is measured simultaneously. The effects of band gap renormalization, carrier saturation, and PL quenching on THz generation are also examined and clarified.

2.2.2 EXPERIMENTS

2.2.2.1 THz Generation in ZnTe Crystals

In these experiments, a long-cavity Ti:sapphire oscillator (Femtosource scientific XL300, Femtolaser) was used for THz wave generation and detection. The characteristics of the laser are described in a way as a central wavelength of 800 nm, a repetition rate of 5.2 MHz and a pulse duration of 70 fs. The pump beam, with a spot diameter of 46 μm, was focused on a 1-mm-thick (110) ZnTe crystal with the resistivity of 100 Ω/cm. A teflon filter was used to block the fundamental light, whereas THz wave transmitting. The transmitted THz wave was collimated and focused on another 0.5-mm-thick (110) ZnTe crystal, which detected the THz wave by free space electro-optical (EO) sampling through a pair of off-axis parabolic mirrors. All of the experiments were performed in a dry nitrogen-purged box. While the power of probing pulses for the EO sampling was kept constant, the pumping power used in this study was varied from 10 to 800 mW by a neutral density (ND) filter, which corresponds to 0.12 ~ 9.26 mJ/cm². The dispersion due to a neutral density (ND) filter was compensated by a compressor inside the laser system. For the sake of analysis convenience, the experimental results were classified into the following three regimes:

DOI: 10.1201/9780429196577-15

1. Relatively low pumping fluences (0.12~0.58 mJ/cm²)
2. Medium pumping fluences (0.58~6.36 mJ/cm²)
3. Relatively high pumping fluences (6.36~9.26 mJ/cm²).

The THz temporal waveforms at various pumping fluences in the range from 0.58 to 9.26 mJ/cm² corresponding to the cases of above (2) and (3) are shown in Figure 2.2.1a–c. As the pumping fluence increases, a delay (shift in time) in the THz waveforms is clearly observed (as shown in Figure 2.2.1a and b). However, this type of time delay in THz temporal waveform of the optical electric field does not continuously increase in the higher pumping fluence regime (as shown in Figure 2.2.1c). The delay in the THz temporal waveforms occurs because of a change in the refractive index, which is described by the relationship: $\Delta t = \Delta n\, d\, c$, where d is the crystal thickness, c is

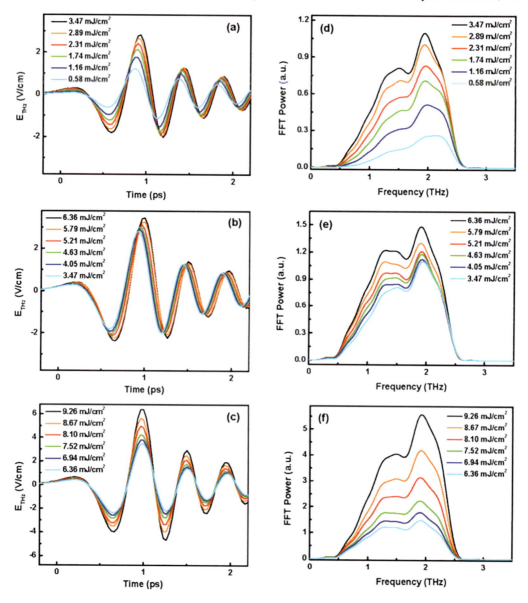

FIGURE 2.2.1 (a)–(c) Temporal waveforms of the THz pulses generated by optical rectification in a ZnTe crystal at various pumping fluences. (d–f) The FFT spectra of the THz waveforms in (a–c).

speed of light, Δn is the change in the refractive index in the THz regime or the optical (fundamental) region. For pumping fluences of 4.63 mJ/cm^2 and 0.58 mJ/cm^2, the time difference, Δt, between the temporal waveforms is approximately 80 fs, which corresponds to a value of $\Delta n = 2.4 \times 10^{-2}$. The possible reasons for the delay in the high pumping fluence regime are discussed in terms of two mechanisms (1) and (2) as follows:

1. High pumping fluences in the crystals may change the refractive index in the optical range (the optical Kerr effect which is one of the third-order optical nonlinear phenomena). However, according to Ref. [9], the change in the refractive index in the optical range is *too small* ($\Delta n \sim 10^{-4}$) to explain the delay in the THz temporal waveforms (in the case of this study, $\Delta n = 2.4 \times 10^{-2}$).

2. Generated photo-excited free carriers change the optical parameters, e.g., the refractive index of ZnTe crystals in the THz range [10]. This suggests that the free carriers created by two-photon absorption cause a change in the refractive index of ZnTe crystals in the THz range. Surprisingly, the THz temporal waveforms are not shifted more at high pumping fluences (>6.36 mJ/cm^2) as shown in Figure 2.2.1c. This demonstrates that the refractive index of ZnTe crystals *does not change* further as the pumping fluence is increased. In other words, the free carriers saturate in the ZnTe crystals and the refractive index in the THz range remains unchanged at pumping fluences of higher than 6.36 mJ/cm^2. This carrier saturation phenomenon is further verified by the photoluminescence results, which are discussed later.

The THz power spectra (Figure 2.2.1d–f) were obtained by fast Fourier transform of the THz temporal waveforms in Figure 2.2.1a–c. The THz spectra generated from ZnTe crystals have two parts (i) and (ii): a low-frequency part (i) at ~1.2 THz and a high-frequency part (ii) at ~1.9 THz (the phase-matched frequency).

The low-frequency part (i) leads (propagates earlier than) the high-frequency part (ii) in time, because of the normal dispersion of THz wave transmitting through the material (ZnTe).

Therefore, the components of the THz temporal waveforms can be described as follows:

①. The main THz pulse corresponds to the low-frequency part of the THz power spectra and

②. The following oscillation corresponds to the high-frequency part (the phase-matching frequency) [11].

In the regime of 0.58 ~ 6.36 mJ/cm^2, the intensity increase rate of the high-frequency part in THz spectra is slower than that for the low-frequency part as the pumping fluence increases, which demonstrates that the growth rate for the trailing part of THz pulse is slower than that of the main part of THz. At pumping fluences greater than 6.36 mJ/cm^2, the spectra do not change as the pumping fluence is increased. This is consistent with an absence of any shift in the THz temporal waveforms (Figure 2.2.1c). Consequently, the THz spectra and the temporal waveforms as a function of pumping fluences are highly correlated with the concentration of free carriers in the ZnTe crystals. The change in the refractive index, Δn, for 800 nm or THz is too small to cause a change in the emitted THz spectra and the temporal waveforms. Instead, free carriers in a ZnTe crystal cause absorption and affect the emitted THz waveforms and spectra.

The quadratic relationship between THz output power and the pumping fluences of less than 0.58 mJ/cm^2 is shown in the inset of Figure 2.2.2. As the pumping fluence increases, the increase in the THz output power does not obey a quadratic relationship when the pumping fluence is larger than 0.58 mJ/cm^2, which is consistent with the results of Hoffmann et al. [4]. This is because the free carriers in the ZnTe crystals, whose concentration increases when the pumping fluence is increased, attenuate the THz output power. Moreover, in the regime of 3.47 ~ 6.36 mJ/cm^2, the THz output power remains almost constant, even if the pumping fluence is increased. Interestingly, in the extremely high pumping fluence regime of above 6.36 mJ/cm², the THz output power is no longer reduced by saturation of the free carriers and there is a quadratic relationship between the pumping fluences and the THz output power, as shown by the solid line in Figure 2.2.2. The saturation of the free carriers does not cause further change in the refractive index of the ZnTe crystals and there is no further shift in the THz temporal waveforms.

2.2.2.2 PHOTOLUMINESCENCE RADIATED FROM ZNTE CRYSTALS

After pumping, the free carriers in ZnTe crystals move from the conduction band to the valence band and emit a Yellow–green photoluminescence (PL), as shown in the inset of Figure 2.2.3. In this study, the PL in the front of the ZnTe was measured using a fiber-coupled spectrometer (Ocean Optics USB4000) with a short pass filter. The two-photon absorption excites free carriers within the entire ZnTe crystal, and the radiated PL must be transmitted and reabsorbed by the ZnTe crystal. The two-photon excited PL spectra represent the band gap edge of the excited ZnTe crystals. At a pumping fluence of 0.58 mJ/cm², the PL is too weak to be isolated from the background noise. When the pumping fluence is increased, a relative red shift in the peak of PL spectra is observed, as shown in Figure 2.2.3a, which indicates the band gap renormalization (BGR) effect that is caused by the screening effect from extra carriers. The repulsion between free carriers results in a smaller band gap, so the relative amount of photo-excited free carriers can be estimated from the shift in the peak of the PL spectra. The red shift in the PL spectrum is usually described by the empirical relationship [12,13]:

$$\Delta E_{PL} = E_g - E_0 = -K\, n^{1/3} \tag{2.2.1}$$

where K is the BGR coefficient and n is the concentration of the free carriers. The $n^{1/3}$ dependence of ΔE_{PL} resembles the prevailing exchange contribution of electron-electron interaction. In the pumping regime of 1.16~6.36 mJ/cm², the monotonic red shift in the PL spectra (Figure 2.2.3a) indicates that the concentration of the free carriers in the ZnTe crystals rises as the pumping fluence increases. This result confirms that a greater number of free carriers results in a greater time delay in the THz temporal waveforms (see Figure 2.2.1) and a greater attenuation of the THz output power (see Figure 2.2.2). At pumping fluences greater than 6.36 mJ/cm², however, the position of the peak in the PL spectra remains almost constant. In other words, the concentration of the free carriers no longer increases in the high pumping fluence regime, from 6.36 to 9.26 mJ/cm², which provides direct evidence of saturation of the free carrier absorption effect in ZnTe crystals.

By integrating various wavelength components of the PL spectra, the integrated intensity of PL signal can be used for discussion. It reveals how many carriers have traveled from the conduction band to the valence band by the photoexcitation pumping light. The dependence of the intensity of the PL signal on pumping fluence is shown in Figure 2.2.3b. At pumping fluences of less than 6.36 mJ/cm², the intensity of the PL signal gradually increases, when the pumping fluence is increased. This indicates an increase in the number of free carriers, which causes a change in the refractive index in the THz range. At pumping fluences larger than 6.36 mJ/cm², all of the results including the time delay in the THz temporal waveforms, the quadratic increase in the THz output power, and the red shift in the PL spectra suggest that the concentration of free carriers remains constant. However, the intensity of the PL signal decreases in this regime. This implies that some free carriers transit back to the ground state associated with luminescence intensity decrease. On the other hand, intense THz waves may induce carrier recombination through the non-radiative relxation, which is termed "The THz quenching effect" [14].

2.2.3 DISCUSSION

To understand the mechanism of the reduction in THz output power, quantitative analysis was made on the time evolution attenuation within a single THz temporal waveform at various pumping fluences. The time-dependent attenuation fraction for a single THz temporal waveform (e.g., Figure 2.2.4a) is given by:

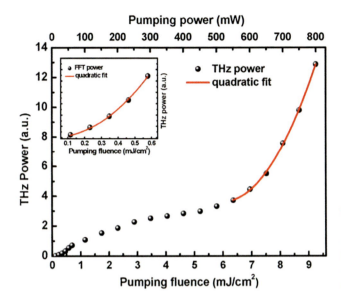

FIGURE 2.2.2 The dependence of THz output power on pumping fluences in a 1-mm ZnTe crystal. The solid line presents the quadratic fit. Inset: an enlargement of the low pumping fluence regime.

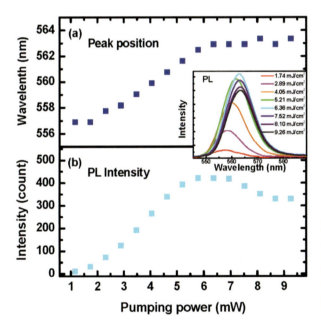

FIGURE 2.2.3 (a) The position of the peak in the PL spectra for a ZnTe crystal as a function of pumping fluences. (b) The PL intensity as an integral of the emission spectra. The intensity spectrum curves with peaks located at 1.74, 2.89, 4.05, 5.21, 6.36, 7.52, 8.10, and 9.26 nm.

$$A(t) = E_{THz}(P_1,t)/[(E_{THz}(P_0,t) \times (P_1/P_0))] \tag{2.2.2}$$

where E_{THz} is the THz temporal waveform, P_0 is a pumping fluence of 0.58 mJ/cm² (without attenuation by free carriers) and P_1 is a higher pumping fluence (with attenuation by free carriers). In Figure 2.2.4b, the symbols show the time-dependent attenuation fraction at a pumping fluence of 4.63 mJ/cm² (the divergent points before 0.25 ps are artificial from calculation), wherein the solid line is a guide for the eyes. The THz attenuation fraction decreases within several hundreds of femtoseconds and then remains constant. The reasons for the appearance of the phenomenon of the time-dependent THz attenuation shown in Figure 2.2.4b can be described in the following way:

1. THz attenuated by free carrier absorption is associated with the number of free carriers and their mobility.
2. A rapid increase in THz attenuation requires that the free carriers take several hundreds of femtoseconds to achieve carrier thermalization in the conduction band (from higher energy levels to low energy levels in conduction band), after two-photon absorption.
3. During the flat portion of THz attenuation, the free carriers require more than 40 ps to relax from the conduction band to the valence band. Because the amount of free carriers does not change during the second and third cycles of the THz temporal waveform, the attenuation by free carriers is almost the same within this period. Thus, a THz wave with a shorter pulse duration (generated by thinner ZnTe crystals) is less attenuated by free carrier absorption at the same pumping fluence as a consequence.

The peak (valley) amplitude of the THz temporal waveforms was also calculated as a function of the pumping fluences. In principle, the peak (valley) amplitude of the THz temporal waveforms should increase linearly, without free carrier absorption. They do not increase linearly, because of free carrier absorption, similarly to previous results. The increase in amplitude of the first cycle of the THz temporal waveforms is more rapid than that in the following cycles, as the pumping fluence increases. Namely, the attenuation in the first cycle of the THz temporal waveforms is smaller than that for the following cycles. The large THz attenuation within the first cycle (~300 fs) verifies the result in Figure 2.2.4b. The subsequent cycles of the THz temporal waveforms demonstrate similar increases for various pumping fluences, which is consistent with the flat THz attenuation after 1 ps, as shown in Figure 2.2.4b. For the high pumping fluences (>6.36 mJ/cm^2), the saturation of free carriers in ZnTe crystals is clearly demonstrated by the analysis of the THz waveform and the PL spectra. This saturation phenomenon of the free carriers results in characteristics that are significant to THz generation in ZnTe crystals at high pumping fluences. They are described as follows in terms of three terms:

1. No further time delay is observed in the THz temporal waveforms because there is no further change in the refractive index of the ZnTe crystals.
2. The THz output power increases quadratically when the pumping fluence is increased.

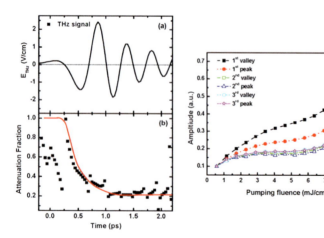

FIGURE 2.2.4 (a) The THz temporal waveform radiated from a ZnTe crystal at a pumping fluence of 4.63 mJ/cm^2. (b) The attenuation fraction of a THz temporal waveform in (a) compared with that at a low pumping fluence of 0.58 mJ/cm^2. The solid line is a guide to the eyes. (c) The amplitude of the peaks and valleys in THz temporal waveforms as a function of the pumping fluences, which is normalized to a value for the pumping fluence of 0.58 mJ/cm^2.

3. There is no shift in the position of the peak in the PL spectra. At high pumping fluences greater than 6.36 mJ/cm^2, thus, the THz electric field strength linearly raises as increasing pumping fluences (Figure 2.2.4c) due to the saturation of the free carrier absorption.

This saturation of free carrier absorption effect (3) has two possibilities.

i. One is the saturation of free carrier concentration. The other one is the reduction of free carrier absorption with smaller carrier mobility or larger effective mass of hot carriers by high THz field strength [15]. For intense THz, the smaller carrier mobility or larger effective mass will cause the change of the refractive index inside crystals and deform the THz waveforms in the time domain.

ii. Also, the spectra of optical pumping pulses and THz pulses after passing through ZnTe crystals gradually shift to lower frequency and become broader in high frequency side as increasing pumping fluences, respectively [16].

However, experimentally we did not observe the broadening of THz spectra as shown in Figure 2.2.1 and the red shift of optical spectra for pumping pulses, especially for high pumping fluence regime (>6.36 mJ/cm^2). In this study, the strongest peak field strength inside the crystal is around 6.3 V/cm as shown in Figure 2.2.1 (THz pulse energy ~1.84 pJ), which is corresponding to the THz fluence of around 0.03 µJ/cm^2. The THz field strength and intensity in this study are much smaller than the reported values (several hundred kV/cm) by D. Turchinovich and M. C. Hoffmann in Ref. [17].

Moreover, the nonlinear plasma response in the crystals would raise the effective mass of hot electrons and modify the curvature of conduction band to lead a relative change of the size of PL spectral bandwidth. Based on our experimental results of the PL spectra in high pumping fluence regime (>6.36 mJ/cm^2), however, we did not observe any changes in the bandwidth and the shift of peak position as shown in the inset of Figure 2.2.3. Only the PL intensity shrinks as increasing the pumping fluences, which is consistent with the results of Ref. [14]. It means no more changes for the band structure of ZnTe crystals such as band gap renormalization induced by the change of the Coulombic interaction of the extensive amount of photogenerated charge carriers. On the other hand, we did not observe the time shift in the THz temporal waveforms as shown in Figure 2.2.1c, which is caused by the intense THz field (several 100 kV/cm) in Ref. [17]. Consequently, we conclude that the quadratic increase of THz output power is due to the saturation of free carrier, and the reason of PL intensity decreasing observed at high pumping fluences of greater than 6.36 mJ/cm^2 in this study would be mainly dominated by the "PL quenching effect". Based on these results, we propose some methods for the reduction of THz attenuation, e.g., shortening the lifetime of carriers (by using another probe light to induce stimulated emission in ZnTe crystals), or changing the optical properties of the ZnTe crystals (by using crystals with different purity [18]).

2.2.4 CONCLUSION

The characteristics of the THz and PL signals emitted from ZnTe crystals over a wide range of pumping fluences are systematically studied in this subsection [19]. In the low pumping fluence regime (0.58~6.36 mJ/cm^2), the concentration of free carriers in the ZnTe crystals increases as the pumping fluence is increased, which results in a band gap renormalization (i.e., a red shift in the PL spectra) in the ZnTe crystals. These free carriers in the ZnTe crystals not only affect the refractive index in the THz range but also cause a time-dependent attenuation of the THz temporal waveforms. In the high pumping fluence regime (above 6.36 mJ/cm^2), the saturation of free carriers in the ZnTe crystal is clearly demonstrated by the unchanged position of the peak in the PL spectra. Therefore, the THz temporal waveforms are not subject to further delay because of the unchanged refractive index in the THz range. Moreover, there is a quadratic increase in the THz output power as the pumping fluence is increased, which is also seen in the results for very low pumping fluences (<0.58 mJ/cm^2). These results demonstrate that there is a potential for intense THz generation in semiconducting nonlinear materials.

The contents of this subsection is obtained by the cooperative research activity of the following people: S. A. Ku, C. M. Tu, W.-C. Chu C. W. Luo, K. H. Wu, A. Yabushita, T. Kobayashi.

REFERENCES

1. B. Ferguson and X.-C. Zhang, "Materials for terahertz science and technology," *Nat. Mater.* **1**(1), 26–33 (2002).
2. E. G. Sun, W. Ji, and X.-C. Zhang, "Two-photon absorption induced saturation of THz radiation in ZeTe," *Proceedings of the Conference on Lasers Electro-Optics, OSA*, 479–480 (2000).
3. V. Y. Gaivoronsky, M. M. Nazarov, D. A. Sapozhnikov, Y. V. Shepelyavyi, S. A. Shkel'nyuk, A. P. Shkurinov, and A. V. Shuvaev, "Competition between linear and nonlinear processes during generation of pulsed terahertz radiation in a ZnTe crystal," *Quantum Electron.* **35**(5), 407–414 (2005).
4. M. C. Hoffmann, K.-L. Yeh, J. Hebling, and K. A. Nelson, "Efficient terahertz generation by optical rectification at 1035 nm," *Opt. Express* **15**(18), 11706–11713 (2007).
5. S. M. Harrel, R. L. Milot, J. M. Schleicher, and C. A. Schmuttenmaer, "Influence of free-carrier absorption on terahertz generation from ZnTe (110)," *J. Appl. Phys.* **107**(3), 033526 (2010).
6. S. Vidal, J. Degert, M. Tondusson, J. Oberlé, and E. Freysz, "Impact of dispersion, free carriers, and two-photon absorption on the generation of intense terahertz pulses in ZnTe crystals," *Appl. Phys. Lett.* **98**(19), 191103 (2011).
7. T. Löffler, T. Hahn, M. Thomson, F. Jacob, and H. G. Roskos, "Large-area electro-optic ZnTe terahertz emitters," *Opt. Express* **13**(14), 5353–5362 (2005).
8. F. Blanchard, L. Razzari, H.-C. Bandulet, G. Sharma, R. Morandotti, J.-C. Kieffer, T. Ozaki, M. Reid, H. F. Tiedje, H. K. Haugen, and F. A. Hegmann, "Generation of 1.5 microJ single-cycle terahertz pulses by optical rectification from a large aperture ZnTe crystal," *Opt. Express* **15**(20), 13212–13220 (2007).
9. A. A. Said, M. Sheik-Bahae, D. J. Hagan, T. H. Wei, J. Wang, J. Young, and E. W. Van Stryland, "Determination of bound-electronic and free-carrier nonlinearities in ZnSe, GaAs, CdTe, and ZnTe," *J. Opt. Soc. Am. B* **9**(3), 405–414 (1992).
10. M. Schall and P. U. Jepsen, "Above-band gap two-photon absorption and its influence on ultrafast carrier dynamics in ZnTe and CdTe," *Appl. Phys. Lett.* **80**(25), 4771–4773 (2002).
11. C.-M. Tu, S. A. Ku, W.-C. Chu, C. W. Luo, J.-C. Chen, and C.-C. Chi, "Pulsed terahertz radiation due to coherent phonon-polariton excitation in <110> ZnTe crystal," *J. Appl. Phys.* **112**(9), 093110 (2012).
12. H. Wang, K. S. Wong, B. A. Foreman, Z. Y. Yang, and G. K. L. Wong, "One-and two-photon-excited timeresolved photoluminescence investigations of bulk and surface recombination dynamics in ZnSe," *J. Appl. Phys.* **83**(9), 4773–4776 (1998).
13. J. D. Ye, S. L. Gu, S. M. Zhu, S. M. Liu, Y. D. Zheng, R. Zhang, and Y. Shi, "Fermi-level band filling and bandgap renormalization in Ga-doped ZnO," *Appl. Phys. Lett.* **86**(19), 192111 (2005).
14. J. Liu, G. Kaur, and X.-C. Zhang, "Photoluminescence quenching dynamics in cadmium telluride and gallium arsenide induced by ultrashort terahertz pulse," *Appl. Phys. Lett.* **97**(11), 111103 (2010).
15. M. C. Hoffmann and D. Turchinovich, "Semiconductor saturable absorbers for ultrafast terahertz signals," *Appl. Phys. Lett.* **96**(15), 151110 (2010).
16. M. Nagai, M. Jewariya, Y. Ichikawa, H. Ohtake, T. Sugiura, Y. Uehara, and K. Tanaka, "Broadband and high power terahertz pulse generation beyond excitation bandwidth limitation via $\chi^{(2)}$ cascaded processes in LiNbO$_3$," *Opt. Express* **17**(14), 11543–11549 (2009).

17. D. Turchinovich and M. C. Hoffmann, "Self-phase modulation of a single-cycle terahertz pulse by nonlinear free-carrier response in a semiconductor," *Phys. Rev. B* **85**(20), 201304 (2012).

18. N. Kamaraju, S. Kumar, E. Freysz, and A. K. Sood, "Influence of two photon absorption induced free carriers on coherent polariton and phonon generation in ZnTe crystals," *J. Appl. Phys.* **107**(10), 103102 (2010).

19. S. A. Ku, C. M. Tu, W.-C. Chu C. W. Luo, K. H. Wu, A. Yabushita, T. Kobayashi, "Saturation of the free carrier absorption in ZnTe crystals," *Opt. Express* **21**(12), 13930–13937 (2013).

2.3 Widely Linear and Non-Phase-Matched Optical-to-Terahertz Conversion on GaSe
Te Crystals

2.3.1 INTRODUCTION

Terahertz (THz) wave technology [1] has been developed in the last several decades, and a large number of useful applications of the THz wave have been developed, such as the measurements of molecular vibrational modes and time-resolved THz spectroscopy, which have led to a new "tera era". Therefore, for practical applications, the development of novel sources and methods in the THz range has become a key issue. A relatively compact and economical avenue for generating coherent THz radiation is taking advantage of optical conversion in nonlinear crystals (e.g., GaSe, LiNbO$_3$, GaP, GaAs, and ZnGeP2) [2–6]. Among these nonlinear crystals, layered ε-GaSe crystals possess many advantages, including a large nonlinear coefficient and significant birefringence, which have been widely used in far-IR generation [3] and mid-IR spectroscopy [7]. Nevertheless, for further application in optics, improvement in optical and mechanical properties of GaSe crystals by doping proper elements is highly desired. High-quality GaSe:Te crystals have been prepared because Te atoms slightly strengthen the mechanical properties [8] and modify the optical and electrical properties noticeably [9,10]. However, GaSe:Te crystals have not yet been studied in THz generation.

This subsection describes a scheme of THz wave generation from GaSe:Te crystals by femtosecond (fs) laser pulses and shows that THz generation efficiency was improved substantially. Compared to the commonly used ZnTe crystals, GaSe:Te crystals can provide widely linear optical-to-THz conversion with an optical pumping fluence of up to 6.9 mJ/cm^2.

2.3.2 EXPERIMENTAL

The p-type pure and Te-doped GaSe single crystals grown by the Bridgman method [11] with five different Te were used in this study. The concentrations of Te in all GaSe:Te crystals were determined by electron probe microanalysis (with an accuracy of 0.01 wt. %) are 0.01, 0.07, 0.38, 0.67, and 2.07 mass %, respectively. On the other hand, a commercial Ti:sapphire oscillator operating at a central wavelength of 800 nm and with pulse duration of 100 fs with a repetition rate of 5.2 MHz was used as a pumping source. Under the scheme of normal incidence (the polarization is along the a–b plane of GaSe crystal sample), the pump beam with 3.5 mJ/cm^2 and diameter of 46 μm was focused on z-cut GaSe:Te crystals to generate THz waves. Any residual 800 nm laser beam was blocked by a Si wafer. The transmitted THz pulses were collected and focused on a 100 μm thick (110)-oriented ZnTe crystal by gold-coated off-axis parabolic mirrors. The electro-optical (EO) sampling technique was subsequently applied to detect the emitted THz fields in the time domain. All experiments were performed in a dry nitrogen-purged box.

DOI: 10.1201/9780429196577-16

2.3.3 RESULTS AND DISCUSSIONS

Typical THz waveforms at various azimuthal angles ϕ were generated from a 0.38 mass % GaSe:Te crystal, as shown in Figure 2.3.1a. Here, ϕ is defined as the angle between the [100] direction of (001) GaSe:Te crystals and the polarization direction of the electric field of the incident optical pulse. Considering EO detection and the 62m point group of the crystals, the electric field of THz generated from GaSe:Te crystals by optical rectification can be derived as

$$E_{\text{THz}} \propto \chi_{22} \cos 2\theta \sin 3\varphi \qquad (2.3.1)$$

where θ is the angle between the c axis of a crystal and the direction of the incident light. The right inset in Figure 2.3.1a shows the intensity of the main peak in THz signals as a function of φ, which can be well described by the φ-dependent relation of Eq. (2.3.1). Therefore, the mechanism of THz generation on the sample GaSe:Te crystals is described in terms of optical rectification, which is different from the difference frequency mixing in ZnGeP2 [6]. The sixfold symmetry of THz intensity is attributed to the hexagonal structure on the (001) plane of GaSe:Te crystals. Moreover, similar 6-fold symmetry observed for all GaSe:Te crystals indicates that the Te atoms doped in GaSe do not destruct the crystal structure of the GaSe matrix.

Figure 2.3.1b unambiguously shows the effect of Te atoms doped in GaSe crystals for THz generation. Nevertheless, the central frequencies (~2.15 THz) of THz radiation are independent of the Te concentration; we found that the THz power suddenly increases once the Te atoms are doped into the GaSe crystals, even for the small Te doping of 0.01 mass %. For 0.38 mass % GaSe:Te crystals,

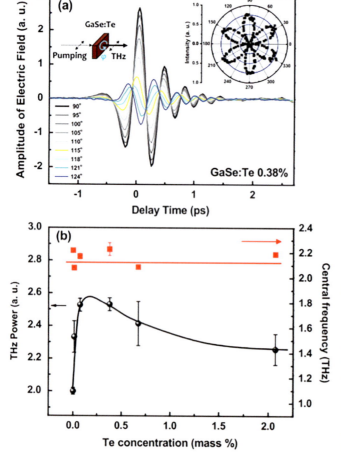

FIGURE 2.3.1 (a) Temporal waveforms of THz radiation at various azimuthal angle φ on a 0.38 mass % GaSe:Te crystal with thickness of 0.30 mm. The right inset illustrates the intensity of the THz main peak as a function of φ. The left inset shows the configuration of THz generation on GaSe:Te crystals. (b) Power and central frequency of THz radiation on various GaSe:Te crystals with thickness of 0.30 mm.

the output power of emitted THz is enhanced by a factor of 21% compared to the GaSe crystals. According to Vodopyanov's studies [12], the optical-to-THz conversion efficiency is proportional to $\chi_{ijk}^{(2)}$. Thus, these results imply that heavy-atom doping, that is, Te, in GaSe crystals can improve its nonlinearity. However, the optical-to-THz conversion efficiency could not be increased further at higher Te-doping level because of increases in several types of structural defects concentration (e.g., polytypes, stacking faults, and dislocation) [9–11].

Figure 2.3.2 shows the Fourier spectrum of the THz radiation from ZnTe, GaSe, and 0.38 mass % GaSe crystals. Inset shows THz power as a function of pump fluences.

In addition, the spectra of THz radiation from GaSe:Te crystals are wider than those of the ZnTe crystals with the same thickness at a high-frequency side. This is caused by the frequency of the transverse optical phonon absorption peak for GaSe centered at 7.1 THz, which is higher than that of ZnTe at 5.3 THz [3,13].

Because of the two-photon absorption of pumping light and the free carrier absorption of THz on ZnTe, the THz output power at first show quadratic dependence in low pump intensity regime and then start to saturate above the pumping fluence of about 3 mJ/cm² [14]. The value of the two-photon absorption coefficient of GaSe (β_{GaSe} 0.558 cm/GW) is smaller than that of ZnTe (βZnTe 4.2 cm/GW) [15], which implies that fewer free carriers are generated in GaSe:Te crystals to diminish the suppression effect in the THz output power due to the two-photon absorption. It is because the second photon is not useful in the generation of THz emitting level population. Therefore, the linear dependence (the originally quadratic dependence is suppressed by two-photon absorption) between the THz output power and pumping fluence started to be observed at some level of pumping. Consequently, at relatively low pump fluences (<4.6 mJ/cm²), the emitted THz power from GaSe:Te crystals is smaller than that from ZnTe crystals. Once the pump fluences are larger than 4.6 mJ/cm², however, the emitted THz power on the GaSe:Te crystals exceeds that on ZnTe crystals considerably because of the larger two-photon absorption of ZnTe. The conversion efficiency of GaSe:Te crystals was significantly higher than that of the GaSe crystal, especially in the high pumping fluence range. Thus, we can expect that GaSe:Te crystals to be a promising material for high-power THz generation, which would be attractive for applications in THz spectroscopy.

Although no significant changes are in the central frequency of THz radiation by fixing the thickness to 0.3 mm on various GaSe:Te crystals (Figure 2.3.1b), the marked redshift of the central frequency is observed by changing the thickness of GaSe:Te crystals, as shown in Figure 2.3.3. Under the slowly varying envelope approximation (which is called SVEA sometimes in literature) [12], the power spectrum of THz radiation can be described as

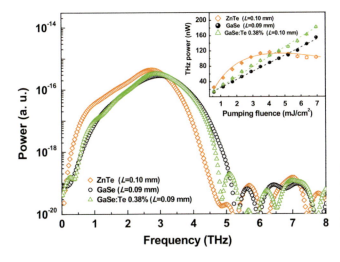

FIGURE 2.3.2 Fourier spectrum of the THz radiation from ZnTe, GaSe, and 0.38 mass % GaSe crystals. Inset, THz power as a function of pump fluences. *L*, thickness of the crystals. The solid lines are guides for the eyes.

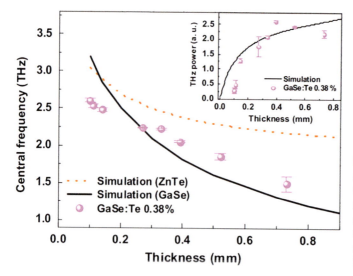

FIGURE 2.3.3 Central frequency of THz radiation on 0.38 mass % GaSe:Te crystals with various thicknesses. Inset, thickness dependence of THz power on 0.38 mass % GaSe:Te. All THz generation experiments were performed by the pumping level of 3.5 mJ/cm².

$$I(\omega_T, L) \alpha\ \omega_T^2 d_{\text{eff}}^2 \tau^2 L^2 \exp\left(-\frac{\tau^2 \omega_r^2}{4}\right) \sin c^2\left(\Delta k L/2\right) \sin c \qquad (2.3.2)$$

where ω_T is the THz angular frequency, L is the thickness of the crystal, $d_{\text{eff}} = (1/2)\chi^{(2)}$, τ is the pulse duration, and the k-vector mismatch is $\Delta k = (n_{\text{THz}} - n_{\text{opt}}^{\text{gr}})\omega_{\text{THz}}/c$. The refractive index in GaSe crystals was obtained from [16], in which $n_{\text{opt}}^{\text{gr}}$ (at 800 nm) = 3.12 and n_{THz} (at 1 THz) = 3.26. For the different thicknesses of the crystals, L, the THz spectrum can be simulated by the above Eq. (2.3.2) and the central frequency of the THz radiation can be further identified from the simulated spectrum. For the thin crystals, the central frequency of THz radiation simulated from Eq. (2.3.2) is dominated by pulse duration. However, the Δk in Eq. (2.3.2) induces a change in the central frequency in the case of thicker crystals. Consequently, the central frequency of THz radiation from thin GaSe and ZnTe crystals is similar, as shown by the solid and dashed lines in Figure 2.3.3. While the crystal thickness increases, the central frequency of THz radiation on ZnTe crystals approaches 1.9 THz, where phase matching is satisfied [17]. However, the central frequencies of THz radiation on GaSe and GaSe:Te crystals decrease gradually to 1.5 THz because no phase matching condition is satisfied in this region. In some manner, we can consider this characteristic is advantageous for tuning the THz central frequency by varying the thickness of GaSe:Te crystals. Moreover, the THz radiation power on 0.38 mass % GaSe:Te crystals increases by 8.8 times while the thickness increases from 0.1 to 0.52 mm (see inset of Figure 2.3.3), which shows excellent agreement with the simulation results. Therefore, the high THz radiation power on GaSe:Te crystals could be obtained simply by using thicker crystals [18].

2.3.4 CONCLUSION

As mentioned above in this subsection [18], it was demonstrated that broadband THz generation with widely linear optical-to-THz conversion on GaSe:Te crystals is realized by non-phase-matched optical rectification. The dopant (Te atoms) in GaSe crystals improves the efficiency of THz generation significantly, especially in the high pumping fluence range. By increasing crystal thickness, the central frequency of THz radiation from GaSe:Te crystals shifts markedly to red, and its power increases. Furthermore, the high-power (>1.36 µW under the pumping of 6.9 mJ/cm² on a 0.52 mm thick 0.38 mass % GaSe:Te crystal) and central-frequency tunable ($\Delta f_c \sim 1.5$ THz) THz radiation can be realized in GaSe:Te crystals. The research described in this subsection was obtained by the collaboration among the following people: W.-C. Chu, S.-A. Ku, H.-J. Wang, C.-W. Luo, Y. M. Andreev, G. Lanskii, and T. Kobayashi [18].

REFERENCES

1. B. Ferguson and X.-C. Zhang, *Nat. Mater.* **1**, 26 (2002).
2. W. Shi, Y. J. Ding, N. Fernelius, and K. Vodopyanov, *Opt. Lett.* **27**, 1454 (2002).
3. Y.-S. Lee, T. Meade, V. Perlin, H. Winful, and T. B. Norris, *Appl. Phys. Lett.* **76**, 2505 (2000).
4. T. Tanabe, K. Suto, J. Nishizawa, K. Saito, and T. Kimura, *Appl. Phys. Lett.* **83**, 237 (2003).
5. K. L. Vodopyanov, M. M. Fejer, X. Yu, J. S. Harris, Y.-S. Lee, W. C. Hurlbut, V. G. Kozlov, D. Bliss, and C. Lynch, *Appl. Phys. Lett.* **89**, 141119 (2006).
6. P. Kumbhakar, T. Kobayashi, and G. C. Bhar, *Appl. Opt.* **43**, 3324 (2004).
7. C. W. Luo, K. Reimann, M. Woerner, T. Elsaesser, R. Hey, and K. H. Ploog, *Phys. Rev. Lett.* **92**, 047402 (2004).
8. A. A. Tikhomirov, Yu. M. Andreev, G. V. Lanskii, O. V. Voevodina, and S. Y. Sarkisov, *Proc. SPIE* **6258**, 625809 (2006).
9. S. Shigetomi and T. Ikari, *J. Appl. Phys.* **95**, 6480 (2004).
10. I. Evtodiev, L. Leontie, M. Caraman, M. Stamate, and E. Arama, *J. Appl. Phys.* **105**, 023524 (2009).
11. S. A. Ku, W.-C. Chu, C. W. Luo, Y. Andreev, G. Lanskii, A. Shaiduko, T. Izaak, V. Svetlichnyi, E. Vaytulevich, K. H. Wu, and T. Kobayashi, *Opt. Express* **20**, 5029 (2012).
12. K. L. Vodopyanov, *Opt. Express* **14**, 2263 (2006).
13. G. Gallot, J. Q. Zhang, R. W. Mcgowan, T. I. Jeon, and D. Grischkowsky, *Appl. Phys. Lett.* **74**, 3450 (1999).
14. M. C. Hoffmann, K.-L. Yeh, J. Hebling, and K. A. Nelson, *Opt. Express* **15**, 11706 (2007).
15. I. B. Zotova and Y. J. Ding, *Appl. Opt.* **40**, 6654 (2001).
16. C.-W. Chen, T.-T. Tang, S.-H. Lin, J. Y. Huang, C.-S. Chang, P.-K. Chung, S.-T. Yen, and C.-L. Pan, *J. Opt. Soc. Am. B* **26**, A58 (2009).
17. A. Nahata, A. S. Weling, and T. F. Heinz, *Appl. Phys. Lett.* **69**, 2321 (1996).
18. W.-C. Chu, S.-A. Ku, H.-J. Wang, C.-W. Luo, Y. M. Andreev, G. Lanskii, and T. Kobayashi, *Opt. Lett.* **37**, 945 (2012).

2.4 THz Emission from Organic Cocrystalline Salt

An Example of 2, 6-Diaminopyridinium-4-Nitrophenolate-4-Nitrophenol

2.4.1 INTRODUCTION

Few-cycle terahertz (THz) electromagnetic pulses have attracted much attention because they have high potential in fundamental studies and practical applications [1]. Inorganic crystals, such as ZnTe, GaSe:Te, used as THz emitters equipped with ultrafast lasers have become popular and high signal-to-noise ratio tabletop THz sources around the world [2–5]. Recently, intense THz pulses have been demonstrated in LiNbO$_3$ by using tilted pulse front excitation, which has extended the applications of intense THz radiation into the nonlinear region [6,7]. On the other hand, due to large second-order nonlinear susceptibilities and low dielectric constants for phase-matching, organic nonlinear optical crystals have received much more attention for THz generation. Some organic crystals such as DAST, DSTMS, and OH1 have been applied to high-field THz generation [8–10]. However, due to low-frequency phonon/resonance absorption in the THz frequency range, the waveforms, and spectra of THz emission from these novel organic crystals are distorted and hence not preferable in some applications. Therefore, ZnTe or LiNbO$_3$ incorporated with Ti:sapphire lasers are still the most popular THz emitters currently at the time of the present paper preparation and expectations of further novel organic materials are strongly hoped to be created.

In this study, we demonstrate THz emission from a novel organic crystal 2,6 diaminopyridinium-4-nitrophenolate-4-nitrophenol (DAP$^+$NP$^-$NP) [11–13]. The field strength of the THz emission from a DAP$^+$NP$^-$NP crystal is comparable with that from a typical THz emitter-ZnTe crystal. Few-cycle THz pulses were observed from a DAP$^+$NP$^-$NP crystal, and both the waveform and spectra from DAP$^+$NP$^-$NP are similar to those from ZnTe.

2.4.2 SAMPLE PREPARATION AND THz EMISSION EXPERIMENTS

The DAP$^+$NP$^-$NP crystals used in this study were synthesized by the slow-evaporation-solution method with ethanol (C$_2$H$_5$OH) as a solvent material [11]. A photo of the used crystal in the present experiment is shown in the inset of Figure 2.4.1a, and the thickness of the sample is ~0.41 mm. We used the XRD θ–2θ scan method to determine the orientation of the used crystal. As shown in Figure 2.4.1a, three diffraction peaks appear at 16.28°, 33.30°, and 50.66°. By comparing with structure in the crystallography database, we confirm that these peaks correspond to the diffracted peaks of (200), (400), and (600), respectively [14]. Therefore, the result of XRD θ–2θ scan suggests that the surface normal of the used crystal is closely along the <100>-direction.

DOI: 10.1201/9780429196577-17

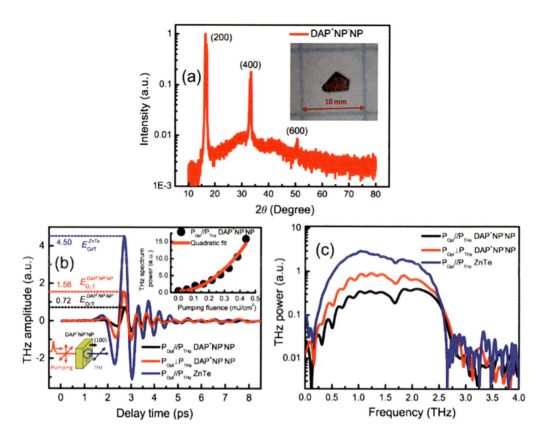

FIGURE 2.4.1 (a) X-ray diffraction θ–2θ scan results of the employed DAP$^+$NP$^-$NP crystal. Compared with the crystallography database, the diffraction peaks at 16.28°, 33.31°, and 50.66° correspond to the diffracted peaks of (200), (400), and (600), respectively [14]. The surface normal of the employed DAP$^+$NP$^-$NP crystal is approximately (100). The inset shows the photo of the employed crystal. (b) THz waveforms from a DAP$^+$NP$^-$NP crystal in the configurations of P_{Opt}/P_{THz} (black) and $P_{Opt}\perp P_{THz}$ (red). The THz waveform from 1-mm <110> ZnTe in the configuration of P_{Opt}/P_{THz} (blue) is also shown as a reference. The inset shows the experimental configuration and the pumping fluence dependence of THz radiation from a DAP$^+$NP$^-$NP crystal (black point) and the quadratic fitting (red line) in P_{Opt}/P_{THz} configuration. (c) The corresponding THz emission spectra.

In the transmission-configuration THz generation experiments, a commercial Ti:sapphire oscillator (Femtosource scientific XL300, FEMTOLASERS Produktions GmbH) operating at a central wavelength of 800 nm was employed as a pumping source, and it produced optical pulses of 100 fs at a repetition rate of 5.2 MHz. The pump beam was focused on the samples with a diameter of about 150 μm and pulse energy of 70.7 nJ to generate THz radiation. The transmitted THz radiation was collimated by two off-axis parabolic mirrors and focused on a 1-mm-thick <110> ZnTe slab for electro-optic sampling. All experiments were performed in a chamber filled with dry nitrogen gas to avoid any deterioration of the nonlinear optical materials for both experimental samples and for pulse laser due to H_2O gaseous molecular contents in the air at room temperature.

2.4.3 RESULTS AND DISCUSSION

Figure 2.4.1b shows the THz temporal waveforms generated from the DAP$^+$NP$^-$NP crystal in different polarization configurations. In the $P_{Opt}//P_{THz}$ configuration, i.e., the polarizations of optical pulses and THz radiation were parallel to each other, a few-cycle THz pulse (black line) can be clearly observed, and the maximum amplitude of electric field is denoted as

DAP+NP−NP $E_{O/T}$. As the polarization of optical pulses was perpendicular to that of THz radiation, in the $P_{Opt} \perp P_{THz}$ configuration, a THz pulse with the same temporal shape (red line *vs.* black line) and an amplitude of about twice larger, i.e., $E_{o \perp T}^{DAP^+NP^-NP}/E_{O/T}^{DAP^+NP^-NP} = 1.56/0.72 = 2.17$, was obtained. In the experiments, we also used a 1-mm-thick <110> ZnTe slab to generate THz radiation, under the same conditions as in P_{Opt}/P_{THz} configuration. As seen clearly in the figure, the amplitude of the electric field from the ZnTe slab is about six times larger than that from the DAP+NP−NP crystal ($E_{o \perp T}^{DAP^+NP^-NP}/E_{O/T}^{DAP^+NP^-NP} = 4.50/0.72 = 6.25$) in the P_{Opt}/P_{THz} configuration.

The FFT spectra of the temporal waveforms are shown in Figure 2.4.1c. We find that the spectra from a DAP+NP−NP crystal (both P_{Opt}/P_{THz} and $P_{Opt} \perp P_{THz}$) show comparable bandwidth to those from 1-mm-thick ZnTe. We also perform pumping fluence-dependent experiments to investigate the mechanism of THz generation from the DAP+NP−NP crystal. As shown in the inset of Figure 2.4.1b, the THz spectrum power in the P_{Opt}/P_{THz} configuration shows quadric dependence on the pumping fluence, which indicates that the THz generation in the DAP+NP−NP crystal is due to the second-order nonlinear optical effect.

To gain more insight into the mechanism of THz generated from a DAP+NP−NP crystal, we rotated the DAP+NP−NP crystal to perform azimuthal angle scan (ϕ-scan) measurements in both P_{Opt}/P_{THz} and $P_{Opt} \perp P_{THz}$ configurations, as shown in Figure 2.4.2c and d, respectively. In the left part of Figure 2.4.2c, the THz power ϕ-scan result reveals sixfold symmetry in the P_{Opt}/P_{THz} configuration. On the other hand, in the left of Figure 2.4.2d, the THz power ϕ-scan result shows twofold symmetry in the $P_{Opt} \perp P_{THz}$ configuration and the polarity also reverses under 180°-rotation.

The symbols of "+" and "−" represent the polarity state of the corresponding THz waveform. Right: the simplified molecular structure of DAP+NP−NP. The symbols of "A" and "D" represent acceptor and donor atoms, respectively. The dashed line represents the direction of the main chain of DAP+NP−NP. The polarization vectors of L (green) and R (brown) represent the THz radiation

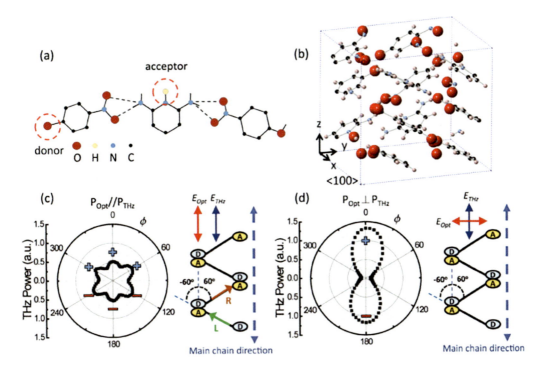

FIGURE 2.4.2 (a) The molecular structure of DAP+NP−NP. O (red), H (yellow), N (blue), C (black), atoms and H-bonds (black dash line) are shown. (b) The crystal structure of DAP+NP−NP [14]. (c) Left: the azimuthal angle scan (ϕ-scan) of THz power in the configuration of P_{Opt}/P_{THz}.

generated in $\phi=300°$ and $60°$, respectively. Left: The azimuthal angle scan (ϕ-scan) of THz power in the configuration of $P_{Opt}\perp P_{THz}$. The maximum THz power was generated along the direction of the main chain of DAP⁺NP⁻NP.

To figure out these interesting ϕ-scan results in DAP⁺NP⁻NP crystals, we propose a model based on the strong second-order-nonlinear effect induced by intramolecular charge transfer between electron donor and acceptor of molecules [15]. The simplified molecular structure of a DAP⁺NP⁻NP crystal in Figure 2.4.2a and b is shown on the right in Figure 2.4.2c and d. The symbols "A" and "D" represent the anionic electron acceptor and the cationic electron donor atoms in Figure 2.4.2a, respectively, while the dashed line represents the direction of the main chain of a DAP⁺NP⁻NP crystal in Figure 2.4.2b [12]. After optical excitation, the nonlinear polarization is induced along the molecular chain between the acceptor and the donor. In the P_{Opt}/P_{THz} configuration, the polarization vectors of "L (green)" and "R (brown)" represent the THz wave generated in $\phi=300°$ and $60°$, respectively. As $\phi=0°$, the induced nonlinear polarizations superimpose and result in a net polarization along the main chain of the DAP⁺NP⁻NP molecules. On the other hand, in the $P_{Opt}\perp P_{THz}$ configuration, the direction of the maximum THz radiation is also along the main chain of a DAP⁺NP⁻NP crystal. All polarities of the measured THz waveform are coincident with the molecular structure of DAP⁺NP⁻NP.

In general, the nonlinear light conversion efficiency in nonlinear optical materials can be brought out significantly through their linear spectra. Therefore, we used a microspectrometer, a spectrometer integrated with a commercial optical-microscope, to study the VIS-NIR transmission spectra at three different positions for the used DAP⁺NP⁻NP crystal, as shown in Figure 2.4.3a. All transmission spectra of these three positions show a cut-off wavelength at ~500 nm and this cut-off wavelength responds to the energy bandgap of the DAP⁺NP⁻NP crystal. The measured transmission spectra in the NIR range show non-negligible differences. At the pumping wavelength, i.e., 800 nm, the measured transmittance at the three positions is 27.0%, 41.5% and 21.8%, which are coincident with the transparencies of these three positions, as shown in the insets of Figure 2.4.3a. This low percentage of transmission indicates that there are some impurities inside the sample and these undesired impurities reduce the THz generation efficiency.

To evaluate the potential of DAP⁺NP⁻NP crystals for high-power THz generation applications, we investigate the effects of thickness and transparency of the samples.

First, for the second-order nonlinear optical process, the amplitude of the electric field of THz radiation is directly proportional to both the crystal thickness L and optical pump intensity I_{Opt}

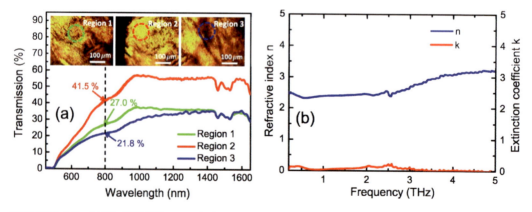

FIGURE 2.4.3 (a) The VIS-NIR transmission spectra at different positions of the employed DAP⁺NP⁻NP crystal. The corresponding photos are shown in the insets and the measured areas (spot size) are ~100 μm in diameter. (b) The refractive index and the extinction coefficient of a DAP⁺NP⁻NP crystal in THz region were obtained by THz time-domain spectroscopy. Compared with Figure 2.4.1c, DAP⁺NP⁻NP does not show significant absorption up to 5 THz, which is coincident with the non-distorted few-cycle THz pulse from a DAP⁺NP⁻NP crystal, as shown in Figure 2.4.1b.

inside of the crystal, i.e., $E_{THz} \propto \times L\, I_{Opt}$ [16]. Second, the undesired impurities inside the samples result in the loss of pumping light and diminish the generated THz intensity (Efficiency is the same since we are claiming that the THz intensity is proportional to the pump intensity). This means the transmittance T of optical pump must be considered, i.e., $I_{Opt} \rightarrow T \times I_{Opt}$. The spot size (~150 μm) of the pump beam is larger than that (~100 μm) used for micro-transmission spectrum. Therefore, we need to consider that the average transmittance $T_{average}$ and the average value of the transmittance at these three positions is 30.1% ((27.0% + 41.5% + 21.8%)/3 = 30.1%). We estimate that the amplitude of the electric field, $E_{O \perp T - Max}^{DAP^+ NP^- NP}$, of THz radiation from DAP⁺NP⁻NP in $P_{Opt} \perp P_{THz}$ configuration would be 5.25 (a.u.)

$$\left(E_{O \perp T - Max}^{DAP^+NP^-NP} = E_{O \perp T - average}^{DAP^+NP^-NP} \times \frac{T_{Max}}{T_{average}} \times \frac{L_{1\,mm}}{L_{0.41\,mm}} = 1.56 \times \frac{41.5\%}{30.1\%} \times \frac{1\,mm}{0.41\,mm} = 5.25 \right)$$

if a highly transparent (in this study, max transmittance T_{Max} = 41.5%) 1-mm-thick DAP⁺NP⁻NP is used. This value would be larger than that of 1-mm-thick ZnTe in P_{Opt} / P_{THz} configuration ($E_{O/T}^{ZnTe} = 4.5$ (a.u.)). Recently, highly transparent DAP⁺NP⁻NP crystals with transmittance of 75% at 800 nm have been reported [13]. This indicates that the amplitude of an electric field twice as large as that from ZnTe could be obtained from a high-quality DAP⁺NP⁻NP crystal. Furthermore, the linear spectrum of DAP⁺NP⁻NP shows higher transmittance around 1000 nm in the NIR range [13]. Compared with the case using 800-nm optical pulses as the pump, we propose that it is possible to use the optical pumping pulse output from a Yb-doped fiber laser to obtain THz radiation with a high electric field. More detailed information about DAP⁺NP⁻NP crystals, such as refractive index and absorption coefficient in the NIR range, is necessary for investigating THz applications in the future. In most organic crystals such as DAST, DSTMS, and OH1, some resonance absorptions in the THz range result in complex waveform and spectra, which is not preferable for applications. We used THz time-domain spectroscopy to analyze the characteristics of the DAP⁺NP⁻NP crystal in the THz range. As shown in Figure 2.4.3b, both the refractive index and extinction coefficient do not show substantial absorbance up to 5 THz in this study. These results are consistent with the observations of non-distorted few-cycle THz pulses from DAP⁺NP⁻NP crystals and indicate that high-quality DAP⁺NP⁻NP crystals have high potential in THz applications.

2.4.4 SUMMARY

In summary, we have examined THz radiation from organic DAP⁺NP⁻NP crystals by using 800-nm optical pulse excitation. The maximum electric field of THz radiation from DAP⁺NP⁻NP is comparable to that from 1-mm ZnTe. This indicates that a high-quality DAP⁺NP⁻NP crystal could be a candidate for low-cost and high-power THz applications [17].

The collaboration research presented in this subsection was conducted by the following people: Chien-Ming Tu, Li-Hsien Chou, Yi-Cheng Chen, Ping Huang, M. Rajaboopathi, Chih-Wei Luo, Kaung-Hsiung Wu, V. Krishnakumar, and Takayoshi Kobayashi [17].

REFERENCES

1. B. Ferguson and X.-C. Zhang, "Materials for terahertz science and technology," *Nat. Mater.* **1**(1), 26–33 (2002).
2. C. M. Tu, S. A. Ku, W. C. Chu, C. W. Luo, J. C. Chen, and C. C. Chi, "Pulsed terahertz radiation due to coherent phonon-polariton excitation in <110> ZnTe crystal," *J. Appl. Phys.* **112**(9), 093110 (2012).
3. S. A. Ku, C. M. Tu, W.-C. Chu, C. W. Luo, K. H. Wu, A. Yabushita, C. C. Chi, and T. Kobayashi, "Saturation of the free carrier absorption in ZnTe crystals," *Opt. Express* **21**(12), 13930–13937 (2013).
4. W.-C. Chu, S. A. Ku, H. J. Wang, C. W. Luo, Y. M. Andreev, G. Lanskii, and T. Kobayashi, "Widely linear and non-phase-matched optical-to-terahertz conversion on GaSe:Te crystals," *Opt. Lett.* **37**(5), 945–947 (2012).

5. S. A. Ku, W.-C. Chu, C. W. Luo, Y. M. Andreev, G. Lanskii, A. Shaidukoi, T. Izaak, V. Svetlichnyi, K. H. Wu, and T. Kobayashi, "Optimal Te-doping in GaSe for non-linear applications," *Opt. Express* **20**(5), 5029–5037 (2012).

6. J. Hebling, K. L. Yeh, M. C. Hoffmann, B. Bartal, and K. A. Nelson, "Generation of high-power terahertz pulses by tilted-pulse-front excitation and their application possibilities," *J. Opt. Soc. Am. B* **25**(7), B6–B19 (2008).

7. A. G. Stepanov, L. Bonacina, S. V. Chekalin, and J.-P. Wolf, "Generation of 30 microJ single-cycle terahertz pulses at 100 Hz repetition rate by optical rectification," *Opt. Lett.* **33**(21), 2497–2499 (2008).

8. C. P. Hauri, C. Ruchert, C. Vicario, and F. Ardana, "Strong-field single-cycle THz pulses generated in an organic crystal," *Appl. Phys. Lett.* **99**(16), 161116 (2011).

9. C. Vicario, B. Monoszlai, and C. P. Hauri, "GV/m single-cycle terahertz fields from a laser-driven large-size partitioned organic crystal," *Phys. Rev. Lett.* **112**(21), 213901 (2014).

10. C. Ruchert, C. Vicario, and C. P. Hauri, "Scaling submillimeter single-cycle transients toward megavolts per centimeter field strength via optical rectification in the organic crystal OH1," *Opt. Lett.* **37**(5), 899–901 (2012).

11. V. Krishnakumar, M. Rajaboopathi, and R. Nagalakshmi, "Studies on vibrational, dielectric, mechanical and thermal properties of organic nonlinear optical co-crystal: 2,6-diaminopyridinium–4-nitrophenolate–4-nitrophenol," *Physica B* **407**(7), 1119–1123 (2012).

12. M. J. Prakash and T. P. Radhakrishnan, "SHG active salts of 4-nitrophenolate with h-bonded helical formations: Structure-directing role of ortho-aminopyridines," *Cryst. Growth Des.* **5**(2), 721–725 (2005).

13. T. Chen, Z. Sun, L. Li, S. Wang, Y. Wang, J. Luo, and M. Hong, "Growth and characterization of a nonlinear optical crystal-2,6-diaminopyridinium 4-nitrophenolate 4-nitrophenol (DAPNP)," *J. Cryst. Growth* **338**(1), 157–161 (2012).

14. http://www.crystallography.net/cod/4505002.html.

15. J. Zyss and D. S. Chemla, "Quadratic nonlinear optics and optimization of the second-order nonlinear optical response of molecular crystals," in *Nonlinear Optical Properties of Organic Molecules and Crystals*, edited by D. S. Chemla, Academic Press (1987).

16. J. Ahn, A. Efimov, R. Averitt, and A. Taylor, "Terahertz waveform synthesis via optical rectification of shaped ultrafast laser pulses," *Opt. Express* **11**(20), 2486–2496 (2003).

17. C-W. Luo, K.-H. Wu, V. Krishnakumar, and T. Kobayashi, "THz emission from organic cocrystalline salt 2, 6-diaminopyridinium-4-nitrophenolate–4-nitrophenol," *Opt. Express* **24**(5) 5039–5044 (2016).

Section 3

CEP (Octave-Span)

3.1 Quasi-Monocyclic Near-Infrared Pulses with a Stabilized Carrier-Envelope Phase Characterized by Noncollinear Cross-Correlation Frequency-Resolved Optical Gating

3.1.1 INTRODUCTION

The broadband phase-matching property of β-BaB$_2$O$_4$ (BBO) crystal in type I noncollinear optical parametric amplifier [1–8] (OPA) and the development of a sophisticated pulse-compression technique (a custom-designed ultra-broadband chirp mirror, an adaptive pulse shaper such as a spatial light modulator, and a deformable membrane mirror) [5–8] facilitated the generation of a pulse as short as 4 fs, covering nearly the full visible spectral range and the near-infrared [8]. Furthermore, the idler from the noncollinear OPA system [9], pumped with second harmonic (SH) of the fundamental Ti:sapphire radiation and seeded with the white-light continuum produced by the same SH, showed the useful phenomenon of self-elimination of pulse-to-pulse carrier-envelope phase slip (CEP$_{slip}$).

Here the CEP slip is explained briefly as follows. The CEP can be defined in time domain and frequency domain for the periodical pulse train. In time domain, the CERP can be defined by the time shift between the peak of the pulse envelope and field peak position measured with the time scale of oscillation as 2π. Repetitive pulse has two frequencies; i.e., repetitive frequency and CEP. Components of the pulse spectrum have equal frequency separation f rep and carrier-envelope frequency. The spectral frequency distribution is given by $f_n = f_{CEO} + n f_{rep}$ with n = integer and $f_{rep} = 1/T_{rep}$ and CEP$_{slip}$ is pulse-after-pulse change of the CEP.

Controlling CEP slip is critical in applications for extreme nonlinear optical phenomena such as high-harmonic generation extending to the soft-x-ray region [10,11] and in precision optical frequency measurement [12]. Many research groups have worked for active stabilization of the CEP of the output from a mode-locked oscillator with a servo-loop feedback system [12–14]. A CEP-sensitive phenomenon was demonstrated first in an experiment on photoelectron emission [15]; recently, CEP dependency was reported in a soft-x-ray spectrum generated with a CEP-stabilized intense laser system [16].

3.1.2 EXPERIMENTAL

The idler spectrum from our noncollinear OPA system as shown below in Figure 3.1.3a has a transform-limited (TL) pulse width of 4.0 fs, which corresponds to 1.2 optical cycles of 3.3 fs with a center wavelength (center of mass in the frequency domain) at 990 nm. Such quasi-monocyclic idler output with no CEP slip can be utilized to study CEP-sensitive phenomena. However, there

DOI: 10.1201/9780429196577-19

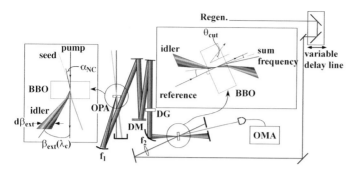

FIGURE 3.1.1 Schematic of the XFROG measurement system with broadband type I SFM. OPA, optical parametric amplifier; DM, deformable membrane mirror with a 19-channel high-voltage driver; DG, BK7 glass plate for dispersion compensation; Regen., fundamental radiation from a Ti:sapphire regenerative amplifier; OMA, optical regenerative analyzer.

is difficulty in characterization of idler pulses with bandwidths of more than an octave by using conventional upconversion or downconversion techniques and in achieving precise pulse compression. In the research reported in this Letter, to overcome this difficulty in characterization, we employed cross-correlation frequency-resolved optical gating [17,18] (XFROG) with broadband sum-frequency mixing (SFM), even taking advantage of the idler's angular dispersion. The idea of broadband parametric upconversion originated with the achromatic phase-matched SH generation of a nanosecond dye laser [19]. Afterward, Szabó and Bor pointed out the importance of noncollinear geometry in a broadband SFM scheme [20], which has been applied in several SH generation and SFM experiments [21–23].

Figure 3.1.1 shows a schematic of our setup for XFROG measurement with broadband type I SFM in a 30-μm-thick BBO crystal. A SH pulse and a SH-originated supercontinuum pulse are introduced into the optical parametric amplifier (OPA) crystal at a noncollinear angle α_{NC}. Spatially and temporally overlapped pulses generate idler radiation. The idler output from the noncollinear OPA has an angular dispersion that fulfills a phase-matching condition among pump, signal, and idler waves. The dispersion outside the BBO crystal $\beta_{ext}\lambda$ is calculated for an OPA noncollinear angle $\alpha_{NC}=3.7°$ and a pump wavelength λ_{pump} 395 nm. After the BBO crystal, the combination of a concave mirror and an off-axis parabolic mirror, of focal lengths of f_1 and f_2, respectively, were used. The relation between the distributed incident angle $\alpha_{idl\text{-}ext}(\lambda)$ to the SFM crystal for XFROG and d $\beta_{ext}(\lambda)$ is $\alpha_{idl\text{-}ext}(\lambda)=\alpha_{idl\text{-}ext}(\lambda_c)\pm Md\,\beta_{ext}(\lambda)/2$, where the magnification factor is $M=f_1/f_2$. Setting the incident angle of the reference $\alpha_{ref\text{-}ext}$ to zero, we determine the wavelength dependency of the exit angle, $\alpha_{sum\text{-}int}\,(\lambda_{sum},\,\theta_{PM})$, of the sum-frequency radiation from the following parallel and perpendicular wave vector matching conditions:

$$k_{sum}^{e} \cos\left[\alpha_{id1\_int}\left(\lambda_{id1},\theta_{PM}\right)\right]-\left\{k_{ref}^{o}+k_{id1}^{o}\cos\left[\alpha_{id1\_int}\left(\lambda_{id1}\right)\right]\right\}=0 \tag{3.1.1}$$

$$k_{id1}^{o} \sin\left[\alpha_{id1\_int}\left(\lambda_{id1}\right)\right]-k_{sum}^{e}\sin\left[\alpha_{sum\_int}\left(\lambda_{sum},\theta_{PM}\right)\right]=0 \tag{3.1.2}$$

Here the subscripts ref, idl, and sum correspond to reference, idler, and sum-frequency radiation, respectively, in the XFROG; θ_{PM} and θ_{cut} are the phase-matching and the cutting angles, respectively, of the SFM crystal with $\theta_{PM}=\theta_{cut}-\alpha_{sum\text{-}int}$ and $\alpha_{idl\text{-}int}=\sin^{-1}$ [sin $(\alpha_{idl\text{-}ext})/n_{idl}^{0}(\lambda_{idl})$]; where $n_{idl}^{0}(\lambda_{idl})$ is the refractive index at the idler wavelength, λ_{idl}. Using these equations, we calculated θ_{PM} and obtained phase-matching gain spectrum $|\eta|^2$ of the SFM interaction by adopting the method of Ref. [19] and using the following expression for acceptance function η:

$$\eta \equiv L^{-1}\int_{0}^{L} d\zeta \exp\left(i\Delta k_{\parallel}\zeta\right)$$

$$= \exp\left(i\Delta k_{\parallel}L/2\right)\mathrm{sinc}\left(\Delta k_{\parallel}L/2\right) \tag{3.1.3}$$

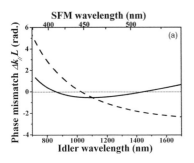 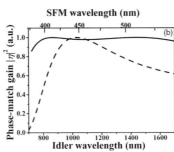

FIGURE 3.1.2 Dependence of (a) Phase mismatch $\Delta k_{\parallel}L$ and (b) Phase-matching gain $|\eta|^2$ on the wavelengths of the idler and the sum frequency calculated with the XFROG in Figure 1 with parameters $f_1=f_2=100$ ($M=1.0$), $\alpha_{\text{idl-int}}(\lambda_c)$ 17.85° for $\lambda_c=1100\,\text{nm}$, and $L=30\,\mu\text{m}$. Solid curves, non-collinear configuration; dashed curves, collinear configuration. The dotted line in (a) represents perfect phase matching.

Here L is the thickness of the SFM crystal, and the phase mismatch ($\Delta k_{\parallel}L$) parallel to the generated wave vector $\boldsymbol{k}_c$ is given by

$$\Delta k_{\parallel}L \equiv \left[k_{\text{sum}}{}^e - k_{\text{ref}}{}^o \cos\left(\alpha_{\text{sum\_int}}\right) - k_{\text{id1}}{}^o \cos\left(\alpha_{\text{id1\_int}} - \alpha_{\text{sum\_int}}\right) \right]L \qquad (3.1.4)$$

In the calculation, we determine M and $\alpha_{\text{idl-int}}(\lambda_c)$ to minimize the angular dispersion $d\alpha_{\text{sum-int}}/d\lambda$ and then select θ_{cos} to produce the maximum SFM bandwidth.

Figure 3.1.2a and b show the dependence of phase mismatch Δk_kL and phase-matching gain $|\eta|^2$, respectively, on the idler and SFM wavelengths under conditions that $f_1=f_2=100$ ($M=1.0$), $\alpha_{\text{idl-int}}(\lambda_c)=17.85°$ for wavelength $\lambda_c=1100\,\text{nm}$, $L=30\,\mu\text{m}$, and $\theta_{\text{cut}}=29.3°$. Counterparts with col-linear configurations that satisfy $\alpha_{\text{idl-int}}(\lambda)=0°$ and consequently $\alpha_{c\text{-int}}=0°$ are also shown for comparison. For the collinear configuration, the phase-matched SFM output power decreases sharply for wavelengths other than that for perfect phase matching ($\Delta k_{\parallel}L=0$), resulting in narrowing of the phase-matching bandwidth (Figure 3.1.2b). Especially in the shorter-wavelength regions, this band narrowing will lead to irreversible loss of information from the measured XFROG trace. In the noncollinear configuration, however, the first-order term of the Taylor-expanded phase mismatch is eliminated by the group-velocity matching condition near the wavelength of 1100 nm in Figure 3.1.2a. As can be observed from Figure 3.1.2b, there is only insignificant reduction of the phase-matching gain–bandwidth in the noncollinear configuration that we used. Hence, with the noncollinear configuration specified above, we have succeeded in characterizing the broadband idler pulses from the noncollinear OPA by XFROG with broadband SFM, utilizing the idler's angular dispersion.

Detailed configuration and specifications for the generation of CEP-locked idler output from a noncollinear OPA are given in Ref. [7]. The generated idler pulse is injected into the BBO crystal for XFROG measurement after bouncing off the flexible mirror and transmission through a 2-mm-thick BK7 (borosilicate) glass plate. The idler radiation of the noncollinear OPA is negatively chirped because the red part of the idler corresponds to the blue part of the positively chirped supercontinuum seed and vice versa. This negative chirp was roughly compensated for by material dispersion of the BK7 glass, and the residual chirp was compensated for by a deformable-membrane mirror (OKO Technologies) with adaptive control of 19 pixels. An optical multichannel analyzer (Ocean Optics S2000) was used to detect the XFROG signal. It is noteworthy to mention that it is impossible to introduce positive chirp by using BK7 glass in a spectral range of >300 nm because of the anomalous property of the glass. However, the residual chirp after the BK7 glass is success-fully compensated for by a deformable-membrane mirror, as shown in Figure 3.1.3. In a XFROG

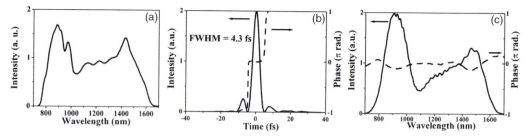

FIGURE 3.1.3 (a) Idler spectrum of the noncollinear OPA system. The TL pulse width is 4.0 ps, with a center wavelength at 900 nm. (b) Pulse shape and (c) Spectrum (solid curves) and the temporal and spectral phases (dashed curves) after retrieval.

measurement with a reference of much narrower spectrum width than the measured pulse (idler), one can estimate the group delay of the idler directly from the XFROG trace obtained (without any retrieval) by calculating the center of gravity of the delay time for each spectral pixel of the optical multichannel analyzer. Feedback to each actuator of the deformable mirror can be made by the group delay obtained as described above. The Ti:sapphire fundamental radiation 790 nm, 120 fs was used as the reference pulse of the XFROG.

An XFROG technique enables a pulse that is shorter than the reference pulse to be measured because the time–frequency contour map of a XFROG trace contains complete information on the measured pulse. This is different from conventional cross-correlation measurement, in which only delay-time-dependent signal intensity is acquired. Compared with SHG FROG pulse diagnostics [24], which may be a candidate for future study, the XFROG method has the advantage of a more intense frequency-conversion signal that uses a strong reference pulse. In addition, XFROG requires a narrower frequency-conversion bandwidth [25] than required in the case of SHG FROG. The frequency-conversion bandwidth limits the thickness of the BBO crystal in each method, so a thicker crystal can be used in XFROG, which is another advantage. These two advantages enable the signal to be acquired more sensitively and efficiently. The reference pulse is separately characterized by the use of an external SHG FROG interferometer.

Figure 3.1.3a and b show the pulse shape and the spectrum along with their phases after retrieval. The TL pulse width that can be calculated from the retrieved spectrum is 4.2 fs, which is close to the value calculated from the noncollinear OPA idler output spectrum in Figure 3.1.3a. This result confirms that the broadband SFM XFROG measurement was performed with full noncollinear OPA idler spectral bandwidth. Our retrieved pulse width (4.3 fs) is also almost the same as the TL pulse width, so the chirp compensation, which corresponds to 1.3 optical cycles in terms of cycle time corresponding to the center wavelength of 970 nm, worked successfully.

3.1.3 CONCLUSION

In this subsection, broadband idler output pulses from a noncollinear OPA has been characterized by using broadband SFM XFROG, taking advantage of the idler's angular dispersion property. By compensating for the residual higher-order dispersion, using adaptive control of a deformable mirror, quasimonocyclic near-infrared pulses with 4.3-fs pulse duration were achieved [26]. Such a sort pulse as short as 4.3fs with CEP stabilize is considered to be very useful for the study of CEP on the laser matter interaction since the effect is reasonable important because the strength of CEP effect is proportional to (CEP induced time shift/ pulse duration). Therefore the effect is 1000 times larger for CEP stabilized 5-fs pulse than hat of 5ps pulse.

REFERENCES

1. G. Cerullo, M. Nisoli, S. Stagira, and S. De Silvestri, *Opt. Lett.* **23**, 1283 (1998).

2. A. Shirakawa and T. Kobayashi, *Appl. Phys. Lett.* **72**, 147 (1998).
3. A. Shirakawa, I. Sakane, and T. Kobayashi, *Opt. Lett.* **23**, 1292 (1998).
4. E. Riedle, M. Beutter, S. Lochbrunner, J. Piel, S. Schenkl, S. Spoerlein, and W. Zinth, *Appl. Phys. B* **71**, 457 (2000).
5. T. Kobayashi and A. Shirakawa, *Appl. Phys. B* **70**, S239 (2000).
6. M. Zavelani-Rossi, G. Cerullo, S. De Silvestri, L. Gallmann, N. Matuschek, G. Steinmeyer, U. Keller, G. Angelow, V. Scheuer, and T. Tschudi, *Opt. Lett.* **26**, 1155 (2001).
7. A. Baltus̆ka, T. Fuji, and T. Kobayashi, *Opt. Lett.* **27**, 306 (2002).
8. A. Baltus̆ka and T. Kobayashi, *Appl. Phys. B* **75**, 427 (2002).
9. A. Baltus̆ka, T. Fuji, and T. Kobayashi, *Phys. Rev. Lett.* **88**, 133901 (2002).
10. A. Apolonski, A. Poppe, G. Tempea, C. Spielmann, T. Udem, R. Holzwarth, T. W. Hänsch, and F. Krausz, *Phys. Rev. Lett.* **85**, 740 (2000).
11. M. Drescher, M. Hentshel, R. Kienberger, G. Tempea, C. Spielmann, G. Reider, P. Corkum, and F. Krausz, *Science* **291**, 1923 (2001).
12. D. J. Jones, S. A. Diddams, J. K. Ranka, A. Stentz, R. S. Windeler, J. L. Hall, and S. T. Cundiff, *Science* **288**, 635 (2000).
13. Y. Kobayashi and K. Torizuka, *Opt. Lett.* **26**, 1295 (2001).
14. F. W. Helbing, G. Steinmeyer, U. Keller, R. S. Windeler, J. Stenger, and H. R. Tulle, *Opt. Lett.* **27**, 194 (2002).
15. G. G. Paulus, F. Grasbon, H. Walther, P. Villoresi, M. Nisoli, S. Stagira, E. Priori, and S. De Silvestri, *Nature* **414**, 182 (2001).
16. A. Baltuska, T. Udem, M. Uiberacker, M. Hentschel, E.̆ Goulielmakis, C. Gohle, R. Holzwarth, V. S. Yakovlev, A. Scrinzi, T. W. Hänsch, and F. Krausz, *Nature* **421**, 611 (2003).
17. S. Linden, H. Giessen, and J. Kuhl, *Phys. Status Solidi B* **206**, 119 (1998).
18. S. Linden, J. Kuhl, and H. Giessen, *Opt. Lett.* **24**, 569 (1999).
19. S. Saikan, D. Ouw, and P. Schäfer, *Appl. Opt.* **18**, 193 (1979).
20. G. Szabó and Z. Bor, *Appl. Phys. B* **58**, 237 (1994).
21. M. Hacker, T. Feurer, R. Sauerbrey, T. Lucza, and G. Szabó, *J. Opt. Soc. Am. B* **18**, 866 (2001).
22. Y. Nabekawa and K. Midorikawa, *Opt. Express* **11**, 324 (2003), http://www.opticsexpress.org.
23. T. Kanai, X. Zhou, T. Sekikawa, S. Watanabe, and T. Togashi, *Opt. Lett.* **28**, 1484 (2003).
24. R. Trebino and D. J. Kane, *J. Opt. Soc. Am. A* **10**, 1101 (1993).
25. A. Baltus̆ka, M. S. Pshenichnikov, D. A. Wiersma, and R. Szipocs, in *Ultrafast Processes in Spectroscopy*, edited by R. Kaarli, A. Freiberg, and P. Saari, Institute of Physics, University of Tartu, Tartu, Estonia, p. 7 (1998).
26. S. Adachi, P. Kumbhakar, and T. Kobayashi, *Opt. Lett.* **29**, 1150 (2004).

3.2 Self-Stabilization of the Carrier-Envelope Phase of an Optical Parametric Amplifier Verified with a Photonic Crystal Fiber

3.2.1 INTRODUCTION

For few-cycle optical pulses [1–5], the maximum electric field amplitude shows a substantial dependence on the carrier-envelope phase (CEP). Therefore, the CEP-stabilized sources of high-intensity femtosecond pulses are important for the field-sensitive nonlinear optical process [6,7]. A high-precision CEP measurement and control of 100-MHz repetition sources have been demonstrated experimentally [8–12]. The stabilization of CEP in a high-energy pulse is still a highly desirable and elusive goal, especially in a power-amplification system at a relatively low repetition rate, such as the kilohertz level until recently [13,14].

The optical parametric process is an important route to generate few-cycle optical pulses. However, for both optical parametric oscillators [15] and super fluorescence-seeded optical parametric amplifiers (OPAs), the CEPs of both the signal and the idler are random because the signal is initiated by quantum noise. Fortunately, the properties of CEP from an externally seeded OPA are different from those of optical parametric oscillators and super fluorescence-seeded OPAs. In a recent Letter by Baltuska et al. [13], it was shown that the CEP of an idler from an externally seeded OPA is automatically stabilized passively when the same pulse source is used for the generation of a white-light seed pulse and for pumping the OPA. This result was also demonstrated experimentally by an OPA pumped with the second harmonic (SH) at 400 nm of a 1-kHz Ti:sapphire chirped-pulse amplifier (CPA) laser [13]. Here we present another experimental proof for the above theoretical prediction, namely, the OPA is pumped by the fundamental of the CPA with a spectrum centered at 786 nm. It was found that a near-infrared optical pulse has self-stabilized CEP.

The mechanism of the CEP self-stabilization in a three-wave-coupling optical parametric process is explained as follows: First, the white-light continuum pulse used as a seed for an OPA is produced by self-phase modulation (SPM) and contains phase c as that of the input pulse, because the SPM process is a purely intensity-dependent nonlinear optical process [16,17]. Furthermore, the supercontinuum (SC) generated with an octave-spanning bandwidth is experimentally utilized in CEP measurement [8,9,11], in which either photonic crystal fiber (PCF) or hollow fiber is used. Second, the CEP of the amplified signal pulse is determined by the seed pulse and is preserved in the amplification process, since the pump is far from resonance and there is no phase-dependent effect. Third, the CEP of the idler pulse accommodates the CEP difference between the pump and the signal pulses. Because the CEP of the pump pulse and that of the signal pulse differ from each other by only a constant value of $\pi/2$ even with the existence of fluctuation c, the CEP of the idler is constant 2p regardless of any pulse-to-pulse fluctuations of either the CPA [18] or the OPA [19].

DOI: 10.1201/9780429196577-20

3.2.2   EXPERIMENTAL

The experimental layout is shown in Figure 3.2.1. The noncollinear optical parametric amplifier (NOPA) is pumped by a 120-fs FWHM pulse of 1.5-mJ energy at 786 nm from a Ti:sapphire regenerative CPA (Thales Laser, Bright) with a 1-kHz repetition rate. Approximately 4% of the energy of the vertically polarized beam is split off for SC generation, whereas the polarization of the main beam with 96% pulse energy is rotated to the horizontal direction by a half-wave plate for pumping the NOPA. The pump beam is telescoped to obtain a high peak intensity (67 GWcm^2 and pump a 2-mm-thick b-barium borate (BBO) crystal (type I, $\theta = 30°$). A variable neutral-density filter and an aperture are used to adjust the intensity of the small fraction to 1 mJ, which is focused into a 2-mm-thick sapphire plate by a 100-mm focal-length lens.

3.2.3   RESULTS AND DISCUSSION

Typically, pulse-to-pulse fluctuations of the CEP can be determined through spectral interference (SI) between the fundamental and its harmonics, referred to as f-to-$2f$ interference [10]. Such a broad spectrum and its SH are easily obtained for an idler pulse from a NOPA with a type I BBO crystal pumped by 400-nm optical pulses [13]. Unfortunately, the NOPA pumped by a 786-nm pulse cannot generate an idler pulse with an octave-wide spectrum. By using an additional method, the spectrum of the idler is extended to measure the CEP drift. Note that the recent invention of PCF has led to the generation of white-light continua that are more than one octave broad [20].

The idler pulses are angle tuned to 1600 nm. When the pump pulse energy is 1.3 mJ, the output energy from a 2-mm-thick BBO crystal is 10 mJ for the idler pulse. The 1600-nm idler pulse is frequency doubled to 800 nm by use of a 50-mm focal-length lens, a 1-mm-thick BBO crystal, and a 100-mm focal-length lens (see Figure 3.2.1b). The 75-mm-long PCF used in our experiments has a 2.0-mm-diameter core surrounded by a cobweb structure of air holes. The period of the hole structure in the cladding is 1.3 μm, and the ratio of the pitch to the hole size is 0.40. The fiber has a small anomalous dispersion of $D \sim 25$ ps/(nm km) at 800 nm, and zero group-velocity dispersion (GVD) at ~765 nm. The SH of the idler is focused into the PCF by a microscope objective lens (20×). The fiber is properly rotated to make the polarization axis of the linearly polarized pulse coincide with one of the principal axes in the fiber. The output pulse (white-light continuum) is focused into a 500-mm-thick BBO again by another microscope objective lens (20×) with the same parameters to generate the SH of the white-light continuum. The spectrum extends from 420 to 980 nm, spanning over a full octave (see Figure 3.2.2a).

The unconverted fraction of the fundamental pulse with horizontal polarization and the vertically polarized SH pulse is directed to an UV polarizer and a color filter after being collimated with a 100-mm focal-length lens, and the transmitted components are measured by a spectrometer

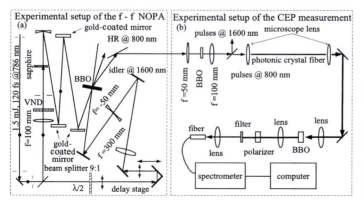

FIGURE 3.2.1 Experimental setup. (a) 786-nm pumped NOPA. l2, half-wave plate at 800 nm; VND, variable neutral-density filter; HR, high-reflection coating. The bidirected arrows indicate the direction of polarization. (b) Experimental setup of the CEP drift measurements.

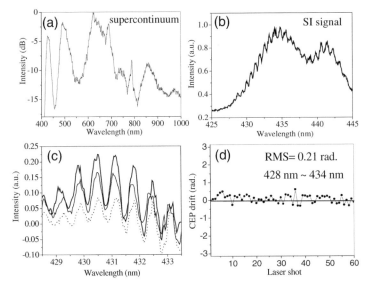

FIGURE 3.2.2 Experimental results of CEP self-stabilization measurement. (a) Spectra of the white-light continuum generated by nonlinear propagation in PCF. (b) SI signal of the white-light continuum and its SH. (c) Part of the SI signal was measured for three successive shots. (d) The stable phase pattern was obtained from the interference of the white-light continuum in PCF and its SH. The result is direct proof of CEP self-stabilization.

(Acton Research, 300i) with an intensified charged-coupled device detector (Princeton Instruments, 7439-0001) (see Figure 3.2.1b). Maximum SI effect between the orthogonally polarized 430-nm pulses is obtained by rotating the UV polarizer (see Figure 3.2.2b). After subtraction of the background by use of a high-pass filter and back Fourier transformation, the SI fringe in the spectral range between 428 and 434 nm is shown for three successive laser shots in Figure 3.2.2c, in which the good stability of the fringe positions from shot to shot is observed even with a relatively large change in intensity. The phase fluctuations are analyzed, and the relative phases are extracted from all recorded SI fringes of 60 shots. The CEP pulse-to-pulse drift measurement results are summarized in Figure 3.2.2d. The CEP of the SH of the idler stays locked within a 6p7-rad range. The rms residual phase drift is 0.21 rad and is attributed to the instability of the pump laser intensity and nonlinear phase noise generated in PCF [21]. Even the intensity of the fringe from 437 to 442 nm is unstable, and the visibility is also decreased; we calculated the CEP drift in this spectrum section (see Figure 3.2.3d-1). The result of rms 0.27 rad is also a direct proof of CEP self-stabilization. When the NOPA is angle tuned to 1400 nm, the bandwidth of the output pulse is 130 nm, which indicates a few-cycle optical pulse with CEP self-stabilized.

To analyze and clarify the mechanism of the fluctuation of fringe in the spectral range from 437 to 442 nm (see Figure 3.2.3c), the wavelength dependence of the fringe visibility is calculated (see Figure 3.2.3b). At the wavelength with maximum visibility, the ratio is assumed to be 1:1 between the fundamental and the SH component. The relative intensity from 428 to 445 nm is shown in Figure 3.2.3a. The measured interference visibility and the theoretical curve are shown in Figure 3.2.3b. It is found that the theoretical simulation is much higher than the experimental data, and there is a gap from 434 to 437 nm in the experimental data. This disagreement is explained as follows: The input pulse has such a high peak power (100 kW) that the SPM can lead to the generation of an ultrabroadband spectrum without the need for other nonlinear effects. In this process, the CEP of the white light is maintained. However, with the pulse injected in the anomalous-dispersion regime, modulational instability occurs as a result of an interplay between SPM and GVD [17]. As a result of modulational-instability gain amplifiers, this process is wavelength dependent and sensitive to input pulse fluctuations (see Figures 3.2.2c and 3.2.3c). Numerical simulations [22], have shown that SC generation can exhibit extreme sensitivity to the fluctuation of input pulse energy. These fluctuations also cause phase and CEP fluctuations [23], which is why the measured CEP drift (0.21 rad) in the section from 437 to 442 nm is relatively larger than the CEP drift (0.27 rad) between

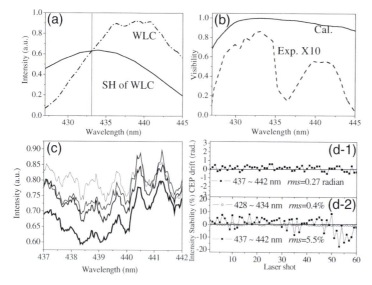

FIGURE 3.2.3 Stimulated Raman effect on the CEP measurement. (a) Relative intensity between the fundamental component and the SH component of the white-light continuum (WLC). (b) Measured interference visibility and theoretical curve. (c) Part of the SI signal was measured for four successive shots. (d-1) Up-phase pattern obtained from the interference of the white-light continuum in PCF and its SH. (d-2) Shot-to-shot intensity fluctuations of spectral fringes at different spectral sections.

428 and 434 nm. Furthermore, even a small amount of light generated in the normal-dispersion regime close to the zero-GVD wavelength generated by SPM, stimulated Raman scattering, non-solitonic radiation, or parametric four-wave mixing can serve as a secondary parametric pump and in turn contribute to the generation of new frequencies and to the formation of the continuum because parametric four-wave mixing is intrinsically phase matched with a broad range of Stokes and anti-Stokes wavelengths that span the whole visible to the near-infrared spectral region [24]. Thus one pump wave creates two new waves with different frequency bands from the spontaneous emission noise [25,26] outside the pump-pulse spectral band. These processes will inevitably add amplified spontaneous emission noise to the shorter and longer wavelength ranges [27], especially in the spectral ranges in which the Stokes and anti-Stokes gain are high, thus inducing the visibility gap from 434 to 437 nm (see Figure 3.2.3b). The amplified quantum noise with a random phase strongly causes coherence degradation, which decreases the visibility of the fringe from 428 to 445 nm (see Figure 3.2.3b).

3.2.4   CONCLUSION

In this subsection, self-stabilization of the CEP of idler pulses from a NOPA was investigated through observation of the SI between an octave-wide supercontinuum and its second harmonic. This SC is generated by nonlinear propagation of the second harmonic of the idler pulse in a PCF [28]. The research output described in the present subsection is performed by the collaboration of X. Fang and T. Kobayashi [28].

REFERENCES

1. T. Wilhelm, J. Piel, and E. Riedle, *Opt. Lett.* **22**, 149 (1997).
2. G. Cerullo, M. Nisoli, S. Stagira, and S. De Silvestri, *Opt. Lett.* **23**, 1283 (1998).
3. A. Shirakawa, I. Sakane, M. Takasaka, and Kobayashi, *Appl. Phys. Lett.* **74**, 2268 (1999).
4. T. Kobayashi and A. Shirakawa, *Appl. Phys. B* **70**, S239 (2000).
5. A. Baltuska and T. Kobayashi, *Appl. Phys. B* **75**, 427 (2002).
6. Ch. Spielmann, C. Kan, N. H. Burnett, T. Brabec, M. Geissler, A. Scrinzi, M. Schnürer, and F. Krausz, *IEEE J. Sel. Top. Quantum Electron.* **4**, 249 (1998).
7. I. P. Christov, M. M. Murnane, and H. C. Kapteyn, *Phys. Rev. Lett.* **78**, 1251 (1997).

8. A. Apolonski, A. Poppe, G. Tempea, C. Spielmann, T. Udem, R. Holzwarth, T. W. Hänsch, and F. Krausz, *Phys. Rev. Lett.* **85**, 740 (2000).
9. S. A. Diddams, D. J. Jones, J. Ye, T. Cundiff, and J. L. Hall, *Phys. Rev. Lett.* **84**, 5102 (2000).
10. D. J. Jones, S. A. Diddams, J. K. Ranka, A. Stentz, R. S. Windeler, J. L. Hall, and S. T. Cundiff, *Science* **288**, 635 (2000).
11. M. Kakehata, H. Takada, Y. Kobayashi, K. Torizuka, Y. Fujihara, T. Homma, and H. Takahashi, *Opt. Lett.* **26**, 1436 (2001).
12. Y. Kobayashi and K. Torizuka, *Opt. Lett.* **25**, 856 (2000).
13. A. Baltuska, T. Fuji, and T. Kobayashi, *Phys. Rev. Lett.* **88**, 133901 (2002).
14. A. Baltuska, Th. Udem, M. Uiberacker, M. Hentschel, E. Goulielmakis, Ch. Gohle, R. Holzwarth, V. S. Yakovlev, A. Scrinzi, T. W. Hänsch, and F. Krausz, *Nature* **421**, 611 (2003).
15. D. C. Edelstein, E. S. Wachman, and C. L. Tang, *Appl. Phys. Lett.* **54**, 1728 (1989).
16. T. Brabec and F. Krausz, *Rev. Mod. Phys.* **72**, 545 (2000).
17. G. P. Agrawal, *Nonlinear Fiber Optics*, 3rd ed., Academic, San Diego, Calif., pp. 46 and 136 (2001).
18. A. Dubietis, R. Danielius, G. Tamosauskas, and A. Piskarskas, *J. Opt. Soc. Am. B* **15**, 1135 (1998).
19. A. Baltuska, T. Fuji, and T. Kobayashi, *Opt. Lett.* **27**, 1241 (2002).
20. J. K. Ranka, R. S. Windeler, and A. J. Stentz, *Opt. Lett.* **25**, 25 (2000).
21. T. M. Fortier, J. Ye, S. T. Cundiff, and R. S. Windeler, *Opt. Lett.* **27**, 445 (2002).
22. A. L. Gaeta, *Opt. Lett.* **27**, 924 (2002).
23. T. M. Fortier, J. Ye, S. T. Cundiff, and R. S. Windeler, *Opt. Lett.* **27**, 445 (2002).
24. J. M. Dudley, L. Provino, N. Grossard, H. Maillotte, R. S. Windeler, B. J. Eggleton, and S. Coen, *J. Opt. Soc. Am. B* **19**, 765 (2002).
25. J. M. Dudley and S. Coen, *Opt. Lett.* **27**, 1180 (2002).
26. K. L. Corwin, N. R. Newbury, J. M. Dudley, S. Coen, S. A. Diddams, K. Weber, and R. S. Windeler, *Phys. Rev. Lett.* **90**, 113904 (2003).
27. H. Kubota, K. R. Tamura, and M. Nakazawa, *J. Opt. Soc. Am. B* **16**, 2223 (1999).
28. X. Fang and T. Kobayashi, *Opt. Lett.* **29**, 1282 (2004).

3.3 Octave-Spanning Carrier-Envelope Phase Stabilized Visible Pulse with Sub-3-fs Pulse Duration

3.3.1 INTRODUCTION

The ability to generate isolated ultrashort pulses with a broad spectrum in the visible region is important for many applications. For example, pump-probe experiments that can resolve molecular vibrations became possible with the development of sub-5-fs noncollinear optical parametric amplifiers (NOPAs) [1–8]. These experiments shed light on many ultrafast photophysical processes, such as exciton self-trapping [9] and breather-soliton formation [10], and photochemical processes such as the transitional states during isomerization in bacteriorhodopsin [11] and Claisen rearrangement [12].

Recently, there have been several reports on the synthesis of short or arbitrary optical pulse trains from Raman sidebands generated by molecular modulation in H2 [13,14]. Even sub-single-cycle optical pulses have been synthesized from seven sidebands in the near-UV-to-near-IR region [14]. However, because the time between pulses is too short (<100 fs) and their extremely high repetition rates prohibit downsampling, these pulses are of little use for investigating chemical or photochemical processes, which typically have time scales of >100 fs.

To our knowledge, the shortest isolated pulse that has been generated in the visible region is a 2.6 fs pulse produced by Matsubara et al. [15]. It was generated by spectral broadening in a hollow fiber and dispersion compensation using a spatial light modulator in a 4-f setup. However, because their output pulse spectrum contained very complicated fine structures that are unsuitable for spectroscopy, it has not been used in any spectroscopic application. In wavelength regions longer than the visible, the shortest pulse is a 4.3 fs pulse produced by our group [16] by compressing from a NOPA idler. This study is an extension of this work to generate a carrier-envelope phase (CEP) stabilized sub-3-fs pulse.

3.3.2 RESULTS AND DISCUSSION

The sub-3-fs pulse is based on a NOPA with a 1-mm thick β-BaB$_2$O$_4$ crystal (BBO) as the gain medium and pumped by the 400 nm second harmonic (width, 70 fs; repetition rate, 5 kHz; energy, 20 µJ; and radius, 50 µm) of a Ti:sapphire regenerative amplifier. As shown in Figure 3.3.1a, the seed of the NOPA is the supercontinuum generated in a CaF$_2$ plate from a fraction (0.6 µJ) of the 400 nm pulse and is injected as the signal, the higher-frequency wave of the parametric amplification. Because the pump and the signal have the same CEPs, the CEP of the newly generated lower-frequency idler (energy of ~0.5 µJ), which is generated as the difference frequency between the pump and the signal, is passively stabilized by the subtraction mechanism of the difference frequency generation in the idler generation process [17]. With slow (~1 s) feedback control to suppress drift, the idler CEP of this system is stable enough to perform optical poling experiments for a few hours [18,19].

The idler output beam with an octave-spanning (800 nm–1.6 µm) spectrum has an angular dispersion of 170 µrad/m, which is similar to the variation in the phase-matching angle with wavelength for the second harmonic generation of the idler in the NOPA BBO.

DOI: 10.1201/9780429196577-21

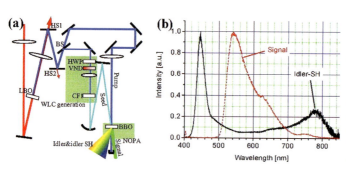

FIGURE 3.3.1 (a) Schematic of NOPA: LBO, LiB_3O_5 crystal for second-harmonic generation. HS, harmonic separator; BS, beam splitter; HWP, half-wave plate; VND, variable neutral-density filter; CF1, calcium fluoride crystal for white-light continuum generation; BBO, β-BaB_2O_4 crystal for parametric amplification. (b) Spectra of idler-SH (solid black curve) and signal (broken red curve).

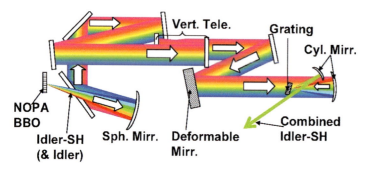

FIGURE 3.3.2 Setup for angular and group-delay dispersion compensation. Sph. Mirr., spherical mirror for collimation; Vert. Tele., telescope for reducing vertical beam width; Cyl. Mirr., cylindrical mirrors for focusing onto grating and collimating after beam combination.

The quasi-satisfactory condition for angularly dispersed achromatic phase-matched frequency doubling [20–22] leads to a considerable amount of the idler second harmonic (idler-SH) generated overlapping onto the idler beam (with different polarization: idler, vertical; idler SH, horizontal). As shown in Figure 3.3.1b, the wavelength range of the idler-SH is 430–800 nm, which is corresponding to ~698–375 THz. The frequency width of over 300 THz.

Figure 3.3.2 shows the optical layout used for the compensation of the angular and group-delay dispersion. We first collimated the diverging fan-shaped idler-SH beam by using a spherical mirror, thereby converting the large angular dispersion into spatial dispersion. The collimated idler-SH with spatial dispersion was then reflected from a deformable mirror [(DM) purchased from Flexible Optical] for high order group-delay dispersion–suppression control. Finally, the spatial dispersion of the idler-SH was converted back into angular dispersion by reflection from a cylindrical mirror, and it was compensated by the diffraction mechanism with the grating.

After adjusting the beam size, the whole idler-SH beam was characterized with a sum-frequency-generation (SFG) cross-correlation frequency-resolved optical gating (XFROG) [23,24] using a 800 nm reference pulse. Compared with SHG-FROG pulse diagnostics [25], a SFG-XFROG can characterize lower energy pulses by employing strong reference pulses; it also requires narrower frequency-conversion bandwidths [26]. An 800 nm reference pulse was split from the same 800 nm regenerative amplifier output that was used to drive the NOPA. It was compressed to 45 fs FWHM using a dual-prism pair and characterized by SHG-FROG using a Michelson interferometer and the same optical setup as described in the following. To avoid geometric smearing in the SFG process, the idler-SH and 800 nm reference pulses were overlapped using a chromium-coated partial mirror and focused by an off-axis parabolic mirror (OAP) onto a 10-µm-thick BBO for an SFG-FROG. The BBO thickness was measured independently. Because the CEP of the 800 nm reference beam varies shot-to-shot randomly and that of the idler-SH is passively stabilized as mentioned earlier in this chapter, interference between them is washed out when integrated over 8000 shots.

Because the idler-SH and 800 nm reference beams were collinearly incidents on the XFROG-BBO, the generated SFG-XFROG signal, the second harmonic of the 800 nm reference beam (ref-SH), the idler-SH beam, and the intense 800 nm reference beam were all collinear with each other. After collimation (i.e., refocusing) by an OAP, the SFG-XFROG signal was filtered in two steps.

1. First, a Glan-Thompson prism was used to transmit a vertically polarized SFG-XFROG signal and ref-SH.
2. Second, a fused-silica prism was used for spectral filtering as follows. The beam spectrally dispersed by the prism was reflected by a spherical mirror and retransmitted through the same prism so that it was recollimated. Because the residual idler-SH and 800 nm reference beams have longer wavelengths than the other beams, they could be completely removed by placing a beam block near the surface of the spherical mirror. The OAP after the XFROG-BBO was adjusted so that the refocused beam was precisely focused on the surface of the spherical mirror. The double-filtered SFG-XFROG signal beam was refocused by another OAP onto the fiber input of a spectrometer (Ocean Optics, USB4000).

Although the above two-step filtering did not reduce the intensity of ref-SH, its intensity was sufficiently stable that we could remove it by simply subtracting the background, of which intensity could be obtained immediately before and/or after the measurement of the target experiment.

A fused-silica wedge was inserted for low-order dispersion control. Fine dispersion control was achieved using the DM, which is an electrostatically driven membrane with 19 electrodes. Note that, in XFROG measurements with a reference beam with a much narrower spectrum than the pulse being measured (idler-SH), the group delay of the idler-SH pulse at some wavelength can be directly estimated without retrieval from the XFROG trace by simply calculating the peak position (assuming a reference pulse gas clean spectral structure) in the delay time for the corresponding spectral pixel of the spectrometer.

The following feedback procedure was used.

1. First, we obtained a SFG-XFROG trace by setting the DM electrode voltages to the middle of their driving ranges.
2. Second, we obtained 19 sets of traces by individually maximizing each of the DM electrode voltages.

Using these sets of data, we tracked the group delay as a function of wavelength, compared 19 driven group-delay spectra with the nondriven group-delay spectrum, and obtained a calibration curve for the group delay as a function of the electrode voltage. The total group-delay variation was assumed to be a linear combination of each electrode drive, and we iteratively minimized the group delay deviation using the calibration curve.

Figure 3.3.3a and b show measured and retrieved XFROG traces, respectively. Figure 3.3.4a and b show the retrieved spectrum and pulse shape along with their phases, respectively. These figures also show the spectrum measured by the spectrometer and the transform-limited (TL) pulse shape

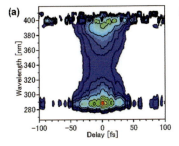
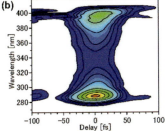

FIGURE 3.3.3 (a) SFG-XFROG trace of idler-SH and (b) retrieved trace.

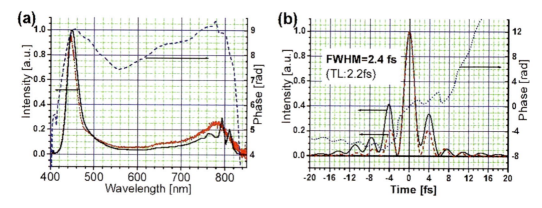

FIGURE 3.3.4   (a) Retrieved and measured spectral pulse shape (solid black curve, retrieved intensity; dotted red curve, measured intensity; broken blue curve, retrieved phase). (b) Retrieved temporal pulse shape (solid black curve, retrieved intensity; dotted blue curve, retrieved phase; broken red curve, intensity of TL pulse).

calculated from the retrieved spectrum. The measured and retrieved spectra agree well with each other. The temporal FWHM of the retrieved pulse was 2.4 fs, which is close to the TL pulse width of 2.2 fs.

3.3.3   CONCLUSION

In this subsection, we have demonstrated compression of CEP-stabilized NOPA second-harmonic visible-NIR output down to 2:4 fs. The total energy of the compressed idler-SH pulse was around 1 nJ, and the energy fraction of the main pulse was 47%. The output spectrum was smooth, making it suitable for spectroscopic applications.

While the energy of the compressed idler-SH pulse is a little too low for the pulse to be used as the pump pulse in pump-probe experiments, there is a higher energy (~1 μJ) signal beam from the same NOPA whose spectrum is shown in Figure 3.3.1b. Because the signal beam has no substantial angular dispersion and its spectrum is sufficiently wide to support a pulse duration of 6 fs, compression using the conventional combination of a negative-chirped mirror pair and a prism pair and used as the pump beam in a pump-probe experiment is feasible. The idler-SH pulse can be used as a probe beam with a wide spectrum in such an experiment.

Because the idler-SH and idler pulses are both CEP stable and simultaneously generated in the same BBO, synthesis of phase-stable pulse by their combination is feasible. Their polarizations can be aligned using a periscope. It was discussed that this approach has the potential of realization of a sub-2 fs pulse generation with a near-two-octave wavelength range of 430 nm–1.6 μm [27]. The research results described in this subsection was obtained by the collaboration of K. Okamura and T. Kobayashi.

REFERENCES

1. G. Cerullo, M. Nisoli, S. Stagira, and S. De Silvestri, *Opt. Lett.* **23**, 1283 (1998).
2. A. Shirakawa and T. Kobayashi, *Appl. Phys. Lett.* **72**, 147 (1998).
3. A. Shirakawa, I. Sakane, and T. Kobayashi, *Opt. Lett.* **23**, 1292 (1998).
4. E. Riedle, M. Beutter, S. Lochbrunner, J. Piel, S. Schenkl, S. Spoerlein, and W. Zinth, *Appl. Phys. B* **71**, 457 (2000).
5. T. Kobayashi and A. Shirakawa, *Appl. Phys. B* **70**, 389 (2000).
6. M. Zavelani-Rossi, G. Cerullo, S. De Silvestri, L. Gallmann, N. Matuschek, G. Steinmeyer, U. Keller, G. Angelow, V. Scheuer, and T. Tschudi, *Opt. Lett.* **26**, 1155 (2001).
7. A. Baltuška, T. Fuji, and T. Kobayashi, *Opt. Lett.* **27**, 306 (2002).

8. A. Baltuška and T. Kobayashi, *Appl. Phys. B* **75**, 427 (2002).
9. A. Sugita, T. Saito, H. Kano, M. Yamashita, and Kobayashi, *Phys. Rev. Lett.* **86**, 2158 (2001).
10. S. Adachi, V. M. Kobryanskii, and T. Kobayashi, *Phys. Rev. Lett.* **89**, 027401 (2002).
11. T. Kobayashi, T. Saito, and H. Ohtani, *Nature* **414**, 531 (2001).
12. I. Iwakura, A. Yabushita, and T. Kobayashi, *Chem. Lett.* **39**, 374 (2010).
13. M. Y. Shverdin, D. R. Walker, D. D. Yavuz, G. Y. Yin, and S. E. Harris, *Phys. Rev. Lett.* **94**, 033904 (2005).
14. W. J. Chen, Z.-M. Hsieh, S. W. Huang, H. Y. Su, C. J. Lai, T. Tang, C. H. Lin, C. K. Lee, R. P. Pan, C. L. Pan, and A. H. Kung, *Phys. Rev. Lett.* **100**, 163906 (2008).
15. E. Matsubara, K. Yamane, T. Sekikawa, and M. Yamashita, *J. Opt. Soc. Am. B* **24**, 985 (2007).
16. S. Adachi, P. Kumbhakar, and T. Kobayashi, *Opt. Lett.* **29**, 1150 (2004).
17. A. Baltuška, T. Fuji, and T. Kobayashi, *Phys. Rev. Lett.* **88**, 133901 (2002).
18. S. Adachi and T. Kobayashi, *Phys. Rev. Lett.* **94**, 153903 (2005).
19. K. Okamura and T. Kobayashi, *Opt. Commun.* **281**, 5870 (2008).
20. S. Saikan, D. Ouw, and F. P. Schäfer, *Appl. Opt.* **18**, 193 (1979).
21. G. Szabó and Z. Bor, *Appl. Phys. B* **50**, 51 (1990).
22. T. Kanai, X. Zhou, T. Sekikawa, S. Watanabe, and Togashi, *Opt. Lett.* **28**, 1484 (2003).
23. S. Linden, H. Giessen, and J. Kuhl, *Phys. Status Solidi B* **206**, 119 (1998).
24. S. Linden, J. Kuhl, and H. Giessen, *Opt. Lett.* **24**, 569 (1999).
25. R. Trebino and D. J. Kane, *J. Opt. Soc. Am. A* **10**, 1101 (1993).
26. A. Baltuška, M. S. Pshenichnikov, D. A. Wiersma, and R. Szipocs, in *Ultrafast Processes in Spectroscopy*, edited by R. Kaarli, A. Freiberg, and P. Saari, Institute of Physics, University of Tartu, Tartu, Estonia, p. 7 (1998).
27. K. Okamura and T. Kobayashi, *Opt. Lett.* **26**, 226 (2011).

3.4 Carrier-Envelope-Phase-Stable, Intense Ultrashort Pulses in Near Infrared

3.4.1 INTRODUCTION

Isolated soft x-ray 50–150 eV attosecond pulses are currently ideal tools to investigate dynamics with atomic temporal resolution and nanometer spatial resolution [1]. Typically, attosecond pulses are produced via high harmonic generation (HHG) in noble gases, driven by intense carrier-envelope phase (CEP) stabilized few-cycle laser pulses at a wavelength around 800 nm [2,3]. To investigate processes with even higher temporal resolution or study the dynamics of core electrons, higher photon energies and shorter attosecond pulses are needed. Both can be obtained using lasers with longer central wavelengths. From the perspective of a single atom's response to the field, increasing wavelength allows for an extension of the cutoff photon energy at the same intensity, described in atomic units by $E_{cut-off} \cong I_p + 3.17(I \lambda^2/16\pi^2 c^2)$ where I is the laser intensity and I_p is the ionization potential of the atom [4]. This comes with the trade-off of reduced photon flux due to increased dispersion of the electron wave packet as it takes more time to re-collide with the parent ion. However, recent work has suggested that this decrease in the single-atom efficiency may be compensated by more favorable phase matching, and thus great interest in long-wavelength sources remains [5–8]. So far, few-cycle, high-intensity IR optical parametric amplifier (OPA) and optical parametric chirped-pulse amplifier (OPCPA) sources have been reported with central wavelengths near 1.8 μm [9,10], 2 μm [11–13], 3 μm [14], and 4 μm [15]. To generate an isolated attosecond pulse by HHG, a sub-two-cycle pulse is highly desirable. Currently, multicycle mid-infrared pulses generated by OPA can be spectrally broadened and compressed to few-cycle durations by self-phase modulation in hollow fibers, but with scalability limited by the transmission efficiency of the fiber. In this Letter, we report a 2.1 μm OPCPA laser system pumped by a Yb:YAG thin-disk amplifier. This system produces 1.2 mJ, CEP-stable, 10.5 fs (1.5 optical cycle) pulses, which are the shortest millijoule level pulses ever generated in this spectral range. It will be an ideal tool for generating isolated attosecond pulses with much higher photon energies. It also can be scaled straightforwardly with pump energy to reach terawatt powers. To achieve millijoule energies and near-single cycle pulses with an OPCPA system, (i) a broad bandwidth seed and (ii) an amplification bandwidth spanning nearly one octave are needed.

To meet the first requirement, we generate a broadband seed by difference-frequency generation (DFG) in a chirped MgO-doped periodically poled lithium niobate (PPLN) crystal, which gives an octave-spanning spectrum by matching the quasi-phase-matching conditions to the evolving envelope of the laser pulse as it passes through the crystal, providing a wider phase matching region than unchirped PPLN due to the increased range of k vectors determined by the varying grating period [16].

Here the structure, property, and functions of PPLN are briefly described in the following.

To achieve high DFG efficiency, the interacting waves need to be phase-matched by choosing the proper poling period. The poling is done in wafer form before the chip is cut and polished. This allows a wide range of possible configurations by designing a poling mask specifically tailored to the application.

DOI: 10.1201/9780429196577-22

3.4.2   EXPERIMENTAL

To obtain the input pulse for the DFG process, a 30 µJ 25 fs 800 nm pulse (4% of the output of a commercial Femtopower Pro multipass Ti:sapphire amplifier) is coupled into a hollow-core fiber filled with 5-bar pressure krypton gas, to be spectrally broadened by means of self-phase-modulation, providing spectral components from ~500 to ~1050 nm. The spectrally broadened pulse is temporally compressed with chirped mirrors, and the low- and high-frequency components of the pulse are subsequently mixed in chirped PPLN with a quasi-phase-matching period changing linearly from 8 to 11 µm, generating a broadband DFG signal centered around 2.1 µm, from <1.5 to >2.8 µm. The DFG process ensures CEP stability [17,18], which is preserved in the subsequent OPA processes.

3.4.3   RESULTS AND DISCUSSION

For broadband amplification of a picojoule-level seed pulse to the millijoule level, an amplification factor of about 10^8 is needed and gain narrowing becomes an issue. The OPA gain spectral narrowing is primarily due to the existence of non-zero phase mismatch ΔkL, where Δk is the wavevector mismatch between signal, pump, and idler, and L is the crystal length and the propagation directions of involved pulses namely, pump, signal, and idler.

$$\Delta k = k_{\mathrm{pump}} - k_{\mathrm{signal}} - k_{\mathrm{idler}}$$

The phase mismatch is a function of the crystal length. With a shorter crystal, the OPA gain bandwidth can be wider, but the gain also depends on the crystal length as proportional to $\alpha \sqrt{IL}$. To support this bandwidth with a nearly constant gain level, shorter crystal lengths and higher pump intensities are needed [19]. The intensity cannot be increased arbitrarily due to the damage threshold of the nonlinear crystals, but this threshold increases with the inverse of the square root of the pulse duration. Therefore, with a shorter pump pulse combined with a thinner crystal, the OPCPA system can achieve more gain bandwidth at the same gain level. With a 1.6 ps pulse, the pump intensity can be more than 50 GW/cm². This pump intensity can support a high gain with a wide bandwidth close to one octave, as shown in Figure 3.4.1.

The full schematic of the OPCPA system is shown in Figure 3.4.2. The pump laser is a home-built, 1.6 ps, 1030 nm Yb:YAG thin-disk regenerative amplifier delivering pulses up to 20 mJ at 3 kHz repetition rate [20]. To optically synchronize the pump pulse and seed pulse, the pump laser is seeded by the same Ti:sapphire oscillator (Femtolasers Rainbow). The oscillator pulse energy within the gain bandwidth of Yb:YAG at 1030 nm, diverted for optical seeding of the regenerative amplifier, is 2 pJ. With 1.6 ps pump pulses, 1–2 mm bulk LiNbO₃ crystals are expected to deliver high parametric gain with an amplification bandwidth approaching an octave. On the other hand, the synchronization of pump and seed pulses is more delicate. To account for this, we used a spectrally resolved cross-correlation technique [21] combined with active stabilization, which allows us to obtain an RMS timing jitter of 24 fs, 1.5% of the pump pulse duration.

To ensure the broadest possible amplification bandwidth, three nearly degenerate OPA stages are used. The signal and pump beams are crossed at a small angle (~3°) to spatially separate them after amplification. 350 µJ of the pump laser pulse is used to amplify the seed with a 400 µm beam diameter (FWHM) in the first stage, a 2 mm thick PPLN crystal with 29.9 µm poling period. The output signal energy is 15 µJ in this stage. Fifteen percent of the pump beam is then used with 2 mm beam diameter in the second stage, which again consists of a 2 mm thick PPLN crystal with 29.9 µm period. The third stage employs a 1.5 mm thick MgO-doped LiNbO₃ crystal. The signal energy is boosted to 1.2 mJ using the remaining 14 m pump pulse with 4 mm beam diameter. Superfluorescence accounts for less than 5% of the total output, as determined by the unseeded output power. PPLN could support much more bandwidth than bulk LiNbO₃ crystal for the same gain but is not currently available with apertures large enough to support the third OPA stage.

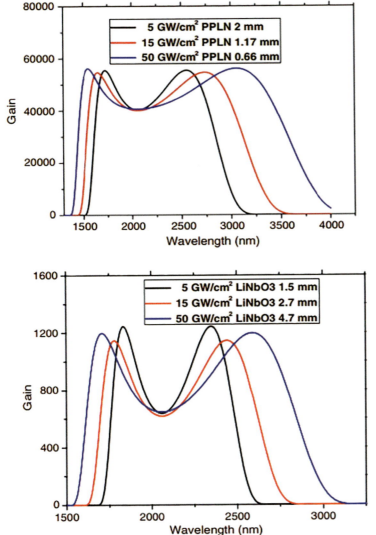

FIGURE 3.4.1 Calculated unsaturated gain profiles of PPLN, and LiNbO$_3$, with pump intensities corresponding to optimal safe operating conditions with 1.6, 25, and 160 ps pump durations (50, 15, and 5 GW/cm^2, respectively) with crystal lengths set such that the gain at the central wavelength is constant. The damage threshold is ~100 GW/cm^2 for LiNbO$_3$ with 1.6 ps pulses at 1030 nm.

To extract the maximum energy from the pump pulse with the bandwidth maintained, the durations of the seed and pump pulses should be matched. In this laser system, the seed pulse is stretched to 700 fs by a Dazzler acousto-optic pulse shaper.

Here we would like to explain very briefly the mechanism of Dazzler. The principle mechanism of Dazzler is based on an interaction between a polychromatic acoustic wave polychromatic optical wave in the bulk block of birefringent crystal. Optical signal in the hundreds of Terahertz range is controlled by tens of Megahertz range.

A 10-mm thick silicon bulk block is placed before the Dazzler to provide positively chirp to the pulse and to match the laser pulse duration to the Dazzler crystal length, to achieve the maximum diffraction efficiency. The amplified IR pulse is recompressed by a 1.5 mm thick silicon wafer using the positive chirp effect. The compressed and amplified pulse of the OPCPA system is characterized by a homemade third harmonic-generation (THG) frequency-resolved optical gating (FROG) apparatus, as shown in Figure 3.4.3. Using the Dazzler, the pulse is compressed to 10.5 fs at FWHM, which is very close to the Fourier limit of 10 fs. At 2.1 μm, this pulse duration corresponds to about 1.5 cycles. Such a short, millijoule-level energy pulse is an ideal tool for producing high photon

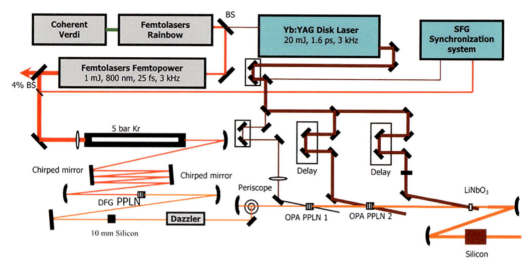

FIGURE 3.4.2   Scheme of the IR OPCPA.

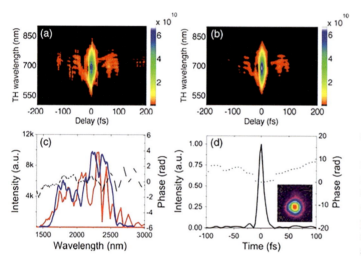

FIGURE 3.4.3   THG FROG measurement results of the compressed 10.5 fs pulse. (a) Measured FROG trace. (b) Retrieved FROG trace. (c) Measured (blue) and retrieved (red) spectral intensity and phase (dashed black). (d) Measured temporal intensity and phase. Inset: measured spatial intensity profile after the third stage.

energy, isolated attosecond pulses via HHG. Furthermore, as shown in the previously reported f-to-$3f$ nonlinear interferometry measurement [11], the CEP of the amplified signal is stable, which is also important for HHG.

3.4.4   CONCLUSION

As described above, we have demonstrated the generation of CEP-stable, 1.5 cycle (10.5 fs), 1.2 mJ pulses at 2.1 μm carrier wavelength and 3 kHz repetition rate, from a broadband OPCPA system pumped by a 1.6 ps short pulse duration laser [22].

This research is the product of collaborative work among the following people: Y. Deng, A. Schwarz, H. Fattahi, M. Ueffing, X. Gu, M. Ossiander, T. Metzger, V. Pervak, H. Ishizuki, T. Taira, T. Kobayashi, G. Marcus, F. Krausz, R. Kienberger, and N. Karpowicz [22].

REFERENCES

1. F. Krausz and M. Ivanov, *Rev. Mod. Phys.* **81**, 163 (2009).
2. E. Goulielmakis, V. S. Yakovlev, A. L. Cavalieri, M. Uiberacker, V. Pervak, A. Apolonski, R. Kienberger, U. Kleineberg, and F. Krausz, *Science* **317**, 769 (2007).

3. I. P. Christov, M. M. Murnane, and H. C. Kapteyn, *Phys. Rev. Lett.* **78**, 1251 (1997).

4. J. L. Krause, K. J. Schafer, and K. C. Kulander, *Phys. Rivulet.* **68**, 3535 (1992).

5. J. Tate, T. Auguste, H. G. Muller, P. Salieres, P. Agostini, and L. F. DiMauro, *Phys. Rev. Lett.* **98**, 013901 (2007).

6. T. Popmintchev, M.-C. Chen, D. Popmintchev, P. Arpin, S. Brown, S. Alisauskas, G. Andriukaitis, T. Balciunas, O. D. Mucke, A. Pugzlys, A. Baltuska, B. Shim, S. E. Schrauth, A. Gaeta, C. Hernandez-Garcia, L. Plaja, A. Becker, A. Jaron-Becker, M. M. Murnane, and H. C. Kapteyn, *Science* **336**, 1287 (2012).

7. V. S. Yakovlev, M. Ivanov, and F. Krausz, *Opt. Express* **15**, 15351 (2007).

8. G. Marcus, W. Helml, X. Gu, Y. Deng, R. Hartmann, T. Kobayashi, L. Strueder, R. Kienberger, and F. Krausz, *Phys. Rev. Lett.* **108**, 023201 (2012).

9. B. E. Schmidt, A. D. Shiner, P Lassonde, J. C. Kieffer, P. B. Corkum, D. M. Villeneuve, and F. Légaré, *Opt. Express* **19**, 6858 (2011).

10. C. Li, D. Wang, L. Song, D. Liu, P. Liu, C. Xu, Y. Leng, R. Li, and Z. Xu, *Opt. Express* **19**, 6783 (2011).

11. T. Fuji, N. Ishii, C. Y. Teisset, X. Gu, T. Metzger, A. Baltuska, N. Forget, D. Kaplan, A. Galvanauskas, and F. Krausz, *Opt. Lett.* **31**, 1103 (2006).

12. X. Gu, G. Marcus, Y. Deng, T. Metzger, C. Teisset, N. Ishii, T. Fuji, A. Baltuska, R. Butkus, V. Pervak, H. Ishizuki, T. Taira, T. Kobayashi, R. Kienberger, and F. Krausz, *Opt. Express* **17**, 62 (2009).

13. K.-H. Hong, S.-W. Huang, J. Moses, X. Fu, C.-J. Lai, G. Cirmi, A. Sell, E. Granados, P. Keathley, and F. X. Kärtner, *Opt. Express* **19**, 15538 (2011).

14. C. J. Fecko, J. J. Loparo, and A. Tokmakoff, *Opt. Commun.* **241**, 521 (2004).

15. G. Andriukaitis, T. Balčiūnas, S. Ališauskas, A. Pugžlys, A. Baltuška, T. Popmintchev, M.-C. Chen, M. M. Murnane, and H. C. Kapteyn, *Opt. Lett.* **36**, 2755 (2011).

16. C. Heese, C. R. Phillips, L. Gallmann, M. M. Fejer, and U. Keller, *Opt. Lett.* **35**, 2340 (2010).

17. A. Baltuska, T. Fuji, and T. Kobayashi, *Phys. Rev. Lett.* **88**, 133901 (2002).

18. T. Fuji, A. Apolonski, and F. Krausz, *Opt. Lett.* **29**, 632 (2004).

19. I. N. Ross, P. Matousek, G. H. C. New, and K. Osvay, *J. Opt. Soc. Am. B* **19**, 2945 (2002).

20. T. Metzger, A. Schwarz, C. Y. Teisset, D. Sutter, A. Killi, R. Kienberger, and F. Krausz, *Opt. Lett.* **34**, 2123 (2009).

21. A. Schwarz, M. Ueffing, Y. Deng, X. Gu, H. Fattahi, T. Metzger, M. Ossiander, F. Krausz, and R. Kienberger, *Opt. Express* **20**, 5557 (2012).

22. Y. Deng, A. Schwarz, H. Fattahi, M. Ueffing, X. Gu, M. Ossiander, T. Metzger, V. Pervak, H. Ishizuki, T. Taira, T. Kobayashi, G. Marcus, F. Krausz, R. Kienberger, and N. Karpowicz, *Opt. Lett.* **37**, 4973 (2012).

Section 4

*Simple NLO Processes
with a Few Colors*

4.1 Three-Photon-Induced Four-Photon Absorption and Nonlinear Refraction in ZnO Quantum Dots

4.1.1 INTRODUCTION

ZnO quantum dots (QDs) with a direct bandgap of 3.37 eV at room temperature and a large exciton binding energy of 60 meV find potential applications in optoelectronics devices [1–17]. Recently, z-scan technique [2] has been extensively used to study the nonlinear optical (NLO) property of such semiconductors [9–13]. At higher power of the incident laser in a material, a strong multiphoton absorption (mPA) process may lead to mPA-assisted excited-state absorption (ESA), i.e., a two-step $m+1$ photon absorption process [1,4,14–19].

There are reports available on three-photon absorption (3PA) and subsequent ESAs in bulk CdS [1], polydiacetylene [14,15], and two-photon absorption (2PA)-induced 3PA in semiconductor QDs [16,17]. However, there is no report so far on the observation of the 3PA-induced four-photon absorption (4PA) via the ESA in semiconductor QDs. Here we have reported the 3PA-induced effective 4PA and the nonlinear refraction (NLR) in ZnO QDs dispersed in methanol, for the first time, to our knowledge, by employing z-scan technique and using Q-switched Nd:YAG laser ($= 1064$ nm, $= 10$ ns, and 10 Hz repetition rate) radiation.

4.1.2 EXPERIMENTAL

The freshly prepared polyvinylpyrrolidone (PVP) capped ZnO QD sample has been synthesized as detailed elsewhere [7]. Figure 4.1.1 depicts the linear optical transmission characteristics of the used sample dispersed in methanol as obtained using an ultraviolet–visible (UV–vis) spectrophotometer (Hitachi U-3010). The excitonic absorption peak is observed at $E_o = 4.7$ eV, which lies much below 3.2 eV, the bandgap of bulk ZnO. The calculated average size (diameter) of the sample from the bandgap shift agrees quite well with the average size of 2.0 ± 0.1 nm as obtained from transmission electron microscope (TEM) micrograph [7]. Insets (a) and (b) of Figure 4.1.1 show pure 4PA and 3PA-induced two-step 4PA processes, respectively. The x-ray diffraction (XRD) pattern is shown in the inset (top) of Figure 4.1.2, which confirms the formation of the wurtzite structure of ZnO [7]. Among the three major XRD peaks that appeared due to (100), (002), and (101) planes, the last one is the most intense, i.e., the orientation of the fastest growth with the lowest surface energy.

4.1.3 RESULTS AND DISCUSSION

For the experimental study of NLO properties of ZnO QDs, a standard z-scan setup [2,6] is made using a Q-switched Nd:YAG laser. A part of the incident laser beam is reflected by a beam splitter to enable monitoring of any fluctuation in the input energy. The transmitted main laser beam is focused by a lens $f = 20$ cm. The beam waist w_0 and the confocal parameter z_o at the focus are thus

DOI: 10.1201/9780429196577-24

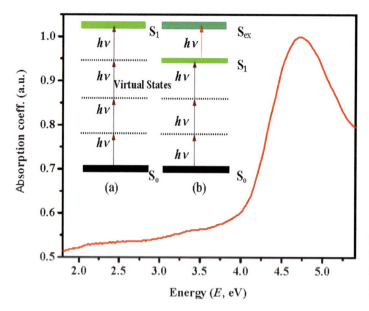

FIGURE 4.1.1 UV–vis absorption characteristics of ZnO nanoparticles. Insets (a) and (b) show pure 4PA and 3PA-induced 4PA via ESA, respectively.

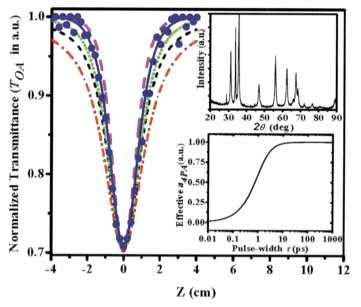

FIGURE 4.1.2 Dashed-dotted (red), short dashed (black), dotted (green), solid (blue), and dashed (magenta) curves are the theoretical normalized OA z-scan transmittance traces obtained for pure 2PA (with $\alpha_2=2.86$ cm/GW), concurrence of 2PA and 3PA (with $\alpha_2=1.2$ cm/GW and $\alpha_3=2.2$ cm³/GW²), pure 3PA (with $\alpha_3=0.71$ cm³/GW²), 3PA induced effective 4PA (with $\alpha_3=0.058$ cm³/GW² and $\alpha_4=0.013$ cm⁵/GW³), and pure 4PA (with $\alpha_4=0.36$ cm⁵/GW³) OA z-scan traces, respectively. Circles are the experimental points. Inset (top) shows XRD pattern of the same QDs. Inset (bottom) shows pulse duration dependence of normalized effective 4PA coefficients [4].

67 m and 1.35 cm, respectively. For a pure mPA process, taking place at a time in a thin sample, the propagation equation of the pulsed laser intensity I is given by

$$\frac{dI}{dz'} = -\alpha_0 I - \alpha_m I^m \qquad (4.1.1)$$

where z' is the propagation length inside the medium and α_0 is the linear and α_m is the m-photon ($m>2$) absorption coefficients. In accordance with the three-level two-step model, simultaneous absorptions of three photons by the absorption of laser pulse promote an electron from S_0 to S_1 (inset

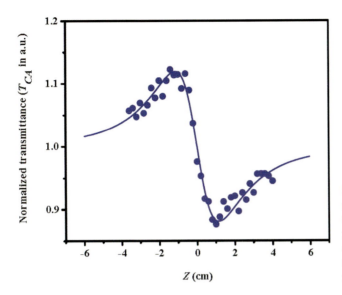

FIGURE 4.1.3 Normalized transmittance trace for CA z-scan taking contribution from pure NLR. The contribution from pure NLR is calculated from the experimental CA and OA transmittance traces and follows the method described in [13].

of Figure 4.1.1). From S_1 the electron may relax to S_0 or experience an upliftment to S_{ex} by absorbing another single photon. The photodynamic of this system is monitored by [4]

$$\frac{\partial I}{\partial z'} = -\sigma_3 p_0 I^3 - \sigma_e p_1 I \tag{4.1.2}$$

where σ_3 and σ_e are the 3PA and ESA cross sections and p_0 and p_1 are the population of the states S_0 and S_1, respectively. The theoretical normalized energy transmittance for a thin sample T (z, p, and q) for the concurrence of the 3PA and excited-state 4PA can be written as [4]

$$T(z,p,q) = \sum_{r=0}^{4} \sum_{s=0}^{4} a_{rs} p^r q^s \tag{4.1.3}$$

where $p = p_0/(1 + z^2/z_0^2)$ and $q = q_0/(1 + z^2/z_0^2)$. The on-axis peak phase shifts owing to the 3PA and 4PA processes are $p_0 = (2\alpha_3 I_0^2 L_{eff})^{1/2}$ and $q_0 = (3\alpha_4 I_0^3 L_{eff})^{1/3}$, respectively, with I_0 being the peak irradiance at the focus. $L_{eff}^{(m-1)} = [1 - \exp\{-(m-1)\alpha_0 L\}]/(m-1)\alpha_0$ is the effective thickness and L is the actual thickness of the sample. The absorption coefficients of pure 3PA and pure 4PA are α_3 and α_4, respectively.

Figure 4.1.2 shows the open-aperture (OA) z-scan transmission trace of the sample with symbols being the experimental points and dashed-dotted (red), short dashed (black), dotted (green), solid (blue), and dashed (magenta) curves are obtained theoretically by using the analytical expressions reported in [4,19] for the cases with pure 2PA, the concurrence of 2PA and 3PA, pure 3PA, 3PA-induced effective 4PA, and pure 4PA OA z-scan traces, respectively. Considering only nonlinear absorption (NLA) processes, the best fitting results give rise to $\alpha_3 = 0.058\,cm^3/GW^2$ and $\alpha_4 = 0.013\,cm^5/GW^3$ with the known values $L = 0.2\,cm$ and $I_0 = 2.5\,GW/cm^2$. The estimated values of NLA coefficients are also summarized in Table 4.1.1.

The used excitation photon energy E_p (1.16 eV) is less than $E_0/3$, giving rise to the concurrence of 3PA and 4PA [16]. Figure 4.1.2 shows that the theoretical curves obtained by using pure 2PA, 3PA, 4PA, and concurrence of 2PA and 3PA fail to fit the experimental data but the curve obtained by considering the 3PA-induced 4PA (via ESA) with Eq. (4.1.3) provides the best fit. So, the dominant and operative mechanism is found to be the 3PA-induced effective 4PA via the ESA. Under the

TABLE 4.1.1

3PA Induced Effective 4PA and NLR Coefficients of ZnO QDs

Sample	3PA$\times10^{18}$ (cm^3/W^3)	Effective 4PA$\times10^{29}$ (cm^5/W^3)	NLR (γ)$\times10^{14}$ (cm^2/W)	NL Index (n_2) 10^{11} (esu)
ZnO QDs	0.058	1.3	1.7	1.1, 1.5[a]
ZnO bulk	0.013[b]	–	0.9[c]	–

[a] Ref. [10].

[b] Ref. [12].

[c] Ref. [11].

steady-state condition, the value of the critical intensity $I_{critical}$ has been estimated [4] to be of the order of 10^8 W/cm^2 by using $I_{critical} = \left(3E_p N_1^c /_3\right)^{1/3}$, where $\alpha_3 = 0.058$ cm^3/GW^2, $= 10$ ns, and critical population of S_1, i.e., $N_1^c = 10^{18}$ cm^{-3} [12]. The calculated value of $I_{critical}$ is lower than that of our used laser intensity, indicating the presence of the ESA [15]. Furthermore, the probability of occurrence of the ESA-induced 4PA as a function of the pulse width, as obtained following the method of Gu and Ji [4], is shown in the inset (bottom) of Figure 4.1.2. It is clearly seen that the probability of occurrence of the ESA-induced 4PA is higher with the nanosecond laser pulses.

Using the closed-aperture (CA) z-scan experiment, the NLR coefficient of the ZnO QD sample is also measured as shown in Figure 4.1.3. The intensity-dependent pure NLR and the nonlinear refractive index n_2 are estimated by using the relations $\Delta\phi_0 = k\gamma I_0 L_{eff}^{(1)}$ and $n_2(esu) = (cn_0/40\pi)\gamma$ [2,13]. Here, $k = 2\pi/\lambda$, $c = 3\times10^{10}$ cm/s, $\lambda = 1064$ nm, and $n_0 = 2.7$. The peak-valley shape indicates the self-defocusing nonlinearity. The estimated values of and n_2 for the sample are shown in Table 4.1.1. The obtained value of n_2 of the ZnO QD is 1.1×10^{-11} esu, which is nearly the same as obtained 1.5×10^{-11} esu previously for ZnO/cetyltrimethylammonium bromide composite nanoparticles [10]. However, the estimated value of n_2 is 2 times larger in comparison to the bulk ZnO [11] and at least 1 order larger than those of ZnS and 1% Mn^{2+}-doped ZnS QDs [13].

The analysis of Kramers–Kronig (K–K) relations predicts a positive NLR as the value of E_p/E_g in the present case is 0.31, which is 0.69 [2]. But the experimental observations are exactly the opposite as had already been seen in ZnO bulk [11,12] and ZnO composites QDs [9,10]. The negative sign in the NLR has also been observed by us in ZnS and doped ZnS QDs at 1064 nm [13]. Ganeev observed such effects in CdS and ZnS nanoparticle-doped zirconium oxide films and its origin was attributed to thermal effect [6]. The value of the thermal nonlinearity γ_T that may arise from light absorption is calculated to be $= 10^{-13}$ cm^2/W, which is only 1 order larger than the obtained value. So, the existing mechanism that contributes to the refractive index variation is the thermal lensing effect owing to the absorption of highly intense nanosecond laser radiation in the investigated QD solution.

4.1.4   CONCLUSION

In conclusion, 3PA-induced effective 4PA through the ESA and self-defocusing NLR have been observed in ZnO QDs with the average size of 2.0 ± 0.1 nm for the first time using 10 ns 1064 nm Nd:YAG laser radiation with the peak intensity of 2.5 GW/cm^2 [20]. Simultaneous presence of the NLA and high nonlinear refractive index of 1.1×10^{-11} esu of ZnO QDs will promote this material as a potential candidate for optoelectronic devices.

The experimental and theoretical research activities described in this subsection is conducted cooperatively by the following people: M. Chattopadhyay, P. Kumbhakar, C. S. Tiwary, A. K. Mitra, U. Chatterjee, and T. Kobayashi [20].

REFERENCES

1. A. Penzkofer and W. Falkenstein, *Opt. Commun.* **16**, 247 (1976).
2. M. Sheik-Bahae, A. A. Said, T. H. Wei, D. J. Hagan, and E. W. Van Stryland, *IEEE J. Quantum Electron.* **26**, 760 (1990).
3. R. L. Sutherland, M. C. Brant, J. E. Rogers, J. E. Slagle, D. G. Maclean, and P. A. Fleitz, *J. Opt. Soc. Am. B* **22**, 1939 (2005).
4. B. Gu and W. Ji, *Opt. Express* **16**, 10208 (2008).
5. B. Gu, J. Wang, J. Chen, Y. Fan, J. Ding, and H. Wang, *Opt. Express* **13**, 9230 (2005).
6. R. A. Ganeev, *J. Opt. A* **7**, 717 (2005).
7. P. Kumbhakar, D. Singh, C. S. Tiwary, and A. K. Mitra, *Chalcogenide Let.* **5**, 387 (2008).
8. T. Kobayashi, *Optoelectron., Devices Technol.* **8**, 309 (1993).
9. L. Irimpan, V. P. N. Nampoori, and P. Radhakrishnan, *J. Appl. Phys.* **103**, 094914 (2008).
10. R. Wang, X. Wu, B. Zou, L. Wang, P. Wu, S. Liu, J. Wang, and J. Xu, *Chin. Phys. Lett.* **15**, 27 (1998).
11. X. J. Zhang, W. Ji, and S. H. Tang, *J. Opt. Soc. Am. B* **14**, 1951 (1997).
12. B. Gu, J. He, W. Ji, and H. T. Wang, *J. Appl. Phys.* **103**, 073105 (2008).
13. M. Chattopadhyay, P. Kumbhakar, C. S. Tiwary, R. Sarkar, A. K. Mitra, and U. Chatterjee, *J. Appl. Phys.* **105**, 024313 (2009).
14. F. Yoshino, S. Polyakov, M. Liu, and G. Stegeman, *Phys. Rev. Lett.* **91**, 063902 (2003).
15. S. Polyakov, F. Yoshino, M. Liu, and G. Stegeman, *Phys. Rev. B* **69**, 115421 (2004).
16. X. B. Feng, G. C. Xing, and W. Ji, *J. Opt. A* **11**, 024004 (2009).
17. G. Xing, W. Ji, Y. Zheng, and J. Ying, *Appl. Phys. Lett.* **93**, 241114 (2008).
18. G. S. He, L. Tan, Q. Zheng, and P. N. Prasad, *Chem. Rev.* **108**, 1245 (2008).
19. B. Gu, X.-Q. Huang, S.-Q. Tan, M. Wang, and W. Ji, *Appl. Phys. B* **95**, 375 (2009).
20. M. Chattopadhyay, P. Kumbhakar, C. S. Tiwary, A. K. Mitra, U. Chatterjee, and T. Kobayashi, *Opt. Lett.* **34**, 3644 (2009).

4.2 Femtosecond Pulses Cleaning by Transient-Grating Process in Optical Media

4.2.1 INTRODUCTION

In ultrafast spectroscopy and laser-matter interaction experiments, clean pulses with a smooth spectrum and temporal shape are vitally important in removing unwanted noise signals introduced by satellite pulses in an uncleaned pulse, which makes the explanations of the experimental results complicated. In laser-matter interaction experiments in the relativistic dominated regime [1,2] in particular, a pulse with high temporal contrast is necessary to prevent unwanted intense subpulses, especially prepulses from generating preplasma before the main pulse, which may block the later arriving main pulse by plasma mirror [3]. To improve the temporal contrast of this highly intense femtosecond pulse, several pulse-cleaning techniques have been developed. These methods include the use of saturable absorbers [4], a nonlinear Sagnac interferometer [5], double chirped pulse amplification (CPA) [6], plasma mirrors [7], polarization rotation [8], cross-polarized wave (XPW) generation [9], and self-diffraction (SD) process [10].

The lowest-order level phenomena of the above-mentioned methods are classified to be third-order nonlinear optical (NLO) phenomena. Among the only-plasma mirrors include real state(s) are involved, while all others can be either "real-state" involved-NLO or "only-virtual-state" involved-NLO processes. Therefore, only the latter group process can be material-relaxation time-unlimited "fast" process, while the formers are limited by material-relaxation time and hence "slow" processes.

By using XPW in BaF_2 crystals, the temporal contrast was improved by about four orders of magnitude [9]. However, the contrast enhancement was limited by the extinction ratio of the polarization discrimination device [9]. Very recently, we demonstrated for the first time the application of a self-diffraction process in a bulk Kerr medium to clean the pulse removing satellite pulses which do not need any polarizer. Through this method, it was proved that both prepulses and postpulses can be cleaned in one picosecond region [10].

As with the SD process, in which two beams are focused into a Kerr bulk medium with a small crossing angle to generate separated SD signals beside the incident beams, the generated signal by the transient-grating (TG) process is also spatially well separated from incident beams, and there is no need for the polarization discrimination device. Both processes are third-order nonlinear optical processes. The TG process is a phase-matched process unlike the SD process; thus, the TG process can have a long nonlinear medium to enhance the signal and can be used in a wide spectral range from ultraviolet (UV) to mid-infrared (MIR) with a broadband bandwidth [11]. Owing to its ability to operate with a long nonlinear medium, the TG process is more sensitive to the input pulse intensity and can be operated with nanojoule pulse energy input [11]. Larger beam angles may also be used in the SD process, reducing the scattered-light background. All these advantages make the TG process another useful process for cleaning ultrashort laser pulses.

In this letter, we demonstrate experimentally the use of the TG process to clean a femtosecond laser pulse. Using this technique in a 0.5-mm-thick fused silica glass plate, the results indicate that the temporal contrast is improved by two orders of magnitude in comparison with the incident

DOI: 10.1201/9780429196577-25

pulses. The laser spectrum is found to be smoothed and broadened simultaneously, and the pulse duration is shortened in this process.

The TG process is also a "third-order" nonlinear process, and the intensity of the TG signal has a cubic dependence on the input intensities I_1, I_2, and I_3. In the time domain, it can be expressed as

$$I_{[sd1]} \propto I_1(t) I_2(t) I_3(t - \tau) \tag{4.2.1}$$

where τ is a delay time between the grating formation pulse pair and the signal pulse to be diffracted. The SD process is a special case of τ being 0. In case the incident intensities are very high, then even higher order nonlinearity than third order, such as fifth-order NLO, may be involved. Especially if the accumulation of the transient grating generated vis photothermal process which is slow in terms of relaxation induced by thermal diffusion it can be higher-order NLO process. The higher-order NLO effects can be either positive or negative, i.e., showing saturation or inverse saturation depending on the mechanisms. It must be also mentioned that the higher-order NLO processes may involve not only such real states but also virtual states or even mixtures of them.

This time-domain expression of Eq. (4.2.1) indicates that the pulse can be cleaned and that the pulse duration of the TG signal will be shortened in the TG process by controlling the delay time τ. The formation and disappearance of the grating in the TG process in non-resonant Kerr media are instantaneous [12]. Therefore, even the satellite pulses in the picosecond range and the weak amplified spontaneous emission will separate the TG process from the main pulse in the time domain. In the frequency domain, the intensity of the generated TG signal can be described as [11]

$$I_{TG}(\omega_{TG}) \propto \left| \iint d\omega_1 d\omega_2 \tilde{E}_1^*(z, \omega_1) \tilde{E}_2(z, \omega_2) \tilde{E}_3(z, \omega_{TG} - \omega_2 + \omega_1) exp\left(i(\omega_2 - \omega_1)\tau\right) \right|^2 \tag{4.2.2}$$

where ω_1, ω_2, ω_3, and ω_{TG} are angular frequencies of three incident beams and the generated TG signal, respectively. The spectral intensity of the TG signal is given by an integral of the spectral intensity of three incident laser pulses. This means that the intensity of the TG signal at every wavelength is the average contribution of the entire spectral region of the incident pulses. As a result, the spectrum of the TG signal is smoothed automatically due to the third-order nonlinear process. Thus, the spectrum of the TG signal together with self-phase modulation and cross-phase modulation is broadened.

4.2.2 EXPERIMENTAL

A commercial Ti:sapphire CPA laser system (Legend USP, Coherent) producing 35-fs, 2.5-mJ maximal pulse energy at 800-nm center wavelength and running at 1-kHz repetition rate was used in the experiment (Figure 4.2.1). Four 1-mm-thick fused silica beamsplitters (BSs) with 80/20, 80/20,

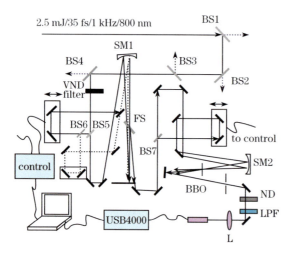

FIGURE 4.2.1 Schematic of the experimental setup. BS1–BS7 are seven 1-mm-thick fused silica beam splitters with 80/20, 80/20, 80/20, 50/50, 67/33, and 50/50 reflection/transmission ratios, respectively. SM1, spherical mirror, $R = -600$ mm; SM2, spherical mirror, $R = -500$ mm; L, focal lens with 50-mm focal length.

80/20, and 50/50 transmission/reflection ratios were used to reduce the pulse energy to about 100 μJ and to introduce prepulses and postpulses through the back-and-front reflections of the BSs. After a variable neutral-density (VND) filter, the laser pulse was split into three beams by a 67/33 and a 50/50 BS. One of the laser beams can tune the time delay through a motor-controlled stage with a 10 nm/step. The three beams were focused with a BOXCARS arrangement [13] using a concave mirror with a 300-mm focal length into a 0.5 mm-thick fused silica (FS) glass plate located after the focal point. The beam diameters on the glass plate were about 400 μm. Transmission pulse energies of beam1, beam2, and beam3 were 32, 37, and 20 μJ, respectively. The pulse energy of the generated TG signal was about 1.2 μJ.

We performed a second-order autocorrelation to measure the temporal contrast which needed much lower incident pulse energy [14]. The measurement procedure was almost the same as our previous work [10]. In the second-order autocorrelation measurement, a 1-mm-thick 50/50 beam splitter was used to split the laser beam. A 170-μm thick beta barium borate (BBO, Type I, $\theta = 29.2°$) crystal was used to generate a sum-frequency (SF) signal. After an aperture, a low-wavelength-pass filter (LPF) cutting at 440 nm, and a 1000-times-decreased natural-density (ND) filter, the SF signal was focused into a fiber and detected using a multi-channel spectrometer (USB4000, Ocean Optics) with 200 ms integration time.

The intensity of the SF signal was obtained by integrating the spectral intensity over the spectral range from 370 to 430 nm to reduce other noises. In the delay time range when the SF signal was very strong to saturate the spectrometer, a 1000-times ND filter was added (for TG signal from −100 to 100 fs, for incident pulse from −1000 to 1000 fs). In this case, the delay time-dependent SF intensities of incident laser pulses and the TG signal were obtained (Figure 4.2.2). In the figure, autocorrelation data of incident laser pulse and TG signal in the delay time show a range of –6 to 6 ps. The data were obtained with a 10-fs delay time step. The temporal contrast of the TG signal was clearly improved by about two orders of magnitude compared with incident pulses. The two peaks in the TG signal autocorrelation at ±1.8 ps were due to the back-and-front reflections of the 170-μm-thick BBO crystal used in the autocorrelator.

4.2.3 RESULTS AND DISCUSSION

The pulse durations of the input pulse and TG signal were measured using second harmonic generation frequency-resolved optical grating SHG-FROG with the same setup as autocorrelation measurement and replacing the beam splitter with a half-reflective mirror. The spectral and temporal profiles, as well as the phase, were retrieved using commercial software (FROG 3.0, Femtosoft Technologies) with a 256×256 grid. The retrieval errors were smaller than 0.005. Figure 4.2.3a shows the retrieved laser spectrum, retrieved spectral phase, and measured spectrum. The measured spectral intensity of the TG signal retrieved spectral intensity profile, and retrieved phase

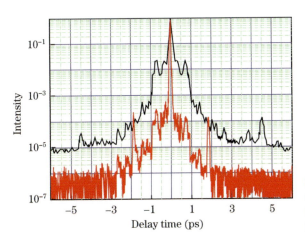

FIGURE 4.2.2 Normalized SAC intensities of the incident pulse (upper curve) and that of the TG signal (lower curve) in the delay time from −6 to 6 ps with a 10-fs/step resolution.

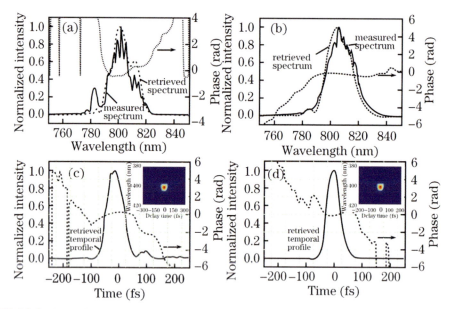

FIGURE 4.2.3   (a) and (b) are the measured spectrum, the retrieved spectrum, and the retrieved spectral phase of the incident pulse and the TG signal, respectively. (c) and (d) are the retrieved temporal profile and the phase of the incident pulse and the TG signal, respectively. The inset patterns in (c) and (d) are the measured 2D SHG-FROG traces of the incident pulse and the TG signal, respectively.

is shown in Figure 4.2.3b. The TG signal spectrum was obviously smoothed and broadened from the input laser spectrum. The center wavelength of the TG signal was red-shifted for several nanometers, which could be due to the crossing phase modulation (XPM) [15]. Even though the input laser spectral phase showed a positive chirp, the obtained TG signal showed a small negative chirp. This is likely due to the XPM process and suitable delay among the incident pulses [10]. The retrieved temporal profile and temporal phase of the input laser pulse and the TG signal are shown in Figure 4.2.3c and d, respectively. Pulse duration of the TG signal was shortened from the input of 74 fs to a compressed 51 fs. The retrieved pulse of the TG signal also showed that the satellite pulses were removed. The two inset patterns shown in Figure 4.2.3c and d are the measured two-dimensional (2D) SHG-FROG traces of incident pulse and TG signal, respectively.

4.2.4   CONCLUSION

In conclusion, we demonstrate femtosecond pulse cleaning using the TG process in Kerr bulk media. Similar to the SD process, there is no need for a polarizer device in this method because the generated TG signal is well separated from the incident laser beams. Simultaneously, the laser spectrum is smoothed and broadened, and the pulse duration is shortened for the generated TG signal in comparison with the incident laser pulse. Although its improvement and efficiency are currently low compared with those of the SD process, the TG process is a phase-matched process, which makes it useful for the optimization of temporal and spectral characteristics by cleaning ultrashort laser pulses in a wide spectral range from UV to MIR with a broadband bandwidth. The research presented here is based on the collaboration among the following people: J. Liu, K. Okamura, Y. Kida, and T. Kobayashi [16].

REFERENCES

1. S. V. Bulanov, T. Zh. Esirkepov, D. Habs, F. Pegoraro, and T. Tajima, *Eur. Phys. J.* **D 55**, 483 (2009).
2. G. A. Mourou, T. Tajima, and S. V. Bulanov, *Rev. Mod. Phys.* **78**, 309 (2006).

3. P. Antici, J. Fuchs, E. d'Humi`eres, E. Lefebvre, M. Borghesi, E. Brambrink, C. A. Cecchetti, S. Gaillard, L. Romagnani, Y. Sentoku, T. Toncian, O. Willi, P. Audebert, and H. P´epin, *Phys. Plasmas* **14**, 030701 (2007).
4. J. Itatani, J. Faure, M. Nantel, G. Mourou, and S. Watanabe, *Opt. Commun.* **148**, 70 (1998).
5. A. Renault, F. Aug´e-Rochereau, T. Planchon, P. ID'Oliveira, T. Auguste, G. Ch´eriaux, and J.-P. Chambaret, *Opt. Commun.* **248**, 535 (2005).
6. M. P. Kalashnikov, E. Risse, H. Sch¨onnagel, and W. Sandner, *Opt. Lett.* **30**, 923 (2005).
7. G. Doumy, F. Qu´er´e, O. Gobert, M. Perdrix, Ph. Martin, P. Audebert, J. C. Gauthier, J. P. Geindre, and T. Wittman, *Phys. Rev. E* **69**, 026402 (2004).
8. D. Homoelle, A. L. Gaeta, V. Yanovsky, and G. Mourou, *Opt. Lett.* **27**, 1646 (2002).
9. A. Jullien, O. Albert, F. Burgy, G. Hamoniaux, J. P. Rousseau, J.-P. Chambaret, F. A. Rochereau, G. Ch´eriaux, J. Etchepare, N. Minkovski, and S. M. Saltiel, *Opt. Lett.* **30**, 920 (2005).
10. J. Liu, K. Okamura, Y. Kida, and T. Kobayashi, *Opt. Express* **18**, 22245 (2010).
11. R. Trebino, *Frequency-Resolved Optical Grating: The Measurement of Ultrashort Laser Pulses*, Kluwer Academic Publishers, Boston (2000).
12. R. Schmid, T. Schneider, and J. Reif, in *Proceedings of Ultrafast Electronics and Optoelectronics*, edited by A. Sawchuk Optical Society of America (2001), UThB3.
13. A. C. Eckbreth, *Appl. Phys. Lett.* **32**, 421 (1978).
14. K. W. DeLong, R. Trebino, J. Hunter, and W. E. White, *J. Opt. Soc. Am. B* **11**, 2206 (1994).
15. J. Liu, Y. Kida, T. Teramoto, and T. Kobayashi, *Opt. Express* **18**, 2495 (2010).
16. J. Liu, K. Okamura, Y. Kida, and T. Kobayashi, *Chin. Opt. Lett.* **9**, 051903(1–3) (2011).

4.3 Non-Degenerate Two-Photon Absorption Enhancement for Laser Dyes by Precise Lock-in Detection

4.3.1 INTRODUCTION

Interest in materials that exhibit high optical nonlinearity has increased dramatically in recent years. In particular, two-photon absorption (TPA) has been exploited in several current technologies, including optical power limiting [1], all-optical shutter [2], two-photon fluorescence microscopy [3–5], and others [6]. Two-photon induced fluorescence (TPIF) [7–9] is the most frequently used method for measuring TPA spectra and cross sections in nonlinear materials, whereas the Z-scan method [10,11] is less commonly employed. However, to acquire broad information on TPA spectra, wide-tunable laser sources such as optical parametric amplifier or dye laser sources are needed [7,12,13]. But, even with a tunable source, the measurement must be repeated for each selected wavelength. As it is often impossible to maintain the experimental conditions throughout the time required to perform the measurement for each wavelength across the full spectral range, experimental errors are introduced and even accumulated. The instability in the peak intensity of the laser pulses also inevitably induces large experimental errors in this time-consuming measurement, a problem that is further enhanced by the nonlinearity of the process. Furthermore, due to the diversification in the pulse duration during the pulse wavelength shifting, the pulse intensity changes for each measurement. In addition, these methods are indirect ways to measure the TPA spectrum, and thus may suffer from other systematic errors related to the intermediate parameters, such as the collection efficiency of emitted fluorescence, fluorescence quantum efficiency, and so on [7]. Finally, the degenerate detection's intensity squared dependence on the excitation beam intensity may lead to a large error during the square calculation. Such inaccuracies may deteriorate the measurements and may explain the quite different TPA cross section values that have been reported when the same method is used to examine some well-known materials [5,7,12,14,15].

In 1999, Belfield et al. reported for the first time a method for measuring the non-degenerate TPA (NDTPA) spectrum using a pump-probe two-beam configuration [16] in which the strong pump was a monochromic IR laser beam, and the weak probe was a white-light super-continuum (WLSC) beam. In this setup, a broad TPA spectrum can be directly measured in a single procedure by the spectral subtraction of the absorbed probe beam from the reference beam. The pump beam intensity has a linear dependence on the TPA cross section (as discussed later), which apparently induces a lower level of error than the intensity squared degenerate detection. However, the acquired spectrum difference is time-dependent due to the overlapping time shift between the pump and probe beams. This is due to the chirping effect occurring during the WLSC generation and the group-velocity dispersion (GVD) effect on the probe pulse propagation time. Kovalenko et al. proved that the chirp could be eliminated by using time-correction procedure when using a supercontinuum probe [17]. However, since the publication of the work by Belfield et al. there have only been a few reports using this setup for a limited number of direct-gap semiconductors and semiconductor quantum dots [18,19].

In the present study, we improve this method in several ways. First, we use chirp mirror pairs to compress the WLSC pulses before propagation in the sample. A very broad WLSC is generated via

DOI: 10.1201/9780429196577-26

complex processes that include self-phase modulation, plasma generation, and stimulated Raman scattering. Due to different contributions of these phenomena to the generation mechanism, the transient refractive index change is wavelength dependent in a very complicated way. Therefore, it is difficult to compress the chirp of WLSC. To minimize the chirp effect in our detector, we replace the spectrometer used in Belfield's setup scheme with a time-resolved two-dimensional induced absorption (ΔA) spectrum map obtained by combining a polychromator with a multi-channel lock-in amplifier (MLA) [16]. This ΔA map is of vital importance for determining the pump and probe pulse overlapping position in time. The MLA not only causes a large improvement of the signal noise ratio (SNR) but also makes it possible to simultaneously measure a broadband difference absorption spectrum with 128 spectral points and a spectral resolution of 0.83 nm. With this high-performance equipment, the TPA profiles for several laser dyes. The TPA spectra thus obtained are confronted with theoretical predictions to further validate our approach.

4.3.2   THEORY

In general, the attenuation of a light beam passing through an optical medium along an axis, designated here as the z-axis, can be expressed by the following phenomenological expression (this case includes linear and nonlinear interactions) [20]:

$$dI(z)/dz = -\alpha I(z) - \beta I^2(z) - \gamma I^3(z) \cdots. \tag{4.3.1}$$

Here, $I(z)$ is the intensity of the incident light beam propagating along the z-axis and α, β, and γ are the one-, two-, and three-photon absorption coefficients of the transmitting medium, respectively. To compare the absorption induced by the two-photon process in the case of an incident laser with a negligible bandwidth to the electronic bandwidths relevant to the two-photon transition including virtual states, we have the following:

$$dI(z)/dz = -\beta I^2 \tag{4.3.2}$$

The non-degenerate condition, in our situation for an intensity I_1 of the WLSC probe beam and I_2 for the pump beam, leads to:

$$dI_1(z)/dz = -\beta_1 I_1 I_2 \tag{4.3.3}$$

$$dI_2(z)/dz = -\beta_2 I_1 I_2 \tag{4.3.4}$$

Here, the two-photon absorption coefficient β is linear and proportional to the imaginary part of the third order nonlinear susceptibility as given by [21]

$$\beta_{1,2} = \frac{8\pi^2 \omega_{1,2}}{c^2 \varepsilon_1^{1/2} \varepsilon_2^{1/2}} \, \text{Im} \, \chi(-\omega_1; \omega_1, -\omega_2, \omega_2) \tag{4.3.5}$$

ε_i denotes the dielectric constant at the angular frequency of ω_i for each incident light, and c is the velocity of light.

The above equation is related with linear transition dipole moments as follows.
Im $\chi^{(3)} = N\pi |M_{fi}|^2 g(\hbar\Delta\omega)(\rho_f - \rho_f)$: imaginary part of the third-order susceptibility,
$M_{fi} = <f|M|i>$: transition dipole moment from an initial state $|i>$ to a final state $<f|$.
$M = \{\Sigma s[\text{er } \mathbf{e}_{1\text{unit}}|s><s|\text{er } \mathbf{e}_{2\text{unit}}|s>/[\hbar(\omega_1 - \omega_{si})] + [\text{er } \mathbf{e}_{2\text{unit}}|s><s|\text{er}<\mathbf{e}_{1\text{unit}}|/[\hbar(\omega_2 - \omega_{si})] + [1\Leftrightarrow 2]\}$

Here, N = number density of molecules with transition dipole, e = elementary charge, $\hbar\Delta\omega$ = detuning energy of photon pair and nondegenerate two-photon absorption-transition relevant excited electronic states $\hbar(\omega_1 + \omega_2) - (E_f - E_i)$, **e**r transition dipole moment, $\mathbf{e}_{1unit}$ unit vector associated with a transition dipole, $|s\rangle$ and $\langle s|$ are intermediate (virtual) stages contributing to the two-photon absorption.

The above equations are describing the non-degenerate two-photon absorption (cross section), it can also be used in the discussion of the Raman scattering cross section.

For input (before the medium) beam intensities labeled I_{10} and I_{20}, and considering the case where $I_{20} \gg I_{10}$, Eqs. (4.3.3) and (4.3.4) can be solved as follows:

$$I_2 \cong I_{20}$$

$$I_1 \cong e^{-\beta_1 I_{20} z} I_{10}$$

(4.3.6), (4.3.7)

Then,

$$\beta_1 I_{20} z = -In\frac{I_1}{I_{10}} = -\frac{1}{\lg e}\lg\frac{I_1}{I_{10}} = \frac{1}{\lg e}\Delta A$$

(4.3.8)

This condition is met in our setup as the intensities at the sample position of the pump and the probe within the spectral resolution are estimated to be 15 and 0.1 GW/cm^2, respectively. From Eq. (4.3.8), it is easy to see that with an intense pump, the TPA coefficient is linear in proportion to the pump beam-induced WLSC probe beam absorbance difference (ΔA). The relationship between the TPA cross section σ in cm^4/GW and TPA coefficient (in cm/GW) is given by [17]

$$\beta_1 = \sigma N_A c \times 10^{-3}$$

(4.3.9)

where N_A is the Avogadro constant and c is the concentration of the sample (in M). From Eqs. (4.3.8) and (4.3.9), we can conclude that the TPA cross section is proportional to the absorbance change ($\sigma \propto \Delta A$). As ΔA can be directly acquired through the pump-probe experiment, the TPA cross section can be calculated from experimental data without interference from other error sources (vide supra). Noteworthy, in this study, all spectra reporting TPA cross sections are based on the transformed TPA wavelength, which is the relative value calculated using ($2/\lambda_{TPA} = 1/\lambda_{pump} + 1/\lambda_{probe}$).

4.3.3 EXPERIMENTAL PROCEDURES

We use pulse energy of about 200 μJ from the Spitfire Ace Laser system (Spectra Physics, 1 kHz repetition rate, central wavelength at 798 nm). Total energy is divided into two using a 10% reflection beam splitter. The transmitted beam affords the pump beam, while the reflected beam is used to generate the WLSC, focusing the fundamental beam on a 2-mm-thick sapphire plate. The strong fundamental part of the beam is blocked by transmitting the beam through a 720 nm short pass dichroic mirror. The WLSC spectral range is from 500 to 730 nm. A pair of chirp mirrors (GVD, 470 ~ 810 nm, −40±20 fs^2) are used to slightly compress the WLSC beam with a triple-pass configuration. Next, the WLSC probe beam and the time-delayed pump beam are both focused using an off-axis parabolic mirror and overlapped at the sample point. After passing through the sample, the beams are collimated by another off-axis parabolic mirror. The pulse duration of the fundamental pump beam is measured as 103 fs using the SHG-FROG method [22]. The spot size of the focused pump beam at the sample surface is measured with a CCD beam profiler (Thorlabs) that has x- and y-axis of 146.1 and 149.9 μm, respectively. For data collection, the WLSC probe beam is focused into a multimode fiber, which is guided to the polychromator, where the signal is captured by a 128-channel MLA. A synchronized chopper with half of the laser repetition frequency is introduced into the fundamental beam path and provides the reference frequency signal for the MLA. The pump beam power is controlled by a wheel gradient-neutral density filter set in the pump path.

Several types of organic fluorescent dyes are used as target samples in this study. They are two xanthene dyes, rhodamine 6G and rhodamine 123 ([6-amino-9-(2-methoxycarbonylphenyl) xanthen3-ylidene]azanium chloride), dissolved in methanol; two coumarins dyes, coumarin 6 (3-(2 benzothiazolyl)-7-diethylamino-coumarin) and coumarin 343 (2,3,6,7-tetrahydro-11-oxo-1H, 5H, 11H-[1]benzopyrano[6,7,8-ij]quinolizine-10-carboxylic acid), dissolved in chloroform; and two oxazine dyes, Nile red (9-diethylamino-5-benzo[a]phenoxazinone) and Nile blue A (basic blue 12), dissolved in chloroform. All the dyes are purchased from Sigma-Aldrich and used without further purification.

4.3.4   RESULTS AND DISCUSSION

A typical result for our TPA spectrum measurement procedure is shown in Figure 4.3.1. The sample is coumarin 6 dissolved in chloroform at a concentration of 23.7 mM. The time-resolved two-dimensional difference absorbance (ΔA) spectra of the probe beam (WLSC) are shown in Figure 4.3.1a. The large positive signal located at the center, around 560 nm, with a delay time of around zero, corresponds to the TPA effect. The black solid line traces the peak of this ΔA map; it shows the zero time position where the pump and probe pulses overlap [17]. The twists in this peak-tracing line are due to the residual chirp of the WLSC and its group velocity dispersion (GVD) in the solvent. The TPA cross section spectrum of coumarin 6 is shown in Figure 4.3.1b. It has a spectral resolution of about 10 nm, which is due to the FWHM of pump spectral shape. For coumarin 6, the result shows a TPA peak located at 652 nm with a cross section of $10^{15} \pm 10^{7}$ GM. This TPA band is in good agreement with the computed NDTPA spectra (Figure 4.3.S2), given the different approximations that cannot be avoided.

The two-dimensional ΔA map for rhodamine 6G dissolved in methanol (concentration: 16.2 mM) is shown in Figure 4.3.2a. Due to losses caused by the stationary absorption of the sample solution, the WLSC probe spectral range is limited to the lowest end at 562 nm. A large positive ΔA signal is observed for the probe wavelength range from 562 to 658 nm. Noticeably, in the same spectral range, some large negative signals appear with delay times longer than 100 fs, and the intensity decreases as the wavelength increases. The comparison of this pattern with its spontaneous fluorescence spectrum confirms that it is the result of the stimulated emission (SE) of rhodamine 6G. This SE can also be observed in the ΔA map of coumarin 6 when the intensity scale is narrowed, although the intensity is too weak to be perceived directly in Figure 4.3.1a.

The TPA is one of the most common third-order nonlinear optical processes, but other third-order nonlinear effects, such as stimulated Raman scattering, have already been well-studied using a similar non-degenerate pump-probe setup [17]. In this work, we stress that a strong stimulated Raman loss [23] (SRL) signal is found in a blank test made of methanol solvent. The details are shown in Figure 4.3.2b. An inverted triangle-shaped signal is located near 645 nm; its intensity distribution has a two-peak structure. These two peaks are caused by the Raman shift of 3000 and 3300 cm^{-1},

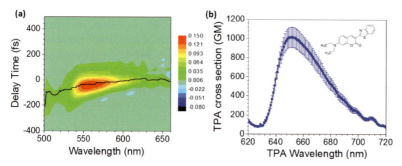

FIGURE 4.3.1 (a) Two-dimensional absorbance change of WLSC through coumarin 6 dissolved in chloroform. (b) TPA cross section spectrum of coumarin 6 in chloroform.

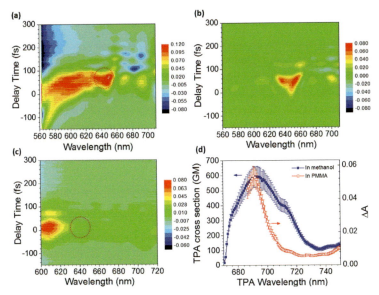

FIGURE 4.3.2 (a) Two-dimensional absorbance change of WLSC through rhodamine 6G dissolved in methanol. (b) Background test for solvent methanol only. (c) Absorbance change of WLSC through rhodamine 6G doped in PMMA film. (d) Measured TPA cross section spectrum of rhodamine 6G in methanol and PMMA.

respectively, which correspond to the $-CH_3$ asymmetric stretch and the $-OH$ stretch. Many common solvents, such as ethanol, DMSO, water, and so on, have similar intense SRL signals in this spectral range. These additional signals must be rigorously accounted for in dye measurements. Carbon tetrachloride (CCl_4) or chloroform ($CHCl_3$) are the recommended solvents for experimental verification as they do not have a large SRL signal around $3000\,cm^{-1}$. Unfortunately, many common dyes such as xanthene dyes do not readily dissolve in such solvents. Therefore, solvents with intense Raman signals in the probe spectral range must sometimes be used, and it is necessary to account for the contribution a Raman solvent makes to the observed TPA spectrum. The comparison of Figure 4.3.2a with Figure 4.3.2b shows that the TPA signal of the sample is contaminated with a Raman loss signal induced in the solvent; the contamination range is marked with a circle of red dots.

It is difficult to distinguish the SRL's contribution to the TPA signal when they are spectrally overlapping, as they are both two-photon processes belonging to comparable third-order nonlinear effects [17,22]. The two incident beams (WLSC (ω_1) and 800 nm (ω_2)) can interact with each other in time overlapping conditions, and both the ω_1 photons and the ω_2 photons can be absorbed during a TPA process. In the same way, when associated with the ω_1 photons, the ω_2 photons can be absorbed (scattered) by the Stokes mechanism or can be amplified by the anti-Stokes mechanism if the vibrational difference level corresponding to the frequency difference ($\omega_1-\omega_2$) is populated. The latter case is not fulfilled in this experiment. Furthermore, coupling between the TPA and Raman processes may occur if the solvent levels in the sample (including at the solvent molecule level) are coherently coupled.

The vibronic states of solvent molecules do not lead to strong coupling with the solute vibronic states due to the lack of strong interaction through hydrogen bonding. Therefore, there is no "intrinsic" interference between the electronic polarization and the solute molecule coherence in the solvent system. However, extrinsic interference is possible. For example, interference may occur if an amplified ω_2 photon created via a stimulated Raman scattering by ω_1 photons is used in the TPA together with ω_1. The Raman gain expected for a ω_2 photon when ω_1 photons exist can be estimated by the Raman correspondence of the solute molecule with a 24.7 M concentration. It is 5×10^{12} photons/cm^{2}s when the number of ω_1 photons is 2×10^{16} photons/cm^{2}s. The loss of the probe light through this process can be completely negligible, and the following equation can be used:

$$\Delta A(\omega) = \Delta A_{TPA}(\omega) + \Delta A_{RS}(\omega) \qquad (4.3.10)$$

Here, ΔA_{TPA} and ΔA_{RS} refer to the TPA and Raman effect, respectively. For the solvent sample, which is a mixture of the solute and solvent, the effect of both should be considered. Especially in this setup, it is expected that the TPA will not create interference, even in the neat solvents, as the absorption edge of the solvent molecules is located much higher (blue side) than half of the shortest edge of the spectrum. However, Raman scattering of the solute scarcely occurs in a relative Raman shift larger than $2000\,\text{cm}^{-1}$. Even if, due to the solute in the sample, the Raman signal contributes, it is expected to have a negligibly low intensity as the concentration of the dye molecules is lower than the concentration of solvent molecules by the factor of 10^{-3}. The concentrations of the solute and solvent in the experiment are calculated to be $24.7\,\text{M}$ and $8.1 \times 10^{-3}\,\text{M}$, respectively. Therefore, from the right part of Eq. (4.3.10), it can be concluded that ΔA_{TPA} is a measure of the solute only, and ΔA_{RS} is a measure of the solvent only.

A polymethyl methacrylate (PMMA) thin film doped with rhodamine 6G is investigated with the expectation of directly removing the solvent interference. PMMA has a large molecular weight and is transparent in the visible and near IR light range; thus, it is expected to be an appropriate matrix to investigate TPA of organic molecules. The PMMA powder and rhodamine 6G are both dissolved in chloroform and mixed with each other; then the mixture is dried and spin-coated on a glass plate. After being stripped from the slide, the film is used for the measurement. The measured ΔA map in Figure 4.3.2c shows that this sample does not have a large ΔA peak around $645\,\text{nm}$ (marked with a red dash circle), which means the Raman background effect is efficiently minimized. However, it is difficult to directly determine the TPA cross section with this doped PMMA film because it is difficult to precisely determine the film thickness and the number of dissolved molecules, as the molecules in the polymer matrix sample are inhomogeneous. This is a serious problem in any kind of spectroscopic measurement of doped molecules in a polymer film. Despite this, the relative values of the TPA cross section recorded at different wavelengths are reliable due to the broadband measurement. This is verified with several scans. Figure 4.3.2b suggests that, for the probe range 580 to $600\,\text{nm}$, the distortion effect from the solvent is negligible. Then, by assuming that rhodamine 6G has the same TPA properties in methanol as in PMMA, we can reconstruct a more reliable TPA cross section spectrum by merging and scaling the result in PMMA to the methanol case. Figure 4.3.2d shows that the measurement starts with the blue round line from $670\,\text{nm}$ and transforms into the red open square line around $690\,\text{nm}$. The optimized TPA peak is located at $691\,\text{nm}$ with a cross section of 596 ± 69 GM. This result is discussed in more detail in the next section.

We use a sample of rhodamine 6G in methanol to study the dependence of the TPA cross section on the pump power and sample concentration. To examine pump power dependence, the sample is fixed at a concentration of $16.2\,\text{mM}$. The pump power is adjusted to between 40 and $840\,\mu\text{W}$ with a neutral density filter. For convenience, we track the ΔA peak values at $620\,\text{nm}$ probe wavelength on each ΔA map acquired by different pump power; they are shown as blue triangle marks in Figure 4.3.3. The linear fitting has a nearly zero intercept, which shows a good agreement with Eq. (4.3.8).

As shown in Figure 4.3.3, there is almost no visible saturation of TPA in this pump power range. For the concentration dependence measurement, the pump power is fixed at $400\,\mu\text{W}$. Several samples with different concentrations are investigated in a similar way: ΔA peak values at $600\,\text{nm}$ and $620\,\text{nm}$ for each concentration are marked with black circles and red round lines in Figure 4.3.3, respectively. The linear fitting for each group shows that the $600\,\text{nm}$ ΔA data has a nearly zero intercept, while the $620\,\text{nm}$ clearly does not. This non-zero intercept evidences the above-mentioned solvent's SRL interference, while zero intercept fitting lines are examples of a purely TPA phenomenon.

Having clarified all these issues, we can further investigate other dyes: rhodamine 123 ($4.2\,\text{mM}$) dissolved in methanol and coumarin 343 ($27.3\,\text{mM}$), Nile red ($0.75\,\text{mM}$), and Nile blue A ($0.68\,\text{mM}$) dissolved in chloroform. The choice for chloroform, despite its low solubility for the chosen solutes, is based on minimum SRL contribution to avoid contamination of the TPA signal. Corresponding TPA cross sections are shown in Figure 4.3.4 (black square dotted line). The rhodamine 6G, rhodamine 123, coumarin 6, coumarin 343, Nile red, and Nile blue A show efficient TPA peaks at 691,

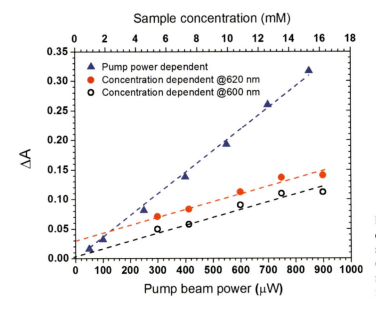

FIGURE 4.3.3 Pump power dependence (blue triangle) and sample concentration dependence (red round and black circle) in rhodamine 6G TPA cross section measurement.

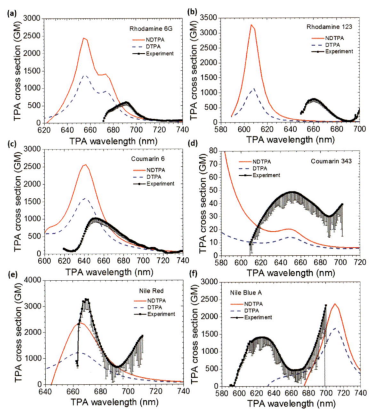

FIGURE 4.3.4 Measured and calculated TPA cross section for (a) Rhodamine 6G, (b) Rhodamine 123, (c) Coumarin, 6 (d) Coumarin 343, (e) Nile red, and (f) Nile blue A. In each figure, black solid line with the error bar is the measured experimental value; calculated DTPA and NDTPA spectrum (800 nm pump) are plotted in blue dashed line and red solid line.

660, 652, 651, 669, and 626 nm, respectively. The corresponded NDTPA cross sections are 596, 776, 1015, 49, 3270, and 1407 GM, respectively.

In general, all the results in this study evidence a larger TPA peak value than those reported using degenerate conditions. Table 4.3.1 compares the rhodamine 6G TPA cross sections obtained

TABLE 4.3.1

Measured Rhodamine 6G TPA Cross Section Value Compared with Previous Literature

References	Wavelength (nm)	TPA Cross Section (GM)	Method	Laser	Pulse Duration
14	694	180±20	NLT[a]	Ruby	15 ps
15	694	355±170	TPIF[b]	Ruby	40 ns
7	690	120	TPIF	Ti:sapphire	100 fs
	700	150			
	720	38			
12	680	55	TPIF	OPA	160 fs
	694	112±12			
	710	94			
This work (include errors)	675.7	337±40	NWLP[c]	WLSC	103 fs[d]
	683.4	486±57			
	699.3	596±69			
	707.4	324±36			
		139±15			

[a] NLT: Nonlinear transmission.
[b] TPIF: Two photon induced fluorescence.
[c] NWLP: Nondegenerate white light probe.
[d] Pump pulse duration.

in this work to previously published values performed using degenerate conditions. The results all agree that the position of the TPA peak is around 690 nm. Obviously, our NDTPA values are larger than the DTPA found in other reports. Noteworthy, ns pulses are known to overestimate TPA cross sections, especially due to excited state absorption (the two-photon absorption can be non-simultaneous). Some studies have reported that the NDTPA is usually enhanced as compared to DTPA [20,25]. This phenomenon can be explained as an intermediate state resonance enhancement (ISRE); one of the photons can have energy close to one of the molecular excitation energies and will achieve (virtual) intermediate state (near) resonance.

To get theoretical confirmation of our experimental results, we have implemented quantum chemical and few states approaches for both linear and nonlinear optical responses of the chromophores of interest. We use density functional theory (DFT) and time-dependent (TD) DFT approaches, as implemented in the Gaussian 03 and 09 packages [26,27]. No simplifications are made for the chemical structures. The properties of interest are related to ground state geometry: that is, geometry optimization and one- and two-photon absorption is related to the electronically excited states (ES). The polarizable continuum model (PCM) as implemented in Gaussian 09 and Gaussian 03 is used to simulate the solvent effects on geometries and optical spectra, respectively. No additional local field corrections are considered [28]. Optical spectra are obtained using the density matrix formalism for nonlinear optical responses as proposed by Tretiak and Chernyak [28,29]. Absolute TPA amplitudes are derived using expression [38] of Ref. [28] for degenerate two-photon absorption (DTPA) considering both diagonal and non-diagonal contributions. For non-degenerate TPA, the TPA cross section is related to the imaginary part of the third-order polarizability $\gamma(-\omega_1; \omega_1,-\omega_2,$ and $\omega_2)$ and the frequency-dependent prefactor ω^2 is replaced by $2\omega_1^2\omega_2/(\omega_1+\omega_2)$ as in Ref. [24], where index 1 refers to the probe beam and index 2 refers to the pump beam. The calculated TPA spectra shown in Figure 4.3.4 are obtained at the TD-B3LYP/6–311+G*//B3LYP/6–311+G* level of theory in conventional quantum chemical notation "single point//optimization level" including up to 20 singlet ES. Given the overall good agreement between the experimental and theoretical OPA band positions (Figure 4.3.S1), NDTPA spectra were computed for the experimental pump wavelength only (1.55 eV/800 nm). The damping factor is introduced to simulate the finite line width Γ in the resonant spectra. The ES structure was further checked at the TD-ω B97×D/6–311+G*// B3LYP/6–311+G* level.

The calculated TPA cross sections are significantly larger than the experimental results shown in Figure 4.3.4. This may be attributed to a number of factors related to the level of theory in use. First, B3LYP, the most suitable exchange-correlation functional in Gaussian 03 for optical properties, is known to overestimate conjugation and thus reduce bond length alternation and overestimate transition dipole moments [30]. For example, if the TPA cross section scales as the fourth power of dipole moment matrix elements, a 15% overestimation of the dipole moments may double the TPA cross section. Next, the choice for the finite line width, which is set the same for all ES, directly affects the TPA amplitudes. When the two-photon excited state is near resonance, the TPA amplitude scales as $1/\Gamma$ (Figure 4.3.S2). In addition to local field corrections such as dynamic contributions [28], other contributing factors include all of those currently considered in predictions of linear optical properties (band shape, amplitude, and position) of solvated chromophores [31]. Specifically, the present solvation model is limited and does not account for state-specific responses [32], or for explicit solvent molecules or counter-ions. Furthermore, vibronic contributions [33] are not simulated. Despite all of these approximations, this level of theory has already proved efficient for rationalizing experimental TPA spectra [28,32]. The OPA spectra are systematically blue-shifted compared to the experimental results. However, compared to the TD-B3LYP/6–31G(d)//HF/6–31G(d) level of theory in a vacuum, both the solvent and a larger basis set improve the agreement between calculated and experimental results. Given the small spectral window investigated experimentally, to be compared with the error in the calculated peak position (vide supra), these TPA spectra reveal a TPA band that is close to that experimentally observed. These bands, which have significant magnitude, are related to one (or several) higher lying excited state(s). Comparison between DTPA and NDTPA calculated spectra (Figure 4.3.4) clearly demonstrates the intermediate state resonance enhancement experimentally observed, which depends on the excited state structure of the particular chromophore. Further intuitive understanding may be gained from few-states models. Besides the larger ISRE for rhodamines as compared to coumarins, the various transition energies taken for the TPA states for Nile Red and Nile blue A (Table 4.3.S2) evidence the strong influence of respective transition energies compared to pump and probe photon energies on the overall enhancement factor.

4.3.5 CONCLUSION

In this study, we improve the method for measuring non-degenerate TPA cross sections by introducing an MLA system into the detection setup. With this system, high resolution and precise chirping correction are achieved using a ΔA map. However, this method still requires further improvement. First, many solvents, such as water and methanol, contribute a strong Raman loss signal with a Raman shift around 3000 is $\sim 3300\,cm^{-1}$. This disturbance can be minimized by selecting an appropriate solvent and by properly correcting for the Raman effect. Second, the TPA spectrum detection range is limited to between 600 and 760 nm, when using 800 nm pumps, as in this study. This problem can be solved by choosing a solvent like chloroform or by changing the pump beam wavelength to a variable range with an OPA setup. Another feasible and easily available solution is to use a much wider WLSC, such as one generated by tapped fiber or photon crystal fiber [34].

This study measures the TPA spectrum for several types of laser dyes including xanthene dyes (rhodamine 6G and rhodamine 123), coumarin dyes (coumarin 6 and coumarin 343), and oxazine dyes (Nile red and Nile blue A). By comparing the results for our analysis of rhodamine 6G to published degenerate measurements, we confirm that TPA is enhanced in non-degenerate cases. We verify this conclusion with theoretical calculations. To the best of our knowledge, this is the first report of measurements for the other laser dyes sampled here. This method makes the measurement of TPA spectrum of materials much more convenient. It contributes to the new TPA materials development and has already been applied by our collaborators [35]. The work presented here is based on collaboration among the following people [36]: B. Xue, C. Katan, J. A. Bjorgaard, and T. Kobayashi.

REFERENCES

1. G. S. He, J. D. Bhawalkar, C. F. Zhao, and P. N. Prasad, *Appl. Phys. Lett.* **67**, 2433 (1995).
2. M. A. M. Versteegh and J. I. Dijkhuis, *Opt. Lett.* **36**, 2776 (2011).
3. W. R. Zipfel, R. M. Williams, and W. W. Webb, *Nat Biotech* **21**, 1369 (2003).
4. W. Denk, *Proc. Natl. Acad. Sci. USA* **91**, 6629 (1994).
5. S. W. Hell, M. Booth, S. Wilms, C. M. Schnetter, A. K. Kirsch, D. J. Arndt-Jovin, and T. M. Jovin, *Opt. Lett.* **23**, 1238 (1998);
6. A. Hayat, A. Nevet, P. Ginzburg, and M. Orenstein, *Semicond. Sci. Tech.* **26**, 083001 (2011)
7. B. Xu and W. W. Webb, *J. Opt. Soc. Am. B* **13**, 481 (1996).
8. M. A. Albota, C. Xu, and W. W. Webb, *Appl. Opt.* **37**, 7352 (1998).
9. Y. Tan, Q. Zhang, J. Yu, X. Zhao, Y. Tian, Y. Cui, X. Hao, Y. Yang, and G. Qian, *Dyes and Pigments* **97**, 58 (2013)
10. A. Nag, A. K. De, and D. Goswami, *J. Phys. B: At. Mol. Opt. Phys.* **42**, 065103 (2009).
11. Y. Xia, Y. Jiang, R. Fan, Z. Dong, W. Zhao, D. Chen, and G. Umesh, *Opt. Laser Technol.* **41**, 700 (2009).
12. N. S. Makarov, M. Drobizhev, and A. Rebane, *Opt. Express* **16**, 4029 (2008).
13. M. Drobizhev, S. Tillo, N. S. Makarov, T. E. Hughes, and A. Rebane, *J. Phys. Chem. B* 113, 855 (2009).
14. P. Sperber and A. Penzkofer, *Opt. Quant. Electron.* **18**, 381 (1986).
15. J. P. Hermann and J. Ducuing, *Opt. Commun.* **6**, 101 (1972).
16. K. D. Belfield, D. J. Hagan, E. W. Van Stryland, K. J. Schafer, and R. A. Negres, *Org. Lett.* **1**, 1575 (1999).
17. S. A. Kovalenko, A. L. Dobryakov, J. Ruthmann, and N. P. Ernsting, *Phys. Rev. A* **59**, 2369–2383 (1999).
18. C. M. Cirloganu, L. A. Padilha, D. A. Fishman, S. Webster, D. J. Hagan, and E. W. Van Stryland, *Opt. Express* **19**, 22951 (2011).
19. L. A. Padilha, J. Fu, D. J. Hagan, E. W. Van Stryland, C. L. Cesar, L. C. Barbosa, C. H. B. Cruz, D. Buso, and A. Martucci, *Phys. Rev. B* **75**, 075325 (2007).
20. G. S. He, L.-S. Tan, Q. Zheng, and P. N. Prasad, *Chem. Rev.* **108**, 1245 (2008).
21. Y. R. Shen, *The Principles of Nonlinear Optics*, J. Wiley, New York, p. 203 (1984).
22. K. W. DeLong, R. Trebino, J. Hunter, and W. E. White, *J. Opt. Soc. Am. B* **11**, 2206 (1994).
23. B. Mallick, A. Lakhsmanna, and S. Umapathy, *J. Raman Spectrosc.* **42**, 1883 (2011).
24. J. M. Hales, D. J. Hagan, E. W. Van Stryland, K. J. Schafer, A. R. Morales, K. D. Belfield, P. Pacher, O. Kwon, E. Zojer, and J. L. Bredas, *J. Chem. Phys* **121**, 3152 (2004).
25. E. Roussakis, J. A. Spencer, C. P. Lin, and S. A. Vinogradov, *Anal. Chem.* **86**, 5937 (2014).
26. Gaussian 03, *Revision D. 02*, M. J. Frisch, G. W. Trucks, H. B. Schlegel, G. E. Scuseria, M. A. Robb, J. R. Cheeseman, J. A. Montgomery, Jr., T. Vreven, K. N. Kudin, J. C. Burant, J. M. Millam, S. S. Iyengar, J. Tomasi, V. Barone, B. Mennucci, M. Cossi, G. Scalmani, N. Rega, G. A. Petersson, H. Nakatsuji, M. Hada, M. Ehara, K. Toyota, R. Fukuda, J. Hasegawa, M. Ishida, T. Nakajima, Y. Honda, O. Kitao, H. Nakai, M. Klene, X. Li, J. E. Knox, H. P. Hratchian, J. B. Cross, V. Bakken, C. Adamo, J. Jaramillo, R. Gomperts, R. E. Stratmann, O. Yazyev, A. J. Austin, R. Cammi, C. Pomelli, J. W. Ochterski, P. Y. Ayala, K. Morokuma, G. A. Voth, P. Salvador, J. J. Dannenberg, V. G. Zakrzewski, S. Dapprich, A. D. Daniels, M. C. Strain, O. Farkas, D. K. Malick, A. D. Rabuck, K. Raghavachari, J. B. Foresman, J. V. Ortiz, Q. Cui, A. G. Baboul, S. Clifford, J. Cioslowski, B. B. Stefanov, G. Liu, A. Liashenko, P. Piskorz, I. Komaromi, R. L. Martin, D. J. Fox, T. Keith, M. A. Al-Laham, C. Y. Peng, A. Nanayakkara, M. Challacombe, P. M. W. Gill, B. Johnson, W. Chen, M. W. Wong, C. Gonzalez, and J. A. Pople, Gaussian, Inc., Wallingford CT (2004).
27. Gaussian 09, *Revision A. 02*, M. J. Frisch, G. W. Trucks, H. B. Schlegel, G. E. Scuseria, M. A. Robb, J. R. Cheeseman, G. Scalmani, V. Barone, B. Mennucci, G. A. Petersson, H. Nakatsuji, M. Caricato, X. Li, H. P. Hratchian, A. F. Izmaylov, J. Bloino, G. Zheng, J. L. Sonnenberg, M. Hada, M. Ehara, K. Toyota, R. Fukuda, J. Hasegawa, M. Ishida, T. Nakajima, Y. Honda, O. Kitao, H. Nakai, T. Vreven, J. A. Montgomery, Jr., J. E. Peralta, F. Ogliaro, M. Bearpark, J. J. Heyd, E. Brothers, K. N. Kudin, V. N. Staroverov, R. Kobayashi, J. Normand, K. Raghavachari, A. Rendell, J. C. Burant, S. S. Iyengar, J. Tomasi, M. Cossi, N. Rega, J. M. Millam, M. Klene, J. E. Knox, J. B. Cross, V. Bakken, C. Adamo,

J. Jaramillo, R. Gomperts, R. E. Stratmann, O. Yazyev, A. J. Austin, R. Cammi, C. Pomelli, J. W. Ochterski, R. L. Martin, K. Morokuma, V. G. Zakrzewski, G. A. Voth, P. Salvador, J. J. Dannenberg, S. Dapprich, A. D. Daniels, Ö. Farkas, J. B. Foresman, J. V. Ortiz, J. Cioslowski, and D. J. Fox, Gaussian, Inc., Wallingford CT (2009).

28. F. Terenziani, C. Katan, E. Badaeva, S. Tretiak, and M. Blanchard-Desce, *Adv. Mat.* **20**, 4641 (2008).

29. S. Tretiak and V. Chernyak, *J. Chem. Phys.* **119**, 8809 (2003).

30. L. Ji, R. M. Edkins, L. J. Sewell, A. Beeby, A. S. Batsanov, K. Fucke, M. Drafz, J. A. K. Howard, O. Moutounet, F. Ibersiene, A. Boucekkine, E. Furet, Z. Liu, J.-F. Halet, C. Katan, and T. B. Marder, *Chem. Eur. J.* 2014(20), 13618–13635.

31. D. Jacquemin and C. Adamo, Computational Molecular Electronic Spectroscopy with TD-DFT, in *Topics in Current Chemistry*, Springer, Berlin/Heidelberg, pp. 1–29 (2015).

32. C. Katan, P. Savel, B. M. Wong, T. Roisnel, V. Dorcet, J.-L. Fillaut, and D. Jacquemin, *Phys. Chem. Chem. Phys.* **16**, 9064–9073 (2014).

33. W. Liang, H. Ma, H. Zang, and C. Ye, *Int. J. Quantum Chem.* **115**, 550–563 (2015).

34. J. Cascante-Vindas, A. Díez, J. L. Cruz, and M. V. Andrés, *Opt. Express* **18**, 14535 (2010).

35. S. Boinapally, B. Huang, M. Abe, C. Katan, J. Noguchi, S. Watanabe, H. Kasai, B. Xue, and T. Kobayashi, *J. Org. Chem.* **79**, 7822 (2014).

36. B. Xue, C. Katan, J. A. Bjorgaard, and T. Kobayashi, *AIP Advances* **5**, 127138 (2015).

Section 5

Multi-Color Involved NLO Processes

5.1 Generation of μJ-Level Multicolored Femtosecond Laser Pulses Using Cascaded Four-Wave Mixing

5.1.1 INTRODUCTION

Over the past decade, μJ-level tunable femtosecond laser pulse system in visible based on three-wave mixing has been well established using noncollinear optical parametric amplifier (NOPA) and has been widely used in pump-probe experiments [1–3]. Recently, four-wave mixing (FWM) has become a hot research spot [4–14]. Ultra-broadband spectrum and ultrashort pulse were generated in various optically transparent media using FWM [4–14]. Tunable visible ultrashort pulse was generated using FWM through filamentation in a gas cell [4]. Few-cycle femtosecond ultrashort pulses in deep UV and mid-IR were also generated by this method [5,7]. However, the energy conversion efficiency was low in a gas medium due to the low nonlinear coefficient. In the case of bulk solid-state media, phase matching will obtain only if the pump beams have a finite crossed angle due to high material dispersion. By using ps pump pulses, high efficiency and high-energy noncollinear four-wave optical parametric amplification in a transparent bulk Kerr medium were observed [9–10]. Multicolored sidebands were generated in a BK7 glass [8], a BBO crystal [6], and a sapphire plate [13] using two crossed femtosecond laser beams.

In our previous report [11], tunable multicolored femtosecond laser pulses were generated simultaneously in a fused silica glass plate. These multicolored femtosecond laser pulses can be used in many experiments, for example femtosecond CARS spectroscopy [15], two-dimensional spectroscopy [16] and some high-intensity laser experiments [17], where two or more femtosecond pulses at different wavelength are needed. However, the pulse energy of sideband was less than 200 nJ, which limited its application in many fields.

In this subsection, μJ-level femtosecond pulses with more than 10 different wavelengths were obtained by optimizing the spectrum and input power of one of the two input beams and improving the spatial and temporal profile of the two input beams. By changing the crossed angle between the two input beams on the glass plate, the wavelengths of the generated pulses were tunable. The properties of the generated sidebands showed that they could be used for multicolor pump-probe experiments and frequency and spatial mode entangled photon generation.

5.1.2 EXPERIMENTAL SETUP

A 1 kHz Ti:sapphire regenerative amplifier femtosecond laser system (Micra+Legend-USP, Coherent) with 40 fs pulse duration and 2.5 W average power was used as a pump source. The laser pulse after the regenerative amplifier was split into several beams using beam splitters. One of the beams (beam1) was spectrally broadened in a 60-cm hollow fiber with a 250 μm inner diameter and that was filled with krypton gas. The broadband spectrum after the hollow fiber was dispersion compensated with a pair of chirped mirrors and a pair of glass wedges. The pulse duration after the hollow fiber compressor was about 10 fs. After passing through a bandpass filter at 700 nm center wavelength with 40 nm bandwidth, beam1 was focused into a 1-mm-thick fused silica glass by a

DOI: 10.1201/9780429196577-28

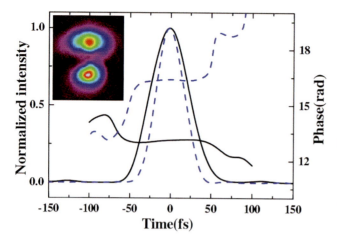

FIGURE 5.1.1 The input pulse duration and phase of beam1 (dashed lines) and beam2 (solid lines). The inset patterns were the two-dimensional beam profile of beam1 (below) and beam2 (upper) on the surface of the fused silica glass.

concave mirror. Another beam (beam2) passed through a delay stage with less than 3 fs resolution. Beam2 was first attenuated by a reflective variable neutral density (VND) filter and was then focused into the fused silica glass by a lens. The third beam (beam3) was used to measure pulse durations by XFROG with the two input beams via sideband generation in a 10-μm-thick BBO crystal.

Here ultrashort pule characterization methods of FROG (frequency-resolved optical gating) and XFROG (cross-correlation frequency-resolved optical gating) are explained briefly as follows.

Frequency-resolved optical gating (FROG) is now a general method for measuring the spectral phase of ultrashort pulse ranging from subfemtosecond to nearly nanosecond in time length. It is based on a spectrally-resolved autocorrelation. XFROG is a method to characterize the spectral phase of weak pulse utilizing an intense pulse with known phase information by cross-correlation.

Figure 5.1.1 depicts a retrieved XFROG trace showing the pulse durations of beam1 and beam2 being 40 ± 3 and 55 ± 3 fs, respectively. The nearly equal pulse durations of the two input pulses indicate them to be temporally well overlapped. The retrieved wavelength-dependent phase shows that beam2 has small positive chirp due to the dispersion of beam splitters and a lens. The spatial mode of the two input beams on the surface of the fused silica glass was measured by a CCD camera (BeamStar FX 33, Ophir Optronics), as shown in the inset of Figure 5.1.1. The beams with elliptical cross sections were obtained by adjusting the beams at the edges of both the lens and the concave mirror. Then, the two beams were aligned to well overlap by monitoring with the CCD. The elliptic beams make the two input beams could be overlapped well in the medium even if there was a crossed angle between them. The glass was not damaged during the whole experiment because the intensity on the surface is one order lower than the optical breakdown threshold intensity of fused silica for femtosecond pulse.

5.1.3 EXPERIMENTAL RESULTS AND DISCUSSION

Multicolored cascaded FWM signals appeared separately in space outside of the two input beams of beam1 and beam2 were synchronously focused on the fused silica glass in both time and space. The photograph on the top of Figure 5.1.2a shows the FWM sideband signals on a white sheet of paper placed about 30 cm after the glass plate. The input powers of beam1 and beam2 were 9 and 20 mW, respectively. An optical fiber was used to pick up different order signals to measure their spectra using a multi-channel spectrometer (USB4000, Ocean Optics). Figure 5.1.2a shows the spectra of the wavelengths of the sidebands from the first-Stokes (S1) to the fourth-order anti-Stokes (AS4) cascaded FWM signals together with the spectra of two input beams when the crossing angle between the two input beams was 1.87°. The spectrum extends from 450 to 1000 nm. The sidebands have a Gaussian profile, and each anti-Stokes spectrum can support a transform-limited

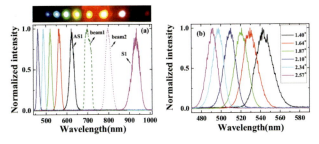

FIGURE 5.1.2 (a) The spectra of sidebands from S1 to AS5 and two input beams from right to left. when the crossing angle between the two input beams was 1.87°. (b) The spectra of AS3 at six different crossing angles of 1.40°, 1.64°, 1.87°, 2.10°, 2.34°, and 2.57° from right to left. The photograph on the top of Figure 5.1.2a shows the sidebands on a sheet of white paper 30 cm after the glass plate when the crossing angle between the two input beams was 1.87°. The first, second, and third ones on the right side are S1, beam2, and beam1, respectively.

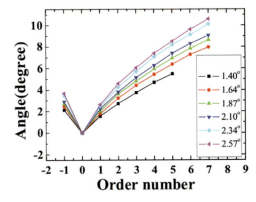

FIGURE 5.1.3 The dependence of the angles between the generated sidebands and beam2 (0 order number) on the order number of the sidebands when the crossed angle between the two input beams were 1.40°, 1.64°, 1.87°, 2.10°, 2.34°, and 2.57° from bottom to top. −1 refers to S1, 1 refers to AS1, and so on.

pulse duration of about 25 fs. Moreover, the wavelengths of the sidebands can be tuned by changing the crossing angle between the two input beams. In the experiment, beam2 was fixed and the crossing angle was changed by changing the position of beam1 on the surface of concave mirror before the fused silica glass. It can be seen from Figure 5.1.2b that the peak wavelength of AS3 can be tuned from 490 to 545 nm by changing the crossing angles from 1.40° to 2.57°. Compared with the previous work [11], the wavelength tunable region was narrower due to the narrower spectral bandwidth of beam2. A white plane was located 30 cm after the fused silica glass to mark the position of sidebands at different crossed angles. In this way, the crossing angles between the generated sidebands and beam2 were recorded, as shown in Figure 5.1.3. The crossing angles between the two neighboring sidebands were decreased as the order number increased for a given crossing angle between the two beams. The crossing angle between each of the sidebands and beam2 was increased with the crossing angle of the incident beams. The slope representing the dependence becomes steeper with the order number increase.

As for the cascaded FWM, the m-order sideband should obey the energy conservation and momentum conservation laws: $(m+1)\omega_1 - m\omega_2 = \omega_{ASm}$ and $k_{ASm} - (k_1 - k_2) = k_{AS(m-1)}$, respectively. Here, 2 and 1 refer to the two input laser beams, ASm refers to the generated m-order anti-Stokes sideband. As for m-order Stokes sideband, the energy conservation and momentum conservation laws were $(m+1)\omega_2 - m\omega_1 = \omega Sm$ and $kSm - (k_2 - k_1) = kS(m-1)$, respectively.

The output power of AS1 and S1 were 1.03 and 1.05 μJ, respectively, when the crossing angle between the input beams was 1.87° and the input power of beam1 and beam2 were 9 and 20 mW, respectively. The output power of sidebands is related to the crossing angle between the two input beams and the order number of sidebands. Figure 5.1.4a shows the dependence of the output power on order number when the crossed angles between the two input beams were 1.40°, 1.87°, 2.10°, and 2.57°. The output powers of S1 and AS1 decreased rapidly with the increase in the crossing

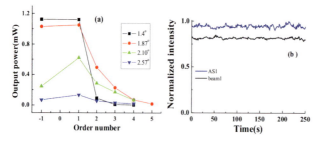

FIGURE 5.1.4 (a) The dependence of the output power on order number when the crossing angles between the two input beams were 1.40° (squares), 1.87° (circle), 2.10° (upward triangle), and 2.57° (downward triangle). −1 refers to S1, 1 refers to AS1, and so on. (b) The power stabilities in terms of standard deviation of AS1 (upper line) and beam1 (lower line) monitored over 4 min, which were determined to be 1.82% RMS and 0.97% RMS, respectively.

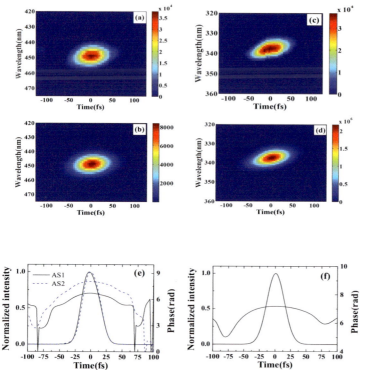

FIGURE 5.1.5 (a) and (b) The measured and the retrieved XFROG traces of S1, respectively; (c) and (d) are the measured and the retrieved XFROG traces of AS2, respectively, for the crossing angle of 1.87°. (e) The recovered intensity profiles extending from −25 to +25 fs. and phase extending from about −75 fs to +75 fs of AS1 (solid line) and AS2 (dashed line) with retrieval errors were 0.011888 and 0.0097389, respectively. (f) The recovered pulse profile extending from −25 to +25 fs and phase extending from about −75 fs to +75 fs of S1 with a retrieval error was 0.0049263.

angle from 1.40° to 2.57°. For a smaller crossing angle between the two input beams, for example 1.40°, the output power of sidebands decreased with the increase in the order number. However, for a larger crossing angle between the two input beams, for example 2.57°, the output power of sidebands decreased slowly with an increase in the order number. The energy conversion efficiency from the input beams to sideband beams in the process was about 10% when the crossing angle between the two input beams was 1.87°. When the input power of beam2 was decreased to 15 mW, the output power of AS1 was about 0.95 µJ. Even in such a situation, the output power of beam2 did not change in the process. These phenomena indicate that the output power of sidebands did not sensitive to the input power of beam2, the same as the previous report. The standard deviation of fluctuation AS1 and beam1 power was monitored over 4 min, which were 1.82% RMS and 0.97% RMS, respectively, as shown in Figure 5.1.4b.

The pulse duration of S1, AS1, and AS2 were measured by generating cross-correlation signals with beam3 and then retrieved the XFROG trace using the commercial FROG software from Femtosoft Technologies. Figure 5.1.5a and b show the measured and the retrieved XFROG traces, respectively, of S1 when the crossed angle was 1.87°. The corresponding traces of AS2 are also

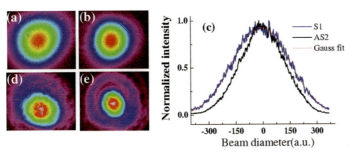

FIGURE 5.1.6 (a) and (b) The two-dimensional spatial modes of S1 and AS3, respectively. (c) The one-dimensional spatial profile of S1 (thick faint curve) and AS3 (thick bold curve). Thin faint line is a Gaussian fit curve of S1. (d) and (e) The two-dimensional spatial mode of beam2 when beam1 was blocked and the input power of beam1 was 20 mW.

shown in Figure 5.1.5c and d. The recovered intensity profiles and phases of AS1 and AS2 are depicted in Figure 5.1.5e with retrieved errors were 0.011888 and 0.0097389, respectively. The recovered pulse duration of AS1 and AS2 were 45±3 and 44±3 fs, respectively. Figure 5.1.5f also shows the recovered pulse profile and phase of S1 with a retrieved error of 0.0049263. The recovered pulse duration was 46±3 fs. All the generated sidebands have similar pulse duration. The retrieved phase shows that there was a chirp in the pulses due to the small positive chirp of input pulses and the dispersion of the glass. Transform limited sidebands may be obtained when the input pulses are small negative chirped with the same absolute value to compensate the dispersion of the glass.

The spatial modes of sidebands were also measured by the CCD camera. All the sidebands have a Gaussian spatial profile even though the two input beams have elliptic cross sections. Figure 5.1.6a and b show the two-dimensional spatial modes of S1 and AS3, respectively. Figure 5.1.6c shows the one-dimensional spatial profile of S1 and AS3. A Gaussian fit curve of S1 is also shown in Figure 5.1.6c indicating the sidebands have a nearly perfect Gaussian profile. The spot size of AS1 with a lens of 700 mm focal length was measured to be less than 1.1 times the diffraction limit. Weak diffraction-like rings around AS1 and AS2 were observed with a screen of white paper. Interestingly, beam2 was focused and its spatial mode changed from elliptic to circular in cross section during the propagating process. Figure 5.1.6d and e show the two-dimensional spatial mode of beam2 when beam1 was shielded and the input power of beam1 was 20 mW, respectively. This spatial mode improvement may be due to the combined effects of FWM, coherent anti-Stokes Raman scattering (CARS), cross-phase modulation (XPM), and self-phase modulation (SPM), which lead to multicolor solitons in a Kerr planar waveguide [18–19].

5.1.4 CONCLUSION

In conclusion, μJ-level multicolored femtosecond pulses were obtained at the same time by a cascaded FWM process: This spatial mode improvement may be due to the combined effects of FWM, coherent anti-Stokes Raman scattering (CARS), cross-phase modulation (XPM), and self-phase modulation (SPM), in a fused silica glass. About 45 fs laser pulse was obtained around 950 nm [20]. The energy conversion efficiency from the input beams to sidebands was about 10%. The properties of the generated sidebands made them suited to various experiments, for example multicolor pump-probe experiment. The work described in this subsection is the collaboration of Jun Liu and Takayoshi Kobayashi [20].

REFERENCES

1. T. Kobayashi, A. Shirakawa, and T. Fuji, "Sub-5-fs transform-limited visible pulse source and its application to real-time spectroscopy," *IEEE J. Select. Topics Quantum Electron* **7**, 525–538 (2001).

2. A. Baltuška, T. Fuji, and T. Kobayashi, "Visible pulse compression to 4 fs by optical parametric amplification and programmable dispersion control," *Opt. Lett.* **27**, 306–308 (2002).

3. G. Cerullo and S. De Silvestri, "Ultrafast optical parametric amplifiers," *Rev. Sci. Inst.* **74**, 1–18 (2003).

4. F. Théberge, N. Aközbek, W. Liu, A. Becker, and S. L. Chin, "Tunable ultrashort laser pulses generated through filamentation in gases," *Phys. Rev. Lett.* **97**, 023904 (2006).

5. T. Fuji, T. Horio, and T. Suzuki, "Generation of 12 fs deep-ultraviolet pulses by four-wave mixing through filamentation in neon gas," *Opt. Lett.* **32**, 2481–2483 (2007).

6. J. Liu, J. Zhang, and T. Kobayashi, "Broadband coherent anti-Stokes Raman scattering light generation in BBO crystal by using two crossing femtosecond laser pulses," *Opt. Lett.* **33**, 1494–1496 (2008).

7. T. Fuji and T. Suzuki, "Generation of sub-two-cycle mid-infrared pulses by four-wave mixing through filamentation in air," *Opt. Lett.* **32**, 3330–3332 (2007).

8. H. Crespo, J. T. Mendonça, and A. Dos Santos, "Cascaded highly nondegenerate four-wave-mixing phenomenon in transparent isotropic condensed media," *Opt. Lett.* **25**, 829–831 (2000).

9. H. Valtna, G. Tamošauskas, A. Dubietis, and A. Piskarskas, "High-energy broadband four-wave optical parametric amplification in bulk fused silica," *Opt. Lett.* **33**, 971–973 (2008).

10. A. Dubietis, G. Tamošauskas, P. Polesana, G. Valiulis, H. Valtna, D. Faccio, P. Di Trapani, and A. Piskarskas, "Highly efficient four-wave parametric amplification in transparent bulk Kerr medium," *Opt. Express* **15**, 11126–11132 (2007), http://www.opticsinfobase.org/abstract.cfm?URI=oe-15-18-11126.

11. J. Liu and T. Kobayashi, "Wavelength-tunable multicolored femtosecond laser pulses generation in a fused silica glass plate," *Opt. Rev.* **17**, 275–2821 (2010).

12. M. Zhi and A. V. Sokolov, "Broadband coherent light generation in a Raman-active crystal driven by two-color femtosecond laser pulses," *Opt. Lett.* **32**, 2251–2253 (2007).

13. J. Liu and T. Kobayashi, "Cascaded four-wave mixing and multicolored arrays generation in a sapphire plate by using two crossing beams of femtosecond laser," *Opt. Express*, **16**, 22119–22125 (2008).

14. H. Crespo and R. Weigand, "Cascaded four-wave mixing technique for high-power few-cycle pulse generation," in *XVI International Conference on Ultrafast Phenomena*, UP (2008) Paper Frilp-5.

15. D. Pestov, R. K. Murawski, G. O. Ariunbold, X. Wang, M. C. Zhi, A. V. Sokolov, V. A. Sautenkov, Y. V. Rostovtsev, A. Dogariu, Y. Huang, and M. O. Scully, "Optimizing the laser-pulse configuration for coherent Raman spectroscopy," *Science*, **316**, 265–268 (2007).

16. R. M. Hochstrasser, "Two-dimensional spectroscopy at infrared and optical frequencies," *PNAS* **104**, 14190–14196 (2007).

17. R. Zgadzaj, E. Gaul, N. H. Matlis, G. Shvets, and M. C. Downer, "Femtosecond pump-probe study of preformed plasma channels," *J. Opt. Soc. Am. B* **21**, 1559–1567 (2004).

18. P. B. Lundquist, D. R. Andersen, and Y. S. Kivshar, "Multicolor solitons due to four-wave mixing," *Phys. Rev. E* **57**, 3551–3555 (1998).

19. G. Fanjoux, J. Michaud, M. Delqué, H. Maillotte, and T. Sylvestre, "Generation of multicolor vector Kerr solitons by cross-phase modulation, four-wave mixing, and stimulated Raman scattering," *Opt. Lett.* **31**, 3480–3482 (2006).

20. J. Liu and T. Kobayashi, "Generation of µJ-level multicolored femtosecond laser pulses using cascaded four-wave mixing," *Opt. Exp.* **17**, 4984–4990 (2009).

5.2 Generation and Optimization of Femtosecond Pulses by Four-Wave Mixing (FWM) Process

5.2.1 INTRODUCTION

Ultrashort laser pulses are powerful tools for laser spectroscopic techniques that are widely used in all fields of science (including chemistry, physics, and biology) and that provide microscopic insights into bulk materials, molecules, and chemical and biochemical reactions [1–4]. Advances in ultrashort-laser-pulse technology have made it possible to generate sub-10-fs pulses with a wavelength of 800 nm and nanojoule pulse energies using Ti:sapphire lasers [5]. Such pulses can be compressed to below 5 fs by using spectral broadening in fibers in combination with dispersion compensation [6]. By using gas-filled hollow fibers or filament compressors, sub-10-fs pulses with high pulse energies can be produced at wavelengths of 800 and 400 nm with kilohertz repetition rates [7–10]. These ultrashort pulses permit the detection of real-time electronic, phonon, and vibrational dynamics in various molecular systems and bulk materials with an extremely high temporal resolution. On the other hand, wavelength-tunable femtosecond pulses are required in various studies of ultrafast phenomena. Over the past decade, wavelength-tunable femtosecond lasers with wavelengths ranging from ultraviolet (UV) to mid-IR have been developed by using three-wave mixing in various nonlinear crystals [11–19]. In particular, wavelength-tunable few cycle pulses have been generated at visible wavelengths using noncollinear optical parametric amplifier (NOPA) based on beta barium borate crystals. These pulses have been widely used in pump–probe experiments [20–24]. Femtosecond mid-IR pulses can be generated by difference frequency generation in nonlinear crystals [17,18,25], and they have been widely used in 2D-IR spectroscopy [26–28]. By achromatic broadband frequency doubling these NOPA pulses, wavelength tunable sub-10-fs pulses can be generated in the spectral region of 275–335 nm [29].

Four-wave mixing (FWM) has recently been investigated in various optically transparent media as a new method for generating tunable ultrashort pulses over an ultrabroad spectral range. Tunable visible ultrashort pulses have been generated by FWM through filament generation in an argon-filled gas cell [30]. In addition, femtosecond pulses in the deep UV (DUV) and mid-IR have been generated by FWM through filamentation in a gas cell [31–34]. Pulses at various UV wavelengths have been generated by cascaded FWM in hollow fibers filled with noble gases [35,36]. Sub-10-fs DUV pulses have also been generated by FWM and third-harmonic generation in gaseous media [37,38].

It was found that ultrabroad spectra and wavelength-tunable ultrashort pulses could be generated in bulk media by FWM if the two pump beams have a finite crossing angle in the medium [39–61]. Wavelength-tunable mid-IR pulses could be obtained in the range of 2.4–12 μm by FWM in CaF_2 and BaF_2 plates [39,40]. A ~30-fs idler pulse at 300 nm was obtained by four-wave optical parametric chirped-pulse amplification in a fused silica plate [41]. In the visible region, spatially separated cascaded FWM multicolored sidebands have been generated in BK7 glass [42–44], fused silica [45–48], and a sapphire plate [49,50]. Up to 15 sidebands can be obtained, and the spectrum of the generated sidebands can extend over more than 1.5 octaves from UV to near-infrared (NIR) [44,45]. Multicolored sidebands have also been observed in many nonlinear crystals [51–61].

DOI: 10.1201/9780429196577-29

This phenomenon has been explained in terms of different-frequency resonant FWM, and it is known as cascaded stimulated Raman scattering or coherent anti-Stokes Raman scattering. By combining these sidebands into a single beam, isolated 25- and 13-fs pulses were obtained in LiNbO$_3$ and KTaO$_3$ crystals, respectively [53,54]. It is expected that these multicolored sidebands with broadband spectrum can be used to generate near-single-cycle pulses [43,44]. These multicolored femtosecond pulses can be conveniently used in multicolored pump–probe experiments. The generated multicolored sidebands contain very similar wavelengths as the emission wavelengths of various fluorescent proteins, such as green fluorescent protein (GFP), cyan fluorescent protein (CFP), and red fluorescent protein (RFP) [62] and some semiconductor quantum dots [63,64]. They could, thus, be used for simultaneous multicolored imaging of biological samples by nonlinear optical microscopy [65–67]. In addition to these FWM processes, by inducing another intense pump pulse, a weak seed pulse can be amplified by noncollinear four-wave optical parametric amplification (FWOPA) in a transparent bulk Kerr medium in the UV and NIR spectral regions [68–74].

This paper presents the research that we have done in which we used FWM to generate and optimize femtosecond laser pulses. It is organized as follows. First, the mechanism of cascaded FWM is presented. Second, we discuss the generation of wavelength-tunable multicolored 15-fs pulses by nondegenerate cascaded FWM in a bulk medium. Self-diffraction (SD) (also known as degenerate cascaded FWM) is described as a powerful method for cleaning femtosecond laser pulses. Then, we describe using FWM in a gas cell or a hollow fiber to generate ultrashort pulses in the UV region. Finally, FWOPA is described and used to simultaneously amplify and compress generated FWM signals. Finally, conclusions and future prospects are given.

5.2.2 CASCADED FWM IN BULK MEDIA

UV pulses were generated by cascaded FWM in a gas-filled hollow fiber with a collinear configuration in 2001 [36]. Bulk media have large dispersions compared with gaseous media. To generate cascaded FWM signals, there should be a small crossing angle between the two incident beams in the medium so as to satisfy the phase-matching condition.

5.2.2.1 PRINCIPLE OF CASCADED FOUR-WAVE MIXING (FWM)

Cascaded FWM processes are schematically depicted in Figure 5.2.1a. The two input beams have wave vectors k_1 and k_2 that have frequencies ω_1 and ω_2 ($\omega_1 > \omega_2$), respectively. Cascaded FWM is deconstructed step by step in Figure 5.2.1b–e. In the first step, two $k_1^{(1)}$ photons interact with a $k_2^{(1)}$ photon to generate a first-order anti-Stokes photon k_{AS1}. A subsequent FWM process among the generated k_{AS1} photon, one $k_2^{(1)}$ photon, and one $k_2^{(1)}$ photon generates a second-order antiStokes photon k_{AS2}. Thus, all the processes are third-order nonlinear FWM processes. Higher-order signals are obtained from the generated lower-order signals; hence, the process is called *cascaded* FWM. The mth-order anti-Stokes sideband has the following phase-matching condition for the different mth-order components:

$$k_{ASm} = k_{AS(m-1)} + k^{(m)} - k_1^{(m)} \approx (m+1)k_1^{(1)} - mk_2^{(1)} \omega_{Asm} \approx (m+1)\omega_1^{(1)} - m\omega_2^{(1)}$$

In each FWM step, k_1 and k_2 photons have the same direction but different frequencies ($\omega_1^{(m)}$ and $\omega_2^{(m)}$) and wave vector magnitudes ($\left|k_1^{(m)}\right|$ and $k_2^{(m)}$). On the Stokes side, the mth-order Stokes sideband will have the following phase-matching condition:

$$k_{s^m} = k_{S(m-1)} + k_2^{(-m)} - k_1^{(-m)} \approx (m+1)k_2^{(-1)} - mk_1^{(-1)}, \quad \omega_{Sm} \approx (m+1)\omega_2^{(-1)} - m\omega_1^{(-1)}$$

It will be a degenerate cascaded FWM process, if $\omega_1 = \omega_2$.

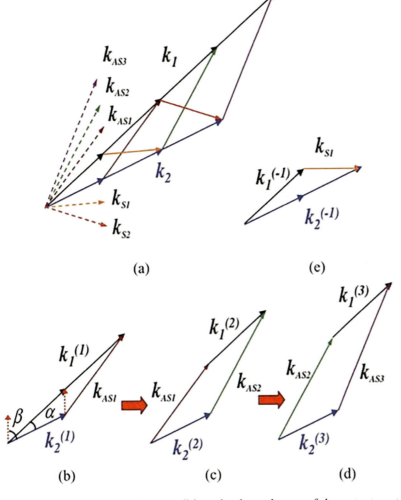

FIGURE 5.2.1 (a) Phase-matching geometry for cascaded FWM. Phase-matching geometries for generating (b) AS1, (c) AS2, (d) AS3, and (e) S1. k_1 and k_2 are the two input beams. The angle α is the crossing angle between the two input beams in the medium [48].

Using the phase-matching condition, the dependences of the output central wavelength and the exit angles of the cascaded FWM signals on the incident crossing angle were calculated. The results showed that the central wavelength of the generated cascaded FWM signals shift to shorter wavelengths as the incident crossing angle increases [48]. In addition to the phase-matching condition, it is also necessary to consider the group velocity delay between the incident pulses and the generated sidebands. The calculation results reveal that using a material with a smaller dispersion and thickness will give a broader output spectrum [48]. However, a thin nonlinear medium has low energy-conversion efficiency. It is because the output of the wave mixing grows exponentially in case the higher-order process including saturation is not taking place and hence the efficiency is limited by the thickness of the nonlinear material.

A numerical simulation of this process revealed the main characteristics of highly nondegenerate cascaded FWM of noncollinear femtosecond pulses in the spatial, spectral, and temporal domains [44].

5.2.2.2 GENERATION OF WAVELENGTH-TUNABLE SELF-COMPRESSED MULTICOLORED PULSES BY NONDEGENERATE CASCADED FWM

According to the phase-matching condition, to generate wavelength-tunable multicolored pulses in a bulk medium by cascaded FWM, the two incident pulses should have different central wavelengths.

In our case, in addition to the Ti:sapphire laser pulse at 800 nm (beam 1), a pulse with a wavelength of around 700 nm was generated by filtering the broadband spectrum after a hollow-fiber compressor (beam 2). The experimental setup is described in detail in [45–48]. Negatively chirped or nearly transform-limited output pulses can be produced by FWM when one pump beam is negatively chirped, and the other is positively chirped [47,48]. This principle can be easily explained. Both chirped input pulses can be written as $E_j(t) \propto \exp\{i[\omega_{j0}t + \varphi_j(t)]\}$, $j=1,2$, where beam 1 is negatively chirped ($\partial^2 \varphi_1(t)/\partial t^2 < 0$), and beam 2 ($\partial^2 \varphi_2(t)/\partial t^2 > 0$) is positively chirped.

The mth-order anti-Stokes signal can be expressed as

$$E_{ASm}(t) \propto \exp\{i[((m+1)\omega_{10} - m\omega_{20})t + (m+1)\varphi_1(t) - m\varphi_2(t))]\}$$

Given $\partial^2 \varphi_1(t)/\partial t^2 < 0$ and $\partial^2 \varphi_2(t)/\partial t^2 > 0$, we obtain

$$\partial^2 \varphi_{ASm}(t)/\partial t^2 = (m+1)\partial^2 \varphi_1(t)/\partial t^2 - m\partial^2 \varphi_2(t)/\partial t^2 < 0$$

This implies that the mth-order anti-Stokes signal is also negatively chirped. Nearly transform-limited pulses will be produced when the negative chirp of the anti-Stokes sidebands just compensates the dispersion of the transparent bulk medium and the phase change in the medium. This phase transfer method has also been used for three-wave mixing [75,76].

The photograph in Figure 5.2.2a shows the generated multicolored sidebands viewed on a white sheet of paper placed about 30 cm after the glass plate. These sidebands are well separated in space. The incident pulse of beam 1 was positively chirped from 35 to 75 fs by passing it through a bulk medium. The pulse of beam 2 was negatively chirped to 45 fs by chirped mirrors. The diameter of beam 1 (beam 2) on the surface of the 1-mm-thick fused silica plate was 250 μm (300 μm) in the vertical direction and 300 μm (600 μm) in the horizontal direction. Beams 1 and 2 had input pulse energies of 24 and 15 μJ.

The spectra of the sidebands from the first-order anti-Stokes (AS1) through to the fifth-order anti-Stokes (AS5) signals extend from 450 to 700 nm when the incident crossing angle was 1.78° (see Figure 5.2.2b). The spectra of AS2 and AS3 can be simultaneously tuned by varying the incident crossing angle in the range of 1.78°–2.78° (see Figure 5.2.2c). The spectrum of AS3 at 1.78° is located between the spectra of AS2 when the crossing angle is between 2.23° and 2.78°, demonstrating that the sideband spectra are continuously tunable with no gap in between. The central wavelength of the generated sideband is different in different bulk media, even at the same incident crossing angle [48].

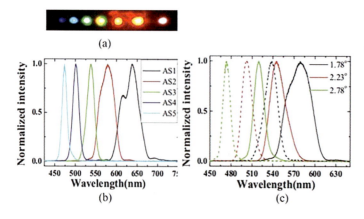

FIGURE 5.2.2 (a) Photograph of sidebands on a sheet of white paper placed 30 cm after the glass plate when the crossing angle between the two input beams is 1.78°. The first, second, and third spots from the right-hand side are beam 2, beam 1, and AS1, respectively. (b) Spectra of the sidebands AS1–AS5 (from right to left) when the crossing angle between the two input beams is 1.78°. (c) Spectra of AS2 (solid curves) and AS3 (dashed curves) for crossing angles of 1.78°, 2.23°, and 2.78° (from right to left correspondingly) [47].

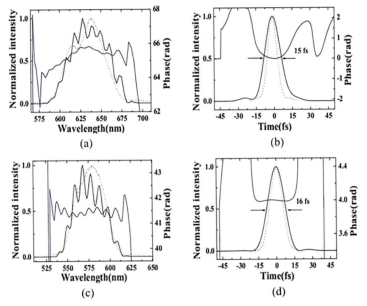

FIGURE 5.2.3 (a) Recovered spectrum (solid black curve), spectral phase (solid thin black curve), and measured spectrum (dotted curve) of AS1. (b) Recovered intensity profile and phase of AS1. The dotted curve is the transform-limited (9 fs) pulse profile of AS1. (c) Recovered spectrum (solid black curve), spectral phase (solid thin curve), and measured spectrum (dotted curve) of AS2. (d) Recovered pulse profile and phase of AS2. The dotted curve is the transform-limited (12 fs) pulse profile of AS1 [47].

We can generate 15-fs AS1 and 16-fs AS2 pulses by simply adding a glass plate to compensate the negative chirp. The pulses were measured using a cross-correlation frequency-resolved optical gating (XFROG) and were retrieved using commercial software (Femtosoft Technologies). A 1-mm-thick CaF_2 plate is used to compensate the dispersion of AS1. The retrieved spectral phase of AS1 still exhibits a clear negative chirp. Figure 5.2.3a–d shows the recovered intensity profiles, spectra, and phases of AS1 and AS2. The spectral phase indicates that both pulses have a small negative chirp. A shorter pulse is expected to be obtained when the negative chirp is completely compensated. AS1 and AS2 have output pulse energies of 0.65 and 0.15 µJ, respectively. Due to self-focusing and FWM, the spatial modes of the sidebands have perfect Gaussian profiles and good beam qualities, as reported earlier [45–48]. Angular dispersion occurs in the generated cascaded FWM beams due to phase matching in the noncollinear parametric process [48].

5.2.2.3 PULSE CLEANING BY DEGENERATE CASCADED FWM

When the two incident pulses have the same central wavelength, the generated cascaded FWM signals will have the same central wavelength as the incident pulses. This process, which is also known as SD, has been extensively used to measure pulse duration using SD-FROG [77]. It has recently been used to smooth laser spectra and clean laser pulses after passing through a hollow-fiber compressor [10,78].

Pulses with high temporal contrast are important for generating plasmas, since they suppress the generation of undesirable preplasma [79,80]. Recently, a third-order nonlinear process, cross-polarized wave generation, has been studied extensively, and it has been used to enhance the temporal contrast of femtosecond laser pulses [81]. Like the cross-polarized wave generation process, SD is a third-order nonlinear process that can also be used to improve the temporal contrast of femtosecond pulses. SD has the advantage that the generated SD signals are spatially separated from the incident beams so that polarization discrimination is not necessary. In a nonresonant electronic Kerr medium, SD occurs over a femtosecond time scale because of inertia-free interaction [82]. Consequently, it can be used to clean even picosecond pulses. The intensity of the first-order SD signal (SD1) can be described by the following equations in both frequency and time domains [77]:

$$I_{sd1}(\omega_{sd1}) \propto \left| \begin{array}{c} \iint d\omega_1\, d\omega_{-1} \chi^{(3)} \tilde{E}_1^*(z,\omega_1) \tilde{E}_{-1}(z,\omega_{-1}) \\ \tilde{E}_1(z,\omega_{sd1}-\omega_{-1}+\omega_1) \\ \sin c\left(\Delta k_z(\omega_{sd1},\omega_1,\omega_{-1})\dfrac{L}{2} \right) \end{array} \right|^2$$

$$I_{sd1} \propto I_1^2(t)\, I_{-1}(t-\tau)$$

where ω_{sd1} is the angular frequency of the SD1 signal and ω_1 and $\omega{-}1$ are those of the two incident beams, Δk_z is the phase mismatch, and L is the path length in the medium. As can be seen, the spectral intensity of the SD1 signal is an integral of the spectral intensity of the two incident pulses. This implies that the SD1 signal intensity for each wavelength component is an average contribution over the whole spectral range of the incident pulses. Therefore, the SD1 signal spectrum is automatically smoothed. The pulse duration of the SD1 signal will be at most √3 shorter than that of the incident pulse [83].

A proof-of-principle experiment was performed with a Ti:sapphire laser. Two incident beams were focused into a 0.5 mm-thick fused silica plate with a crossing angle of about 1.5°. The silica plate was located about 20 mm behind the focal point. Both beams had $1/e^2$ diameters of about 360 μm on the glass plate. The transmission pulse energies of the two incident beams after the glass plate, beam_1 and beam_−1, were 40 and 51 μJ, respectively. The SD1 signals generated in addition to beam_1 and beam_−1 had pulse energies of about 5 and 6 μJ. The energy-conversion efficiency from the input laser beams to the two SD1 signals was about 12%. Due to the low pulse energy and low power of the generated SD signals, we performed a second-order autocorrelation (SAC) measurement to measure the temporal contrast. This measurement requires a much lower incident pulse energy and is more sensitive to low-energy noise than other measurement techniques that use third-order nonlinear processes because it involves only one second-order nonlinear process [77,84]. The pulse durations of the input pulse and the SD signal were measured using a second harmonic generation FROG (SHG-FROG).

Figure 5.2.4 shows that the pulse is cleaned, even in 1 ps, and that extraneous components are removed, while the main pulse remains. For the incident pulse, the SAC peak intensity of around ±0.7 ps is about 1.2×10^{-2} of the main pulse. The SAC of the SD1 signal has a small peak at the same delay that is less than 1.2×10^{-6} of the main pulse, which is four orders of magnitude smaller and is less than the cube of 1.2×10^{-2} (i.e., 1.7×10^{-6}). The pulse is self-compressed in this process;

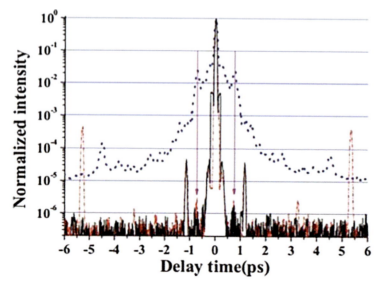

FIGURE 5.2.4 SAC intensity of the incident pulses (dotted curve) and SAC intensity of the SD1 signal when the incident beams were incident at the Brewster angle (solid black curve) and perpendicular (dash-dotted curve) to the glass plate for delay times from −6 to 6 ps and with a 5-fs/step resolution [78].

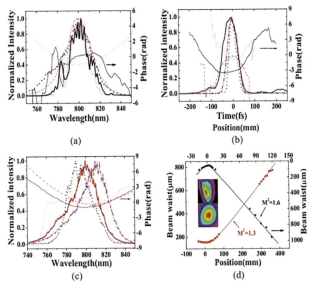

FIGURE 5.2.5 (a) Measured spectrum (thick curve) and retrieved spectral phase (thin curve) of the incident pulse (solid black curve) and the SD1 signal (dotted curve). The thin dash-dotted blue and magenta curves show the retrieved spectra of the SD1 and the incident pulse, respectively. (b) Retrieved temporal profile (thick curve) and temporal phase (thin curve) of the incident pulse (solid black curve), and the SD1 signals at delay times of 0 fs (dash-dotted curve) and + 33 fs (dotted curve). (c) Measured spectra (thick curve) and retrieved spectral phase (thin curve) of the SD1 signals for delay times of 0 fs (solid curve), −33 fs (dash-dotted thin curve), and + 33 fs (dotted blue curve). (d) M2 and 2-D beam profiles of the incident beam (squares; upper pattern at Position 0 mm) and the SD1 signal (circles; lower pattern at zero Position.) [78].

in addition, self-focusing also improves the temporal contrast. In nonresonant electronic Kerr media, self-focusing is instantaneous and it has a power threshold due to competition with diffraction. The intensity-dependent self-focusing effect increases the main pulse intensity, whereas amplified spontaneous emission and noise peaks are not enhanced. Consequently, the intensity-enhanced main pulse has a much-improved temporal contrast. When the glass plate is located after the focal point, the generated SD1 signal has a smaller divergence angle than the scattered light due to self-focusing; this reduces the noise of the scattered light.

Due to the convolution effect, the spectrum of the SD1 signal is smoother and broader than the input laser spectrum (see Figure 5.2.5a).

The pulse duration of the SD1 signal for a zero delay time was shortened from 75 to 54 fs relative to the input pulse (see Figure 5.2.5b). The retrieved temporal and spectral phases of the SD1 signal were found to be smoothed with some positive chirp. In the medium, self-phase modulation (SPM) and cross-phase modulation (XPM) accompany SD. The peak wavelength of the SD1 signal was shifted about ±10 nm at a delay time of ±33 fs (the positive sign indicates that beam_1 is ahead of beam_−1) due to XPM and the small-frequency chirps of the incident pulses (see Figure 5.2.5c) [74]. The retrieved spectral phase also shows that the reduction or enhancement of the chirp rates depends on the sign of the delay time for the same SD1 signal (see Figure 5.2.5c). Using suitable delay times and chirps of the incident pulse will induce self-compression of the SD1 signal to a nearly transform-limited pulse. As shown in Figure 5.2.5b, the pulse duration was shortened to 39 fs, which is close to its transform-limited pulse duration of 33 fs.

As in cascaded FWM experiments [45–48] and pulse compression experiments in bulk media [85], the spatial profile and beam quality of the SD1 signal were proved in this SD process compared with the input laser beam due to spatial filtering induced by self-focusing in the medium. The 2-D beam profile of the SD1 signal is improved from an asymmetric incident beam to a nearly symmetric Gaussian beam (see inset of Figure 5.2.5d). M^2 of the SD1 beam was also improved from 1.6 of the input laser beam to 1.3.

Here I would like to add some comments on cascaded NLO processes. Ordinary cascaded NLO phenomena are considered from lower order to higher order as described before in this paper. However, the cascade to the other direction can also take place. Simple representation of 2nd order + 2nd order like $\omega + \omega = 2\omega$ (2nd order). $2\omega + \omega = 3\omega$ (2nd order), and so on. It looks like the 2nd order + 2nd oruder results in the 3rd order (third harmonic generation: THG). One example of cascade down is $\omega + \omega = 2\omega$ (2nd order). $2\omega − \omega = \omega$ (2nd order) even though new frequency radiation is not generated in this process, it can change the phase of ω radiatin like in Kerr effect. Therefore, this process also looks like third order even without any new frequency.

5.2.3   UV PULSE GENERATION BY FWM IN HOLLOW FIBER

This section discusses using degenerate FWM in a gas to generate ultrashort DUV pulses. Unlike bulk media, gases cannot be optically damaged making them suitable media for frequency conversion of intense pulses. Since a gas medium has a much lower density than a bulk medium, a long interaction length or input pulses with high peak intensities are necessary to achieve a high frequency-conversion efficiency. There are two ways to realize a long interaction length: filamentation or using a hollow waveguide. The former method forms a filament by spatially and temporally overlapping two input laser pulses (i.e., the pump and idler pulses) [30,31,34]. The laser beams propagate with a constant beam size in the filament so that a long interaction length is realized. An energy-conversion efficiency from the pump to the signal of 4% and generation of a DUV pulse with an energy of 20 μJ has been demonstrated [31]. In this approach, input pulses with high peak intensities are necessary to form a fil-ament. The signal-generation stability depends on the stability of the filament. The latter method uses a hollow waveguide to achieve a long interaction length, which is determined by the waveguide length [37] and is independent of the input pulse parameters. Unlike free-space propagation, the waveguide has a negative group-velocity dispersion (GVD) that is used to cancel the positive GVD induced by the gas medium. This allows the phase-matching condition to be satisfied for a high energy-conversion efficiency [36,37]. FWM in a hollow waveguide is thus applicable for energy conversion of laser pulses with low peak intensities as well as laser pulses with high peak intensities. Another advantage of this method is the high beam quality of the signal pulse generated in the waveguide.

FWM in a gas has been utilized to generate ultrashort DUV pulses. FWM in a filament has been used to generate 12-fs DUV pulses [31] and FWM in a hollow waveguide can produce 8-fs DUV pulses [37] after passing through a dispersion compensator that has a grating compressor. Wavelength-tunable UV laser pulses can also be generated by combining FWM with NOPA [86]. Since chirped mirrors are not available in the DUV region, dispersion compensation in this wave-length region must be performed using a grating compressor or a prism compressor. Since both compressors produce large third-order dispersion, it may be necessary to use two of the following three elements, a grating compressor, a prism compressor, and a deformable mirror, to compensate third- or higher-order dispersion to generate a pulse shorter than 8 fs. However, this requires a com-plex setup, which leads to a large energy loss.

Wojtkiewicz et al. proposed a method for achieving chirped pulse FWM [87,88]. In this scheme, chirped input pulses are used to generate a negatively chirped signal pulse. The chirped signal pulse can be compressed by propagation through a transparent medium so that no external pulse compres-sor is required. Precise dispersion compensation is possible by selecting an appropriate transparent medium, such as magnesium fluoride that has no appreciable high-order GVD in the DUV region [87–90]. Since the input pulses are chirped and they have lower peak intensities than the corre-sponding transform-limited pulses, a hollow fiber is suitable for chirped-pulse FWM.

Our group has recently generated a sub-10-fs DUV pulse by using a self-phase-modulated pulse as the input idler for chirped-pulse FWM [91]. Using a broadband idler supporting a sub-10-fs trans-form-limited pulse duration leads to the generation of a DUV pulse with a pulse duration shorter than 10 fs. The following section describes the principle of broadband chirped-pulse FWM using a self-phase-modulated idler pulse. After presenting a scheme for this process, the properties of the experimentally generated sub-10-fs DUV pulses are discussed.

5.2.3.1   CHIRPED-PULSE FWM IN A GAS-FILLED HOLLOW WAVEGUIDE

In degenerate FWM with energy conservation $\omega_{sig}=2\omega_{pump}-\omega_{idler}$, using a negatively chirped pump pulse and a positively chirped idler pulse produces a negatively chirped signal pulse [87,88]. Figure 5.2.6 shows a schematic diagram of this process (it is similar to that shown in [89].

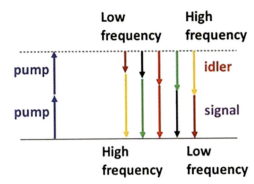

Low frequency   High frequency

pump

pump

idler

signal

High frequency   Low frequency

FIGURE 5.2.6 Energy diagram for chirped-pulse FWM.

In addition, the figure shows that the negative frequency chirp in the signal is due to the positive frequency chirp in the idler. The idler frequency increases with time, whereas the signal frequency decreases with time. A negative frequency chirp in the pump pulse leads to a negative frequency chirp in the signal pulse, which can be explained in a similar manner as the relation between the frequency chirps of the idler and the signal.

Because chirped input pulses are used, SPM and XPM, which broaden the signal spectrum [37], hardly occur so that the signal pulse bandwidth is mainly determined by the input pulse bandwidths. To generate a sub-10-fs DUV pulse by chirped-pulse FWM, it is necessary to use a broadband idler or a pump pulse that supports a sub-10-fs transform-limited pulse duration [90].

Signal pulse generation by FWM in a gas-filled hollow fiber generated by input pulses propagating along the z-axis is expressed by [92]

$$\frac{\partial \varepsilon_s}{\partial z} = i D_s \varepsilon_s + i \left(\frac{\omega_s}{c} \right) n_2 \, T_s \left\{ \left[|\varepsilon_s|^2 + 2|\varepsilon_p|^2 + 2|\varepsilon_i|^2 \right] \varepsilon_s + \varepsilon_p^2 \varepsilon_i^* \exp(-i\Delta\beta z) \right\} \tag{5.2.1}$$

$$D_S = \frac{-\alpha_S}{2} - i \left(\frac{\beta_S^{(2)}}{2} \right) \left(\frac{\partial^2}{\partial t^2} \right) + i \left(\frac{\beta_S^{(3)}}{6} \right) + \dots \tag{5.2.2}$$

where β_s is the propagation constant of the signal pulse inside the hollow fiber [93], the asterisks indicate complex conjugates, ω_s is the angular frequency of the electric field, and n_2 is the nonlinear refractive index of the core medium. The phase mismatch is given by $\Delta\beta = \beta_i + \beta_s - 2\beta_p$ and the higher order dispersion $\beta_S^{(n)} = \partial^n \beta / \partial \omega^n |_{\omega=\omega_k}$ In (5.2.2), α_S is the loss constant of the gas-filled hollow waveguide at the signal frequency [93]. $T_s = \{1 + (i/\omega_s)(\partial/\partial t)\}$ contains the effect of self-steeping, A_{eff} is the effective core area, and c is the speed of light in vacuum [94,95]. The complex amplitudes of the signal, idler, and pump pulses are, respectively, ε_s, ε_i and ε_p in (5.2.1), which is expressed in a frame of reference propagating with the signal group velocity.

For chirped-pulse FWM with input pulses having low peak intensities and a low-density gas medium, the terms related to SPM and XPM in (5.2.1) may be dropped. By assuming an input pump energy that is sufficiently low to neglect the pump depletion, a negligibly small GVD due to the low gas density to satisfy phase matching [37], and a negligibly low propagation loss, the propagation of the signal pulse generated by FWM is expressed as

$$\varepsilon_S (t,z) = \left(\frac{z n_2 \omega_S}{c A_{\text{eff}}} \right) A(t)_p^2 \, A(t)_i \, \exp \left[i \left(2\phi_p (t) - \phi_i (t) + \frac{\pi}{2} \right) \right] \tag{5.2.3}$$

In deriving (5.2.3), $\varepsilon_{p,i}(t) = A_{p,i}(t)\exp[i\phi_{p,i}(t)]$ was substituted into the coupled Eq. (5.2.1), where $A_{p,i}$ (t) and $\phi_{p,i}(t)$ are, respectively, the real amplitudes and phases of the pump and idler. Differentiating the phase term in (5.2.3) gives the time evolution of the instantaneous signal frequency, which is expressed as $2d\phi_p(t)/dt - d\phi_i(t)/dt$. This explains the relation between the frequency chirp in the input pulses and the chirp in the signal, as discussed earlier.

5.2.3.2 BROADBAND CHIRPED-PULSE FWM

To generate a broadband signal pulse by chirped-pulse FWM, a broadband pump pulse or a broadband idler pulse is necessary. For this purpose, a self-phase-modulated pulse is used as an input pulse for chirped-pulse FWM. By using SPM in a hollow waveguide filled with a noble gas [96], it is possible to expand the spectral width of a 30-fs pulse generated by a commercial Ti:sapphire chirped-pulse amplifier to a spectral width that supports 5-fs transform-limited pulses. However, a self-phase modulated pulse contains nonlinear phase distortion in its temporal profile, which is related to the spectral intensity and phase distribution of the pulse. Specifically, SPM induces phase modulation within the temporal profile of a laser pulse that varies proportionally with the temporal intensity distribution of the laser pulse [94]. The pulse contains a positively chirped central region that is suitable for chirped-pulse FWM, but it also contains negatively chirped leading and trailing edges. Such nonlinear frequency modulation produces dips in the spectrum. When a self-phase-modulated pulse is used as the input for chirped-pulse FWM, the nonlinear phase distortion in the SPM idler may be transferred to the signal pulse, resulting in a nonlinear temporal phase and a non-uniform signal spectrum. However, this does not matter, if the pump pulse is much shorter (i.e., less than about 50%) than the idler pulse. Since a signal pulse is generated only in the temporal region in which the two input pulses overlap, by employing a short pump pulse, it is possible to generate a signal pulse that results from only the interaction between the central region of the self-phase-modulated idler and the pump pulse. In this case, the idler pulse behaves as if it is a linearly chirped broadband pulse.

When the amplitudes of the input idler and pump pulses are Gaussian functions and the pump pulse is at least two times shorter than the idler pulse, the signal becomes a linearly (negatively) chirped Gaussian pulse (see Figure 5.2.7) and a smooth spectral shape is obtained. A nearly transform-limited pulse duration is available for the signal after compensating the negative frequency chirp, provided that a positive group-delay dispersion (GDD) is added to the signal without adding substantial high-order GVD. Such dispersion compensation is realized using a transparent medium (e.g., magnesium fluoride) whose absorption wavelength is far from the signal wavelength.

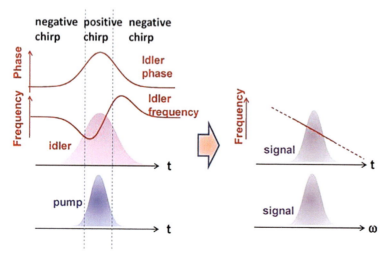

FIGURE 5.2.7 Frequency sweep of a SPM idler with respect to time for a pump pulse that is shorter than the idler. The shorter pump produces a linearly chirped signal and a smooth spectral profile.

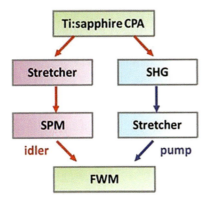

FIGURE 5.2.8 Scheme for broadband chirped-pulse FWM. SHG, second harmonic generation; CPA, chirped-pulse amplifier.

5.2.3.3 PRACTICAL ISSUES IN BROADBAND CHIRPED-PULSE FWM

Figure 5.2.8 shows an example of an experimental setup for demonstrating the afore-described broadband chirped-pulse FWM. A femtosecond NIR pulse from a Ti:sapphire chirped-pulse amplifier is split into two pulses. One is used for generating a broadband idler and the other is used for generating a pump pulse by frequency doubling the NIR pulse. Before the idler is used as the input for FWM, its spectral width is expanded by SPM in a gas-filled hollow waveguide. The near-UV (NUV) pulse generated through frequency doubling is stretched by a prism pair (prism compressor) that induces a negative frequency chirp.

The NUV pump pulse duration is lengthened in this process, but it remains shorter than the broadband idler pulse duration. Finally, the SPM idler and the negatively chirped pump pulses are spatially and temporally synthesized in a gas-filled hollow waveguide to generate a signal pulse by FWM. Filamentation in a gas cell [30,31] could also be used when the input pulse energies are above about 1 mJ.

As discussed earlier, the pump pulse should be shorter than the idler to generate a signal pulse with a smooth spectral shape. Two procedures are considered for achieving this: pulse broadening in a transparent medium before or after spectral broadening by SPM. The self-phase modulated idler has a broad spectral width, and consequently, the temporal profile of the self-phase modulated idler is substantially distorted by high-order dispersion in a transparent medium. Moreover, the phase of the self-phase modulated pulse is nonlinear in terms of its temporal evolution. The positive chirp in the central region is enhanced by the positive GVD in the transparent medium, while the negative chirps in the leading and trailing edges are compensated and produce intense spikes due to the GVD. This affects the temporal profile and, thus, the spectrum and spectral phase of the signal generated by FWM. Furthermore, the temporal width of the positively chirped region of the self-phase modulated idler is not effectively expanded in this case.

On the other hand, pulse broadening of the idler before spectral broadening is useful. In this case, the positive frequency chirp induced by the transparent medium and the positive frequency chirp induced by SPM both contribute to generate a positive chirp. This leads to a wide positively chirped region within the idler pulse duration because of the extended pulse duration of the idler prior to SPM. The positive chirp due to the transparent medium, which is produced before SPM, partially cancels the negative chirp at the leading and trailing edges of the idler pulse induced by SPM. Since the pulse duration of the idler is already broadened by GVD in the transparent medium, it is not necessary to significantly increase the pulse duration after spectral broadening. Distortion in the temporal profile of the broadband idler can be minimized in this case, making it possible to generate a signal with smooth temporal and spectral profiles. It also minimizes the high-order spectral phase in the signal, resulting in the generation of a single ultrashort pulse after dispersion compensation.

A prism compressor can be used for negatively chirping a NUV pump pulse with a narrow spectral width whose transform-limited pulse duration is longer than about 30 fs. A prism compressor generally induces a large negative third-order dispersion as well as a negative group-delay dispersion. This distorts the temporal profile and the phase of a pulse when the pulse has a wide bandwidth that can support a pulse duration shorter than 10 fs. However, this does not matter for narrowband pulses since the nonlinear spectral phase distortion due to the third-order spectral phase distortion lies outside the narrow bandwidth. Chirped mirrors can also be used to negatively chirp a NUV pulse. Although chirped mirrors for NUV pulses have oscillations in their dispersion curves, their effect may not be significant for a narrowband pulse that has a transform-limited pulse duration longer than 30 fs.

By spectrally broadening the idler rather than the NUV pump and by prechirping the idler pulse prior to spectral broadening, it is possible to generate a nearly linearly chirped signal pulse. This results in minimal distortion of the phases and temporal profiles of the input pulses, and, thus, gives signal pulses with a smooth temporal profile and phase. The experimental scheme discussed above is used in [91].

5.2.3.4 SUB-10-FS DUV PULSES GENERATED BY BROADBAND CHIRPED-PULSE FWM

The energy-conversion efficiency from the pump to the signal is determined by the input pump and idler intensities. Since the response is nonlinear with respect to the pump intensity, the conversion efficiency of the signal is much more sensitive to the pump intensity than that of the idler. Because chirped input pulses are used, chirped-pulse FWM has a lower energy conversion efficiency than FWM which uses transform-limited input pulses [31,37]. Therefore, chirped-pulse FWM is not useful for generating high-energy DUV pulses when only a low-energy pump pulse is available. In the case of a chirped pump pulse with a pulse energy of about 100 µJ, the pulse energy of a signal pulse generated by broadband FWM is limited to 300 nJ [91]. Increasing the pulse energy (and thus the intensity) of the pump pulse will significantly increase the signal pulse energy. Using a 1-mJ pump pulse, it is expected to be possible to generate a signal pulse with an energy of several tens of microjoules [88,90,92].

It is possible to generate a broadband signal pulse with a spectrum extending from 260 to 290 nm using broadband chirped pulse FWM [91]. The signal spectrum differs from the idler and pump pulse spectra. For example, in an experiment, the idler spectrum contained two deep dips, whereas the signal spectrum had only one deep dip (see Figure 5.2.9) [91]. This difference in spectral shapes is related to the different pump and idler pulse durations, as discussed above. Contrary to the

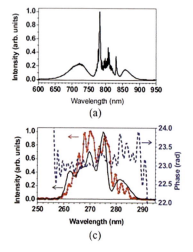

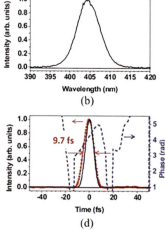

FIGURE 5.2.9 Spectra of the input (a) NIR and (b) NUV, and output DUV [solid line in (c)] pulses measured with spectrometers. The spectrum (solid line with filled circles) and spectral phase (broken line) measured by the SD-FROG are shown in (c), while the corresponding temporal profile (solid line with filled circles) and phase (broken line) are shown in (d). The inverse Fourier transform of the solid line in (c) is indicated by the solid line in (d) [91].

discussion, the spectral shape of the signal is not a single peaked one like a Gaussian. This is due to the pump pulse not being sufficiently shorter than the idler pulse. A signal pulse with a Gaussian-shaped spectrum can be generated, if a larger chirp than that in the experiment is added to the idler prior to spectral broadening by SPM [92].

A broadband negatively chirped DUV pulse is generated in broadband chirped-pulse FWM. The pulse can be easily compressed by a normal GVD induced by a transparent medium. Unlike when using prism and grating compressors, third-order dispersion is not induced in this case, and the pulse energy loss is small. For example, in an experiment, a DUV pulse generated by broadband chirped-pulse FWM had a negative GDD of 265 fs^2 [91]. Some of the GDD was compensated by the positive GDD induced by the magnesium fluoride output window of the hollow-fiber chamber, which generated FWM. The residual negative GDD in the DUV pulse was compensated by propagation in the air [91]. SD-FROG measurements are useful for optimizing the path length in air. In a FROG, the edge of an aluminum mirror is used for beam splitting, which results in negligible pulse broadening [37]. The spectral change of the SD signal with respect to the time delay of a replica is sensitive to the GDD of a pulse to be measured [97]. In the experiment, only three SD-FROG measurements were used to optimize the path length and, thus, the positive GDD induced by the air [91]. The DUV pulse duration was compressed to less than 10 fs. The same pulse compression can also be realized by varying the path length of the pulse in a pair of wedges made of MgF$_2$ or CaF$_2$ thin plates.

The DUV pulse duration can be precisely compressed in broadband chirped-pulse FWM. The pulse compression procedure is relatively simple, as discussed above. The transform-limited pulse duration estimated from an experimentally measured spectrum is 8 fs, while the corresponding measured pulse duration after dispersion compensation is 9.7 fs [91]. These pulse durations differ by only 24%. The compressed signal pulse has a smooth temporal profile (see Figure 5.2.9c). It contains almost no satellite pulses; the energy of the satellite pulses is less than 5% of the total energy. In broadband chirped-pulse FWM, a linearly chirped DUV pulse is obtained by using a pump pulse that is shorter than the idler pulse. The linearly chirped DUV pulse has a smooth temporal profile and, thus, has almost no high-order spectral phase distortion. An ultrashort DUV pulse with a single pulse structure can be generated by compensating the negative GDD in the DUV pulse without inducing a substantial high-order dispersion using a transparent medium.

5.2.4 FWOPA IN BULK MEDIA

We have demonstrated using FWM to generate and clean femtosecond laser pulses in both gases and bulk media. Another FWM process, FWOPA, has recently been used to amplify ultrashort pulses in a glass plate at different wavelengths [69–74]. This method is particularly useful in the UV region, due to the lack of suitable nonlinear crystals in this region. Here, we demonstrate that a weak laser pulse can be simultaneously amplified and compressed by another intense laser pulse [74].

The principle of this method is schematically illustrated in Figure 5.2.10a. An intense pump beam and a weak seed beam are focused onto a glass plate with a crossing angle α. When the pump and seed pulses are synchronous in time and overlap in space in a transparent bulk medium, the seed pulse spectrum will be broadened due to XPM in the medium induced by the intense pump pulse. Furthermore, the weak seed pulse will be simultaneously amplified when the crossing angle α satisfies the phase-matching condition for FWM [68–74].

According to the phase-matching condition, the crossing angle α_{in} in the medium is given by $\cos \alpha_{in} [(2k_p)^2 + k_s^2 - k_i^2] / 4k_p k_s$ terms of the wavenumber k, where the subscripts p, s, and i indicate the pump, seed, and idler beams, respectively. Figure 5.2.10b and its inset show phase-matching curves for the crossing angle in air α as a function of the seed wavelength for fused silica and CaF$_2$, respectively. The pump pulse has typical wavelengths of 800 and 400 nm for fused silica and CaF$_2$, respectively. There is a broad phase-matching bandwidth around 500 nm (250 nm) when the crossing angle α is about 3.1° (6.4°), as shown in Figure 5.2.10b.

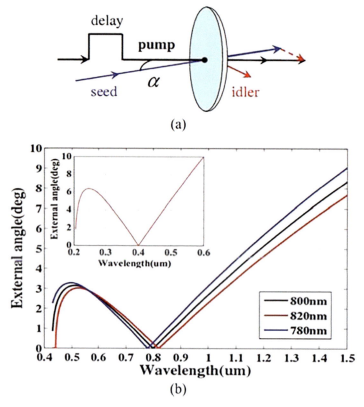

FIGURE 5.2.10 (a) Schematic showing the experimental setup. α is the crossing angle. (b) Phase-matching curves for the crossing angle α as a function of the seed wavelength for fused silica (CaF_2, inset) when the pump pulse was fixed at wavelengths of 780, 800, and 820 nm depicted by curves from top to bottom at 1 um-wavelength position. (400 nm, inset) [74].

The intense pump pulse at 800 nm simultaneously spectrally broadens and amplifies the incident seed pulse AS1 at 620 nm when the crossing angle α is around 2.80°±0.05°. Figure 5.2.11a shows the spectral profile and intensity of the amplified pulse as a function of the delay time t_{ps}. The 400-nJ incident pulse was amplified to 1.1 µJ with a 140 µJ pump pulse. Cascaded FWM signals were simultaneously generated around 500 nm with 250 nJ, as shown in the photograph in the inset of Figure 5.2.11a. Figure 5.2.11b shows the output energy of the seed pulse as a function of the pump intensity at a delay time of 0 fs. Much higher output energies are expected to be obtained when a cylindrical lens is used for focusing [69–71]. The thin glass plate ensured that the phase-matching spectral bandwidth was broad. Consequently, broadband amplification still occurred around 620 nm.

A quasi-linear chirp can be induced across the weak seed pulse when the pump pulse is much wider than it [98]. The phase induced by XPM can also be compensated by using a pair of chirped mirrors. Furthermore, XPM-based compressors are more flexible than SPM ones because the phase can be tuned by the pump pulse. After passing through the chirped mirrors for four bounces ($-40\,fs^2$/bounce), the pulse was compressed from 22.6 to 12.6 fs (see Figure 5.2.11c), which is close to the calculated transform-limited pulse duration of 10.5 fs. Figure 5.2.11d shows the incident laser spectrum, the broadened and retrieved spectrum, and the spectral phase of the compressed pulse. Several pulses with different wavelengths can be simultaneously guided into this system; the system will then simultaneously amplify them and broaden their spectra [74].

5.2.5 CONCLUSION AND PROSPECTS

FWM has been used to generate and optimize ultrashort laser pulses. Self-compressed 15-fs multi-colored pulses are simultaneously generated from UV to NIR by cascaded FWM in a bulk medium. The wavelengths of the generated multicolored pulses can be tuned by varying the incident-crossing angle. Studies on degenerate and nondegenerate cascaded FWM reveal that the generated cascaded

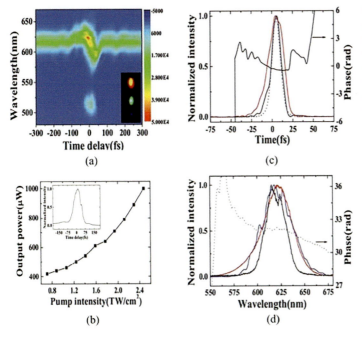

FIGURE 5.2.11 (a) Spectral profile and intensity of the output seed beam (AS3) as a function of the delay time t_{ps} when the crossing angle α was $2.80\pm0.05°$. (b) Dependence of the output energy of the output seed pulse on the pump intensity. The curve in the inset is the temporal profile of the pump pulse. (c) Retrieved temporal profiles of the incident seed pulse (solid faint line extending from ~–40 fs to ~+30 fs) and the compressed output seed pulse (solid black line extending from ~–20 fs to ~+20 fs), and the transform-limited pulse of the broadened spectrum (dotted line). (d) Retrieved spectrum (solid faint line) and spectral phase (dotted line) of the output seed pulse. Measured spectra of the incident seed pulse (solid line) and output seed pulse (smooth solid faint line extending from ~560 nm to ~660 nm.) [74].

signals have broader and smoother spectra, shorter and cleaner pulses, and improved beam profiles and spatial quality. These outstanding characteristics make cascaded.

FWM signals are useful for multicolor time-resolved spectroscopy and multicolor nonlinear optical microscopy. The high temporal contrast of cascaded FWM signals makes them suitable for seed pulses in background-free petawatt lasers. Broadband DUV pulses are generated using self-phase modulated pulses as the idler in chirped-pulse FWM. The DUV pulse is negatively chirped and can be easily compressed by a transparent medium to below 10 fs. The temporal profile of the compressed pulse is smooth and useful for ultrafast spectroscopic applications in the DUV region.

The FWOPA method was expected to be used to simultaneously amplify and compress laser pulses in a bulk medium. This method is expected to amplify DUV pulses pumped by 400-nm pulses and compress DUV pulses to below 5 fs in the near future [99]. The research products represented in this subsection are based on the cooperative activity of the following people: T. Kobayashi, J. Liu, and Y. Kida [99].

REFERENCES

1. A. H. Zewail, "Femtochemistry: Atomic-scale dynamics of the chemical bond," *J. Phys. Chem. A* **104**, 5660–5694 (May 2000).
2. J. Shah, *Ultrafast Spectroscopy of Semiconductors and Semiconductor Nanostructures*, 1st ed., Springer, Berlin, Germany (1999).
3. T. Kobayashi, T. Okada, K. A. Nelson, and S. De Silvestri, *Ultrafast Phenomena XIV*, 1st ed., Springer, New York (2004).
4. T. Kobayashi, A. Shirakawa, and T. Fuji, "Sub-5-fs transform-limited visible pulse source and its application to real-time spectroscopy," *IEEE J. Sel. Topics Quantum Electron* **7**(4), 525–538 (Jul. 2001).
5. L. Xu, Ch. Spielmann, F. Krausz, and R. Szipocs, "Ultrabroadband ring" oscillator for sub-10-fs pulse generation," *Opt. Lett.* **21**, 1259–1261 (Aug. 1996).

6. A. Baltuska, Z. Wei, M. S. Pshenichnikov, and D. A. Wiersma, "Optical pulse compression to 5 fs at a 1-MHz repetition rate," *Opt. Lett.* **22**, 102–104 (Jan. 1997).

7. M. Nisoli, S. De Silvestri, O. Svelto, R. Szipocs, K. Ferencz, C. Spielmann, S. Sartania, and F. Krausz, "Compression of high-energy laser pulses below 5 fs," *Opt. Lett.* **22**, 522–524 (Apr. 1997).

8. S. Bohman, A. Suda, T. Kanai, S. Yamaguchi, and K. Midorikawa, "Generation of 5.0 fs, 5.0 mJ pulses at 1 kHz using hollow-fiber pulse compression," *Opt. Lett.* **35**, 1887–1889 (Jun. 2010).

9. C. P. Hauri, W. Kornelis, F. W. Helbing, A. Heinrich, A. Couairon, A. Mysyrowicz, J. Biegert, and U. Keller, "Generation of intense, carrier envelope phase-locked few-cycle laser pulses through filamentation," *Appl. Phys. B* **79**, 673–677 (Sep. 2004).

10. J. Liu, K. Okamura, Y. Kida, T. Teramoto, and T. Kobayashi, "Clean sub-8-fs pulses at 400 nm generated by a hollow fiber compressor for ultraviolet ultrafast pump-probe spectroscopy," *Opt. Express* **18**, 20645–20650 (Sep. 2010).

11. A. Shirakawa, I. Sakane, and T. Kobayashi, "Pulse-front-matched optical parametric amplification for sub-10-fs pulse generation tunable in the visible and near infrared," *Opt. Lett.* **23**, 1292–1294 (Aug. 1998).

12. G. Cerullo, M. Nisoli, S. Stagira, and S. De Silvestri, "Sub-8-fs pulses from an ultrabroadband optical parametric amplifier in the visible," *Opt. Lett.* **23**, 1283–1285 (Aug. 1998).

13. A. Shirakawa, I. Sakane, M. Takasaka, and T. Kobayashi, "Sub 5 faviiable pulse generation by pulse-front-matched noncollinear optical parametric amplification," *Appl. Phys. Lett.* **74**, 2268–2270 (Apr. 1999).

14. A. Baltuska, T. Fuji, and T. Kobayashi, "Visible pulse compression to 4 fs by optical parametric amplification and programmable dispersion control," *Opt. Lett.* **27**, 306–308 (Mar. 2002).

15. G. Cerullo and S. De Silvestri, "Ultrafast optical parametric amplifiers," *Rev. Sci. Instrum.* **74**, 1–18 (Jan. 2003).

16. D. Brida, G. Cirmi, C. Manzoni, S. Bonora, P. Villoresi, S. De Silvestri, and G. Cerullo, "Sub-two-cycle light pulses at 1.6 μm from an optical parametric amplifier," *Opt. Lett.* **33**, 741–743 (Apr. 2008).

17. R. A. Kaindl, M. Wurm, K. Reimann, P. Hamm, A. M. Weiner, and M. Woerner, "Generation, shaping, and characterization of intense femtosecond pulses tunable from 3 to 20 μm," *J. Opt. Soc. Amer. B* **17**, 2086–2094 (Dec. 2000).

18. C. Heese, C. R. Phillips, L. Gallmann, M. M. Fejer, and U. Keller, "Ultrabroadband, highly flexible amplifier for ultrashort midinfrared laser pulses based on aperiodically poled Mg:LiNbO₃," *Opt. Lett.* **35**, 2340–2342 (Jul. 2010).

19. I. Kozma, P. Baum, S. Lochbrunner, and E. Riedle, "Widely tunable sub 30 fs ultraviolet pulses by chirped sum frequency mixing," *Opt. Express* **11**, 3110–3115 (Dec. 2003).

20. T. Kobayashi, T. Saito, and H. Ohtani, "Real-time spectroscopy of transition states in bacteriorhodopsin during retinal isomerization," *Nature* **414**, 531–534 (Nov. 2001).

21. G. Cerullo, D. Polli, G. Lanzani, S. De Silvestri, H. Hashimoto, and R. J. Cogdell, "Photosynthetic light harvesting by carotenoids: Detection of an intermediate excited states," *Science* **298**, 2395–2398 (Dec. 2002).

22. S. Adachi, V. M. Kobryanskii, and T. Kobayashi, "Excitation of a breather mode of bound soliton pairs in trans-polyacetylene by sub-5-fs optical pulses," *Phys. Rev. Lett.* **89**, 027401-1–027401-4 (Jul. 2002).

23. P. Kukura, D. W. McCamant, S. Yoon, D. B. Wandschneider, and R. A. Mathies, "Structural observation of the primary isomerization in vision with femtosecond-stimulated Raman," *Science* **310**, 1006–1009 (Nov. 2005).

24. D. Polli, P. Altoe, O. Weingart, K. M. Spillane, C. Manzoni, D. Brida, G. Tomasello, G. Orlandi, P. Kukura, R. A. Mathies, M. Garavelli, and G. Cerullo, "Conical intersection dynamics of the primary photoisomerization event in vision," *Nature* **467**, 440–443 (Sep. 2010).

25. J. A. Gruetzmacher and N. F. Scherer, "Few-cycle mid-infrared pulse generation, characterization, and coherent propagation in optically dense media," *Rev. Sci. Instrum.* **73**, 2227–2236 (Jun. 2002).

26. Y. S. Kim and R. M. Hochstrasser, "Applications of 2D IR spectroscopy to peptides, proteins, and hydrogen-bond dynamics," *J. Phys. Chem. B* **113**, 8231–8251 (Dec. 2009).

27. R. M. Hochstrasser, "Two-dimensional spectroscopy at infrared and optical frequencies," *Proc. Nat. Acad. Sci. USA* **104**, 14190–14196 (Jul. 2007).

28. J. M. Anna, M. J. Nee, C. R. Baiz, R. McCanne, and K. J. Kubarych, "Measuring absorptive two-dimensional infrared spectra using chirped pulse upconversion detection," *J. Opt. Soc. Amer. B* **27**, 382–393 (Mar. 2010).

29. P. Baum, S. Lochbrunner, and E. Riedle, "Tunable sub-10-fs ultraviolet pulses generated by achromatic frequency doubling," *Opt. Lett.* **29**, 1686–1688 (Jul. 2004).

30. F. Theberge, N. Ak´ozbek, W. Liu, A. Becker, and S. L. Chin, "Tunable¨ ultrashort laser pulses generated through filamentation in gases," *Phys. Rev. Lett.* **97**, 023904-1–023904-4 (Jul. 2006).

31. T. Fuji, T. Horio, and T. Suzuki, "Generation of 12 fs deep-ultraviolet pulses by four-wave mixing through filamentation in neon gas," *Opt. Lett.* **32**, 2481–2483 (Sep. 2007).

32. T. Fuji and T. Suzuki, "Generation of sub-two-cycle mid-infrared pulses by four-wave mixing through filamentation in air," *Opt. Lett.* **32**, 3330–3332 (Nov. 2007).

33. P. Zuo, T. Fuji, and T. Suzuki, "Spectral phase transfer to ultrashort UV pulses through four-wave mixing," *Opt. Express* **18**, 16183–16192 (Jul. 2010).

34. M. Beutler, M. Ghotbi, F. Noack, and I. V. Hertel, "Generation of sub-50 fs vacuum ultraviolet pulses by four-wave mixing in argon," *Opt. Lett.* **35**, 1491–1493 (May 2010).

35. G. C. Durfee III, S. Backus, M. M. Murnane, and C. H. Kapteyn, "Ultrabroadband phase-matched optical parametric generation in the ultraviolet by use of guided waves," *Opt. Lett.* **22**, 1565–1567 (Oct. 1997).

36. L. Misoguti, S. Backus, C. G. Durfee, R. Bartels, M. M. Murnane, and H. C. Kapteyn, "Generation of broadband VUV light using third-order cascaded processes," *Phys. Rev. Lett.* **87**, 013601-1–013601-4 (Jul. 2001).

37. C. G. Durfee, S. Backus, H. C. Kapteyn, and M. M. Murnane, "Intense 8-fs pulse generation in the deep ultraviolet," *Opt. Lett.* **24**, 697–699 (May 1999).

38. U. Graf, M. Fieß, M. Schultze, R. Kienberger, F. Krausz, and E. Goulielmakis, "Intense few-cycle light pulses in the deep ultraviolet," *Opt. Express* **16**, 18956–18963 (Nov. 2008).

39. H. Okamoto and M. Tatsumi, "Generation of ultrashort light pulses in the mid-infrared (3000–800 cm^{-1}) by four-wave mixing," *Opt. Commun.* **121**, 63–68 (Nov. 1995).

40. H. K. Nienhuys, P. C. M. Planken, R. A. van Santen, and H. J. Bakker, "Generation of mid-infrared pulses by $\chi^{(3)}$ difference frequency generation in CaF_2 BaF_2," *Opt. Lett.* **26**, 1350–1352 (Sep. 2001).

41. J. Darginavicius, G. Tamoauskas, A. Piskarskas, and A. Dubietis, "Generation of 30-fs ultraviolet pulses by four-wave optical parametric chirped pulse amplification," *Opt. Express* **18**, 16096–16101 (Jul. 2010).

42. H. Crespo, J. T. Mendonc¸a, and A. Dos Santos, "Cascaded highly nondegenerate four-wave-mixing phenomenon in transparent isotropic condensed media," *Opt. Lett.* **25**, 829–831 (Jun. 2000).

43. R. Weigand, J. T. Mendonca, and H. Crespo, "Cascaded nondegenerate four-wave mixing technique for high-power single-cycle pulse synthesis in the visible and ultraviolet ranges," *Phys. Rev. A* **79**, 0638381–063838-5 (Jun. 2009).

44. J. L. Silva, R. Weigand, and H. Crespo, "Octave-spanning spectra and pulse synthesis by non-degenerate cascaded four-wave mixing," *Opt. Lett.* **34**, 2489–2491 (Aug. 2009).

45. J. Liu and T. Kobayashi, "Wavelength-tunable multicolored femtosecond laser pulse generation in fused silica glass," *Opt. Lett.* **34**, 1066–1068 (Apr. 2009).

46. J. Liu and T. Kobayashi, "Generation of uJ-level multicolored femtosecond laser pulses using cascaded four-wave mixing," *Opt. Express* **17**, 4984–4990 (Mar. 2009).

47. J. Liu and T. Kobayashi, "Generation of sub-20-fs multicolor laser pulses using cascaded four-wave mixing with chirped incident pulses," *Opt. Lett.* **34**, 2402–2404 (Aug. 2009).

48. J. Liu and T. Kobayashi, "Cascaded four-wave mixing in transparent bulk media," *Opt. Commun.* **283**, 1114–1123 (2010).

49. J. Liu and T. Kobayashi, "Cascaded four-wave mixing and multicolored arrays generation in a sapphire plate by using two crossing beams of femtosecond laser," *Opt. Express* **16**, 22119–22125 (Dec. 2008).

50. J. Liu, T. Kobayashi, and Z. G. Wang, "Generation of broadband two dimensional multicolored arrays in a sapphire plate," *Opt. Express* **17**, 9226–9234 (May 2009).

51. M. Zhi and A. V. Sokolov, "Broadband coherent light generation in a Raman-active crystal driven by two-color femtosecond laser pulses," *Opt. Lett.* **32**, 2251–2253 (Aug. 2007).

52. M. Zhi and A. V. Sokolov, "Broadband generation in a Raman crystal driven by a pair of time-delayed linearly chirped pulses," *New J. Phys.* **10**, 025032-1–025032-12 (Feb. 2008).

53. E. Matsubara, T. Sekikawa, and M. Yamashita, "Generation of ultrashort optical pulses using multiple coherent anti-Stokes Raman scattering in a crystal at room temperature," *Appl. Phys. Lett.* **92**, 071104-1–071104-3 (Feb. 2008).

54. E. Matsubara, Y. Kawamoto, T. Sekikawa, and M. Yamashita, "Generation of ultrashort optical pulses in the 10 fs regime using multicolor Raman sidebands in $KTaO_3$," *Opt. Lett.* **34**, 1837–1839 (Jun. 2009).

55. H. Matsuki, K. Inoue, and E. Hanamura, "Multiple coherent antiStokes Raman scattering due to phonon grating in $KNbO_3$ induced by crossed beams of two-color femtosecond pulses," *Phys. Rev. B* **75**, 024102-1–024102-4 (Jan. 2007).

56. K. Inoue, J. Kato, E. Hanamura, H. Matsuki, and E. Matsubara, "Broadband coherent radiation based on peculiar multiple Raman scattering by laser-induced phonon grating in TiO$_2$," *Phys. Rev. B* **76**, 041101(R)–041101-4 (Jul. 2007).

57. E. Matsubara, K. Inoue, and E. Hanamura, "Violation of Raman selection rules induced by two femtosecond laser pulses in KTaO$_3$," *Phys. Rev. B* **72**, 134101-1–134101-5 (Oct. 2005).

58. J. Takahashi, E. Matsubara, T. Arima, and E. Hanamura, "Coherent multistep anti-Stokes and stimulated Raman scattering associated with third harmonics in YFeO$_3$ crystals," *Phys. Rev. B* **68**, 155102-1–155102-5 (Oct. 2003).

59. J. Takahashi, M. Keisuke, and Y. Toshirou, "Raman lasing and cascaded coherent anti-Stokes Raman scattering of a two-phonon Raman band," *Opt. Lett.* **31**, 1501–1503 (May 2006).

60. M. Zhi, X. Wang, and A. V. Sokolov, "Broadband coherent light generation in diamond driven by femtosecond pulses," *Opt. Express* **16**, 12139–12147 (Aug. 2008).

61. J. Liu, J. Zhang, and T. Kobayashi, "Broadband coherent anti-Stokes Raman scattering light generation in BBO crystal by using two crossing femtosecond laser pulses," *Opt. Lett.* **33**, 1494–1496 (Jul. 2008).

62. G. Patterson, R. Day, and D. Piston, "Fluorescent protein spectra," *J. Cell Sci.* **114**, 837–838 (May 2001).

63. M. Bruchez Jr., M. Moronne, P. Gin, S. Weiss, and A. P. Alivisatos, "Semiconductor nanocrystals as fluorescent biological labels," *Science* **281**, 2013–2016 (Sep. 1998).

64. X. Michalet, F. F. Pinaud, L. A. Bentolila, J. M. Tsay, S. Doose, J. J. Li, G. Sundaresan, A. M. Wu, S. S. Gambhir, and S. Weiss, "Quantum dots for live cells, in vivo imaging, and diagnostics," *Science* **307**, 538–544 (Jan. 2005).

65. S. Shrestha, B. E. Applegate, J. Park, X. Xiao, P. Pande, and J. A. Jo, "High-speed multispectral fluorescence lifetime imaging implementation for in vivo applications," *Opt. Lett.* **35**, 2558–2560 (Aug. 2010).

66. H. Kobayashi, M. Ogawa, R. Alford, P. L. Choyke, and Y. Urano, "New strategies for fluorescent probe design in medical diagnostic imaging," *Chem. Rev.* **110**, 2620–2640 (May 2009).

67. C. W. Freudiger, W. Min, B. G. Saar, S. Lu, G. R. Holtom, C. He, J. C. Tsai, J. X. Kang, and X. S. Xie, "Label-free biomedical imaging with high sensitivity by stimulated Raman scattering microscopy," *Science* **322**, 1857–1861 (Dec. 2008).

68. A. Penzkofer and H. J. Lehmeier, "Theoretical investigation of noncollinear phase-matched parametric four-photon amplification of ultrashort light pulses in isotropic media," *Opt. Quantum Electron.* **25**, 815–844 (Nov. 1993).

69. H. Valtna, G. Tamosauskas, A. Dubietis, and A. Piskarskas, "High-energy˘ broadband four-wave optical parametric amplification in bulk fused silica," *Opt. Lett.* **33**, 971–973 (May 2008).

70. A. Dubietis, G. Tamosauskas, P. Polesana, G. Valiulis, H. Valtna, D. Fac-˘cio, P. Di Trapani, and A. Piskarskas, "Highly efficient four-wave parametric amplification in transparent bulk Kerr medium," *Opt. Express* **15**, 11126–11132 (Sep. 2007).

71. J. Darginavicius, G. Tamo˘sauskas, G. Valiulis, and A. Dubietis, "Broad-˘ band four-wave optical parametric amplification in bulk isotropic media in the ultraviolet," *Opt. Commun.* **282**, 2995–2999 (Jul. 2009).

72. A. Dubietis, J. Darginavicius, G. Tamo˘sauskas, G. Valiulis, and A. Piskarskas, "Generation and amplification of ultrashort UV pulses via parametric four-wave interactions in transparent solid-state media," *Lithuanian J. Phys.* **49**, 421–431 (Apr. 2009).

73. D. Faccio, A. Grun, K. P. Bates, O. Chalus, and J. Biegert, "Optical¨ amplification in the near-infrared in gas-filled hollow-core fibers," *Opt. Lett.* **34**, 2918–2920 (Oct. 2009).

74. J. Liu, Y. Kida, T. Teramoto, and T. Kobayashi, "Simultaneous compression and amplification of a laser pulse in a glass plate," *Opt. Express* **18**, 4665–4672 (Feb. 2010).

75. S. Shimizu, Y. Nabekawa, M. Obara, and K. Midorikawa, "Spectral phase transfer for indirect phase control of sub-20-fs deep UV pulses," *Opt. Express* **13**, 6345–6353 (Aug. 2005).

76. M. T. Seidel, S. Yan, and H.-S. Tan, "Mid-infrared polarization pulse shaping by parametric transfer," *Opt. Lett.* **35**, 478–480 (Feb. 2010).

77. R. Trebino, *Frequency-Resolved Optical Grating: The Measurement of Ultrashort Laser Pulses*, Kluwer, Norwell, MA (2000).

78. J. Liu, K. Okamura, Y. Kida, and T. Kobayashi, "Temporal contrast enhancement of femtosecond pulses by a self-diffraction process in a bulk Kerr medium," *Opt. Express* **18**, 22245–22254 (Oct. 2010).

79. G. A. Mourou, T. Tajima, and S. V. Bulanov, "Optics in the relativistic regime," *Rev. Mod. Phys.* **78**, 309–371 (Apr./Jun. 2006).

80. S. V. Bulanov, T. Esirkepov, D. Habs, F. Pegoraro, and T. Tajima, "Relativistic laser-matter interaction and relativistic laboratory astrophysics," *Eur. Phys. J. D* **55**, 483–507 (Nov. 2009).

81. A. Jullien, O. Albert, F. Burgy, G. Hamoniaux, J.-P. Rousseau, J.-P. Chambaret, F. A. Rochereau, G. Che´riaux, J. Etchepare, N. Minkovski, and S. M. Saltiel, "10^{-10} temporal contrast for femtosecond ultraintense lasers by cross-polarized wave generation," *Opt. Lett.* **30**, 920–922 (Apr. 2005).

82. T. Schneider, D. Wolfframm, R. Mitzner, and J. Reif, "Ultrafast optical switching by instantaneous laser-induced grating formation and self-diffraction in barium fluoride," *Appl. Phys. B* **68**, 749–751 (Nov. 1999).

83. A. Jullien, L. Canova, O. Albert, D. Boschetto, L. Antonucci, Y. H. Cha, J. P. Rousseau, P. Chaudet, G. Cheriaux, J. Etchepare, S. Kourtev, N. Minkovski, and S. M. Saltiel, "Spectral broadening and pulse duration reduction during cross-polarized wave generation: Influence of the quadratic spectral phase," *Appl. Phys. B* **87**, 595–601 (Jun. 2007).

84. K. W. DeLong, R. Trebino, J. Hunter, and W. E. White, "Frequency resolved optical gating with the use of second-harmonic generation," *J. Opt. Soc. Amer. B* **11**, 2206–2215 (Nov. 1994).

85. X. W. Chen, Y. X. Leng, J. Liu, Y. Zhu, R. X. Li, and Z. Z. Xu, "Pulse self-compression in normally dispersive bulk media," *Opt. Commun.* **259**, 331–335 (Mar. 2006).

86. A. E. Jailaubekov and S. E. Bradforth, "Tunable 30-femtosecond pulses across the deep ultraviolet," *Appl. Phys. Lett.* **87**, 021107-1–021107-3 (Jul. 2005).

87. J. Wojtkiewicz and C. G. Durfee, "Hollow-fiber OP-CPA for energetic ultrafast ultraviolet pulse generation," in *Proc. Conf. Lasers and Electro Optics*, California, pp. 423–424 (2002).

88. J. Wojtkiewicz, K. Hudek, and C. G. Durfee, "Chirped-pulse frequency conversion of ultrafast pulses to the deep-UV," in *Proc. Conf. Lasers and Electro-Optics*, Maryland, pp. 186–188 (2005).

89. P. Tzankov, O. Steinkellner, J. Zheng, A. Husakou, J. Herrmann, W. Freyer, V. Petrov, and F. Noack, "Generation and compression of femtosecond pulses in the vacuum ultraviolet by chirped-pulse four-wave difference frequency mixing," in *Proc. Conf. Lasers and Electro-Optics*, California, pp. 1–2 (2006).

90. I. Babushkin and J. Herrmann, "High energy sub-10fs pulse generation in vacuum ultraviolet using chirped four wave mixing in hollow waveguides," *Opt. Express* **16**, 17774–17779 (Oct. 2008).

91. Y. Kida, J. Liu, T. Teramoto, and T. Kobayashi, "Sub-10fs deep-ultraviolet pulses generated by chirped-pulse four-wave mixing," *Opt. Lett.* **35**, 1807–1809 (Jun. 2010).

92. Y. Kida and T. Kobayashi, "Generation of sub-10-fs ultraviolet Gaussian pulses," *J. Opt. Soc. Amer. B* **28**, 139–148 (Jan. 2011).

93. E. J. Marcateli and R. Schmeltzer, "Hollow metallic and dielectric waveguides for long distance optical transmission and lasers," *Bell Syst. Tech. J.* **43**, 1783–1809 (Jul. 1964).

94. J.-C. Diels and W. Rudolph, *Ultrashort Laser Pulse Phenomena*, 2nd ed., Academic Press, San Diego (2006).

95. G. P. Agrawal, *Nonlinear Fiber Optics*, 4th ed., Academic Press, San Diego (2006).

96. M. Nisoli, S. D. Silvestri, and O. Svelto, "Generation of high energy 10fs pulses by a new pulse compression technique," *Appl. Phys. B* **68**, 2793–2795 (May 1996).

97. D. J. Kane and R. Trebino, "Characterization of arbitrary femtosecond pulses using frequency-resolved optical gating," *IEEE J. Quantum Electron.* **29**(2), 571–579 (Feb. 1993).

98. M. Spanner, M. Y. Ivanov, V. Kalosha, J. Hermann, D. A. Wiersma, and M. Pshenichnikov, "Tunable optimal compression of ultrabroadband pulses by cross-phase modulation," *Opt. Lett.* **28**, 749–751 (May 2003).

99. T. Kobayashi, J. Liu, and Y. Kida, "Generation and optimization of femtosecond pulses by four-wave mixing process," *IEEE J. of Selected topics in Quantum Electron.* **18**(1), 54–65 (Jan./Feb. 2012).

5.3 Tunable Multicolored Femtosecond Laser Pulses Generation by Using Cascaded Four-Wave Mixing (CFWM) in Bulk Materials

5.3.1 INTRODUCTION

Tunable, ultrashort laser pulses in different spectral ranges are powerful tools with applications in scientific research including ultrafast time-resolved spectroscopy [1–7], nonlinear microscopy [8–14], and laser micro-machining [15–18]. In the case of ultrafast time-resolved spectroscopy, which is widely used in the investigation of electronic and vibrational dynamics in molecules, the absorption peaks vary from sample to sample, and some of the molecular dynamics under investigation take place in less than 100 fs. As a result, sub-20 fs pulses with a time resolution high enough to observe real-time vibrational quantum beat and that are wavelength tunable in a wide range will play a key role. Nonlinear microscopy, such as two- or three-photon and second- or third-harmonic generation (SHG/THG) microscopy, are technologies widely used in biological research. Two-photon microscopy can be used in tissue imaging with a depth of several 100 μm [8,9], and three-photon microscopy can image to a depth of 1.4 mm [10]. SHG/THG microscopy can be used to image some biological tissues without the need for fluorescent proteins or staining with dyes and can achieve imaging depths of several 100 μm due to its use of long excitation wavelengths [11–14]. The pump laser sources used in nonlinear microscopy have pulse widths of ~100 fs or shorter, and a visible to middle-IR spectral range [8–14]. Some spectroscopy and microscopy experiments, such as the multicolored pump-probe experiment [19], two-dimensional spectroscopy [20], or multicolored nonlinear microscopy [21–23] require ultrashort pulses with several colors.

Conventionally used ultrashort laser sources have a spectral range of 650–950 nm (Ti:sapphire laser), 1000–1100 nm (Yb-/Nd-doped solid-state laser or fiber laser), or 1550 nm (Er-doped fiber laser). Great efforts have been made to extend the spectral range using nonlinear processes [24–40]. Optical parametric amplifier (OPA) and optical parametric oscillator (OPO) technologies are among the most successful methods for generating μJ-level pulses with spectral ranges from UV to mid-IR [32–35]. Spectrally tunable few-cycle pulses can be generated using a noncollinear optical parametric amplifier (NOPA) [36–40]. Commercial NOPA setups are available from several companies, although the price is still too high for many research groups. Pulses with broadband spectra from visible to IR (known as supercontinuum white light) also can be generated through filamentation in gases, bulk media, or fibers [41–45], although there are problems with the stability of the supercontinuum laser pulses [46–48].

Recently, four-wave mixing (FWM) has been studied as a new method for the generation of ultrashort pulses, including few-cycle pulses, with a spectral range from deep-UV (DUV) to mid-IR [49–66]. Among these results, the multicolored laser pulses that can be generated using cascaded four-wave mixing (CFWM) in transparent bulk materials are particularly attractive, due to their ultra-broadband spectral range, large wavelength tunable range and compact configurations [54–66]. Multicolored laser pulses generated by CFWM were first shown in semiconductor lasers in the

DOI: 10.1201/9780429196577-30

1980s [67]. Highly efficient multicolored (>4 colors) signals were generated in a nearly degenerate intracavity FWM experiment in a GaAs/GaAlAs semiconductor laser with a dye laser as the pump source for both the semiconductor laser and the FWM process. This was used as a method for the quantitative determination of the third-order nonlinear optical susceptibility of the semiconductor. Eckbreth then generated multicolored light (>4) with a coherent anti-Stokes Raman scattering process in several gases, and the light was used for hydrogen-fueled scramjet applications [68]. Harris and Sokolov showed that more than 13 sidebands with a spectral range from 195 nm to 2.94 μm were generated in D_2 gas by using a Raman process [69]. In 2000, Crespo and his co-workers reported multicolored (>11 colors) sideband generation using a cascaded highly nondegenerate FWM process in common glass [54]. Since then, studies have been conducted using other materials, such as sapphire plate [55,56], BBO crystal [57,58], fused silica glass [59], CaF_2 [60], BK7 glass [60], and diamond [61,62]. These studies have carefully investigated the mechanism and characteristics of multicolored laser pulses. The phase-matching condition of CFWM has also been discussed and used to explain the generation of multicolored sidebands with two noncollinear pump laser pulses [60]. Our experiment has shown that more than 15 spectral upshifted sidebands and two spectral downshifted pulses can be obtained with a spectral width broader than 1.8 octaves, covering the range from UV to near-IR [55–57,59,60]. The spectra of the multicolored sidebands can also be tuned in the broadband spectral range by adjusting the cross-angle of the two pump beams or simply by replacing the nonlinear media [59,60]. The pulse duration of different sidebands can be shorter than 50 fs without any extra dispersion compensation components [55–57,59,60], and sub-20 fs pulses can be obtained when the pump pulse chirp is carefully optimized [63]. Weigand and his co-workers also tried to recombine and synthesize all the sidebands and found that few-cycle visible-UV pulses were feasible [64,65]. The pulse energy of the first sideband can be higher than 1 μJ, with an energy conversion efficiency of around 10% [66]. A low pump threshold for multicolored sideband generation was reported when materials with high nonlinear refractive indices, such as diamond [62] or nanoparticle-doped materials [70], were used as the nonlinear medium in the experiment. A compact experimental setup for multicolored laser pulse generation was also constructed [62]. Aside from the one-dimensional multicolored sidebands discussed above, a two-dimensional (2-D) multicolored sideband array can be generated when the pump intensity is increased in various materials such as a sapphire [55,56], diamond [62], and BBO [71,72]. Characteristically, more than 10 arrays could be generated with pump energies of several to several tens of μJ [55,56,62]. CFWM, together with beam breakup due to ellipticity of pump beams or anisotropy of nonlinear media, are thought to be the main mechanisms behind this new phenomenon [62,71]. However, simulations based on the nonlinear Schrödinger equation are still needed for the phenomenon to be fully understood.

The remainder of this chapter is organized as follows. In Section 5.3.2, the theoretical analysis for multicolored pulse generation is presented. In Section 5.3.3, the characteristics of multicolored pulses are shown. The experimental setups are shown in Section 5.3.3.1. In Section 5.3.3.2, the spectral characteristics are introduced, i.e., the spectral range of the sidebands, the spectral width of each sideband, and the wavelength tunability of each sideband. The characterization of pulses is described in Section 5.3.3.3. Then, the pulse energy/output power and power stability are given in Section 5.3.3.4. Multicolored pulse generation with a low pump threshold is discussed in Section 5.3.3.5. In Section 5.3.4, 2-D multicolored sideband arrays are introduced and discussed. Finally, conclusions and some prospects for future research directions are given in Section 5.3.5. This article is written as a summary of recent publications reported by the authors.

5.3.2 THEORETICAL ANALYSIS

5.3.2.1 FWM PROCESS

FWM was found in the first decade of the laser epoch, and it has rapidly developed in the last 20 years. FWM is a third-order optical parametric process, in which four waves interact with each other through third-order optical nonlinearity [73]. Three waves form a nonlinear polarization at the frequency of the fourth wave during the FWM process. The wave functions of the four waves can be expressed as:

$$E_j(\mathbf{r},t) = A_j(r)\exp[i(\mathbf{k}_j \mathbf{r} - \omega_j t)] \quad (j = 1, 2, 3, 4) \tag{5.3.1}$$

where ω_j and $\mathbf{k}_j$ are frequencies and wave vectors of the four waves, and $A(r_j) = |A_j(r)|\exp[i\,(\phi(r))]$ is the complex amplitude.

There are two possible roadmaps of the FWM process that satisfy the conservation of photon energies and momenta. The phase-matching condition or conservation of photon energies and momenta can be written as:

(i) $\omega_4 = \omega_1 + \omega_2 + \omega_3$, $k_4 = \mathbf{k}_1 + \mathbf{k}_2 + \mathbf{k}_3$; (ii) $\omega_4 + \omega_3 = \omega_1 + \omega_2$, $\mathbf{k}_4 + \mathbf{k}_3 = \mathbf{k}_1 + \mathbf{k}_2$

The case (i) involves THG and third-order sum frequency generation. We are more interested in the FWM in case (ii), where the nonlinear polarization at frequency ω_4 can be written as:

$$P^{(3)}(\omega_4) = 3\varepsilon_0 \chi_{\mathrm{eff}}^{(3)} E_1(\omega_1) E_2(\omega_2) E_3 {}^*(\omega_3)\exp\left(i\,\Delta\mathbf{k}\cdot\mathbf{r}\right) \tag{5.3.2}$$

where $\chi_{\mathrm{eff}}^{(3)}$ is the effective third-order nonlinear optical susceptibility, and $k = \mathbf{k}_1 + \mathbf{k}_2 + \mathbf{k}_3 - \mathbf{k}_4$ is the wave vector phase-mismatching in the process. By solving the coupled-wave equations for FWM shown as follows:

$$dE_4(\omega_4)/dr = \left(i\omega_4 / \left(2\varepsilon_0 cn(\omega_4)\right)\right) P^{(3)}(\omega_4)\exp\left[-i\,\Delta\mathbf{k}\cdot\mathbf{r}\right] \tag{5.3.3}$$

we can get the optical field, $E_4(\omega_4)$, of the newly generated signal.

5.3.2.2 CFWM PROCESS

The theoretical analysis of CFWM processes for multicolored laser pulse generation is given in [60]. The schematic of CFWM processes is shown in Figure 5.3.1a. Two vectors, $\mathbf{k}_1$ and $\mathbf{k}_2$, correspond to the two input beams with frequencies of ω_1 and ω_2 ($\omega_1 > \omega_2$) respectively. The mth-order anti-Stokes (spectrally blue-shifted) and Stokes (spectrally red-shifted) sidebands are marked as ASm and Sm ($m = 1, 2, 3, \ldots$). Figure 5.3.1b–e shows the phase-matching geometries for generating the first three anti-Stokes sidebands (AS1, AS2, AS3) and the first Stokes sideband (S1). Based on these phase-matching geometries, the phase-matching condition for the mth-order anti-Stokes sideband can be written as: $k_{\mathrm{ASm}} = k_{\mathrm{AS}}(m-1) + k_1^{(m)} - k_2^{(m)} \approx (m+1)k_1^{(1)} - mk_2^{(1)}$, $\omega_{\mathrm{ASm}} \approx (m+1)\omega_1^{(1)} - m\omega_2^{(1)}$. Since the two input beams are never single-frequency lasers, $k_1^{(m)}$ and $k_2^{(m)}$ are used instead of $\mathbf{k}_1$ and $\mathbf{k}_2$.

The values of $\omega_1^{(m)}$, $\omega_2^{(m)}$, $\left|k_1^{(m)}\right|$, and $\left|k_2^{(m)}\right|$ may be different for every step of the m FWM processes. Similarly, with $k_1^{(-m)}$ and $k_2^{(-m)}$ used instead of $\mathbf{k}_1$ and $\mathbf{k}_2$, the mth-order Stokes sideband will have the following phase-matching condition: $k_{\mathrm{Sm}} = k_{S(m-1)} + k_2^{(-m)} - k_1^{(-m)} \approx (m+1)k_2^{(-1)} - mk_1^{(-1)}$, $\omega_{\mathrm{Sm}} \approx (m+1)\omega_2^{(-1)} - m\omega_1^{(-1)}$. As the lower-order signals will participate in the generation of adjacent higher-order signals as pump pulses, this process is called CFWM.

Based on the phase-matching condition expressed above, the output parameters, such as wavelength and output angle, of the generated sidebands can be calculated to explain and inform experimental work. In our experiments, the wavelength range of the two pump beams were 660–740 nm (Beam 1) and 800 nm (Beam 2). The nonlinear medium was assumed to be fused silica plate with a

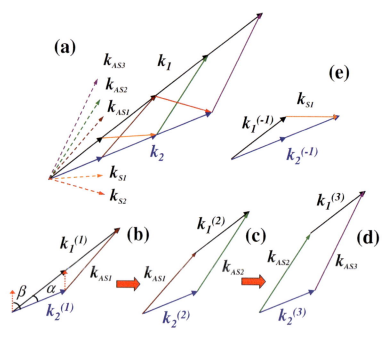

FIGURE 5.3.1 Schematic of multicolored sidebands generation using CFWM process. The phase-matching geometries for (a) AS1–AS3 and S1. (b) AS1. (c) AS2. (d) AS3, and (e) S1 [60].

thickness of 1 mm. The simulations were performed under these conditions. The wavelength dependence of Beam 1 for optimal phase-matching on the order number at different cross-angles is shown in Figure 5.3.2a. To fulfill the phase-matching condition, the wavelength of Beam 1 should redshift for higher-order anti-Stokes sidebands. The wavelengths of generated sidebands for different cross-angles are shown in Figure 5.3.2b, which clearly shows that the wavelengths of same order sidebands can be tunable by changing the cross-angle of the two pump beams, and the tuning range covered the wavelength gap between adjacent sidebands. The exit angles of the generated sidebands are plotted against the order number at different cross-angles in Figure 5.3.2c. The difference in exit angle between the multicolored sidebands was large enough for easy separation, even for adjacent sidebands. The dependence of the exit angle on the center wavelength of the generated sidebands at different cross-angles in different materials is shown in Figure 5.3.2d. The phase mismatching for the first four anti-Stokes sidebands at two different angles, 1.87° and 2.34°, is shown in Figure 5.3.2e. The increase of the slope of the curves with the order numbers means the reduction of the gain bandwidth for the sidebands, which was confirmed by our experiment. The calculations based on the phase-matching condition of CFWM agree with the experimental results, which will be given in the next section.

5.3.3   EXPERIMENTAL CHARACTERISTICS OF MULTICOLORED PULSES

5.3.3.1   EXPERIMENTAL SETUPS

Various experimental setups for multicolored laser pulse generation have been reported in the literature. The main differences between these setups are the methods for preparing the two pump laser beams. Crespo and his coworkers used two femtosecond pulses from a dye-laser amplifier system, with Beam 1 (561 nm, 40 fs) and Beam 2 (618 nm, 80 fs), and a pulse energy of 20 μJ for each beam [54].

Zhi used two OPA systems, pumped with a commercial Ti:sapphire amplifier [61]. The SHG signals of the signal and idler pulses from the two OPAs were used as the pump pulses for the generation of multicolored sidebands. The central wavelength and pulse energy of the two pump beams were 630 nm/1–3 μJ and 584 nm/1–3 μJ. There were also some other differences, including the Ti:sapphire amplifier pulse, and the supercontinuum generated in bulk materials [58].

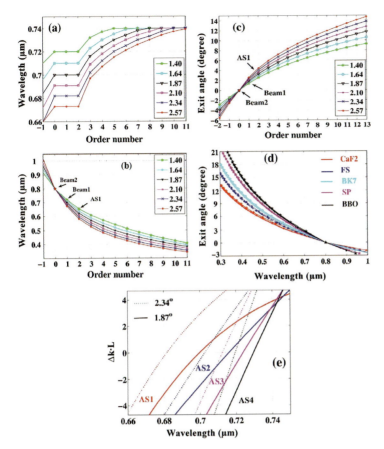

FIGURE 5.3.2 Calculated output parameters of generated sidebands from right to left. Dependence of (a) Central wavelength of Beam 1 for minimum phase-mismatching. (b) The wavelength and (c) The exit angles of generated sidebands on the order number at different crossing angles. Angles of 1.40, 1.64, 1.87, 2.10, 2.34, 2.57 are marked by lines composed of star, open circle, downward triangle, square, cross, filled circle. (d) Dependence of exit angle of the generated sidebands on the center wavelength at three different cross-angles, 1.40°, 1.87°, and 2.57°, in different nonlinear media. Curves from the top to the bottom at 0.5 um are for BBO, SP, BK7, SF, CaF2. (e) Phase mismatching of the sidebands from AS1 to AS4 at 1.87° and 2.34° in 1 mm fused silica [60].

We have used two experimental setups for multicolored pulse generation. As shown in Figure 5.3.3, we used Type-1 experimental setup for most of our work. The pump source was a 1 kHz Ti:sapphire regenerative amplifier laser system (35 fs/2.5 mJ/1 kHz/800 nm, Micra+Legend-USP, Coherent, Santa Clara, CA, USA). The pump laser was split into four beams for different uses. One beam (Beam 1), with energy of 300 μJ, was focused into a krypton-gas-filled hollow-core fiber with inner and outer diameters of 250 μm and 3 mm, and a length of 60 cm. The spectrum of Beam 1 broadened to a range extending from 600 to 950 nm after transmission through the hollow-core fiber, while the pulse energy decreased to about 190 μJ, due to coupling and propagation loss. A pair of chirped mirrors and two glass wedges were applied to compensate for the chirp of Beam 1 with broadband spectrum. A nearly transform-limited pulse, with a pulse duration of ~10 fs, was obtained by changing the bounce times on the chirped mirrors and the insertion of the glass wedges. Negatively and positively chirped pulses also can be obtained for different experiments. Beam 1 was then spectrally filtered with band-pass filters (BPF) short-wavelength-pass filters (SPF), or long-wavelength-pass filters (LPF) in different experiments. A concave mirror with a focal length of 600 cm was used to focus Beam 1 into the nonlinear medium (G1). Beam 2 was focused into the nonlinear medium by a lens with a focal length of 1 m. The fourth beam (Beam 4) was used to characterize the generated multicolored pulses with the cross-correlation frequency-resolved optical gating (XFROG) technique [74] in a 10 μm-thick BBO crystal.

Figure 5.3.4 shows the schematic of Type-2 setup, which was used for the generation of low-threshold multicolored sidebands and 2-D multicolored arrays in a diamond plate. Another Ti:sapphire laser system (35 fs/2.5 mJ/1 kHz/800 nm, Spitfire ACE, Spectra-Physics) was used as the pump source, and a beam with pulse energy of 150 μJ was used in the experiment. A BK7 glass plate with a thickness of 3 mm was used to spectrally broaden the laser pulse using a self-phase

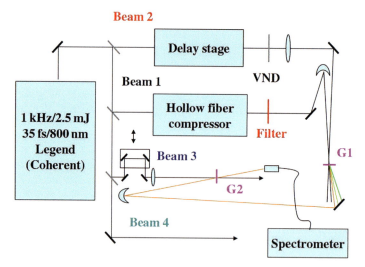

FIGURE 5.3.3 Type-1 experimental setup. VND, variable neutral-density filter; G1, nonlinear medium for multicolored sidebands generation; G2, nonlinear medium for pulse measurement with an X-FROG system [74].

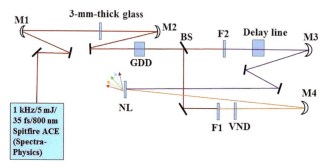

FIGURE 5.3.4 Type-2 experimental setup. Focal lengths of concave mirrors, M1, M3, and M4 are 500 mm, while that of M2 is 250 mm; GDD, chirped mirrors; F1, longpass filter with a cut-off wavelength of 800 nm; F2, short-pass filter with a cut-off wavelength of 800 nm; BS, beamsplitter; VND, variable neutral-density filter; NL, nonlinear medium for multicolored pulse generation.

modulation (SPM) process. Then, a pair of chirped mirrors (GDD) was used to compensate for the dispersion induced by the BK7 glass and other components. Then, the laser beam was split into two parts, Beam 1 and Beam 2, with a beamsplitter (BS). Beam 1 first propagated through a short-pass filter (F1, cut-off wavelength of 800 nm) and was then focused into the nonlinear medium by a concave mirror (M4) with a focal length of 500 mm. Beam 2 was spectrally filtered with a long-pass filter (F2, cut-on wavelength of 800 nm), and then focused into the nonlinear medium by another concave mirror (M3) with a focal length of 500 mm. The beam diameters of both Beam 1 and Beam 2 were ~300 μm on the 1 mm thick diamond plate. As no hollow fiber or gas chamber is used, the Type-2 experimental setup occupied half the space of a Type-1 setup.

In the experiment, the diameters of the incident beams at the position of the nonlinear media were measured using a CCD camera (BeamStar FX33, Ophir Optronics: Jerusalem, Israel). The pulses were characterized using the XFROG or SHG-FROG technique and retrieved with a commercial software package (FROG 3.0, Femtosoft Technologies: Chennai, India). The spectra of the pulses were measured with a commercial spectrometer (USB4000, Ocean Optics: Dunedin, FL, USA). To avoid optical damage, the intensities of the two pump beams on the surface of nonlinear media surface were set at least one order of magnitude lower than the damage threshold for all the media used. Neither damage nor supercontinuum generation was observed in any experiment. CCD devices convert or manipulate an electrical signal into some kinds of output, including sets of digital valued ensemble. Thus CCD can catch visual information to convert into an image or motional image like video images. They are in this sense digital cameras. The captured images are transferred to the camera's memory system to record them as electronic data. A CCD camera imaging plane is composed of light-sensitive pixel elements with high quality, low noise image with a dynamic range of higher than 3×10^5.

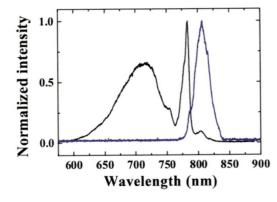

FIGURE 5.3.5 The spectra of the two pump beams, Beam 1 (black curve with a broad peak around 720 nm) and Beam 2 (blue curve with a sharp peak at 810 nm) [75].

5.3.3.2 SPECTRA AND WAVELENGTH TUNING OF MULTICOLORED SIDEBANDS

5.3.3.2.1 Tuning the Wavelength of Sidebands by Changing Cross-Angle

The Type-1 experimental setup was applied, and a short-pass filter with cut-off wavelength of 800 nm was used to eliminate the red components of Beam 1. As a result, the spectra of the two pump pulses were as in Figure 5.3.5. The input powers of Beam 1 and Beam 2 were 11 and 19 mW, respectively.

Multicolored sidebands were obtained when the input beams overlapped well both spatially and temporally, as shown in Figure 5.3.6a. These sidebands were in the same plane as the pump beams but separated with different exit angles. Figure 5.3.6b depicts the spectra of the lowest four-order red-shifted sidebands, AS1–AS4, when the pump beam cross-angle was 2.1°. The spectral width decreased with an increase in the order number of the sidebands, as was predicted by the calculations, shown in Figure 5.3.2e. Figure 5.3.6c shows the central wavelength of AS2 with different pump beam cross-angles. The central wavelength of AS2 shifted from 500 to 625 nm, and the cross-angle increased from 1.5° to 2.5°. This tuning range successfully surpassed the wavelength gap between AS1 and AS3, as shown in Figure 5.3.6b. Therefore, it was possible to tune the wavelength continuously by simple angle tuning, without a gap between the neighboring order sidebands. The spectra of AS8–AS15 are shown in Figure 5.3.6d. The spectra of S1 and S2 are shown in Figure 5.3.6e. The whole wavelength range obtained by angle tuning of all sidebands covered the near UV-visible-near IR range from 360 nm to 1.2 μm, corresponding to more than 1.8 octaves. These broadband spectra and large tunability are very attractive and useful from the viewpoints of application and creating simple tuning mechanisms.

5.3.3.2.2 Tuning the Wavelength of Sidebands by Changing Nonlinear Media

The phase-matching condition for CFWM is $k_{Sm} = k_{S(m-1)} + k_2^{(-m)} - k_1^{(-m)} \approx (m+1)k_2^{(-1)} - mk_1^{(-1)}$, $\omega_{Sm} \approx (m+1)\omega_2^{(-1)} - m\omega_1^{(-1)}$, as discussed in Section 5.3.2. The wave vectors k can be written as $k = n \times k_0$, where n is the linear refractive index of the nonlinear medium and k_0 is the wave vector in vacuum. This means that the refractive index (dispersion curve) of the medium will also influence the phase-matching conditions. The refractive index (dispersion curve) can be adjusted by replacing the medium with different optical properties. Figure 5.3.7 shows the spectra of AS1 and AS3 for nonlinear media (CaF_2 plate, fused silica plate, BK7 glass plate, sapphire plate, and BBO crystal) with a fixed pump beam cross-angle of 1.8°. By changing the media, the central wavelength of AS1 could be tuned from 640 to 610 nm, while the central wavelength of AS3 was adjusted correspondingly from 490 to 560 nm. The spectrum of AS3 in the BBO crystal overlapped with the spectrum of AS1 in the CaF_2 crystal, which means that spectral gaps between neighboring sidebands can be bridged simply by replacing the nonlinear medium.

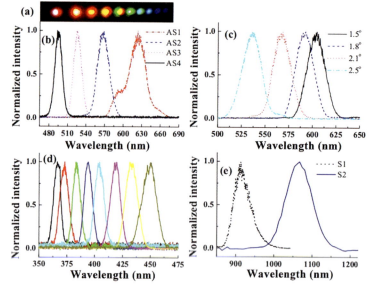

FIGURE 5.3.6 (a) Photograph of the first ten anti-Stokes sidebands on a white paper set 1 m far from the nonlinear medium. (b) The spectra of AS1–AS4 with cross-angle of 2.1° for two pump beams. (c) The spectra of AS2 with different cross-angles. Spectra of (d) AS8–AS15 from right to left and (e) Two Stokes signals S1 and S2 with cross-angle of two pump beams set as 1.5° [59].

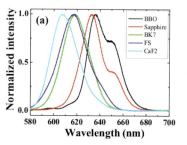
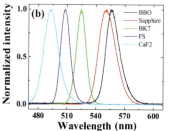

FIGURE 5.3.7 Spectra of (a) AS1 emitted from BBO, Sapphire BK7, FS, CaF2 are depicted from right to left and (b) AS3 of five different materials with cross-angle of 1.8° [60].

5.3.3.3 TEMPORAL CHARACTERISTICS OF MULTICOLORED PULSES

The pulse durations of Beam 1 and Beam 2 were measured to be 40 and 55 fs, respectively. The characteristics of the sidebands, S1, AS1, and AS2, which were generated using these two pump pulses are shown in Figure 5.3.8. The pulse durations of AS1, AS2, and S1 were calculated to be 45, 44, and 46 fs, respectively. The retrieved phase showed that these pulses were all positively chirped, and that the positive chirp induced by material dispersion of the nonlinear medium prevented shorter pulses from being obtained.

As discussed in [63], chirped pump pulses can be used for pre-compensation of the positive chirp of the sidebands, resulting in even shorter pulse durations. The principle of the process can be explained in the following way. In the CFWM process, the m-th-order anti-Stokes sideband has the phase matching condition: $k_{ASm} = k_{AS(m-1)} + k_1 - k_2 = (m+1)k_1 - mk_2$, $\omega_{ASm} \approx (m+1)\omega_1 - m\omega_2$. The m-th ($m > 0$) order anti-Stokes signal can be expressed as follows, if the electric field of two incident pulses is given as:

$$E_j(t) \propto \exp\left\{i\left[\omega_{j0}t + \phi_j(t)\right]\right\}, j = 1, 2$$

$$E_{ASm}(t) \propto \exp\left\{i\left[\left((m+1)\omega_{10} - m\omega_{20}\right)t + \left((m+1)\phi_1(t) - m\phi_2(t)\right)\right]\right\}$$

(5.3.4)

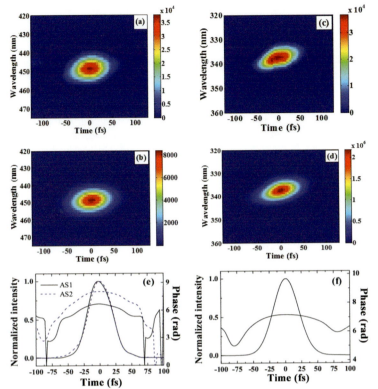

FIGURE 5.3.8 (a) Measured and (b) Retrieved XFROG traces of S1. (c) Measured and (d) Retrieved XFROG traces of AS2 when the cross-angle was 1.87°. Retrieved temporal intensity profiles and phases of (e) AS1 (solid line), AS2 (dashed line) and (f) S1 [66].

If Beam 1 is negatively chirped $\left(\partial^2\phi_1(t)/\partial t^2 < 0\right)$ and Beam 2 is positively chirped $\left(\partial^2\phi_2(t)/\partial t^2 > 0\right)$, we can obtain:

$$\partial^2\phi_{ASm}(t)/\partial t^2 = (m+1)\partial^2\phi_1(t)/\partial t^2 - m\partial^2\phi_2(t)/\partial t^2 < 0 \qquad (5.3.5)$$

This means that the m-th order blued-shifted field, E_{ASm}, can be negatively chirped. As such, a nearly transform-limited pulse can be achieved, if the negative chirp of the ASm field is precisely adjusted to correctly compensate for the dispersion induced by the nonlinear media and other optical components used in the processes of pulse generation and characterization. By this method, the pulse durations of AS1 and AS2 were compressed to 15 and 16 fs, respectively, as shown in Figure 5.3.9. Further optimization of the dispersion, including higher-order dispersion, is needed to obtain truly transform-limited pulses.

5.3.3.4 OUTPUT POWER/ENERGY OF MULTICOLORED PULSES

Table 5.3.1 shows the average power of AS1–AS3 obtained with five different bulk media. The external cross-angle of the two pump beams was 1.8°, while the input powers of Beam 1 and Beam 2 were set at 6.5 and 25 mW respectively. CaF_2 had the highest AS1 output power, and the lowest AS3 output power of all five media. Conversely, in BBO crystal, the powers of the sidebands decreased the most rapidly with an increasing order number. This phenomenon can be explained by the different phase-matching conditions and dispersion properties of the five materials.

Figure 5.3.10a shows the power dependence of AS1 on the power of Beam 1, with the power of Beam 2 fixed at 19 mW and the cross-angle set at 1.8°. The output power of AS1 was sensitive to the pump power with a low pump rate, and saturation occurred when the power of Beam 1 increased to

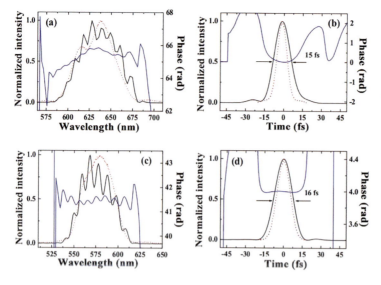

FIGURE 5.3.9 (a) Measured (dotted line), retrieved (black) spectral intensity profile and retrieved spectral phase (blue) of AS1. (b) Retrieved temporal intensity profile (black), temporal phase (blue), and calculated transform-limited temporal intensity profile (dotted line) of AS1. (c) Measured (dotted), retrieved (black) spectral intensities and retrieved spectral phase (blue) of AS2. (d) Retrieved temporal intensity profile (black), phase (blue), and calculated transform-limited temporal intensity profile (red) of AS2 [63].

TABLE 5.3.1

The Output Power of AS1–AS3 with Five Commonly Used Third-Order Nonlinear Media

µW	CaF$_2$	Fused Silica	BK7	Sapphire Plate	BBO
AS1	480	700	715	750	780
AS2	210	315	295	210	135
AS3	125	90	60	40	10

Note. The external cross-angle of two pump beams is 1.8°, while the input powers of Beam 1 and Beam 2 are 7 and 25 mW, respectively

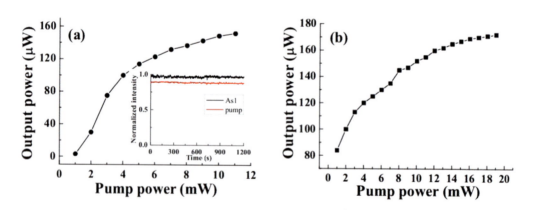

FIGURE 5.3.10 The power dependence of AS1 on (a) Beam 1 with power of Beam 2 fixed at 19 mW, and (b) Beam 2 with power of Beam 1 fixed at 11 mW. Power stabilities of AS1 (top) and Beam 1 (bottom) in 20 min are shown as the insertion of (a) [59].

about 11 mW. Similarly, when the power of Beam 1 was set to 11 mW and Beam 2 had a high pump power, saturation of the output power of AS1 appeared, as shown in Figure 5.3.10b. This saturation may have helped to obtain sidebands with high stability. The power stability of AS1 and Beam 1 were 0.95% and 0.62% in RMS, respectively, as shown in the inset of Figure 5.3.10a.

By optimizing the spatial and temporal overlap of the two pump beams, the maximum pulse energy of S1 and AS1 reached was higher than 1 µJ [66]. An even higher output power was achieved by enlarging the pump beam size and increasing the pump power.

5.3.3.5 Multicolored Sidebands Generated with Low Threshold

The polarization at frequency ω_{AS1} in the FWM process generating AS1 can be written as:

$$P^{(3)}(\omega_{AS1}) \alpha \chi_{eff}^{(3)} E^2(\omega_1) E^*(\omega_2) \tag{5.3.6}$$

According to the coupled-wave equations in the FWM process, the optical field and polarization at frequency ω_{AS1} had the following relationship:

$$\frac{dE(\omega_{AS1})}{dz} = \frac{i\omega_{AS1}}{2\varepsilon_0 c n_{AS1}} P^{(3)}(\omega_{AS1}) e^{-i\Delta kz}$$

Here, Δk is the wave vector mismatch in the FWM process. From Eqs. (5.3.5) and (5.3.6), it shows that the AS1 intensity becomes higher, following the proportionality relation with the squared absolute nonlinear optical susceptibility $\left(\left|\chi_{eff}^{(3)}\right|^2\right)$ of the material used.

Based on this, diamond, the nonlinear optical susceptibility of which is ~5 times larger than that of sapphire and ~10 times larger than that of CaF_2 [73], was used in the experiment to obtain multicolored sidebands with a low threshold.

The experiment was performed with the Type-2 setup shown in Section 5.3.3.1. The spectra of two pump beams, Beam 1, and Beam 2, are depicted in Figure 5.3.11. The two spectral positions were adjusted by tuning the angle between input beams and filters. The retrieved temporal intensity profiles and phases of two pump pulses are shown in Figure 5.3.12, where the Beam 1 and Beam 2 pulse durations are 81 and 47 fs, respectively.

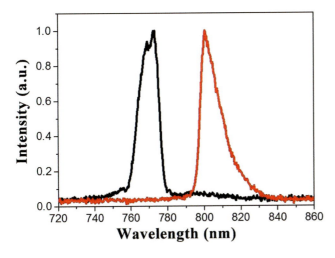

FIGURE 5.3.11 The spectra of Beam 1 (black left), Beam 2 (red right) [62].

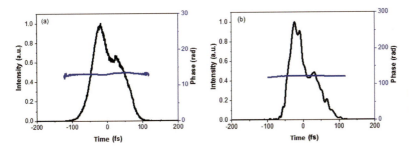

FIGURE 5.3.12 The retrieved intensity profile and phase of (a) Beam 1. (b) Beam 2 [62].

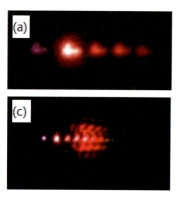

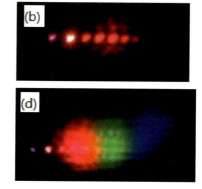

FIGURE 5.3.13 Multicolored pattern generated with different pump levels. (a) Beam 1: 0.855 mW, Beam 2: 0.410 mW. (b) Beam 1: 0.855 mW, Beam 2: 0.856 mW. (c) Beam 1: 0.855 mW, Beam 2: 1.121 mW. (d) Beam 1: 0.855 mW, Beam 2: 1.970 mW [62].

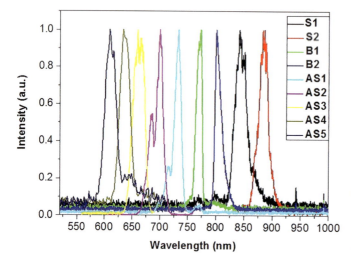

FIGURE 5.3.14 The normalized spectra of two pump beams, Beam 1 and Beam 2, and several sidebands AS5–AS1 (from left most to right), S2, S1 from rightmost to the next [62].

The average power of Beam 1 was set to 0.855 mW, and the average power of Beam 2 was continuously changed by a VND. Figure 5.3.13 shows the multicolored sidebands at different pump levels. The intensities of Beam 1 and Beam 2 on the diamond plate in Figure 5.3.13a were calculated to be 14.9×10^9 and 12.3×10^9 W/cm^2 respectively. These were much lower than the threshold intensities obtained previously for multicolored sideband generation in a fused silica plate, of 60×10^9 and 8×10^9 W/cm^2 [70]. This low pump threshold for multicolored sideband generation is important in the context of an actual experiment because pump lasers with high repetition frequencies, i.e., several 100 kHz to several MHz, inevitably have a low pulse energy when used with a conventional amplifier system. Compared to the multicolored sidebands generated with a 1 kHz amplifier, pulses with a higher repetition frequency are more useful in nonlinear microscopy. Low repetition frequencies make the image frame times unconventionally long and lead to high noise levels. Figure 5.3.13b–d shows the 2D structure obtained by increasing the power of Beam 2, a detailed discussion of which will be given in the next section.

Figure 5.3.14 shows the spectra of generated multicolored sidebands obtained under pump rates of 0.855 mW and 0.856 mW for Beam 1 and Beam 2, respectively. The normalized spectra, AS1–AS5, of the two pump beams, S1 and S2, are shown in Figure 5.3.14. The spectral width of these sidebands was broader than 10 nm, which means that a pulse duration of <100 fs was achievable. The spectra of these sidebands could also be continuously tuned by adjusting the cross-angle of Beam 1 and Beam 2.

TABLE 5.3.2

Output Power of AS1–AS5 and S1 with Average Power of 0.855 and 0.856 mW for Beam 1 and Beam 2, Respectively [62]

Sidebands	AS1	AS2	AS3	AS4	AS5	S1
Power (µW)	34.0	6.1	2.5	1.3	0.8	31.4

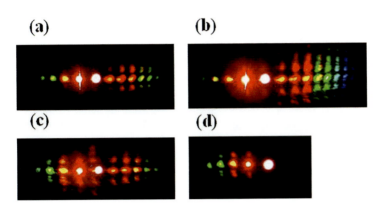

(a) **(b)**

(c) **(d)**

FIGURE 5.3.15 Photographs of the multicolored arrays generated with (a) Pulse energy of beam 2 of 220 µJ. (b) Pulse energy of beam 2 of 250 µJ. (c) Time delay of two pump beams of 7 fs and pulse energy of beam 2 of 250 µJ. (d) A short-pass filter cut-off wavelength at 820 nm inserted in the Beam 1 path [55].

Table 5.3.2 shows the output power of AS1–AS5 and S1 when the average power of Beam 1 and Beam 2 were set at 0.855 mW and 0.856 mW. The conversion efficiency was about 1.84%, 1.99%, 0.36%, 0.15%, 0.08% and 0.05% for S1, AS1, AS2, AS3, AS4 and AS5 respectively. The power of AS1–AS4 could be increased by increasing the pump rate, as shown in Figure 5.3.13c and d.

5.3.4   2-D MULTICOLORED SIDEBANDS ARRAYS

Zeng and his coworkers first observed 2-D multicolored arrays in 2006 in a quadratic nonlinear medium (BBO crystal) with two closely-overlapped femtosecond laser beams from Ti:sapphire amplifier and its SHG signal [71]. The cause of the 2-D pattern was thought to be the cascaded quadratic nonlinear process, together with spatial breakup of the quadratic spatial solitons induced by ellipticity of the input beams. The 2-D structure could also be suppressed by another weak SHG beam.

Zhi also generated 2-D multicolored arrays in a diamond plate with three pump beams [61], attributed to the interaction of two different sets of cascaded stimulated Raman scattering processes.

We observed a similar structure in a cubic nonlinear medium, sapphire plate, with only two pump beams in 2008 [55,56]. 2-D multicolored arrays were generated when pump energy was increased. Figure 5.3.15 shows the 2-D multicolored arrays generated under various conditions. The 2-D multicolored arrays could be controlled by changing the intensity, delay, or polarization of one input beam.

We performed another experiment to study the characteristics of the 2-D multicolored arrays in detail. The schematic of this experiment is shown in Figure 5.3.16a. The two pump beams had a cross-angle of 1.8°, and a beam size of 300 µm in a sapphire plate. Stable 2-D multicolored arrays were generated when Beam 1 and Beam 2 overlapped in time and space in the sapphire plate, as shown in Figure 5.3.16b. Spatially well-separated multicolored sidebands with >10 columns and rows were observed. The columns were approximately normal to the center row while the rows adjacent to the center row were not parallel to each other. The 2-D multicolored array sidebands are defined as $B_{m,n}$ for convenience, as shown in Figure 5.3.16c. $B_{0,0}$ and $B_{-1,0}$ stand for the two pump beams, Beam 1 and Beam 2, respectively.

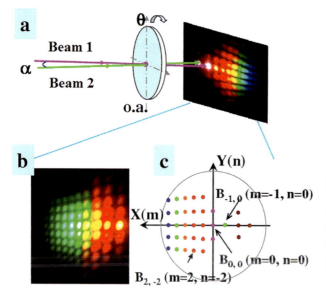

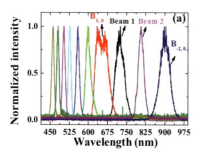

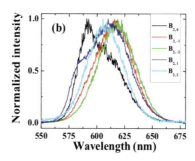

FIGURE 5.3.16 (a) Schematics of the generation of 2-D multicolored arrays. (b) A photograph of the 2-D multicolored arrays generated in sapphire plate. (c) 2-D multicolored arrays are defined as $B_{m,n}$, in which $B_{0,0}$ and $B_{-1,0}$ refer to Beam 1 and Beam 2, respectively [56]. o.a., optical axis.

FIGURE 5.3.17 The spectra of sidebands on (a) The center row $B_{m,0}$; and (b) The second column $B_{2,n}$ [56].

The spectra of the sidebands on the center row $B_{m,0}$ and the second column $B_{2,n}$ were measured, as shown in Figure 5.3.17. The sidebands on the central row were generated through a CFWM process, which is almost the same as discussed in previous sections. The center wavelengths between neighboring spots on the same column were approximately the same, as shown in Figure 5.3.17b. A more accurate experiment using pump beams with narrower spectra will help to confirm these characteristics.

The powers of Beam 1 and Beam 2 were set to 0.1 and 25 mW, respectively. The measured powers of some sidebands at this pump rate are shown in Figure 5.3.18a. The powers of sidebands on the rows of

$B_{m,0}$, $B_{m,1}$, $B_{m,-1}$, $B_{m,2}$, $B_{m,-2}$ are shown as the star, red circle, black square, green triangle, and blue triangle, respectively. We found that the power of the sidebands on $B_{m,1}$ and $B_{m,-1}$ were approximately the same as the value of m. The sidebands in $B_{m,2}$ and $B_{m,-2}$ also had this property, showing that the power distribution had mirror symmetry with the central line of $B_{m,0}$. The power dependence of three sidebands, $B_{1,0}$, $B_{2,0}$, and $B_{2,1}$, on the input power of Beam 2 is shown in the inset of Figure 5.3.18a.

During the experiment, the power of Beam 1 was amplified from 0.1 to 0.17 mW, which means that the FWM process could also be used for parametric amplification. The power stability of several sidebands in different arrays, measured with a photodiode, is shown in Figure 5.3.18b. The stabilities are all in the range of 0.5%–2% in RMS within 200 s.

In 2013, two-dimensional multicolored arrays were observed with a low pump rate in a diamond plate, as shown in Figure 5.3.13. The experimental setup was the same as the Type-2 setup shown in

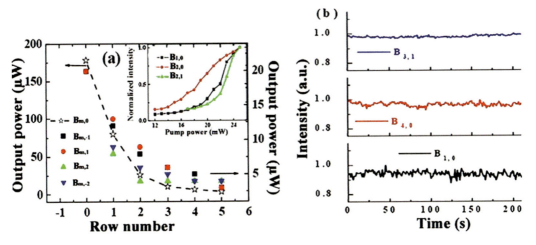

FIGURE 5.3.18 (a) The output power of sidebands on the $B_{m,0}$, $B_{m,1}$, $B_{m,-1}$, $B_{m,-2}$, $B_{m,2}$, with pump power of 0.1 and 25 mW for Beam 1 and Beam 2, respectively. The inset shows the power dependence of three different sidebands, $B_{1,0}$, $B_{2,0}$, and $B_{2,1}$, on power of Beam 2. (b) The power stabilities of three sidebands $B_{3,1}$, $B_{4,0}$ and $B_{1,0}$ [56].

Section 5.3.3.1. Increasing the pulse energy of Beam 2 to 0.856 μJ created 2-D multicolored arrays, as shown in Figure 5.3.13b. The threshold energy was much lower than that for a sapphire plate, i.e., <2 μJ for a diamond plate and 25 μJ for a sapphire plate [56,62]. More sidebands, lines and brighter arrays were observed when the energy of Beam 2 was increased further, as shown in Figure 5.3.13c and d.

The mechanism for the generation of 2D structure is not yet fully clear, although we have given some explanations based on the CFWM and beam breakup [62]. Detailed simulations of spontaneous breakup of elliptical laser beams were performed by Majus and his coworkers [76]. They attributed this breakup to multistep four-wave and parametric amplification of certain components occurring in the spatial spectrum of the self-focusing laser beam. Interestingly, beam breakup was observed even with a near circular (the ellipticity $e = 1.09$) input beam when the input power was ~20 times larger than P_{cr}, which is defined as follows [77]:

$$P_{cr} = 3.77\lambda^2 / (8\pi n n_2) \tag{5.3.7}$$

Here, λ is the laser wavelength in vacuum, n is the refractive index, and n_2 is the nonlinear refractive index. The pump beams have an ellipticity of ~1.2 in the experiment, due to asymmetric focusing with several concave mirrors. The summation of the peak powers of the two input beams is about ~60 P_{cr} ($P_{cr} = 0.4$ MW for diamond plate) and would be large enough for beam breakup [76].

The combination of this beam breakup and CFWM is the main cause of this 2-D multicolored structure.

Beam breakup can also occur when the pump beams are circular. Dergachev has investigated the interaction of two noncollinear femtosecond laser filaments in sapphire both numerically and experimentally [78]. The simulation was based on the nonlinear Schrödinger equation. The experiment was performed with a Ti:sapphire amplifier with a pulse width of 120 fs. The two incident beams, which are all from the same laser source, have a cross-angle of 4.64°. Because of the asymmetric distributions of refractive index changes in the nonlinear material due to the Kerr effect of the pump beams, additional "hot points" or plasma channels arise in the plane oriented perpendicular to the pulse propagation plane with input pulse power above 10 P_{cr} (corresponding to a pulse energy of 3~4 μJ). Clear beam breakup was observed. The energy distribution of light spots at the output end can be adjusted by tuning the phase between the two input beams.

We are working on simulations based on the nonlinear Schrödinger equation including FWM and other nonlinear processes, such as optical Kerr effect and multiphoton absorption, which are required for full understanding of this phenomenon.

5.3.5   CONCLUSION AND PROSPECTS

In conclusion, we have investigated multicolored sideband generation based on CFWM both theoretically and experimentally. Analysis and computer simulation including the effect of phase-matching in FWM reasonably explain the reported experimental results.

The main characteristics of multicolored sidebands obtained in our experiment can be summarized as follows.

1. Tunability in a wide spectral region

    Fifteen spectral up-shifted pulses and two spectral down-shifted emissions were obtained simultaneously in a spectra domain that spanned more than 1.8 octaves. The wavelengths of the sidebands could be tuned from near-ultraviolet to near-infrared by adjusting the crossing angle between the two input beams or by replacing the nonlinear bulk medium.
2. Ultrashort pulse width

    The pulse width of the sidebands remained nearly unchanged for the Stokes and anti-Stokes pulses. Nearly transform-limited compressed pulses as short as 15 fs could be obtained when one of the two input beams was properly negatively chirped and the other was positively chirped.
3. High output energy

    The pulse energy of the sideband could be increased to 1 µJ, power stability better than 1% RMS. We expect that an even higher output power could be generated by increasing the pump energy and expanding the spot sizes of the two pump beams on the optical medium to avoid saturation.

We have reported the generation of multicolored sidebands consisting of 2-D arrays and provided some possible explanations. Beam breakup and CFWM are responsible for this interesting phenomenon. Careful investigation using both simulation and experiment is still needed for complete understanding of this new phenomenon [79].

Future studies into multicolored sidebands extending in the visible and near-IR spectral regions, which are generated with pump lasers with MHz repetition rates, would be useful in numerous applications such as nonlinear microscopy [79]. Pulse energy of 1 µJ for each pump pulse has been confirmed to be enough energy for multicolored sideband generation in a diamond plate. This energy can be reduced further, if the pump beams are tightly focused on a medium with higher third-order nonlinearity. For example, CdSSe-nanoparticle-doped glass has 5.6 times larger third-order susceptibility than a diamond plate, even at wavelengths far from its resonant frequency [73].

REFERENCES

1. M. Dantus, M. J. Rosker, and A. H. Zewail, "Real-time femtosecond probing of "transition states" in chemical reactions," *J. Chem. Phys.* **87**, 2395 (1987).
2. R. M. Bowman, M. Dantus, and A. Zewail, "Femtosecond transition-state spectroscopy of iodine: From strongly bound to repulsive surface dynamics," *Chem. Phys. Lett.* **161**, 297–302 (1989).
3. A. Douhal, S. K. Kim, and A. H. Zewail, "Femtosecond molecular dynamics of tautomerization in model base pairs," *Nature* **378**, 260–263 (1995).
4. V. Hertel and W. Raldoff, "Ultrafast dynamics in isolated molecules and molecular clusters," *Rep. Prog. Phys.* **69**, 1897 (2006).

5. T. Kobayashi, T. Saito, and H. Ohtani, "Real-time spectroscopy of transition states in bacteriorhodopsin during retinal isomerization," *Nature* **414**, 531–534 (2001).

6. T. Kobayashi and Y. Kida, "Ultrafast spectroscopy with sub-10 fs deep-ultraviolet pulses," *Phys. Chem. Chem. Phys.* **14**, 6200–6210 (2012).

7. C. T. Middleton, K. L". Harpe, C. Su, Y. K. Law, C. E. C. Hernandez, and B. Kohler, "DNA excited-state dynamics: From single bases to double Helix," *Annu. Rev. Phys. Chem.* **60**, 217–239 (2009).

8. W. Denk, W. J. H. Strickler, and W. W. Webb, "Two-photon laser scanning fluorescence microscopy," *Science* **248**, 73–76 (1990).

9. F. Helmchen and W. Denk, "Deep tissue two-photon microscopy," *Nat. Methods* **2**, 932–940 (2005).

10. N. G. Horton, K. Wang, D. Kobat, C. W. Clark, F. W. Wise, C. B. Schaffer, and C. Xu, "*In vivo* three-photon microscopy of subcortical structures within an intact mouse brain," *Nat. Photon.* **7**, 205–209 (2013).

11. P. J. Campagnola and L. M, Loew, "Second-harmonic imaging microscopy for visualizing biomolecular arrays in cells, tissues and organisms," *Nat. Biotechnol.* **21**, 1356–1360 (2003).

12. P. J. Campagnola, H. A. Clark, and W. A. Mohler, "Second-harmonic imaging microscopy of living cells," *J. Biomed. Opt.* **6**, 277–286 (2001).

13. D. Débarre, W. Supatto, A. M. Pena, A. Fabre, T. Tordjmann, L. Combettes, M. C. Schanne-Klein, and E. Beaurepair, "Imaging lipid bodies in cells and tissues using third-harmonic generation microscopy," *Nat. Methods* **3**, 47–53 (2006).

14. Y. Barad, H. Eisenberg, H. M. Horowitz, and Y. Silberberg, "Nonlinear scanning laser microscopy by third harmonic generation," *Appl. Phys. Lett.* **70**, 922 (1997).

15. R. R. Gattass and E. Mazur, "Femtosecond laser micromachining in transparent materials," *Nat. Photon.* **2**, 219–225 (2008).

16. X. Liu, D. Du, and G. Mourou, "Laser ablation and micromachining with ultrashort laser pulses," *IEEE J. Quant. Electron.* **33**, 1706–1716 (1997).

17. C. B. Schaffer, A. Brodeur, J. F. Garcia, and E. Mazur, "Micromachining bulk glass by use of femtosecond laser pulses with nanojoule energy," *Opt. Lett.* **26**, 93–95 (2001).

18. M. Huang, M. F. L. Zhao, Y. Cheng, N. S. Xu, and Z. Z. Xu, "Origin of laser-induced near-subwavelength ripples: Interference between surface plasmons and incident laser," *ACS Nano* **3**, 4062–4070 (2009).

19. R. Zgadzaj, E. Gaul, N. H. Matlis, G. Shvets, and M. C. Downer, "Femtosecond pump-probe study of preformed plasma channels," *J. Opt. Soc. Am. B* **21**, 1559–1567 (2004).

20. R. M. Hochstrasser, "Two-dimensional spectroscopy at infrared and optical frequencies," *Proc. Natl. Acad. Sci. USA* **104**, 14190–14196 (2007).

21. K. W. Dunn, R. M. Sandoval, K. J. Kelly, P. C. Dagher, G. A. Tanner, S. J. Atkinson, R. L. Bacallao, and B. A. Molitoris, "Functional studies of the kidney of living animals using multicolor two-photon microscopy," *Am. J. Physiol. Cell Physiol.* **283**, 905–916 (2002).

22. E. Sahai, J. Wyckoff, U. Philippar, J. E. Segall, F. Gertler, and J. Condeelis, "Simultaneous imaging of GFP, CFP and collagen in tumors *in vivo* using multiphoton microscopy," *BMC Biotechnol.* **5**, 14 (2005).

23. P. Mahou, M. Zimmerley, K. Loulier, K. S. Matho, G. Labroille, X. Morin, W. Supatto, J. Livet, D. Débarre, and E. Beaurepaire, "Multicolor two-photon tissue imaging by wavelength mixing," *Nat. Methods* **9**, 815–818 (2012).

24. P. A. Frank, A. E. Hill, C. W. Peters, and G. Weinreich, "Generation of optical harmonics," *Phys. Rev. Lett.* **7**, 118–120 (1961).

25. F. Seifert, J. Ringling, F. Noack, V. Petrov, and O. Kittelmann, "Generation of tunable femtosecond pulses to as low as 172.7 nm by sum-frequency mixing in lithium triborate," *Opt. Lett.* **19**, 1538–1540 (1994).

26. J. Liu, Y. Kida, T. Teramoto, and T. Kobayashi, "Generation of stable sub-10 fs pulses at 400 nm in a hollow fiber for UV pump-probe experiment," *Opt. Express* **18**, 4664–4672 (2010).

27. P. Baum, S. Lochbrunner, and E. Riedle, "Tunable sub-10-fs ultraviolet pulses generated by achromatic frequency doubling," *Opt. Lett.* **68**, 2793–2795 (2004).

28. B. Zhao, Y. Jiang, K. Sueda, N. Miyanaga, and T. Kobayashi, "Sub-15 fs ultraviolet pulses generated by achromatic phase-matching sum-frequency mixing," *Opt. Express* **17**, 17711–17714 (2009).

29. N. Aközbek, A. Iwasaki, A. Becker, S. L. Chin, and C. M. Bowden, "Third-harmonic generation and self-channeling in air using high-power femtosecond laser pulses," *Phys. Rev. Lett.* **89**, 143901 (2002).

30. P. Tzankov, O. Steinkellner, J. Zheng, M. Mero, W. Freyer, A. Husakou, I. Babushkin, J. Herrmann, and F. Noack, "High-power fifth-harmonic generation of femtosecond pulses in vacuum ultraviolet using a Ti: Sapphire laser," *Opt. Express* **15**, 6389–6395 (2007).

31. J. J. Macklin, J. D. Kmetec, and C. L. Gordon, "High-order harmonic generation using intense femtosecond pulses," *Phys. Rev. Lett.* **70**, 766 (1993).

32. J. A. Giordmaine and R. C. Miller, "Tunable coherent parametric oscillation in LiNbO$_3$ at optical frequencies," *Phys. Rev. Lett.* **14**, 973–976 (1965).

33. D. C. Edelstein, E. S. Wachman, and C. L. Tang, "Broadly tunable high repetition rate femtosecond parametric oscillator," *Appl. Phys. Lett.* **54**, 1728 (1989).

34. G. M. Gale, M. Cavallari, T. J. Driscoll, and F. Hasche, "Sub-20-fs tunable pulses in the visible from an 82-MHz optical parametric oscillator," *Opt. Lett.* **20**, 1562–1564 (1995).

35. K. C. Burr, C. L. Tang, M. A. Arbore, and M. M. Fejer, "Broadly tunable mid-infrared femtosecond optical parametric oscillator using all-solid-state-pumped periodically poled lithium niobate," *Opt. Lett.* **22**, 1458–1460 (1997).

36. T. Wilhelm, J. Piel, and E. Riedle, "Sub-20 fs tunable across the visible from blue-pumped single-pass nonlinear parametric converter," *Opt. Lett.* **22**, 1494–1496 (1997).

37. G. Cerullo, M. Nisoli, S. Stagira, and S. De Silvestri, "Sub-8-fs pulses from an ultrabroadband optical parametric amplifier in the visible," *Opt. Lett.* **23**, 1283–1285 (1998).

38. A. Shirakawa, I. Sakane, and T. Kobayashi, "Pulse-front-matched optical parametric amplification for sub-10-fs pulse generation tunable in the visible and near infrared," *Opt. Lett.* **23**, 1292–1294 (1998).

39. K. Okamura and T. Kobayashi, "Octave-spanning carrier-envelope phase stabilized visible pulse with sub-3-fs pulse duration," *Opt. Lett.* **36**, 226–228 (2011).

40. A. Shirakawa, I. Sakane, M. Takasaka, and T. Kobayashi, "Sub-5-fs visible pulse generation by pulse-front-matched noncollinear optical parametric amplification," *Appl. Phys. Lett.* **74**, 2268–2270 (1999).

41. P. B. Corkum, C. Rolland, and T. Srinivasan-Rao, "Supercontinuum generation in gases," *Phys. Rev. Lett.* **57**, 2268 (1986).

42. J. Kasparian, R. Sauerbrey, D. Mondelain, S. Niedermeier, J. Yu, J. P. Wolf, Y. B. André, M. Franco, B. Prade, and S. Tzortzakis et al., "Infrared extension of super continuum generated by femtosecond terawatt laser pulses propagating in the atmosphere," *Opt. Lett.* **25**, 1397–1399 (2000).

43. V. P. Kandidov, O. G. Kosareva, I. S. Golubtsov, W. Liu, A. Becker, N. Akozbek, C. M. Bowden, and S. L. Chin, "Self-transformation of a powerful femtosecond laser pulse into a white-light laser pulse in bulk optical media (or supercontinuum generation)," *Appl. Phys. B* **77**, 149–165 (2003).

44. W. J. Wadsworth, A. O. Blanch, J. C. Knight, T. A. Birks, T. P. Martin Man, and P. S. J. Russell, "Supercontinuum generation in photonic crystal fibers and optical fiber tapers: A novel light source," *JOSA B* **19**, 2148–2155 (2002).

45. C. Xia, M. Kumar, O. P. Kulkarni, M. N. Islam, F. L. Terry, J. M. J. Freeman, M. Poulain, and G. Mazé, "Mid-infrared supercontinuum generation to 4.5 μm in ZBLAN fluoride fibers by nanosecond diode pumping," *Opt. Lett.* **31**, 2553–2555 (2006).

46. C. Dunsby, P. M. P. Lanigan, J. McGinty, D. S. Elson, J. R. Isidro, I. Munro, N. Galletly, F. McCann, B. Treanor, and B. Önfelt et al., "An electronically tunable ultrafast laser source applied to fluorescence imaging and fluorescence lifetime imaging microscopy," *J. Phys. D* **37**, 3296–3303 (2004).

47. X. Gu, L. Xu, M. Kimmel, E. Zeek, P. O'Shea, A. P. Shreenath, and R. Trebino, "Frequency-resolved optical gating and single-shot spectral measurements reveal fine structure in microstructure-fiber continuum," *Opt. Lett.* **27**, 1174–1176 (2000).

48. J. M. Dudley, G. Genty, and S. Coen, "Supercontinuum generation in photonic crystal fiber," *Rev. Mod. Phys.* **78**, 1135–1184 (2006).

49. T. Fuji, T. Horio, and T. Suzuki, "Generation of 12 fs deep-ultraviolet pulses by four-wave mixing through filamentation in neon gas," *Opt. Lett.* **32**, 2481–2483 (2007).

50. H. Okamoto and M. Tatsumi, "Generation of ultrashort light pulses in the mid-infrared (3000–800 cm^{-1}) by four-wave mixing," *Opt. Commun.* **121**, 63–68 (1995).

51. T. Fuji and T. Suzuki, "Generation of sub-two-cycle mid-infrared pulses by four-wave mixing through filamentation in air," *Opt. Lett.* **32**, 3330–3332 (2007).

52. Y. Kida, J. Liu, T. Teramoto, and T. Kobayashi, "Sub-10 fs deep-ultraviolet pulses generated by chirped-pulse four-wave mixing," *Opt. Lett.* **35**, 1807–1809 (2010).

53. J. He and T. Kobayashi, "Generation of sub-20 fs deep-ultraviolet pulses by using chirped-pulse four-wave mixing in CaF$_2$ plate," *Opt. Lett.* **38**, 2938–2940 (2013).

54. H. Crespo, J. T. Mendonça, and A. Dos Santos, "Cascaded highly nondegenerate four-wave-mixing phenomenon in transparent isotropic condensed media," *Opt. Lett.* **25**, 829–831 (2000).

55. J. Liu and T. Kobayashi, "Cascaded four-wave mixing and multicolored arrays generation in a sapphire plate by using two crossing beams of femtosecond laser," *Opt. Express* **16**, 22119–22125 (2008).
56. J. Liu, T. Kobayashi, and Z. G. Wang, "Generation of broadband two-dimensional multicolored arrays in a sapphire plate," *Opt. Express* **17**, 9226–9234 (2009).
57. J. Liu, J. Zhang, and T. Kobayashi, "Broadband coherent anti-Stokes Raman scattering light generation in BBO crystal by using two crossing femtosecond laser pulses," *Opt. Lett.* **33**, 1494–1496 (2008).
58. W. Liu, L. Zhu, and C. Fang, "Observation of sum-frequency-generation-induced cascaded four-wave mixing using two crossing femtosecond laser pulse in a 0.1 mm beta-barium-borate crystal," *Opt. Lett.* **37**, 3783–3785 (2012).
59. J. Liu and T. Kobayashi, "Wavelength-tunable multicolored femtosecond laser pulse generation in fused silica glass," *Opt. Lett.* **34**, 1066–1068 (2009).
60. J. Liu and T. Kobayashi, "Cascaded four-wave mixing in transparent bulk media," *Opt. Comm.* **283**, 1114–1123 (2010).
61. M. Zhi, X. Wang, and A. V. Sokolov, "Broadband coherent light generation in diamond driven by femtosecond pulses," *Opt. Express* **16**, 12139–12147 (2008).
62. J. He, J. Du, and T. Kobayashi, "Low-threshold and compact multicolored femtosecond laser generated by using cascaded four-wave mixing in a diamond plate," *Opt. Comm.* **290**, 132–135 (2013).
63. J. Liu and T. Kobayashi, "Generation of sub-20-fs multicolor laser pulses using cascaded four-wave mixing with chirped incident pulses," *Opt. Lett.* **34**, 2402–2404 (2009).
64. R. Weigand, J. T. Mendonca, and H. Crespo, "Cascaded nondegenerate four-wave mixing technique for high-power single-cycle pulse synthesis in the visible and ultraviolet ranges," *Phys. Rev. A* **79**, 063838 (2009).
65. J. L. Silva, R. Weigand, and H. Crespo, "Octave-spanning spectra and pulse synthesis by non-degenerate cascaded four-wave mixing," *Opt. Lett.* **34**, 2489–2491 (2009).
66. J. Liu and T. Kobayashi, "Generation of μJ-level multicolored femtosecond laser pulses using cascaded four-wave mixing," *Opt. Express* **17**, 4984–4990 (2009).
67. R. Nietzke, P. Fenz, W. Elsässer, and E. O. Göbel, "Cascaded fourwave mixing in semiconductor laser," *Appl. Phys. Lett.* **51**, 1298–1300 (1987).
68. A. C. Eckbreth, T. J. Anderson, and G. M. Dobbs, "Multi-color CARS for Hydrogen-fueled scramjet applications," *Appl. Phys. B* **45**, 215–223 (1988).
69. A. V. Sokolov, D. R. Walker, D. D. Yavuz, G. Y. Yin, and S. E. Harris, "Raman generation by phased and antiphased molecular states," *Phys. Rev. Lett.* **85**, 562–565 (2000).
70. H. Zhang, H. Liu, J. Si, W. Yi, F. Chen, and X. Hou, "Low threshold power density for the generation of frequency up-converted pulses in bismuth glass by two crossing chirped femtosecond pulses," *Opt. Express* **19**, 12039–12044 (2011).
71. H. Zeng, J. Wu, H. Xu, and K. Wu, "Generation and weak beam control of two-dimensional multicolored arrays in a quadratic nonlinear medium," *Phys. Rev. Lett.* **96**, 083902 (2006).
72. W. Liu, L. Zhu, and C. Fang, "*In-situ* weak-beam and polarization control of multidimensional laser sidebands for ultrafast optical switching," *Appl. Phys. Lett.* **104**, 111114 (2014).
73. R. W. Boyd, *Nonlinear Optics*, 3rd ed., Elsevier, Singapore (2010).
74. J. Liu, Y. Kida, T. Teramoto, and T. Kobayashi, "Simultaneous compression and amplification of a laser pulse in a glass plate," *Opt. Express* **18**, 2495–2502 (2010).
75. J. Liu and T. Kobayashi, "Generation and amplification of tunable multicolored femtosecond laser pulses by using cascaded four-wave mixing in transparent bulk media," *Sensors* **10**, 4296–4341 (2010).
76. D. Majus, V. Jukna, G. Valiulis, and A. Dubietis, "Generation of periodic filament arrays by self-focusing of highly elliptical ultrashort pulsed laser beams," *Phys. Rev. A* **79**, 033843 (2009).
77. A. Dubietis, G. Tamosauskas, G. Fibich, and B. Ilan, "Multiple filamentation induced by input beam ellipticity," *Opt. Lett.* **29**, 1451–1453 (2004).
78. A. A. Dergachev, V. N. Kadan, and S. A. Shlenov, "Interaction of noncolinear femtosecond laser filaments in sapphire," *Quant. Electron.* **42**, 125–129 (2012).
79. J. He, J. Liu, and T. Kobayashi, "Tunable multicolored femtosecond pulse generation using cascaded four-wave mixing in bulk materials," *Appl. Sci.* **4**, 444–467 (2014).

5.4 Mechanism Study of 2-D Laser Array Generation in a YAG Crystal Plate

5.4.1 INTRODUCTION

Recently, generation of ultrabroadband spectrum and ultrashort pulse through four-wave mixing (FWM) process induced by the third-order susceptibility has attracted considerable interest. When two crossed femtosecond beams overlapping in time and space were synchronized in a piece of BK7 glass [1], a 1-D broadband multicolored side band array occurred, which was also observed in fused silica [2,3] and certain crystals such as $PbWO_4$ [4], $LiNbO_3$ [5], $KNbO_3$ [6], TiO_2 [7], $KTaO_3$ [8], and BBO [9]. In addition, 2-D multicolored arrays were obtained in a quadratic nonlinear crystal [10] and sapphire plate [11,12].

The generation of 1-D multicolored sideband array can be explained by a cascaded four-wave mixing process, and the upshifted and downshifted sidebands correspond to the phase matching condition [1,3]. But the explanation of the two-dimensional multicolored array generation using two crossed femtosecond laser beams in a sapphire plate or a quadratic nonlinear crystal hasn't been cleared up to now. In this chapter, we explored the mechanism of the 2-D multicolored arrays generated in the process of cascaded four-wave-mixing process based on numerical simulation, considering cross-phase modulation (XPM) and self-focusing under the pump of two noncollinear cross-overlapped femtosecond beams. In addition, the threshold input laser beam power for the two crossing beams splitting in a YAG crystal plate has been studied numerically.

5.4.2 NUMERICAL SIMULATION MODEL

We performed numerical simulation of the multicolored array generation in a YAG crystal plate by using two crossing femtosecond laser pulses according to the experimental conditions approximately [3,11–13]. The schematic of the experiment is shown in Figure 5.4.1.

Femtosecond laser beam1 and beam2 were synchronously focused on a 1 mm-thick YAG crystal plate as depicted in Figure 5.4.1a. The YAG crystal plate was placed around the focus. The crossing angle between the two incident beams was about 1.87°. Figure 5.4.1b shows the cross section of the two elliptical beams on the surface of the YAG crystal plate, and the diameter was about 500 μm for both beam1 and beam2 in the experiment. The two input laser beams were horizontal

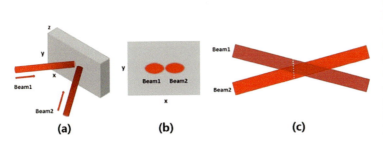

FIGURE 5.4.1 Schematics of the simulation process. (a) Two laser beams were synchronously focused into a 1 mm-thick sapphire plate with a crossing angle of 1.87°. (b) The cross section of the two beams on the surface of the sapphire plate. (c) Longitudinal distribution of the two crossing laser beams in the $y=0$ plane.

DOI: 10.1201/9780429196577-31

polarization with the repetition frequency of 1 kHz. The peak wavelength of beam1 and beam2 was 800 and 700 nm, respectively. The pulse duration of beam1 was 35 fs with an average power of 7 mW, while the pulse duration of beam2 was 60 fs. When the average power of beam2 was set to be 25 mW in the experiment which corresponded to a peak power of about 167 MW, the two beams started to split.

Since we mainly focus on the spatial distribution of the two crossing femtosecond laser beams in the YAG crystal, the temporal aspects of the propagation can be ignored. This processing method has been demonstrated to be valid by previous studies [14,15]. Then we represent beam1 and beam2 as:

$$E_1 = A_1 \exp(-i(\omega_1 t - k_1 z)) \quad \text{and} \quad E_2 = A_2 \exp(-i(\omega_2 t + k_2 z)) \tag{5.4.1}$$

where $A1$ and $A2$ denote the amplitude of the light field. ω_1 and ω_2 represent the frequency of beam1 and beam2. k_1 and k_2 denote the wave number of beam1 and beam2, respectively.

The total field within the nonlinear medium is given by:

$$E = E_1 + E_2 \tag{5.4.2}$$

This field produces a third-order nonlinear polarization within the medium, given by:

$$P = \varepsilon_0 \chi^{(3)} E^3 \tag{5.4.3}$$

With slowly varying amplitude approximation, we assume that the laser beam1 obeys the wave equation in the form:

$$2ik_1 \frac{\partial A}{\partial z} + \Delta_\perp A = -u_0 \omega_1 P^{\text{NL}}(\omega_1) \tag{5.4.4}$$

Then we can obtain $P^{\text{NL}}(\omega_1)$ from Eqs. (5.4.1)–(5.4.3) as follows:

$$P^{\text{NL}}(\omega_1) = \varepsilon_0 \chi^{(3)}(\omega_1, \omega_1, -\omega_1, \omega_1)\left[3A_1^2 A_1^*\right]_0 + \chi^{(3)}(\omega_1, \omega_2, -\omega_2, \omega_1)\left[6A_1 A_2^* A_1\right] \tag{5.4.5}$$

We introduce Eq. (5.4.5) into Eq. (5.4.4) and bring in the defocusing of plasma. Then the equation of the laser beam1 propagation in the nonlinear medium can be expressed as:

$$2ik_1 \frac{\partial A_1}{\partial z} + \Delta_\perp A_1 + \frac{2k_1^2}{n_1} \Delta n A_1 = 0 \tag{5.4.6}$$

where k_1 and n_1 denote the wave number and refractive index of beam1 propagating in the sapphire plate respectively. Δn corresponds to the intensity-dependent refractive index:

$$\Delta n = n_2 I_1 + 2n_2 I_2 - \alpha(I_1 + I_2)^m \tag{5.4.7}$$

The first item on the right hand of the above equation denotes the optical Kerr effect that induced nonlinear refractive index and n_2 is 7×10^{-16} cm^2/W [16]. The second item refers to the nonlinear refractive index of beam1 induced by beam2 and reflects the cross-phase modulation process. The third item corresponds to the plasma defocusing effect-induced nonlinear refractive index, and m is chosen to be 6, which is approximately the effective nonlinearity order of multiphoton ionization rate following representation [17]. Here, α denotes an empirical parameter which gives rise to a clamped intensity of 5×10^{13} W/cm^2 in our simulation [17]. Note that different methods have been considered to take into account the counteracting effect to the self-focusing, such as saturable nonlinear refractive index [18] or multi-photon absorption associated with plasma generation [19]. Saturable nonlinear refractive index model has the same effect as Eq. (5.4.7) without specifying the underlying physical mechanism which balances the self-focusing. Multi-photon absorption is crucial for long propagation distance. Thus, plasma defocusing plays a dominant role to balance the self-focusing, since in our case the thickness of our sample is much shorter than those used in [19].

Here I would like to mention two NLO phenomena relevant to self-focusing as follows.

i. Self-defocusing: In the case of negative nonlinear refractive index, lateral lase intensity distribution provides concave thermal lens effect with lower refractive index than peripheral.
ii. Temporal cross-phase modification: In a collinear (copropagating) pump-probe configuration with intense pump and weak probe with different wavelength the central (peak) part of the probe pulse is affected by the change of refractive index (assuming to be positive namely increase in the refractive index) results in the slower phase and group velocity. This effect may distort the probe temporal shape skewed (peak shifted) to the reverse direction to the propagation direction.

Similarly, the wave equation of beam2 propagating in the sapphire plate can be written as:

$$2ik_2 \frac{\partial A_2}{\partial z} + \Delta_\perp A_2 + \frac{2k_2^2}{n_2}\left[n_2I_2 + 2n_2I_1 - \alpha\left(I_1 + I_2\right)^6\right]A_2 = 0 \tag{5.4.8}$$

Since in the experiment described by [11–13], two beams are centered at different wavelengths, i.e., 700 and 800 nm, we carried out numerical simulation of the two crossing femtosecond laser beams with the central wavelength of 700 and 800 nm, respectively based on the Eqs. (5.4.6) and (5.4.8) synchronously. It is worth mentioning if two beams are centered at the same wavelength, stimulated Raman scattering as observed in [11–13] will be suppressed. Besides, strong interference will take place between two beams with identical wavelength, resulting in intensity fringes. For the sake of saving computation time, during the simulation process, we reduced the diameters of the two incident beams on the surface of the YAG crystal plate to 1/5 of those in the experiment and the input powers of the two beams were reduced to 1/25 in order to ensure the peak intensity remain the same. For beam1, the beam widths were 88 μm and 57 μm in the horizontal and vertical directions. The beam widths of beam2 were 100 μm and 60 μm in the horizontal and vertical directions, respectively. The input powers of beam1 and beam2 were set to be 12 P_{cr} and 15 P_{cr} respectively. P_{cr} refers to the critical power for self-focusing and is defined as follows:

$$P_{cr} = 3.77\lambda^2/(8\pi nn_2) \tag{5.4.9}$$

where λ the laser wavelength, n_2 is the nonlinear refractive index and n refers to the refractive index.

5.4.3   RESULTS AND DISCUSSION

The simulation result is shown in Figure 5.4.2. The elliptical initial profile of beam1 and beam2 is given in Figure 5.4.2a and b, respectively. Figure 5.4.2c shows the incoherent superposition of two laser beams' intensity distribution in the horizontal plane of $y=0$ when the two crossed laser beams propagate through a 1-mm YAG crystal plate. Multiple strips could be clearly observed as shown in Figure 5.4.2c. Each strip indeed represents one filament induced by the dynamic interplay between the Kerr effect-induced self-focusing and plasma defocusing [17–19]. At the output plane of the crystal, multiple filaments are observed as multiple laser spots. Therefore, Figure 5.4.2d–i depicts the evolution of the beam profile at different propagation distances until 2-D laser array is observed. In detail, Figure 5.4.2d represents the transverse intensity distribution at $z=0.5$ mm, which indicates the self-focusing process of the two elliptical laser beams in the YAG crystal. The two laser beams

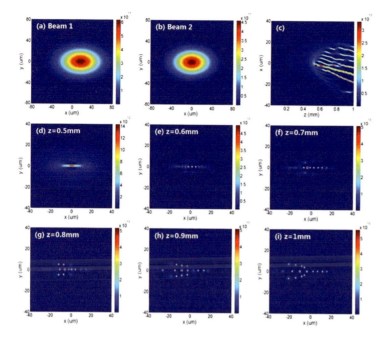

FIGURE 5.4.2 Simulation result with two crossing elliptical laser beams. (a) and (b) Pattern of input beam 1 and beam 2. (c) Longitudinal laser intensity distribution in the $y=0$ plane and transverse intensity distribution at (d) $z=0.5$ mm, (e) $z=0.6$ mm, (f) $z=0.7$ mm, (g) $z=0.8$ mm, (h) $z=0.9$ mm, (i) $z=1$ mm.

intersect at a distance of 0.5 mm in the plate as shown in Figure 5.4.2d. Then both two beams start to split. Figure 5.4.2e shows that a one-dimensional array has been formed at the distance of 0.6 mm. With the increase of the propagation distance, beam1 and beam2 start to split in the vertical direction as shown in Figure 5.4.2f. Figure 5.4.2g indicates that a two-dimensional array with three rows and two columns has appeared apparently. Further, with the increase of the propagation distance, more columns have come up as shown in Figure 5.4.2h and i.

As the spatial asymmetric distribution of the initial input pulse could lead to the beam breakup [18,19], the ellipticity of the initial laser beams plays an important role in the formation of two-dimensional array. More importantly, XPM significantly enhances the asymmetry of the pulse's phase front when two beams cross each other at an angle. As indicated in Figure 5.4.1c, across the dotted white line, which is parallel to the x-axis, the upper part of the beam2 suffers stronger phase modulation than the bottom part because of XPM. However, this asymmetry does not occur along the y-axis. The same phenomenon happens to beam1 as well. Hence, due to XPM, the cylindrical symmetry of the pulse phase will be broken no matter if the initial beam profile is elliptic or not. One would expect that the enhanced asymmetry will lower the power required to generate a 2D laser array. Then we can conclude that the formation of two-dimensional array was induced by both cross-phase modulation (XPM) and cylindrical symmetry breaking in the initial beam profile.

In addition, we have studied the threshold initial power for the laser beams splitting in a YAG crystal plate by using two crossing femtosecond laser pulses based on simulation. We carried out the same simulation process with different input power of beam2. The input peak power of beam1 was set to be 8.5 P_{cr} and the input peak power of beam2 was set as 5, 10, and 15 times the critical power for self-focusing, respectively.

In Figure 5.4.3, the laser intensity distributions obtained in a YAG crystal by using two crossing laser beams are depicted for different input power values of laser beam2. Figure 5.4.3a–c illustrates the longitudinal intensity distribution at three different input peak powers of beam2, while Figure 5.4.3d–f shows the corresponding laser intensity cross-section patterns at the propagation distance $z=1$ mm on the exit surface of the crystal plate.

When the input laser power of beam2 is 5 times the critical power for self-focusing, it can be seen that only a single spot has been formed on the cross section at the propagation distance of

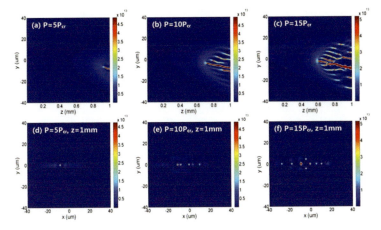

FIGURE 5.4.3 Simulated longitudinal laser intensity distribution in the $y=0$ plane and cross section with two crossing elliptical laser beams when the input laser power of beam 2 is set as (a) and (d) $P=5\,P_{cr}$, (b) and (e) $P=10\,P_{cr}$, (c) and (f) $P=15\,P_{cr}$.

1 mm. Only single filament is found in Figure 5.4.3a, while multiple filaments are generated in Figure 5.4.3b and c similar to Figure 5.4.2c. As the input laser power of beam2 increases to 10 P_{cr}, it is clear that the two laser beams start to split at the distance of $z=0.8$ mm and one-dimensional array has been formed on the exit surface of the YAG crystal plate. When the input laser power of beam2 is 15 P_{cr}, it can be seen that the number of spots in the x direction increases, and the beams start to split in the y direction.

Finally, we conclude that the threshold input power of beam2 for the laser beam splitting is about 10 P_{cr} which corresponds to a peak power of about 181 MW. This result keeps consistent with that in the experiment [11–13]. According to the previous discussion, it could be foreseen that if one could further enhance the asymmetry of XPM-caused phase modulation, such as increasing the crossing angle or the ellipticity of the initial beam profile, offset the beam in the y-axis and so on, the threshold power required to generate 2D laser array could be even lower.

5.4.4 CONCLUSION

In summary, we have reproduced a two-dimensional laser array by using two crossing elliptical laser beams in a YAG plate based on numerical simulation considering cross-phase modulation (XPM) and self-focusing [20]. We concluded that both XPM and the cylindrical symmetry breaking in the initial beam profile contribute to the generation of two-dimensional laser array. In addition, we have studied the threshold input laser beam power for the two crossing beams splitting in a YAG crystal plate. Our study could be valuable in various applications, such as 2-D all-optical switching devices or multicolor pump-probe experiments. The contents described in this subsection are based on the research activity based on the collaboration among the following people: Tao Zeng, Jinping He, Takayoshi Kobayashi, and Weiwei Liu [20].

REFERENCES

1. H. Crespo, J. T. Mendonça, and A. Dos Santos, "Cascaded highly nondegenerate four-wave-mixing phenomenon in transparent isotropic condensed media," *Opt. Lett.* **25**(11), 829–831 (2000).
2. J. Liu and T. Kobayashi, "Wavelength-tunable, multicolored femtosecond-laser pulse generation in fused-silica glass," *Opt. Lett.* **34**(7), 1066–1068 (2009).
3. J. Liu and T. Kobayashi, "Generation of microJ-level multicolored femtosecond laser pulses using cascaded four-wave mixing," *Opt. Express* **17**(7), 4984–4990 (2009).
4. M. Zhi and A. V. Sokolov, "Broadband coherent light generation in a Raman-active crystal driven by two-color femtosecond laser pulses," *Opt. Lett.* **32**(15), 2251–2253 (2007).
5. E. Matsubara, T. Sekikawa, and M. Yamashita, "Generation of ultrashort optical pulses using multiple coherent anti-Stokes Raman scattering in a crystal at room temperature," *Appl. Phys. Lett.* **92**(7), 071104 (2008).

6. H. Matsuki, K. Inoue, and E. Hanamura, "Multiple coherent anti-Stokes Raman scattering due to phonon grating in $KNbO_3$ induced by crossed beams of two-color femtosecond pulses," *Phys. Rev. B* **75**(2), 024102 (2007).

7. K. Inoue, J. Kato, E. Hanamura, H. Matsuki, and E. Matsubara, "Broadband coherent radiation based on peculiar multiple Raman scattering by laser-induced phonon grating in TiO_2," *Phys. Rev. B* **76**(4), 041101 (2007).

8. E. Matsubara, K. Inoue, and E. Hanamura, "Violation of Raman selection rules induced by two femtosecond laser pulses in $KTaO_3$," *Phys. Rev. B* **72**(13), 134101 (2005).

9. J. Liu, J. Zhang, and T. Kobayashi, "Broadband coherent anti-Stokes Raman scattering light generation in BBO crystal by using two crossing femtosecond laser pulses," *Opt. Lett.* **33**(13), 1494–1496 (2008).

10. H. Zeng, J. Wu, H. Xu, and K. Wu, "Generation and weak beam control of two-dimensional multicolored arrays in a quadratic nonlinear medium," *Phys. Rev. Lett.* **96**(8), 083902 (2006).

11. J. Liu and T. Kobayashi, "Cascaded four-wave mixing and multicolored arrays generation in a sapphire plate by using two crossing beams of femtosecond laser," *Opt. Express* **16**(26), 22119–22125 (2008).

12. J. Liu, T. Kobayashi, and Z. Wang, "Generation of broadband two-dimensional multicolored arrays in a sapphire plate," *Opt. Express* **17**(11), 9226–9234 (2009).

13. J. P. He, J. Liu, and T. Kobayashi, "Tunable multicolored femtosecond pulse generation using cascaded fourwave mixing in bulk materials," *Appl. Sci* **4**(3), 444–467 (2014).

14. A. Dubietis, G. Tamošauskas, G. Fibich, and B. Ilan, "Multiple filamentation induced by input-beam ellipticity," *Opt. Lett.* **29**(10), 1126–1128 (2004).

15. W. Liu and S. L. Chin, "Abnormal wavelength dependence of the self-cleaning phenomenon during femtosecond-laser-pulse filamentation," *Phys. Rev. A* **76**(1), 013826 (2007).

16. F. Silva, D. R. Austin, A. Thai, M. Baudisch, M. Hemmer, D. Faccio, A. Couairon, and J. Biegert, "Multi-octave supercontinuum generation from mid-infrared filamentation in a bulk crystal," *Nat. Commun.* **3**, 807 (2012).

17. L. Sudrie, A. Couairon, M. Franco, B. Lamouroux, B. Prade, S. Tzortzakis, and A. Mysyrowicz, "Femtosecond laser-induced damage and filamentary propagation in fused silica," *Phys. Rev. Lett.* **89**(18), 186601 (2002).

18. A. Dubietis, G. Tamošauskas, G. Fibich, and B. Ilan, "Multiple filamentation induced by input-beam ellipticity," *Opt. Lett.* **29**(10), 1126–1128 (2004).

19. D. Majus, V. Jukna, G. Valiulis, and A. Dubietis, "Generation of periodic filament arrays by self-focusing of highly elliptical ultrashort pulsed laser beams," *Phys. Rev. A* **79**(3), 033843 (2009).

20. T. Zeng, J. He, T. Kobayashi, and W. Liu, "Mechanism study of 2-D laser array generation in a YAG crystal plate," *Opt. Express* **23**(15), 19092–19097 (2015).

Section 6

Broadband Ultrashort Pulse Generation

6.1 Broadband Coherent Anti-Stokes Raman Scattering Light Generation in BBO Crystal by Using Two Crossing Femtosecond Laser Pulses

6.1.1 INTRODUCTION

In the past decade, collinear high-order stimulated Raman scattering (RS) [1] and high-order harmonic generation (HHG) [2] have been extensively studied to generate ultrabroadband spectra to generate subfemtosecond light pulses. Although subfemtosecond pulses have been generated by means of HHG extensively [3], RS still offers an attractive alternative to obtain ultrashort pulse owing to its higher conversion efficiency than HHG. A 1.6 fs ultrashort pulse has been generated through Fourier synthesis of several discrete Raman sidebands of cooled D_2 gas [4]. Recently, a similar RS phenomenon was found even in solid-state materials by using two crossing femtosecond laser beams, such as $YFeO_3$ [5], $SrTiO_3$ [6], $KTaO_3$ [7], $LiNbO_3$ [8], $KNbO_3$ [9], and TiO_2 [10], all at room temperature. As many as 20 anti-Stokes (AS) and 2 Stokes (S) coherent beams were generated in a lead tungstate $PbWO_4$ very recently [11], and this kind of RS could be selectively excited by using a pair of time-delayed linearly chirped pulses [12].

Here, we report broadband high-order coherent anti-Stokes RS generation in a BBO crystal. This is very interesting because BBO crystal is the most frequently used crystal in few-cycle ultrashort femtosecond pulse generation in visible by means of noncollinear optical parametric amplification [13–15].

6.1.2 EXPERIMENTAL

A femtosecond laser system Micra+Legend −USP produces 2.5 mJ, 40 fs, and 1 kHz pulses centered around 800 nm. The laser pulses were divided into two beams by a beam splitter. One beam (beam 1) was spectrum broadened in a hollow fiber and then was dispersion compensated with a chirped-mirror pair and a pair of glass wedges. The other beam (beam 2) passed through a delay stage with better than 3 fs resolution. The two laser beams were attenuated by a variable neutral-density (ND) filter and then focused into a 2-mm-thick BBO (type I, $\theta=21°$, $\phi=0°$) crystal. The laser polarizations of both beams were parallel to the optic axis of BBO. The spectra of different order CARS signal were measured by a spectrometer (USB4000) through an optical fiber attached to an arm on a moveable stage normal to the diffracted signal beam.

6.1.3 RESULTS AND DISCUSSION

The laser spectrum after the hollow fiber extended from 660 to 900 nm. The pulse before the crystal was positively chirped to about 100 fs duration owing to the variable ND filter and the glass wedge

DOI: 10.1201/9780429196577-33

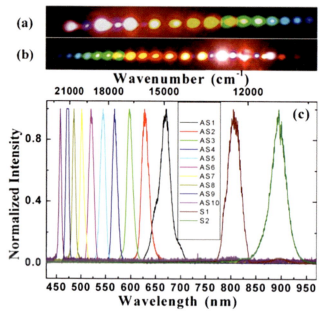

FIGURE 6.1.1 Photograph of the sidebands on a white sheet of paper behind the BBO crystal when (a) Beam1 works as a pump and (b) Beam2 works as a pump. (c) Spectra of sidebands from Figure 1a; A Sm $m=1$–10 refers to the mth-order anti-Stokes spectrum, Left panel of (c) AS10-AS1 from leftmost to right. Right part panel of (c) Stokes S1 and S2 from left to right, and Sn $n=1$, 2 refers to the mth-order Stokes spectrum.

pair. At first, the crystal was placed before the focal point of a lens. The beam diameters of both beams on the crystal were measured to be about 0.8 mm. The pulse energies on the BBO crystal were 40 J (beam 2) and 54 J (beam 1). The angle between the two beams in the air was 1.75° and nearly normal to the surface of the crystal. When the two laser beams were well overlapped temporally and spatially, multiple bright sidebands were generated on both sides of the two input beams. A photograph of sidebands light on a white sheet of paper placed behind the BBO crystal is shown at the top of Figure 6.1.1a. As many as 15 AS sidebands and 2S sidebands were generated, and they were well separated spatially. The spectra of different order RS signal were also measured and shown in Figure 6.1.1c. The spectrum of the sidebands can extend from the ultraviolet to the infrared with more than 1 octave. When the delay time of beam2 was tuned to less than 20 fs, multiple AS sidebands moved to the side of beam2 (Figure 6.1.1b). In this case, beam2 was used as a pump. The wavelengths of the same order sidebands were shifted when the AS sidebands emit on the different sides of the two input beams. The frequency spacings between two neighboring sidebands were also different between the signals shown in Figure 6.1.1a and b.

Dependence of the sideband spectra on the crossing angle was studied by changing the direction of beam1 with a fixed direction of beam2. In the experiment, two beams were set at five different crossing angles in the air: 1.53°, 1.75°, 2.18°, 2.62°, and 3.05°. The spectra of the sidebands and the conversion efficiency were found to be changed by varying the crossing angle. This phenomenon was also reported recently by Zhi and Sokolov [11]. The brightest sideband signals were observed when the crossing angle was set at 1.75°. As the crossing angle was decreased, the sidebands became close to each other both in frequency and space. A weak continuous line was generated when the crossing angle was reduced to smaller than 1.0°. The frequency spacing between the two neighboring sidebands increased gradually with the crossing angle. The spectrum of AS1 at different crossing angles is shown in the inset of Figure 6.1.2. The spectrum of AS1 was shifted to high frequency when the crossing angle was increased. We could see that the center wavelength of the sidebands can be tuned in a large bandwidth simply by changing the crossing angle. As the sideband order increased, the frequency separation between the two neighboring sidebands gradually decreased, as shown in Figure 6.1.2. The frequency spacing between AS1 and AS2 was about 1117 cm^{-1} for 2.18° crossing angle, which is gradually decreasing to 691 cm^{-1} for AS8 and AS9.

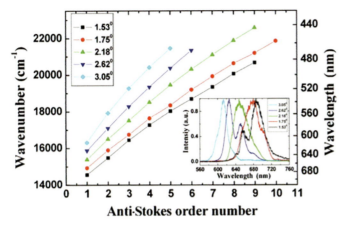

FIGURE 6.1.2 Center wavenumber (wavelength) of different sidebands varies with the anti-Stokes order number when the crossing angle between two input beams is 1.53°, 1.75°, 2.18°, 2.62°, and 3.05°. The inset shows the spectra of the first-order sidebands when the crossing angles between the two beams are 1.53°, 1.75°, 2.18°, 2.62°, and 3.05°.

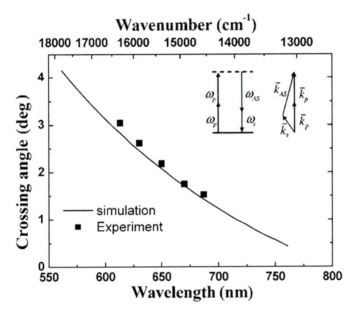

FIGURE 6.1.3 Dependence of peak wavelength of the first-order AS sidebands on the crossing angle of two input beams. The crossing angle was measured in the air. Squares, experimental results. Curve: calculated results by using the phase-matching condition.

In the CARS process, the laser beams should obey the energy conservation. Photon energy and photon momentum conservation laws are obtained by multiplying $\hbar$ in the relations among the frequencies and wavevectors, $\omega_{AS} = 2\omega_p - \omega_s$ and $2k_p - k_s - k_{AS} = 0$, respectively, as shown in Figure 6.1.3.

Here, s and p (seed and pump) refer to the two input laser beams, while AS refers to the generated sidebands. Using the two conservation laws, the dependence of the AS1 wavelength on the crossing angle was calculated, the results being shown in Figure 6.1.3. In the calculation, we fix the wavelength of s at 800 nm and vary the wavelength of p light from 660 to 780 nm to accord with the two input beams under the experiment condition. It can be seen that the calculated results agree very well with the experiment results. Therefore, frequency dependence on the angle can be explained in terms of phase matching. High conversion efficiency is obtained when the crossing angle is around 2°. It is because the frequency difference between the p and s in the calculation is close to the broad Raman line at 1547 cm^{-1} [16], and it is enhanced by the difference-frequency resonance as in the ordinary Raman process. The observed frequencies of the Raman shift are different from the Raman shift measured by the conventional Raman spectrum [16]. This is probably because the four-wave mixing process is taking place being associated with optical phonon mode, which satisfies the phase-matching condition [9,10].

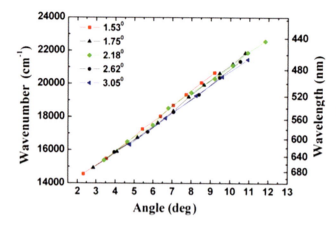

FIGURE 6.1.4 Relationship between the center wavenumber (wavelength) of different sidebands and the emitting angle of different sidebands. The crossing angles between the two input beams of which direction is normal to the crystal surface are 1.53°, 1.75°, 2.18°, 2.62°, and 3.05°.

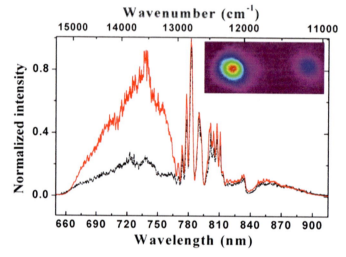

FIGURE 6.1.5 Spectrum of the pulse after the hollow fiber compressor (solid curve) and the spectrum of the pulse after the BBO crystal with CARS effect (dotted curve). The inset is the photograph showing the spatial profile of the second-order (left) and the third-order (right) sidebands.

The dependence of the center wavenumber and wavelength of different order AS on the emitting angle at different crossing angles is plotted in Figure 6.1.4. As can be seen, the slope for center wavenumber to output angle is almost constant, which is about $843 \, cm^{-1}$/deg. This means that the emitting angle of sidebands is not related to the AS number but to the center wave nubber of the sidebands. This linear relationship between the emitting angle and the wavelength of sidebands makes it possible to synthesize these sidebands by using dispersion optics that has been recently realized by Matsubara et al. [8].

When the crystal was located on the focal point of the lens, the beam diameter was measured to be about 200 m. When the energy of each incident laser pulse was 3 J, bright sidebands were also observed. The spatial profile of the sidebands was also measured by using a CCD camera behind the crystal. Figure 6.1.5 shows a spatial profile of the AS2 and AS3 sideband signals. The figure also shows the spectrum of the pump laser with and without the seed. The conversion efficiency of pump into the sidebands was higher than 30%, as also can be seen from the spectral intensity of the pump pulse with and without the seed in Figure 6.1.5. There was almost no dependence of the spectrum of the first-order AS sidebands on the phase-match angle within ±15°. There was no sideband generation when the polarization of one input beam was rotated 90°.

6.1.4 CONCLUSION

We have observed a broadband CARS signal generation in a most frequently used nonlinear crystal BBO pumped by using two crossing femtosecond laser pulses. More than 1 octave spectrum from ultraviolet to infrared was obtained in this way. The energy conversion efficiency in this kind of nonlinear process is about 30% in the experiment [17]. The phase matching in crystal plays an important role in the CARS process. This phenomenon will extend the possible application of BBO crystal in a new device. Subfemtosecond light may be able to be obtained by synthesizing this kind of sideband in the future. The work presented here in this subsection is performed cooperatively by the following people: Jun Liu, Jun Zhang, Tadayoshi Kobayashi [17].

REFERENCES

1. A. V. Sokolov, M. Y. Shverdin, D. R. Walker, D. D. Yavuz, A. M. Burzo, G. Y. Yin, and S. E. Harris, *J. Mod. Opt.* **52**, 285 (2005).
2. M. Hentschel, R. Kienberger, Ch. Spielmann, G. A. Reider, N. Milosevic, T. Brabec, P. Corkum, U. Heinzmann, M. Drescher, and F. Krausz, *Nature* **414**, 509 (2001).
3. R. Kienberger, E. Goulielmakis, M. Uiberacker, A. Baltuska, V. Yakovlev, F. Bammer, A. Scrinzi, Th. Westerwalbesloh, U. Kleineberg, U. Heinzmann, M. Drescher, and F. Krausz, *Nature* **427**, 817 (2004).
4. M. Y. Shverdin, D. R. Walker, D. D. Yavuz, G. Y. Yin, and S. E. Harris, *Phys. Rev. Lett.* **94**, 033904 (2005).
5. J. Takahashi, E. Matsubara, T. Arima, and E. Hanamura, *Phys. Rev. B* **68**, 155102 (2003).
6. J. Takahashi, M. Keisuke, and Y. Toshirou, *Opt. Lett.* **31**, 1501 (2006).
7. E. Matsubara, K. Inoue, and E. Hanamura, *Phys. Rev. B* **72**, 134101 (2005).
8. E. Matsubara, T. Sekikawa, and M. Yamashita, *Appl. Phys. Lett.* **92**, 071104 (2008).
9. H. Matsuki, K. Inoue, and E. Hanamura, *Phys. Rev. B* **75**, 024102 (2007).
10. K. Inoue, J. Kato, E. Hanamura, H. Matsuki, and E. Matsubara, *Phys. Rev. B* **76**, 041101(R) (2007).
11. M. Zhi and A. V. Sokolov, *Opt. Lett.* **32**, 2251 (2007).
12. M. Zhi and A. V. Sokolov, *New J. Phys.* **10**, 025032 (2008).
13. A. Baltuska, T. Fuji, and T. Kobayashi, *Opt. Lett.* **27**, 306 (2002).
14. A. Shirakawa, I. Sakane, M. Takasaka, and T. Kobayashi, *Appl. Phys. Lett.* **74**, 2268 (1999).
15. A. Shirakawa, I. Sakane, and T. Kobayashi, *Opt. Lett.* **23**, 1292 (1998).
16. J. L. You, G. C. Jiang, H. Y. Hou, Y. Q. Wu, H. Chen, and K. D. Xu, *Chin. Phys. Lett.* **19**, 205 (2002).
17. J. Liu, J. Zhang, and T. Kobayashi, *Opt. Lett.* **33**, 1894 (2008).

6.2 Generation of Broadband Two-Dimensional Multicolored Arrays in a Sapphire Plate

6.2.1 EXPERIMENTAL

When an ultrashort pulse propagates in a nonlinear bulk medium, various nonlinear optical phenomena such as supercontinuum light generation [1], optical solitons [2], optical parametric generation [3], and X-wave generation [4] have been observed. When two ultrashort pulses that overlap in space and time propagate in a nonlinear bulk medium, much more interesting nonlinear optical phenomena have appeared. Optical parametric amplification based on a nonlinear crystal is well established and has been used for many scientific experiments [5]. A 1-D broadband multicolored sideband array was observed when two crossed femtosecond beams were synchronized in BK7 glass [6], sapphire [7] and certain crystals, for example, $PbWO_4$ [8], $LiNbO_3$ [9], $KNbO_3$ [10], TiO_2 [11], $KTaO_3$ [12], and BBO [13]. However, all these phenomena occurred in one dimension only. There were only a few experiments based on 2-D nonlinear optical phenomena. 2-D arrays of transverse patterns were experimentally observed in the multifilament interactions [14,15], and 2-D solitons were the subject of considerable research over the last decade. Two-dimensional discrete solitons [16] and surface solitons [17] were observed in nonlinear photonic lattices. Regular 2-D multicolored transverse arrays were also observed when two crossed femtosecond laser beams were synchronized in a quadratic nonlinear crystal [18] or a sapphire plate [7]. However, 2-D nonlinear optical phenomena have not been widely studied to date.

In this chapter, we reported the generation of stable broadband 2-D multicolored arrays in a sapphire plate using two crossed femtosecond laser beams overlapping in time and space. The properties of the spectrum, spatial mode, power stability, and pulse duration of the 2-D muliticolored arrays were studied in detail. These stable 2-D multicolored arrays were sensitive to the orientation of the optical axis of the sapphire plate with respect to the plane of polarization of the incident beams. 2-D discrete multicolored soliton-like spots were observed.

6.2.2 EXPERIMENTAL SETUP

A 1 kHz Ti:sapphire regenerative amplifier fs laser system (Micra+Legend-USP, Coherent) with 40 fs pulse duration and 2.5 W average output power was used as a pump source. The laser pulse after the regenerative amplifier system was split into three beams using beam splitters. One of the beams (beam_1) was spectrally broadened in a hollow fiber with a 250 μm inner diameter and 60 cm in length that was filled with krypton gas. The broadband spectrum after the hollow fiber was dispersion compensated with a pair of chirped mirrors and a pair of glass wedges. The pulse duration after the hollow fiber compressor was about 10 fs. After passing through a bandpass filter at 720 nm, beam_1 was focused into a 2-mm-thick sapphire plate by a concave mirror. Another beam (beam_2) passed through a delay stage with less than 3 fs resolution. Beam_2 was first attenuated by a variable neutral density (VND) filter and was then focused into the sapphire plate by a lens.

DOI: 10.1201/9780429196577-34

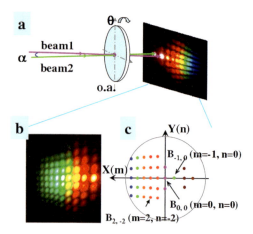

FIGURE 6.2.1 (a) Schematics of the experimental setup for multicolored array generation. α is the crossing angle between the two input beams, beam_1 and beam_2. θ is the rotation angle of the sapphire plate. (b) A photograph of the 2-D multicolored array on a UV light sensitive plate. (c) Definition of 2-D multicolored arrays, where $B_{0,0}$ and $B_{-1,0}$ refer to two incident beams, beam_1 and beam_2, respectively.

The third beam (beam_3) was used to generate correlation signals with the input beams and the generated sidebands in a 10-μm-thick BBO crystal, which was used to measure pulse durations.

Schematics of the experimental setup for multicolored array generation are shown in Figure 6.2.1a. The sapphire plate was cut on the [0001] plane and there were two orthogonal optical axes in the plane of the sapphire plate. The plane formed by the two orthogonal crystal axes was set normal to the two input beams. The two input beams had perpendicular polarizations and coincided with one of the crystal axes. The diameters of both incident beams on the surface of sapphire plate were 300 μm, as initially measured by a CCD camera (BeamStar FX 33, Ophir Optronics). The crossing angle α between the two input beams was 1.8°. When beam_1 and beam_2 were synchronously focused on the sapphire plate in both time and space, stable separate 2-D transverse multicolored array signals at different wavelengths were generated. The polarizations of the multicolored arrays were the same as those of the input beams as tested using a film polarizer. Figure 6.2.1b shows a photograph of the 2-D multicolored arrays on a UV light-sensitive plate placed about 20 cm after the sapphire plate. More than ten quasi-periodic columns and more than ten rows of multicolor signals can be seen, well separated from each other in space. The columns were approximately normal to the center row and the rows adjacent to the center row were not parallel to the center row. For convenience, the 2-D multicolored array signals were defined to be $B_{m,n}$, as shown in Figure 6.2.1c, where $B_{0,0}$ and $B_{-1,0}$ refer to two incident beams, beam-1 and beam-2, respectively. There were also two sidebands, $B_{-2,1}$ and $B_{-2,-1}$, beside the first-order Stokes sideband $B_{-2,0}$. The divergence angle of the sidebands was measured using a paper 50 cm after the sapphire plate to mark the position of each sideband. Neighboring spots on the same column have nearly the same crossing angle. However, the angle between two neighboring signals was decreased from 2.2° to 0.7° in the x direction (row direction, Figure 6.2.1c) and from 1.7° to 1.0° in the y direction (column direction, Figure 6.2.1c) as the sidebands changed from column $B_{-2,n}$ to the $B_{7,n}$ column.

6.2.3   EXPERIMENTAL RESULTS AND DISCUSSION

The spectra of array signals on the center row $B_{m,0}$ were measured using a multichannel spectrometer (USB4000, Ocean Optics), as shown in Figure 6.2.2a. For clarity, not all sideband spectra are shown. A broadband spectrum from 400 nm to 1.2 μm with more than 1.5 octaves was generated [7]. These generated sidebands were explained to be the result of a cascaded FWM process, which was recently reported in detail [6,7]. The spectra were tunable by changing the crossing angle α between the two input beams, and also by changing the center wavelength of the bandpass filter. The maximum difference between peak wavelengths of the side spots and the center spot on the same column was about 20 nm, as shown in Figure 6.2.2b.

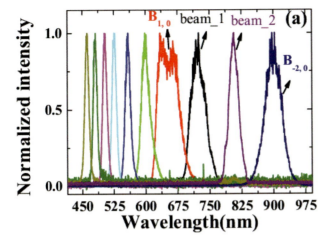

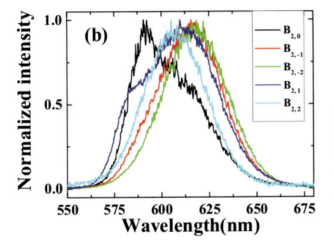

FIGURE 6.2.2 (a) The spectra of array signals on the center row $B_{m,0}$, where beam_1 and beam_2 are two incident beams. (b) The spectra of array signals on the second column $B_{2,n}$. B20 has leftmost relatively sharp peak. Relatively broad peaks from left(most faint line) to right are corresponding to B22, B21, B2-1, B2-2.

FIGURE 6.2.3 (a) The spatial profiles of $B_{0,0}$, $B_{1,0}$, and $B_{4,1}$ in one dimension. The inset patterns are spatial modals of $B_{0,0}$, $B_{1,0}$, and $B_{4,1}$ from bottom to top, measured by a CCD camera. (b) The retrieved XFROG pulse trace and phase of the $B_{1,0}$ with a retrieved error of 0.01022. The retrieved pulse duration is 35 ± 3 fs. The inset pattern is the measured XFROG trace.

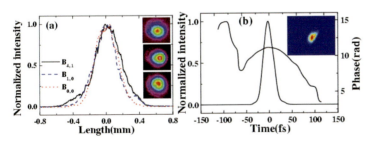

The spatial profiles of different signals in the arrays were measured using a CCD camera. Figure 6.2.3a shows spatial profiles of $B_{0,0}$, $B_{1,0}$, and $B_{4,1}$ in two dimensions and one dimension. The spatial profile changed from Gaussian ($B_{0,0}$) to a sech^2 profiles ($B_{4,1}$). The pulse duration of the sidebands was measured by cross-correlation with beam_3. Figure 6.2.3b shows the retrieved XFROG pulse trace and the phase of $B_{1,0}$ with a retrieved error of 0.01022. The retrieved pulse duration was 35 ± 3 fs. The retrieved phase shows that there was some chirp in the pulse due to the positively chirped input pulses and the dispersion of the glass. The pulse duration of $B_{1,0}$ was even shorter than the two incident pulses, the cross-correlation width of which with beam_3 were 82 ± 5 fs and 84 ± 5 fs for beam_1 and beam_2, respectively.

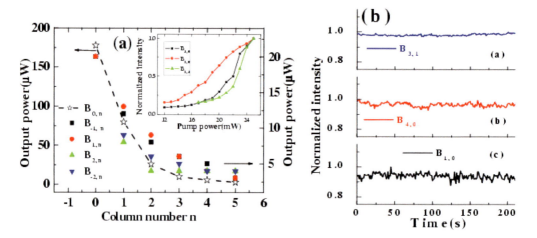

FIGURE 6.2.4 (a) The output power of array signals on the 0, ±1, and is ±2 rows when the power of the two incident beams, beam_1 and beam_2, were 0.1 and 25 mW, respectively. Only the signals in the center row are marked with star symbols and dashed line, as shown on the left. The inset figure shows the dependence of the output power of different sidebands on the input power of beam_2. (b) The monitored power stabilities of $B_{4,0}$, $B_{3,1}$, and $B_{1,0}$ were for 200 s. They were 1.25% RMS, 0.63% RMS, and 1.84% RMS, respectively.

We measured the powers of some array signals when the power of the two incident beams, beam_1 and beam_2, were 0.1 and 25 mW, respectively, as shown in Figure 6.2.4a. The signal power decreased rapidly with increasing column order for signals on the center row $B_{m,0}$, as shown by the star symbols in Figure 6.2.4a. The difference in output power between the side spots and the center spot on the same column decreased continuously as the row order increased. For signals on column 5 ($B_{5,n}$), signal power on the center row $B_{5,0}$ was even smaller than that on rows ±1 and ±2. The dependence of the different sideband output power on the input power of beam_2 is shown in the inset of Figure 6.2.4a. The increase in signal power with higher column or row order was delayed, but more rapid than that of lower order columns or rows. In this case, the power of beam_1 was very low, and saturation did not take place. The output power of beam_1 was amplified from 0.1 to 0.17 mW in the experiment. The power stabilities of different array signals were monitored by a Si power sensor, as shown in Figure 6.2.4b. The stability fluctuation of $B_{1,0}$ was about 1.84% RMS when measured over 200 s. Interestingly, the stabilities of the high-order sidebands were much better than that of the first-order, especially for the sidebands beside the beam on the center row. The stability fluctuations of $B_{4,0}$, $B_{3,1}$, and $B_{4,1}$ were 1.25% RMS, 0.63% RMS, and 0.97% RMS over 200 s, respectively.

It can be concluded from the sech^2 spatial profile and excellent power stability properties that the array signals at the higher order were spatial soliton-like spots. These discrete multicolored spatial soliton-like sidebands were generated due to the combined effects of the cascaded third-order nonlinear processes of FWM, coherent anti-Stokes Raman scattering (CARS), cross-phase modulation (XPM), and self-phase modulation (SPM) [19–20].

We observed the multicolored arrays to be sensitive to the rotation of the sapphire plate in the plane normal to the input beams. The brightest and largest number of array signals appeared when the plane of the polarization of the two input beams coincided with one of the crystal axes, as shown in Figure 6.2.5a. It was found that there were four angles 0°, 90°, 180°, and 270° at which the sidebands were brightest, and four angles 45°, 135°, 225°, and 315° at which the sidebands were weakest, as shown in Figure 6.2.5b. The periodic array did not appear continuously but appeared aperiodically, i.e., when the angle was rotated 6°, 14°, 0°, −6°, −11°, −13°, −16°, and −19° and further. At other angles, these regular arrays were replaced by a noise pattern, as shown in Figure 6.2.5c. The rotation angle of the sapphire plate also affected the position of the array signals. Figure 6.2.5d and e showed photographs when the sapphire plate was rotated by −16° and 14°, respectively. It can be seen that

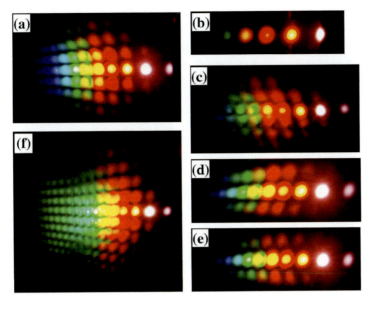

FIGURE 6.2.5　Photographs of 2-D multicolored arrays on a sheet of white paper when (a) The plane of polarization of the two input beams coincided with one of the crystal axes. (b) The sapphire plate rotated for 45°. (c) Noise pattern. (d) and (e) Show photographs with the sapphire plate rotated by −16° and 14°, respectively. (f) A photograph of the 2-D multicolored arrays on a UV light-sensitive plate when the input power of beam_2 was increased to 27 mW.

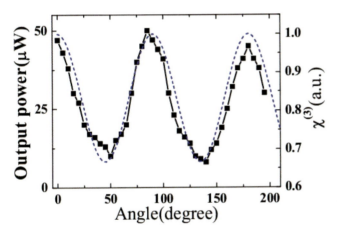

FIGURE 6.2.6　The dependence of $B_{1,0}$ output power (square symbols) and $|\chi^{(3)}(\theta)|$ (dashed line) of the sapphire plate on the rotation angle θ of the sapphire plate.

the column line was tilted a different direction when the sapphire plate was rotated. The spectrum of the tilted signal was slightly narrower in the shorter wavelength region when the signal was tilted in the high order direction, and vice versa. If the power of beam_2 was increased to 27 mW, bright stable "fish-like shaped" multicolored arrays were observed, as shown in Figure 6.2.5f.

Even though the temporal evolution of the 2D array of the multicolored array has not been studied, it is considered that higher order spot grows are delayed from the lower one.

The dependence of the $B_{1,0}$ output power on the rotation angle θ of the sapphire plate was measured, as shown in Figure 6.2.6. Here we only show the evolution in the rotation angle region from 0° to 210°. Clearly, the output power changed periodically with the rotation of the sapphire plate. This periodic evolution was because of the periodic dependence of $\chi^{(3)}(\theta) \propto \cos^2(2\theta)+1$ of the sapphire plate on the rotation angle θ, as shown by the dashed line in Figure 6.2.6. The peak wavelength of the sidebands shifted continuously to a shorter wavelength by about 20 nm when the sapphire plate was rotated from 0° to 45° due to a phase matching condition. The polarization of the sidebands also changed with rotation of the sapphire plate. To detect the polarization rotation of the sidebands, a thin film polarizer was located normal to the sideband beams and placed 50 cm after the sapphire plate. The polarizations of the multicolored arrays were the same as those of the input beams when the plane of polarization of the input beams coincided with one of the crystal axes. Very little orthogonally polarized light was detected. As the sapphire plate was rotated from 0° to 90°, the orthogonally polarized light continuously increased from 0° to 45° and decreased from 45° to 90°, and light with parallel polarization showed the opposite effect. At 45°, the sidebands were the weakest, and the orthogonally polarized light had power equal to the parallel polarized light. We rotated the thin film polarizer to minimize the intensity of light passing through the film polarizer. The rotation angle of the thin film polarizer β was in accordance with the rotation angle of the sapphire plate θ, and had the same angle when the sapphire plate was rotated from 0° to 45° and was $90° - \theta$ when the sapphire plate was rotated from 45° to 90°. This rotation angle dependence phenomenon was also recently observed in supercontinuum generation process [21].

A half-wave plate or a quartz-wave plate can also be used in the path of one of the input beams to safely control the multicolored arrays. The polarization, intensity, and position of the multicolored arrays can be controlled by rotating the sapphire plate. Note that this phenomenon did not appear when a fused silica glass was used as the medium, due to its symmetric structure. This phenomenon was very easily repeated in the experiment, and the sapphire plate was not damaged over the course of the experiment.

6.2.4   CONCLUSION

In conclusion, an interesting 2-D nonlinear optical phenomenon was observed. Broadband femto-second 2-D multicolored arrays were generated in a sapphire plate. These stable 2-D multicolored arrays can be controlled by rotating the sapphire plate, a half-wave plate, or a quarter wave plate. 2-D multicolored discrete solitons were expected to obtain by this method [22]. The properties of the spectrum, spatial modal, power stability, and pulse duration of the 2-D multicolored arrays show they could be used in various applications, for example, 2-D all-optical switching devices or multi-colored pump-probe experiments. The contents of this suction are obtained by collaborative work of Jun Liu, Takayoshi Kobayashi, and Zhiguang Wang [22].

REFERENCES

1. A. Saliminia, S. Chin, and R. Vallée, *Opt. Express* **13**, 5731–5738 (2005).
2. M. Segev and G. Stegeman, *Phys. Today* **51**, 43–48 (1998).
3. C. G. Durfee III, S. Backus, M. M. Murnane, and H. C. Kapteyn, *Opt. Lett.* **22**, 1565–1567 (1997).
4. C. Conti, S. Trillo, P. Di Trapani, G. Valiulis, A. Piskarskas, O. Jedrkiewicz, and J. Trull, *Phys. Rev. Lett.* **90**, 170406 (2003).
5. T. Kobayashi, A. Shirakawa, and T. Fuji, *IEEE J. Select. Topics Quantum Electron* **7**, 525–538 (2001).
6. H. Crespo, J. T. Mendonça, and A. Dos Santos, *Opt. Lett.* **25**, 829–831 (2000).
7. J. Liu and T. Kobayashi, *Opt. Express* **16**, 22119 (2008).
8. M. Zhi and A. V. Sokolov, *Opt. Lett.* **32**, 2251–2253 (2007).
9. E. Matsubara, T. Sekikawa, and M. Yamashita, *Appl. Phys. Lett.* **92**, 071104 (2008).
10. H. Matsuki, K. Inoue, and E. Hanamura, *Phys. Rev. B* **75**, 024102 (2007).
11. K. Inoue, J. Kato, E. Hanamura, H. Matsuki, and E. Matsubara, *Phys. Rev. B* **76**, 041101(R) (2007).
12. E. Matsubara, K. Inoue, and E. Hanamura, *Phys. Rev. B* **72**, 134101 (2005).
13. J. Liu, J. Zhang, and T. Kobayashi, *Opt. Lett.* **33**, 1494–1496 (2008).
14. D. Faccio, A. Dubietis, G. Tamosauskas, P. Polesana, G. Valiulis, A. Piskarskas, A. Lotti, A. Couairon, and P. Di Trapani, *Phys. Rev. A* **76**, 055802 (2007).
15. D. Kip, M. Soljacic, M. Segev, E. Eugenieva, and D. N. Christodoulides, *Science* **290**, 495–498 (2000).
16. J. W. Fleischer, M. Segev, N. K. Efremidis, and D. N. Christodoulides, *Nature* **422**, 147–150 (2003).
17. X. Wang, A. Bezryadina, Z. Chen, K. G. Makris, D. N. Christodoulides, and G. I. Stegeman, *Phys. Rev. Lett.* **98**, 123903 (2007).
18. H. Zeng, J. Wu, H. Xu, and K. Wu, *Phys. Rev. Lett.* **96**, 083902 (2006).
19. P. B. Lundquist, D. R. Andersen, and Y. S. Kivshar, *Phys. Rev. E* **57**, 3551–3555 (1998).
20. G. Fanjoux, J. Michaud, M. Delqu'e, H. Mailotte, and T. Sylvestre, *Opt. Lett.* **31**, 3480–3482 (2006).
21. V. Kartazaev and R. R. Alfano, *Opt. Commun.* **281**, 463–468 (2008).
22. J. Liu and T. Kobayashi, *Opt. Lett.* **34**, 1066–1068 (2009).

Section 7

NLO Materials

7.1 Sellmeier Dispersion for Phase-Matched Terahertz Generation in ZnGeP$_2$

7.1.1 INTRODUCTION

In recent years, among researchers, significant interest has been seen in the development of systems for the generation and detection of coherent tunable terahertz THz radiation. THz radiation has several potential applications in, e.g., space communications, time-domain far-infrared spectroscopy, THz ranging, and THz imaging [1]. Generation of such radiation is possible by various techniques, namely, by optical rectification in semiconductors with femtosecond laser radiation sources [2,3], from photoconductive antennas [4], and by nonlinear optical NLO frequency-conversion techniques, such as optical parametric oscillation and difference-frequency mixing DFM in various NLO crystals [5–9]. Zinc germanium diphosphide ZnGeP$_2$, ZGP is one of the most promising crystals for NLO applications [5–10] because of its high second-order nonlinearity ($d_{36} = 75$ pm/V), low absorption in the infrared, and wide THz spectral transmission [7,8]. Since this crystal in addition has a high thermal conductivity relative to other infrared crystals, it is attractive for use in high-average-power nonlinear optics [5–10]. The generation of THz radiation in this crystal was reported by Boyd et al. [7] and also by Apollonov et al. [8] using CO$_2$ lasers as input pump radiation sources. Recently, Shi and Ding [9] reported the generation of continuously tunable coherent THz radiation in this crystal by phase-matched DFM between the 1.064-m Nd:YAG laser radiation and the tunable idler radiations of a Nd: YAG-pumped optical parametric oscillator. They used an annealed sample of ZGP crystal, which has a smaller absorption coefficient of 1.52 cm^{-1} than the 5.63 cm^{-1} for an unannealed sample at 1.064 m. However, no difference was found in the absorption property of the two crystal samples at other wavelengths, including the THz range covering 66.5–300 m. It was noted by Shi and Ding [9] that the absolute value for the refractive index of the crystal sample at 1.064 m is not affected by the annealing process. The change in absorption in the crystal only at 1.064 m caused by the annealing process is due to the reduction of the impurity density of the crystal without modification of the phonon modes [9].

However, it is observed that the experimental phase-matching data of Ref. [9] differ substantially, by 5°–18°, from those calculated from the Sellmeier dispersion relations of Bhar et al. [10] which provide the best fit. Shi and Ding [9] noted this discrepancy and speculated only qualitatively that the dispersion of the crystal had been modified by the annealing process only at 1.064 m, and so the experimental phase-matching data are different from the results calculated from the Sellmeier dispersion relations of Bhar et al. [10] which were formulated with the measured refractive indices of an unannealed crystal sample. Shi and Ding [9] also speculated that the dispersion of the crystal in the THz range has a negligible effect on the determination of the phase-matching angle. However, it is shown here that the dispersion of the crystal in the THz region has a significant role in determining the phase-matching angle of the considered NLO process. We also find that the refractive indices of the crystal in the relevant THz spectral region cannot be obtained appropriately with the Sellmeier dispersions of Bhar et al. [10] which were formulated with mid-infrared refractive indices. An earlier, separate dispersion relation for determining the ordinary indices of the ZGP crystal in the THz spectral region was formulated by Boyd et al. [7] to calculate the phase-matching angle for the generation of THz radiation by

DOI: 10.1201/9780429196577-36

DFM between different CO_2 laser wavelengths. Apollonov et al. [8] also previously reported some refractive indices of ZGP crystal at THz wavelengths. It is observed that with these [7,8] refractive-index dispersion data for the THz region, the theoretical fit to the experimental phase-matching data [9] is improved to some extent but cannot be explained properly over the entire THz wavelength region. However, an appropriate Sellmeier dispersion is formulated here to determine the refractive index of an ordinary polarized ray in the ZGP crystal in the THz region. With the Sellmeier dispersion formulated here, the experimental phase-matching data of Shi and Ding [9] as well as those of Boyd et al. [7] are explained satisfactorily over the entire THz range.

7.1.2 DERIVATION OF THE SELLMEIER DISPERSION

The Sellmeier dispersion relation model [11] for the refractive index of NLO crystals is very useful for the calculation of NLO properties. Here, the Sellmeier dispersion for the ordinary refractive index of ZGP crystal in the THz spectral range is formulated. The ordinary refractive indices n_o of ZGP crystal in the THz wavelength region were measured earlier by Boyd et al. [7] 66.7 m and Apollonov et al. [8] 100 m. The former fitted a Sellmeier dispersion with their measured indices, while the latter did not. Based on the measured data of Boyd et al. [7] and Apollonov et al. [8] in the shorter and longer wavelengths, respectively, we have formulated a new Sellmeier dispersion relation, Eq. (7.1.1), for n_o of ZGP crystal, for the determination of the phase-matching properties for the generation of THz radiation. The formulated Sellmeier dispersion relation is

$$n_o(l)^2 = 10.93904 + 0.60675l^2/(l^2 - 1600) \qquad (7.1.1)$$

where $(\lambda > 60\,\mu m)$ is in micrometers.

In Figure 7.1.1, the dependence of n_o on the THz wavelength is shown, and curves 1, 2, 3, and 4 correspond to Refs. [7,8,10] and our Eq. (7.1.1), respectively. The variation of n_o obtained from the Sellmeier dispersion for n_o in Ref. [10] has also been shown in Figure 7.1.1 for comparison with the measured data of Boyd et al. [7] and Apollonov et al. [8] From Figure 7.1.1 it is clearly observed that the values of n_o that are obtained from Ref. [10] are different from the measured data [7,8]. The Sellmeier dispersion of Ref. [10] was obtained earlier, with the measured refractive indices covering only the 0.64–12 µm wavelength range, so it is not surprising that this Sellmeier dispersion [10] does not reproduce the refractive indices of the crystal properly in the THz region. In the following sections, it will be shown that our formulated Sellmeier dispersion relation, given by Eq. (7.1.1), provides a satisfactory explanation of the experimental phase-matching data [7,9] for the generation of THz radiation in this crystal by DFM, whereas the other data [7,8] for the THz refractive-index dispersion fail.

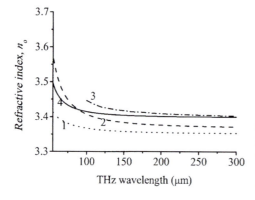

FIGURE 7.1.1 Dispersion of the ordinary refractive index n_o of the ZnGeP$_2$ crystal in the THz spectral range. Curves 1, 2, 3, and 4 are obtained from the dispersion data of Refs. [7,8,10], and Eq. (7.1.1) of the present paper, respectively.

7.1.3 GENERATION OF TERAHERTZ RADIATION WITH A ND:YAG LASER

By collinear phase-matched DFM between the ordinary- (*o*) polarized 1.064 µm (λ_1) Nd:YAG laser radiation and the extraordinary (*e*) polarized tunable ($\lambda_2 > 1.064$ µm) idler radiation of an 1.064-µm pumped optical parametric oscillator, the generation of continuously tunable THz (λ_3)radiation is possible in ZGP crystal. Two different DFM configurations, *oe–o* and *oe–e*, were used [9]; in the first case, the polarization of the generated THz radiation is *o*, whereas for the latter one it is *e*. The experimental and computed dependence of the external phase-matching angle (θ_{ext}) with λ_3 for *oe–o* and *oe–e* DFM configurations are shown in Figures 7.1.2 and 7.1.3, respectively. In Figures 7.1.2 and 7.1.3, the refractive indices at the input pump wavelengths have been calculated by use of the Bhar et al. [10] dispersion equations, whereas the ordinary refractive indices at the generated THz wavelengths have been taken from Refs. [7,8,10] for the computed curves 1, 2, and 3, respectively. Curves 4 of Figures 7.1.2 and 7.1.3 are obtained with our Eq. (7.1.1). In the case of the *oe–e* configuration, the phase-matching angle depends on both the ordinary and extraordinary indices at the generated THz wavelength, unlike for the *oe–o* configuration, for which the phase-matching angle depends only on the ordinary index at the generated THz wavelength. Therefore, to obtain curves 2, 3, and 4 in Figure 7.1.3, the extraordinary index has been calculated as in Ref. [12] and with the corresponding ordinary index from Refs. [7,8] and Eq. (7.1.1), respectively.

However, for curve 1, both the ordinary and the extraordinary indices for the input pump as well as for the generated THz wavelengths have been calculated from Ref. [10]. The vertical and horizontal error bars in Figures 7.1.2 and 7.1.3 show the probable errors in the experimental data [9].

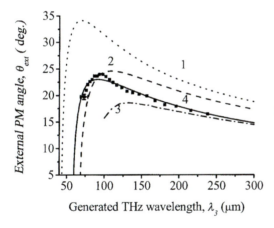

FIGURE 7.1.2 External phase-matching angle θ_{ext} versus λ_3 for *oe–o* DFM. THz ordinary refractive indices are obtained from Refs. [7,8,10], and our Eq. (7.1.1) for curves 1, 2, 3, and 4, respectively. Ref. [10] has been used for the calculation of the refractive indices at the input pump wavelengths. The horizontal and vertical error bars show the uncertainties in the experimental data (squares) [9].

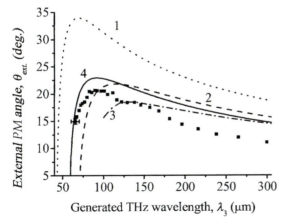

FIGURE 7.1.3 External phase-matching angle θ_{ext} versus λ_3 for *oe–e* DFM. For curves 2, 3, and 4, the ordinary refractive index n_o at a generated THz wavelength λ_3 has been obtained from Refs. [7,8], and Eq. (7.1.1), respectively, and the corresponding extraordinary THz index n_e has been calculated as in Ref. [12]. For curve 1, both n_o and n_e for each λ_3 have been calculated with the Sellmeier dispersion of Ref. [10], while those for the input pump wavelengths have been calculated from Ref. [10] for all curves. The horizontal and vertical error bars show the uncertainties in the experimental data (squares) [9].

In Figure 7.1.2, it can be seen that the experimental phase-matching data [9] differ from those curves 1 calculated with the Bhar et al. [10] Sellmeier dispersion over the entire wavelength region. This Sellmeier dispersion [10] is not able to reproduce the experimental phase-matching data of Ref. [9] because, as was shown in Section 7.1.2, the Sellmeier dispersion of Ref. [10] could not reproduce the refractive indices of the crystal in the relevant THz spectral region. Curve 2 in Figure 7.1.2 shows that the experimental data also cannot be explained with the Boyd et al. [7] dispersion, except for small improvements at wavelengths shorter than 100 m.

On the other hand, curve 3 in Figure 7.1.2 shows that the fit to the experimental data is improved with the Apollonov et al. [8] refractive index data only in the longer THz wavelength region. Moreover, no refractive-index data are available in Ref. [8] for 100-m wavelengths. However, in Figure 7.1.2 it is observed that the calculated phase-matching data curve 4 with our formulated Sellmeier dispersion Eq. (7.1.1) excellently matches the experimental data of *oe–o* DFM [9] over the entire wavelength region. In the case of the *oe–e* DFM, for which the phase-matching curves are shown in Figure 7.1.3, it is observed that curve 4, obtained with the Sellmeier dispersion Eq. (7.1.1), fits the experimental phase-matching data well, particularly, in the shorter-wavelength region, whereas results curves 1, 2, and 3 calculated with other dispersion data [7,8,10] deviate from the experiment over a broad spectral region.

In the case of *oe–e* DFM in the longer-wavelength region there are still some discrepancies between the experimental data and our calculated results curve 4; however, these might be within the uncertainty in the experimental data. For example, for the generation of ~157 µm radiation by *oe–e* DFM the deviation in the external phase-matching angle is ~2.2°, which corresponds to an internal angle deviation of only ~0.6°.

7.1.4   GENERATION OF TERAHERTZ RADIATION WITH CO_2 LASERS

The generation of THz radiation in ZGP crystal is also possible by DFM between different wavelengths of two CO_2 laser sources [7,8]. The experimental data, together with the computed phase-matching characteristics of ZGP crystal for the generation of THz radiation from CO_2 laser radiations, are shown in Figure 7.1.4. Figure 7.1.4 corresponds to the *eo–o* DFM configuration, i.e., in this case, the input radiations are orthogonally *e* and *o* polarized CO_2 laser lines, and the generated THz radiation is *o* polarized.

To yield the theoretical curves in Figure 7.1.4, the refractive indices at the input pump wavelengths have been calculated with the Bhar et al. [10] dispersion equations, whereas the refractive indices at the generated THz wavelengths have been taken from Refs. [7,8,10], and our Eq. (7.1.1) for curves 1, 2, 3, and 4, respectively. In Figure 7.1.4 the wavelength of the *e*-polarized radiation₂ 9.588 m, and that of the *o*-polarized radiation₂ is varied to obtain the generation of phase-matched

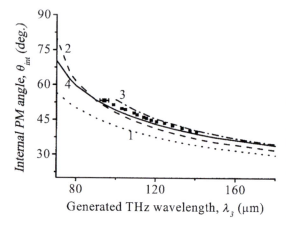

FIGURE 7.1.4 Internal phase-matching angle θ_{int} versus λ_3 for *eo–o* DFM between CO_2 laser radiations. Computed curves 1, 2, 3, and 4 are based on THz ordinary index data from Refs. [7,8,10], and our Eq. (7.1.1), respectively, while Ref. [10] has been used for the calculation of the refractive indices at the input wavelengths. The error bar shows the uncertainty in the experimental data (squares) [7].

wavelength-tunable o-polarized THz radiation. It may be noted that in the present case the input radiation sources are CO_2 lasers that are line tunable, and so the wavelength of the generated THz radiation is not continuously tunable, unlike the cases presented above in Section 7.1.3. The phase-matching configuration considered in Figure 7.1.4 is the same as that of forward-wave phase-matching in Ref. [7]. Unlike in Figures 7.1.2 and 7.1.3, the internal phase-matching angle (θ_{int}) has been plotted in Figure 7.1.4 to show similarity with Ref. [7]. Figure 7.1.4 shows that our Sellmeier dispersion Eq. (7.1.1) provides the best to the experimental data [7].

Deviations of the experimental data from the calculated results with the available THz Sellmeier dispersion of Boyd et al. [7] are within ~$2.5°$–$2.9°$. On the other hand, the deviation of the experimental data from the calculated results with our formulated Sellmeier dispersion Eq. (7.1.1) are within $1.0°$–$1.5°$, the same as the reported uncertainty in obtaining the original phase-matching data, because of the refractive-index uncertainty [7]. It may also be noted that refractive-index dispersion data of Apollonov et al. [8] also provide a better fit to the experimental phase-matching data, particularly in the longer-wavelength region; however, the deviations at shorter wavelengths are larger, and there is no refractive-index data available for wavelengths shorter than $100\,\mu m$.

7.1.5 DISCUSSION

Now we again consider Figures 7.1.2 and 7.1.3 for some detailed discussion. As was mentioned above, the measured data of Shi and Ding [9] deviate as widely as $5°$–$18°$ from their data calculated with the Bhar et al. [10] Sellmeier dispersion relations. In the experiment [9] an annealed sample of ZGP crystal had been used to reduce $1.064\,m$ absorption in the crystal. Shi and Ding [9] speculated that through the annealing process the dispersion property and the phase-matching angle of the crystal at $1.064\,m$ had been modified. They also presumed that the dispersion of the crystal in the THz domain had a negligible effect on the phase-matching angle. Here we have found that the origin of the phase-matching discrepancy in Ref. [9] is the use of the inappropriate Sellmeier dispersion relations for the calculation [9] of the refractive-index dispersion of the crystal in the THz range. As was mentioned above, the dispersion equations of Bhar et al. [10] were constructed with measured refractive indices up to only $12\,\mu m$ in the long-wavelength range, and hence it is not surprising that these dispersion relations cannot reproduce the refractive index dispersion of the crystal in the THz region as shown in Section 7.1.2 with $12\,\mu m$. It is clearly observed from curves 2 and 3 of Figures 7.1.2 and 7.1.3 that, when separate dispersion data from Refs. [7,8] are used for the THz region, the calculated phase-matching data reverse their trend of variation in the shorter wavelength region and also that the absolute values of the phase-matching angles are reduced substantially, coming closer to the experimental data [9]. However, it is observed from Figures 7.1.2 and 7.1.3 that the experimental data of Shi and Ding [9] over the whole THz region considered can be explained neither by the THz dispersion data of Apollonov et al. [8] nor by that of Boyd et al. [7] On the other hand, the calculated results shown as curves 4 in Figures 7.1.2 and 7.1.3 with our formulated Sellmeier dispersion Eq. (7.1.1) satisfactorily explain the measured phase-matching data throughout the entire wavelength region.

The importance of the THz dispersion for determining the phase-matching angle can also be seen in Figure 7.1.4. It is observed in Figure 7.1.4 that the calculated phase matching data (curve 1) for Bhar et al. [10] dispersion differ as widely as ~$5°$–$10°$ from the measured data [7]; the deviations are almost similar to the DFM cases with Shi and Ding [9]. However, it has been seen in Section 7.1.4 that our calculation results curve 4 in Figure 7.1.4 with the Sellmeier dispersion Eq. (7.1.1) satisfactorily explain the experimental phase-matching data of Boyd et al. [7].

7.1.6 CONCLUSION

In conclusion, we have presented a Sellmeier dispersion of ZGP crystal, which is used to obtain the refractive-index dispersion of the crystal in the THz spectral range. We have also presented the phase-matching characteristics of the crystal for the generation of widely tunable THz radiation by phase-matched DFM with Nd:YAG and CO_2 lasers as fundamental radiation sources. The computed results, with our formulated Sellmeier dispersion Eq. (7.1.1) and other available THz refractive index dispersion data [7,8], have been compared with the available experimental data [7,9]. Our formulated Sellmeier dispersion Eq. (7.1.1) explains well the phase-matching data of Boyd et al. [7] and the results computed with Eq. (7.1.1) also provide a satisfactory explanation of phase-matching discrepancies as large as 5°–18° reported recently in Ref. [9]. Here, we note that the refractive index and birefringence of the material under consideration, particularly near the band-edge region, are subject to variations from sample to sample owing to differences in stoichiometry and impurity concentrations [6,11]. However, the Sellmeier dispersion of Bhar et al. [10] for near-infrared and our Eq. (7.1.1) for the THz range presented here provide a good reproduction of the currently available data for nonlinear experiments for THz generation in the crystal samples used [7,9]. The presented Sellmeier dispersion in the paper [13] was claimed to be useful for determining the NLO properties of this material for different NLO applications in the THz spectral region. The work presented in this subsection was conducted cooperatively among the following people: Pathik Kumbhakar, Takayoshi Kobayashi, and Gopal C. Bhar,

REFERENCES

1. B. Ferguson and X.-C. Zhang, "Materials for terahertz science and technology," *Nat. Mater.* **1**, 26–33 (2002).
2. A. Bonavalet, M. Joffre, J.-L. Martin, and A. Migus, "Generation of ultrabroadband femtosecond pulses in the mid-infrared by optical rectification of 15 fs light pulses at 100 MHz repetition rate," *Appl. Phys. Lett.* **67**, 2907–2909 (1995).
3. A. Nahata, A. S. Weling, and T. F. Heinz, "A wideband coherent terahertz spectroscopy system using optical rectification and electro-optic sampling," *Appl. Phys. Lett.* **69**, 2321–2323 (1996).
4. M. S. Tani, M. Herrmann, and K. Sakai, "Generation and detection of terahertz pulsed radiation with photoconductive antennas and its application to imaging," *Meas. Sci. Technol.* **13**, 1739–1745 (2002).
5. K. L. Vodopyanov, F. Ganikhanov, J. P. Maffetone, I. Zwieback, and W. Ruderman, "ZnGeP$_2$ optical parametric oscillator with 3.8–12.4 m," *Opt. Lett.* **25**, 841–843 (2000).
6. G. D. Boyd, E. Buehler, and F. G. Storz, "Linear and nonlinear optical properties of ZnGeP$_2$ and CdSe," *Appl. Phys. Lett.* **18**, 301–304 (1971).
7. G. D. Boyd, T. J. Bridges, C. K. N. Patel, and E. Buehler, "Phase-matched submillimeter wave generation by difference frequency mixing in ZnGeP$_2$," *Appl. Phys. Lett.* **21**, 553–555 (1972).
8. V. V. Apollonov, A. I. Gribenyukov, V. V. Korotkova, A. G. Suzdal'tsev, and Yu. A. Shakir, "Subtraction of the CO$_2$ laser radiation frequencies in a ZnGeP$_2$ crystal," *Sov. J. Quantum Electron.* **26**, 469–470 (1996).
9. W. Shi and Y. J. Ding, "Continuously tunable and coherent terahertz radiation by means of phase-matched difference frequency generation in zinc germanium phosphide," *Appl. Phys. Lett.* **83**, 848–850 (2003).
10. G. C. Bhar, L. K. Samanta, D. K. Ghosh, and S. Das, "Tunable parametric crystal oscillator," *Sov. J. Quantum Electron.* **17**, 860–861 (1987).
11. G. C. Bhar, "Refractive index interpolation in phasematching," *Appl. Opt.* **15**, 305–307 (1976).
12. The extraordinary index n_e $n_o n$; n 0.0397, which is the long-infrared birefringence, taken from Ref. [6].
13. P. Kumbhakar, T. Kobayashi, and G. C. Bhar, "Sellmeier dispersion for phase-matched terahertz generation in ZnGeP$_2$," *Appl. Opt.* **43**, 3324–3328 (2004).

7.2 Broadband Sum-Frequency Mixing (SFM) in Some Recently Developed Nonlinear Optical Crystals

7.2.1 INTRODUCTION

Broadly tunable ultrafast laser radiation has demanded considerable attention due to its several applications in spectroscopy as well as in the generation of higher harmonics and attosecond pulses [1–8]. However, the development of a high-power, broadly tunable laser source in the IR region having an ultrashort pulse width remains a challenging task. One of the promising possibilities is the idler radiation from the noncollinear optical parametric amplifier NOPA system, pumped with the second harmonic of a Ti:sapphire laser and seeded with the white light continuum produced by the same second harmonic. Such idler radiation shows the useful phenomenon of self-elimination of pulse-to-pulse carrier-envelope-phase CEP slip [5–7]. This is a quite significant advantage since controlling CEP slip is important for applications in different nonlinear optical NLO phenomena such as high-harmonic generation and in precision optical frequency measurement [5,7,8]. Until now, BBO-BaB$_2$O$_4$ is the only crystal that has been used extensively for different types of NLO device applications including those in the femtosecond regime. Broadband ultrashort pulses as short as 4 fs in the visible to near-infrared NIR region have been generated [1–8] in BBO. Several borate group crystals, namely, CLBO CsLiB$_6$O$_{10}$, LB4 Li$_2$B$_4$O$_7$, KABO K$_2$Al$_2$B$_2$O$_7$, KBBF KBe$_2$BO$_3$F$_2$, LBO Li$_2$B$_3$O$_5$, and NYAB Nd$_x$Y$_{1-x}$Al$_3$BO$_{34}$, have been developed relatively recently with improved NLO properties [9–18]. It has already been demonstrated that the generation of ultrabroadband visible radiation through NOPA is possible in all these newly developed crystals [10,11].

The group delay walk-off limits the bandwidth, and hence this is the main constraint for broadband generation of shorter pulses. Such a deleterious effect was overcome for the first time by Szabo and Bor [4]. They pointed out the advantages of noncollinear angularly dispersed geometry in sum-frequency mixing SFM for obtaining the broadband spectrum. By employing such an advantageous configuration, a broadband UV pulse below 20 fs was generated by Nabekawa et al. [1] in BBO. The potentiality of BBO crystal has also been demonstrated for characterization of the octave-spanned CEP-locked angularly dispersed idler radiation of the Ti:sapphire second-harmonic i.e., 395 nm-pumped NOPA through the cross-correlation frequency-resolved optical grating XFROG technique [5].

We present here, for the first time, the broadband phase matching characteristics of seven different borate group crystals, namely, BBO, CLBO, KABO, LB4, KBBF, LBO, and NYAB, for type I noncollinear SFM between a radiation with a fixed wavelength 790 nm, the Ti:sapphire fundamental wavelength and the broadband angularly dispersed idler radiation from a 395-nm-pumped broadband NOPA for the generation of a sum-frequency SF wave covering the 397- to 539-nm region of the visible spectrum. Moreover, we show that KBBF crystal offers some additional advantages as compared to other borates. Finally, we demonstrated that even through simple conventional noncollinear type I SFM, in a very thin crystal of about 5-m thickness, it is possible to characterize an ultrabroadband optical pulse covering the 0.4- to 1.6-m region [12,13].

DOI: 10.1201/9780429196577-37

7.2.2 SCHEMATIC OF THE EXPERIMENTAL ARRANGEMENT

Figure 7.2.1a depicts the schematic of the experimental setup of the noncollinear interaction geometry of the broadband SFM between 790 nm and the broadband idler from a NOPA. Figure 7.2.1b shows the schematics of the principle of the non-collinear angularly dispersed broadband sum frequency mixing geometry. The basic experimental setup for the field reconstruction of an ultrashort laser pulse by means of the XFROG technique is shown [19] in Figure 7.2.1c.

In the XFROG technique, one can easily measure extremely complex pulses by the spectrally resolved SFM between the unknown field $E(t)$ with the use of a fully characterized intense reference field $E_{ref}(t)$ in a nonlinear crystal (depicted as C2 in Figure 7.2.1) with $\chi^{(2)}$ nonlinearity as a function of delay. The fields E_{ref} and E should be of comparable duration, with an allowance of an order of magnitude difference between them. The SFM signal field is proportional to the product of Et and $E_{ref}(t-\tau)$, i.e., $E_{sum}(t) \propto E(t)E_{ref}(t-\tau)$. The form the of E_{sum} field shows that the reference pulse acts as a gate for the pulse, hence the acronym XFROG. The signal E_{sum} is then spectrally analyzed. The resulting spectrum is the key quantity. It contains enough information to reconstruct the amplitude and phase of the pulse $E(t)$.

Here, the pump radiation of NOPA is Ti.sapphire second-harmonic radiation, i.e., 395 nm. The seed radiation is the white light continuum WLC generated in a sapphire plate pumped by the same 395-nm radiation. For a type I interaction, the pump radiation is e-polarized and the signal and idler radiations are both o-polarized. The noncollinear angle between pump and signal idler radiations is $\alpha(\beta)$ and that between signal and idler radiation is ψ. The details of the NOPA have been described elsewhere [5–7]. To achieve broadband amplification of white light seed radiation was set to an optimum value [3,11] (α_{opt}). This value of α_{opt} is to be determined by a zero group velocity mismatch GVM_{s-i} between signal and idler wavelength in the NOPA, which can generate octave-spanned angularly dispersed broadband idler radiation [5]. The typical pulse energy pulse duration of the pump, 790 nm, and its second-harmonic generation SHG 395-nm radiations are 400 and 100 µJ (150 fs), respectively. The transform-limited pulse duration of the o-polarized idler radiation after appropriate pulse compression is typically ~4.2 fs and the pulse energy is ~1 J [5]. Note here that out of the seven crystals considered in this paper, LBO is the only biaxial crystal, and we considered phase-matching in the xy plane of this crystal. To satisfy the type I phase matching condition in the LBO crystal in the case of SFM, two input radiations are both polarized along the

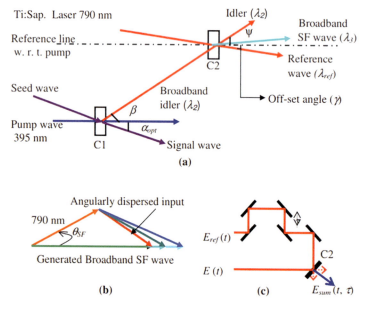

FIGURE 7.2.1 (a) Schematic of the experimental setup to be used for broadband SFM with a reference beam (790 nm, Ti:sapphire laser) and the angularly dispersed broadband idler radiation from NOPA. (b) Noncollinear angularly dispersed broadband SFM geometry. (c) XFROG experimental setup.

z-axis and the generated radiation is polarized and propagates in the xy plane. For NOPA in LBO, the pump wave is polarized and propagates in the xy plane, the signal and idler are both polarized along the z-axis, and the phase-matching angle is measured with the x-axis.

For the generation of broadband visible laser radiation and subsequent characterization of the broadband NOPA idler radiation, the noncollinear angular geometry (top portion of Figure 7.2.1a) is used [5]. Here, a reference wave $\lambda_{ref} = 790\,nm$ is SF mixed with the angularly dispersed NOPA idler radiation$_2$ that varies between 0.8 and 1.6 m through type I phase-matched interaction. This generates angularly collimated broadly tunable visible radiation λ_3 extending from 0.397 to 0.529 m.

7.2.3   THEORETICAL BACKGROUND OF PHASE MATCHING AND BROADBAND SFM

The phase-matching condition equivalent to the momentum conservation for type I SFM between two ordinary waves in a negative uniaxial crystal is given as [10–14]

$$k_{SF}^e(\theta_{SF}) = k_{ref}^o + k_2^o \tag{7.2.1a}$$

The superscripts o and e represent the ordinary and extraordinary polarizations of the interacting waves. The requirement of energy conservation in the process is given by

$$\omega_{SF} = \omega_{ref} + \omega_2 \tag{7.2.1b}$$

Here, $k_{SF}^e(\theta_{SF})$, k_{ref}^o, and k_2^o are the wave vectors of the extraordinary polarized SF wave and two ordinary polarized input waves, respectively. The frequencies of SF wave are ω_{SF}, ω_{ref}, and ω_2 are the frequencies of the SF wave and two input waves with wavelengths SF, ref, and 2, respectively. From Eqs. (7.2.1a) and (7.2.1b), one obtains an analytical expression of the phase matching angle θ_{SF} as

$$\theta_{SF} = \cos^{-1}\left\{\left[\left(k_{SF}^e/Y\right)^2 - 1\right] / \left[\left(k_{SF}^e/k_{SF}^o\right)^2 - 1\right]\right\}^{0.5} \tag{7.2.2}$$

Here, $Y = [(k_{ref}^o)^2 + (k_2^o)^2 + 2k_{ref}^o k_2^o \cos\psi]$ can be defined as the average wave vector of the two interacting beams in the SF mixing and is along the generated SF wave direction.

Also $k_{SF}^e = 2\pi n_{SF}^e \lambda_{SF}$, $k_{ref}^o = 2\pi n_{ref}^o / \lambda_{ref}$, and $k_2^o = 2n_2^o / \lambda_2$, where n_{SF}, n_{ref}, and n_2 are the refractive indices of a crystal at the λ_{SF}, λ_{ref}, and λ_2 wavelengths, respectively. The noncollinear angles between different interacting radiations are designated as ψ, β, and γ, as shown in Figure 7.2.1. The dependence of noncollinear angle on$_2$ determined by the phase-matching condition is described elsewhere [5,10,11]. The offset angle is set in such a way that apart from helping to achieve a broader spectrum in SFM, it facilitates automatic separation of the interacting radiations. The magnitude of the wave vector mismatch in the interaction is given by

$$\Delta k = k_{SF}^e(\theta_{SF}) - \sqrt{\left[\left(k_{ref}^o\right)^2 + \left(k_2^o\right)^2 + 2k_{ref}^o k_2^o \psi\right]^{1/2}} \tag{7.2.3}$$

From Eq. (7.2.3), we get

$$\frac{\partial \Delta k}{\partial \lambda_2} = \frac{\partial k_{SF}^e(\theta_{SF})}{\partial \lambda_2} - \frac{1}{2}\left[\left(k_{ref}^o\right)^2 + \left(k_2^o\right)^2 + 2k_{ref}^o k_2^o \cos(\psi)\right]^{-1/2}$$
$$\times \left[2k_2^o \frac{\partial k_2^o}{\partial \lambda_2} + 2k_{ref}^o \frac{\partial k_2^o}{\partial \lambda_2}\cos(\psi) - 2k_{ref}^o k_2^o \sin(\psi)\frac{\partial \psi}{\partial \kappa_2}\right] \tag{7.2.4}$$

It is clear from Eqs. (7.2.1a), (7.2.1b), and (7.2.2–7.2.4) that perfect phase matching in an SFM interaction will take place, in general, only for a given set of values of λ_{ref}, λ_2, and λ_{SF}. However, to characterize the broadband NOPA idler radiation or for the generation of broadband visible radiation, the phase-matching condition is required to be satisfied for a wider variation of the input wavelengths. In our case, we want to vary wavelength$_2$, i.e., the wavelength of the NOPA idler radiation.

The relation between $\partial \Delta k / \partial \lambda_2$ and $\partial \Delta k / \partial \lambda_{SF}$ is $\partial \Delta k / \partial \lambda_{SF} = (\lambda_2^2 / \lambda_{SF}^2)(\partial \Delta k / \partial \lambda_2)$. Hence, by controlling the value of $\partial \Delta k / \partial \lambda_2$, i.e., by incorporating the angular dispersion $(\partial \psi / \partial \lambda_2)$ into the NOPA idler, it is possible to achieve phase matching for the generation of the SF wave in a wider wavelength region.

7.2.4   RESULTS AND DISCUSSION

Figure 7.2.2 shows the type I SFM phase-matching characteristics of seven different borate crystals with the broadband noncollinear interaction geometry, as shown in Figure 7.2.1a. Phase-matching angles for type I SFM between the fixed reference wave at 790 nm and the angularly dispersed idler radiation (λ_2) covering 0.8–1.6 μm are calculated using Eqs. (7.2.1a) and (7.2.1b) keeping $\gamma = 4°$ Curves marked as 1, 2, 3, 4, 5, 6, and 7 correspond to BBO, CLBO, LB4, KBBF, KABO, LBO, and NYAB crystals, respectively. From Figure 7.2.2, we clearly see that the wavelength dependences of the phase-matching angle for all of the crystals are small. The phase-matching angle difference in the full spectral range from 0.8 to 1.6 m is smaller than 3° in all cases, except in the LBO and NYAB crystals. Therefore, by utilizing the angular dispersion i.e., I_2 in$_2$ radiation one can easily generate broadband SFM in the visible region and thus the broadband idler radiations are characterized, which is difficult to be performed otherwise.

Under plane-wave approximation, the energy E_{SF} of the generated λ_{SF} radiation can be given as [12,15]

$$E_{SF} = \frac{E_2 \lambda_2}{\lambda_{SF}} \tan h^2 \left[\frac{52.2 L^2 d_{eff}^2 E_{ref}}{n_{ref}^o n_2^o n_{SF}^e (\theta_{SF}) T \lambda_2 \lambda_{SF} \pi^2} \right]^{1/2}$$
$$\times \sin c^2 \left(\Delta k l / 2 \right)$$

(7.2.5)

Here, T is the pulse duration; L is the length of the SFM crystal; d_{eff} is the effective coupling coefficient; and E_{ref} and E_2 are the energies of the λ_{ref} and λ_2 beams, respectively. The refractive indices of the λ_{ref} and λ_2, and λ_{SF} beams are described as n_{ref}^o, n_2^o, and n_{SF}^e (θ_{SF}), respectively, in Eq.

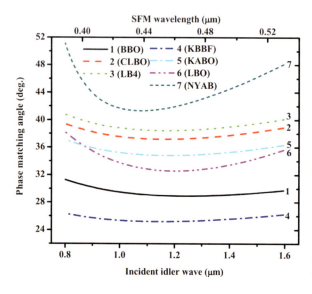

FIGURE 7.2.2 Ultrabroadband phase-matching in type I SFM with a fixed monochromatic beam 790 nm and broadband NIR wavelengths 0.8–1.6. Curves marked as 1, 2, 3, 4, 5, 6, and 7 are for BBO, CLBO, LB4, KBBF, KABO, LBO, and NYAB crystals, respectively. Here the offset angle is considered as 4°.

(7.2.5). While writing Eq. (7.2.5), we neglected the absorption losses in the SFM crystal and we have taken the radii of all the interacting radiations are equal to r. The term $\sin c^2(\Delta kL/2)$ is proportional to the phase-matching gain of the SFM process and it characterizes the dependence of the wave vector mismatch Δk on the conversion efficiency. Figure 7.2.3a shows the variations of Δk versus λ_2 and λ_{SF} for the SFM interaction in all the crystals. The values of Δk were calculated using Eq. (7.2.3) and thickness of each of crystal was taken as 5 μm. The curves of Figure 7.2.3a indicate that the value of ΔkL in KBBF crystal is the smallest, and it is $\Delta kL < 0.1$ rad in the whole wavelength region from 0.8 to 1.6 μm. Figure 7.2.3b depicts the phase-matching gain spectrum, and it clearly shows that it is possible to characterize the full one-octave-spanned idler radiation in each of the seven borate crystals. It is observed from Figure 7.2.3a that the gain spectrum is modulated by a large amount in the case of the NYAB crystal; whereas, in the KBBF crystal the phase-matching gain spectrum is completely flat, and it offers the best performance. This facilitates the generation of broadband visible radiation without any modulation in the output spectrum.

The conversion efficiency of the SFM process is defined as $\eta = E_{SF}/(E_{ref}E_2)^{1/2}$ and the estimated values are 0.990%, 0.065%, 0.003%, 0.154%, 0.046%, 0.160%, and 0.066%, respectively, for the BBO, CLBO, LB4, KBBF, KABO, LBO, and NYAB crystals, for example, for SFM between 0.79 and 1.32 μm idler radiation for the generation of 0.494-m visible radiation with the reference 0.79 μm

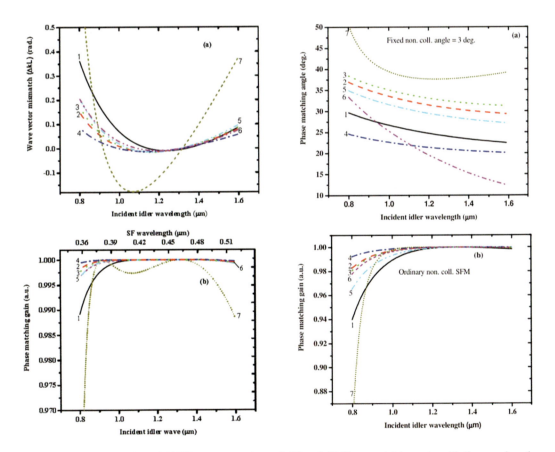

FIGURE 7.2.3 Variation of (a) Wave vector mismatch kL and (b) Phasematching gain with the wavelength of the generated SF wave also with the idler wavelength for type I SFM with noncollinear angularly dispersed geometry in seven newly developed crystals. Curves marked as 1, 2, 3, 4, 5, 6, and 7 are for BBO, CLBO, LB4, KBBF, KABO, LBO, and NYAB crystals, respectively. Here the reference beam is set at 790 nm and idler wavelengths come from the NOPA covering the 0.8- to 1.6-μm region.

and idler radiation 1.32 μm energy of 10 and 1 μJ, respectively. The thickness of each crystal was set as 5 μm. The value of the KBBF crystal is smaller than that of the BBO and LBO crystals, but it is larger than all other five crystals. The calculated results show that BBO is more efficient than KBBF also for the fifth-harmonic generation of 213 nm from 1064 nm Nd:YAG laser radiation by type I SFM. Therefore, in terms of conversion efficiency, BBO remains the most efficient crystal among the borate group of crystals. However, the damage threshold, i.e., the maximum value of reference beam intensity ($I_{ref} = E_{ref}/T$) that can be used in the KBBF crystal is almost three times larger than that of the BBO crystal.

We can see from Eq. (7.2.5) that the conversion efficiency η is proportional to the intensity of the reference radiation (I_{ref}). Therefore, by utilizing the higher values of I_{ref}, the disadvantage of the KBBF crystal for its smaller value can be compensated to some extent. Note here that the refractive indices and their dispersions for all the crystals were calculated using the published reports [9,15,20].

7.2.5   BROADLY TUNABLE CONVENTIONAL SFM IN A THIN CRYSTAL

Finally, we considered the conventional noncollinear type I SFM for the generation of broadly tunable visible laser radiation by SFM between 790-nm reference and tunable collinear 0.8- to 1.6-μm IR waves in all the considered potential borate crystals.

Figure 7.2.4a and b show the phasematching characteristics and the phase-matching gain, respectively, of all the considered crystals when the conventional SFM is considered. In Figure 7.2.4a and

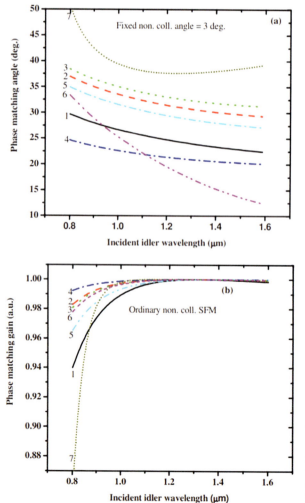

FIGURE 7.2.4   (a) Phase-matching characteristics. (b) Variation of acceptance power spectrum is for conventional type I SFM in seven different borate crystals. Curves marked as 1, 2, 3, 4, 5, 6, and 7 are for BBO, CLBO, LB4, KBBF, KABO, LBO, and NYAB crystals, respectively, in both graphs.

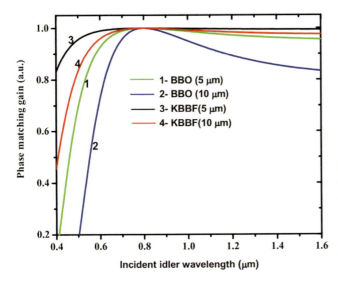

FIGURE 7.2.5 Phase-matching gain of BBO crystal for ordinary collinear type-I SFM of 790-nm radiation with broadband tunable input radiation covering 0.4–1.6 mm. Curves marked as 1 (3) and 2 (4) are for BBO (KBBF) crystals of 5- and 10-mm thicknesses, respectively.

b curves marked as 1, 2, 3, 4, 5, 6, and 7 correspond to BBO, CLBO, LB4, KBBF, KABO, LBO, and NYAB crystals, respectively. In the present case of SFM, the wavelength of one of the interacting beams is fixed at 790 nm and the other one is tuned in the range of 0.8–1.6 m but not angularly dispersed in contrast to the case as discussed in the previous sections. In this case, also the phase-matching angle and phase matching gain are calculated with a fixed value of noncollinear angle of 3° between the two input radiations. Figure 7.2.4b shows that with a very thin crystal of 5-µm thickness, it is possible to characterize the whole one-octave-spanned idler radiation in each crystal, and the SFM gain spectrum is the broadest for KBBF.

Figure 7.2.5 shows a case of conventional type I collinear SFM in BBO and KBBF crystals. One of the interaction wavelengths is 790 nm, while the other input wavelength varies in an even wider wavelength region covering 0.4–1.6 m. In Figure 7.2.5, curves 1 and 2 correspond to BBO crystal of 5 and 10 m thicknesses and curves 3 and 4 correspond to KBBF crystal of the same thicknesses, respectively. From Figure 7.2.5 it is clear that a 5-m-thick KBBF crystal can characterize a very wide band radiation covering the 0.4- to 1.6-m region through type I SFM with the Ti:sapphire laser fundamental 790-nm radiation.

7.2.6 CONCLUSION

We presented for the first time the phase-matching characteristics of the seven different borate group crystals BBO, CLBO, LB4, KABO, KBBF, LBO, and NYAB for the generation of broadly tunable visible pulses by type I SFM [21]. Here we applied the angular chirped pump technique, where angularly dispersed broadband idler radiation is SF mixed with the Ti:sapphire fundamental radiation of 790 nm. This broadband SFM technique can be a useful tool for the characterization of octave-spanned CEP-locked idler pulses generated from a Ti:sapphire second-harmonic pumped broadband NOPA. We also presented that through a simple conventional SFM in a very thin KBBF crystal of thickness 5 m a large phase-matching bandwidth can be achieved, and thus, it is possible to characterize a very wide band radiation covering the 0.4- to 1.6-µm region through type I SFM with 790 nm, the Ti:sapphire laser fundamental radiation. A comparison of the NLO performances of the inorganic borate group crystals studied in this paper can be made with the organic aromatic compounds, such as MNA and NPP [15]. These organic crystalline materials have several advantageous properties, such as larger figure of merit, low-cost growth process, and large birefringence, that make them suitable for use in different NLO frequency converters. However, these organic crystalline materials have significant drawbacks that limit their application in nonlinear optics. These materials are hygroscopic and extremely soft so that their surfaces must be protected with coatings [15]. The inorganic borate group potential materials that are studied in this paper are

TABLE 7.2.1

NLO Properties of Several Borate Crystals for Broadband SFM

Name of the Crystal (Crystal Symmetry)	Transparency (nm)	SFM Conversion Efficiency (n in %)[a]	θ_{pm} (°) (Broadband SFM)[a]	θ_{pm} (°) (Ordinary SFM with Noncolinear angle = 3°)	LDT (GW/cm²)[b]	Chemical Stability	Commercial Availability
BBO (3 m)	189–3500	0.99	29.0	23.86	13	Hygroscopic	Available
CLBO (42 m)	180–2750	0.07	37.42	30.58	26	Highly hygroscopic	Available
LB4 (4 mm)	160–330	0.003	38.65	32.21	40	Non hygroscopic	Available
KBBF (32)	155–3780	0.15	25.38	20.83	>40	Non hygroscopic	Not available
KABO (32)	180–3600	0.05	35.01	28.50	>15	Non hygroscopic	Not available
LBO (mm²)	160–2600	0.16	32.84	16.84	>10	Non hygroscopic	Available
NYAB (32)	325–2300	0.07	43.34	37.68	>0.6	Non hygroscopic	Available

[a] $\lambda_{ref} = 0.79\ \mu m$, $\lambda_2 = 1.320\ \mu m$, and $\lambda_{SF} = 0.4942\ \mu m$.
[b] At 10-ns, 1064-nm Nd:YAG laser wavelength.

nonhygroscopic, except for the CLBO and BBO crystals. The optical damage threshold of MNA and NPP are also lower than that of the potential borate crystals considered in this paper [15]. Therefore, these crystalline materials are suitable for usual practical applications in comparison to the organic crystals NPP and MNA.

Several NLO characteristics of all these crystals are summarized in Table 7.2.1 for easy comparison. From Table 7.2.1 we see that the KBBF crystal has several favorable characteristics, such as a high laser damage threshold, nonhygroscopic nature, and the shortest SHG cut-off. Thus, the KBBF crystal is the best candidate among all the available borate crystals for characterization of CEP-locked octave spanned broadband angular-dispersed idler radiation generated from a broadband NOPA. However, due to difficulties in growth, properly cutting, and polishing KBBF crystals are yet to be available commercially. The research described in this subsection [21] was performed in collaboration with the following people: M. Chattopadhyay, P. Kumbhakar, and T. Kobayashi [21].

REFERENCES

1. Y. Nabekawa and K. Midorikawa, "Broadband sum frequency mixing using noncollinear angularly dispersed geometry for indirect phase control of sub-20-femtosecond UV pulses," *Opt. Express* **11**, 324–338 (2003).
2. G. M. Gale, M. Cavallari, T. J. Driscoll, and F. Hache, "Sub-20-fs tunable pulses in the visible from an 82-MHz optical parametric oscillator," *Opt. Lett.* **20**, 1562–1564 (1995).
3. E. Riedle, M. Beutter, S. Lochbrunner, J. Piel, S. Schenkl, S. Spörlein, and W. Zinth, "Generation of 10 to 50 fs pulses tunable through all of the visible and the NIR," *Appl. Phys. B: Lasers Opt.* **71**, 457–465 (2000).
4. G. Szabo and Z. Bor, "Frequency conversion of ultrashort pulses," *Appl. Phys. B* **58**, 237–241 (1994).
5. S. Adachi, P. Kumbhakar, and T. Kobayashi, "Quasi-monocyclic near-infrared pulses with a stabilized carrier-envelope phase characterized by noncollinear cross-correlation frequency-resolved optical gating," *Opt. Lett.* **29**, 1150–1152 (2004).
6. T. Kobayashi and A. Shirakawa, "Tunable visible and near-infrared pulse generator in a 5 fs regime," *Appl. Phys. B: Lasers Opt.* **70**, S239–S246 (2000).
7. A. Baltuska and T. Kobayashi, "Adaptive shaping of two-cycle visible pulses using a flexible mirror," *Appl. Phys. B: Lasers Opt.* **75**, 427–443 (2002).
8. A. Baltuska, T. Udem, M. Uiberacker, M. Hentschel, E. Goulielmakis, C. Gohle, R. Holzwarth, V. S. Yakovlev, A. Scrinzi, T. W. Hänsch, and F. Krausz, "Attosecond control of electronic processes by intense light fields," *Nature (London)* **421**, 611–615 (2003).
9. C. Chen, Z. Lin, and Z. Wang, "The development of new borate based UV nonlinear optical crystals," *Appl. Phys. B: Lasers Opt.* **80**, 1–25 (2005).
10. P. Kumbhakar and T. Kobayashi, "Ultrabroad-band phase matching in two recently grown nonlinear optical crystals for the generation of tunable ultrafast laser radiation by type-I noncollinear optical parametric amplification," *J. Appl. Phys.* **94**, 1329–1338 (2003).
11. P. Kumbhakar and T. Kobayashi, "Nonlinear optical properties of $Li_2B_4O_7$ LB_4 crystal for the generation of tunable ultra-fast laser radiation by optical parametric amplification," *Appl. Phys. B: Lasers Opt.* **78**, 165–170 (2004).
12. G. C. Bhar, U. Chatterjee, A. M. Rudra, P. Kumbhakar, R. K. Route, and R. S. Feigelson, "Generation of tunable 187.9–196-nm radiation in -Ba_2BO_4," *Opt. Lett.* **22**, 1606–1608 (1997).
13. G. C. Bhar, P. Kumbhakar, U. Chatterjee, A. M. Rudra, and A. Nagahori, "Widely tunable deep ultraviolet generation in CLBO," *Opt. Commun.* **176**, 199–205 (2000).
14. T. Kojima, S. Konno, S. Fujikawa, K. Yasui, K. Yoshizawa, Y. Mori, T. Sasaki, M. Tanaka, and Y. Okada, "20-W ultraviolet-beam generation by fourth-harmonic generation of an all-solid-state laser," *Opt. Lett.* **25**, 58–60 (2000).
15. D. N. Nikogosyan, *Nonlinear Optical Crystals: A Complete Survey*, Springer, New York (2006).
16. F. Yang et al., "Theoretical and experimental investigations of nanosecond 177.3 nm deep-ultraviolet light by second harmonic generation in KBBF," *Appl. Phys. B* **96**, 415–422 (2009).
17. J. T. Lin, "Non-linear crystals for tunable coherent sources," *Opt. Quantum Electron.* **22**, S283–S313 (1990).

18. D. A. Hammons, M. Richardson, B. H. T. Chai, A. Chin, and R. Jollay, "Scaling diode-pumped Nd^3 and Yb^3-doped YCa_4OBO_{33} YCOB self-frequency doubling lasers," *Proc. SPIE* **3945**, 50–62 (2000).

19. S. Linden, H. Giessen, and J. Kuhl, "XFROG—A new method for amplitude and phase characterization of weak ultrashort pulses," *Phys. Status Solidi* **206**, 119–124 (1998).

20. Z. D. Luo, J. T. Lin, A. D. Jiang, Y. C. Huang, and M. W. Qui, "Features and applications of a new self-frequency-doubling laser crystal—NYAB," *Proc. SPIE* **1104**, 132–141 (1989).

21. M. Chattopadhyay, P. Kumbhakar, and T. Kobayashi, "Broadband sum-frequency mixing in some recently developed nonlinear optical crystals," *Opt. Engineering* **48**, 124201 (2009).

7.3 Optimal Te-Doping in GaSe for Nonlinear Applications

7.3.1 INTRODUCTION

The optical properties of GaSe have been successfully used to generate coherent radiation in the mid-infrared and down to the terahertz (THz) frequency range [1,2]. Several unique properties of GaSe are associated with its layered structure. The basic 4-fold layer consists of two monoatomic sheets of Ga sandwiched between two monoatomic sheets of Se. The strong covalent interaction within these basic layers and the Van-der-Waals-type weak bonding between these basic layers render GaSe a highly anisotropic material. GaSe crystals are negative uniaxial crystals and belong to the point group of $62^-\, m$. Four polymorphic modifications were identified in the GaSe compound. The atom arrangement in one layer is the same for all four modifications; however, layer stacking can be classified by the noncentrosymmetric δ, ε, and γ or the centrosymmetric β modifications [3]. In general, GaSe crystals grown by the conventional Bridgeman technique are ε-polytype. Because of the layer structure, GaSe crystals exhibit considerable anisotropic absorption at short-wavelength edges of visible and THz ranges, and at the long-wavelength edge of the mid-IR range. The strongly anisotropic absorption ($\alpha_e > \alpha_o$) at the short-wavelength visible range is related to anisotropic band structure and the selection rules for the optical absorption in layered GaSe crystals [4,5]. Additionally, the layer structure of GaSe results in low (almost zero by Mohs scale) hardness, and the crystals can be easily cleaved along planes parallel with the atomic layers, which hamper large-area applications.

GaSe is an excellent matrix material for doping with various elements. An original ε-polytype structure of GaSe was strengthened by doping; the physical properties responsible for the frequency conversion efficiency were also modified [6–13]. Although the THz generation and optical properties in GaSe have been examined in detail [2,14,15], few studies focused on the optical properties in the THz range and the THz generation by doped GaSe crystals. The optical properties were studied experimentally in S-doped GaSe (GaSe:S) [16], GaSe:In [17], GaSe:Er [10,18], GaSe:Al [12], and GaSe:Te [17,19,20]. This indicates that the optical properties and THz generation efficiency of GaSe are strongly doping dependent. However, no systematic studies were conducted to find the optimally doped GaSe-based system for both mid-IR and THz applications. To our knowledge, no simple and reliable methods can be applied to identify the optimal doping in GaSe crystals for THz applications.

Among the physical properties, the far-infrared absorption of a non-linear crystal is a practical limitation on optical rectification and down-conversion. The far infrared absorption of a crystal is usually attributed to infrared-active phonon modes or their combination modes. For instance, Chen et al. [15] experimentally studied the effect of phonon mode of rigid layer mode $E'^{(2)}$ centered at 0.596 THz on the refractive index and THz generation efficiency in GaSe. In addition, the phonon mode of $E''^{(2)}$ at 1.78 THz was found in GaSe:S crystals [16], which demonstrated that the doping of S suppressed the rigid layer mode $E'^{(2)}$ and caused the $E''^{(2)}$ mode. However, the doping-dependent evolution between two modes of $E'^{(2)}$ and $E''^{(2)}$ remains unclear.

This paper reports the growth of centimeter-sized ingots with ε-GaSe:Te (nominal 0.05, 0.1, 0.5, 1, and 3 mass % in the charge composition) single crystals for non-linear applications. The strong correlation between the structure and the phonon modes in various Te-doped GaSe crystals was observed for the first time by measuring the absorption spectra in mm-long samples. As Te concentration increased, the absorption peak of rigid layer mode $E'^{(2)}$ markedly rose and subsequently shrunk with the simultaneous appearance of phonon mode $E''^{(2)}$ at 1.77 THz in heavy Te-doping. This study also proposed that the evolution of $E'^{(2)}$ and $E''^{(2)}$ modes can be used to identify the

DOI: 10.1201/9780429196577-38

lattice structure and the optical quality in Te-doped GaSe crystals, which was experimentally confirmed by the study of THz generation efficiency at various Te-doping levels. The THz generation efficiency from ~0.3-mm-thick Te-doped samples with 0.07 mass % was over 20% higher than that in a pure GaSe crystal with same thickness and was consistent with the highest absorption peak of the rigid layer mode E$'^{(2)}$. Finally, the doping-dependent evolution of the rigid layer mode E$'^{(2)}$ may be used as a criterion for identifying the optimal doping in GaSe or other crystals.

7.3.2 CRYSTAL GROWTH AND CHARACTERIZATION

7.3.2.1 GROWTH TECHNOLOGY

The GaSe:Te crystals were prepared according to the following processes. Initially, the polycrystalline materials with 120–150 g were synthesized in a two-zone horizontal furnace by using high purity (99.9999%) gallium (Ga), selenium (Se), and 99.9% tellurium (Te). The synthesis was performed in sealed quartz ampoules, which were evacuated to 10^{-5} Torr. Weighted charges of Ga and Se were placed in the boats located at hot and cold ends of the ampoule. A GaSe crystal was synthesized through three sequential stages with different temperature profiles over the ampoule, as described elsewhere [11]. Chemical reaction of the reagents up to GaSe formation was produced in the first stage by interacting between the vapor from the sublimation of Se at 690°C and Ga melt at 970°C; that is, the GaSe compound was synthesized in the reaction ampoule under selenium vapor pressure. In the second stage, the melt further homogenized at 1000°C through diffusion. In the final stage, the melt was cooled for 36 h to form a large block and homogeneous GaSe ingot. For GaSe:Te crystals, Te with 0.05, 0.1, 0.5, 1, and 3 mass % were added into the boat with gallium during synthesis. The temperature gradient at the crystallization front was 10°C cm^{-1}, and the crystal pulling rate was 10 mm/day.

Figure 7.3.1 shows the typical microscopic image of the pure and Te-doped GaSe single crystals.

The concentration of Te in GaSe:Te crystals was determined as 0.01, 0.07, 0.38, 0.67, and 2.07 mass % by electron probe micro-analyzer (EPMA, JEOL JXA-8800M).

An electron probe micro-analyzer (EPMA) is an instrument to analyze the elements constructing the materials. The method is based on X-ray analysis generated by the irradiation of electron beams onto the target materials. EPMA has evolved into an instrument that can handle the elemental analysis of sub-micron areas as well as observation, analysis, and image analysis for areas as large as 10 cm². This instrument is used for basic research in a diverse range of fields, such as steel, minerals, semiconductors, ceramics, textiles, medical and dental materials, medicine, and biology as well as application research.

In addition, the EPMA results show the homogenous distribution of Te in all of the samples used in this study. The crystal structures of all samples were recognized as the ε-polytype of $\overline{6}2m$ point group by the X-ray diffraction patterns. The samples were prepared by cleaving the middle part with the most homogeneity of an as-grown ingot parallel to the z-cut layer and were used without

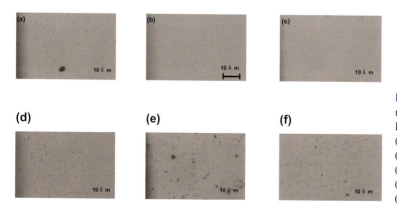

FIGURE 7.3.1 Confocal microscopic images of (a) Pure GaSe. (b) GaSe:Te (0.01 mass %). (c) GaSe:Te (0.07 mass %). (d) GaSe:Te (0.38 mass %). (e) GaSe:Te (0.67 mass %). (f) GaSe:Te (2.07 mass %) crystals.

additional treatment. The same samples or their parts were also used in various studies of this work. Additionally, the microhardness of GaSe:Te crystals was approximately 10 kg/mm² which is higher than the value of ~8 kg/mm² in GaSe crystals.

7.3.2.2 OPTICAL PROPERTIES

Mid-IR transmission spectra were recorded by FTIR VERTEX 70v (Bruker Optics Corp.) spectrometer with an operation wavelength range of 8000–375 cm⁻¹ and spectral resolution of 0.16 cm⁻¹. Thickness of the GaSe sample examined in this measurement was 0.89 mm. The thickness of the 0.01, 0.07, 0.38, 0.67, and 2.07 mass % Te-doped GaSe samples were 1.14, 1.00, 1.00, 0.98, and 0.87 mm, respectively. The typical mid-IR transmission spectra are presented in Figure 7.3.2. The features with strong low-frequency absorption and the position of phonon mode in GaSe remain the same for Te-doped GaSe. This implies that Te doping is unavailable for changing the optical quality in the mid-IR range.

Mid-IR absorption coefficients were estimated from Figure 7.3.2 and also determined in local point-to-point measurements by using low-power CO_2 laser ($\phi \sim 1.0$ mm) at 9.6 µm to minimize the influence of the microscopic surface and bulk defects on the measurement results. Subsequently, the point measurement data were applied to calibrate transmission spectra recorded by the mid-IR spectrophotometer. Because of the low mid-IR absorptivity, the presence of surface and bulk micro-defects, and the drifts of zero and 100% levels in spectra, we estimated only the upper limit of the o-wave absorption coefficients and were unable to quantitatively characterize the optical quality as a function of doping levels in these six GaSe:Te crystals. The average absorption coefficients α in pure GaSe and GaSe:Te (0.01, 0.07, 0.38 mass %) crystals estimated from calibrated absorption spectra were within 0.1–0.2 cm⁻¹. For a 0.67 mass % GaSe:Te crystal, the average absorption coefficient was estimated as 0.3–0.4 cm⁻¹, which can be reduced to 0.2 cm⁻¹ by measuring the local point. However, because of the noticeable precipitates in a 2.07 mass % GaSe:Te crystal, as shown in Figure 7.3.1f, the higher absorption coefficients with a fluctuating range from ≤1 cm⁻¹ to a few cm⁻¹ were obtained in the local points. Moreover, the strongly anisotropic absorption ($\alpha_e > \alpha_o$) in GaSe crystals does not change in Te-doped GaSe crystals.

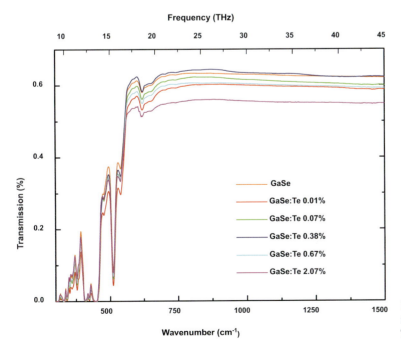

FIGURE 7.3.2 IR transmission spectra of various Te-doped GaSe crystals.

The optical properties in the range of 0.4–2.0 THz were measured by a homemade THz time-domain spectroscopy (TDS) system, as described elsewhere [16]. The THz pulses were generated by a homemade biased 5×5×1 mm InP photoconductive switch under the pumping of 50 fs pulses from Ti:sapphire laser (central wavelength: 800 nm, pulse repetition rate: 80 MHz). The generated THz pulses propagated through four off-axis parabolic mirrors and focused on a 1-mm-thick (110) ZnTe crystal, which was used to probe the THz pulse waveform by the free-space electro-optic (EO) sampling technique. The entire experimental setup was placed in an airtight enclosure purged with dry nitrogen and maintained at a relative humidity of <3.0% to avoid the strong absorption of the water vapor in the THz range. In this study, the THz beam was normally incident to the crystal surfaces.

It was necessary to consider the lattice vibration contribution to the free carriers to investigate the complex dielectric function of GaSe:Te crystals in the THz range. According to the combined Drude-Lorentz model, the total complex dielectric function $\tilde{\varepsilon}(\omega)$ [21] is given by

$$\tilde{\varepsilon}(\omega) = \varepsilon(\infty) + \sum_{j=1}^{J} \frac{S_j \omega_{TO_j}^2}{\omega_{TO_j}^2 - \omega^2 - i\Gamma_j \omega} - \frac{\omega_p^2}{\omega(\omega + i\langle\tau\rangle^{-1})} \quad (7.3.1)$$

where S_j is the strength of the oscillator, ω_{TO_j} is the frequency of the transverse optical phonon, Γ_j is the phonon relaxation rate, ω_p is the plasma frequency, and $\langle\tau\rangle$ is the average momentum relaxation time for free carriers. The first term of the right-hand side, $\varepsilon(\infty)$, is the high-frequency dielectric constant related to bound electrons; the second term describes the contribution of optical phonons; and the third term is the contribution from free electrons or plasmons. From the Drude-Lorentz model approximation, the complex dielectric function is given by the following equation:

$$\tilde{\varepsilon}(\omega) = (n(\omega) + ik(\omega))^2 = \varepsilon(\infty) + \frac{i\hat{\sigma}}{\omega\varepsilon_o} \quad (7.3.2)$$

where the complex conductivity $\hat{\sigma}(\omega)$ can be written as $\hat{\sigma}(\omega) = \sigma_r(\omega) + i\sigma_i(\omega)$. The real part of conductivity $\sigma_r(\omega) = 2n\kappa\omega\varepsilon_0 (\kappa = c\alpha/2\omega)$ can be obtained from Eqs. (7.3.1) and (7.3.2), as follows:

$$\sigma_r(\omega) = \frac{\varepsilon_0 \omega_p^2 \langle\tau\rangle^{-1}}{[\omega^2 + \langle\tau\rangle^{-2}]} + \sum_{j=1}^{J} \frac{\varepsilon_0 S_j \Gamma_j \omega_{TO_j}^2 \omega^2}{(\omega_{TO_j}^2 - \omega^2)^2 + \Gamma_j^2 \omega^2} \quad (7.3.3)$$

By using Eq. (7.3.3), we fit the experimental data of the real part conductivity (see Figure 7.3.3) with the free-space permittivity of $\varepsilon_0 = 8.854 \times 10^{-12}$ F/m. All the fitting parameters are summarized in Table 7.3.1.

By theoretical fitting in Figure 7.3.3, the absorption phonon modes can be described precisely by a number of physical parameters, as listed in Table 7.3.1. For instance, the oscillator strength S_j, an estimation of the intermolecular interaction, is Te-doping dependent in GaSe crystals and has a maximal value for 0.07 mass % GaSe:Te.

Because of the low absorptivity in the THz range, the presence of surface and bulk microdefects, and the drifts of zero and 100% levels in spectra, we estimated only the upper limit of o-wave absorption coefficients and were unable to quantitatively characterize the optical quality as a function of doping levels in these six GaSe:Te crystals and identify the optimal doping level. The average absorption coefficients in the THz range were <5 cm^{-1} of GaSe:Te with 0, 0.01, 0.07, and 0.38 mass %. Furthermore, the higher absorption coefficients of 7.7 cm^{-1} and ≥9.7 cm^{-1} were obtained from 0.67 and 2.07 mass % GaSe:Te crystals, respectively.

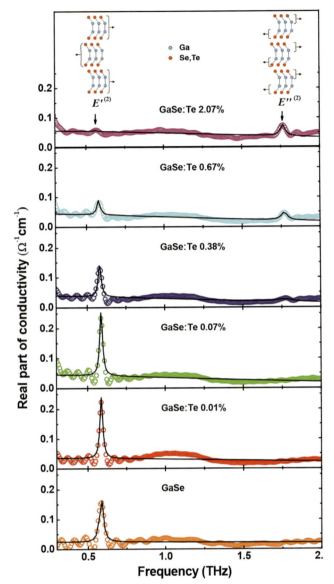

FIGURE 7.3.3 O-wave real part of conductivity spectra in GaSe:Te crystals. Points: experimental data. Solid lines: fitting by Eq. (7.3.3). The inset in the upper figure shows the vibrational displacements of the rigid layer mode $E'^{(2)}$ and two-atom sub-layer mode $E''^{(2)}$ in a primitive layer.

TABLE 7.3.1

Fitting Parameters for Eq. (7.3.3)

Te Concentration (mass %)	0	0.01	0.07	0.38	0.67	2.07
Phonon frequency ω_{TO1} (THz)	0.584	0.586	0.587	0.580	0.577	0.564
Phonon relaxation rate Γ_1 (THz)	0.030	0.022	0.022	0.024	0.026	0.017
Oscillator strength S_1 (a.u.)	0.146	0.150	0.154	0.085	0.037	0.004
Phonon frequency ω_{TO2} (THz)	–	–	–	1.77	1.77	1.76
Phonon relaxation rate Γ_2 (THz)	–	–	–	0.121	0.046	0.038
Oscillator strength S_2 (a.u.)	–	–	–	0.0037	0.0043	0.0050

7.3.2.3   THz Generation via Optical Rectification

Here, we describe THz generation in pure and Te-doped GaSe crystals through the optical rectification method by a commercial Ti:sapphire oscillator with a pulse duration of ~70 fs and repetition
rate of 5.2 MHz. By using a 100-μm-thick (110)-oriented ZnTe crystal as EO detector, the temporal
waveforms of the THz field from GaSe:Te crystals were measured using the EO sampling technique,
as shown in Figure 7.3.4a.

All specimens were fabricated with an almost fixed thickness (approximately 0.3 mm) to compare
the THz generation efficiency. The fast Fourier transformation (FFT) of the temporal profile of THz
pulse was used to obtain the central frequency and output power of THz radiation, as illustrated in
the inset of Figure 7.3.4a. Consequently, the power and the central frequency of the generated THz
pulses on all GaSe:Te crystals were extracted, as shown in Figure 7.3.4b. The central frequency of
THz emission was independent of the Te-doping in GaSe crystals. However, the Te dopant in GaSe
crystals considerably improved the THz output power. The highest conversion efficiency, over 20%
higher than that of a pure GaSe crystal, was found in a 0.07 mass % Te-doped GaSe crystal.

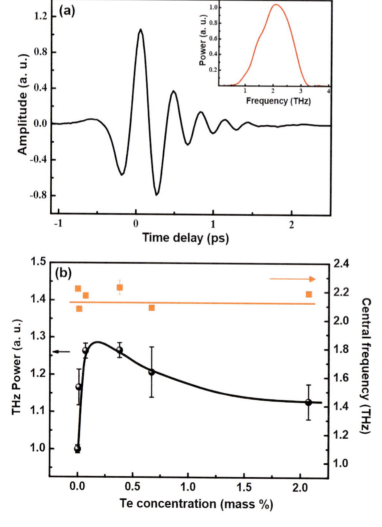

FIGURE 7.3.4   (a) Temporal
waveform of THz radia-
tion on a 0.30-mm GaSe:Te
(0.07 mass %) crystal. The
inset illustrates the power
spectra from the fast Fourier
transform of the temporal
waveform. (b) The central
frequency and the THz generation power at various Te
concentrations.

7.3.3   DISCUSSION

Crystals grown from the charge compositions of GaSe with nominal 0.05, 0.1, 0.5, 1, and 3 mass % of Te were identified as the noncentrosymmetric ε-GaSe:Te (0.01, 0.07, 0.38, 0.67, and 2.07 mass %) crystals, which are useful for non-linear applications. The crystal structure of the heavy Te-doped (nominal 5 and 10 mass %) GaSe crystals was polycrystalline and unsuitable for nonlinear applications. A small 0.02-THz shift of the central frequency of the phonon mode $E'^{(2)}$ (in Table 7.3.1) was observed only in GaSe:Te with 2.07 mass % crystals. This indicates that the slight doping of Te has almost no influence on the lattice parameters of GaSe crystals. The low absorptivity in slightly Te-doped GaSe limited us to resolve the small changes of the absorption coefficients in THz and mid-IR ranges with maximal transparency, as well as to identify optimal Te-doping in GaSe crystals.

Furthermore, the noticeable changes in the absorption peak intensities of the phonon modes $E'^{(2)}$ and $E''^{(2)}$ were observed with various Te doping levels, as shown in Figure 7.3.3. For $E'^{(2)}$ at ~0.584 THz, the so-called rigid layer mode was formed because the GaSe layers vibrated as rigid units and no relative displacement occurred between the Ga and Se atoms within a basic layer, as illustrated in the inset of Figure 7.3.3. However, if the layers vibrate with the relative displacement between two Ga-Se sub-layers within four-atom rigid layers (see the inset of Figure 7.3.3), the frequency of phonon mode would be higher than the rigid layer mode, that is, the $E''^{(2)}$ mode at 1.77 THz, as shown in Figure 7.3.3 [22]. First, we focused on the intensity evolution of absorption peaks as increasing the concentration of Te. For slight Te-doping, the absorption peak of the rigid layer mode $E'^{(2)}$ gradually increased until it reached a maximal value at Te-doping of 0.07 mass %. This indicates that the lattice structure in GaSe was improved by decreasing the number of point- and layer-stacking defects with Te-doping [8,23], which correlates with the clear images in Figure 7.3.1b and c, respectively. Consequently, the optical quality of GaSe:Te was also considerably improved by Te-doping. However, the shrink of the absorption peak of the rigid layer mode $E'^{(2)}$ reveals that the improvement of the lattice structure and optical quality in GaSe cannot be further extended by higher Te doping levels. This degradation of optical quality in Te-doped GaSe crystals can be caused by increasing structural defects (polytypism, stacking faults, dislocations) [20,23,24], defect complexes [13,24], exciton-phonon and exciton-impurity interactions [25], intensive interlayer (interstitial) species [23,26], and formation of strained regions [25]. For heavy Te-doped GaSe crystals, the rigid layer mode $E'^{(2)}$ disappears, which is consistent with the results of [20] and the observations in GaSe:S crystals [16]. As the rigid layer mode $E'^{(2)}$ shrinks, the phonon mode $E''^{(2)}$ at 1.77 THz, which was also observed in GaSe:S crystals [16], gradually increased in intensity in conjunction with the Te doping level. No distinct reason was found to explain the increase of $E''^{(2)}$ mode; however, interlayer intercalation of larger-size Te atoms [13,25] led to the formation of the local strained regions that bond hard Se-Ga sub-layers.

The results of the THz pulse generation in GaSe:Te (0, 0.01, 0.07, 0.38, 0.67 and 2.07 mass %) through optical rectification in Figure 7.3.4 reveal the strong correlation between the intensity of the absorption peak of phonon mode $E'^{(2)}$ and THz generation efficiency, that is, the optical quality in Te-doped GaSe crystals. The THz generation efficiency in 0.07 mass % GaSe:Te crystals was over 20% higher than that in pure GaSe crystals. Further improvement in the generation efficiency is possible by carefully tuning the Te-doping level between 0.07 and 0.38 mass %, optimizing the crystal length because of the so-called "scaling up" effect [8], and improving surface quality. It is also proposed that the higher damage threshold can be reached due to higher optical quality in optimally doped GaSe crystals if the surface defects can be suppressed to allow higher pump intensity.

Because of a narrow window for the optimal concentration of Te doping in GaSe crystals, which were not considered or occasionally omitted, no optimally Te-doped GaSe have been reported in previous studies. However, the optimal doping in GaSe crystals is crucial in the application of sub-cm and cm-sized crystals in long-pulse frequency conversion at mid-IR and down-conversion into THz range [15] as it proceeds from the "scaling up" effect [8]. This study is the first to demonstrate the evolution of E′$^{(2)}$ and E″$^{(2)}$ phonon modes measured by THz TDS to identify the optical quality and the optimal doping levels in doped GaSe crystals.

7.3.4 CONCLUSION

Centimeter-sized Te-doped GaSe ingots were grown from the charge compositions of GaSe with nominal 0.05, 0.1, 0.5, 1, and 3 mass % of Te, which were identified as ε-GaSe:Te (0.01, 0.07, 0.38, 0.67, and 2.07 mass %) single crystals and suitable for non-linear applications [27]. We found a strong correlation between the intensity of the rigid layer mode E′$^{(2)}$ at ~0.584 THz and the optical quality in Te-doped GaSe crystals. This study was the first to use this correlation as a sensitive mean for determining the optical quality in doped GaSe, and as an efficient tool for determining the optimal doping level. This was further confirmed by the THz generation experiments. The Te doping of approximately 0.07 mass % was identified as the optimal doping for THz generation through optical rectification, resulting in 20% improvement in generation efficiency. Further improvement in the efficiency can be achieved by fine-tuning the doping levels and the crystal length because of the so-called "scaling up" effect, by canceling the surface defects and increasing pump intensity [27].

The research work was conducted cooperatively among the cooperatively among the following people: Shin An Ku, Wei-Chen Chu, Chih Wei Luo, Yu. M. Andreev, Grigory Lanskii, Anna Shaidukoi, Tatyana Izaak, Valery Svetlichnyi, Kaung Hsiung Wu, and T. Kobayashi [27].

REFERENCES

1. V. G. Dmitriev, G. G. Gurzadyan, and D. N. Nikogosyan, *Handbook of Nonlinear Optical Crystals*, Springer, Berlin (1997), pp. 166–169.
2. R. Huber, A. Brodschelm, F. Tauser, and A. Leitenstorfer, "Generation and field-resolved detection of femtosecond electromagnetic pulses tunable up to 41 THz," *Appl. Phys. Lett.* **76**, 3191–3193 (2000).
3. N. C. Fernelius, "Properties of gallium selenide single crystal," *Prog. Cryst. Growth Charact.* **28**, 275–353 (1994).
4. W. Y. Liang, "Optical anisotropy in GaSe," *J. Phys. C: Solid State Phys.* **8**, 1769–1768 (1975).
5. R. Le Toullec, N. Piccioli, M. Mejatty, and M. Balkanski, "Optical constants of ε-GaSe," *Nuovo Cimento* **38B**, 159–167 (1977).
6. A. A. Tikhomirov, Y. M. Andreev, G. V. Lanskii, O. V. Voevodina, and S. Y. Sarkisov, "Doped GaSe nonlinear crystals," *Proc. SPIE*. 6258, 64–72 (2006).
7. K. R. Allakhverdiev, R. I. Guliev, E. Y. Salaev, and V. V. Smirnov, "An investigation of linear and non-linear optical properties of GaS_xSe_{1-x} crystals," *Sov. J. Quantum Electron.* **12**, 947–949 (1982).
8. D. R. Suhre, N. B. Singh, V. Balakrishna, N. C. Fernelius, and F. K. Hopkins, "Improved crystal quality and harmonic generation in GaSe doped with indium," *Opt. Lett.* **22**, 775–777 (1997).
9. N. B. Singh, D. R. Suhre, W. Rosch, R. Meyer, M. Marable, N. C. Fernelius, F. K. Hopkins, D. E. Zelmon, and R. Narayanan, "Modified GaSe crystals for mid-IR applications," *J. Cryst. Growth* **198**, 588–592 (1999).
10. Y.-K. Hsu, C.-W. Chen, J. Y. Huang, C.-L. Pan, J.-Y. Zhang, and C.-S. Chang, "Erbium doped GaSe crystal for mid-IR applications," *Opt. Express* **14**, 5484–5491 (2006).
11. Z.-S. Feng, Z.-H. Kang, F.-G. Wu, J.-Y. Gao, Y. Jiang, H.-Z. Zhang, Y. M. Andreev, G. V. Lanskii, V. V. Atuchin, and T. A. Gavrilova, "SHG in doped GaSe:In crystals," *Opt. Express* **16**, 9978–9985 (2008).
12. Y.-F. Zhang, R. Wang, Z.-H. Kang, L.-L. Qu, Y. Jiang, J.-Y. Gao, Y. M. Andreev, G. V. Lanskii, K. Kokh, A. N. Morozov, A. V. Shaiduko, E. Vinnik, and V. V. Zuev, "$AgGaS_2$- and Al-doped GaSe for IR application," *Opt. Commun.* **284**, 1677–1681 (2011).
13. Z. Rak, S. D. Mahanti, K. C. Mandal, and N. C. Fernelius, "Doping dependence of electronic and mechanical properties of $GaSe_{1-x}Te_x$ and $Ga_{1-x}In_xSe$ from first principles," *Phys. Rev. B* **82**, 155203 (2010).
14. Y. J. Ding and W. Shi, "Widely tunable monochromatic THz sources based on phase-matched difference frequency generation in nonlinear-optical crystals: A novel approach," *Laser Phys.* **16**, 562–570 (2006).
15. C.-W. Chen, T.-T. Tang, S.-H. Lin, J. Y. Huang, C.-S. Chang, P.-K. Chung, S.-T. Yen, and C.-L. Pan, "Optical properties and potential applications of ε-GaSe at terahertz frequencies," *J. Opt. Soc. Am. B* **26**, A58–A65 (2009).

16. Z.-W. Luo, X.-A. Gu, W.-C. Zhu, W.-C. Tang, Y. M. Andreev, G. Lanskii, A. Morozov, and V. Zuev, "Optical properties of GaSe:S crystals in terahertz frequency range," *Opt. Precision Eng. (in Chinese)* **19**, 354–359 (2011).

17. K. C. Mandal, S. H. Kang, M. Choi, J. Chen, X.-C. Zhang, J. M. Schleicher, C. A. Schmuttenmaer, and N. C. Fernelius, "III-VI chalcogenide semiconductor crystals for broadband tunable THz sources and sensors," *IEEE J. Sel. Top. Quant. Electron* **14**, 284–288 (2008).

18. C.-W. Chen, Y-K. Hsu, J. Y. Huang, C.-S. Chang, J.-Y. Zhang, and C.-L. Pan, "Generation properties of coherent infrared radiation in the optical absorption region of GaSe crystal," *Opt. Express* **14**, 10636–10644 (2006).

19. S. Y. Sarkisov, V. V. Atuchin, T. A. Gavrilova, V. N. Kruchinin, S. A. Bereznaya, Z. V. Korotchenko, O. P. Tolbanov, and A. I. Chernyshev, "Growth and optical parameters of GaSe:Te crystals," *Russ. Phys. J.* **53**, 346–352 (2010).

20. G. B. Abdullaev, K. R. Allakhverdiev, S. S. Babaev, E. Yu. Salaev, M. M. Tagyev, L. K. Vodopyanov, and L. V. Golubev, "Raman scattering from $GaSe_{1-x}Te_x$," *Solid State Commun.* **34**, 125–128 (1980).

21. B. L. Yu, F. Zeng, V. Kartazayev, R. R. Alfano, and K. C. Mandal, "Terahertz studies of the dielectric response and second-order phonons in a GaSe crystal," *Appl. Phys. Lett.* **87**, 182104 (2005).

22. H. Yoshida, S. Nakashima, and A. Mitsuishi, "Phonon Raman spectra of layer compound GaSe," *Phys. StatusSolidi B* **59**, 655–666 (1973).

23. I. Evtodiev, L. Leontie, M. Caraman, M. Stamate, and E. Arama, "Optical properties of p-GaSe single crystals doped with Te," *J. App. Phys.* **105**, 023524 (2009).

24. G. M. Mamedov, M. Karabulut, H. Ertap, O. Kodolbas, O. Oktu, and A. Bacoglu, "Exciton photoluminescence, photoconductivity and absorption in $GaSe_{0.9}Te_{0.1}$ alloy crystals," *J. Lumin.* **129**, 226–230 (2009).

25. E. A. Meneses, N. Jannuzzi, J. R. Freitas, and A. Gouskov, "Photoluminescence of layered $GaSe_{1-x}Te_x$ crystals," *Phys. Stat. Sol. B* **78**, K35–K38 (1976).

26. S. Shigetomi and T. Ikari, "Optical and electrical characteristics of p-GaSe doped with Te," *J. Appl. Phys.* **95**, 6480–6482 (2004).

27. S.-A. Ku, W.-C. Chu, C.-W. Luo, Y. M. Andreev, G. Lanskii, A. Shaidukoi, T. Izaak, V. Svetlichnyi, K. H. Wu, and T. Kobayashi, "Optimal Te-doping in GaSe for non-linear applications" *Opt. Express* **20**, 5029–5037 (2012).

7.4 Widely Linear and Non-Phase-Matched Optical-to-Terahertz Conversion on GaSe
Te Crystals

7.4.1 INTRODUCTION

Despite that terahertz (THz) waves [1] have been developed only in the past few decades, a large number of useful applications of the THz wave have been developed, such as the measurements of molecular vibrational modes and time-resolved THz spectroscopy, which have led to a new "tera era". Therefore, for practical applications, the development of novel sources and methods in the THz range has become a key issue. A relatively compact and economical avenue for generating coherent THz radiation is taking advantage of optical conversion in nonlinear crystals (e.g., GaSe, LiNbO$_3$, GaP, GaAs, and ZnGeP$_2$) [2–6]. Among these nonlinear crystals, layered ε-GaSe crystals possess a number of advantages, including a large nonlinear coefficient and significant birefringence, which have been widely used in far-IR generation [3] and mid-IR spectroscopy [7]. Nevertheless, for further application in optics, improvement in the optical and mechanical properties of GaSe crystals by doping proper elements is highly desired. High-quality GaSe:Te crystals have been prepared because Te atoms slightly strengthen the mechanical properties [8] and modify the optical and electrical properties noticeably [9,10]. However, GaSe:Te crystals have not yet been studied in THz generation.

This subsection describes THz wave generation from GaSe:Te crystals by femtosecond (fs) laser pulses and shows that THz generation efficiency was improved substantially. Compared to the commonly used ZnTe crystals, GaSe:Te crystals can provide widely linear optical-to-THz conversion with an optical pumping fluence of up to 6.9 mJ/cm^2.

7.4.2 EXPERIMENTAL

The p-type pure and Te-doped GaSe single crystals grown by the Bridgman method [11] were used in this study.

There are most popular crystallization methods. One is recrystallization from a supersaturated solution of materials, which are typically organic molecules. The other is a Bridgman method used mainly for inorganic material. The Bridgman method uses a crucible containing the melt. It is moved in a vertical temperature gradient toward cooler regions of the furnace. For the production of single crystals, solidification generally begins at the bottom of the crucible where a seed is located.

The concentrations of Te in all GaSe:Te crystals were determined by electron probe microanalysis (with an accuracy of 0.01 wt. %) are 0.01, 0.07, 0.38, 0.67, and 2.07 mass %, respectively. On the other hand, a commercial Ti:sapphire oscillator operating at a central wavelength of 800 nm and with a pulse duration of 100 fs with a repetition rate of 5.2 MHz was used as a pumping source. Under the scheme of normal incidence (the polarization is along the a–b plane), the pump beam with

DOI: 10.1201/9780429196577-39

3.5 mJ/cm^2 and diameter of 46 μm was focused on z-cut GaSe:Te crystals to generate THz waves. Any residual 800 nm laser beam was blocked by a Si wafer. The transmitted THz pulses were collected and focused on a 100 μm thick (110)-oriented ZnTe crystal by gold-coated off-axis parabolic mirrors. The electro-optical (EO) sampling technique was subsequently applied to detect the emitted THz fields in the time domain. All experiments were performed in a dry nitrogen-purged box.

7.4.3   RESULTS AND DISCUSSION

Typical THz waveforms at various azimuthal angles ϕ were generated from a 0.38 mass % GaSe:Te crystal, as shown in Figure 7.4.1a. Here, ϕ is defined as the angle between the [100] direction of (001) GaSe:Te crystals and the polarization of the incident optical pulse. Considering EO detection and the 62 m point group of the crystals, the electric field of THz generated from GaSe:Te crystals by optical rectification can be derived as

$$E_{THz} \propto x_{22} \cos 2\theta \sin 3\varphi \qquad (7.4.1)$$

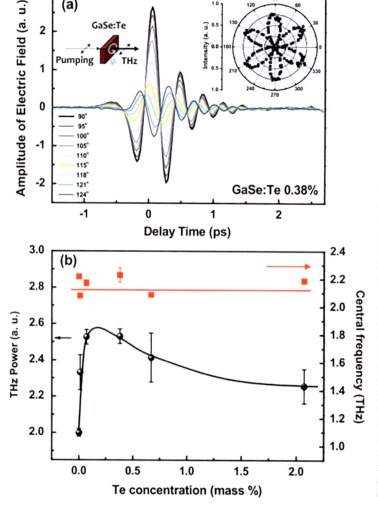

FIGURE 7.4.1 (a) Temporal waveforms of THz radiation at various azimuthal angle ϕ on a 0.38 mass % GaSe:Te crystal with a thickness of 0.30 mm. The right inset illustrates the intensity of the THz main peak as a function of ϕ. The left inset shows the configuration of THz generation on GaSe:Te crystals. Corresponding temperature of the curve is assigned in such a way that at 0 ps signal peak is the maximum and absolute values of the peaks are decreasing with increasing temperature. (b) Power and central frequency of THz radiation on various GaSe:Te crystals with a thickness of 0.30 mm.

where θ is the angle between the c axis of a crystal and the direction of the incident light. The right inset in Figure 7.4.1a shows the intensity of the main peak in THz signals as a function of ϕ, which can be well described by the ϕ-dependent relation of Eq. (7.4.1). Therefore, the THz generation mechanism on GaSe:Te crystals is optical rectification, which is different from the difference frequency mixing in ZnGeP$_2$ [6]. The 6-fold symmetry of THz intensity is attributed to the hexagonal structure on the (001) plane of GaSe:Te crystals. Moreover, similar 6-fold symmetry observed for all GaSe:Te crystals indicates that the Te atoms doped in GaSe do not destroy the crystal structure of the GaSe matrix.

Figure 7.4.1b unambiguously shows the effect of Te atoms doped in GaSe crystals for THz generation. Nevertheless, the central frequencies (~2.15 THz) of THz radiation are independent of the Te concentration; we found that the THz power suddenly increases once the Te atoms are doped into the GaSe crystals, even for the small Te doping of 0.01 mass %. For 0.38 mass % GaSe:Te crystals, the output power of emitted THz is enhanced by a factor of 21% compared to the GaSe crystals. According to Vodopyanov's studies [12], the optical-to THz conversion efficiency is proportional to x_{ijk}^2. Thus, these results imply that heavy-atom doping, that is, Te, in GaSe crystals can improve its nonlinearity. However, the optical-to-THz conversion efficiency could not increase further at higher Te-doping level because of increases in structural defects (e.g., polytypism, stacking faults, and dislocation) [9–11].

To describe the THz wave generated on GaSe:Te crystals more clearly, Figure 7.4.2 shows a comparison of THz Fourier spectra from a GaSe crystal, a 0.38 mass % GaSe:Te crystal, and a commonly used ZnTe crystal. For GaSe and 0.38 mass % GaSe:Te crystals, the spectra of the emitted THz pulse cover the range of 0.5–5 THz. Additionally, the spectra of THz radiation from GaSe:Te crystals are wider than those of the ZnTe crystals with the same thickness at a high-frequency side. This is caused by the frequency of the transverse optical phonon absorption peak for GaSe centered at 7.1 THz, which is higher than that of ZnTe at 5.3 THz [3,13].

The inset in Figure 7.4.2 shows the average output power of THz radiation calibrated by a liquid He-cooled bolometer as a function of pumping fluences on ZnTe, GaSe, and GaSe:Te crystals. Because of the two-photon absorption of pumping light and the free carrier absorption of THz on ZnTe, the THz output power saturates above the pumping fluence of 3 mJ/cm^2 [14]. However, the value of the two-photon absorption coefficient of GaSe (βGaSe 0.558 cm/GW) is smaller than that of ZnTe (βZnTe 4.2 cm/GW) [15], which implies that fewer free carriers are generated in GaSe:Te crystals to diminish the suppression effect in the THz output power. Therefore, the linear dependence (the originally quadratic dependence is suppressed by two-photon absorption) between the

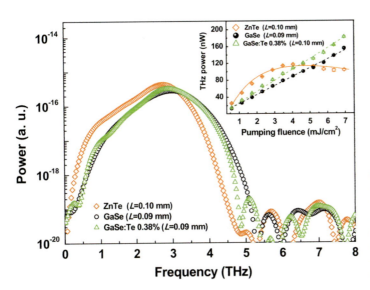

FIGURE 7.4.2 Fourier spectra of the THz radiation from ZnTe, GaSe, and 0.38 mass % GaSe:Te crystals. Inset, THz power as a function of pump fluences. L, thickness of the crystals. The solid and dashed lines are guides for the eyes.

THz output power and pumping fluence is clearly observed. Consequently, at relatively low pump fluences (<4.6 mJ/cm²), the emitted THz power from GaSe:Te crystals is smaller than that from ZnTe crystals. Once the pump fluences are larger than 4.6 mJ/cm², however, the emitted THz power on the GaSe:Te crystals exceeds that on ZnTe crystals considerably because of the larger two-photon absorption of ZnTe. The conversion efficiency of GaSe:Te crystals was significantly higher than that of the GaSe crystal, especially in the high pumping fluence range. Thus, we can expect that GaSe:Te crystals to be a promising material for high-power THz generation, which would be attractive for applications in THz spectroscopy.

Although no significant changes are in the central frequency of THz radiation by fixing the thickness to 0.3 mm on various GaSe:Te crystals (Figure 7.4.1b), the marked redshift of the central frequency is observed by changing the thickness of GaSe:Te crystals, as shown in Figure 7.4.3.

Under the slowly varying envelope approximation [12], the power spectrum of THz radiation can be described as

$$\mathrm{L} \propto \omega_T^2 d_{\mathrm{eff}}^2 \tau^2 L^2 \exp\left(-\frac{\tau^2 \omega_\tau^2}{4}\right) \mathrm{sin}\, c^2\left(\Delta k L / 2\right) \qquad (7.4.2)$$

where ω_T is the THz angular frequency, L is the thickness of the crystal, $d_{\mathrm{eff}}=(1/2)^{(2)}$, τ is the pulse duration, and the k-vector mismatch is $\Delta k = (n_{\mathrm{THz}} - n_{\mathrm{opt}}^{\mathrm{gr}})\omega_T/c$. The refractive index in GaSe crystals was obtained from [16], in which $n_{\mathrm{opt}}^{\mathrm{gr}}$ (at 800 nm) = 3.12 and n_{THz} (at 1 THz) = 3.26. For the different thicknesses of the crystals, L, the THz spectrum can be simulated by Eq. (7.4.2) and the central frequency of the THz radiation can be further identified from the simulated spectrum. For the thin crystals, the central frequency of THz radiation simulated from Eq. (7.4.2) is dominated by pulse duration. However, the Δk in Eq. (7.4.2) leads to the change in central frequency on thicker crystals. Consequently, the central frequency of THz radiation from thin GaSe and ZnTe crystals is similar, as shown by the solid and dashed lines in Figure 7.4.3. While the crystal thickness increases, the central frequency of THz radiation on ZnTe crystals approaches 1.9 THz, where phase matching is satisfied [17]. However, the central frequencies of THz radiation on GaSe and GaSe:Te crystals decrease gradually to 1.5 THz because of no phase matching condition being satisfied in this region. In some manner, we can consider this characteristic advantageous for tuning the THz central frequency by varying the thickness of GaSe:Te crystals. Moreover, the THz radiation power on 0.38

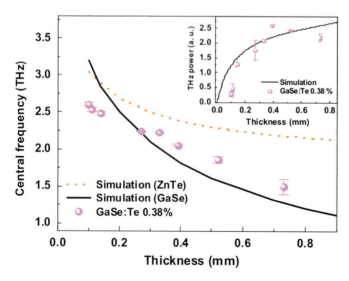

FIGURE 7.4.3 Central frequency of THz radiation on 0.38 mass % GaSe:Te crystals with various thicknesses. Inset, thickness dependence of THz power on 0.38 mass % GaSe:Te. All THz generation experiments were performed at the pumping level of 3.5 mJ/cm².

mass % GaSe:Te crystals increases by 8.8 times while the thickness increases from 0.1 to 0.52 mm (see inset of Figure 7.4.3), which is excellently consistent with the simulation results. Therefore, the high THz radiation power on GaSe:Te crystals could be obtained simply by using thicker crystals.

7.4.4 CONCLUSION

In conclusion, we demonstrated broadband THz generation with widely linear optical-to-THz conversion on GaSe:Te crystals by non-phase-matched optical rectification. The dopant-Te atoms-in GaSe crystals improve the efficiency of THz generation significantly, especially in the high pumping fluence range. By increasing crystal thickness, the central frequency of THz radiation from GaSe:Te crystals shifts markedly to red, and its power increases. Furthermore, the high-power (>1.36 μW under the pumping of 6.9 mJ/cm^2 on a 0.52 mm thick 0.38 mass % GaSe:Te crystal) and central-frequency tunable (Δf_c ~1.5 THz) THz radiation can be realized in GaSe:Te crystals [18].

The research presented in this subsection was performed collaboratively among the following people: Wei-Chen Chu, Shin An Ku, Harn Jiunn Wang, Chih Wei Luo, Yu. M. Andreev, Grigory Lanskii, and Takayoshi Kobayashi [18].

REFERENCES

1. B. Ferguson and X.-C. Zhang, *Nat. Mater.* **1**, 26 (2002).
2. W. Shi, Y. J. Ding, N. Fernelius, and K. Vodopyanov, *Opt. Lett.* **27**, 1454 (2002).
3. Y.-S. Lee, T. Meade, V. Perlin, H. Winful, and T. B. Norris, *Appl. Phys. Lett.* **76**, 2505 (2000).
4. T. Tanabe, K. Suto, J. Nishizawa, K. Saito, and T. Kimura, *Appl. Phys. Lett.* **83**, 237 (2003).
5. K. L. Vodopyanov, M. M. Fejer, X. Yu, J. S. Harris, Y.-S. Lee, W. C. Hurlbut, V. G. Kozlov, D. Bliss, and C. Lynch, *Appl. Phys. Lett.* **89**, 141119 (2006).
6. P. Kumbhakar, T. Kobayashi, and G. C. Bhar, *Appl. Opt.* **43**, 3324 (2004).
7. C. W. Luo, K. Reimann, M. Woerner, T. Elsaesser, R. Hey, and K. H. Ploog, *Phys. Rev. Lett.* **92**, 047402 (2004).
8. A. A. Tikhomirov, Yu. M. Andreev, G. V. Lanskii, O. V. Voevodina, and S. Y. Sarkisov, *Proc. SPIE* **6258**, 625809 (2006).
9. S. Shigetomi and T. Ikari, *J. Appl. Phys.* **95**, 6480 (2004).
10. I. Evtodiev, L. Leontie, M. Caraman, M. Stamate, and E. Arama, *J. Appl. Phys.* **105**, 023524 (2009).
11. S. A. Ku, W.-C. Chu, C. W. Luo, Y. Andreev, G. Lanskii, A. Shaiduko, T. Izaak, V. Svetlichnyi, E. Vaytulevich, K. H. Wu, and T. Kobayashi, *Opt. Express* **20**, 5029 (2012).
12. K. L. Vodopyanov, *Opt. Express* **14**, 2263 (2006).
13. G. Gallot, J. Q. Zhang, R. W. Mcgowan, T. I. Jeon, and D. Grischkowsky, *Appl. Phys. Lett.* **74**, 3450 (1999).
14. M. C. Hoffmann, K.-L. Yeh, J. Hebling, and K. A. Nelson, *Opt. Express* **15**, 11706 (2007).
15. I. B. Zotova and Y. J. Ding, *Appl. Opt.* **40**, 6654 (2001).
16. C.-W. Chen, T.-T. Tang, S.-H. Lin, J. Y. Huang, C.-S. Chang, P.-K. Chung, S.-T. Yen, and C.-L. Pan, *J. Opt. Soc. Am. B* **26**, A58 (2009).
17. A. Nahata, A. S. Weling, and T. F. Heinz, *Appl. Phys. Lett.* **69**, 2321 (1996).
18. W.-C. Chu, S. A. Ku, H. J. Wang, C. W. Luo, Y. M. Andreev, G. Lanskii, and T. Kobayashi. *Opt. Lett.* **37**, 945 (2012).

Section 8

NLO Processes in Time-Resolved Spectroscopy

8.1 Elimination of Coherence Spike in Reflection-Type Pump-Probe Measurements

8.1.1 INTRODUCTION

The pump-probe technique has been applied to time-resolved measurements over the past decades. In order to understand the reaction mechanisms or the electronic structure of the excited states in media, measuring the characteristics of lifetime is widely adopted by scientists. Moreover, this pump-probe technique is also of increasing interest in solid state physics, where it is used for studying metals, semiconductors, superconductors, and other materials. However, this technique gives rise to coherence interference, i.e., the so-called coherence spike or coherence artifact, around zero delay time between pump and probe pulses. In 1981, Vardeny and Tauc [1] first proposed that the coherence artifact mostly refers to a pump polarization coupling term appearing when pump and probe overlap, which was also confirmed by the spectral hole burning of Cruz et al. [2]. Then, Eichler et al. further explained it by diffraction from a transient grating induced by the interference of the pump and probe beams [3]. They provided very important basic principles for the generation of the coherence artifact. Indeed, this coherence artifact could be found in most time-resolved spectroscopy (the transient reflectivity change $\Delta R/R$ or the transient transmissivity change $\Delta T/T$) [4–7] and disturbs the analysis of relaxation dynamics to determine the amplitude of signal near zero delay and relaxation time from the trace of $\Delta R/R$ or $\Delta T/T$. For instance, Wang et al. [4] revealed that the controversy in lifetime of p-like excited state of the hydrated electron is due to the existence of a coherence spike at zero delay time in pump-probe spectroscopic kinetics traces. After removing this spike effect, they could obtain the intrinsic lifetimes of the two incompletely relaxed states in bulk water are 180 ± 30 and 545 ± 30 fs. Besides, the reflection-type pump-probe measurements for the numerous solid-state materials also suffer from this serious problem. Therefore, how to unambiguously distinguish the true pump-probe signal of materials from the annoying coherence spike or obtain the coherence-spike-free pump-probe signal is indeed the key issue in the time-resolved femtosecond spectroscopy. In this chapter, we report the systematical studies for the origin of the coherence spike in reflection-type pump-probe measurements and further demonstrate the effective methods for removing it.

8.1.2 EXPERIMENTS

The experiments were performed with a femtosecond Ti:sapphire laser of Micra-10 (Coherent) pumped at 532 nm from an Nd:YAG laser (Verdi, Coherent). A beam splitter reflected 50% of light in a pump channel, whereas the remnant was transmitted and served as a probe. Both pump and probe beams passed through two acousto-optic modulators (AOM).

AOM is an optical device which modulates light intensity transmitting through the device. It is composed of AO crystal and high-voltage supply. AO crystal is crystalline material with high AO coefficient which does not have spatial inversion symmetry. By providing acoustic wave the refractive index of AO crystal in the modulator is modulated at the acoustic wave frequency. This modulation is utilized for the time-resolved reflection spectral intensity as explained in the following.

DOI: 10.1201/9780429196577-41

However, only one in the pump beam was driven by the RF driver and modulated the pump beam at 87 kHz. After travelling through a delay stage, a half-wave ($\lambda/2$) plate, and a polarizer, the pump beam was focused by a 200-mm lens on the surface of a sample of 125 μm in diameter. The $\lambda/2$ plate and polarizer allowed us to adjust the intensity and polarization (electric field, E) of the pump beam (both needed for intensity control). On the other hand, the probe beam only passed through the $\lambda/2$ plate and the polarizer after the AOM and focus on the surface of the sample with 84 μm in diameter by the 150 mm lens. The powers of the pump and probe beams were 32 and 3 mW, respectively. The best spatial overlap of pump and probe beams on the sample was realized by monitoring with a CCD camera. The reflection of the probe beam was received by the photodiode. However, since the variation due to the reflectivity change of the samples was very small, typically between 10^{-5} and 10^{-7}, it was very difficult to detect by the photodiode directly under the noisy background such as laser noise, electric noise, and mechanical vibration. Usually, the lock-in technique is used to reduce such background noise. To eliminate the high-level noise in the audio frequency, the small signal was modulated at 87 kHz where the noise was smaller and the gain of the narrow-band amplifier was maxima. The narrower bandwidth could be adjusted by the longer time constant of the phase detector. However, too long time constant would obscure the signal. The typical value used in this experiment was 1 s. As mentioned above, the AOM modulation was commonly applied to the pump pulse train to detect the signal due merely to change in the reflected probe intensity induced by the pump pulses. Then, the probe pulse train was also modulated by the reflectivity change of the samples on a constant intensity of the probe pulse train, i.e., a AC signal $\Delta I(t)$ was added to a DC signal $I_0(t)$. This signal was detected by the photodiode and sent to the lock-in amplifier, which was phase-locked to the AOM. The lock-in amplifier only extracted the AC signal $\Delta I(t)$, of which frequency was exactly equal to the modulation frequency in-phase with the AOM. By varying the delay time (t) between pump and probe pulses, $\Delta I(t)$ would change as a function of delay time. Therefore, the temporal evolution of the reflectivity change (ΔR) could be measured in the reflection-type pump-probe measurements.

8.1.3 RESULTS AND DISCUSSION

In order to investigate the coherence spike in the reflection-type pump-probe measurements, we chose the popular semiconductor (100) InP as a test sample, which has been well studied in ultrafast dynamics [8]. The typical $\Delta R/R(t)$ of InP is shown in Figure 8.1.1. After the excitation of a pump pulse, the $\Delta R/R$ rapidly grows and then decays to a thermal equilibrium state. Besides, the most dramatic variation is the strong coherence interference (spike) around zero delay time marked by the dashed circle in Figure 8.1.1. This coherence spike appears as the second-order (intensity) autocorrelation trace of pump and probe beams. Namely, this can only be observed at the temporal

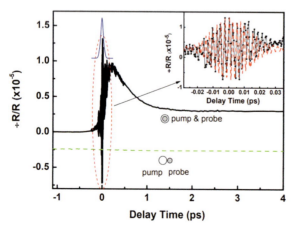

FIGURE 8.1.1 The transient reflectivity change of InP. The thin line represents the second-order autocorrelation trace. The dashed line indicates that the transient reflectivity change under the separation of pump and probe spots on the surface of samples. The inset shows the coherence spike on an enlarged scale. The dashed line in the inset is the first-order interferometric autocorrelation trace. All of the polarization configurations are pump ⊥ (the polarization of pump pulses is perpendicular to the incident plan) and probe (the polarization of probe pulses is parallel to the incident plane). All of the measurements were performed at $\theta_{pump}=1°$, $\theta_{probe}=7°$, and the interval of delay time $\Delta t=0.33$ fs.

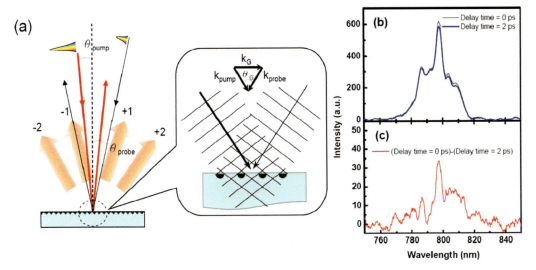

FIGURE 8.1.2   (a) Schematics of a reflection-type pump-probe measurement and the generation of a transient grating in materials. The thick lines represent the pump beat. The thin lines represent the probe beam. The thick arrows represent the diffracted light at various orders ($m=0, +1, -1, \ldots$) $\theta_G=\theta_{pump}+\theta_{probe}$. (b) The spectra were measured at $\theta=8°$ (from the normal of samples). (c) The spectrum was obtained by subtracting the spectrum with a thick line from the spectrum with a thin line in (b).

overlapping region of pump and probe pulses. Through the precisely delay-time scanning with the resolution of 0.33 fs, a periodic oscillation could be clearly observed at zero delay time as shown in the inset of Figure 8.1.1.

Compared with the result of the first-order interferometric autocorrelation curve (the dashed line in the inset of Figure 8.1.1), which was directly measured at the position of the samples, the periodic oscillation is due to the interference between pump and probe pulses. However, under the standard pump-probe setup, e.g. the configuration in Figure 8.1.2a, the detector only receives the probe pulses and no pump pulses. There are two possibilities of interfering with pump pulses. One is due to the scattering from the surface of samples. The other is due to the diffraction from the transient grating in materials. The coherence spike disappears while the pump and probe spots are slightly separated in space as shown by the dashed line in Figure 8.1.1. This implies that the coherence spike cannot be simply explained by the scattering due to the surface roughness.

Figure 8.1.2a sketches the generation of a transient grating in the reflection-type pump-probe experiments. Both pump and probe pulses with different propagation directions overlap on the surface of a sample. If the delay time between pump and probe pulses is around zero, they produce an interference pattern on the sample. The modulation of the interference pattern causes a periodical change in the refractive index (caused by the bleaching of the interband absorption [9]) [3]. The transient grating vector is given by (Eq. 8.1.1)

$$\mathbf{k}_G = \mathbf{k}_{pump} - \mathbf{k}_{probe} \tag{8.1.1}$$

where k_{pump} and k_{probe} are the propagation vectors of pump and probe pulses, respectively.

According to the grating Eq. (8.1.2):

$$d\,(\mathrm{Sin}\,\theta_d - \mathrm{Sin}\,\theta_{pump}) = m\lambda \tag{8.1.2}$$

where $d=2\pi/k_G$ is the period of the transient grating, θ_{pump} is the incident angle of a pump beam, m is the order of interference, and θ_m is the diffraction angle of the m-th order. The pump pulse could be

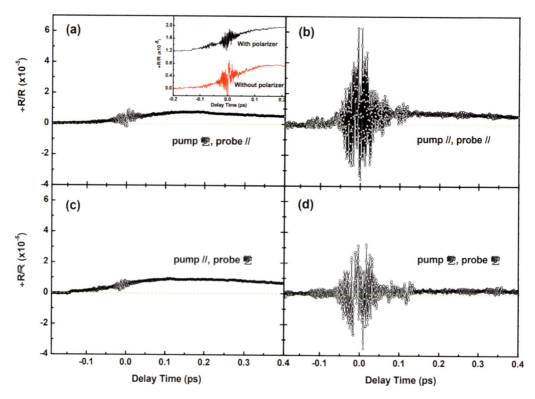

FIGURE 8.1.3 The coherence spike in $\Delta R/R$ under four polarization configurations (a) pump $\perp$ (the polarization of pump pulses was perpendicular to the incident plane) and probe $\parallel$ (the polarization of probe pulses is parallel to the incident plane), (b) pump $\parallel$ and probe $\parallel$, (c) pump $\parallel$ and probe $\perp$, (d) pump $\perp$ and probe $\perp$. The inset of (a) is that the measurements were performed with and without the polarizer in front of the detector. All of the measurements were performed at $\theta_{pump}=1°$, $\theta_{probe}=7°$, and the interval of delay time $\Delta t=0.33$ fs.

diffracted by this transient grating as shown with the thick arrows in Figure 8.1.2a. At $\theta=8°$ (from the surface normal of the sample) around $m=+1$, the spectra of the scattering light have been measured at the delay time=0 and 2 ps. Figure 8.1.2c shows the difference between the thin line with a delay time 0 ps and the thick line with delay time 2 ps in Figure 8.1.2b. This additional spectrum around the diffractive angle with $m=+1$ indicates that the existence of the transient grating at zero delay time. Once the pump pulse is diffracted into the detector, the interference between the pump pulse and probe pulse will be observed. For instance, the optical path of diffracted light with $m=-1$ is just collinear with the optical path of the probe pulses and then leads to coherence interference around the zero delay time. This implies that the coherence spike is unambiguously caused by the diffracted light of pump pulses due to the transient grating of the samples, which is consistent with the theoretical results of Eichler et al. in transmission-type pump-probe experiments [3].

In principle, the interference is expected not to take place between the pump and probe pulses with perpendicular polarization. However, the coherence spike was still observed in Figure 8.1.1 which was performed with pump $\perp$ (the polarization of pump pulses is perpendicular to the incident plane) and probe (the polarization of probe pulses is parallel with the incident plane). This maybe simply assigned to the imperfect linear polarization of light even the polarizer with extinction ratio >10^4:1 we used in this experiment. Additionally, we added one polarizer in front of the detector to exclude the pump pulse of which polarization is perpendicular to the polarization of the probe pulse or eliminate the coherence spike. As you can see in the inset of Figure 8.1.3a, unexpectedly, the coherence spike was still observed. When the polarizations of both pump and probe pulses are set to be parallel with the incident plane (pump, probe), the extremely large coherence spike appears

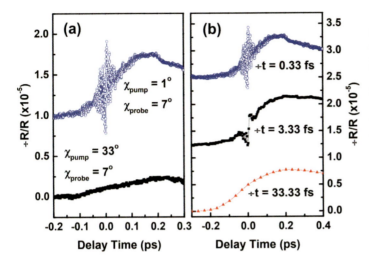

FIGURE 8.1.4 (a) The transient reflectivity changes with various incident angles of pump and probe beams. All of the measurements were performed at the interval of delay time $\Delta t=0.33$ fs. (b) The transient reflectivity changes with various intervals of delay time Δt. All of the polarization configurations are pump $\perp$ and probe.

around zero delay time as shown in Figure 8.1.3b. Similarly, this situation also happens in the case of the polarizations of both pump and probe pulses set to be perpendicular to the incident plane (pump$\perp$, probe$\perp$ in Figure 8.1.3d), but it is smaller than that in Figure 8.1.3b. This is because of the weaker modulation strength in the interference pattern due to the polarization of pump and probe pulses which are parallel to the incident plane in Figure 8.1.3d. Thus, the coherence spike in the case of pump //probe (Figure 8.1.3b and d) is much larger than that in the case of pump $\perp$ probe (Figure 8.1.3a and c).

Could we eliminate the coherence spike in this reflection-type pump-probe measurement? According to the above analysis, the coherence spike is mainly induced by the transient grating due to the interference between pump and probe pulses. Therefore, one possible way is to reduce the transient grating in the sample. For instance, the coherence spike completely disappears with $\theta_{pump}=33°$ and $\theta_{probe}=7°$ in Figure 8.1.4a. When the incident angles of the pump and probe beams are increased, the size of the transient grating vector k_G will increase according to Eq. (8.1.1). Then, the period (d) of transient grating becomes shorter to increase the diffraction angle θ_m. Further, the grating Eq. (8.1.2) will be failed to be satisfied with a larger incident angle of the pump beam. Thus, the pump pulses cannot be diffracted by the transient grating established by the interference between pump and probe pulses and no coherence spike in $\Delta R/R$. In some experiments, small θ_{pump} and θ_{probe} are needed to be set for the angle-resolved ultrafast spectroscopy in anisotropic materials such as high-Tc superconductor, $YBa_2Cu_3O_7-\delta$, by varying the polarizations of pump and probe pulses [10–12]. If the θ_{pump} and θ_{probe} must be fixed in small angle, the coherent spike in $\Delta R/R$ could be removed by increasing the interval of delay time. As shown in Figure 8.1.4b, the coherence spike gradually vanishes with increasing the interval of delay time Δt. For $\Delta t=33.3$ fs, there is no coherence spike in $\Delta R/R$.

Generally, the coherence spike is not welcome in the pump-probe measurements. On the other hand, however, this coherence spike can give us information about the characteristics of the pulses we used. In the inset of Figure 8.1.1, the oscillation of the coherence spike is almost equal to the results of standard first-order autocorrelation measurements. This means that the characteristics of pulses, i.e., the coherent length or bandwidth can be directly estimated from the coherence spike.

8.1.4 SUMMARY

In summary, we have demonstrated the origin of coherence spike in the reflection-type pump-probe measurement. The strength of coherence spike changes with various setups in the pump-probe measuring system, such as the polarization of pump and probe pulses, incident angle of pump and probe beams, and the interval of the delay time. Moreover, two effective methods to eliminate the coherence spike in the reflection-type pump-probe measurement have been suggested and demonstrated [13]. One is to utilize the experimental configuration with the large incident angles of pump and probe beams which cause the grating equation to fail to be satisfied. The other way is to increase the interval of delay time Δt during measurements.

The work presented in this subsection is the product of the collaboration of the following people: C. W. Luo, Y. T. Wang, F. W. Chen, H. C. Shih, and T. Kobayashi [13].

The research presented in this section is reproduced and adapted with permission from Ref. [13]©The Optical Society. *Opt. Exp.* **17**, 11321–11327 (2009), was conducted by collaborative work of the following people: C. W. Luo, Y. T. Wang, F. W. Chen, H. C. Shih, and T. Kobayashi.

REFERENCES

1. Z. Vardeny and J. Tauc, "Picosecond coherence coupling in the pump and probe technique," *Opt. Commun.* **39**, 396–400 (1981).
2. C. H. B. Cruz, J. P. Gordon, P. C. Becker, R. L. Fork, and C. V. Shank, "Dynamics of spectral hole burning," *IEEE J. Quantum Electron.* **24**, 261–266 (1988).
3. H. J. Eichler, D. Langhans, and F. Massmann, "Coherence peaks in picosecond sampling experiments," *Opt. Commun.* **50**, 117–122 (1984).
4. C.-R. Wang, T. Luo, and Q.-B. Lu, "On the lifetimes and physical nature of incompletely relaxed electrons in liquid water," *Phys. Chem. Chem. Phys.* **10**, 4463–4470 (2008).
5. C. Chudoba, E. T. J. Nibbering, and T. Elsaesser, "Site-specific excited-state solute-solvent interactions probed by femtosecond vibrational spectroscopy," *Phys. Rev. Lett.* **81**, 3010–3013 (1998).
6. U. Conrad, J. Gdde, V. Jhnke, and E. Matthias, "Ultrafast electron and magnetization dynamics of thin Ni and Co films on Cu(001) observed by time-resolved SHG," *Appl. Phys. B* **68**, 511–517 (1999).
7. M. V. Lebedev, O. V. Misochko, T. Dekorsy, and N. Georgiev, "On the nature of coherent artifact," *J. Exp. Theoret. Phys.* **100**, 272–282 (2005).
8. Y. Kostoulas, L. J. Waxer, I. A. Walmsley, G. W. Wicks, and P. M. Fauchet, "Femtosecond carrier dynamics in low-temperature-grown indium phosphide," *Appl. Phys. Lett.* **66**, 1821–1823 (1995).
9. H. J. Eichler, P. Gunter, and D. W. Pohl, *Laser-Induced Dynamic Gratings*, Springer-Verlag, Berlin (1986).
10. C. W. Luo, M. H. Chen, S. P. Chen, K. H. Wu, J. Y. Juang, J.-Y. Lin, T. M. Uen, and Y. S. Gou, "Spatial symmetry of the superconducting gap of $YBa_2Cu_3O_7$–δ obtained from femtosecond spectroscopy," *Phys. Rev. B* **68**, 220508-1(R)–220508-4 (2003).
11. C. W. Luo, P. T. Shih, Y.-J. Chen, M. H. Chen, K. H. Wu, J. Y. Juang, J.-Y. Lin, T. M. Uen, and Y. S. Gou, "Spatially resolved relaxation dynamics of photoinduced quasiparticles in underdoped $YBa_2Cu_3O_7$–δ," *Phys. Rev. B* **72**, 092506-1–092506-4 (2005).
12. C. W. Luo, C. C. Hsieh, Y.-J. Chen, P. T. Shih, M. H. Chen, K. H. Wu, J. Y. Juang, J.-Y. Lin, T. M. Uen, and Y. S. Gou, "Spatial dichotomy of quasiparticle dynamics in underdoped thin-film $YBaCu_3O_{7-\delta}$ superconductors," *Phys. Rev. B* **74**, 184525-1–184525-4 (2006).
13. C. W. Luo, Y. T. Wang, F. W. Chen, H. C. Shih, and T. Kobayashi, "Eliminate coherent spike in reflection-type pump-probe measurements," *Opt. Express* **17**, 11321–11327 (2009).

8.2 Vibrational Fine Structures Revealed by the Frequency-to-Time Fourier Transform of the Transient Spectrum in Bacteriorhodopsin

8.2.1 INTRODUCTION

Bacteriorhodopsin (bR) is a photoactive retinoid protein with a proton-pumping function that generates a pH gradient across the cellular membrane, using solar energy to synthesize ATP. This pumping involves a change in a local structure of the protein near the retinal molecule. The conformational change in the protein is initiated by the optically triggered trans-cis photoisomerization, which is closely related to the photoisomerization of retinal from 11-cis to the all-trans configuration in rhodopsin, which is the functional pigment in the visual sensor process [1]. The absorption of a photon by the chromophore molecule in the retinal causes isomerization around the $C_{13}=C_{14}$ double bond followed by photocycle with several intermediates that have distinct spectra. Because of its interesting functionality, bR has been widely studied both theoretically [2,3] and experimentally [4–10]. An ultrafast switch that exploits its ultrafast photoisomerization has also been implemented [11], although the mechanism remains controversial. Ruhman et al. determined experimentally that locked-retinal that is contained in bR undergoes a similar photoinduced spectral change to that of ordinary bR [12–14]. On the basis of theoretical models of the photoisomerization mechanism, the groups of Olivucci [2,15] and Schulten [3,16] proposed "a two-state, two-mode model" and "a three-state model", respectively.

Ultrafast dynamics of chemical reactions have been studied by using ultrafast spectroscopy, which has demonstrated the transient existence of intermediate species during the reactions [17,18]. Ultrafast spectroscopy is a method that complements the electron diffraction method presented by Miller et al. [19] and Zewail [20] and the X-ray diffraction methods proposed by Anfinrud et al. [21]. Instantaneous vibrational amplitude detection with a subfemtosecond resolution, which is demonstrated in this work, provides a much higher time resolution than electron or X-ray diffraction. Additionally, the method can be used to make measurements of amorphous and liquid-phase materials, to which X-ray or electron diffraction cannot easily be applied. In the authors' earlier work [22], we detected ultrafast changes in the frequencies of in-plane and out-of-plane bending modes caused by a structural change in real time during photoisomerization of retinal molecule in bR. Femtosecond stimulated Raman spectroscopy with high time and spectral resolutions were employed to observe time-dependent conformational changes [23,24], but the method does not yield information on the vibrational phase, which can be obtained in the pump-probe measurement.

As was discussed in a work published in 2007 [12], time-resolved measurement of absorbance change involves wave packets on both potential surfaces of the electronic ground state and the excited state. Detailed observations made in the authors' recent investigation [24] on the photoexcited dynamics revealed that the primary process of the photoisomerization begins with the activation of the C=N stretching mode of the highest frequency, which decays in 30 fs, followed by that of

DOI: 10.1201/9780429196577-42

the C=C stretching mode, which has a slightly lower frequency. The C=C mode is modulated at frequencies of 142 ± 35, 105 ± 3, 158 ± 3, and $150\pm18\,cm^{-1}$ in wavelength ranges of 505–530 (*S*), 540–600 (*M*), 610–630 (*L$_1$*), and 635–663 nm (*L$_2$*), respectively, corresponding to the excited state (*H* state), ground state, *I* intermediate, and *J* intermediate, respectively. In the first three ranges (*S*, *M*, *L$_1$*), the lifetimes were <100, 1050 ± 180, and $270\pm30\,fs$ (with a rise time of $160\pm20\,fs$), respectively.

In this work, pump-probe measurements of bR were made by using ultrashort visible laser pulses, to obtain simultaneously ultrafast time-resolved absorbance change (Δ*A*) spectra at 128 wavelengths. The Δ*A* spectra revealed a vibrational progression hidden in a featureless spectrum of induced absorption and stimulated emission. This observation can be made because of the localization of the wave packet along the potential multimode hyper surfaces. The transition energy of the induced absorption or stimulated emission corresponds to a localized point (space) on the hyper surface, which is visited by the wave packets with fixed phases. By taking the first and second derivatives of time-resolved Δ*A* spectra covering 128 wavelength points, a detailed discussion of the relevant wavepacket motion could be made.

8.2.2 EXPERIMENTAL SECTION

A noncollinear optical parametric amplifier (NOPA) [25–27] was adopted as a light source in the pump-probe experiment, as described elsewhere [22,28,29]. Several key features of the system are briefly described below. The output pulse from the NOPA was compressed by using a compressor that was composed of a pair of prisms and chirp mirrors. The pulse duration was sub-10 fs and the spectral range covered 520–750 nm, within which the spectral phase was almost constant, resulting in the Fourier transform-limited pulses. The pulse energies of the pump and probe were about 10 and 1 nJ, respectively. A 128-channel lock-in amplifier was employed as a phase-sensitive broadband detector. All of the experiments were conducted at room temperature 293 (1 K).

8.2.3 RESULTS AND DISCUSSION

Figure 8.2.1 presents the laser spectrum and the absorption spectrum of the bR sample. The absorbed photon distribution spectrum, also displayed in Figure 8.2.1, was obtained from the pump laser spectrum and the absorption probability spectrum, which was calculated from the absorption spectrum.

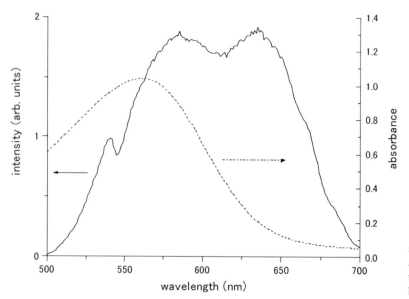

FIGURE 8.2.1 Laser spectrum (solid curve) and absorption spectrum (dashed curve) of bacteriorhodopsin.

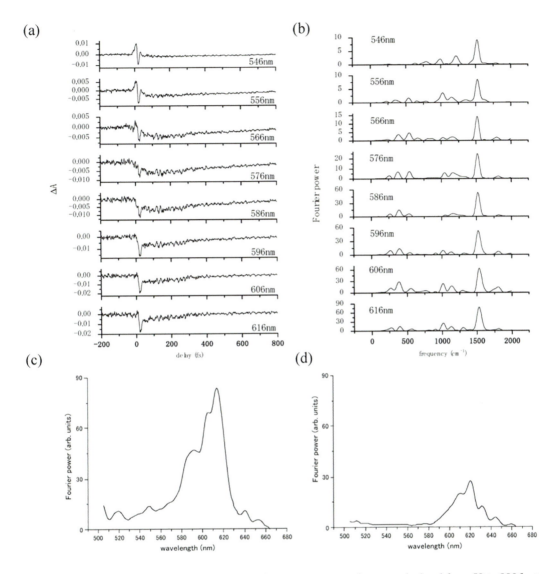

FIGURE 8.2.2   (a) Real-time traces and (b) Fourier power spectra of traces calculated from 50 to 800 fs at eight typical wavelengths. Fourier power spectra of the (c) 1527 and (d) 1008 cm^{-1} modes.

The pump-probe experiment on the bR sample was performed with a probe delay time of −400 to 800 fs. The change in absorbance caused by the pump pulse was probed as a function of the probe delay. The curve of the absorbance change versus delay is called the real-time (vibration) trace and the curve of the absorbance change versus probe wavelength is called the vibration real-time spectrum.

Panels a and b of Figure 8.2.2 present the real-time traces and Fourier power spectra of the traces, calculated from 50 to 800 fs, respectively, at eight typical wavelengths. The Fourier transform was performed from a time other than 0 fs, to eliminate any effect of interference between the probe pulse and the scattered pump pulse close to zero delay. Even though the pulse duration was sub-10 fs, the spatial coherence was very high and the multiply scattered pump pulse interfered with the probe pulse. Therefore, probe delays shorter than 50 fs had to be eliminated to prevent this effect of coherence. The power spectra in Figure 8.2.2b include two intense peaks at 1527 and 1008 cm^{-1}. Recently, time-resolved Raman spectra of bacteriorhodopsin were reported by the groups of Mathies and of Mizutani in 2009 [30,31]. Figure 8.2.2b resembles Raman spectra of the early delay time range shown in their reports. Therefore, Fourier power spectra shown in Figure 8.2.2b are thought to be mainly reflecting the J intermediate, which has an isomerized structure. Panels c and d of Figure 8.2.2 display the relative powers of the oscillations at 1527 and 1008 cm^{-1}, respectively, in the real-time traces measured at various wavelengths.

Figure 8.2.3a shows the time-resolved spectrum of the sample integrated over the range from 400 to 800 fs. It is referred to as the ΔA spectrum below. Previous study by Heller has theoretically predicted that fine structure on the absorption spectrum reflects a time-correlation function, which is periodically modulated by wave packet motion on the potential energy surface [32]. The relation between optical absorption intensity $I(\omega)$ and time-correlation function of the transition dipole moment was obtained as follows [33]:

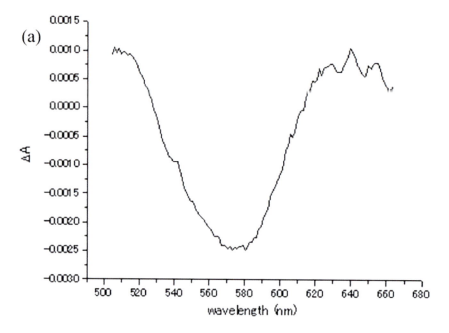

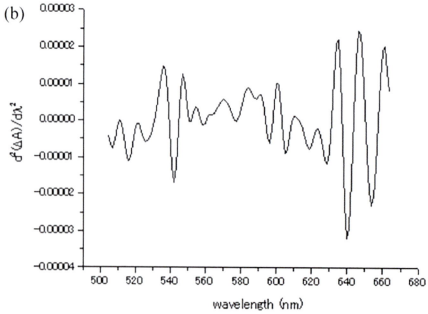

FIGURE 8.2.3
(a) Time-resolved spectrum of sample integrated from 400 to 800 fs and (b) its second derivative.

$$\int_{-\infty}^{\infty} \frac{I(\omega)}{\omega} = \sum_{u} \sum_{v} \rho_u \left\langle \left\langle \left\{ iu \middle| \mu(0) \middle| fv \right\} \cdot \left\{ fv \middle| \mu(t) \middle| iu \right\} \right\rangle_{\text{ah}} \right\rangle_{\text{slv}}$$

$$\equiv \left\langle \left\langle \mu_{\text{if}}(0) \cdot \mu_{\text{if}}(t) \right\rangle \right\rangle_0$$

where $\mu_{\text{if}}(0) = \left(i \middle| \mu \middle| f \right)$ and $\mu_{\text{if}}(t) = \left(i \middle| \mu(t) \middle| f \right) \cdot \middle| iu \right\}$ and $\middle| fv \}$ denote the initial and final vibronic states. μ is the dipole moment operator written by

$$\mu = e \sum_{a} r_a$$

where a is the ath electron of a solute molecule. ρ_u is the state density of the iu state. Kakitani et al. have applied this method for the study of the stationary absorption spectrum of bR [34].

As is seen in Figure 8.2.3a, the time-resolved absorption change ΔA observed in this work shows a fine structure of $\sim$150 cm^{-1} period in the spectral region where ΔA has a positive value dominated by the effect of induced absorption. The period of 150 cm^{-1} agrees with the reported values of the torsion period around C$_{12}$=C$_{13}$ [22,35,36]. Therefore, the observed fine structure is thought to be reflecting anharmonicity of the torsion mode in the higher excited state, which is the final state of the induced absorption process. The spectral features are thus observed by freezing the wavepacket motion in the excited state revealing the large displacement along the potential curve corresponding to the torsion not only in the lowest excited state but also in the higher excited state, which is located higher than the lowest excited state by 16000 cm^{-1}. The result is very reasonable because the instability of the lowest excited state to trigger the ultrafast torsion relevant to the photoisomerization is due to the change of the bond order and steric hindrance of the methyl group attached to C$_{11}$. To go into deeper discussion, the calculation of the second derivative of the ΔA spectrum was performed. Peaks and valleys in the second derivative of the spectrum correspond to the hidden valleys and peaks in the ΔA (we call it the zeroth derivative) spectrum. Hence, the second derivative of the ΔA spectrum in Figure 8.2.3b has a fine structure that has various peaks and valleys. We confirmed that the spectral structure is reproducible in all four different scans of data. The structure that is evident in the relative powers of the oscillations at 1527 and 1008 cm^{-1} (see Figure 8.2.2c and d) has peaks or valleys at close to the same photon energies as the ΔA spectrum and its second derivative. Table 8.2.1 presents the frequencies of the peaks and valleys in the Fourier power spectra, the ΔA spectrum, and its second derivative.

The ΔA signal reflects the modulation of electronic transition probability by wave packet motion on a potential energy surface. When the time resolution of the measurement is much better than the period of wave packet motion associated with molecular vibrations of several tens of femtoseconds, the observed ΔA spectrum at each delay step reflects the spectral structure of electronic states with vibration modes. As was reported in our previous work [24] the dynamics of bR is dominated by the J intermediate in the time region from 400 to 800 fs and the spectral region from 610 to 660 nm where the peaks and valleys were observed above in the time-resolved ΔA spectrum and Fourier power spectra. Since the time resolution of this work is better than 10 fs, the wave packet on the potential energy surface can be assumed to be motionless at a localized location on the surface corresponding to each probe delay in the time-resolved measurements. Therefore, the fine structures on the time-resolved ΔA spectra are thought to reflect the vibrational levels of the J intermediate, which was observed by freezing the position of the wave packet in the electronic state.

The distances between neighboring peaks or valleys are 153$\pm$12 or 309$\pm$27 cm^{-1}, respectively. Accordingly, peaks (valleys) correspond to the vibrational progression in the time-resolved spectrum that is associated with the excited state absorption (stimulated emission). Valley signals, associated with bleaching that is induced by ground-state depletion, are thought to be negligible, because of the absence of sufficiently intense fluorescence at this low probe frequency (long wavelength).

TABLE 8.2.1

Peak and Valley Frequencies in Fourier Power Spectra of 1527 and 1008 cm⁻¹ Modes, in the ΔA Spectra, and in $d^2(\Delta A)/d\lambda^2$

1527 cm⁻¹		1008 cm⁻¹		ΔA		$d^2(\Delta A)/d\lambda^2$	
Peak	Valley	Peak	Valley	Peak	Valley	Peak	Valley
			15180		15140	15166	
15296		15325		15301			15286
	15444		15533		15456	15451	
15625		15655					
	15748		15625	15620			15620
		15841			15791	15755	
			15904	15893			15884
					16010	15984	
				16054			16070
		16129			16151	16135	
				16185			16212
			16326		16304	16291	
16293		16393		16348			16366
	16460				16456	16443	
16528				16501			16526

Even though the peak (valley) positions may not be very precise, the frequency separations are almost the same, indicating that they still can be considered to represent the vibrational progression, which is sensitively revealed by the Fourier power spectrum. The frequency separations among these peaks and valleys are independent of the delay, indicating that the spectral shape of these valleys and peaks does not arise from spectral interference between the scattered pump and the probe. The spectral spacings of 153 and 309 cm⁻¹ may be assigned to the fundamental and double frequencies of the twisting motion associated with photoisomerization that had an observed period of around 200 fs [22,35,36].

Panels a and b of Figure 8.2.4 present the peak frequencies and valley frequencies, respectively, of ΔA spectra over a range of delays in steps of 1 fs. Panels c and d of Figure 8.2.4 show those of the second derivatives of the ΔA spectra. The standard deviations of the peak (valley) frequencies and of the frequency difference between the frequencies of the neighboring peaks (valleys) were calculated by using the data in panels a–d of Figure 8.2.4, and displayed in panels e–h of Figure 8.2.4, respectively. The range of delays used to calculate the standard derivation was from 400 to 800 fs, to avoid any interference between the probe pulse and the scattered pump pulse. A mean frequency of neighboring peaks was used as the abscissa in the plot of the standard deviation of the difference between frequencies. If the neighboring peaks (valleys) move independently of each other, then the standard deviation of the difference between the frequencies of the neighboring peaks (valleys) should be $2^{1/2}$ times that of the peak (valley) frequencies themselves as expected from the Gaussian random statistics. However, the ratio of the former to the latter was in fact observed to be 0.90 ± 0.24, which is much smaller than $2^{1/2}$. Therefore, neighboring peaks (valleys) move in a synchronized way, verifying that vibrational progression appears as the spectral structure, which is revealed by data.

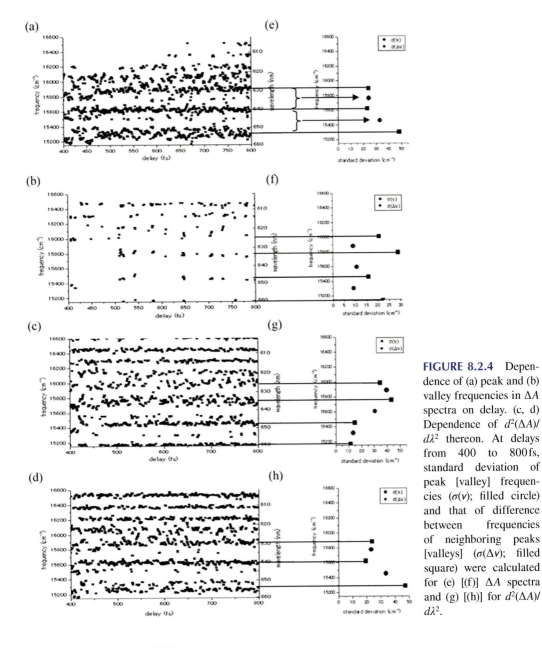

FIGURE 8.2.4 Dependence of (a) peak and (b) valley frequencies in ΔA spectra on delay. (c, d) Dependence of $d^2(\Delta A)/d\lambda^2$ thereon. At delays from 400 to 800 fs, standard deviation of peak [valley] frequencies ($\sigma(v)$; filled circle) and that of difference between frequencies of neighboring peaks [valleys] ($\sigma(\Delta v)$; filled square) were calculated for (e) [(f)] ΔA spectra and (g) [(h)] for $d^2(\Delta A)/d\lambda^2$.

8.2.4 CONCLUSIONS

In conclusion, for the first time, this subsection has clarified that the vibrational progression that is hidden in a featureless spectrum of induced absorption and stimulated emission is found by using the Fourier power spectrum of molecular vibrational modes, because of the localization of the wave packet along the potential multimode hyper surfaces. The transition energy of the induced absorption or stimulated emission corresponds to a localized point (space) on the hyper surface, which the wave packets visit with fixed phases.

The advantages of this real time-resolved vibrational spectroscopy over conventional time-resolved vibration spectroscopy are the capability of even the time course of vibrational periods of mode and can further discuss the initial phase of each vibrational mode. These are our future works published sometime later in this chapter.

The research presented in this subsection was performed by the collaboration of A. Yabushita and T. Kobayashi [37].

REFERENCES

1. A. P. Shreve and R. A. Mathies, *J. Phys. Chem.* **99**, 7285–7299 (1995).
2. R. Gonza´les-Luque, M. Garavelli, F. Bernardi, M. Merchan, M. A. Robb, and M. Olivucci, *Proc. Natl. Acad. Sci. U.S.A.* **97**, 9379–9384 (2000).
3. W. Humphrey, H. Lu, I. Logonov, H. J. Werner, and K. Schulten, *Biophys. J.* **75**, 1689–1699 (1998).
4. S. Schenki, F. van Mourik, G. van der Zwan, S. Haacke, and M. Chergui, *Science* **309**, 917–920 (2005).
5. J. Herbst, K. Heyne, and K. R. Diller, *Science* **297**, 822–825 (2002).
6. G. Haran, K. Wynne, A.-H. Xie, Q. He, M. Chance, and R. M. Hochstrasser, *Chem. Phys. Lett.* **261**, 389–395 (1996).
7. F. Gai, K. C. Hasson, J. C. McDonald, and P. A. C. Anfinrud, *Science* **279**, 1886–1891 (1998).
8. Q. Zhong, S. Ruhman, and M. I. Ottolenghi, *J. Am. Chem. Soc.* **118**, 12828–12829 (1996).
9. L. Song and M. A. El-Sayed, *J. Am. Chem. Soc.* **120**, 8889–8890 (1998).
10. M. Du and G. R. Fleming, *Biophys. Chem.* **48**, 101–111 (1993).
11. S. Roy, C. P. Singh, and K. P. J. Reddy, *Curr. Sci.* **83**, 623–626 (2002).
12. A. Kahan, O. Nahmias, N. Friedman, M. Sheves, and S. Ruhman, *J. Am. Chem. Soc.* **129**, 537–546 (2007).
13. B.-X. Hou, N. Friedman, M. Ottolenghi, M. Sheves, and S. Ruhman, *Chem. Phys. Lett.* **381**, 549–555 (2003).
14. A. C. Terentis, Y. Zhou, Y. G. H. Atkinson, and L. Ujj, *J. Phys. Chem. A* **107**, 10787–10797 (2003).
15. M. Olivucci, A. Lami, and F. Santoro, *Angew. Chem., Int. Ed.* **44**, 5118–5121 (2005).
16. S. Hayashi, E. Tajkhorshid, and K. Schulten, *Biophys. J.* **85**, 1440–1449 (2003).
17. T. S. Rose, M. J. Rosker, and A. H. Zewail, *J. Chem. Phys.* **88**, 6672–6673 (1988).
18. J. C. Polanyi and A. H. Zewail, *Acc. Chem. Res.* **28**, 119–132 (1995).
19. V. I. Prokhorenko, A. M. Nagy, S. A. Waschuk, L. S. Brown, R. R. Birge, and R. J. D. Miller, *Science* **313**, 1257–1261 (2006).
20. R. Srinivasan, J. S. Feenstra, S. T.; Park, S. Xu, and A. H. Zewail, *Science* **307**, 558–563 (2005).
21. F. Schotte, J. Soman, J. S. Olson, M. Wulff, and P. A. Anfinrud, *J. Struct. Biol.* **147**, 235–246 (2004).
22. T. Kobayashi, T. Saito, and H. Ohtani, *Nature* **414**, 531–534 (2001).
23. D. W. McCamant, P. Kukura, and R. A. Mathies, *Appl. Spectrosc.* **57**, 1317–1323 (2003).
24. A. Yabushita and T. Kobayashi, *Biophys. J.* **96**, 1–15 (2009).
25. A. Shirakawa, I. Sakane, and T. Kobayashi, *Opt. Lett.* **23**, 1292–1295 (1998).
26. A. Shirakawa, I. Sakane, M. Takasaka, and T. Kobayashi, *Appl. Phys. Lett.* **74**, 2268–2270 (1999).
27. A. Baltuska, T. Fuji, and T. Kobayashi, *Opt. Lett.* **27**, 306–308 (2002).
28. T. Kobayashi, A. Shirakawa, H. Matsuzawa, and H. Nakanishi. *Chem. Phys. Lett.* **321**, 385–393 (2000).
29. N. Ishii, E. Tokunaga, S. Adachi, T. Kimura, H. Matsuda, and T. Kobayashi, *Phys. Rev. A* **70**, 023811 (2004).
30. S. Shim, J. Dasgupta, and R. A. Mathies, *J. Am. Chem. Soc.* **131**, 7592–7597 (2009).
31. M. Mizuno, M. Shibata, J. Yamada, H. Kandori, and Y. Mizutani, *J. Phys. Chem. B* **113**, 12121–12128 (2009).
32. H. J. Heller, *J. Chem. Phys.* **68**, 3891–3896 (1978).
33. T. Kakitani, *J. Phys. Soc. Jpn.* **55**, 993–1010 (1986).
34. R. Akiyama, T. Kakitani, Y. Imamoto, Y. Shichida, and Y. Hatano, *J. Phys. Chem.* **99**, 7147–7153 (1995).
35. X. Chen and V. S. Batista, *J. Photochem. Photobiol., A* **190**, 274–282 (2007).
36. A. B. Mayers, R. A. Harris, and R. A. Mathies, *J. Chem. Phys.* **79**, 603–613 (1983).
37. A. Yabushita and T. Kobayashi, *J. Phys. Chem. B* **114**, 4632–4636 (2010).

Section 9

Low Dimensional (D) Materials

Section 9.1

0D

9.1.1 Superior Local Conductivity in Self-Organized Nanodots on Indium-Tin-Oxide Films Induced by Femtosecond Laser Pulses

9.1.1.1 INTRODUCTION

Indium-tin-oxide (ITO) is an important transparent conducting oxide (TCO). ITO films have been widely used as transparent electrodes in optoelectronic devices, such as solar cells [1] and organic light-emitting devices (OLEDs) [2], because of their high electrical conductivity ($\sim 10^{-4}$ Ω–cm), coupled with their high transmission ($\sim 90\%$) in the visible range [3,4]. In particular, the surface properties of ITO films, such as their electron affinity and work function, also play a key role in establishing the characteristics of OLEDs, owing to their direct contact with the organic materials, as a hole injection layer [4,5]. ITO has also been extensively used as a good ohmic contact material in GaN-base light emitting diodes (LEDs), because ITO shows excellent ohmic behavior in terms of high surface current, which is evidenced by the excellent surface conductivity [6,7].

In general, a thermal annealing process ($\geq 200°C$) is frequently used to change the crystallinity of ITO from amorphous (a-ITO) to crystalline (c-ITO), which results in a diminution of resistivity and an increase in transparency [8]. However, this conventional method of thermal annealing at high temperatures does not work for flexible devices due to their inherently poor thermal stability and the constraint of a low glass transition temperature (T_g), for flexible substrates based on plastic materials. To overcome the constraint of avoiding a high-temperature processing step for flexible polymer substrates, therefore, ultraviolet (UV) lasers, such as KrF and XeCl excimer lasers with nanosecond pulses, have been reported to anneal ITO films and modify the crystallinity, without a marked rise in the sample temperature [9–11].

Recently, material processing by femtosecond (fs) laser irradiation has attracted a great deal of attention because a fs pulse of energy can be precisely and rapidly transferred to the film, without thermal effects [12–14]. This is so-called femtosecond laser annealing (FLA). Nonthermal melting in semiconductors, using FLA, exhibits great potential to solve the problem of the thermal budget, in annealing [15,16]. Pan et al. reported near-infrared femtosecond laser-induced crystallization in amorphous silicon [16]. Very recently, a femtosecond laser was also used to pattern a-ITO films, using the crystallization effect [17]. The high fluences (~ 100–1000 mJ/cm^2), close to the ablation threshold of a-ITO under a focused fs laser beam, destroyed the a-ITO films, causing many micro-cracks and nanogratings, due to the thermal cycling effect [17]. On the other hand, laser-induced periodic surfaces structures (LIPSS) or ripples, with typical areas of $5\,\mu m^2$

DOI: 10.1201/9780429196577-45

to several tens of μm^2, have been observed for various materials, under pulsed laser illumination near their ablation thresholds [18–20]. At present, there are a few publications on the modification of ITO films over a large area, using fs laser pulses at low fluence. In addition, the interaction between the fs laser pulses and the ITO film and their impact on the electrical and optical properties are still unclear.

In this study, a large-area (over $200 \times 200\,\mu m$ in the center of the beam spot), periodic ripple structure, composed of self-organized nanodots on the surface of ITO films, was induced with low-fluence fs laser pulses (0.1–0.3 mJ/cm^2), without scanning. The resistivity, carrier mobility, carrier concentration and especially the surface conductivity of ITO films are significantly changed, due to the laser-induced periodic surface structures (LIPSS), while the optical transmittance is unaffected. The cause of this significant enhancement of the surface electrical properties of the ITO films, with FLA, has been identified via local composition inspection, using Auger electron spectroscopy (AES), and the chemical bonding state analysis, using X-ray photoelectron spectroscopy (XPS).

9.1.1.2 EXPERIMENTS

ITO thin films with a thickness of 30 nm and resistivity of $\sim 4 \times 10^{-2}\ \Omega$–cm (O$_2$/(Ar+O$_2$) flow ratio is around 0.047) were deposited without optimization on the glass substrates (1×1 cm), by magnetron sputtering deposition at 1000 W power. The ITO target (58×15 cm) was composed of In$_2$O$_3$ with 10 wt% SnO$_2$. After the deposition of ITO films, at room temperature, these samples were then irradiated, using a regenerative, amplified Ti:sapphire laser (Legend USP, Coherent), with 800 nm wavelength, 100 fs pulse duration, $\sim$0.5 mJ pulse energy and 5 kHz repetition rate. The diameter of the laser beam was adjusted to $\sim$14 mm, to ensure full exposure for a sample size of 1 cm × 1 cm.

The morphology of the ITO films was examined using a scanning electron microscope (SEM) (HITACHI-S2500 JSM-6500F). The thicknesses of the ITO film, before and after laser pulse irradiation, were determined by surface contour measurement (KOSAKA ET4000A), using a vertical resolution of 0.1 nm. The resistivity, carrier concentration and Hall mobility of the ITO films were measured by Hall measurements, using the Van der Pauw technique (Bio-Rad Microscience HL5500). The refractive index of the ITO film was measured with an n&k Analyzer 1280 (n&k Technology, Inc.). The topography and surface current distribution were further analyzed using a current sensing-atomic force microscope (CSAFM, Agilent 5500) with a Cr/Pt-coated CSAFM tip (ContE-G type, BudgetSensors, force constant: 0.2 N/m). All CSAFM images reported here were measured with a tip bias of 0.1 V, in a scan area of $10 \times 10\,\mu m$ (in the center of 1 cm × 1 cm ITO films). The optical transmission measurements were performed using an UV-visible-near-IR spectrophotometer (Shimadzu SolidSpec-3700).

The local compositions of the as-deposited and fs laser irradiated ITO films were examined by Auger electron spectroscopy (AES, ULVAC-PHI 700). Due to the spot size yielded by an integral electron gun of under 5 KeV operating voltage, the data collected by a cylindrical mirror analyzer had a spatial resolution of $\sim$30 nm. The chemical bonding in and composition of the as-deposited and fs laser irradiated ITO films were determined using X-ray photoelectron spectroscopy (XPS, PHI Quantera AES 650), with a monochromatic Al $K\alpha$ source, at 1486.7 eV. The position of all XPS peaks was calibrated by using the binding energy of 84.0 eV, in Au. For all data in this study, the diameter of analyzed spot was 100 μm, for pass energy of 15 keV. The peak fitting of the O 1s spectra was performed using the Gaussian-Lorentzian function, using an XPS peak fitting program (XPSPEAK4.1 from Dr. R. M. Kwok), and subsequent quantification of the XPS data was obtained by using the peak areas and experimental sensitivity factors. All the O $1s$ fitted peaks were calculated using 70% Gaussian-30% Lorentzian function and the same full width at half maximum (FWHM). The convolution of Gaussian and Lorentzian is called Voigt function named after Woldemar Voigt (German mathematical physicist).

9.1.1.3 RESULTS AND DISCUSSION

Following normal-incidence irradiation with fs laser pulses, a periodic structure was clearly observed on the surface of ITO films, as shown in the scanning electron microscope (SEM) images of Figure 9.1.1.1a–f.

SEM is a type of an electron microscope which produces images of a sample by scanning the surface with a focused beam of electrons. The electrons interact with atoms in the sample, producing various signals that contain information about the surface topology and composition of the sample. The electron beam is raster scanned and the position of the beam is combined with the intensity of the detected signal to produce an image.

Under the same fluence of 0.1 mJ/cm^2, the periodic surface structure evolves noticeably, with an increasing number of pulses (N). For $N=5 \times 10^3$ shots (Figure 9.1.1.1b), only a few small dots appear on the surface of ITO films. For $N \geq 2.5 \times 10^4$ shots (Figure 9.1.1.1c–f), however, the laser-induced periodic structure is clearly observed on the surface of ITO films. As shown in the inset of Figure 9.1.1.1f, the ripple structure is composed of many sub-micron, in-line dots and the size of the self-organized dots is 20–500 nm. The long axis of the periodic ripple pattern is perpendicular to the direction of the lasers polarization, as represented by the arrow in Figure 9.1.1.1f. The laser light is probably scattered and diffracted by the grains, or defects and these scattered waves interfere with each other, to induce the subsequent enhancement of the local field [21]. Thus, the mixture of the dotted and ripple structures on the surface of ITO films, which is similar to our previous results for ordered YBCO array structures [22], is presumably formed by the solidification of melted dot patterns, under conditions of constructive interference and minimized surface energy.

The spacing of the laser-induced periodic surface structures (LIPSS) on the surface of ITO films was estimated by 2D-Fourier transformation, as represented by the red square in Figure 9.1.1.1f.

From the positions of satellite peaks, we obtained three kinds of periodicity in the laser-induced ripples, i.e., 798 ± 15, 420 ± 14, and 230 ± 15 nm. For the case of larger ripple spacing with 798 ± 15 and 420 ± 14 nm, it can be easily explained by classical scattering model [21]:

$$\Lambda = \frac{\lambda}{1 \pm \sin \theta} \qquad (9.1.1.1)$$

where Λ is the ripple spacing, λ is the laser wavelength and θ is the incident angle of the laser beam onto the target. However, the shorter ripple spacing of 230 ± 15 nm, which is much smaller than the laser wavelength of 800 nm, cannot be predicted by the classical scattering model with Eq. (9.1.1.1). Thus, due to the scale of ripple spacing with ~200 nm, it may be caused by the second harmonic

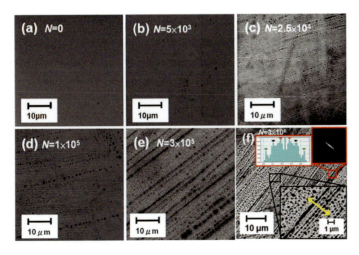

FIGURE 9.1.1.1 (a)–(f) Show the SEM images of periodic surface structures induced by 800 nm fs laser pulses at a fluence of 0.1 mJ/cm^2 and with various pulse numbers ($N = 0$, 5×10^3, 2.5×10^4, 1×10^5, 3×10^5, and 3×10^6, respectively). The black-square inset shows the enlarged surface features at corresponding locations of (f). The red-square inset shows the 2D Fourier-transformed pattern and its cross-section profile at corresponding locations of (f). The arrow indicates the direction of the laser polarization.

generation (SHG) with a shorter wavelength of 400 nm around the surface of ITO film. It has been reported that sub-wavelength ripple structures have been observed on the surface of InP, GaP and GaAs semiconductors [23] and Si [24], after fs laser multiple-pulse irradiation in the transparency region. It has been postulated that these nanostructures on the surface of films are induced by the harmonic generation from the near-surface region of films [23,24]. In our case, the sub-wavelength ripple with ~200 nm was indeed observed in high-intensity regions especially in the center of the laser Gaussian beam as shown in Figure 9.1.1.1f, which implies the high possibility of SHG owing to the surface asymmetry [24]. However, the formation mechanism of this sub-wavelength ripple in ITO films is still unclear. Nevertheless, the effect of self-organized nanodots, near the surface, on the electrical and optical properties and the morphology of self-organized nanodots and their formation will be further examined in the following sections.

Figure 9.1.1.2 shows the carrier concentration, carrier mobility and resistivity of ITO films, as a function of the number of pulses, from 0 to 3×10^6 shots, at a fluence of 0.1 mJ/cm². For total shots (N) less than 1000, the carrier concentration of fs laser-treated ITO films is almost the same as that of the as-deposited ITO films. Upon further increasing the number of laser shots, to 3×10^6, the carrier concentration rises noticeably, from $\sim 1 \times 10^{19}$ to $\sim 1.6 \times 10^{19}$ cm⁻³, and is linearly dependent on the number of shots in the semi-logarithmic plot. In contrast, the carrier mobility is correspondingly reduced, from 12.3 to 10.2 cm²/V-s, i.e., a 17% reduction, after fs laser irradiation with 3×10^6 shots, at a fluence of 0.1 mJ/cm². It is believed that the fs laser interference in the reaction duration of each laser pulse is too weak to clearly detect the slight changes in carrier concentration, using the Hall measurement. However, when the accumulated number of pulses exceeds the threshold energy of 1×10^3 shots, the thickness of the laser-irradiated area becomes thick enough to distinctly show the change in the carrier concentration, using Hall measurement. The resistivity is relatively less sensitive to the number of pulses, showing a 14% reduction (4.3×10^{-2} to 3.7×10^{-2} Ω–cm), from 0 to 3×10^6 shots, which can be simply attributed to an increase in carrier concentration.

The thicknesses of the ITO film before and after irradiation were 30 ± 1.5 nm and showed no noticeable variation, for repeated surface contour measurement. This strongly implies that the laser fluences used in this study were much less than the ablation threshold energy of ITO films, which is reported to be greater than 100 mJ/cm² [17]. Thus, the effect of thickness on the electrical properties can be eliminated.

Figure 9.1.1.3 shows the optical transmittance in the as-deposited ITO film and fs laser-treated ITO films, for various numbers of shots ($N=5 \times 10^5$ and 5×10^6 shots), at a fluence of 0.1 mJ/cm². Compared

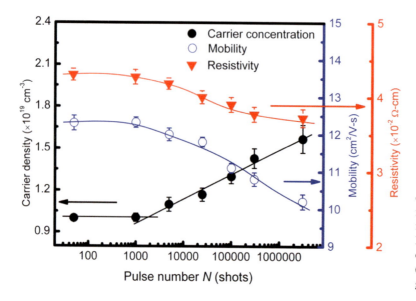

FIGURE 9.1.1.2 The carrier concentration, mobility, and resistivity in the fs laser-treated ITO films as a function of the pulse numbers. (The solid lines are a guide to the eyes.)

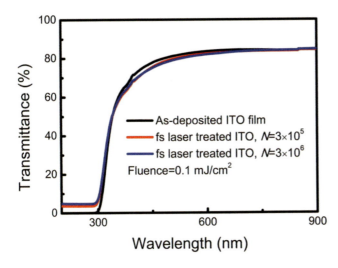

FIGURE 9.1.1.3 The transmittance in the fs laser treated ITO films as a function of wavelengths with various pulse numbers ($N = 0$, 5×10^5 and 5×10^6, respectively) at a fluence of 0.1 mJ/cm^2.

with an as-deposited ITO film, the optical transmittance of fs laser-treated ITO films is about the same, regardless of the number of pulses. Although the electrical properties of the fs laser-treated ITO films are noticeably changed, for different number of pulses, N, due to the formation of self-organized nanodots on the surface, the fs laser-treated ITO films still maintain their transparency, in the visible to near-infrared (NIR) range. These results indicate that fs laser annealing represents a new way to modify the electrical properties of ITO films while retaining their high optical transmittance.

The reaction depth of fs laser pulses in ITO films is limited to the top surface, which is unambiguously demonstrated by the current-sensing AFM (CSAFM) measurement. Figure 9.1.1.4a1–c1 show the topographic images of ITO films, for various numbers of pulses, at a fluence of 0.1 mJ/cm^2, and their corresponding surface current images are illustrated in Figure 9.1.1.4a2 and c2. For an as-deposited ITO film, the surface roughness is around 0.4 nm (Figure 9.1.1.4a1) and the surface current, for all of the measured area of $10 \times 10\,\mu$m, is around 0.3 pA in rms (RMS) (Figure 9.1.1.4a2). After pulsed fs laser irradiation, both the surface roughness (Figure 9.1.1.4b1–c1) and the surface current of ITO films (Figure 9.1.1.4b2–c2) increase, as the number of pulses increases. For $N = 3 \times 10^5$ shots, the surface roughness is 3.4 nm, and the surface current is 7.2 pA in rms (RMS). In addition, some larger dots appear on the surface of the ITO films (Figure 9.1.1.4b1) and many white spots are noted in the surface current image (Figure 9.1.1.4b2).

When the number of pulses is increased to $N = 3 \times 10^6$, the dots on the surface of ITO films become larger and even form a regular ripple pattern, with a roughness of 4.2 nm. Meanwhile, the corresponding surface current increases significantly, to 10 pA in rms (RMS), presumably due to the nanodots with high conductivity illustrated in Figure 9.1.1.4c2, which corresponds precisely to the nanodot pattern in the topographic image of Figure 9.1.1.4c1. The height of a bright spot is around 3–5 nm, as shown by the cross-sectional analysis in Figure 9.1.1.4d1, and its corresponding current is about 10 pA, as shown in Figure 9.1.1.4d2.

In short, the local conductivity of ITO films is remarkably enhanced, i.e., the surface current in the nanodots is ~30 times higher than that of the as-deposited ITO film. However, such a significant reduction in resistivity is not due to the phase transition, because the grazing incident XRD of fs laser-treated ITO films reveals no improvement in the crystallinity. Thus, the composition of the nanodots on the top of the fs laser-treated ITO films is further examined in the following section, to determine the cause of the significant increase in the local surface conductivity of the ITO films after femtosecond laser irradiation.

The first derivative (dN/dE) of the AES peaks, including In(MNN), Sn(MNN), and O(KLL) of an as-deposited ITO film and a fs laser-treated ITO film are shown in Figure 9.1.1.5. For an as-deposited ITO film (bottom-black curve, in Figure 9.1.1.5a), three dN/dE signals can be assigned

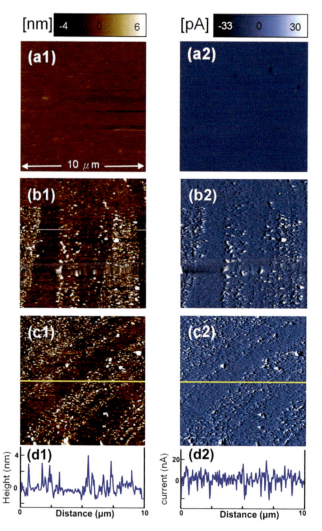

FIGURE 9.1.1.4 (a1)–(c1) The topographic images and (a2)–(c2) Their corresponding surface current images of ITO films induced by 800 nm fs laser pulses with various pulse numbers ($N = 0$, 3×10^5 and 3×10^6, respectively) at a fluence of 0.1 mJ/cm². (d1) Cross-section analysis on the height along the solid line in the AFM image (c1). (d2) Cross-section analysis on the current along the solid line in the CAFM image (c2).

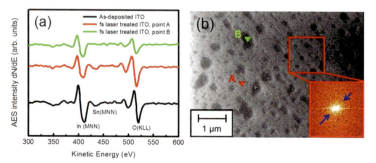

FIGURE 9.1.1.5 (a) The first derivative (dN/dE) of AES peaks, In(MNN), Sn(MNN), and O(KLL) as measured for the as-deposited ITO and fs laser-treated ITO films. Point A (outside of dot) and point B (inside of dot) correspond to the spots as marked in the SEM topview image (b) of a fs laser-treated ITO film. The red-square inset shows the 2D Fourier transformed pattern at corresponding locations of (b).

to the In(MNN) at a kinetic energy of 410 eV [25], Sn(MNN) at a kinetic energy of 433 eV [25,26] and O(KLL) at a kinetic energy of 519 eV [25,27], respectively. For a fs laser-treated ITO film, the *dN/dE* signals of In(MNN), Sn(MNN) and O(KLL) at point A in Figure 9.1.1.5b are slightly reduced, as compared with those of an as-deposited ITO film. However, the *dN/dE* signals at point B in Figure 9.1.1.5b which is located where the dots induced by the fs laser pulses, drop significantly. The reduction in signal is most significant for the *dN/dE* signal of Sn(MNN). This implies that the composition of ITO films is indeed changed by fs laser annealing, particularly at the positions of the dots.

Based on the disappearance of Sn and the noticeable reduction in In and O, the composition of the self-organized nanodots on the fs laser-treated ITO films deviates from the stoichiometry of an as-deposited ITO film. This suggests that the surface of an ITO film irradiated by fs laser pulses changes its composition, from In_2O_3:SnO_2 to InO_x-like. However, the electrical properties of In_2O_3, or InO_x films are inferior to that of ITO films [28,29]. Further examination of the peak intensities of as-deposited and fs laser-treated ITO films reveals that the *dN/dE* peak intensity ratio, for O(KLL) to In(MNN), changed from 1.7 to 1.3 (~23% reduction), after fs laser irradiation. Therefore, the oxygen content may play a key role in inducing such superior local conductivity in the nanodots of fs laser-treated ITO films. Additionally, the periodicity of the ripple seems to be smeared in high-magnification SEM images due to the ripple being formed by individual dots and too few structures (dots) were included in the image as shown in Figure 9.1.1.5b. Although the period in high-magnification SEM images is hardly recognized by the eyes, the anisotropic 2D Fourier-transformed pattern (the inset of Figure 9.1.1.5b) indicates the intrinsic feature of periodicity in laser-induced ITO ripples.

The electronic structures of oxides and Indium in fs laser-treated ITO films, determined by X-ray photoelectron spectroscopy (XPS) analysis, provides further information on chemical bonding, which may identify the cause of the superior surface conductivity of fs laser-treated ITO films.

A schematic representation of as-deposited ITO surface composition, based on Donley's model [30], is shown in Figure 9.1.1.6a. Figure 9.1.1.6d–g shows the O 1*s* XPS spectra of fs laser-treated ITO films, for various numbers of pulses. For an as-deposited ITO film (Figure 9.1.1.6b), the O 1*s* XPS spectrum can be fitted by three peaks, which are attributed to the In_2O_3-like oxygen, at 529.6±0.1 eV [30,31], oxygen that is adjacent to the oxygen-deficient sites, at 531.0±0.1 eV [30,31], in addition to the hydroxide and/or oxy-hydroxide peak, at 532.0±0.1 eV [27,30,31]. Compared with a reference sample of In_2O_3 powder, in Figure 9.1.1.6b, the peak for oxygen (at 531.0±0.1 eV) that is adjacent to the oxygen-deficient sites in an as-deposited ITO film (Figure 9.1.1.6c) increases, owing to the formation of oxygen vacancies, during the thin film sputtering process [27].

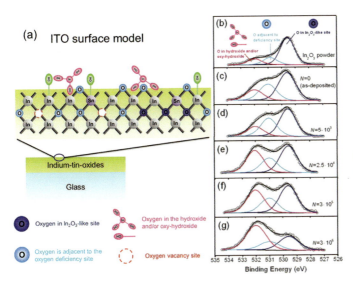

FIGURE 9.1.1.6 (a) Schematic representation of as-deposited ITO surface composition based on Donley's model [30]. (b) The O 1*s* XPS spectra of In_2O_3 powder and fs laser-treated ITO films with various pulse numbers ($N = 0$, 5×10^3, 2.5×10^4, 10^5, 3×10^5, and 3×10^6, respectively).

TABLE 9.1.1.1

The Relative Magnitude of Three Fitting Peaks in XPS O 1s Spectra (Figure 9.1.1.6) for Various Pulse Numbers

Samples	O in In$_2$O$_3$ Like Site and Adjacent to Deficient Site	O in Hydroxide and/or Oxy-Hydroxide
In$_2$O$_3$ powder	85.2%	14.8%
As-deposited ITO film, $N=0$	80.9%	19.1%
$N=5 \times 10^3$	79.5%	20.5%
$N=1 \times 10^5$	65.0%	35.0%
$N=3 \times 10^5$	61.5%	39.5%
$N=3 \times 10^6$	47.8%	52.2%

× The relative magnitude of the fitted peaks in O 1s spectra.

It is worth emphasizing that there is a dramatic change in the relative intensities of these three peaks, as the number of pulses increases, in fs laser-treated ITO films, which is quantitatively summarized in Table 9.1.1.1. During the FLA process, the oxygen atoms in the In$_2$O$_3$-like sites, adjacent to the oxygen-deficient sites, may be vaporized by the breaking of the In-O (bond strength ~3.31 eV) and Sn-O (5.53 ± 0.13 eV) bonds [32], via possible absorption of multiple photons, especially at the positions of nanodots, due to the higher energy associated with the constructive interference. As N is increased, from 0 to 3×10^6, the oxygen atoms in the In$_2$O$_3$like sites, adjacent to the oxygen-deficient sites, are significantly reduced, from 80.9% to 47.8%, in the fs laser-treated ITO films. These removed oxygen atoms may further form dangling bonds, on the top of the surface, leading to a gradual increase in the intensity of oxygen signals in the OH group, as N increases.

In addition, Figure 9.1.1.7 shows the XPS spectra of In $3d_{5/2}$, for In metals, In$_2$O$_3$ powders and fs laser-treated ITO films, for various numbers of pulses. For the In metals, the In $3d_{5/2}$ peak located at a lower binding energy of 443.7 eV corresponds to the In^o bonding state of In-In bonds [27]. For the In$_2$O$_3$ powders, however, the In $3d_{5/2}$ peak located at the higher binding energy of 444.6 eV

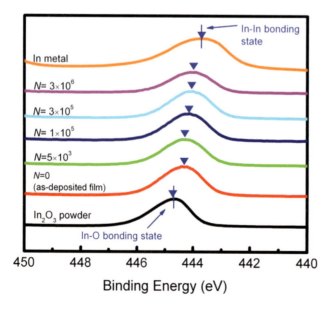

FIGURE 9.1.1.7 The In $3d_{5/2}$ XPS spectra of In$_2$O$_3$ powder, In metal, and fs laser-treated ITO films with various pulse numbers ($N = 0$, 5×10^3, 2.5×10^4, 10^5, 3×10^5, and 3×10^6, respectively).

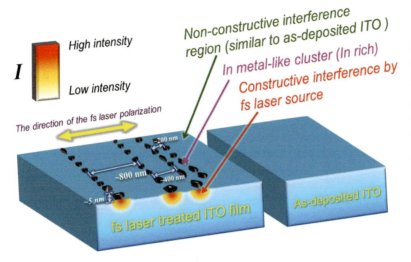

FIGURE 9.1.1.8 A schematic illustration for the formation of self-organized nano-dots induced by the constructive interference of fs laser at the near-surface region. The dot is composed of In-rich clusters with a height ~5 nm as a result of In-O bonding breaking into In–In under local-field enhancement.

corresponds to the In^{3+} bonding state of In_2O_3 [31,33]. Hence, the In $3d_{5/2}$ peak of 444.4 eV for an as-deposited ITO film demonstrates related to the valence states of In_2O_3. As the number of pulses increases, the In $3d_{5/2}$ peak gradually shifts, from the In-O bonding state to the In-In bonding state. In particular, for $N = 3 \times 10^6$, the peak of In $3d_{5/2}$ is located at 443.9 eV, which is almost equal to the binding energy of In–In bonds.

This strongly indicates that the appearance of In metal-like clusters in the fs laser-treated ITO films, particularly inside the self-organized nanodots, causes the high conductivity in the CSAFM measurements. Similar oxygen deficiencies (SiO_{2-x}) with periodical distribution, have been reported by Shimotsuma et al. [34], for SiO_2 glass under fs laser irradiation. Recently, Loeschner et al. also demonstrated a related phenomenon in aggregated metal nanoparticles that strongly interact via dipolar forces by using fs laser pulses [35].

Upon closer inspection of the curve in Figure 9.1.1.4d1, it can be observed that the nanodots are bulges on the surface of the ITO films. This can be attributed to the formation of In clusters with a bond length of 3.82 ~ 3.84 A, which is longer than the bond length of In-O (2.12° ~ 2.21 A) [36], leading to an effective volume increase for the nanodots. If one In-O bond breaks° and an In-In bond forms, the increase in the bond length is approximately 72%. The breaking ratio of the In-O bonds in the dots is around 23% based on the AES results. Consequently, we can estimate that the effective increase in the volume of the nanodots, for a 30 nm ITO film, is approximately 5 nm, which is consistent with the roughness results, shown in Figure 9.1.1.4d1. To summarize the findings for In clusters and superior local surface conductivity, the formation of self-organized nanodots induced by constructive interference of fs laser, and the size and height of those nanodots are schematically illustrated in Figure 9.1.1.8 and compared with those for the as-deposited ITO film. This study's observation of self-organized nanodots with superior local surface conductivity may be of significant interest for applications such as nanolithography, nanophotoelectrons and nanomechanics, in large-area nanotechnology.

9.1.1.4 CONCLUSION

We report the formation of periodic structures with self-organized nanodots on the surface of ITO films, after fs laser pulse irradiation. This periodic ripple microstructure, which is composed of self-organized nanodots, of 20–500 nm size, over an area of 200×200 μm, can be directly fabricated using single-beam fs laser irradiation, without scanning. The multi-periodic spacing of ~800, ~400, and ~200 nm observed in the laser-induced ripple of ITO films can be attributed to the interference between the incident fs lasers. The in-line, sub-micron dots are presumably formed by the solidification of melted dot patterns, under conditions of constructive interference and minimized surface energy. After irradiation with fs laser pulses, the electrical properties of ITO films are noticeably modified, e.g., the carrier concentration is increased and the carrier mobility is reduced, while a high optical transmittance is retained. Moreover, the much increased surface current is found to be $\sim$30 times higher than that of as deposited ITO films, according to CSAFM measurements. From AES and XPS analysis, it is deduced that the self-organized nanodots contain fewer oxygen atoms, i.e., 25%$\sim$30% reduction in the dot regions, owing to the breaking of In-O bonds. In addition, the In $3d_{5/2}$ XPS spectra results further show that the In-In bonding state gradually appears in the fs laser-treated ITO films, as the number of pulses increases. Therefore, the much greater local conductivity in the self-organized nanodots originates from the formation of In metal-like clusters, which, in turn, leads to an effective increase in the volume of the nanodots, with budges of height $\sim$5 nm, for a 30-nm-thick ITO film. The work in this subsection was conducted cooperatively among the following people: Chih Wang, Hsuan-I Wang, Wei-Tsung Tang, Chih-Wei Luo, Takayoshi Kobayashi, and Jihperng Leu [37].

REFERENCES

1. M. Al-Ibrahim, H. K. Roth, and S. Sensfuss, "Efficient large-area polymer solar cells on flexible substrates," *Appl. Phys. Lett.* **85**, 1481–1483 (2004).
2. H. Liu and R. Sun, "Laminated active matrix organic light-emitting devices," *Appl. Phys. Lett.* **92**, 063304-1–063304-3 (2008).
3. H. Kim, C. M. Gilmore, A. Pique, J. S. Horwitz, H. Mattoussi, H. Murata, Z. H. Kafafi, and D. B. Chrisey, "Electrical, optical, and structural properties of indiumtinoxide thin films for organic light-emitting devices," *J. Appl. Phys.* **86**, 6451–6461 (1999).
4. C. Guillen and J. Herrero, "Structure optical and electrical properties of indium tin oxide thin films prepared by´ sputtering at room temperature and annealed in air or nitrogen," *J. Appl. Phys.* **101**, 073514-1–073514-7 (2007).
5. C. C. Wu, C. I. Wu, J. C. Sturm, and A. Kahn, "Surface modification of indium tin oxide by plasma treatment: An effective method to improve the efficiency, brightness, and reliability of organic light emitting devices," *Appl. Phys. Lett.* **70**, 1348–1350 (1997).
6. J. K. Sheu, Y. K. Su, G. C. Chi, P. L. Koh, M. J. Jou, C. M. Chang, C. C. Liu, and W. C. Hung, "High-transparency Ni/Au ohmic contact to *p*-type GaN," *Appl. Phys. Lett.* **74**, 2340–2342 (1999).
7. R. H. Horng, D. S. Wuu, Y. C. Lien, and W. H. Lan, "Low-resistance and high-transparency Ni/indium tin oxideohmic contacts to *p*-type GaN," *Appl. Phys. Lett.* **79**, 2925–2927 (2001).
8. M. Gross, A. Winnacker, and P. J. Wellmann, "Electrical, optical and morphological properties of nanoparticle indium-tin-oxide layers," *Thin Solid Films* **515**, 8567–8572 (2007).
9. H. Hosono, M. Kurita, and H. Kawazoe, "Excimer laser crystallization of amorphous indium-tin-oxide thin films and application to fabrication of Bragg gratings," *Thin Solid Films* **351**, 137–140 (1999).
10. G. Legeay, X. Castel, R. Benzerga, and J. Pinel, "Excimer laser beam/ITO interaction: From laser processing to surface reaction," *Phys. Stat. Sol. (C)* **5**, 3248–3254 (2008).
11. J. G. Lunney, R. R. O'Neill, and K. Schulmeister, "Excimer laser etching of transparent conducting oxides," *Appl. Phys. Lett.* **59**, 647–649 (1991).
12. H. M. van Driel, J. E. Sipe, and J. F. Young, "Laser-induced periodic surface structures on solids: A universal phenomenon," *Phys. Rev. Lett.* **49**, 1955–1958 (1982).
13. J. F. Young, J. S. Preston, H. M. van Driel, and J. E. Sipe, "Laser-induced periodic surface structure. II. Experiments on Ge, Si, Al and brass," *Phys. Rev. B* **27**, 1155–1172 (1983).
14. B. C. Stuart, M. D. Feit, A. M. Rubenchik, B. W. Shore, and M. D. Perry, "Laser-induced damage in dielectrics with nanosecond to subpicosecond pulses," *Phys. Rev. Lett.* **74**, 2248–2251 (1995).
15. A. Rousse, C. Rischel, S. Fourmaux, I. Uschmann, S. Sebban, G. Grillon, P. Balcou, E. Frster, J. P. Geindre, and P. Audebert, "Non-thermal melting in semiconductors measured at femtosecond resolution," *Nature* **410**, 65–68 (2001).
16. J. M. Shieh, Z. H. Chen, B. T. Dai, Y. C. Wang, A. Zaitsev, and C. L. Pan, "Near-infrared femtosecond laser induced crystallization," *Appl. Phys. Lett.* **85**, 1232–1234 (2004).
17. C. W. Cheng, W. C. Shen, C. Y. Lin, Y. J. Lee, and J. S. Chen, "Fabrication of micro/nano crystalline ITO structures by femtosecond laser pulses," *Appl. Phys. A* **101**, 243–248 (2010).

18. M. Huang, F. Zhao, Y. Cheng, N. Xu, and Z. Xu, "Origin of laser-induced near-subwavelength ripples: Interference between surface plasmons and incident laser," *ACS Nano* **3**, 4062–4070 (2009).

19. Q. Z. Zhao, S. Malzer, and L. J. Wang, "Formation of subwavelength periodic structures on tungsten induced byultrashort laser pulses," *Opt. Lett.* **32**, 1932–1935 (2007).

20. X. Jia, T. Q. Jia, Y. Zhang, P. X. Xiong, D. H. Feng, Z. R. Sun, J. R. Qiu, and Z. Z. Xu, "Periodic nanoripples in the surface and subsurface layers in ZnO irradiated by femtosecond laser pulses," *Opt. Lett.* **35**, 1248–1250 (2010).

21. G. Zhou, P. M. Fauchet, and A. E. Siegman, "Growth of spontaneous periodic surface structures on solids during laser illumination," *Phys. Rev. B* **26**, 5366–5381 (1982).

22. C. W. Luo, C. C. Lee, C. H. Li, H. C. Shih, Y.-J. Chen, C. C. Hsieh, C. H. Su, W. Y. Tzeng, K. H. Wu, and J. Y. Juang, "Ordered YBCO sub-micron array structures induced by pulsed femtosecond laser irradiation," *Opt. Express* **16**, 20610–20616 (2008).

23. A. Borowiec and H. K. Haugen, "Subwavelength ripple formation on the surfaces of compound semiconductors irradiated with femtosecond laser pulses," *Appl. Phys. Lett.* **82**, 4462–4464 (2003).

24. R. L. Harzic, D. Dorr, D. Sauer, M. Neumeier, M. Epple, H. Zimmermann, and F. Stracke, "Large-area, uniform, ¨ high-spatial-frequency ripples generated on silicon using a nanojoule-femtosecond laser at high repetition rate," *Opt. Lett.* **36**, 229–231 (2011).

25. J. A. Chaney and P. E. Pehrsson, "Work function changes and surface chemistry of oxygen, hydrogen, and carbon on indium tin oxide," *Appl. Surf. Sci.* **180**, 214–226 (2001).

26. D. Briggs and M. P. Seah, *Practical Surface Analysis*, John Wiley and Sons, New York (1993).

27. F. Zhu, C. H. A. Huan, K. Zhang, and A. T. S. Wee, "Investigation of annealing effects on indium tin oxide thin films by electron energy loss spectroscopy," *Thin Solid Films* **359**, 244–250 (2000).

28. M. Mizuhashi, "Electrical properties of vacuum-deposited indium oxide and indium tin oxide films," *Thin Solid Films* **70**, 91–100 (1980).

29. S. Noguchi and H. Sakata, "Electrical properties of undoped In_2O_3 films prepared by reactive evaporation," *J. Phys. D: Appl. Phys.* **13**, 1129–1134 (1980).

30. C. Donley, D. Dunphy, D. Paine, C. Carter, K. Nebesny, P. Lee, D. Alloway, and N. R. Armstrong, "Characterization of indium-tin oxide interfaces using X-ray photoelectron spectroscopy and redox processes of a chemisorbed probe molecule: Effect of surface pretreatment conditions," *Langmuir* **18**, 450–457 (2002).

31. T. Szor¨enyi, L. D. Laude, I. Bert´oti, Z. Kntor, and Z. S. Geretovszky, "Excimer laser processing of indiumtinoxide´ films: An optical investigation," *J. Appl. Phys.* **78**, 6211–6219 (1995).

32. D. R. Lide, *CRC Handbook of Chemistry and Physics*, Taylor and Francis, Boca Raton, Florida (2003–2004).

33. J. C. C. Fan and J. B. Goodenough, "X-ray photoemission spectroscopy studies of Sn-doped indium-oxide films," *J. Appl. Phys.* **48**, 3524–3531 (1977).

34. Y. Shimotsuma, P. G. Kazansky, J. Qiu, and K. Hirao, "Self-organized nanogratings in glass irradiated by ultrashort light pulses," *Phys. Rev. Lett.* **91**, 247405-1–247405-8 (2003).

35. K. Loeschner, G. Seifert, and A. Heilmann, "Grating like nanostructures in polymer films with embedded metal nanoparticles induced by femtosecond laser irradiation," *J. Appl. Phys.* **108**, 073114–073123 (2010).

36. I. Tanaka, M. Mizuno, and H. Adachi, "Electronic structure of indium oxide using cluster calculations," *Phys. Rev. B* **56**, 3536–3539 (1997).

37. C. Wang, H.-I. Wang, W.-T. Tang, C.-W. Luo, T. Kobayashi, and J. Leu, "Superior local conductivity in self-organized nanodots on indium-tin-oxide films," *Opt. Express* **19**, 24286–2297 (2011).

9.1.2 Observation of an Excitonic Quantum Coherence in CdSe Nanocrystals

9.1.2.1 INTRODUCTION

Recent observations of electronic and/or vibronic coherences [1–4] in biological light-harvesting complexes by ultrafast multidimensional spectroscopy have led to speculation that such phenomena are exploited to boost energy transfer efficiencies in photosynthesis [5]. Quantum coherence between electronic states manifest themselves as periodic oscillations of the electronic density with time, in which the modulation frequencies scale with the energy differences between the participating eigenstates [6].

The most common concept about "coherence" is the properties of wave. Because particle-matterduality is introduced to explain "strange properties of quantum mechanics" it was accepted in terms of de Broglie wave. However, even more curious "spooky" phenomenon discussed by Albert Einstein is the entanglement. Here this chapter is not going to discuss the problem of entanglement. Just the quantum "coherence" of exciton quasi "particle." Even though exciton is a particle they can interfere among them.

Motivated by fundamental scientific interest and potential applications, studies of electronic coherences have also been extended to a variety of nanoscale artificial light-harvesting systems [7–11]. Indeed, theoretical studies have put forth the possibility of harnessing electronic quantum coherences to enhance the output of solar cells [12].

Among the multitude of artificial light-harvesting systems, semiconductor nanocrystals [13,14], also known as quantum dots (QDs), stand out due to their desirable optical properties [15] and relatively well-established synthetic procedures [16]. The latter allows exquisite control over the size and shape and hence the photophysical properties of these nanocrystals. High incident photon-to-current conversion efficiencies of 8.55% have been demonstrated [17] by solar cells that incorporate QDs as the photosensitizer [18]. While the excited-state dynamics [19,20] and the coherent phonon phenomena [21,22] of semiconductor nanocrystals have been actively investigated, it is only in recent years that excitonic quantum coherence has been studied in CdSe QDs [23,24]. Two-dimensional electronic spectroscopy (2DES) performed on zinc-blende CdSe QDs at ambient temperature reveals a coherent superposition between $1S_e1S_{3/2}$ and $1S_e2S_{3/2}$ excitonic states, for which a dephasing time of 15 fs is found [23].

More recent 2DES measurements elucidate multilevel quantum coherences with dephasing times that extend to ~100 fs [24]. These studies did not address the excitonic decoherence mechanism, for which physical insight is all the more critical given the disparate dephasing time scales reported. In addition, the possibility of steering coherent phonon wave packet dynamics by the excitonic coherence remains unexplored.

9.1.2.2 EXPERIMENTAL

Here, femtosecond optical pump–probe spectroscopy is employed to investigate coherent excitonic motion associated with the $1S_e1S_{3/2}$–$1S_e2S_{3/2}$ excitonic superposition in wurtzite CdSe QDs. In contrast with zinc blende CdSe QDs, it is noteworthy that excitonic coherences in the thermodynamically

DOI: 10.1201/9780429196577-46

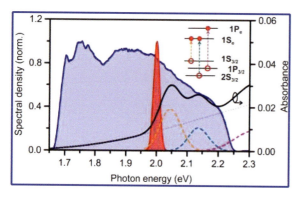

FIGURE 9.1.2.1 Linear absorption spectrum of the CdSe QD thin-film sample (black line) and the spectra of the broadband (blue line) and narrowband (red line) laser pulses. The absorption spectrum can be fit to a sum of optical transitions to the three lowest excitonic states (dashed lines), in addition to Rayleigh scattering (dotted line). The spectrum of the broadband laser pulse excites predominantly the transitions to the two lowest-energy excitonic states. The inset shows the excitonic level diagram denoted by the optical transitions observed in the absorption spectrum.

more stable wurtzite form of CdSe have so far eluded detection [25]. Spectral signatures of excitonic coherence are clearly discerned from our low-temperature optical pump–probe data, from which an ultrafast charge migration that is mediated by excitonic quantum coherence is reconstructed. Results from temperature-dependent measurements are suggestive of decoherence induced by exciton-acoustic phonon scattering, although the dominant contribution to decoherence is found to be temperature-independent. Finally, the presence of excitonic coherence is found to suppress exciton-LO-phonon coupling, while the exciton-LA-phonon coupling is enhanced. These observations are supported by semiclassical ab initio molecular dynamics (AIMD) simulations.

9.1.2.3 RESULTS AND DISCUSSION

The optical absorption spectrum of the CdSe QD thin-film sample collected at 77 K is shown in Figure 9.1.2.1. The spectrum reveals well-resolved peaks at 2.04, 2.14, and 2.32 eV, which correspond to transitions to the $1S_e1S_{3/2}$, $1S_e2S_{3/2}$, and $1P_e1P_{3/2}$ 19 excitonic states, respectively; note that the background rising toward the high-energy side of the spectrum is due to Rayleigh scattering by the thin-film sample. The optical absorption spectrum of the CdSe QDs in toluene solution at 295 K reveals a band edge of 1.99 eV (see Supporting Information), which suggests a mean diameter of 6.4 nm for the CdSe QDs [26]. The mean diameter inferred from the band-edge absorption energy is in good agreement with that obtained from transmission electron microscopy, from which an average diameter of 6.1 ± 0.4 nm is measured (see Supporting Information).

Photoexcitation of the sample by transform-limited, broadband laser pulses of 6 fs duration and spectral range of 550–750 nm results in the formation of a coherent superposition of the $1S_e1S_{3/2}$ and $1S_e2S_{3/2}$ excitonic states. The normalized differential transmission $\Delta T/T$ spectra obtained at 77 K show positive features, which correspond to ground-state bleaching and stimulated emission from the $1S_e1S_{3/2}$ and $1S_e2S_{3/2}$ excitonic states, as well as negative features, which can be assigned to excited-state absorption to the biexciton manifold [19] (see Supporting Information). Temporal oscillations in the time-resolved $\Delta T/T$ spectra can be assigned to coherent longitudinal-optical (LO) and longitudinal-acoustic (LA) phonons, with frequencies of 208 and 18 cm^{-1}, respectively. Inspection of the $\Delta T/T$ signal at short time delays ($t < 100$ fs) reveals an additional high-frequency, albeit short-lived oscillatory component (Figure 9.1.2.2a), which is suggestive of excitonic quantum coherence.

Further analysis of the early time oscillatory signal is performed on a time trace obtained at a probe wavelength in the band-edge transition region where the amplitude of the coherent LO phonon is a minimum. In this way, the contribution of the coherent LO phonon to the signal can be neglected. The resultant time traces obtained at 77, 100, 120, and 140 K show that the early time oscillation becomes more rapidly damped with temperature (Figure 9.1.2.2b). To see this trend, we note that the secondary maximum of the $\Delta T/T$ signal at 40 fs time delay, apparent at 77 K (see arrow in the top panel of Figure 9.1.2.2b), becomes indiscernible at 140 K. Furthermore, the appearance of the time trace at 295 K (bottom panel of Figure 9.1.2.2b) is qualitatively different from

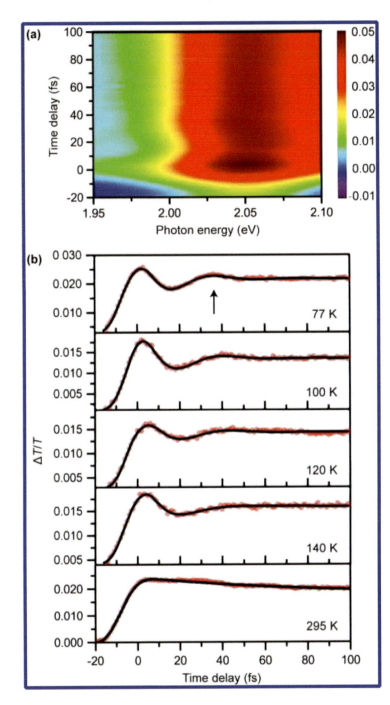

FIGURE 9.1.2.2 (a) Contour plot of the differential transmission spectra collected as a function of time delay following excitation of 6.1 nm diameter wurtzite-type CdSe QDs at 77 K. The data reveals strongly damped, high-frequency oscillations that are due to excitonic quantum coherence. (b) Time-resolved differential transmission signal was collected in the region of the band-edge transition for temperatures of 77, 100, 120, 140, and 295 K (top to bottom). The solid lines for the 77–140 K time traces are fits to Eq. (9.1.2.1). The $\Delta T/T(t)$ time trace collected at 295 K is fit to a convolution of the instrument response function with exponential decay and an offset. The arrow in the top panel denotes the secondary maximum of the $\Delta T/T$ signal that is apparent at 77 K.

those recorded between 77 and 140 K: the monotonically decaying time trace collected at 295 K is consistent with the population dynamics of an incoherent ensemble of CdSe QDs. The absence of coherent dynamics at ambient temperature suggests decoherence within the <10 fs time resolution of the experiment, in agreement with the results of previous 2DES measurements [25]. To extract the damping times, the time traces are fit to the function

$$S(t) = \sqrt{\frac{4 \ln 2}{\pi \Delta_{\mathrm{IRF}}^2}} \exp\left(-\frac{4 \ln 2 \, t^2}{\Delta_{\mathrm{IRF}}^2}\right) \Theta(t) \times \left[A_1 + A_2 \cos\left(\omega t + \varphi\right) \exp\left(-t / \tau\right)\right] \quad (9.1.2.1)$$

TABLE 9.1.2.1

Parameters Obtained from the Fit of the Early-Time Periodic Oscillation to Eq. (9.1.2.1)

Temperature (K)	Frequency ω (cm^{-1})	Phase ϕ (π rad)	Damping Time τ (fs)	Dephasing Time T_{12}^* (fs)
77	851 ± 17	0.14 ± 0.02	14.7 ± 1.2	15.8 ± 1.5
100	756 ± 21	0.07 ± 0.02	14.1 ± 1.0	15.1 ± 1.2
120	753 ± 63	0.18 ± 0.05	13.8 ± 3.8	14.7 ± 4.2
140	750^a	0.13 ± 0.04	11.4 ± 2.4	11.9 ± 2.6

[a] Fixed to allow the fit to the oscillation frequency at 140 K was converge.

which is a convolution of a damped oscillation atop a step function with a normalized Gaussian instrument response function of fwhm Δ_{IRF}. In the expression, $\Theta(t)$ is the Heaviside function with amplitude A_1, and A_2, ω, ϕ, and τ are the amplitude, frequency, phase, and damping time of the oscillation, respectively. The fits to the time traces are shown in Figure 9.1.2.2b and the fit parameters ω, ϕ, and τ are summarized in Table 9.1.2.1.

The frequencies of the oscillations observed at $t < 100$ fs are ~750–850 cm^{-1} for the range of temperatures employed in the experiments. In the absence of phonon modes with such high frequencies, the origin of the short-lived oscillatory component can be attributed to coherent excitonic dynamics. This assignment is bolstered by the following observations. First, the measured oscillation frequency coincides with the $\Delta E \approx 730$–750 cm^{-1} energy separation between the $1S_e 1S_{3/2}$ and $1S_e 2S_{3/2}$ excitonic states determined for the CdSe QD sample over the same temperature range (see Supporting Information). The good agreement between ΔE and ω strongly suggests that the observed oscillations originate from excitonic quantum beats. Second, the retrieved oscillation phases for all temperatures are ~0 rad, which implies that the exciton density distribution starts its oscillation from an extremum, as one would intuitively expect for the excitation of coherent superposition by transform-limited laser pulses [27,28].

A coherent superposition of excitonic states encodes the motion of exciton density. In the present work, a superposition of the $1S_e 1S_{3/2}$ and $1S_e 2S_{3/2}$ excitonic states yields a hole radial wave packet, described by the time-dependent wave function

$$\Psi(r,t) = c_{1s}(t)\Psi_{1s}(r)\exp(-iE_{1s}t/\hbar) + c_{2s}(t)\psi_{2s}(r)\exp(-iE_{2s}t/\hbar) \qquad (9.1.2.2)$$

where $\psi_{1s}(r)$ and $\psi_{2s}(r)$ are the $1S_{3/2}$ and $2S_{3/2}$ hole radial wave functions with coefficients $c_{1s}(t)$ and $c_{2s}(t)$, respectively, and E_{1s} and E_{2s} are the associated eigenenergies. The hole wave functions are obtained from solving the Luttinger Hamiltonian that includes an additional spherical confinement potential [29]. The coefficients $c_{ns}(t)$ ($n=1,2$) are related to the fractional populations $f_{ns}(t)$ of the $1S_e nS_{3/2}$ states by $c_{ns}(t) = [f_{ns}(t)]^{1/2}$, where $f_{1s}(t) + f_{2s}(t) = 1$; the fractional populations are in turn determined by the spectral overlap between the sample optical absorption spectrum and the laser spectral density (see Supporting Information). Note that, in principle, $c_{1s}(t)$ and $c_{2s}(t)$ are time-dependent due to the decay of the $1S_e 2S_{3/2}$ excited state to the $1S_e 1S_{3/2}$ band-edge state with a time constant of 245 fs [20], as well as the further relaxation of the $1S_e 1S_{3/2}$ state to the ground state and/or to trap states [30]. The corresponding motion of the radial hole density is given by the expression

$$|\Psi(r,t)|^2 = c_{1s}^2(t)|\Psi_{1s}(r)|^2 + c_{2s}^2(t)|\Psi_{2s}(r)|^2$$

$$+ 2c_{1s}(t)c_{2s}(t)\Psi_{1s}(r)\Psi_{2s}(r)\cos\left[(E_{2s}-E_{1s})t/\hbar\right]\exp(-t/T_{12}) \qquad (9.1.2.3)$$

where a phenomenological damping term with time constant T_{12} has been introduced to account for the decoherence between the $1S_{3/2}$ and $2S_{3/2}$ hole states. In the limit that the population dynamics are slow compared to the decoherence time, i.e., $T_{12} \ll T_{1s}$, T_{2s}, where T_{1s} and T_{2s} are the population decay time constants of the $1S_{3/2}$ and $2S_{3/2}$ hole states, respectively, $c_{1s}(t)$ and $c_{2s}(t)$ can be assumed to be time independent. That is, $c_{1s}(t) = c_{1s}(0)$ and $c_{2s}(t) = c_{2s}(0)$, where $c_{1s}(0)$ and $c_{1s}(0)$ are determined by the initial excitation conditions to be $c_{1s}(0) = 0.849$ and $c_{2s}(0) = 0.528$ (see Supporting Information). In this limit, the experimentally measured damping time τ corresponds to T_{12}.

It is important to note that observations of coherent dynamics by ensemble-averaged pump–probe measurements are complicated by inhomogeneous dephasing [31]. In the present work, inhomogeneity of the optical response arises primarily from the finite size dispersity of the CdSe QD sample. In a recent 2DES study, the influence of size dispersion was effectively eliminated by analyzing the dephasing of the zero quantum coherence at specific coherence energy, thereby yielding the decoherence time for only a narrow subset of QD sizes [24]. Here, we account for inhomogeneous dephasing by considering the normal distribution of oscillation frequencies $\omega = (E_{2s} - E_{1s})/\hbar$ that arises from the size-dependence of E_{2s} and E_{1s} (see Supporting Information) [32]. According to our estimates, the experimentally measured $T_{12} = 14.7 \pm 1.2$ fs at 77 K corresponds to a homogeneous dephasing time of $T_{12}^* = 15.8 \pm 1.5$ fs (see Table 9.1.2.1). This result is supported by AIMD simulations performed on a 1.3 nm-diameter $Cd_{33}Se_{33}$ model cluster, which yield a decoherence time of 17 fs at 77 K for the $1S_e1S_{3/2}$–$1S_e2S_{3/2}$ excitonic superposition. While the use of a QD with smaller radius R in the simulations constitutes an approximation, given that higher acoustic phonon frequencies [21] ($\omega_a \sim R^{-1}$) and stronger electron–phonon coupling via the deformation potential [33] ($S \approx R^{-2}$) would predict shorter decoherence times, we note that the computed $1S_e1S_{3/2}$–$1S_e2S_{3/2}$ energy gap of the model QD (0.08 eV) is similar to the experimental value of 0.09 eV. As a result, the amplitude of the phonon-induced fluctuations for the model QD is expected to be commensurate with that of the experimental system [34]. According to linear response theory [35], the computed dephasing time should therefore be directly comparable to the experimental results.

The experimental data can be used to reconstruct the time evolution of the hole radial distribution function (Figure 9.1.2.3).

The radial distribution function that initially peaked at a radius of 1.04 nm moves to 1.76 nm in 22 fs. The corresponding charge migration rate of 0.33 Å/fs is comparable to some of the fastest electron transfer rates inferred for strongly coupled electron donor–acceptor systems [36–38]. In the present work, the observed ultrafast charge migration is driven solely by excitonic quantum

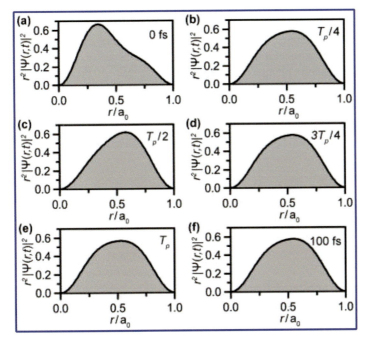

FIGURE 9.1.2.3 Radial distribution functions $r^2|\Psi(r, t)|^2$ of the hole density reconstructed from the experimental data collected at 77 K for the time delays (a) 0 fs, (b) $T_p/4$, (c) $T_p/2$, (d) $3T_p/4$, (e) T_p, and (f) 100 fs, where $T_p = h/(E_{2s}-E_{1s})$ is the classical orbital period. For the CdSe QD studied here, T_p corresponds to 44 fs. The plot at 100 fs is representative of the asymptotic hole density. The radius of the QD a_0 is 3.05 nm in the present work.

coherence without the involvement of nuclear motion. Furthermore, because the radial distribution function at the moment of coherent photoexcitation is governed by the relative phases of the $1S_e1S_{3/2}$ and $1S_e2S_{3/2}$ excitonic states, and as phase coherence is lost, this initial radial distribution function asymptotically evolves into that given by the relative populations of the $1S_e1S_{3/2}$ and $1S_e2S_{3/2}$ states, the rate of charge migration can be increased simply by reducing the decoherence time. In the present system, for example, the initial and asymptotic hole densities are peaked at 1.04 and 1.67 nm, respectively. Hence, a shortened decoherence time of 1 fs would yield a charge migration rate of 2 Å/fs. A corollary to this point is that coherent charge migration can be expected as long as the initial and asymptotic charge density distributions are different, even when the decoherence time is ultrashort.

Examining the temperature-dependence of the excitonic decoherence provides insight into the decoherence mechanism. In bulk semiconductors, carrier decoherence occurs via carrier–carrier and carrier–phonon scattering [39,40]. In the case of semiconductor QDs, three-pulse photon-echo measurements reveal optical dephasing rates that scale linearly with sample temperature T [41,42]. In the limit $k_BT \gg \hbar\omega_a$, where k_B is the Boltzmann constant and ω_a is the frequency that characterizes the quasi-continuum of acoustic phonons; the linear temperature dependence suggests exciton–phonon scattering involving low-frequency, incoherent acoustic phonons as the dominant dephasing mechanism. Such linear scaling has also been observed, for example, in the case of GaAs quantum wells [43], carbon nanotubes [44], and dye molecules in the condensed phase [45]. Within experimental error, our temperature-dependent T_{12}^* values follow the linear relation $1/T_{12}^* (T)=\Gamma_{12}(T)=\Gamma_{12}(0)+aT$, where $\Gamma_{12}(0)=45\pm8$ ps^{-1} is the temperature-independent offset and $a=0.22\pm0.09$ ps^{-1} K^{-1} is the slope (Figure 9.1.2.4). It is evident that $\Gamma_{12}(0)$ dominates the decoherence rates that are obtained in the 77–140-K range, with the temperature-dependent term aT accounting for only ~30% of the measured decoherence rate. This result suggests that decoherence of the $1S_e1S_{3/2}$–$1S_e2S_{3/2}$ excitonic superposition in the 77–140-K temperature range is only partially induced by acoustic phonons. Possible origins of $\Gamma_{12}(0)$ include exciton–exciton scattering between the two excitonic states that comprise the superposition [46], as well as scattering that involves surface defects [42]. The former could be enhanced by the complex exciton fine structure of the wurtzite CdSe QDs [47], whereas the latter is conceivable for the ligand-capped QDs examined here. Finally, while we caution against the direct comparison between optical dephasing rates and intraband dephasing rates, which is not meaningful [48], we note that the slope a obtained from our experiments is ~4× larger than the value of $a\approx0.06$ ps^{-1} K^{-1} obtained for the optical dephasing rates of similar-sized CdSe QDs [42]. The origin of this discrepancy is unknown and requires a systematic study over a wider temperature range.

While incoherent acoustic phonons are found to participate in the decoherence of the $1S_e1S_{3/2}$–$1S_e2S_{3/2}$ excitonic superposition, the experimental data also reveals the influence of the excitonic superposition on the behavior of the coherent phonons of CdSe QDs. Two types of coherent phonon modes are known to exist in CdSe quantum dots [22]: the coherent LO phonon (208 cm^{-1}) and the

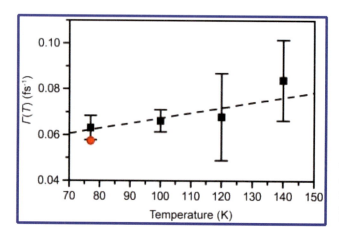

FIGURE 9.1.2.4 Measured excitonic decoherence rates (black squares) exhibit a linear dependence on temperature with an offset. The decoherence rate computed by AIMD simulations at 77 K is also shown (red circle).

coherent LA phonon (18 cm⁻¹). In the present work, the $\Delta T/T$ time traces reveal that broadband (Figure 9.1.2.5a) and narrowband (Figure 9.1.2.5b) excitation predominantly launch the coherent LA and LO phonons, respectively. Further insight into the effect of the $1S_e1S_{3/2}$–$1S_e2S_{3/2}$ superposition on the coherent phonon dynamics can be obtained from analyzing the first-moment time trace $\langle\Omega^{(1)}(t)\rangle$ computed about the band-edge transition (Figure 9.1.2.5c).

The first moment $\langle\Omega^{(1)}\rangle$ of a differential transmission spectrum is related to the energy gap between the bands which are optically coupled by the probe pulse [49]. Considering the contributions from both LO and LA phonons, $\langle\Omega^{(1)}(t)\rangle$ can be fit into the expression

$$\langle\Omega^{(1)}(t)\rangle = A_{LO}\cos(\omega_{LO}t + \varphi_{LO})\exp(-t/\tau_{LO}) + A_{LA}\cos(\omega_{LA}t + \varphi_{LA})\exp(-t/\tau_{LA}) \quad (9.1.2.4)$$

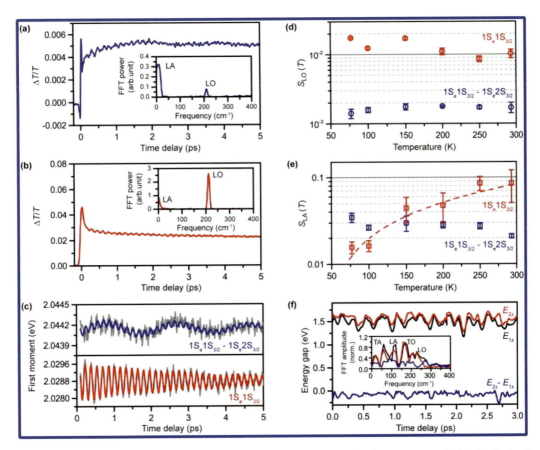

FIGURE 9.1.2.5 (a) $\Delta T/T$ signal as a function of time delay for a probe photon energy of 1.98 eV, obtained following the phase-coherent excitation of the $1S_e1S_{3/2}$ and $1S_e2S_{3/2}$ states. The inset shows the FFT power spectrum, which reveals oscillation frequencies that can be assigned to the LA and LO phonons of the CdSe QD. (b) $\Delta T/T$ signal as a function of time delay for a probe photon energy of 2.01 eV, obtained following the selective excitation of the $1S_e1S_{3/2}$ state. The inset shows the FFT power spectrum, which reveals oscillation frequencies that can be assigned to the LA and LO phonons of the CdSe QD. (c) The spectral first moment computed about the band-edge transition for both broadband coherent excitation (top panel) and narrowband state-selective excitation (bottom panel) of the CdSe QD sample at 77 K. Note the different spans of the vertical scales. (d) Huang–Rhys factor S_{LO} was obtained at different temperatures for the LO phonon. (e) Huang–Rhys factor S_{LA} was obtained at different temperatures for the LA phonon. The increase in S_{LA} with temperature, observed with narrowband excitation, is described by a linear fit (dashed line). (f) AIMD trajectories of the phonon-induced fluctuations of the E_{1s} (black) and E_{2s} (red) energy gaps, as well as the difference E_{2s}–E_{1s} (blue). The inset shows the FFT amplitudes of the energy gaps. The peaks at 60, 120, 170, and 230 cm⁻¹ can be assigned to the TA, LA, TO, and LO phonon, respectively [81].

where A_{LO}, ω_{LO}, φ_{LO}, and τ_{LO} (A_{LA}, ω_{LA}, φ_{LA}, and τ_{LA}) correspond to the amplitude, frequency, phase, and damping time of the LO (LA) phonon, respectively (see Supporting Information). To clarify the influence of coherent excitonic motion on coherent phonon dynamics, the first-moment time traces obtained with narrowband, state-selective excitation to the $1S_e1S_{3/2}$ state are also recorded (Figure 9.1.2.5c). The $\langle\Omega^{(1)}(t)\rangle$ traces obtained under the two different excitation conditions reveal qualitative differences: excitation of the $1S_e1S_{3/2}$–$1S_e2S_{3/2}$ superposition drives predominantly the coherent LA phonon, whereas state-selective excitation mostly yields the coherent LO phonon.

The observed suppression of the coherent LO phonon and the enhancement of the coherent LA phonon can be further quantified by computing their corresponding Huang–Rhys factors S_i (i=LO, LA) [50]. The Huang–Rhys factor characterizes the exciton–phonon coupling strengths and can be extracted from the $\langle\Omega^{(1)}(t)\rangle$ oscillation amplitude by the relation [51] $A_i=2\omega_iS_i$. Several observations can be made about the S_{LO} and S_{LA} values measured over the temperature range of 77–295 K (Figure 9.1.2.5b and c). First, it is evident that simultaneous excitation of the $1S_e1S_{3/2}$ and $1S_e2S_{3/2}$ states as compared to state-selective excitation of only the $1S_e1S_{3/2}$ state leads to a one order-of magnitude suppression of S_{LO} over the entire temperature range of 77–295 K. Second, broadband excitation of the $1S_e1S_{3/2}$ and $1S_e2S_{3/2}$ states yields relatively temperature invariant S_{LA} values, whereas a linear increase in S_{LA} with temperature is observed for excitation of only the $1S_e1S_{3/2}$ state. The latter observation can be rationalized in terms of a linearly increasing phonon occupation number in the electronic ground state with temperature, which in turn launches an excited-state vibrational wave packet with a larger nuclear displacement amplitude upon photoexcitation [52]. From the experimental data, it can be deduced that the intrinsic Huang–Rhys factor for the LA phonon, accessed in the low-temperature limit and therefore independent of the phonon occupation number, is larger for coherent excitation of the $1S_e1S_{3/2}$ and $1S_e2S_{3/2}$ states than for selective excitation of the $1S_e1S_{3/2}$ state.

AIMD simulations reveal that the LO-phonon-induced modulation of the E_{1s} and E_{2s} gaps occur in phase, signifying similarly signed electron-LO phonon coupling matrix elements for the $1S_e1S_{3/2}$ and $1S_e2S_{3/2}$ states (Figure 9.1.2.5d). In this case, the energy difference $\Delta E=E_{2s}-E_{1s}$, which encodes the coherent excitonic superposition, exhibits suppressed LO phonon oscillations (Figure 9.1.2.5d inset). In agreement with experiment, the simulation results show that phase-coherent excitation of the $1S_e1S_{3/2}$ and $1S_e2S_{3/2}$ states leads to a ~10-fold suppression of the LO phonon mode. These results mirror qualitatively the previously observed suppression of the radial breathing mode (RBM) coherent phonon following the simultaneous excitation of the E_{11} and E_{22} transitions of single-walled carbon nanotubes by broadband, few-cycle laser pulses [53]. However, the AIMD simulations predict the suppression of the coherent LA phonon with simultaneous excitation of the $1S_e1S_{3/2}$ and $1S_e2S_{3/2}$ states, even though the experimental results point to an enhancement. This contradiction between experiment and theory suggests the direct involvement of excitonic motion in driving the LA phonon, an effect that is not considered in the AIMD simulations. Intuitively, the ultrafast radial charge migration that is associated with the $1S_e1S_{3/2}$–$1S_e2S_{3/2}$ excitonic superposition impulsively alters the electronic potential along the radial direction, which in turn triggers atomic motion along the radial coordinate, i.e., the coherent LA phonon is launched. The observed temperature independence of S_{LA} can be attributed to the persistence of coherent charge migration even at elevated temperatures. Similar launching of coherent phonons by ultrafast charge transfer has been observed [54,55]. In the resonant coupling regime, Bloch oscillations in semiconductor quantum wells have been shown to drive coherent LO phonons adiabatically [56].

The direct observation of coherent valence electron motion represents one of the holy grails in femtochemistry and attosecond physics [27,57]. Compared to coherent exciton migration in photosynthetic light-harvesting complexes [1–4] or to charge migration that has been predicted for ionized molecules [58–60], the relative simplicity of the electronic structure of QDs makes them an attractive platform for visualizing coherent electron motion. Pioneering investigations of excitonic coherences in zinc blende-type CdSe QDs by 2DES spectroscopy, however, yielded largely disparate compositions of the excitonic superpositions and decoherence times despite similar experimental conditions [23,24]. In the present work, optical pump–probe spectroscopy performed on a highly monodisperse sample of wurtzite-type CdSe QDs reveals unambiguous spectral signatures of quantum coherence between the $1S_e1S_{3/2}$ and $1S_e2S_{3/2}$ exciton states. The high signal-to-noise ratio afforded by our experimental data allows the first reconstruction of ultrafast charge migration in a nanoscale system that is driven solely by excitonic quantum coherence. We note that the charge migration distance can be controlled by spectral shaping of the excitation laser pulse. For example, photoexcitation of an equal population of the $1S_e1S_{3/2}$ and $1S_e2S_{3/2}$ excitonic states would extend the inner and outer turning points of the radial wave packet to 0.97 and 1.98 nm, respectively. Unlike conventional donor–acceptor charge transfer, however, it is important to note that the charge migration observed herein involves the center-of-mass motion of the hole distribution within a single CdSe nanocrystal, which preserves its overall electrical neutrality at all times.

9.1.2.4 CONCLUSION

The valence radial wave packet that is observed in this work is reminiscent of atomic Rydberg radial wave packets that were previously generated with picosecond pulses [61]. Unlike the numerous revivals exhibited by Rydberg wave packets [62], however, the QD excitonic superposition is found to decohere within a fraction of the classical orbit period. Variable temperature measurements reveal that the decoherence originates predominantly from electronic factors, exciton–exciton scattering, exciton fine structure, and defect scattering, rather than the presence of the phonon bath. This observation suggests that efforts to achieve extended decoherence times should focus on engineering the excitonic structure of QDs and minimizing the number of defect sites. Another possible avenue for further exploration is to investigate how the decoherence of the excitonic superposition is affected by the presence of coherent phonons that are simultaneously generated by the excitation pulse.

Through the observation of coherent LO and LA phonons in the present work, we have elucidated the effect of excitonic coherence and the associated ultrafast charge migration on the behavior of the phonons. This result paves way for the coherent control of atomic motion [63,64] via the optical manipulation of valence electron densities [65]. In addition, when applied to donor–acceptor motifs in which the CdSe QD serves as either the hole donor or acceptor [66,67], the ultrafast charge migration that occurs within the CdSe nanocrystal can potentially be harnessed to gate charge transfer on ultrashort time scales.

Methods: Sample Preparation. Colloidal wurtzite-type CdSe QDs are synthesized following the literature procedure [16] before they are dispersed in a poly(methyl methacrylate) (PMMA) matrix and spin-coated onto a 1.5 mm-thick fused silica window. The average diameter of the QDs is 6.1 nm with 6% rms dispersity, as determined by transmission electron microscopy (see Supporting Information). The narrow size dispersity, further evidenced by the well-resolved features in the absorption spectrum (Figure 9.1.2.1), is critical in allowing the observation of spectral signatures of excitonic quantum coherence.

Detailed Information of Experimental Optical Pump–Probe Spectroscopy. The optical pump–probe setup employs few-cycle pulses in the visible and pulseto-pulse measurements of the differential transmission spectra. The details of the apparatus can be found in Ref. [68]. For the study reported herein, multiphoton intrapulse interference phase scan (MIIPS) is incorporated to characterize and compensate for the residual high-order dispersion of the broadband laser pulses [68], thereby furnishing transform-limited ∼6 fs pulses in the 550–750 nm spectral range for experiments (see Supporting Information). The typical excitation fluence is 0.4 mJ/cm² and the corresponding average number of excitons [19] per QD is $\langle N \rangle \approx 0.3$. Fluence-dependence measurements confirm that the $\Delta T/T$ signal is linear in the range of excitation fluences employed in the experiments (see Supporting Information). Narrowband pump pulses are produced by inserting a 10 nm bandpass dielectric interference filter into the broadband pump beam. Pump and probe pulses are orthogonally polarized to suppress contributions from scattering and coherent artifacts, which could otherwise obfuscate the short-lived excitonic coherence signal. In addition, coherent artifacts from the PMMA matrix are eliminated by subtracting the measured response of a pure PMMA sample from the signal of the QD sample (see Supporting Information) [69]. Accurate determination of time zero is accomplished via linear spectral interferometry between pump and probe pulses (see Supporting Information) [70].

Ab Initio Molecular Dynamics Simulations. The $Cd_{33}Se_{33}$ cluster with a diameter of 1.3 nm was constructed using bulk wurtzite lattice. Recent experiments [71] have shown that such "magic" size cluster is one of the smallest stable CdSe QDs that support a crystalline-like core [72,73]. These properties make $Cd_{33}Se_{33}$ an excellent model for studying electronic and vibrational properties of semiconductor QDs. The cluster geometry was optimized using ab initio density functional theory with a plane wave basis, as incorporated in the Vienna ab initio simulation package (VASP) [74]. The PBE functional [75] with projector-augmented-wave (PAW) pseudopotentials [76] was employed in a converged plane wave basis. The simulations were performed in a periodically replicated cubic cell with at least 8 Å of vacuum between QD replicas. The fully optimized structure was then heated to the desired temperatures with repeated velocity rescaling. Three picosecond-long microcanonical MD trajectories were generated using the Verlet algorithm with the 1 fs time step and Hellmann–Feynman forces. The decoherence time was obtained with the semiclassical optical response formalism [35], which allows one to use the MD simulation. The pure-dephasing time is associated with fluctuations of the energy levels due to coupling of the electronic degrees of freedom to phonons. The fluctuations in the energy levels are best characterized in terms of correlation functions. The pure-dephasing function is defined as,

$$D(t) = \exp(i\omega t)\left\langle \exp\left(\frac{i}{h}\int_0^t \Delta E(\tau)d\tau\right)\right\rangle \qquad (9.1.2.5)$$

where the angular brackets denote thermal averaging. The dephasing function can be approximated using the second-order cumulant expansion as,

$$D(t) = \exp(-g(t)) \qquad (9.1.2.6)$$

Where

$$g(t) = \frac{1}{h^2}\int_0^t d\tau_1 \int_0^{\tau_1}\left\langle \Delta E(\tau)\Delta E(0)\right\rangle d\tau_2 \qquad (9.1.2.7)$$

The method based on the cumulant expansion shows better numerical convergence than the direct expression Eq. (9.1.2.5), which involves averaging of an oscillating function. Both direct and cumulant methods have shown excellent agreement with experiments for several systems [77–79]. The data reported here are based on the cumulant expansion. The research presented in this subsection was performed collaboratively among the following people: Shuo Dong, Dhara Trivedi, Sabyasachi Chakrabortty, Takayoshi Kobayashi, Yinthai Chan, Oleg V. Prezhdo, and Zhi-Heng Loh [80].

SUPPORTING INFORMATION

The Supporting Information is available free of charge on the ACS Publications website.

Sample characterization by TEM and variable-temperature UV/vis spectroscopy, details of laser pulse compression and data processing to remove artifacts and determine time-zero, fluence-dependence measurements, time-resolved differential transmission spectra collected up to 5 ps time delay, estimation of size dispersity-induced inhomogeneous dephasing, parameters used in the construction of the hole wave functions, and fit parameters obtained from the time-domain analysis of the spectral first moment (PDF).

REFERENCES

1. E. Collini, C. Y. Wong, K. E. Wilk, P. M. G. Curmi, P. Brumer, and G. D. Scholes, *Nature* **463**, 644–669 (2010).
2. G. Panitchayangkoon, D. Hayes, K. A. Fransted, J. R. Caram, E. Harel, J. Wen, R. E. Blankenship, and G. S. Engel, *Proc. Natl. Acad. Sci. U.S.A.* **107**, 12766–12770 (2010).
3. E. Romero, R. Augulis, V. I. Novoderezhkin, M. Ferretti, J. Thieme, D. Zigmantas, and R. van Grondelle, *Nat. Phys.* **10**, 677–683 (2014).
4. G. D. Scholes, G. R. Fleming, A. Olaya-Castro, and R. Grondelle, *Nat. Chem.* **3**, 763–774 (2011).
6. G. C. Schatz and M. A. Ratner, *Quantum Mechanics in Chemistry*, Dover Publications, Mineola, NY (2002).
7. C. A. Rozzi, S. M. Falke, N. Spallanzani, A. Rubio, E. Molinari, D. Brida, M. Maiuri, G. Cerullo, H. Schramm, J. Christoffers, and C. Lienau, *Nat. Commun.* **4**, 7 (2013).
8. A. Halpin, J. M. Johnson Philip, R. Tempelaar, R. S. Murphy, J. Knoester, L. C. Jansen Thomas, and R. J. D. Miller, *Nat. Chem.* **6**, 196–201 (2014).
9. S. M. Falke, C. A. Rozzi, D. Brida, M. Maiuri, M. Amato, E. Sommer, A. De Sio, A. Rubio, G. Cerullo, E. Molinari, and C. Lienau, *Science* **344**, 1001–1005 (2014).
10. J. Yuen-Zhou, D. H. Arias, D. M. Eisele, C. P. Steiner, J. J. Krich, M. G. Bawendi, K. A. Nelson, and A. Aspuru-Guzik, *ACS Nano* **8**, 5527–5534 (2014).
11. E. Cassette, R. D. Pensack, B. Mahler, and G. D. Scholes, *Nat. Commun.* **6**, 6086 (2015).
12. K. E. Dorfman, D. V. Voronine, S. Mukamel, and M. O. Scully, *Proc. Natl. Acad. Sci. U.S.A.* **110**, 2746–2751 (2013).
13. M. L. Steigerwald and L. E. Brus, *Acc. Chem. Res.* **23**, 183–188 (1990).
14. A. P. Alivisatos, *Science* **271**, 933–937 (1996).
15. V. I. Klimov, *Nanocrystal Quantum Dots*, 2nd ed., CRC Press, Boca Raton, FL (2010).
16. C. B. Murray, D. J. Norris, and M. G. Bawendi, *J. Am. Chem. Soc.* **115**, 8706–8715 (1993).
17. A. C.-H. M. Chuang, P. R. Brown, V. Bulovic, and M. G. Bawendi, *Nat. Mater.* **13**, 796–801 (2014).
18. P. V. Kamat, *J. Phys. Chem. Lett.* **4**, 908–918 (2013).
19. V. I. Klimov, *J. Phys. Chem. B* **104**, 6112–6123 (2000).
20. P. Kambhampati, *Acc. Chem. Res.* **44**, 1–13 (2011).
21. G. Cerullo, S. De Silvestri, and U. Banin, *Phys. Rev. B: Condens. Matter Mater. Phys.* **60**, 1928–1932 (1999).
22. D. M. Sagar, R. R. Cooney, S. L. Sewall, E. A. Dias, M. M. Barsan, I. S. Butler, and P. Kambhampati, *Phys. Rev. B: Condens. Matter Mater. Phys.* **77**, 235321 (2008).
23. D. B. Turner, Y. Hassan, and G. D. Scholes, *Nano Lett.* **12**, 880–886 (2012).
24. J. R. Caram, H. Zheng, P. D. Dahlberg, B. S. Rolczynski, G. B. Griffin, A. F. Fidler, D. S. Dolzhnikov, D. V. Talapin, and G. S. Engel, *J. Phys. Chem. Lett.* **5**, 196–204 (2014).
25. G. B. Griffin, S. Ithurria, D. S. Dolzhnikov, A. Linkin, D. V. Talapin, and G. S. Engel, *J. Chem. Phys.* **138**, 014705 (2013).
26. W. W. Yu, L. Qu, W. Guo, and X. Peng, *Chem. Mater.* **15**, 2854–2860 (2003).
27. E. Goulielmakis, Z.-H. Loh, A. Wirth, R. Santra, N. Rohringer, V. S. Yakovlev, S. Zherebtsov, T. Pfeifer, A. M. Azzeer, M. F. Kling, S. R. Leone, and F. Krausz, *Nature* **466**, 739–737 (2010).
28. J. Li, Z. Nie, Y. Y. Zheng, S. Dong, and Z.-H. Loh, *J. Phys. Chem. Lett.* **4**, 3698–3703 (2013).
29. K. E. Knowles, E. A. McArthur, and E. A. Weiss, *ACS Nano* **5**, 2026–2035 (2011).
30. A. L. Efros, *Phys. Rev. B: Condens. Matter Mater. Phys.* **46**, 7448–7458 (1992).
31. K. M. Pelzer, G. B. Griffin, S. K. Gray, and G. S. Engel, *J. Chem. Phys.* **136**, 164508 (2012).
32. D. J. Norris and M. G. Bawendi, *Phys. Rev. B: Condens. Matter Mater. Phys.* **53**, 16338–16346 (1996).
33. T. Takagahara, *Phys. Rev. Lett.* **71**, 3577–3580 (1993).
34. A. V. Akimov and O. V. Prezhdo, *J. Phys. Chem. Lett.* **4**, 3857–3864 (2013).
35. S. Mukamel, *Principles of Nonlinear Optical Spectroscopy*, Oxford University Press, New York (1995).
36. J. B. Asbury, E. Hao, Y. Q. Wang, H. N. Ghosh, and T. Q. Lian, *J. Phys. Chem. B* **105**, 4545–4557 (2001).
37. G. Benkö, J. Kallioinen, J. E. I. Korppi-Tommola, A. P. Yartsev, and V. Sundström, *J. Am. Chem. Soc.* **124**, 489–493 (2002).
38. W. R. Duncan, W. M. Stier, and O. V. Prezhdo, *J. Am. Chem. Soc.* **127**, 7941–7951 (2005).
39. F. Rossi and T. Kuhn, *Rev. Mod. Phys.* **74**, 895–950 (2002).
40. J. Shah, *Ultrafast Spectroscopy of Semiconductors and Semiconductor Nanostructures*, Springer, Berlin, Heidelberg (2013).

41. R. W. Schoenlein, D. M. Mittleman, J. J. Shiang, A. P. Alivisatos, and C. V. Shank, *Phys. Rev. Lett.* **70**, 1014–1017 (1993).

42. D. M. Mittleman, R. E. Schoenlein, J. J. Shiang, V. L. Colvin, A. P. Alivisatos, and C. V. Shank, *Phys. Rev. B: Condens. Matter Mater. Phys.* **49**, 14435–14447 (1994).

43. L. Schultheis, A. Honold, J. Kuhl, K. Kohler, and C. W. Tu, *Phys. Rev. B: Condens. Matter Mater. Phys.* **34**, 9027–9030 (1986).

44. M. W. Graham, Y.-Z. Ma, A. A. Green, M. C. Hersam, and G. R. Fleming, *J. Chem. Phys.* **134**, 034504 (2011).

45. C. J. Bardeen, G. Cerullo, and C. V. Shank, *Chem. Phys. Lett.* **280**, 127–133 (1997).

46. M. R. Salvador, P. Sreekumari Nair, M. Cho, and G. D. Scholes, *Chem. Phys.* **350**, 56–68 (2008).

47. C. Y. Wong and G. D. Scholes, *J. Phys. Chem. A* **115**, 3797–3806 (2011).

48. S. Mukamel, *J. Phys. Chem. A* **117**, 10563–10564 (2013).

49. W. T. Pollard, S. Y. Lee, and R. A. Mathies, *J. Chem. Phys.* **92**, 4012–4029 (1990).

50. K. Huang and A. Rhys, *Proc. R. Soc. London, Ser. A* **204**, 406–423 (1950).

51. M. Lax, *J. Chem. Phys.* **20**, 1752–1760 (1952).

52. A. T. N. Kumar, F. Rosca, A. Widom, and P. M. Champion, *J. Chem. Phys.* **114**, 701–724 (2001).

53. Z. Nie, R. Long, J. Li, Y. Y. Zheng, O. V. Prezhdo, and Z.-H. Loh, *J. Phys. Chem. Lett.* **4**, 4260–4266 (2013).

54. P. Tyagi, R. R. Cooney, S. L. Sewall, D. M. Sagar, J. I. Saari, and P. Kambhampati, *Nano Lett.* **10**, 3062–3067 (2010).

55. L. Dworak, V. V. Matylitsky, M. Braun, and J. Wachtveitl, *Phys. Rev. Lett.* **107**, 247401 (2011).

56. T. Dekorsy, A. Bartels, H. Kurz, K. Kohler, R. Hey, and K. Ploog, *Phys. Rev. Lett.* **85**, 1080–1083 (2000).

57. F. Krausz and M. Ivanov, *Rev. Mod. Phys.* **81**, 163–234 (2009).

58. A. L. Kuleff, J. Breidbach, and L. S. Cederbaum, *J. Chem. Phys.* **123**, 044111 (2005).

59. F. Remacle and R. D. Levine, *Proc. Natl. Acad. Sci. U.S.A.* **103**, 6793–6798 (2006).

60. S. Lünnemann, A. L. Kuleff, and L. S. Cederbaum, *J. Chem. Phys.* **129**, 104305 (2008).

61. A. Ten Wolde, L. D. Noordam, A. Lagendijk, and H. B. V. van den Heuvell, *Phys. Rev. Lett.* **61**, 2099–2101 (1988).

62. J. A. Yeazell, M. Mallalieu, and C. R. Stroud, *Phys. Rev. Lett.* **64**, 2007–2010 (1990).

63. A. Assion, T. Baumert, M. Bergt, T. Brixner, B. Kiefer, V. Seyfried, M. Strehle, and G. Gerber, *Science* **282**, 919–922 (1998).

64. E. M. Grumstrup, J. C. Johnson, and N. H. Damrauer, *Phys. Rev. Lett.* **105**, 257403 (2010).

65. F. Lepine, M. Y. Ivanov, and M. J. Vrakking, *J. Nat. Photonics* **8**, 195–204 (2014).

66. R. Costi, A. E. Saunders, and U. Banin, *Angew. Chem., Int. Ed.* **49**, 4878–4897 (2010).

67. K. Wu, H. Zhu, and T. Lian, *Acc. Chem. Res.* **48**, 851–859 (2015).

68. Z. Nie, R. Long, L. Sun, C. Huang, J. Zhang, Q. Xiong, D. W. Hewak, Z. Shen, O. V. Prezhdo, and Z.-H. Loh, *ACS Nano* **8**, 10931–10940 (2014).

69. A. L. Dobryakov, S. A. Kovalenko, and N, P. Ernsting, *J. Chem. Phys.* **119**, 988–1002 (2003).

70. C. Dorrer, N. Belabas, J. P. Likforman, and M. Joffre, *J. Opt. Soc. Am. B* **17**, 1795–1802 (2000).

71. A. Kasuya, R. Sivamohan, Y. A. Barnakov, I. M. Dmitruk, T. Nirasawa, V. R. Romanyuk, V. Kumar, S. V. Mamykin, K. Tohji, B. Jeyadevan, K. Shinoda, T. Kudo, O. Terasaki, Z. Liu, R. V. Belosludov, V. Sundararajan, and Y. Kawazoe, *Nat. Mater.* **3**, 99–102 (2004).

72. A. Puzder, A. J. Williamson, F. Gygi, and G. Galli, *Phys. Rev. Lett.* **92**, 217401 (2004).

73. S. Kilina, S. Ivanov, and S. Tretiak, *J. Am. Chem. Soc.* **131**, 7717–7726 (2009).

74. G. Kresse and J. Furthmüller, *Phys. Rev. B: Condens. Matter Mater. Phys.* **54**, 11169–11186 (1996).

75. J. P. Perdew, K. Burke, and M. Ernzerhof, *Phys. Rev. Lett.* **77**, 3865–3868 (1996).

76. G. Kresse and D. Joubert, *Phys. Rev. B: Condens. Matter Mater. Phys.* **59**, 1758–1775 (1999).

77. B. F. Habenicht, H. Kamisaka, K. Yamashita, and O. V. K. Prezhdo, *Nano Lett.* **7**, 3260–3265 (2007).

78. A. B. Madrid, K. Hyeon-Deuk, B. F. Habenicht, and O. V. Prezhdo, *ACS Nano* **3**, 2487–2494 (2009).

79. Z. Guo, B. F. Habenicht, W.-Z. Liang, and O. V. Prezhdo, *Phys. Rev. B: Condens. Matter Mater. Phys.* **81**, 125415 (2010).

80. S. Dong, D. Trivedi, S. Chakrabortty, T. Kobayashi, Y. Chan, O. V. Prezhdo, and Z.-H. Loh, *Nano Lett.* **10**, 6875–6882 (2015).

81. F. Widulle, S. Kramp, N. M. Pyka, A. Göbel, T. Ruf, A. Debernardi, R. Lauck, and M. Cardona, *Phys. B* **263–264**, 448–451 (1999).

Section 9.2

1D CNT

9.2.1 Coherent Phonon Generation in Semiconducting Single-Walled Carbon Nanotubes Using a Few-Cycle Pulse Laser

9.2.1.1 INTRODUCTION

Single-walled carbon nanotubes (SWNTs), with unique mechanical, electronic and optical properties, enable groundbreaking applications in nanoelectronics and photonics. Their one-dimensionality provides a playground for studying the dynamics of confined electrons and phonons and their interplay [1–4]. Recently, many efforts have been made to investigate the coherent lattice vibrations (phonon) in SWNTs by coherent phonon (CP) spectroscopy via femtosecond pump–probe techniques [5–14]. Extensive studies were made on the mechanism of the spectroscopic appearance of the CP generation. It is argued that the coupling of the phonon modes to the electronic structure results in the modulations of the difference absorbance [5–14]. The vibrations of radial breathing mode (RBM) were explained as ultrafast modulations of optical constant at frequency ω_{RBM} due to band gap (E_g) oscillations induced by the change of the diameter Φ_d ($E_g \propto 1/\Phi_d$). It is claimed that the photon energy dependence of the CP signal shows a derivative-like behavior [6–8]. However, as described later discussion in the present paper the agreement of the probe wavelength dependence of the RBM amplitude is not good enough to support the explanation. The problem in the previous papers discussing the origin of the real-time vibration (phonon) signal associated with the modulation of the electronic (excitonic) transitions based on the experiments was caused by a relatively smaller number of probe energies, which made the argument on the mechanism ambiguous due to the limited the precise acquisition of complete dependence signal of vibrational amplitudes on probe photon energy.

Here, we report on the use of resonant 7.1-fs visible pulses to generate and detect CPs in SWNTs. We probed 128 different wavelengths simultaneously by using a detection system composed of a polychromator and a multichannel lock-in amplifier. We separately observed four semiconducting chiral systems without ambiguity and obtained abundant data points of the probe photon energy dependence of the phonon amplitudes for RBMs, which allow an in-depth study of the origin of the CP generation in SWNTs further. Since the process of pump–probe experiment is the third-order nonlinear process, the effects of the real and imaginary parts of the third-order susceptibility on the modulation of the probed absorbance change are fully discussed. The probe photon energy-dependent amplitude profiles are discussed with respect to the mechanism of the modulation of excitonic transition probability.

DOI: 10.1201/9780429196577-48

9.2.1.2 EXPERIMENTAL DETAILS

The SWNT sample was prepared by CoMoCat method [15–17]. The pump and probe light sources were from the non-collinear parametric amplifier (NOPA). The pump source of this system is a commercially supplied regenerative amplifier (Spectra Physics, Spitfire). The central wavelength, pulse duration, power of the output, and repetition rate of this amplifier were 800 nm, 50 fs, 740 mW, and 5 kHz, respectively. With the use of a compression system composed of a pair of prisms and a pair of chirped mirrors, the system supported a pulse with a pulse duration of 7.1 fs with constant spectral phase, indicating that the pulses are nearly Fourier-transform limited. The energy of the pump is 32 nJ. The probe pulse energy is five times weaker than the pump pulse.

The polarization of the pump and probe beams are parallel to each other. In the pump–probe experiment, signal was spectrally dispersed with a polychromator (JASCO, M25-TP) over 128 photon energies (wavelengths) from 1.71 to 2.36 eV (722–524 nm). It was detected by 128 sets of avalanche photodiodes and lock-in amplifiers with a reference from an optical chopper intersecting the pump pulse at the 2.5-kHz repetition rate. Details about our 7.1-fs pump-probe experimental setup and working principle of the techniques are described elsewhere [18,19].

9.2.1.3 RESULTS AND DISCUSSION

9.2.1.3.1 STATIONARY ABSORPTION SPECTRUM OF THE SAMPLE AND LASER SPECTRUM

Figure 9.2.1.1a shows the stationary absorption spectrum of SWNTs with the relevant chirality assignments [15]. The assignments of some absorption bands might be uncertain since they might be shared by more than one type of tube because of their finite spectral widths. The broadband visible laser spectrum is resonant with the second exciton transitions (E_{22}) of the tubes in the 1.712.36 eV range. For the analysis of the real-time data in the latter part of this paper, the absorption spectrum was fitted by the sum of five dominant Voigt functions, which are the convolution of the Gaussian and Lorentzian functions at 2.17, 2.08, 1.91, 1.84, and 1.78 eV corresponding to (6,5), (6,4), (7,5), (8,3), and (9,1) tubes, respectively. The properties of tubes (9,1) will not be discussed here due to their weak absorption. The absorbed laser spectrum for different chiralities is also calculated, as shown in Figure 9.2.1.1b, which is defined later by the difference in the spectrum between the probe light before and after passing through the sample. The application of the absorbed spectrum will be further discussed later for successful fitting the experimental probe photon energy-dependent amplitude profiles.

9.2.1.3.2 TWO-DIMENSIONAL (2D) REAL-TIME SPECTRA AND EXACT CHIRALITY ASSIGNMENT

Two-dimensional (2D) difference absorption ΔA (E_{pr}, t) was represented in Figure 9.2.1.2a, showing clear oscillations in ΔA amplitude with time as stripe-like structures. The Fourier transform (FT) of the ΔA time traces is shown in Figure 9.2.1.2b. This plot identifies two well-known dominant vibrational modes [5]: RBMs at 250–350 cm⁻¹ (95–133 fs) and G modes at 1587 cm⁻¹ (21 fs), generated by impulsive excitation with pulse duration (7.1 fs) much smaller than vibrational periods. The vibrations with frequencies from 360 to 1525 cm⁻¹ are disregarded here since they are too weak to be resolved. The probe photon energy dependence of vibrational amplitudes is well displayed for RBMs. According to the results previously reported, the four dominant symmetric double-peak structures, as indicated by the two-way arrows, should follow the first derivative of electronic resonances [6–8]. The middle dip for each structure should correspond to the E_{22} transitions for relevant chiralities since the oscillation becomes minimal at resonance. And then, bearing in mind that the Raman shifts for different RBMs should theoretically correspond to their phonon frequencies in the CP spectra, chirality assignments for four chiralities (6,4), (6,5), (7,5), and (8,3) can be achieved

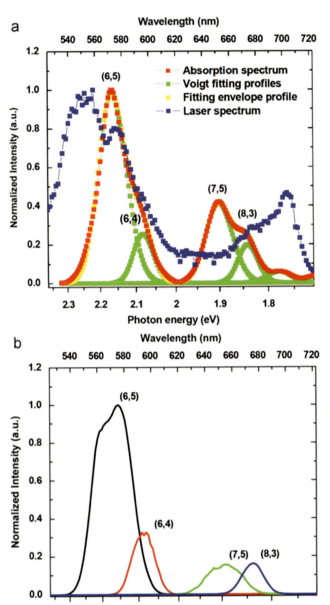

FIGURE 9.2.1.1 (a) Laser spectrum (blue line) and stationary visible absorption spectrum of SWNTs (red line) after subtracting the weak background and its Voigt fitting profiles (green lines), illustrating individual E_{22} absorption components in 1.72.4 eV spectral range. The chirality assignments are shown together. (b) Absorbed spectra, defined by the difference in the spectrum between the probe light before and after passing through the sample, for different chiralities. (For interpretation of the references to color in this figure legend, the reader is referred to the web version of this article.)

exactly, since only these specific type of tubes can fulfill the above frequency and resonance conditions simultaneously for each mode. For the G mode vibrations, however, the amplitude profiles overlap together for different chiralities and cannot be distinguished because the axial G^+-mode is known to be insensitive to the diameter and chirality of SWNTs [5].

9.2.1.3.3 PROBE PHOTON ENERGY DEPENDENT AMPLITUDE PROFILES

In previous reports studying the CPs in SWNTs, the probe photon energy (or wavelength) dependence of the RBM amplitudes was compared with the first-derivative of the relevant absorptions and it was claimed that the close resemblance between the amplitude profiles to the derivative is the

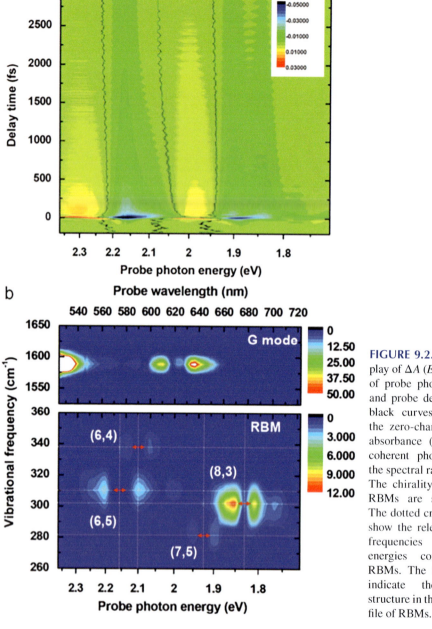

FIGURE 9.2.1.2 (a) 2D display of $\Delta A (E_{pr}, t)$ as functions of probe photon energy, E_{pr}, and probe delay time, t. The black curves are related to the zero-change lines in the absorbance ($\Delta A = 0$). (b) 2D coherent phonon spectra in the spectral range of 1.72.4 eV. The chirality assignments of RBMs are shown together. The dotted crisscrossing lines show the relevant vibrational frequencies and resonance energies corresponding to RBMs. The two-way arrows indicate the double-peak structure in the amplitude profile of RBMs.

verification of the wavepacket motion [6–8]. However, the results of fitting analyses are far from the sufficient agreement and the exact mechanism of the wavepacket motion has not been investigated in detail.

It is well known that the wavefunction of the electronic state can be factorized into the electronic part and vibrational part under the Born–Oppenheimer approximation, $\psi(Q,q) = \phi(Q,q)\chi(Q)$, where $\phi(Q,q)$ and $\chi(Q)$ represent wavefunctions of the electrons and nuclei, respectively. Usually Franck–Condon approximation given below is good enough to formulate the time dependence of the transition probability with the frequency of CP in case without anharmonicity. Following the Franck–Condon principle, dipole transition under the Condon approximation is given by

$$\left\langle \Psi_1(Q,q) \middle| \mathrm{eq} \middle| \Psi_2(Q,q) \right\rangle_{Q,q}$$

$$\approx \left\langle \left\langle \Phi_1(Q,q) \middle| \mathrm{eq} \middle| \Phi_2(Q,q) \right\rangle_q \chi(Q) \right\rangle_Q \qquad (9.2.1.1)$$

$$\approx \left\langle \Phi_1(Q_0,q) \middle| \mathrm{eq} \middle| \Phi_2(Q_0,q) \right\rangle_q \left\langle \chi(Q) \middle| \chi(Q) \right\rangle_Q$$

The Franck–Condon factor, $< \chi(Q) | \chi(Q) >_Q = < \sum_i c_i \chi_i(Q) \middle| \sum_j c_j \chi_j(Q) >_Q$ gives the time dependence of the transition probability with the frequency of CP in case without anharmonicity. Then the change of the transition probability due to the wave-packet motion cannot be simply described by the smooth motion of the wave-packet along the corresponding normal mode [20]. The observed spectral shift described by the derivative-type dependence caused by small smooth shift is due to phase modulation of the probe light by the wavepacket motion. This is in principle related to the real part of the third-order susceptibility on the absorption difference. In this case at fixed probe energy, this effect translates into an amplitude modulation via a probe pulse spectral change induced by the cross-phase modulation mechanism [21,22]. This mechanism in effect results from the refractive index change caused by the deformation of the lattice/molecular configuration during the coherent lattice/molecular vibrations. The refractive index change then introduces the linear shift of the probe energy. Therefore the signal can be simply approximated with the spectral (linear) shift of the probe pulse induced by cross-phase modulation (XPM), and the probe energy dependence of phonon amplitude follows the first-derivative of the electronic resonance (denoted by derivative-type hereafter), in consistent with the first-derivative analysis as reported recently [6–8]. The meaning of XPM is described as follows.

Pump laser vibronically excites the system (in this case CNT). Due to Franck–Condon principle, the molecular configuration excited is deviated from the most stable structure in the ground state and hence starts to vibrate along the vibronic hypersurface of this vibronic (both excited in electronic state and vibration level) state. The real-time change of the vibronic spectrum of the state is coupled to the refractive index change (modulation) corresponding to the modulational phase change (cross phase change, XPM).

The probe photon energy dependence of the vibrational amplitude of RBMs for three chiralities is plotted in Figure 9.2.1.3a–c, which is obtained by cutting through the 2D FT power spectra at the relevant central frequencies and then taking the square roots. We firstly performed the fitting of the amplitude profiles (black lines) with the first derivative of the relevant absorption components (gray lines) for different chiralities, as shown in the top panels in Figure 9.2.1.3. Each of the corresponding absorption components is obtained by the spectral deconvolution of the stationary absorption spectrum by Voigt fitting in Figure 9.2.1.1. Similar to the results reported in literature [8], the typical two-peak structure, which is expected by the first-derivative dependence, cannot be well fitted, even after slightly modifying the absorption components by second-derivative to improve the disagreement between the absorption spectra and the fitted spectra. There are always substantial deviations in the valley-dip positions in the spectra of the probe photon energy dependence and the first derivative of the absorptions, and the line shapes between the derivatives and the amplitudes cannot agree well with each other. A similar phenomenon was also reported for chirality (6,5) in highly enriched samples when the amplitude dependence was fitted by the first derivative profile [8]. Considering that the line shape of the laser spectrum can affect the actual acquisition of laser energy, we further compared the amplitudes with the derivative of the absorbed laser spectrum (red lines), defined later as the spectral distribution of the

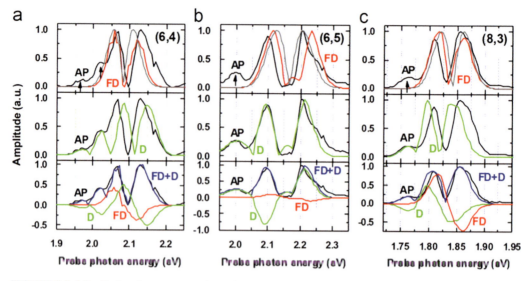

FIGURE 9.2.1.3 Probe photon energy-dependent RBM amplitudes (AP, black lines) for (a) (6,4), (b) (6,5), and (c) (8,3) tubes, fitted with the first-derivative type (FD, gray lines), the absorbed laser spectra (FD, red lines), the difference-type only (D, green lines), and the sum of FD and D (FD+D, blue lines) with relevant contributions, respectively. The arrows indicate the side bands. In the bottom columns, the original FD (red lines) and D (green lines) lines are plotted together before taking absolute value to show their corresponding contributions to the vibrational amplitudes. The amplitude profile of (7,5) is not studied here since it is very weak.

laser photon absorbed by the sample. In this case, it seems the line-shape fitting is improved, but large deviations in the valley-dip positions in the probe-dependent spectra remain substantially large. Especially, the side bands, as indicated by the arrows in Figure 9.2.1.3, are already outside of the range of the resonance energy distributions. They cannot be well fitted in any case, as just inferred from the first-derivative calculation. More interestingly, we found the energy difference between the side bands and each one of the peaks in the double-band structure is always of an integral number of the relative RBM frequencies. This observation brought us to further take into account the Raman interaction contributions between the probe pulse and the coherent vibrations [21,22], as discussed below.

Because of the Kramers–Kronig relations, the effect of the imaginary part (denoted by difference-type hereafter) is also taking place, which is essentially the Raman gain/loss process induced by the energy exchange between the CPs and the probe optical field, according to the relation, $\Delta A(\omega, \tau) \propto -\mathrm{Im}[P^{(3)}(\omega, \tau)/E_{\mathrm{probe}}(\omega, \tau)]$, where $P^{(3)}(\omega, \tau)$ is the nonlinear polarization induced by pump and probe pulses, and $E_{\mathrm{probe}}(\omega, \tau)$ is the electronic field of the optical probe pulse. In this process, the probe optical field is alternately deamplified and amplified in corresponding transitions depending on the phase change of the vibrations. So the FT power spectral shapes are considered to depend on the spectral distribution of the laser photons absorbed by the sample. For quantitative discussion, the probe photon energy dependence of vibrational amplitude $V(\omega_{\mathrm{probe}})$ can be phenomenologically described by

$$V\left(\omega_{\mathrm{probe}}\right) = C_S \left| a\left(\omega_{\mathrm{probe}}\right) - a\left(\omega_{\mathrm{probe}} - \omega_v\right) \right| \qquad (9.2.1.2.1a)$$

$$V_{\mathrm{AS}}\left(\omega_{\mathrm{probe}}\right) = C_s \left| a\left(\omega_{\mathrm{probe}}\right) - a\left(\omega_{\mathrm{probe}} + \omega_v\right) \right| \qquad (9.2.1.2.1b)$$

$$a\left(\omega_{\mathrm{probe}}\right) = L\left(\omega_{\mathrm{probe}}\right)\left(1 - 10^{-A\left(\omega_{\mathrm{probe}}\right)}\right) \qquad (9.2.1.2.2)$$

where C_S and C_{AS} are proportionality constant, corresponding to the cases of pump/Stokes and pump/anti-Stokes interactions, respectively, $\alpha(\omega_{probe})$ is the absorbed laser spectrum at ω_{probe}, the optical frequency of probe light, ω_v is the molecular vibration frequency, $L(\omega_{probe})$ is the laser spectrum and $A(\omega_{probe})$ is the absorbance of the sample at ω_{probe}. The interaction between the CP and probe optical field can be between the first (anti-) Stokes beam and the absorbed beam and also can be between higher-order (anti-) Stokes beams. Based on the analysis above, the amplitude profiles can be fitted with the absolute difference between the absorbed photon energy distribution and the distribution shifted by the amount of vibration frequency (denoted by difference type). The difference-type fitting is displayed in the middle panels of Figure 9.2.1.3. In the fitting, the π-phase jump between the neighboring contributions was taken into account. Including the Raman contributions, the fitting is obviously improved very much, except for those of the valley-dip positions, which thus means the good fitting cannot be achieved only with the mechanism of either difference-type or derivative-type mechanism. Considering that the effect of the real part can play roles simultaneously with the imaginary part, we continued to perform the fitting with the sum contributions of the two types of mechanisms with adjustable parameters of the relative amount of contributions. As shown in the bottom panels of Figure 9.2.1.3, we finally achieved the fitting perfectly.

The sum contributions (blue lines) of the two types of mechanisms for different RBMs is demonstrated by adjusting the parameters of the relative amount of contributions from different fitting origins, derivative type (red lines) and difference-type (green lines), before taking absolute values. Thus, in contrast to the results reported in literature recently [6–8], in which the intensity of the RBM mode was poorly fitted to the first-derivative of the corresponding absorption of the chiral species due to spectral change associated with the wavepacket motion, the fitting results here indicate that the real and the imaginary parts of the third-order susceptibility can both contribute to the modulation of the absorbance change, and the latter is also a dominant contributor to the modulation of the difference absorbance.

9.2.1.4   CONCLUSION

We separately observed four RBM systems without ambiguity in CoMoCat grown SWNTs by using a 7.1 fs broadband visible pulse and an ultrahigh sensitive detection system [23]. The amplitude profiles of RBMs can be fitted perfectly by the sum of the first derivative of the absorption due to the real part of the susceptibility and the difference absorption due to the imaginary part induced by a Raman gain/loss process with adjustable contributions. The imaginary part can also be a dominant contributor in the modulation of difference absorbance. The Raman process is explained in terms of the energy exchange between coherent phonons and probe optical field. The research described here in this subsection was conducted cooperatively by the following people: T. Kobayashi, Z. Nie, J. Du, H. Kataura, Y. Sakakibara, and Y. Miyata [23].

REFERENCES

1. S. Iijima, *Nature (London)* **354**, 56 (1991).
2. M. S. Dresselhaus, G. Dresselhaus, and P. Avouris (Eds.), *Carbon Nanotubes: Synthesis, Structure, Properties and Applications*, Springer, Berlin (2001).
3. M. Dresselhaus, "Carbon nanotubes: Synthesis, structure, properties," in *Physical Properties of Carbon Nanotubes*, edited by R. Saito, G. Dresselhaus, and M. Dresselhaus, World Scientific, Singapore (2003).
4. P. Harris, *Carbon Nanotubes and Related Structures: New Materials for the Twenty-First Century*, Cambridge University Press, Cambridge, England (1999).
5. A. Gambetta, C. Manzoni, E. Menna, M. Meneghetti, G. Cerullo, G. Lanzani, S. Tretiak, A. Piryatinski, A. Saxena, and R. L. Martin, and A. R. Bishop, *Nat. Phys.* **2**, 515 (2006).
6. Y.-S. Lim, K.-J. Yee, J. H. Kim, E. H. Hároz, J. Shaver, J. Kono, S. K. Doorn, R. H. Hauge, and R. E. Smalley, *Nano Lett.* **6**, 2696 (2006).

7. J.-H. Kim, K.-J. Han, N.-J. Kim, K.-J. Yee, Y.-S. Lim, G. D. Sanders, C. J. Stanton, L. G. Booshehri, E. H. Haŕoz, and J. Kono, *Phys. Rev. Lett.* **102**, 037402 (2009).

8. L. Luer, C. Gadermaier, J. Crochet, T. Hertel, D. Brida, and G. Lanzani, *Phys. Rev. Lett.* **102**, 127401 (2009).

9. G. D. Sanders, C. J. Stanton, J.-H. Kim, K.-J. Yee, Y.-S. Lim, E. H. Haŕoz, L. G. Booshehri, J. Kono, and R. Saito, *Phys. Rev. B* **79**, 205434 (2009).

10. K. Kato, K. Ishioka, M. Kitajima, J. Tang, R. Saito, and H. Petek, *Nano Lett.* **8**, 3102 (2008).

11. K. Makino, A. Hirano, K. Shiraki, Y. Maeda, and M. Hase, *Phys. Rev. B* **80**, 245428 (2009).

12. Y.-S. Lim, J.-G. Ahn, J.-H. Kim, K.-J. Yee, T. Joo, S.-H. Baik, E. H. Haroz, L. G. Booshehri, and J. Kono, *ACS Nano* **4**, 3222 (2010).

13. J. Wang, M. W. Graham, Y. Ma, G. R. Fleming, and R. A. Kaindl, *Phys. Rev. Lett.* **104**, 177401 (2010).

14. Z. Zhu, J. Crochet, M. S. Arnold, M. C. Hersam, H. Ulbricht, D. Resasco, and T. Hertel, *J. Phys. Chem. C* **111**, 3831 (2007).

15. M. S. Arnold, S. I. Stupp, and M. C. Hersam, *Nano Lett.* **5**, 713 (2005).

16. S. M. Bachilo, L. Balzano, J. E. Herrera, F. Pompeo, D. E. Resasco, and R. B. Weisman, *J. Am. Chem. Soc.* **125**, 11186 (2003).

17. Y. Miyata, K. Yanagi, Y. Maniwa, T. Tanaka, and H. Kataura, *J. Phys. Chem. C* **112**, 15997 (2008).

18. T. Kobayashi, M. Yoshizawa, U. Stamm, M. Taiji, and M. Hasegawa, *J. Opt. Soc. Am. B* **7**, 1558 (1990).

19. A. Baltuŝka, T. Fuji, and T. Kobayashi, *Opt. Lett.* **27**, 306 (2002).

20. T. Kobayashi, Z. Wang, and I. Iwakura, *New J. Phys.* **10**, 065009 (2008).

21. N. Ishii, E. Tokunaga, S. Adachi, T. Kimura, H. Matsuda, and T. Kobayashi, *Phys.Rev. A* **70**, 023811 (2004).

22. T. Kobayashi and Z. Wang, *IEEE J. Quant. Electron.* **44**, 1232 (2008).

23. T. Kobayashi, Z. Nie, J. Du, H. Kataura, Y. Sakakibara, and Y. Miyata, *J. Lumin.* **133**, 157–161 (2011).

9.2.2 Electronic Relaxation and Coherent Phonon Dynamics in Semiconducting Single-Walled Carbon Nanotubes with Several Chiralities

9.2.2.1 INTRODUCTION

Extensive studies have been carried out on carbon nanotubes (CNTs) following their discovery by Iijima in 1991 [1]. Nanostructured carbon materials include fullerenes, peapods, graphene, and CNTs. Single-walled carbon nanotubes (SWNTs) with one-dimensional nanostructures have unique mechanical, electronic, and optical properties [2–4]. In particular, depending on their chirality, they can exhibit either metallic or semiconducting characteristics [5]. In addition, their one-dimensional nature, which is similar to materials such as conjugated polymers [6,7], provides a playground for studying the dynamics of confined electrons and phonons [8–11]. Similar to conjugated polymers, SWNTs exhibit unique vibronic and exciton-phonon couplings [12,13]. Theoretical and experimental studies have revealed a variety of phonon-assisted peaks, suggesting strong exciton-phonon coupling [14–18], which is often at the heart of many important phenomena in condensed matter physics. Although exciton-phonon coupling in SWNTs is usually studied by Raman spectroscopy [19–23], it is only a sensitive probe of ground-state vibrations. Recently, efforts have been made to investigate coherent lattice vibrations (phonons) in SWNTs by coherent phonon (CP) spectroscopy via femtosecond pump-probe techniques, which can also enable direct measurement of time-domain CP dynamics in excited states [24].

There are two methods of studying the vibrational dynamics in condensed matter. One is in the frequency domain, and the other is in the time domain. The former techniques are more common and include time-resolved Raman scattering or infrared absorption. The latter technique is real-time vibrational spectroscopy, which is used to obtain information about the amplitudes of vibration through modulation of the electronic transition probabilities. Information can also be obtained about the vibronic coupling strengths and dynamics of the vibrational modes in both the ground and excited electronic states. In this manner, the relaxation of electronic states and the dynamics of vibrational levels can both be studied using the same experimental equipment under exactly the same conditions. The real-time vibrational spectroscopy has been used for many different molecular, bimolecular, polymer, and biopolymer systems [25–27].

CP dynamics in SWNTs have been studied by several groups using the impulsive excitation method [24,28–36]. However, there are the following three important subjects(Of course s is needed.) that need to be addressed.

1. The first is the mechanism of the CP generation, which is related to coupling between phonon modes and the electronic structure, resulting in modulation of the probed difference absorbance. Using a tunable laser with a 50-fs pulse duration, Lim et al. observed

DOI: 10.1201/9780429196577-49

radial breathing mode (RBM) vibrations [28]. The oscillations in the probe transmittance were found to be the result of ultrafast modulation of the optical constants at a frequency ω_{RBM} due to bandgap (E_{gap}) oscillations coupled with periodical change (modulation) of the SWNT diameter d ($E_{gap} \propto 1/d$). It was claimed that the photon energy dependence of the CP signal shows the first-derivative behavior. Subsequently, Kim et al. [29] and Luer et al. [32] drew similar conclusions in their investigations of chiral SWNTs using other experimental schemes. These papers suggested that the CPs induced by femtosecond visible or near-infrared pulses are in the ground states. However, impulsive excitation can also generate CPs in excited states when the femtosecond pulse is resonant with the electronic transition [24,28–38].

2. The second concern is that there has been no discussion of the effects of modulation of the phase of the light field, as opposed to that of the molecular vibrations, due to the periodic change in the refractive index of the material. Rapid modulation of the refractive index by molecular vibrations can lead to a spectral shift, leading to spectral changes in the probe due to molecular phase modulation (MPM), which is similar in concept to self-phase modulation [39] and cross-phase modulation [40].

3. The third issue is that energy exchange between CPs and the optical field of the probe has not been fully considered. Probe photons with energies higher and lower than some specific spectral component can interact simultaneously with CPs, resulting in a complicated dependence of the CP signal on the probe energy. For example, in the study of Lim et al. in 2006 [28] the probe photon energy dependence of the CP signal was investigated using broad probe bandwidths (25 nm or more), and the spectral shape was compared with the absolute value of the first derivative of the absorption spectrum. Although they concluded that the two spectral shapes were similar, their results showed that the CP amplitude profile was much steeper than the derivative-based profile, explained in terms of a band-gap oscillation that induces a shift in the transition absorption spectrum. The separations between the two inflection points and between the two peaks in the CP amplitude are calculated to be about 40 and 70 meV, respectively, using Figure 9.2.2.5 of the paper [28]. The two inflection points and the two derivative peaks should be coincident if the spectral shift is the origin of the oscillating signal. In fact, the two values differ by a factor of more than 1.7 times, and thus the real-time traces cannot simply be explained by the spectral shift mechanism. In 2009, Kim et al. [29] studied CNTs by selectively exciting CPs for RBMs in SWNTs using a pulse shaping technique and measured the probe photon energy dependence of the amplitudes in the same way as Lim et al.

From Figure 9.2.2.4 in Ref. [29], it can be easily seen that the full width at half-maximum (FWHM) of the photoluminescence excitation profile corresponding to the absorption spectrum is ~55 meV, while that of the reconstructed spectrum calculated from the CP excitation profile is ~125 meV, which is more than twice as large. In the same year, Luer̈ et al. [32] used sub-10-fs visible pulses and a broadband detector to obtain a nearly continuous spectrum of the probe wavelength dependence of the CP amplitude and phase, based on the time-dependent wave-packet theory of Kumar et al. [41,42]. The positions of the low- and high-energy peaks and the valley bottom of the modulation depth were found to be 564, 589, and 575 nm, respectively, compared with values of 562, 579, and 571 nm obtained from simulations, as shown in Figure 9.2.2.5a of their paper [30]. The half-widths of the CP amplitude spectrum and of the first-derivative profiles were estimated to be 25 and 17 nm, respectively, from the data of their paper [30]. Thus, the deviation is as large as almost 50%, which is particularly noticeable if the data are displayed on top of each other in the same graph.

The solution to all of these problems is to clarify the mechanism of CP generation in SWNTs. This will aid in the development of new applications of CNTs in optical devices such as capillary containers for small molecules.

In addition, the bond length change associated with CPs results in photoinduced reactions, such as the opening and fragmentation of the end caps of CNTs [43–45]. Such fragmentation is induced by structural transformations triggered by femtosecond laser excitation [43–45]. Dumitrica et al. [43,44] demonstrated the possibility of selectively opening a nonequilibrium cap of a CNT after femtosecond excitation. Ultrafast bond weakening and simultaneous excitation of two CP modes with different frequencies, localized in the spherical caps and in the cylindrical nanotube body, are responsible for the selective cap opening. The nanotube radius initially increases due to the induced RBM deformation triggered by the exciton-phonon coupling and then decreases followed by the oscillatory behavior. When the radius becomes larger than that of the enlarged spherical cap at the end of the CNT, bond breaking takes place, causing the cap to open. This photoinduced cap opening is a serious problem for CNT applications because of photostability and robustness issues, as well as for applications involving encapsulation and decapsulation of chemicals in CNTs. The initial phase of the coherent vibrations, particularly the vibrations of RBMs, affects the cap opening. The importance of coherent vibrational modes has been discussed for molecular systems in solution or solid materials including CNTs. For solutions, the mechanism of coherent vibration during a chemical reaction was studied for various molecules [46,47].

In the present paper, we report a detailed pump-probe study of CPs in SWNTs using sub-10-fs visible pulses. Although many chiral systems coexist in the sample, we can distinguish four different semiconducting systems because of the large bandwidth of the short pulse, extending from the visible to the near-infrared, and because of our sensitive broadband detection system. Consequently, we obtain abundant information about the probe photon energy dependence of the phonon amplitudes of the RBMs, leading to an understanding of CP generation. The probe photon amplitude profiles are analyzed in terms of the modulation of the excitonic transition probabilities. Since the signal from the pump-probe experiment is generated in a third-order nonlinear (NL) process, the real and imaginary parts of the NL susceptibility can both play roles. The effects of the CPs on the difference absorbance are fully modeled. Furthermore, given the importance of the structural response to the impulsive photoexcitation for chiral-selective end functionalization of SWNTs, the electronic origin of tube structural changes is discussed in terms of the first moment of the electronic transition energy associated with the difference absorption. The effect of the refractive index changes associated with the CP vibrations is clarified.

9.2.2.2 EXPERIMENT

9.2.2.2.1 ULTRAFAST SPECTROSCOPY

The pump and probe beams are generated in a noncollinear parametric amplifier (NOPA) [48,49], excited in turn by a commercial regenerative amplifier (Spectra Physics Spitfire). The central wavelength, pulse duration, output power, and repetition rate of the regenerative amplifier are 800 nm, 50 fs, 740 mW, and 5 kHz, respectively. The output from the NOPA spans the spectral range of 1.71–2.37 eV (524–723 nm). Using a compression system composed of a pair of prisms and chirped mirrors, 7.1-fs pulses with a constant spectral phase, indicating that they are nearly Fourier-transform (FT) limited, are generated. The pulse energies of the pump and probe are 32 and 6 nJ, respectively. The polarization directions of the pump and probe beams are parallel to each other.

In the experiment, the signal is spectrally dispersed using a polychromator (Jasco M25-TP) into 128 photon energies between 1.71 and 2.37 eV. These beams are detected by 128 sets of avalanche photodiodes and lock-in amplifiers with a reference from an optical chopper intersecting the pump pulse at a 2.5 kHz repetition rate. The experiment is performed at room temperature (293 ± 1 K).

9.2.2.2.2 SAMPLE PREPARATION

A CoMoCAT synthesis is performed using a silica support (Sigma-Aldrich SiO_2 with a 6-nm average pore size and a Brunauer-Emmett-Teller [BET] surface area of $480 \, m^2 g^{-1}$) and a bimetallic catalyst prepared from cobalt nitrate and ammonium heptamolybdate precursors [50–52]. The total metallic loading in the catalyst is 2 wt%, with a Co:Mo molar ratio of 1:3. Before exposure to the Co feedstock, the catalyst is heated to 500°C in a flow of gaseous H_2 and further heated to 750°C in flowing He. A Co-disproportionation reaction is used to produce SWNTs in a fluidized bed reactor under a flow of pure Co at five atmospheres. The SWNTs grown by this method remain mixed with the spent catalyst, containing the silica support and the Co and Mo species. To eliminate the silica from the mixture, the solid product is suspended in a stirred 20% HF solution for 3 h at (a is needed. TK) 25°C. The suspension is then filtered through a 0.2-μm polytetrafluoroethylene (PTFE) membrane and washed with deionized water to neutralize it. PTFE was invented by a researcher in Du Pont (a company of chemicals) in the USA. The name of the polymer was registered as "Teflon" in 1941. This polymer is chemically very stable and used in varieties of applications.

Next, the solid product is added to an aqueous solution containing the surfactant sodium dodecylbenzene sulfonate at twice the concentration of its critical micelle level, and it is then ultrasonically agitated for 1 h using a Fisher Scientific Model 550 homogenizer (with a 550 W output). This creates a stable suspension of individual and bundled nanotubes. The suspension is centrifuged for 1 h at 72600 g to separate metallic catalyst particles and suspended tube bundles from the lower-density surfactant-suspended nanotubes. Only a small fraction of the product becomes deposited at the bottom of the centrifuge tube. Finally, the supernatant liquid, enriched in individual surfactant-suspended SWNTs, is withdrawn and adjusted to a pH of between 8 and 9 for spectral analysis.

We have measured the morphologies of SWNTs studied in this work using atomic force microscope (AFM) (SII NanoTechnology, S-image). The average length was 330 nm with a standard deviation of 160 nm, and the heights were about 1 nm, which is close to the diameter of individual SWNT. The average length was slightly shorter than the original length of CoMoCAT written in the specification, but the shortening should not give serious effects on the optical properties of SWNTs. Although the width of each SWNT was not well known due to the limited horizontal resolution of AFM, the low height means that SWNTs are thought to be well dispersed to almost individuals or at least very thin bundles. Details are described in the Supplemental Material [53]. The samples are then used without further fractionation or extraction.

9.2.2.3 RESULTS AND DISCUSSION

9.2.2.3.1 STATIONARY ABSORPTION SPECTRUM

Figure 9.2.2.1a plots the stationary absorption spectrum of the SWNTs with the relevant chirality assigned. There are more than 15 absorption peaks and shoulders in the spectral range of 1.22–2.67 eV. In the energy range of our laser spectrum [1.71–2.36 eV, see Figure 9.2.2.1b], the absorption spectrum has three peaks at 2.17, 1.91, and 1.78 eV, which are assigned to the (6,5), (7,5), and (9,1) chiral systems, respectively. There are also two shoulders near 2.10 and 1.85 eV, which are attributed to the (6,4) and (8,3) chiral systems, respectively. This assignment is made based on the published relationship between the transition energy and chiral index set (n, m) [50,54–58]. The broadband laser spectrum in the visible region, indicated by the dashed line in Figure 9.2.2.1b, is resonant with the second excitonic transitions (E_{22}) of the semiconducting tubes, avoiding any possible excitation of metallic tubes (M_{11}) at higher energies [59] or the first exciton transition (E_{11}) at lower energies. The absorption spectrum is fit to the sum of five Voigt functions (shown in gray), which are convolutions of Gaussian and Lorentzian functions, peaking at 2.17, 2.08, 1.91, 1.84, and 1.78 eV, which correspond to the (6,5), (6,4), (7,5), (8,3), and (9,1) tubes, respectively. The assumption of symmetric Voigt spectral shapes is reasonable because of the much smaller width of the theoretically expected van Hove singularity than that of the observed spectra [60,61]. The spectral shapes and peak photon energies obtained in the simulation are slightly uncertain because of the overlap with neighboring peaks. However, the overall spectral positions are estimated to be accurate to less than 5 meV based on the noise level and simulations. Since the (9,1) tubes exhibit only weak absorption, despite the

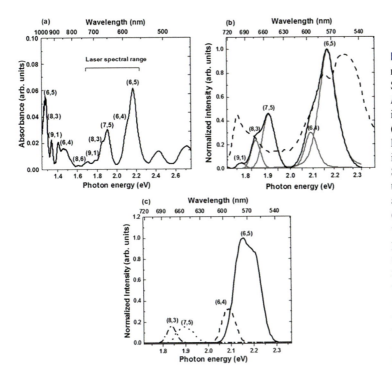

FIGURE 9.2.2.1 (a) Stationary absorption spectrum of SWNTs, showing the E_{11} and E_{22} transitions in visible and infrared (1.2–2.7 eV) range. (b) Laser spectrum (dashed line) and stationary visible absorption spectrum of SWNTs (solid line) after subtracting the weak background and its Voigt-fitting profiles (gray lines), illustrating individual E_{22} absorption components in the 1.71–2.36 eV spectral range. The chirality assignments are shown together. (c) Absorbed laser spectra, defined by the difference in the spectrum between the probe light before and after passing through the sample, for different chiralities.

high laser intensity in this spectral region, the stationary absorption spectrum and pump-probe results focus on the other four chiral systems, namely (6,5), (6,4), (7,5), and (8,3).

Table 9.2.2.1 lists the fitted widths of the Gaussians and Lorentzians used in the Voigt functions. It is assumed that these widths are due to inhomogeneous and homogeneous broadening, respectively. Accordingly, the homogeneous and inhomogeneous widths of each component can be estimated, as listed in Table 9.2.2.1. The four chiral systems have similar inhomogeneous widths of 45–85 meV, while the homogeneous widths vary from 4 to 26 meV, corresponding to a dephasing time of 17–120 fs. The shorter dephasing times are for the broader 26.3 and 22.8 meV homogeneous widths for the (6,5) and (6,4) chiral systems, respectively, which have higher transition energies than the other systems. These shorter dephasing times probably arise from the phase relaxation induced by the interaction between higher vibrational levels that are almost resonant with neighboring chiral systems of lower electronic energy. Given that the lineshape of the laser spectrum can affect the

TABLE 9.2.2.1

Observed Absorption Band Width and Band Center and Fitted Parameter with a Voigt Function

Chirality	Center (eV)	FWHM (meV)	σ (meV)	γ (meV)	F_g (meV)	f_L (meV)	f_v (meV)	L_G (fs)	L_L (fs)
(6,5)	2.17	89.9	36.1	13.1	84.9	26.3	97.4	21.5	17.2
(6,4)	2.09	64.6	19.2	11.4	45.1	22.8	58.1	40.4	19.9
(7,5)	1.91	69.9	27.4	1.88	64.4	3.96	66.7	28.2	114
(8,3)	1.84	49.2	19.9	1.87	46.9	3.76	48.9	38.9	121

σ (meV), Gaussian width; γ (meV), Lorentzian width; f_G (meV), FWHM of the Gaussian profile; F_L (meV), FWHM of the Lorentzian profile; f_v (meV), FWHM of the voigt profile.

amount of laser energy absorbed, the "absorbedphotonenergyspectrum" [25,26] is calculated as shown in Figure 9.2.2.1c. It is defined as the difference in the spectrum of the probe light before and after passing through the sample. The application of the absorbed photon energy spectrum will be discussed later when considering the successful fitting of the experimental results for the dependence of the vibrational amplitude on the probe photon energy, referred to as the probe amplitude spectrum for CPs.

9.2.2.3.2 ELECTRONIC RELAXATION AND THERMALIZATION OF EXCITED POPULATION

Figure 9.2.2.2a graphs the two-dimensional (2D) difference absorption spectra A plotted against the probe photon energy E_{pr} and the delay time t. The striped oscillatory structures parallel to the time axis represent the modulation of the difference absorbance $\delta\Delta A(E_{pr}, t)$ by CPs. Here δ is the modulation of the absorbance change $\Delta A(E_{pr}, t)$ due to CPs at a specific probe photon energy E_{pr} and delay time t induced by the spectral shift and transition probability change, which give rise to the horizontal and vertical modulation, respectively. Figure 9.2.2.2b plots time-resolved spectra integrated over 200-fs delay-time steps at ten center-probe delay times ranging from 50 to 1850 fs. There are four prominent bleaching bands composed of three peaks and one shoulder, nearly coincident with the relevant E_{22} transitions for the (6,5), (6,4), (7,5), and (8,3) chiral systems in Figure 9.2.2.1. In addition, there are three isosbestic points near 2.21, 2.02, and 1.94 eV at delay times longer than 250 fs, as indicated by the small squares. Figure 9.2.2.2c expands the spectra near the isosbestic points over a 0–400 fs range with an integration step of 50 fs. The crossing points between neighboring time-resolved spectra are within ± 0.03 eV of the average photon energies, except between 0 and 50 fs. Therefore, the relaxation after 200 fs can be described using a simple two-state model composed of a single intermediate state or excited state after photoexcitation, where a conversion from one state to the other is occurring. This behavior can be explained in terms of intraband relaxation from the E_{22} excitonic state to the E_{11} excitonic state, followed by a slower decay from E_{11} to the ground state. Moreover, Figure 9.2.2.2a shows that the zero-crossing photon energy rapidly changes just after excitation until a probe delay time of 300 fs, after which the change slows down. The mean rates of change for the crossing points before and after 300 fs are 82 and 7.7 meV/ps, respectively.

Figure 9.2.2.3 displays the electronic decay dynamics of transient absorbance changes probed at ten different probe photon energies. The relaxation can be fit in all cases by the sum of two exponential functions plus a long lifetime component. The two resulting time constants are listed in Table 9.2.2.2. When the probe photons are resonant with the tube absorption, the population signal exhibits an initially rapid partial recovery of the photobleaching and photoinduced absorption (τ_1 ranging from 50 to 90 fs), followed by slower recovery and growth processes (τ_2 ranging from 600 to 850 fs) akin to previous results ($\tau_1 < 100$ fs and $\tau_2 \approx 1$ ps) obtained using longer pulses for less dispersed samples [24,32]. Both decay times are shorter for (6,5) and (6,4) than for (7,5) and (8,3) chiral systems, similar to the reduction of the homogeneous dephasing times discussed above.

The decrease in the absolute value of A can either be due to bleaching recovery or to the growth of induced absorption. Positive and negative absorbance changes have been previously observed in two-color pump-probe experiments by Lauret et al. [62] The relaxation that occurred over a long period of 1 ps was fit using a power law-decay $t^{-0.5}$. This was explained in terms of a random walk along a one-dimensional CNT chain followed by disappearance due to trapping by defects [36]. However, such a random-walk description is only valid after a few instances of hopping.

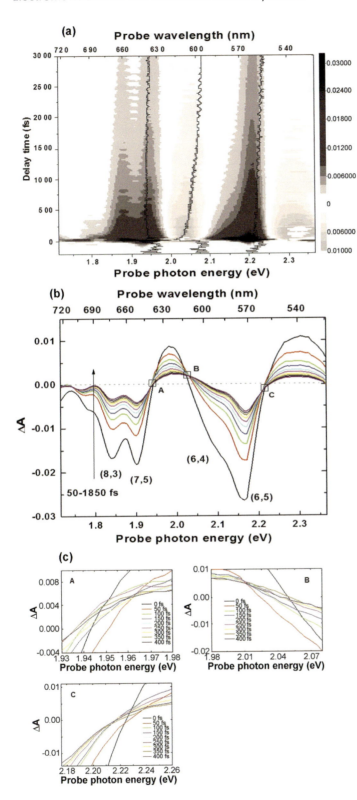

FIGURE 9.2.2.2 (a) 2D display of $\Delta A(E_{pr}, t)$ as functions of probe photon energy, E_{pr}, and probe delay time, t. The black solid curves are the zero-change lines in the absorbance ($\Delta A = 0$). (b) Time-resolved A spectra at the delay time points from 50 to 1850 fs with a 200-fs integration step. (c) The enlargements of the three isosbestic point regions, shown in the three insets of (b), at around (a) 2.21 eV, (b) 2.03 eV, and (c) 1.93 eV with a 50-fs integration step.

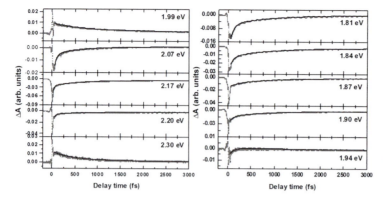

FIGURE 9.2.2.3 Electronic decay dynamics of transient absorbance change, probed at 10 probe photon energies (gray dotted lines). The electronic relaxations are expressed by two-order exponential functions plus very long lifetimes (black solid lines). The relevant fitting parameters are shown in Table 9.2.2.3.

TABLE 9.2.2.2

Fitting Parameters with the Two-Exponential Function for the Time Traces Displayed in Figure 9.2.2.3

Hv_{Probe} (eV)	Chirality	a_1	τ_1 (fs)	a_2	τ_2 (fs)	c
1.81	(8,3)	−0.017	70.7	−0.006	838	−7E-5
1.84	(7,5)	−0.016	91.2	−0.010	877	−0.002
1.90	(7,5)	−0.014	87.0	−0.012	730	−0.003
2.07	(6,4)	−0.023	53.4	−0.006	602	8E-4
2.17	(6,5)	−0.017	96.5	−0.016	796	−0.005
2.20	(6,5)	−0.012	62.8	−0.003	730	−0.002

9.2.2.3.3 FT SPECTRA AND CHIRALITY ASSIGNMENTS

Eight typical real-time traces in the photon energy range of 1.77–2.21 eV are displayed in Figure 9.2.2.4a. Figure 9.2.2.4b plots the corresponding FT spectra. The calculations are performed on the data after 50 fs to avoid interference near zero time delay. The FT plots show two vibrational modes due to RBMs at ~300 cm^{-1} (~100 fs) and to G modes at 1587 cm^{-1} (21 fs) [24,31,32], generated by the impulsive excitation for a pulse duration (less than 10 fs) much shorter than all of the vibrational periods. Other vibrational modes are too weak to be resolved.

A 2D map of the Fourier power plotted against the probe photon energy and the vibrational frequency is presented in Figure 9.2.2.5. Four RBMs are evident with vibrational frequencies of about 337, 310, 301, and 282 cm^{-1}. The assignment of chirality using only the electronic absorption or Raman spectra is often ambiguous due to spectral congestion [54–58]. In contrast, the intensity of the four dominant double-peaked structures indicated by the double-headed arrows in Figure 9.2.2.5 corresponds to the first derivative of the electronic absorption spectrum for the different types of tubes [28,29,32]. The central dip in each structure represents the E_{22} transition since the oscillation is minimal at resonance [32]. However, as will be discussed later, some corrections must be made to this assertion.

The Raman shifts for different RBMs theoretically correspond to the vibrational frequencies of the FT of their CP profiles [54,55]. Consequently, the dips at the intersections of the horizontal and vertical lines in Figure 9.2.2.5 correspond to the relevant vibrational frequency and electronic resonance transition energy. Using this relationship, the chirality can be assigned for the (6,4), (6,5), (7,5), and (8,3) systems, as indicated in Figure 9.2.2.5, because only these tubes can simultaneously fulfill these two conditions. Other kinds of tubes may have weak absorption lines in the laser spectral range, but they cannot be efficiently excited. Using broadband high-sensitivity multichannel

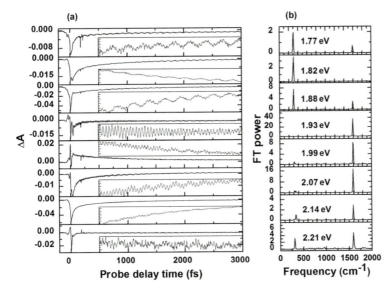

FIGURE 9.2.2.4 (a) Real-time traces of the absorbance changes from −200 to 3000 fs at eight typical probe photon energies and (b) their corresponding FT power spectra. The insets in (a) are the enlarged real-time traces in 500–1500 fs range showing the pronounced vibrations.

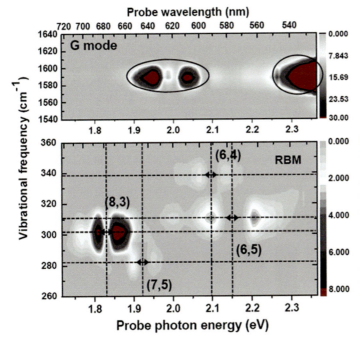

FIGURE 9.2.2.5 2D CP spectra in the spectral range of 1.71–2.36 eV. The chirality assignments of RBMs are shown together. In the bottom panel, the dotted criss-crossing lines show the relevant vibrational frequencies and resonance energies corresponding to RBMs. The two-way arrows indicate the double-peak structure in the amplitude profile of RBMs. The circles in the top panel show the main features observed in the G mode.

lock-in detectors, the power spectra of the four systems can be uniquely distinguished for RBMs, even though their absorption spectra overlap as shown in Figure 9.2.2.1, without the need for time-consuming or complex pulse-shaping techniques [29]. Therefore, this method is advantageous for the simultaneous analysis of a sample containing many chiral systems. However, the dip position and the peak of the absorption spectrum do not exactly coincide, as will be discussed later. In contrast, for G-mode vibrations, the amplitude profiles for different chiralities overlap and cannot be distinguished. The frequency of the axial G-mode is insensitive to the diameter and chirality of SWNTs.

9.2.2.3.4 CP AMPLITUDES OF CHIRAL SYSTEMS

The CP amplitude $Q(t)$ of CNTs induced impulsively by ultrashort optical pulses is approximately represented by

$$Q(t) = Q_e(t) + Q_v(t) \cos(\omega_v t + \varphi) \tag{9.2.2.1}$$

where

$$Q_e(t) = Q_{ex} - Q_{gr} \quad \text{for} \quad t > 0, \text{ and}$$
$$Q_e(t) = 0 \qquad \text{for} \quad t < 0 \tag{9.2.2.2}$$

Here $Q_e(t)$ is the difference in the equilibrium position of the phonon coordinate between the electronic ground state Q_{gr} and the exciton state Q_{ex}, and it has a short rising part due to the finite pump pulse width. The second term on the right-hand side of Eq. (9.2.2.1) is the product of the envelope vibrational amplitude $Q_v(t)$ and the oscillation with a frequency ω_v and an initial phase φ. The time dependence of $Q_v(t)$ is due to homogeneous and inhomogeneous vibrational dephasing. Figure 19 in the article by Sanders et al. [30] shows that $0 > Q_e(t)$ and $|Q_e(t)| > Q_v > 0$ for positive times t. The phases were predicted such that $\varphi = 0$ for E_{11} transitions in mod 2 CNTs and for E_{22} transitions in mod 1 CNTs, and $\varphi = \pi$ for E_{11} transitions in mod 1 CNTs and for E_{22} transitions in mod 2 CNTs.

The four E_{22} CP intensities $I_{(n,m)}$ are plotted in terms of their chirality in Figure 9.2.2.6. Here, the CP intensities are normalized by the absorbed light intensity, given by the overlap of the laser spectrum and the corresponding Voigt components for the chiral systems in Figure 9.2.2.1b. The intensities $I_{(n,m)}$ of the (n, m) systems are in the order $I_{(8,3)} > I_{(7,5)} > I_{(6,4)} > I_{(6,5)}$. This trend is consistent with that predicted by a new microscopic theory describing the generation and detection of CPs in SWNTs, developed by Sanders et al. [30]. For the RBM modes, they argue that (i) the E_{22} CP intensity decreases with increasing chiral angle $\theta_{(n,m)}$ for a given value of $(2n+m)$; (ii) as the family index increases, the E_{22} CP intensity increases if $(n-m)$ mod 3 equals 1 or 2; and (iii) the CP intensities of mod 1 tubes are considerably weaker than those of mod 2 tubes. In our case, $I_{(8,3)}$ is expected to be more intense than $I_{(7,5)}$ because $\theta_{(8,3)} < \theta_{(7,5)}$. Both of these tubes have the

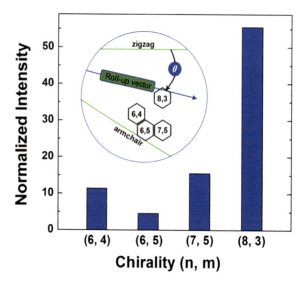

FIGURE 9.2.2.6 Comparison of the four E_{22} CP intensities between different chiralities (n, m) considering their chirality difference. The inset shows the position of the tubes studied in a chirality map.

same family index of $2n+m=19$. Second, tubes (6,4), (7,5), and (8,3) are all mod 2, and $I_{(6,4)}$ is the smallest among the three tubes because the family index of 16 for (6,4) is smaller than that for (7,5) and (8,3). Moreover, according to the new theory, since tube (6,5) is in the mod 1 group and the other three tubes belong to the mod 2 group, $I_{(6,5)}$ should be the smallest among the four tube types.

9.2.2.3.5 RAMAN PROCESSES IN A CLASSICAL MODEL

To understand the origin of the oscillatory signal in the real-time traces, we first consider a classical model involving coupled equations describing CPs for two pulsed beams at different frequencies. When an ultrashort pulse with a broad spectrum is used for pumping, two sets of electric fields are involved: E_L and E_S, and E_L and E_{AS}. The difference between ω_L and ω_S corresponds to the molecular vibrational frequency ω_v. The stimulated Raman process is described by the following set of coupled differential equations in terms of the normal coordinate Q of the molecular vibration [63]

$$\frac{\partial}{\partial x'} E_S = k_{IS} E_L Q^* \tag{9.2.2.3}$$

$$\frac{\partial}{\partial x'} E_L = k_{IS} E_S Q + k_{1A} E_{AS} Q^* \tag{9.2.2.4}$$

$$\frac{\partial}{\partial x'} E_{AS} = k_{1A} E_L Q \tag{9.2.2.5}$$

$$\left(\frac{\partial}{\partial t'} + \frac{1}{T_{2v}} \right) Q = k_{2S} E_L E_S^* + k_{2A} E_{AS} E_L^* \tag{9.2.2.6}$$

Here, T_{2v} is the vibrational phase (transverse) relaxation time including inhomogeneity. The constants κ_{1A} and κ_{1S} are the coupling strengths of the laser and the Stokes field with CPs, respectively, while the constants κ_{2A} and κ_{2S} are those between the laser and the anti-Stokes field and between the laser and the Stokes field, respectively. In these equations, it is assumed that there are no spectrally overlapped components at $\omega_L-\omega_v$ nor at $\omega_L+\omega_v$, which correspond to the downshifted frequency of the Stokes field E_S and to the upshifted anti-Stokes field E_{AS}, respectively. Equations (9.2.2.3)–(9.2.2.5) determine the spatial variation of the fields E_A, E_L, and E_{AS} and of the coherent vibrational amplitude Q in a sample along the propagation direction of the laser beam. Equation (9.2.2.6) describes the dynamics of the phonons driven by the pair of fields E_L and E_S and by the pair E_L and E_{AS}. After the generation of a CP by a pump pulse, the probe pulse interacts with the coherent vibration. In this process, the coherent vibration with an amplitude Q interacts with the fields E_L and E_S (or E_{AS}), as expressed in Eqs. (9.2.2.3)–(9.2.2.5). Thus, these equations allow the spectral dynamics in the time-resolved spectrum to be clearly understood. However, they do not consider the electronic and molecular vibrational energy structure. To take the vibrational energy of a CP into account, a coherent vibration can be modeled as a harmonic oscillator. Furthermore, quantization of the CP coupled exciton needs to be considered. In the following, the quantization of electronic and vibrational excitations is described using a semiclassical model in which the optical field remains classical. This model is appropriate because the pump laser field is strong and the effect of the vacuum of the phonon field inducing spontaneous phonon emission by the Raman scattering process can be neglected. The probe is much weaker than the pump, but it is still strong enough for the signal to neglect any detectable spontaneous Raman or electronic emissions, which are emitted over a wide solid angle of 4π steradians and have a dipolar angular dependence.

9.2.2.3.6 RAMAN AND RAMAN-LIKE PROCESSES IN A SEMICLASSICAL MODEL

The laser field is treated classically, whereas the CNT is modeled quantum mechanically using a diagram scheme. The wave packet can be generated either by simultaneous coherent excitation of the vibronic polarization of several vibrational levels $|e_{2v}>$ in the electronic or excitonic excited states or by impulsively stimulated Raman scattering in the ground state. The former involves a split-excited state, which can be represented in terms of the V-type interaction. Analogously, the latter case requires a split-ground state, expressed as the Λ-type interaction discussed later. Since more than one vibrational level can be excited, at least two levels in both the ground and excited states must be taken into account. As displayed in Figure 9.2.2.7b, we include three levels in the ground state, $|g_0>$, $|g_v>$, and $|g_{2v}>$ with energies E_0, E_0+E_v, and E_0+2E_v, respectively, and two levels in the excited state, $|e_0>$ and $|e_v>$ with energies E_e and E_e+E_v, respectively. The third electronic state, $|e_{2v}>$, which is not necessarily the second excited state but could instead be a higher excited state, E_0+2E_v, respectively, and two levels in the excited state, and with energies E_e and E_e+E_v, respectively. The third electronic state, which is not necessarily the second excited state but could instead be a higher excited state, is located near an energy of $2E_e$. In the V-type interaction, two light fields with photon energies of E_e and E_e+E_v interact with the vibrational coherent polarization in the excited states, whereas in the -type interaction, two fields with photon energies of E_0 and E_0+E_v interact with the vibrational polarization in the ground states.

Figure 9.2.2.8 shows the relevant double-sided Feynman diagrams for NL processes in a four-level system, corresponding to the processes induced by ground state vibrational coherence due to a stimulated Raman process. The two fundamental NL processes A and B are depicted in Eqs. (9.2.2.1) and (9.2.2.2). Processes A and A describe the amplitude distributions at low and high energies, respectively, whereas processes B and B correspond to vibrational signals at intermediate energy. Diagrams (1), (3), (5), (7), (9), and (12) are due to the NL process A/A', while the others are due to B/B'. Feynman diagrams (3)–(6) and (9)–(12) involve a vibrational quantum number of 2.

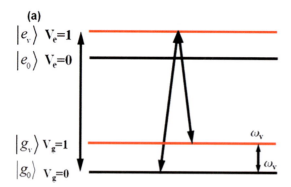

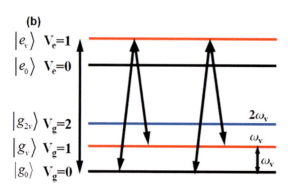

FIGURE 9.2.2.7 (a) Three-level systems and parameters interacting in the Λ-type configurations. (b) Five-level systems and parameters interacting in the Λ-type configuration. $|g_o\rangle$ and $|e_o\rangle$ are the ground electronic state and the lowest electronic excited state with energy E_0 and E_e, respectively. $|g_v\rangle$ and $|g_{2v}\rangle$ are the vibrational levels of quantum number v and $2v$ in the ground electronic states with energy E_0+E_v and E_0+2E_v, respectively. $|e_v\rangle$ is the vibrational level of quantum number v in the electronic excited state with energy E_e+E_v.

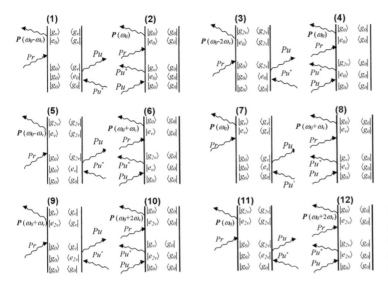

FIGURE 9.2.2.8 Double-sided Feynman diagrams of the two nonlinear processes corresponding to ground state bleaching.

Before discussing the origins of CPs in SWNTs in detail, we first consider the mechanism for the appearance of the signals. According to the 2D FT power spectrum in Figure 9.2.2.5, the power of RBMs is weak at photon energies corresponding to the center of the "absorbed spectrum" for the different chiral systems. Although the exact chirality is difficult to assign for G-mode vibrations, the vibrational spectra have two main components, which are outlined in Figure 9.2.2.5 with a circular ring (upper panel). They display a similar probe photon energy dependence to that previously reported for a highly purified sample CNT sample with a (6,5) chirality [32]. Here, as discussed previously in Refs. [26,27], a similar method is used to analyze the ground state NL processes A and B. The phase relations are

$$\phi_{A'}(\omega_v,\omega) = \arctan\left(\frac{-\omega + \omega_0 + \omega_v}{\gamma^2}\right) \tag{9.2.2.7}$$

$$\phi_{B'}(\omega_v,\omega) = \arctan\left(\frac{\omega - \omega_0}{\gamma_2}\right) \tag{9.2.2.8}$$

$$\phi_{A}(\omega_v,\omega) = \arctan\left(\frac{\omega - \omega_0 + \omega_v}{\gamma_2}\right) \tag{9.2.2.9}$$

$$\phi_{B}(\omega_v,\omega) = \arctan\left(\frac{-\omega + \omega_0}{\gamma_2}\right) \tag{9.2.2.10}$$

The frequencies ω_0 and ω_v correspond to the electronic 0–0 transition and vibrational frequency of the system, respectively. As expressed in Eqs. (9.2.2.8) and (9.2.2.10), the phases of processes B and B are opposite, which suggests that the intensity of NL process B will decrease if the intensity of process B increases. Since both lie in the same spectral range, the result of this antiphase relation is that the vibrational spectra, due to these two processes, will cancel each other. A similar phenomenon has been observed for impulsively excited coherent vibrations in chlorophyll [64].

Consequently, the vibrational spectra vanish at the center of the "absorbed spectrum," where intense resonant third-order NL interactions are expected to occur. The degree of cancellation depends on the intensity distribution of the probe spectrum since the modulation varies with the probe spectral intensities at frequencies $\omega_0 - \omega_v$ and $\omega_0 + \omega_v$. If these two intensities are the same, they will exactly cancel each other. When the probe spectrum is broad enough, energy will be exchanged. The modulation at the probe frequency $\omega_0 - \omega_v$ driven by the resonantly coupled spectral components can trigger oscillations near $\omega_0 - 2\omega_v$. Likewise, oscillations can be induced at $\omega_0 + 2\omega_v$. Higher-order spectral variations can also occur through cascade processes. However, modulations at $\omega_0 \pm n\omega_v$ are not due to high-order molecule-field interactions induced by potential anharmonicities such as overtones. (Here we refer to the CNT system as a molecule for simplicity.) On the contrary, the phenomenon arises due to consecutive interactions between the probe and field at different photon energies, even for a harmonic molecular potential. This analysis assumes coherent vibrations in the ground states. Since the interband relaxation occurs with a time constant on the order of 40 fs, the vibrational wave packet contributing to the modulation of the electronic state is attributed to the ground state [65].

9.2.2.3.7 PROBE PHOTON ENERGY DEPENDENCE OF THE VIBRATIONAL AMPLITUDES

The probe photon energy dependence of the vibrational amplitude profiles (black lines) of RBMs for the three chiralities (6,4), (6,5), and (8,3) is plotted in Figure 9.2.2.9a–c. The amplitude profile for (7,5) is not shown since it is not strong enough for detailed analysis. Several groups have studied RBM amplitude profiles [28,29,32]. In their work, the probe photon energy dependence of the amplitudes was compared with the first derivative of the absorption spectra. It was argued that the resemblance of the amplitude profiles to the first derivative implies wave-packet motion. This type of analysis is hereafter referred to as a DER-type analysis. However, the physical mechanism does not support that conclusion.

Panels (2) in Figure 9.2.2.9 show a fit of the probe amplitude profiles with the first derivatives of the relevant stationary absorption components (gray lines) for different chiralities. The absorption for each chiral system is obtained from a spectral deconvolution of the stationary absorption spectrum using the Voigt function in Figure 9.2.2.1b. The double-peak structure associated with a

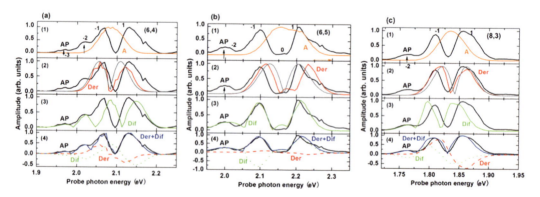

FIGURE 9.2.2.9   Probe photon energy dependence of normalized RBM amplitude profiles (AP, black lines) for (a) (6,4), (b) (6,5), and (c) (8,3) tubes, fitted with the absorbed laser spectra (A, orange lines), first-derivative of the stationary absorptions (gray lines, panels 2) and absorbed laser spectra (DER, red lines); the absolute difference between the absorbed photon energy distribution by the sample and the distribution shifted by the relevant RBM frequency (DIF, green lines); and the sum of DER and DIF (DER+DIF, blue lines) with relevant contributions, respectively. In the top panels, the arrows indicate the side bands, and the numbers represent the peak number in amplitude profiles. The original DER (dashed red lines) and DIF (dotted green lines) lines in the bottom panels are plotted together before taking absolute value to show their corresponding contributions to the vibrational amplitudes. The amplitude profile of (7,5) is not shown here since it is very weak.

first-derivative dependence [28,29,32] does not produce a good fit. Even after modifying the absorption components to take into account the errors in fitting the stationary absorption with the analytic function, the disagreement between the calculated derivative function and the probe photon energy dependence remains large. There are substantial deviations in the positions of the valley in the spectra, and even the line shapes of the absolute values of the derivatives and the vibrational amplitudes do not match each other.

It is assumed that no absorption saturation is introduced by the pump laser, as verified by the linear dependence of the signal on the pump intensity. The NL macroscopic polarization $\tilde{P}_A^{(3)}(\omega,\tau)$ is induced by three fields, two from the pump pulse $E_{pu}(\omega, \tau)$ and one from probe pulse $E_{pr}(\omega, \tau)$. It can be represented by the following equation as a function of the probe optical frequency ω and the pump-probe delay time τ [26,27]

$$\tilde{P}_A^{(3)}(\omega,\tau) = \chi_i(\omega)\tilde{E}_{pu}^*\tilde{E}_{pr}\alpha I(\omega_0)\tilde{E}_{pr}I_{pu}(\omega) \qquad (9.2.2.11)$$

The polarization is generated by a Raman gain/loss process associated with energy exchange between the coherent vibrations and the probe optical field, according to the relation $\Delta A(\omega, \tau) \propto -\text{Im}[P^{(3)}(\omega, \tau)/E_{probe}(\omega, \tau)]$ [26,27]. In this process, the probe optical field is alternately deamplified and amplified, depending on the phase change of the vibrations. Thus, the probe photon dependence of the FT power is dependent on the spectral distribution of the laser photons absorbed by the sample. In that case,

$$\Delta A_A(\omega,t) \propto \int d(\omega_0 - \omega_v)\exp(-t/\tau_{2v}) \times \cos(\omega_v t + \phi_A)I(\omega_0)L_{pu}(\omega) \cong \delta(\omega)\cos(\omega_v t + \phi_A)$$

$$(9.2.2.12)$$

We next consider a phenomenological analysis. Since the pump laser is resonant with several chiral species, we first separate the spectra associated with each of these species. For quantitative discussion, the phenomenological description of the probe photon energy dependence $\Delta A(\omega_{probe})$ of the vibrational amplitude can be expressed as [66]

$$\Delta A(\omega_{probe} = \omega_{AS}) = C_{1AS}a(\omega_{probe}) - a(\omega_{probe} - \omega_v)| \qquad (9.2.2.13)$$

$$\Delta A(\omega_{probe} = \omega_{AS}) = C_{1AS}a(\omega_{probe}) - a(\omega_{probe} - \omega_v)| \qquad (9.2.2.14)$$

$$a(\omega_{probe}) = L(\omega_{probe})(1 - 10 - A_{\omega probe}) \qquad (9.2.2.15)$$

Here C_{1S} and C_{1AS} are proportionality constants, ω_{probe} is the frequency of the probe, ω_v is the molecular vibration frequency, $L(\omega)$ is the laser spectrum, $a(\omega)$ is the absorbed laser spectrum (which is the frequency distribution of photons being absorbed by the sample, shown by orange lines in panels (1) of Figure 9.2.2.9), and $A(\omega)$ is the absorbance of the sample at a frequency ω. Equations (9.2.2.14) and (9.2.2.15) correspond to pump/Stokes and pump/anti-Stokes interactions, respectively.

The above calculation is based on the assumption that the imaginary part of the third-order susceptibility $\chi_i^{(3)}$ corresponding to the Raman interaction can be written as

$$\chi_i^{(3)}(-\omega_2 : \omega_1, -\omega_1, \omega_2) = C_2(a(\omega_2) - a(\omega_1 = \omega_2 \pm \omega_v)) \qquad (9.2.2.16)$$

Here C_2 is a proportionality constant, the plus and minus signs correspond to the cases of pump/Stokes and pump/anti-Stokes interactions, respectively, and ω_1 and ω_2 are the components of the probe spectrum. The sideband peaks should occur at frequencies corresponding to the difference between the vibrational frequencies [25–27].

The energy difference between the sidebands, as numbered in panels (1) of Figure 9.2.2.9a, is always found to be close to an integer multiple of the relative RBM frequencies. This observation suggests the presence of overtones in the Raman interactions between the probe pulse and the coherent vibrations. However, these overtones are not induced by anharmonicity but are instead due to cascaded Raman processes, as discussed below. Because the spectral distribution of the pump and probe lasers is broad, energy exchange can take place between coherent lattice or molecular vibrations and the probe optical field. The interaction can be between the first Stokes beam and the probe beam (the so-called laser used in the discussion of Raman interaction) or between the first Stokes beam and the second (or higher) Stokes beams via the coherent vibrations.

As a result, the probe photon energy dependence can be compared with the absolute value of the difference between the absorbed photon energy spectrum and the distribution shifted by the vibrational frequency, as expressed in Eq. (9.2.2.16). Hereafter, this analysis is referred to as a DIF-type analysis. Fitting of the amplitude profiles with the difference between the shifted and unshifted absorbed probe energy distributions were carried out for Raman ground-state and Raman-like excited-state interactions. They included higher order contributions with adjustable parameters (due, for example, to the phonon amplitudes). The results are plotted in panels (3) of Figure 9.2.2.9, corresponding to RBMs for (6,4), (6,5), and (8,3) tubes. It can be seen that the fits are still not good.

9.2.2.3.8 FITTING THE AMPLITUDE SPECTRUM WITH CONTRIBUTIONS FROM THE REAL AND IMAGINARY PARTS OF THE THIRD-ORDER SUSCEPTIBILITY

In this section, the fit to the probe photon energy dependence of the vibrational amplitude is improved by combining the contributions from the DIF- and DER-type analyses. In previous papers [24,28–36], based on the DER analysis, the modulation was explained in terms of a sinusoidal electronic energy modulation due to an RBM-induced change in the diameter of a CNT [24,28–36]. In the following, the mechanism involved in the DER-type contribution is more fully discussed.

A change in the refractive index is induced by the MPM process due to the DER contribution to the modulation [67]. This MPM produces a periodic shift in the probe spectrum. This, in turn, modulates the time-resolved spectrum composed of ground-state absorption bleaching, stimulated emission, and induced absorption. This effect results from the change in refractive index caused by the deformation of the lattice and molecular structure during the coherent vibrations, which the electronic distribution instantaneously follows [26]. The index change introduces a modulation of the probe frequency because of the change in the phase of the probe field, whose time derivative is the optical frequency. Consequently, the signal can be approximated as a spectral shift of the probe pulse induced by a kind of "cross-phase modulation (XPM)" [40], and the probe energy dependence of the phonon amplitude follows the first derivative of the electronic resonance.

By adding the contributions from the DIF and DER processes, excellent fits are finally obtained, as shown in panels (4) of Figure 9.2.2.9. They show the absolute value of the sum (blue lines) of the contributions of the two types of mechanisms for different RBMs with an adjustable relative contribution used as the fitting parameter for DER (dashed lines) and DIF (dotted lines) analyses, before taking absolute values of the sum. We adjusted the relative intensity of different order of phonon peaks taking into account of the differences in the signs of amplitudes and add them to the DER profile to make the fitting line shape match the amplitude profile. At first we start with an apparently reasonable ratio to fit with the experimental results of the vibrational amplitude profile. After this, we take the sum (over probe photon energy data points) of the squared values of the difference between the fitted and observed values. Then tried to minimize the value by changing the ratios by small amount one after the other to read the minimum deviation. Finally, we take the absolute

TABLE 9.2.2.3

The Corresponding Parameters Used for the Fitting of the RBM Amplitude Profiles, Including Absorption Band Center (cm⁻¹), Shift Amounts (cm⁻¹) for first-Derivative Fitting, Absorption Band Width (cm⁻¹), FD-Type and D-Type Contributions to the Amplitudes (%), and the Stokes (−) and Anti-Stokes (+) Order of Relevant Peaks in Amplitude Profiles

Chirality (n,m)	DER-Type Contribution (%)	Band Center (cm⁻¹)	Shift Amount (cm⁻¹)	Band Width (cm⁻¹)	Shift/ Width	DIF-Type Contribution (%)	Peak Number and Its Stokes (−) and Anti-Stokes (+) Order			
(6.4)	~45%	16840	~30	455.78	0.066	~55%	+1	−1	−2	−3
(6.5)	~10%	17510	~29	801.92	0.036	~90%	1	0	−1	−2
(8.3)	~58%	14821	~22	327.92	0.067	~42%	1	−1	−2	/

The zero-order stokes peaks represent the absorbed laser beam.

value. The fitting profiles of DER and DIF types are already shown in panel (4) of Figure 9.2.2.9 before taking absolute values. For the weight of the DIF and the DER, the fraction (percent) of the contributions from DIF and DER are listed in Table 9.2.2.3.

In the cases of the (6,4) and (6,5) tubes, the contributions from the DER analysis are smaller than those from the DIF analysis. In particular, for the (6,5) tubes, the DIF contribution is 90%. The size of the contribution is reversed for the (8,3) tubes, for which more than half is due to the DIF process. The reason for the higher DIF contribution for the (8,3) tubes is the narrower spectral bandwidth of this system compared with the others. The FWHM of the stationary spectra for the (6,4), (6,5), and (8,3) tubes is found in Figure 9.2.2.1 to be 64.6, 89.9, and 49.2 meV, respectively, as listed in Table 9.2.2.1. The order of the widths matches the DIF contributions. It can be concluded that both the DIF and DER analyses make significant contributions to the probe photon energy dependence of the absorbance change, in contrast to previous reports [28,29,32]. The physical mechanisms for the DER and DIF contributions are respectively the real and imaginary parts of the third-order susceptibility. Table 9.2.2.3 lists the contributing Stokes and anti-Stokes bands for three different chiral systems. Because of the non-uniform probe spectral distribution, the (6,5) and (8,3) tubes show only anti-Stokes sidebands up to second order, while the (6,4) tubes exhibit both Stokes and anti-Stokes bands. The relative sizes of the DER contributions are further discussed below.

From the fitted probe photon energy distributions of the DIF analysis, the Huang-Rhys (HR) factors could be obtained. Since the amplitude of the 0–0 transition energy is reduced by the interference of the Stokes and anti-Stokes interactions, a correct value may not be obtained for the (6,5) case. Instead, the intensity ratios between 0 and 1 and between 1 and 2 were used for the determination of the HR factors for the (6,4) and (6,5) systems, respectively. In the case of the (8,3) tubes, the distributions over 0, 1, and 2 were all used. The HR factors were found to be 0.26 ± 0.05, 0.32 ± 0.05, and 0.75 ± 0.04 for (6,4), (6,5), and (8,3), respectively.

The trends of HR factors can be explained by the stiffness difference of the three SWNT structures. As illustrated in Figure 9.2.2.10, the expansion in the armchair structure is associated with C–C bond expansion, whereas in the zigzag structure it is associated with C–C–C bending, which is easier than the stretching mode, resulting in a large HR factor. Based on the preceding discussion, the HR factors are predicted to be in the order $(6,5) < (6,4) < (8,3)$ from Figure 9.2.2.6, since (8,3) is the closest to a zigzag structure among the three, while (6,5) is closest and (6,4) is second closest to an armchair structure. The order of the (6,4) and (6.5) tubes is reversed in this prediction, but the numerical discrepancy is small.

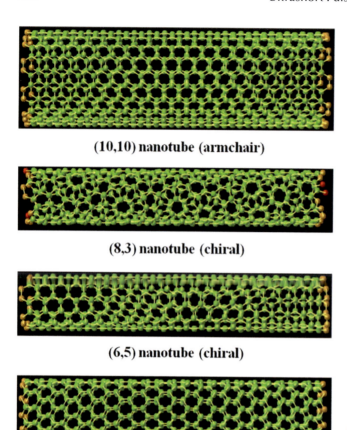

(10,10) nanotube (armchair)

(8,3) nanotube (chiral)

(6,5) nanotube (chiral)

(0,10) nanotube (zig-zag)

FIGURE 9.2.2.10 Side views of CNTs with zigzag, chiral, and armchair structures.

9.2.2.3.9 SIZE AND MEANING OF THE CONTRIBUTION FROM THE REAL PART OF THE THIRD-ORDER SUSCEPTIBILITY

The probe photon energy dependence of the vibrational amplitude exhibits substantial DER contributions. The derivative is the result of a small shift in the spectrum. The size of the shift can be calculated from the derivative with respect to the energy of the original function. However, as shown in Figure 9.2.2.9 and in the data presented in Refs. [28,29], the fits are poor. Therefore, they cannot be reliably used to evaluate the amount of shift. Instead, the shift can be more accurately determined by the moment method. The first moment is defined by

$$M_1(t) = \int_{w1}^{w2} w\Delta A(w)dw \Big/ \int_{w1}^{w2} \Delta A(w)dw \qquad (9.2.2.17)$$

Here, ω is the probe frequency and $\Delta A(\omega))$ is the difference absorption spectrum due to induced absorption, stimulated emission, and bleaching.

There are three advantages to this moment method compared with the derivative method. The first is that the moment calculation is insensitive to the noise in the signal while the derivative calculation is very sensitive to it. As a result, there is no ambiguity arising from the derivative fitting to the energy dependence of the vibrational amplitude. The second advantage is due to the separation between the apparent transition energy and transition amplitude modulations. The third is the elimination of the contributions of the Raman and Raman-like interactions from wave packets in the ground and excited states. They do not contribute to the first moment if an appropriate integration range is selected, because the Raman contributions are determined by the zeroth moment [66].

The time-dependent modulation of the difference absorption due to the MPM of each tube can be evaluated by performing a first-moment calculation on the corresponding absorption band in the A spectrum as a function of sampling time. Figure 9.2.2.11a graphs $M_1(t)$ time traces for four spectral components corresponding to the absorption associated with different chiralities. In these time traces, the slow dynamics have been subtracted to remove the slowly varying contributions from electronic relaxation. After performing a fast FT (FFT), as displayed in Figure 9.2.2.11b, two dominant vibrational modes appear in the RBM and G-modes. The vibrational frequencies for the two modes are in good agreement with those in the 2D display of the CP spectra in Figure 9.2.2.5; no signal overlap from different chiralities can be observed for RBMs in each time trace. Although there are weak beat patterns in the traces, indicating the presence of a slight superposition of RBM vibrations with different frequencies but similar decay times, tubes of specific chirality dominate the contribution to the first-moment signals.

The results initially appear to indicate that the energy of the E_{22} transition periodically oscillates at the RBM and G-mode frequencies [28,29]. However, this spectral shift, or equivalently the spectral dependence of the DER amplitude, is due to the refractive index modulation by the molecular vibrations. Molecular vibrational modes, such as the RBM, induce changes in the size of the coefficients of the vibrational wave functions with only a few quantum numbers, but represent changes in the vibrational wave function coupled to the electronic wave function so that the coefficient changes with time. The dynamical process may be stated such that the CNT is a linear combination of several virtual tubes with different diameters proportional to the square root of the vibrational quantum number, $\sqrt{v}$. The vibrational quantum numbers have coefficients determined by the Franck-Condon factors, where the laser spectrum corresponds to the energy $E_{22} + v\hbar\omega_{vib}$ with electronic energy E_{22} and vibrational energy $v\hbar\omega_{vib}$, where $v = 0$, 1, and 2. Thus, the diameter does not change continuously as proposed in previous papers [28,29] but instead varies discretely as a linear combination of several different diameters.

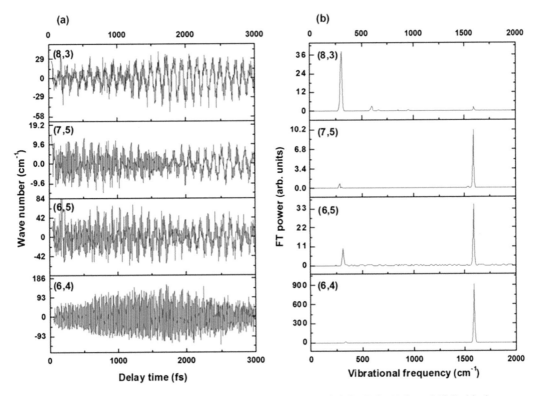

FIGURE 9.2.2.11 (a) The dynamics of the first moment, $M1(t)$, of (8,3), (7,5), (6,5), and (6,4) chiral systems tracked as a function of the pump-probe delay time. (b) The FFT power spectra of the first moment $M1(t)$.

9.2.2.3.10 RBMs Studied by the Moment Calculation

The method by which the CPs modulate the time-resolved spectrum is next discussed, after clarifying that the CP packet wave function is given by a linear combination of eigenmodes.

One might wonder whether the refractive index change can be expressed as a linear combination of the vibronic wave function, such that the refractive index would have a steplike structure. In fact, even though the change in the spectral shape of the probe field induced by CPs is separated by the vibrational frequency, the phase modulation spectrum is smooth and continuous because of the characteristics of the Kramers-Kronig relations. The nonresonant contribution can dominate the refractive index spectrum of the relevant chiral system in the resonant region, together with a significant contribution from other chiral species having nearby resonant peaks. This behavior was already observed in a one-dimensional system of J-aggregated molecules [66]. Accordingly, the spectral shift due to the refractive index modulation is so small that the index changes can be considered to be linear with respect to the probe spectral changes. Therefore, the MPM mechanism can be depicted in a classical way using $Q(t)$, as shown in Figure 9.2.2.12. The displacement $Q(t)$ of the radial axis of a CNT from its equilibrium position is not described by a single sharp line but instead by several (two or three) lines with some probability distribution. This is because the wave function of the CP is expressed by the linear combination of wave functions of several vibrational quantum number. The potential curves in the ground and excited states along the CNT diameter have minima at the equilibrium positions in the ground state, as illustrated in Figure 9.2.2.13. It shows schematically the generation of wave-packet motion in the ground state. In the excited state, the lifetime (50–90 fs) is not long enough to establish a well-defined frequency of the breathing mode (~100 fs), along which coordinate the wave packet moves for a long time enough.

The corresponding time evolution of the refractive index and resulting probe frequency shift are also shown in Figure 9.2.2.12. The lines, as mentioned above, may have some distribution along the ordinate $\Delta n(t)$ and $\Delta\omega(t)$ axes with single peaks, but they are not separated as in the DIF case. Because of the Kramers-Kronig relations, if an energy shift takes place, the spectra of both the absorption and refractive index move in the same direction. The electronic distribution follows the nuclear motion and hence the instance of the maximum diameter the electronic density is lowest along the periphery of the tube. This corresponds to the turning point of the nuclear wave packet on

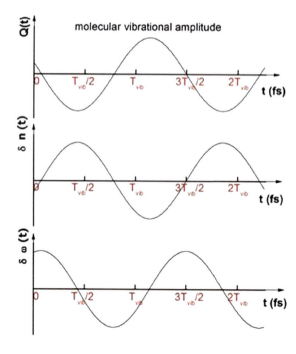

FIGURE 9.2.2.12 Classical views of displacement coordinate $Q(t)$ of a CNT tube from the equilibrium state as a function of time in the ground state, and corresponding time evolution in the refractive index, $\Delta n(t)$ and resulting probe frequency shift $\Delta\omega(t)$ as a function of time. T_{vib} is vibrational period of CP.

the ground-state potential curve. Then the refractive index becomes smallest because of minimum polarizability at the optical frequency. Hence $\delta n(t)$ shows π out of phase shift from that of $Q(t)$, as shown in Figure 9.2.2.12. Since the instantaneous frequency $(\omega + \delta\omega)$ of optical field is given by the time derivative of electric field, $\delta\omega$ has $\pi/2$ shift from $\delta n(t)$, as depicted in the figure.

The NL refractive index due to the molecular vibration is determined by $\delta n = n_2 I$, where I is the peak intensity of the focused pump laser on the sample surface at the relevant transition energy of the corresponding chiral system. The values of δn for the four chiral systems were calculated using the relation $\delta\omega/\omega_E \approx \delta n/n$ in terms of the standard deviation of the probe frequency modulation $(\delta\omega)$ and of the electronic transition frequency (ω_E). The resulting values for the effective NL refractive index of the four modes due to MPM are listed in Table 9.2.2.4, together with the pump intensity I at the relevant transition energy. Here again, the value for (8,3) tubes is much larger than that for (6,5) tubes, because the former have a much larger DER contribution (58%) than the latter (10%), according to Table 9.2.2.3. The effective NL refractive index due to the RBM was determined to be in the range 0.2–$3.1 \times 10^{-17}\,\mathrm{cm^2/W}$ for each system.

Finally, consider the three differences between the present discussion and the theory by Sanders et al. [30]. The first is that Sanders treated the system in terms of a displacement potential upon excitation and thus it does not include Raman processes.

This fact is evident in Figures 9.2.2.14 and 9.2.2.19 of their paper, in which the CP vibration appears in the form of a cosine function with a delay due to the finite pulse duration. Accordingly,

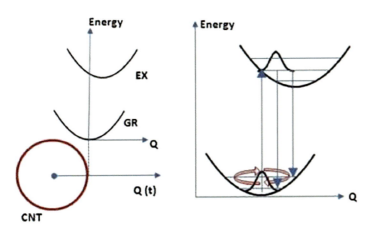

FIGURE 9.2.2.13 Schematics of the potential curve along the displacement with the origin of the equilibrium position of the radial coordinate starting from the symmetric center (not the origin) of the CNT. The arrows schematically show the motion of the wave packet along the potential curve with schematic vibrational levels.

TABLE 9.2.2.4

Vibration-Induced Nonlinear Refractive Index Change Estimated by Analyzing the First-Moment Dynamics

Chirality (n,m)	ω_E (cm^{-1})	ω_{RBM} [ϕ (Phase π)] cm^{-1} $\Delta\omega_R$	(RBM) (cm^{-1})	$\Delta\omega_R/\omega_E$ (RBM) (10^{-4})	I (10^{10}W/cm^2)	n_2 (RBM) (10^{-17} cm^2/w)
(6,4)	16,855	337 (0.48)	10.93	6.48	0.99	0.66
(6,5)	17,462	310 (0.07)	17.22	9.86	4.71	0.21
(7,5)	15,349	282	12.14	7.91	0.71	1.11
(8,3)	14,892	301 (0.05)	30.60	20.5	0.66	3.09

ω_E, transition energy; Ω_r, ω_{RBM}, RBM frequency Ω_g, $\Delta\omega_R$, the root mean square magnitude of the frequency modulation of RBM; $\Delta\omega_G$, the root mean square of the frequency modulation of G-mode; N_2, nonlinear refractive index determined by the equations of refractive index change due to molecular vibration; $\Delta n = N_2 i$ and $\Delta\omega/\Omega_{r(G)} \sim \Delta n/N$ and I Is the peak intensity of focused pump laser on the sample surface at the relevant transition energy of the corresponding chiral system.

their assignment of the signal is in terms of Raman generation of a wave packet in the excited state, but it excludes Raman generation in the ground state. The second difference is that they described the CP dynamics semiclassically as seen in the same figures with a continuous displacement associated with the CP motion. Third, the real part of the nonlinearity is not included in their analysis. In addition, the effect of signal modulation due to MPM, which typically appears in impulsive excitations, is not included.

9.2.2.4   CONCLUSIONS

CP and electronic relaxation dynamics of SWNTs were investigated for the four chiral systems (6,4), (6,5), (7,5), and (8,3) in CoMoCat-grown ensembles, using a sub-10-fs broad band visible pulse and a high-sensitivity detection system based on a 128-channel lock-in amplifier. From the measured data on the probe energy-dependent amplitudes of the RBMs, it was found that the imaginary and real parts of the third-order susceptibility both play important roles in the modulation of the difference absorbance. The amplitude profiles of RBMs can be accurately fit by the sum of the first derivative of the absorption due to the real part of the third-order susceptibility and the difference absorption due to the imaginary part of the susceptibility induced by a Raman process, with the relative contributions being adjustable. The imaginary part, given by a DIF contribution in the fit, arises from energy exchange between CPs and the probe optical field. The size of the HR factors obtained from DIF fits to the (6,4), (6,5), and (8,3) systems are estimated to be 0.26, 0.32, and 0.75, respectively; they depend on the stiffness of the CNT structure. The real part, resulting from a refractive index change due to molecular vibrations, manifests itself as DER dependence due to the MPM process. The NL refractive index for each chiral system is determined to be in the range of $0.2–3.1 \times 10^{-17}\,\mathrm{cm^2/W}$.

The research presented here in this subsection is a collaborative work of the following people: T. Kobayashi, Z. Nie, J. Du, K. Okamura, H. Kataura, Y. Sakakibara, Y. Miyata [67].

REFERENCES

1. S. Iijima, *Nature (London)* **354**, 56 (1991).
2. M. S. Dresselhaus, G. Dresselhaus, and P. Avouris (Eds.), *Carbon Nanotubes: Synthesis, Structure, Properties and Applications*, Springer, Berlin (2001).
3. M. Dresselhaus, *Carbon Nanotubes: Synthesis, Structure, Properties, and Applications*, Springer, New York (2001).
4. R. Saito, G. Dresselhaus, and M. Dresselhaus, *Physical Properties of Carbon Nanotubes*, World Scientific, Singapore (2003).
5. P. Harris, *Carbon Nanotubes and Related Structures: New Materials for the Twenty-First Century*, Cambridge University Press, Cambridge, England (1999).
6. T. Kobayashi, M. Yoshizawa, U. Stamm, M. Taiji, and M. Hasegawa, *J. Opt. Soc. Am. B* **7**, 1558 (1990).
7. T. Kobayashi, *J-aggregates*, World Scientific, Singapore (1996); B. J. LeRoy, S. G. Lemay, J. Kong, and C. Dekker, *Nature* **432**, 371 (2004).
8. H. Htoon, M. J. O'Connell, S. K. Doorn, and V. I. Klimov, *Phys. Rev. Lett.* **94**, 127403 (2005).
9. V. Perebeinos, J. Tersoff, and P. Avouris, *Phys.Rev.Lett.* **94**, 027402 (2005).
10. O. A. Dyatlova, C. Koehler, E. Malic, J. Gomis-Bresco, J. Maultzsch, A. Tsagan-Mandzhiev, T. Watermann, A. Knorr, and U. Woggon, *Nano Lett.* **12**, 2249 (2012).
11. V. Shank, R. Yen, R. L. Fork, J. Orenstein, and G. L. Baker, *Phys. Rev. Lett.* **49**, 1660 (1982).
12. S. Adachi, V. M. Kobryanskii, and T. Kobayashi, *Phys. Rev. Lett.* **89**, 027401 (2002).
13. Y. Yin, A. G. Walsh, A. N. Vamivakas, S. B. Cronin, D. E. Prober, and B. B. Goldberg, *Phys. Rev. B* **84**, 075428 (2011).
14. S. Reich, C. Thomsen, and J. Robertson, *Phys. Rev. Lett.* **95**, 077402 (2005).
15. C. D. Spataru, S. Ismail-Beigi, L. X. Benedict, and S. G. Louie, *Phys. Rev. Lett.* **92**, 077402 (2004).
16. F. Wang, G. Dukovic, L. E. Brus, and T. F. Heinz, *Science* **308**, 838 (2005).
17. A. Srivastava, H. Htoon, V. I. Klimov, and J. Kono, *Phys. Rev. Lett.* **101**, 087402 (2008).
18. A. Jorio, R. Saito, J. H. Hafner, C. M. Lieber, M. Hunter, T. McClure, G. Dresselhaus, and M. S. Dresselhaus, *Phys. Rev. Lett.* **86**, 1118 (2001).

19. A. Jorio, M. A. Pimenta, A. G. S. Filho, R. Saito, G. Dresselhaus, and M. S. Dresselhaus, *New J. Phys.* **5**, 139 (2003).
20. S. K. Doorn, D. A. Heller, P. W. Barone, M. L. Usrey, and M. S. Strano, *Appl. Phys. A: Mater. Sci. Process* **78**, 1147 (2004).
21. J. Jiang, R. Saito, G. G. Samsonidze, A. Jorio, S. G. Chou, G. Dresselhaus, and M. S. Dresselhaus, *Phys. Rev. B* **75**, 035407 (2007).
22. J. Jiang, R. Saito, K. Sato, J. S. Park, G. G. Samsonidze, A. Jorio, G. Dresselhaus, and M. S. Dresselhaus, *Phys. Rev. B* **75**, 035405 (2007).
23. C. M. Gambetta, E. Menna, M. Meneghetti, G. Cerullo, G. Lanzani, S. Tretiak, A. Piryatinski, A. Saxena, R. L. Martin, and A. R. Bishop, *Nat. Phys.* **2**, 515 (2006).
24. T. Kobayashi, J. Zhang, and Z. Wang, *New J. Phys.* **11**, 013048 (2009).
25. T. Kobayashi and Z. Wang, *IEEE J. Quant. Electron.* **44**, 1232 (2008).
26. N. Ishii, E. Tokunaga, S. Adachi, T. Kimura, H. Matsuda, and T. Kobayashi, *Phys. Rev. A* **70**, 023811 (2004).
27. Y. S. Lim, K.-J. Yee, J. H. Kim, E. H. Haroz, J. Shaver, J. Kono, S. K. Doorn, R. H. Hauge, and R. E. Smalley, *Nano Lett.* **6**, 2696 (2006).
28. J. H. Kim, K.-J. Han, N. J. Kim, K. J. Yee, Y. S. Lim, G. D. Sanders, C. J. Stanton, L. G. Booshehri, E. H. Haroz, and J. Kono, *Phys. Rev. Lett.* **102**, 037402 (2009).
29. G. D. Sanders, C. J. Stanton, J. H. Kim, K. J. Yee, Y. S. Lim, E. H. Haroz, L. G. Booshehri, J. Kono, and R. Saito, *Phys. Rev. B* **79**, 205434 (2009).
30. K. Kato, K. Ishioka, M. Kitajima, J. Tang, R. Saito, and H. Petek, *Nano Lett.* **8**, 3102 (2008).
31. L. Luer, C. Gadermaier, J. Crochet, T. Hertel, D. Brida, and G. Lanzani, *Phys. Rev. Lett.* **102**, 127401 (2009).
32. K. Makino, A. Hirano, K. Shiraki, Y. Maeda, and M. Hase, *Phys. Rev. B* **80**, 245428 (2009).
33. Y. S. Lim, J. G. Ahn, J. H. Kim, K. J. Yee, T. Joo, S. H. Baik, E. H. Haroz, L. G. Booshehri, and J. Kono, *ACS Nano* **4**, 3222 (2010).
34. J. Wang, M. W. Graham, Y. Z. Ma, G. R. Fleming, and R. A. Kaindl, *Phys. Rev. Lett.* **104**, 177401 (2010).
35. Z. Zhu, J. Crochet, M. S. Arnold, M. C. Hersam, H. Ulbricht, D. Resasco, and T. Hertel, *J. Phys. Chem. C* **111**, 3831 (2007).
36. W. T. Pollard, S.-Y. Lee, and R. A. Mathies, *J. Chem. Phys.* **92**, 4012 (1990).
37. W. T. Pollard, H. L. Fragnito, J.-Y. Bigot, C. V. Shank, and R. A. Mathies, *Chem. Phys. Lett.* **168**, 239 (1990).
38. R. R. Alfano and S. L. Shapiro, *Phys. Rev. Lett.* **24**, 592 (1970).
39. M. N. Islam, L. F. Mollenauer, R. H. Stolen, J. R. Simpson, and H. T. Shang, *Opt. Lett.* **12**, 625 (1987).
40. T. N. Kumar, F. Rosca, A. Widom, and P. M. Champion, *J. Chem. Phys.* **114**, 701 (2001).
41. T. N. Kumar, F. Rosca, A. Widom, and P. M. Champion, *J. Chem. Phys.* **114**, 6795 (2001).
42. T. Dumitrica, M. E. Garcia, H. O. Jeschke, and B. I. Yakobson, *Phys. Rev. Lett.* **92**, 117401 (2004).
43. T. Dumitrica, M. E. Garcia, H. O. Jeschke, and B. I. Yakobson, *Phys. Rev. B* **74**, 193406 (2006).
44. H. O. Jeschke, A. H. Romero, M. E. Garcia, and A. Rubio, *Phys. Rev. B* **75**, 125412 (2007).
45. I. Iwakura, A. Yabushita, and T. Kobayashi, *J. Am. Chem. Soc.* **131**, 688 (2009).
46. I. Iwakura, A. Yabushita, and T. Kobayashi, *Chem. Phys. Lett.* **501**, 567 (2010).
47. A. Shirakawa, I. Sakane, M. Takasaka, and T. Kobayashi, *Appl. Phys. Lett.* **74**, 2268 (1999).
48. A. Baltuska, T. Fuji, and T. Kobayashi, *Opt. Lett.* **27**, 306 (2002).
49. M. S. Arnold, S. I. Stupp, and M. C. Hersam, *Nano Lett.* **5**, 713 (2005).
50. S. M. Bachilo, L. Balzano, J. E. Herrera, F. Pompeo, D. E. Resasco, and R. B. Weisman, *J. Am. Chem. Soc.* **125**, 11186 (2003).
51. Y. Miyata, K. Yanagi, Y. Maniwa, T. Tanaka, and H. Kataura, *J. Phys. Chem. C* **112**, 15997 (2008).
52. See Supplemental Material at http://link.aps.org/supplemental/10.1103/PhysRevB.88.035424 for sample morphology.
53. S. M. Bachilo, M. S. Strano, C. Kittrell, R. H. Hauge, R. E. Smalley, and R. B. Weisman, *Science* **298**, 2361 (2002).
54. R. Bruce Weisman and S. M. Bachilo, *Nano Lett.* **3**, 1235 (2003).
55. C. Fantini, A. Jorio, M. Souza, M. S. Strano, M. S. Dresselhaus, and M. A. Pimenta, *Phys. Rev. Lett.* **93**, 147406 (2004).
56. H. Telg, J. Maultzsch, S. Reich, F. Hennrich, and C. Thomsen, *Phys. Rev. Lett.* **93**, 177401 (2004).
57. J. Maultzsch, H. Telg, S. Reich, and C. Thomsen, *Phys. Rev. B* **72**, 205438 (2005).
58. Y. Kim, N. Minami, and S. Kazaoui, *Appl. Phys. Lett.* **86**, 073103 (2005).

59. W. G. Wildoer, L. C. Venema, A. G. Rinzler, R. E. Smalley, and C. Dekker, *Nature (London)* **391**, 59 (1998).
60. T. W. Odom, J. L. Huang, P. Kim, and C. M. Lieber, *Nature (London)* **391**, 62 (1998).
61. J. S. Lauret, C. Voisin, G. Cassabois, C. Delalande, Ph. Roussignol, O. Jost, and L. Capes, *Phys. Rev. Lett.* **90**, 057404 (2003).
62. G. W. Chantry, *The Raman Effect: Principles*, edited by A. Anderson, Vol. **1**, Marcel Dekker, New York, pp. 287–342 (1971).
63. J. Du, K. Nakata, Y. Jiang, E. Tokunaga, and T. Kobayashi, *Biophys. J.* **101**, 995 (2011).
64. C. Manzoni, A. Gambetta, E. Menna, M. Meneghetti, G. Lanzani, and G. Cerullo, *Phys. Rev. Lett.* **94**, 207401 (2005).
65. Y. Wang and T. Kobayashi, *Chem. Phys. Chem.* **11**, 889 (2010).
66. N. Zhavoronkov and G. Korn, *Phys. Rev. Lett.* **88**, 203901 (2002).
67. T. Kobayashi, Z. Nie, J. Du, K. Okamura, H. Kataura, Y. Sakakibara, and Y. Miyata, *Phys. Rev. Lett.* **88**, 035424 (2013).

SUPPLEMENTAL MATERIAL

SUPPORTING INFORMATION: SAMPLE MORPHOLOGY

We have measured morphologies of SWCNTs used in this work using AFM (SII S-image). Thin films were fabricated on a Si substrate with thermally oxidized 200 nm thick SiO_2 layer. To take the AFM image, we ensured the adsorption of SWCNTs to the substrate, the SiO_2 surface was modified with 3-aminopropyltriethoxysilane (APTES, Sigma Aldrich) which has affinity to SWCNTs. Then, the SWCNT solution was spin coated on the substrate. The substrate was then washed in methanol to remove the surfactant. Morphology of the SWCNTs were observed by an atomic force microscopy (AFM, SII, S-image). Average length was 330 nm with standard deviation of 160 nm, and the heights were about 1 nm, which is close to the diameter of individual SWCNT. The average length was slightly shorter than the original length of CoMoCAT written in the specification but the damage should not be serious to affect the optical properties of SWCNTs. Although the width of each SWCNT was not well known due to the limited horizontal resolution of AFM, the low height means that SWCNTs are thought to be well dispersed to almost individuals or at least very thin bundles.

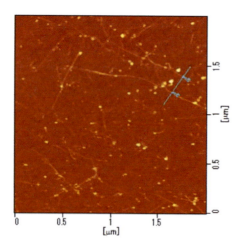

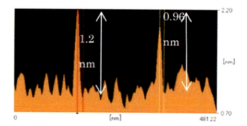

9.2.3 Coherent Phonon Coupled with Exciton in Semiconducting Single-Walled Carbon Nanotubes Using a Few-Cycle Pulse Laser

9.2.3.1 INTRODUCTION

Coherent phonon (CP) dynamics in single-walled carbon nanotubes (SWCNTs) have been studied by several groups, the impulsive excitation method using short pulses [1–10]. However, there are still some important concerns which need to be addressed, especially the mechanism of CP generation, which is related to coupling between the phonon modes and the excitonic state. Using a tunable laser with a 50-fs pulse duration, Lim et al. observed radial breathing mode (RBM) vibrations [2]. The oscillations in the probe transmittance were found to be the result of ultrafast modulation of the optical constants at a frequency ω_{RBM} due to band gap (E_{gap}) oscillations induced by changes in the SWCNT diameter d, which follows the relation, $E_{\mathrm{gap}} \propto 1/d$. It was claimed that the photon energy dependence of the CP signal shows a first derivative-like behavior. Although they concluded that the two spectral shapes were similar, their results showed that the CP amplitude profile was much steeper than the derivative-based profile, explained in terms of a band-gap oscillation that induces a shift in the transition absorption spectrum. The separations between the two inflection points and between the two peaks in the CP amplitude are calculated to be about 40 and 70 meV, respectively, using Figure 9.2.3.5 of the paper [2]. The two inflection points and the two derivative peaks should be coincident if the small spectral shift is the origin of the oscillating signal. The two values differ by a factor of more than 1.7 times, and thus the real-time traces cannot simply be explained by the spectral shift mechanism.

In this chapter, we report a detailed pump–probe study of CPs in SWCNTs using sub-10-fs visible pulses and 128-channel lock-in amplifier. The latter can cover the absorption spectral range of several chiral systems with high sensitivity to the difference absorption. Thanks to the sensitivity and broad spectral coverage; it has become possible to clearly compare the probe wavelength dependence and the derivatives of the absorption spectra of several carbon nanotubes (CNTs). The effects of the CPs on the difference absorbance are fully modeled. The probe photon amplitude profiles are analyzed in terms of the modulation of the excitonic transition probabilities. Through an analysis, it was found that the Raman interaction and molecular phase modulation, corresponding to the imaginary and real parts of the third-order susceptibility, are found to be the origins of the probe wavelength dependence. It is reasonable since the signal from the pump–probe experiment is generated in a third-order nonlinear optical process; both the real and imaginary parts of the nonlinear susceptibility can play roles at the same time because of the Kramers–Kronig relations.

9.2.3.2 EXPERIMENT

The SWCNT sample was prepared by the CoMoCat method [11–13]. Details about our sub-5-fs pump–probe experimental setup and working principle of the techniques are described elsewhere [14,15].

9.2.3.3 RESULTS AND DISCUSSION

9.2.3.3.1 ELECTRONIC RELAXATION AND THERMALIZATION OF EXCITED POPULATION

Figure 9.2.3.1a shows the two-dimensional (2D) difference absorption spectra ΔA plotted against the probe photon energy E_{pr} and the delay time t. The striped oscillatory structures parallel to the time axis represent the modulation of the difference absorbance $\delta\Delta A(E_{pr}, t)$ by CPs. Here δ is the modulation of the absorbance change $\Delta A(E_{pr}, t)$ due to CPs at a specific probe photon energy E_{pr} and delay time t induced by the spectral shift and transition probability change, which give rise to the horizontal and vertical modulations, respectively. Figure 9.2.3.1b plots time-resolved spectra integrated over 200-fs delay-time steps at 10 center probe delay times ranging from 50 to 1850 fs. There are four prominent bleaching bands composed of three peaks and one shoulder, nearly coincident with the relevant E_{22} transitions for the (6,5), (6,4), (7,5), and (8,3) chiral systems. In addition, there

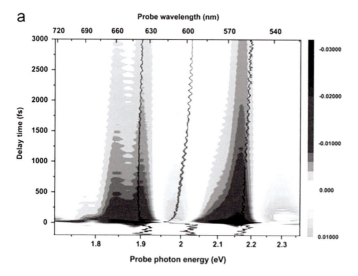

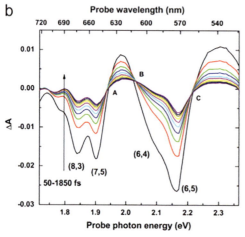

FIGURE 9.2.3.1   (a) Two-dimensional difference absorption spectrum. The black solid curves represent the zero-change lines in the absorbance ($\Delta A=0$). (b) Time-resolved ΔA spectra at the delay time points from 50 to 1850 fs with a 200-fs integration step.

are three isosbestic points near 2.21, 2.02, and 1.94 eV at delay times longer than 250 fs, as indicated by the small squares. The crossing points between neighboring time-resolved spectra are within 70.03 eV of the average photon energies, except between 0 and 50 fs. Therefore, the relaxation after 200 fs can be described using a simple two-state model composed of a single intermediate state or excited state after photo-excitation, where a conversion from one state to another is occurring. This behavior can be explained in terms of intraband relaxation from the E_{22} excitonic state to the E_{11} excitonic state, followed by a slower decay from E_{11} to the ground state.

9.2.3.3.2 FOURIER-TRANSFORM (FT) SPECTRA AND CHIRALITY ASSIGNMENTS

A 2D map of the Fourier power plotted against the probe photon energy and the vibrational frequency is presented in Figure 9.2.3.2. The FT plots show two vibrational modes due to RBMs at 300 cm^{-1} (100 fs) and to G modes at 1587 cm^{-1} (21 fs) [1,5,6], generated by the impulsive excitation for a pulse duration (less than 10 fs) much shorter than all the vibrational periods. Other vibrational modes are too weak to be resolved. Four RBMs are evident with vibrational frequencies of about 337, 310, 301, and 282 cm^{-1}. The assignment of chirality using only the electronic absorption or Raman spectra is often ambiguous due to spectral congestion. In contrast, the intensity of the four dominant double peaked structures indicated by the double-headed arrows in Figure 9.2.3.2 corresponds to the first derivative of the electronic absorption spectrum for the different types of tubes [2,3,6]. The central dip in each structure represents the E_{22} transition since the oscillation is minimal at resonance [6]. However, as will be discussed later, some corrections must be made to this assertion.

The Raman shifts for different RBMs theoretically correspond to the vibrational frequencies of the FT of their CP profiles [16,17]. Consequently, the dips at the intersections of the horizontal and vertical lines in Figure 9.2.3.2 correspond to the relevant vibrational frequency and electronic resonance transition energy. Using this relationship, the chirality can be assigned for the (6,4), (6,5), (7,5), and (8,3) systems, as indicated in Figure 9.2.3.2, because only these tubes can simultaneously fulfill these two conditions. Other kinds of tubes may have weak absorption lines in the laser spectral range, but they cannot be efficiently excited. Using broadband high sensitivity multichannel lock-in detectors, the power spectra of the four systems can be uniquely distinguished for RBMs,

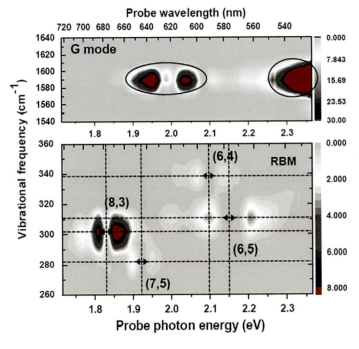

FIGURE 9.2.3.2 2D CP spectra in the spectral range of 1.71–2.36 eV. The chirality assignments of RBMs are shown together. In the bottom panel, the dotted crisscrossing lines show the relevant vibrational frequencies and resonance energies corresponding to RBMs. The two-way arrows indicate the double-peak structure in the amplitude profile of RBMs. The elliptic lines in the top panel display the main features observed in the G mode.

even though their absorption spectra overlap without the need for time-consuming or complex pulse-shaping techniques [3]. Therefore, this method is advantageous for the simultaneous analysis of a sample containing many chiral systems. However, the dip position and the peak of the absorption spectrum do not exactly coincide, as will be discussed later. In contrast, for *G*-mode vibrations, the amplitude profiles for different chiralities overlap and cannot be distinguished. The frequency of the axial *G*-mode is insensitive to the diameter and chirality of SWCNTs and is thus not reflected in the location of the signal in Figure 9.2.3.2.

9.2.3.3.3 FITTING THE AMPLITUDE SPECTRUM WITH CONTRIBUTIONS FROM THE REAL AND IMAGINARY PARTS OF THE THIRD-ORDER SUSCEPTIBILITY

In this section, the fit to the probe photon energy dependence of the vibrational amplitude is improved by combining the contributions from the difference (DIF)-type and derivative (DER)type analyses. In the previous papers [2–10], based on the DER analysis, the modulation was explained in terms of a sinusoidal electronic energy modulation due to an RBM-induced change in the diameter of a CNT [2–10]. In the following, the mechanism involved in the DER-type contribution is more fully discussed.

A change in the refractive index is induced by the molecular phase modulation (MPM) process due to the DER contribution to the modulation [18]. This MPM produces a periodic shift in the probe spectrum. This in turn modulates the time-resolved spectrum composed of ground-state absorption bleaching, stimulated emission, and induced absorption. This effect results from the change in refractive index caused by the deformation of the lattice and molecular structure during the coherent vibrations [19], which the electronic distribution instantaneously follows. The index change introduces a modulation of the probe frequency because of the change in the phase of the probe field, whose time derivative is the optical frequency. Consequently, the signal can be approximated as a spectral shift of the probe pulse induced by a kind of "cross-phase modulation" [20], and the probe energy dependence of the phonon amplitude follows the first derivative of the electronic resonance.

By adding the contributions from the DIF and DER processes, excellent fits are finally obtained, as shown in panels (4) of Figure 9.2.3.3. They show the absolute value of the sum (blue lines) of the contributions of the two types of mechanisms for different RBMs with an adjustable relative contribution used as the fitting parameter for DER (dashed lines) and DIF (dotted lines) analyses, before taking absolute values of the sum. We adjusted the relative intensity of different order of phonon peaks taking into account the differences in the signs of amplitudes and adding them with the DER profile to make the fitting line shape match the amplitude profile. First, we start with the apparently reasonable ratio to fit with the experimental results of the probe photon dependence of the vibrational amplitude, and then take the sum (over probe photon energy data points) of the squared values of the difference between the fitted and observed and try to minimize the value by changing the ratio by small amount to each of the minimum deviation value. Finally, we take the absolute value. The fitting profiles of DER and DIF types are already shown in panels (4) of Figure 9.2.3.3 before taking absolute values.

The physical mechanisms for the DER and DIF contributions are respectively the real and imaginary parts of the third-order susceptibility. Because of the non-uniform probe spectral distribution, the (6,5) and (8,3) tubes show only anti-Stokes sidebands up to second order, while the (6,4) tubes exhibit both Stokes and antiStokes bands. The relative sizes of the DER contributions are further discussed below.

9.2.3.4 CONCLUSIONS

CP and electronic relaxation dynamics of SWCNTs were investigated in CoMoCat grown ensembles for the four chiral systems (6,4), (6,5), (7,5), and (8,3). It was found that both the imaginary and real parts of the third-order susceptibility play important roles in the modulation of the

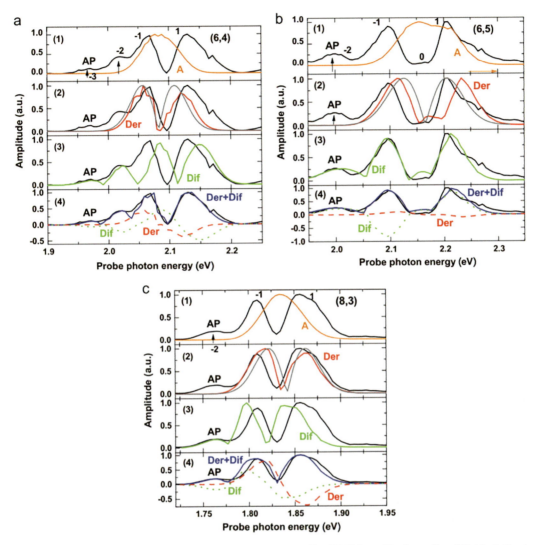

FIGURE 9.2.3.3 Probe photon energy dependence of normalized RBM amplitude profiles (AP, black lines) for (a) (6,4), (b) (6,5) and (c) (8,3) tubes, fitted with the absorbed laser spectra (A, thin orange lines) in panels (1), first-derivative of the stationary absorptions (thin gray lines, panels (2)) and absorbed laser spectra (DER, medium width red lines in columns (2)), the absolute difference between the absorbed photon energy distribution by the sample and the distribution shifted by the relevant RBM frequency (DIF, medium thick green lines in columns (3)), and the sum of DER and DIF (DER+DIF, medium thick blue lines in columns (4)) with relevant contributions, respectively. In panels (1) and (2), the arrows indicate the side bands, and the numbers represent the peak number in amplitude profiles defined from the center band being number 0. The original DER (dashed red lines) and DIF (dotted green lines) lines in the bottom panels (4) are plotted together before taking absolute values to show their corresponding contributions to the vibrational amplitude spectra.

difference absorbance. The research presented in this subsection was conducted cooperatively by the following people: Takayoshi Kobayashi, Zhaogang Nie, Juan Du, and Bing Xue [21].

ACKNOWLEDGMENTS

The authors would like to acknowledge Drs. H. Kataura, Y. Sakakibara, and Y. Miyata for providing us the samples.

REFERENCES

1. C. M. Gambetta, E. Menna, M. Meneghetti, G. Cerullo, G. Lanzani, S. Tretiak, A. Piryatinski, A. Saxena, R. L. Martin, and A. R. Bishop, *Nat. Phys.* **2**, 515 (2006).
2. Y. S. Lim, K.-J. Yee, J. H. Kim, E. H. Hároz, J. Shaver, J. Kono, S. K. Doorn, R. H. Hauge, and R. E. Smalley, *Nano Lett.* **6**, 2696 (2006).
3. J. H. Kim, K.-J. Han, N. J. Kim, K. J. Yee, Y. S. Lim, G. D. Sanders, C. J. Stanton, L. G. Booshehri, E. H. Hároz, and J. Kono, *Phys. Rev. Lett.* **102**, 037402 (2009).
4. G. D. Sanders, C. J. Stanton, J. H. Kim, K. J. Yee, Y. S. Lim, E. H. Hároz, L. G. Booshehri, J. Kono, and R. Saito, *Phys. Rev. B* **79**, 205434 (2009).
5. K. Kato, K. Ishioka, M. Kitajima, J. Tang, R. Saito, and H. Petek, *Nano Lett.* **8**, 3102 (2008).
6. L. Lüer, C. Gadermaier, J. Crochet, T. Hertel, D. Brida, and G. Lanzani, *Phys. Rev. Lett.* **102**, 127401 (2009).
7. K. Makino, A. Hirano, K. Shiraki, Y. Maeda, and M. Hase, *Phys. Rev. B* **80**, 245428 (2009).
8. Y. S. Lim, J. G. Ahn, J. H. Kim, K. J. Yee, T. Joo, S. H. Baik, E. H. Hàroz, L. G. Booshehri, and J. Kono, *ACS Nano* **4**, 3222 (2010).
9. J. Wang, M. W. Graham, Y. Z. Ma, G. R. Fleming, and R. A. Kaindl, *Phys. Rev. Lett.* **104**, 177401 (2010).
10. Z. Zhu, J. Crochet, M. S. Arnold, M. C. Hersam, H. Ulbricht, D. Resasco, and T. Hertel, *J. Phys. Chem. C* **111**, 3831 (2007).
11. M. S. Arnold, S. I. Stupp, and M. C. Hersam, *Nano Lett.* **5**, 713 (2005).
12. S. M. Bachilo, L. Balzano, J. E. Herrera, F. Pompeo, D. E. Resasco, and R. B. Weisman, *J. Am. Chem. Soc.* **125**, 11186 (2003).
13. Y. Miyata, K. Yanagi, Y. Maniwa, T. Tanakka, and H. Kataura, *J. Phys. Chem. C* **112**, 15997 (2008).
14. T. Kobayashi, M. Yoshizawa, U. Stamm, M. Taiji, and M. Hasegawa, *J. Opt. Soc. Am. B* **7**, 1558 (1990).
15. A. Baltuska, T. Fuji, and T. Kobayashi, *Opt. Lett.* **27**, 306 (2002).
16. S. M. Bachilo, M. S. Strano, C. Kittrell, R. H. Hauge, R. E. Smalley, and R. B. Weisman, *Science* **298**, 2361 (2002).
17. R. Bruce Weisman and S. M. Bachilo, *Nano Lett.* **3**, 1235 (2003).
18. N. Zhavoronkov and G. Korn, *Phys. Rev. Lett.* **88**, 203901 (2002).
19. T. Kobayashi and Z. Wang, *IEEE J. Quant. Electron.* **44**, 1232 (2008).
20. M. N. Islam, L. F. Mollenauer, R. H. Stolen, J. R. Simpson, and H. T. Shang, *Opt. Lett.* **12**, 625 (1987).
21. T. Kobayashi, N. Zhaogang, and J. Du, *J. Lumin.* (Selected papers from DPC'13) **152**, 11–14 (2013).

9.2.4 Real-Time Spectroscopy of Single-Walled Carbon Nanotubes for Negative Time Delays by Using a Few-Cycle Pulse Laser

9.2.4.1 INTRODUCTION

Single-walled carbon nanotubes (SWNTs) have unique mechanical, electronic, and optical properties due to their one-dimensional nanostructures and are promising candidates as devices for nanoelectronics and photonics [1–3]. The peculiar optical properties of one-dimensional systems have been studied extensively in various other materials such as conjugated polymers [4,5] and J-aggregates [6], providing a playground for studying the dynamics of confined electrons and phonons and their interplay [4,7]. Advances in optical studies such as Raman scattering and photoluminescence excitation spectroscopies have led to definitive assignments of spectral features to specific chirality classes given by a set of two integers (n, m) [8–11]. Recent theoretical and experimental studies have illuminated the importance of pronounced excitonic effects in interband optical processes and have revealed a variety of phonon-assisted peaks, suggesting strong exciton–phonon coupling [11–15], which is at the heart of many phenomena in SWNTs. Significant efforts have been made to investigate the coherent phonon dynamics in SWNTs using the pump–probe technique in the femtosecond time regime, i.e., coherent phonon spectroscopy or real-time vibrational spectroscopy [16–25]. However, the most relevant spectroscopic analyses by far have generally been performed in the positive time range, with the pump pulse coming before the probe to perturb the absorption spectrum of the medium, which is subsequently probed after a set time delay. In comparison with Raman spectroscopy, which is relevant to ground-state vibrations [26], real-time spectroscopy is much more suitable for studying the excited-state electronic relaxation and vibrational dynamics of materials under the same excitation and environmental conditions. However, the coherent phonon vibrations observed in the femtosecond pump/probe experiment include wave packet motion effects in both the ground and excited states, which make it difficult to ascribe the signals to either of the two states [27,28]. To date, only a limited number of experiments have overcome this difficulty by using chirped-pulse excitation [29,30] and detailed analysis of the delay-time dependence of pump-dump pulses in stimulated Raman scattering [31].

In contrast to investigation using positive time ranges, investigation of the real-time traces for negative time delays, whereby the probe pulse precedes the pump pulse, can provide information on vibronic polarization modulated by phonons only in the excited states [28]. Previous reports have discussed the experimental results observed in the negative time range in terms of perturbed free-induction decay and coherent coupling [32–37]. The difference absorption spectrum has been calculated for a two-level system and has already been applied to molecular systems [38,39]. Theoretically, in this coherent regime, it is the probe that "deposits" polarization in the medium, and this subsequently relaxes over time. The pump then "probes" the decay by perturbing the probe polarization, although no pump energy is absorbed by the sample medium, and the signal

DOI: 10.1201/9780429196577-51

is detected in the direction of the probe beam [32,33]. The perturbed free-induction decay term, which occurs only in the negative time range, can be used to determine the electronic phase relaxation time and to study the vibrational phase relaxation dynamics in excited electronic states [28]. Recently, we reported the experimental observation of absorbance changes for several molecular systems for negative time delays [28,40]. This negative-time measurement is a powerful method for studying coherent phonon vibration in excited states without the effect of wave packet motion in the ground states under the same experimental condition as the real population relaxation associated with vibrational dynamics, including both the ground state and excited state. In addition, knowledge of electronic dephasing dynamics is very important for elucidating the properties of excited states and the dynamics of optical nonlinear processes, which offer information on the response dynamics of various device applications such as optical switching and optical signal processing [41,42]. The time sequences of two fields of the pump and one from the probe are the same pulse ordering as that denoted by type S_{III} configuration in the two-dimensional spectroscopy [43,44]. In this paper, we report what is to the best of our knowledge the first observation of the vibrational and electronic coherence dynamics of SWNTs in negative time delays using a pump–probe technique with an extremely short (7.1 fs) and broad bandwidth pulse in the visible spectrum and using an ultrahigh-sensitivity broadband detection system composed of a polychromator and a multichannel lock-in amplifier, which can detect 128 different wavelengths simultaneously. Vibrational and electronic phase relaxation dynamics studied under the same conditions are elucidated and compared with the results reported in the positive time range.

9.2.4.2 EXPERIMENTAL METHOD

9.2.4.2.1 Pump–Probe Experiment

The details of the ultrafast pump–probe experiment were described in our previous papers [28,45]. In short, the main light source was a 7.1-fs pulse from a noncollinear parametric amplifier (NOPA) developed by our group [46,47]. The spectrum of the output pulse from the NOPA extended from 520 to 725 nm. The focus areas of the pump and probe pulses were about 100 and 75 μm^2, respectively, with a common center on the surface of the sample film. The polarizations of the pump and probe pulses were parallel to each other. In the pump–probe experiment, the signal was spectrally dispersed with a polychromator (JASCO, M25-TP) over 128 photon energies (wavelengths) from 1.71 to 2.37 eV (from 722 to 524 nm). The pump–probe step size is 1 fs. All measurements were performed at room temperature (295 ± 1 K).

9.2.4.2.2 Sample Preparation

The details of the sample preparation were described in our previous paper. In short, a CoMoCat synthesized sample was prepared using silica support and a bimetallic catalyst with a Co:Mo molar ratio of 1:3 [48–50].

9.2.4.3 RESULTS AND DISCUSSION

9.2.4.3.1 Stationary Absorption Spectrum

Figure 9.2.4.1 **shows the stationary absorption spectrum of swnts without further** processes of fractionation of separation and extraction. More than 15 peaks and shoulders can be distinguished in the spectrum. Their assignment was made according to the relation between the transition energy and chiral index as set in the literature [8]. The broadband visible laser pulse with photon energy in the 1.7–2.4 eV range was resonant with the second exciton transitions (E_{22}) in the tubes. The chirality assignments of some absorption bands and shoulders around 1.46, 1.7, 2.4, and 2.7 eV in the spectrum were not made because they might have been shared by more than one type of tube.

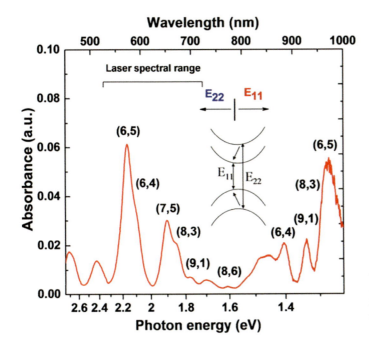

FIGURE 9.2.4.1 Stationary absorption spectrum of SWNTs, showing the E_{11} and E_{22} transitions in visible and infrared (1.2–2.7 eV) range, and the relevant chirality assignments.

9.2.4.3.2 Two-Dimensional (2D) Real-Time Vibration Spectra

In the pump–probe experiment for a simplified two-level system, the differential absorbance $\Delta A(\omega)$, proportional to the differences of the imaginary parts of the susceptibilities, can be given by

$$\Delta A(\omega) : \mathrm{Im}\left\{ \frac{f_2(\omega)}{e_{pr}(\omega)} F\left[E_{pr}(t) N_{pu}^{(2)}(t) + E_{pu}(t)\left[F_1(t) \otimes \left(E_{pu}(t) E_{pu}^{*}(t) \right) \right] \right] \right.$$
$$\left. - E_{pu}(t)\left[F_1(t) \otimes \left(E_{pu}^{*}(t) \right) \right] \right]$$

(9.2.4.1)

where Im stands for the imaginary part, $\otimes$ denotes convolution, F[A(t)] is the Fourier transform of A(t), $E_{pu}(t)$, and $E_{pr}(t)$ are the pump and probe fields, respectively, and $N_{pu}^{(2)}$ represents the population change induced by the pump field. Definitions of the remaining terms are given elsewhere [32]. The three terms contained on the right side of Eq. (9.2.4.1) represent the level population (LP) term, pump polarization coupling (PPC) term, and perturbed free-induction decay (PFD) term, respectively. Reports published recently have mostly focused on LP, the first term [16–25], which appears only after the onset of the pump pulse, corresponding to the positive time region. The PPC term arises from the coherent coupling between pump field-induced polarization and the probe field appearing only when the probe and pump overlap. In contrast to the first term, the third term, PFD, represents the case in which the probe pulse comes earlier than the pump pulse, contributing only to the negative time delay, and there is no temporal overlap between them.

The 2D differential absorption $\Delta A(E_{pr}, t)$ as a function of probe–photon energy E_{pr} and pump–probe delay time t from 0 to −200 fs is displayed in Figure 9.2.4.2a. The real-time traces show sharp and intense peaks (or spikes) around zero delay time and noticeable slow-decay signals with a finite size of ΔA delaying by as long as −150 fs. The peaks are caused by PPC when the pump and probe temporally overlap with each other, and the slow-decay signals are due to the PFD process. Because the coupling term is given by the convolution function of the pump and probe pulse and provides no contribution outside of this function, it is effective only when the pump pulse overlaps the probe pulse in time close to the zero time range on the order of the pump pulse duration (7.1 fs) [32,33].

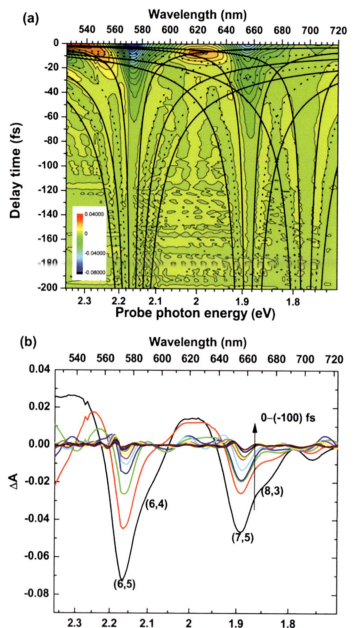

FIGURE 9.2.4.2 (a) 2D display of $\Delta A(E_{pr}, t)$ as functions of probe photon energy, E_{pr}, and pump–probe delay time, t. The solid and dotted lines are simulated traces of fringe peak and valley positions respectively. (b) Probe photon energy dependences of ΔA at the delay times from 0 to -100 fs with a 10-fs step. The blue dashed lines and the arrows are drawn as a guide to the eye, showing the peak position shifts with time passage.

Similar to the positive time results [16–18], there are also some periodic structures in the time traces. For negative time delays, the preceding probe pulse at first generates macroscopic electronic polarization in the sample, and then later, the coming intense pump field interferes with the probe polarization, resulting in the formation of a grating. PFD, the third term in Eq. (9.2.4.1), is generated by another electronic field component of the pump pulse to be diffracted into the probe direction, satisfying the causality.

In the present case, the polarization generated by the probe pulse is not due to a pure electronic transition but to a vibronic transition. Therefore, the observed vibrations are theoretically due to the vibronic transition between the vibration levels in the ground electronic states and the vibronic excited states. The excited-state vibrational modes can still be observed without a population in the excited state in the negative time range because the vibronic polarization is modulated by lattice (or molecular) vibrations in the excited states when the vibrational mode is at the ground level in the ground electronic state [28]. The advantages of this method are summarized in the following way. In the case of positive time, the signal can both be due to the ground state or due to excited state and because of this ambiguity, it is difficult to discuss clearly the vibrational dynamics. However, in the case of negative time experiment, only the excited vibronic coherence can be studied. Another advantage is the ability of the study of the electronic excited-state coherence relaxation and vibrational coherence decay under the same condition of laser and samples. Usually, the study of electronic and vibration requires two different sets of lasers and detection systems. Also by tuning the pump laser using a spectral filter, coherence phenomena in some spectral ranges can be studied with a bandwidth in a specific range at the expense of time resolution.

In the case of SWNTs, the main absorption spectra in near-infrared and visible range originate from the electronic transition from the highest valence band (v_1) to the lowest conduction band (c_1) with E_{11} of about 8000 cm^{-1} and that from the second highest valence band (v_2) to the second lowest conduction band (c_2) with energy E_{22} of about 16000 cm^{-1} [51,52]. The laser spectrum used in the present study is extending in the range of 520–720 nm (corresponding to 19200–13800 cm^{-1}) as described in our previous paper [45]. Therefore vibronic coherence above-discussed is due to the electronic (vibronic coupling) connecting between the ground state and E_{22} state, which is the excitation from v_2 to c_2 in a SWNT. The electronic coupling and/or vibronic coupling between E_{11} and E_{22} excitons are not possible. The coupling including the ground state, E_{11}, and E_{22} is also not possible, and hence electronic coherences between E_{11} and E_{22} are not excited.

Figure 9.2.4.2b depicts the differential absorption (ΔA) lines obtained by intersecting the 2D plot of ΔA in Figure 9.2.4.2a at delay times from 0 to −100 fs with a 10-fs step. Three features are shown in this figure. First is the prominent "bleaching" located at 2.18, 1.88, and 1.77 eV, which corresponds to the peak wavelength of the absorption (bleaching) spectra of (6,5), (7,5), and (9,7). There is an unclear structure at ~2.08 eV, probably due to the shoulder from the spectrum of (6,4). The second feature is the hyperbolic contour curves asymptotically approaching these peaks. Third is the quasi-periodic line features present parallel to the zero-delay time line of the horizontal axis. The third feature is buried by the hyperbolic appearing contour at the shorter negative delay time, and they start to be clear at longer negative delay than about −80 fs.

The first feature, bleaching, appears in the same positions as in the positive delay time data [16,18]. The location of the intense bleaching is consistent with that of the E_{22} transitions in Figure 9.2.4.1. Brito Cruz et al. [32] and Joffre et al. [53] also reported the existence of intense bleaching. It can be explained in the following way. The amplitude of the grating formed by the interference between the probe polarization and pump field is reduced when the probe field is virtually absorbed during interaction with the sample, even though it is not really absorbed in the sense of true energy transfer from the probe light to the sample. The virtual electronic energy oscillates between the sample electronic state and probe field, as in the case of Rabi oscillation. The diffracted pump field intensity is then reduced because of the smaller grating modulation amplitude. Then, the amplitude of the PFD signal has a similar spectral dependence to the electronic population observed in the positive delay time [45].

The second feature of the hyperbolic structure can be explained in terms of frequency domain interference between the pump field at zero delay and probe-induced polarization of which fringe separation in the spectral interference pattern is inversely proportional to the separation between the two. The negative time transmittance change is given by [53]

$$\Delta_r T(\omega,t) \alpha I_p(\omega - \omega_0) \exp(-t/T_2)$$

$$\times \left[\cos((\omega - \omega_0)t) - T_2(\omega - \omega_0)\sin((\omega - \omega_0)t) \right] / \left[(\omega - \omega_0)^2 + (1/T_2)^2 \right] \quad (9.2.4.2)$$

Here, $I_p(\omega)$ is the probe light spectrum, T_2 is the electronic dephasing time, and ω_0 and ω are resonance frequency and probe light frequency, respectively. The right-hand side is transformed into the following equations:

$$\Delta_r T(\omega, t) \alpha I_p(\omega - \omega_0) T_2^2 \exp(-t/T_2)\left[\cos((\omega - \omega_0)t) + \theta\right] \qquad (9.2.4.3)$$

$$\theta = \tan^{-1}(T_2(\omega - \omega_0)) \qquad (9.2.4.4)$$

Equations (9.2.4.2) and (9.2.4.3) show that the coherent transient spectrum has a sinusoidal modulation with a modulation period given by $2\pi/(\omega-\omega_0)$ with the initial phase (at zero delay time) of θ. It maintains the form of exponentially decaying Lorentzian function with a half-width of half-maximum of $(1/T_2)$. Because of the (co)sinusoidal modulation, the modulation can appear several times (in the frequency domain) as seen in the range of 2.35–2.18 eV, where the transient absorbances in both negative time and positive time ranges are most intense. There is a complex structure in the range of 2.07 eV because of the coherent perturbed decay interference between the converging signal to 2.18 and ~2.08 eV. There are two more converging frequencies at 2.89 and 1.78 eV. As these modulated spectra are at the center of the resonance frequency of ω_0 in the coherent transient spectrum, this initial phase at the frequency is 0π, but at near resonance frequency ω a small phase shift determined by Eq. (9.2.4.4) is expected. Using the equation we plotted the fringe peak and valley positions with solid and dotted lines, respectively, in Figure 9.2.4.2a, and the theoretical expectation is found to be satisfied.

The third feature of the coherent transient spectrum is vibrational oscillation. As the limited life of the electronic phase decay signals in which molecular vibration is providing modulation, it is difficult to determine the vibrational frequency precisely. It decays not only because of vibrational phase decay (including both homogeneous and inhomogeneous decay) but also because of electronic phase decay. In this way, the decay is different from that in the positive decay range, which is caused by both vibrational phase decay and electronic population decay. Therefore, it disappears even before the lattice system comes into equilibrium due to vibrational population decay.

As previously described, the PPC and PFD terms can both contribute to the differential absorption signals for negative time delays. The ΔA line at zero time is governed by the PPC term, which occurs only around zero delay, whereas for delays of longer than 10 fs, most of the signal should be due to the PFD term. In theory, the PPC term is proportional to the pump-induced polarization present at the time the probe is coexisting with the pump at the sample, whereas the PFD term is proportional to the remaining probe-induced polarization. This occurs because the presence of the pump field modifies the otherwise free decay of the polarization that is "deposited" by the probe field [32,33]. For the PPC term, with our ultrabroad optical pulses, it is possible to excite polarizations in a band of states that are resonant with the pump energy. Because the PFD term represents the process of the relaxation of macroscopic polarization corresponding to the transition between the excited vibronic state and the ground state, the sign of ΔA and its vibrational and electronic decay should also be dependent on the relevant vibronic state belonging to the electronic excited states.

In addition, as shown in Figure 9.2.4.2b, there is a perceptible transition energy variation for the two strongest peak positions at ~2.16 eV (572 nm) and ~1.92 eV (650 nm) with the passage of time. This could occur for two reasons: the transition energy could be modulated by coherent phonon vibrations, as will be described later, or the macroscopic electronic coherent polarization dephasing time could be very different for different chiralities with absorption bands adjacent to each other. Then, their peak positions would be affected over time due to the overlap of the different decay components.

9.2.4.3.3 Electronic Phase Relaxation Time

We can use two methods to measure the electronic coherence decay time. The first method uses Eq. (9.2.4.3), and the second method uses the coefficient of the modulation term on the left side of Eq. (9.2.4.2).

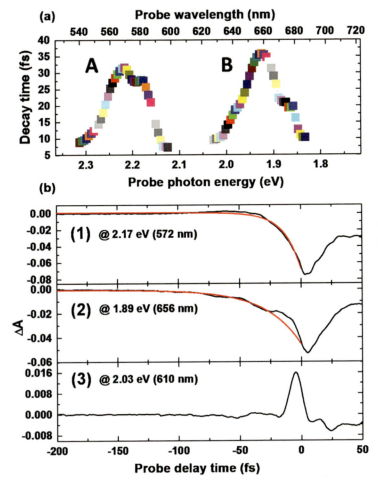

FIGURE 9.2.4.3 (a) Probe photon energy dependence of electronic dephasing time estimated by single-exponential function fitting. (b) Negative time traces (black lines) corresponding to the two maximum values in ranges A and B in (a), and their fitting (red lines) with single exponential function. A good fitting by single-exponential for line (3) cannot be achieved because in this case the pump polarization coupling effect is too large and dominating the total signal for the decay time to be determined precisely.

In the negative time range, the PFD term increases with the electronic dephasing time constant, T^{el}_2, as described in Eq. (9.2.4.2), and shuts off quickly for times later than the peak of the pump pulse because it is only responsive to the presence of the pump following the probe [32,33]. The vibrational dephasing time T^{vib}_2 is mostly on the order of picoseconds in many molecules, whereas the T^{el}_2 of medium-size molecules at room temperature is typically as short as a few tens of femtoseconds. Pulses that are at least shorter than 10 fs are thus required to study the electronic and vibrational dephasing times simultaneously [41]. The appearance of the slow PFD process and oscillatory structures discussed above is proof that T^{el}_2 is much longer than the pump duration (7.1 fs).

The dephasing time of the electronic coherence under the experimental conditions was determined by assuming that the time traces followed the first-exponential decay in spectral ranges where the interference contribution was weak relative to that of the PFD. The results are displayed in Figure 9.2.4.3a. Second-order exponential fitting was also performed because there were spectral overlaps for adjacent chiralities. However, the two time constants obtained were very close to each other and almost equal to the first-exponential fittings. The single-exponential fitting was thus approximately successful here, which also indicates that the dephasing times were very close for different chiralities in this sample.

Figure 9.2.4.3a shows that the decay time constant is also probe photon energy dependent, similar to the behavior of PFD as the same function. There could be two reasons for this. First, the PFD could be the process of the relaxation of macroscopic polarization corresponding to the transition between the excited vibronic states and ground states, as described above. Then, the sign of ΔA and its decay would be dependent on the relevant vibronic state belonging to the same electronic excited states. Because the PFD term would be proportional to the unperturbed population differences in this scenario, the initial decay intensity would be different according to the absorption cross-section of different transitions. Second, the PPC and PFD terms could have mixed contributions at negative time near the zero delay time. The sum of the two contributions could be either constructive or destructive depending on the sign of the two terms. Because the coherent term should provide no contribution outside the convolution function of pump and probe pulse, the decay could be steeper than that predicted based on T^{el}_2 owing to the dominant contribution of the coherent term with a minor contribution of the PFD term, especially when the sign of the PFD term is opposite that of the PPC term. Thus in some spectral ranges, it would be difficult to determine the exact time constants because of the mixed contributions of the two terms near the zero time delay, as represented by line 3 in Figure 9.2.4.3b. Therefore, the longest decay time is considered to be minimally affected by the PPC term and is consequently closest to the true electronic dephasing time.

The single-exponential fittings of the dephasing time corresponding to the two maxima in ranges A and B in Figure 9.2.4.3a are demonstrated by lines 1 and 2 in Figure 9.2.4.3b, respectively. The maximum phase relaxation time is 32 ± 1 fs in range A and $36 + 1$ fs in range B. The chiral systems of (6,5) and (7,5) mainly contribute to ranges of A and B, respectively. The spectral decomposition reported in our previous paper [45] gave the results of spectral half-widths of (6,5) and (7,5), which are 89.9 and 63.9 meV, respectively. If we assume that the widths are dominantly determined by homogeneous contribution only, the electronic dephasing time T^{el}_2 of the coherence between the ground state and the lowest excited state is determined to be in the range of 60–40 fs. They are 40.5, 59.7, 55.0, and 55.7 fs for (6,5), (6,4), (7,5), and (8,3), respectively. The electronic (polarization) dephasing time is substantially longer than the pump duration $T_{pu} = 7.1$ fs $\left(T^{el}_2 \gg T_{pu} \right)^{32}$. This is the prerequisite that needs to be fulfilled for a noticeable signal to be observed for negative delays in the case of the simple two-level system. In addition, the decay lines in Figure 9.2.4.3b are not following pure exponential decays. Clearly visible oscillatory modulation is superimposed on the exponential decay lines. These periodic oscillations are consistent with those in Figure 9.2.4.2a, indicating the modulation of the vibronic polarization by lattice vibrations in excited states.

Here, we discuss the various components that contribute to the electronic dephasing time of the four chiral systems, (6,5), (6,4), (7,5), and (8,3), obtained by the negative time experiment, which is listed in Table 9.2.4.1.

The observed dephasing time $T_2 = 1/k_2$ includes the contributions of the population decay $\tau_1 = 1/k_1$ of the $E_{22}45$ exciton as determined in a previous paper, the inhomogeneous broadening $T^*_2 = 1/k^*_2$, and pure dephasing $T^0_2 = 1/k^0_2$ using the following equation:

$$\frac{1}{T_2} = \frac{1}{2\tau_1} + \frac{1}{T^*_2} + \frac{1}{T^0_2}, \quad k_2 = (1/2)k_1 + k^*_2 + k^0_2 \tag{9.2.4.5}$$

TABLE 9.2.4.1

Spectral Widths of the Chiral Components Separated by Simulation and Widths Obtained from the Electronic Dephasing Time by the Negative Time Method

Chirality	Probe Wavelength (nm)	Probe Photon Energy (eV)	Fwhm of Absorption Spectrum (meV) [45]	Dephasing Time from Negative Time Measurement (fs)	Calcd Fwhm from Dephasing Time (meV)	$1/k^*_2 + k^0_2$ (fs)	$k^*_2 + k^0_2$ (10^{13} s^{-1})
(6,5)	569.5	2.18	90	31.6	262	47.0	2.13
(6,4)	580.4	2.14	65	27.8	198	58.0	1.72
(7,5)	658.6	1.88	64	35.7	232	60.5	1.65
(8,3)	675.8	1.83	49	19.4	426	26.7	3.75

Using the equation, we can obtain the value of $k_2^* + k_2^0$, as shown in Table 9.2.4.1. It is difficult to separate the effects of inhomogeneous broadening $T_2^* = 1/k_2^*$ and pure dephasing $T_2^0 = 1/k_2^0$ from the calculation. However, we can tell that the electronic decoherence time of the E_{22} exciton is longer than 47, 58, 61, and 27 fs in (6,5), (6,4), (7,5), and (8,3), respectively.

From the dephasing lifetime values, the width of each spectrum can be calculated, as listed in the table. The table shows that the spectra of several of the chiral systems heavily overlap but the spectral widths of the components can be separately obtained using the relation $T_2 \Delta \omega = \pi$ between the electronic dephasing time T_2 and the full width at half maximum (fwhm) $\Delta \omega$ of the band if the absorption band is mainly determined by homogeneous broadening. Even in the case of a mixture of homogeneous and inhomogeneous broadening, the factor does not change much. This result suggests that the estimation of the spectral widths of bands that overlap with each other can be separated by the negative time method. In our previous paper [45], we attempted to separate the spectral widths of chiral components by simulation with the Voigt function. The widths obtained by this simulation method suffered from the overlapping tail of the absorption spectrum. It is clear from the spectra of the sample after removing the broad basal widths, the estimated widths were found to be about three times narrower than those shown in Table 9.2.4.1. The corrected values of the fwhm of the bands are then nearly equal to those obtained from the dephasing time except for (8,3), which is only a shoulder of (7,5).

9.2.4.3.4 FOURIER TRANSFORM POWER SPECTRA AND PROBE PHOTON ENERGY-DEPENDENT AMPLITUDES

As previously described, excited-state vibrational modes can still be observed without a population excitation in the negative time range because the electric polarization can be modulated by lattice vibrations in excited states. To find the vibrational signals, the time traces in the negative and positive time ranges were compared by Fourier transform (FT), as shown in Figure 9.2.4.4. The FT was performed after removing the slowly varying exponential decay contributions, and the initial 40-fs time range near the zero-time delays was omitted to avoid the effect of the initial mixed contributions from the PPC term. The positive time FT plots in Figure 9.2.4.4a2, calculated from 40 to 200 fs and from 40 to 3000 fs, identify two well-known dominant vibrational modes [16,22], the radial breathing mode (RBM) with a frequency of ~320 cm^{-1} (period of ~104 fs) and the G mode of ~1587 cm^{-1} (~21 fs), generated by the impulsive excitation with sub-5-fs laser pulses. Other vibrational modes are too weak to be resolved. In contrast, for the FT calculated for negative time traces in both Figure 9.2.4.4a1 and b1, obvious RBM signals appear in both the positive and negative time ranges, whereas the G-mode vibration at ~1587 cm^{-1} in Figure 9.2.4.4a1 and b1 is too weak to be resolved well in the negative time range.

There are bumps around 1300 and 1000 cm^{-1}, respectively in parts a1 and b1, but they are probably the noise being covered by the broadened FT regions due to the short time range of the PFD signal limited by the short electronic dephasing time. The high-frequency vibration of the G mode with a period of ~21 fs should be able to be fully resolved in the negative time range, which is long enough for the lattice to vibrate nearly ten times. Therefore, the most significant distinction in the FT power spectrum between the positive and negative time ranges is the intensity ratio between G mode and RBM mode. The high-frequency G-mode vibration is much weaker in the negative time due to exciton state than in the positive time range due to the ground state, and the former is too weak to appear clearly in the FT power spectrum for negative delays.

To further substantiate the vibrational signals described above, the 2D FT power spectra calculated in the negative time range, −40 to −200 fs, are shown in Figure 9.2.4.5a. For comparison, a part of the 2D FT power spectra of positive time range traces in the 40–3000-fs region is displayed in Figure 9.2.4.5b. The vibrational frequencies below 1000 cm^{-1} shown in Figure 9.2.4.5a and b are due to the low-frequency RBM vibrations. They are observed in the spectral ranges corresponding to the

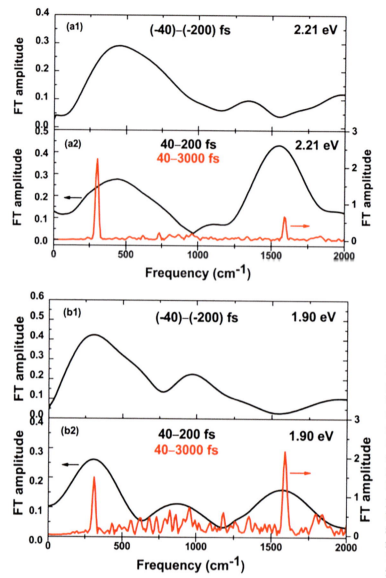

FIGURE 9.2.4.4 (a1) FT amplitude spectra calculated for the real-time traces probed at 2.21 eV from −40 fs to −200 fs. (a2) FT amplitude spectra calculated for the real-time traces probed at 2.21 eV from 40 to 200 fs and from 40 to 3000 fs. (b1) FT amplitude spectra calculated for the real-time traces probed at 1.90 eV from −40 to −200 fs. (b2) FT amplitude spectra calculated for the real-time traces probed at 1.90 eV from 40 to 200 fs and from 40 to 3000 fs.

absorption spectra of the chiral system shown in Figure 9.2.4.1. The most significant distinction in the spectra between the positive and negative time ranges is the intensity ratio between G mode and RBM mode. The high-frequency G-mode vibration is too weak to be resolved for negative delays, as already shown in Figure 9.2.4.4, even though it is very intense in the positive time range and its vibrational period (∼21 fs) is short enough to vibrate many more times than RBMs (∼100 fs). As the negative time measurement is a powerful method for studying excited-state dynamics without the effect of wave packet motion in ground states, the observations above indicate that the RBM is intense in the exciton state while G mode is very weak.

To see better the RBM signal, the magnified images of spectral ranges A and B are shown in Figure 9.2.4.5c and d, respectively. In good contrast with previous positive time studies, the dependence on the probing photon energy is still clearly displayed for the RBMs. The modes show a clear dip around 2.16 eV (572 nm) for range A and 1.92 eV (650 nm) for range B, which correspond to the energies of the second excitonic transitions in the (6,5) and (7,5) tubes, respectively, and two

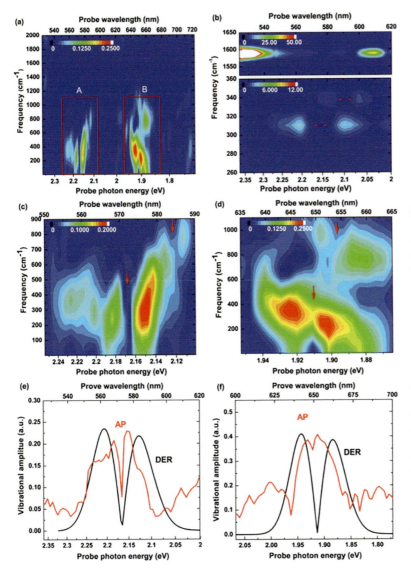

FIGURE 9.2.4.5 (a) 2D FT power spectra of negative time traces from −40 to −500 fs in a spectral range of 1.7–2.4 eV. (b) Part of 2D FT power spectra of positive time traces as a comparison from 40 to 3000 fs in the range of 1.7–1.99 eV. (c) and (d) are the zoom-ins of A and B parts in (a). (e) Comparison of probe photon energy dependence of amplitude profiles (AP, red line) and the derivative (DER, black lines) profiles of (6,5) and (7,5) chiralities, respectively. The amplitude profiles were calculated by taking the square root of the FT powers.

further lobes at higher and lower energies, implying that the vibration becomes minimal at the resonance transition peak. Analogous to those in the positive time range, the vibrational properties were interpreted using modulation spectroscopy [17,22,45]. In our previous paper, we discussed the probe–photon energy-dependent amplitude profiles in terms of the energy exchange between coherent lattice vibration and the probe optical field and phase modulation induced by molecular phase modulation [27,54].

The mechanisms of the appearance of dips in Figure 9.2.4.5e and f are described in our previous paper. The essential points of the probe wavelength dependence are summarized as follows.

The molecular vibration changes electronic distribution, which causes susceptibility change described by nonlinear susceptibility (NS). The real part of NS induces the refractive index change, which changes the probe spectrum by molecular phase modulation. Because of this, the amplitude dependence of the vibration due to this mechanism depends on the first derivative of the absorbed photon energy spectrum as seen in Figure 9.2.4.5c. The imaginary part of NS provides energy loss and gain of probe light through the stimulated Raman processes. It gives bumps on both sides of the most intensely interacting probe wavelength in the Stokes and anti-Stokes sides, which can be seen in Figure 9.2.4.5f. The wavelength dependence are discussed in our previous paper for positive delay time, where the S/N is higher [45]. Figure 9.2.4.5e and f show the amplitude profiles and the first derivative of the relevant Gaussian components resulting from spectral deconvolution of the stationary absorption spectrum for different chiralities. To obtain the amplitude profiles, the transverse cutting lines through the 2D plot of the FT power spectra from 100 to $350\,cm^{-1}$ were first averaged, and then the square root of the resulting profile was taken. To discuss in the same way as mentioned in our previous paper [54], we compared the profile with the first derivative (DER) of the absorbed photon energy spectra of chiral systems of (6,5) and (7,5) around 572 and 652 nm, respectively. It can be seen that the RBM amplitude profiles are close to the DER spectra. However, as in the previous paper discussing the positive time signal, there are substantial deviations in both cases. Especially the existence of sidebands in both higher and lower energy sides is clearly seen. This is due to the Raman interaction inducing energy exchange between two spectral regions through coherent vibration. It results in the appearance of the sidebands separated by the vibration frequency to both red- and blue-shifted directions forming difference (DIF)-type contribution as discussed in the previous paper [45]. The fitting in the present case of the probe photon energy dependence of the vibrational amplitude with the sum of the DIF and DER is difficult because of low S/N due to short time integration of Fourier in the negative time range. Therefore, it was not made to show the fitting using the sum of them.

So far, the low-frequency vibrational modes at ~$320\,cm^{-1}$ have been assigned to RBMs. Usually, exact chirality assignment can be further achieved for RBMs by combining the resonance conditions described above with the mode frequencies of the vibrations in the 2D coherent phonon spectra [17,19,45]. However, compared to the electronic population decay time T^{el}_1 determined in the positive time range, the negative time data provide a much shorter electronic dephasing time T^{el}_2, which limits the precision of the vibrational frequency together with the vibrational dephasing time by the FT. As a result, the mode patterns are severely distorted as seen in the 2D plots shown in Figure 9.2.4.5c and d. The most distinct distortion is that the central frequencies of the two lobes on both sides of the resonance position in the fingerprint belonging to the same vibrational mode are shifted away from their original values with respect to the frequency axis. The distortion of the fingerprint signal pattern makes it difficult to assign chirality. There are at least four overlapping absorption peaks that resonate with the laser spectrum, but resonant dips of only two appear in the 2D FT plot. The vibrational modes with resonance dips at 2.16 eV (572 nm) and 1.92 eV (650 nm) can be ascribed to the (6,5) and (7,5) tubes with vibrational frequencies close to their Raman shifts at 310 and $285\,cm^{-1}$, respectively. The vibrational modes in the upper right corner of Figure 9.2.4.5c and d represent two small dips around 2.12 eV (584 nm) and 1.89 eV (653 nm), respectively, which are near the resonance positions of (6,4) at 590 nm and (8,3) at 670 nm. However, their average vibrational frequencies of more than $600\,cm^{-1}$ are much larger than their own Raman shifts at around 334 and $301\,cm^{-1}$, respectively. Some other vibrational modes, such as D modes [21], have also been reported below $2000\,cm^{-1}$, but their vibrational amplitudes are usually much weaker than those of RBMs and their vibrational frequencies are higher than $1000\,cm^{-1}$. Therefore the possibility of the D modes can be ruled out, and they may probably be due to the overtones of RBMs of (6,4) and (8,3) appearing only in the exciton state as discussed made before using Figure 9.2.4.4.

9.2.4.4   CONCLUSIONS

In this subsection, we studied the vibrational and electronic coherence relaxation dynamics for negative time delays in CoMoCat-grown SWNTs using a 7.1-fs laser pulse [55]. The real-time traces in negative time provide information on the electronic dephasing time and also the vibrational phase relaxation dynamics in excited electronic states. The electronic phase relaxation time was found to be in the range of 30–40 fs for this sample. The most interesting result is that the vibrations of the RBM could be detected, whereas those of the G mode are too weak to be resolved in the negative time range. This implies that the RBMs in the negative time are mostly due to wave packets generated in the exciton state. The dynamics of vibrational RBMs are studied from the negative time traces under experimental conditions that were identical to those of the vibronic dephasing process. Our findings indicate that the probe photon energy-dependent amplitude profiles of RBMs follow a first derivative behavior and difference type, analogous to the results in the positive time range. The research described in this subsection was conducted by the following people: Takayoshi Kobayashi, Zhaogang Nie, Bing Xue, Hiromichi Kataura, Youichi Sakakibara, and Yasumitsu Miyata [55].

REFERENCES

1. P. Harris, *Carbon Nanotubes and Related Structures: New Materials for the Twenty-First Century*, Cambridge University Press, Cambridge, England (1999).
2. M. S. Dresselhaus, G. Dresselhaus, and P. Avouris, *Carbon Nanotubes: Synthesis, Structure, Properties and Applications*, Springer, Berlin, Germany (2001).
3. M. Dresselhaus, *Carbon Nanotubes: Synthesis, Structure, Properties, and Applications*, Springer, New York (2001).
4. R. Saito, G. Dresselhaus, and M. Dresselhaus, *Physical Properties of Carbon Nanotubes*, World Scientific, Singapore (2003).
5. T. Kobayashi, M. Yoshizawa, U. Stamm, M. Taiji, and M. Hasegawa, "Relaxation dynamics of photoexcitations in polydiacetylenes and polythiophene," *J. Opt. Soc. Am. B* **7**, 1558–1578 (1990).
6. T. Kobayashi, *J-aggregates*, World Scientific, Singapore (1996).
7. S. Reich, C. Thomsen, and J. Maultzsch, *Carbon Nanotubes: Basic Concepts and Physical Properties*, Wiley-VCH, Berlin, Germany (2004).
8. S. M. Bachilo, M. S. Strano, C. Kittrell, R. H. Hauge, R. E. Smalley, and R. B. Weisman, "Structure-assigned optical spectra of single walled carbon nanotubes," *Science* **298**, 2361–2366 (2002).
9. C. Fantini, A. Jorio, M. Souza, M. S. Strano, M. S. Dresselhaus, and M. A. Pimenta, "Optical transition energies for carbon nanotubes from resonant Raman spectroscopy: Environment and temperature effects," *Phys. Rev. Lett.* **93**, 147406 (2004).
10. H. Telg, J. Maultzsch, S Reich, F. Hennrich, and C. Thomsen, "Chirality distribution and transition energies of carbon nanotubes," *Phys. Rev. Lett.* **93**, 177401 (2004).
11. J. Maultzsch, H. Telg, S. Reich, and C. Thomsen, "Radial breathing mode of single-walled carbon nanotubes: Optical transition energies and chiral-index assignment," *Phys. Rev. B* **72**, 205438 (2005).
12. S. Reich, C. Thomsen, and J. Robertson, "Exciton resonances quench the photoluminescence of zigzag carbon nanotubes," *Phys. Rev. Lett.* **95**, 077402 (2005).
13. C. D. Spataru, S. Ismail-Beigi, L. X. Benedict, and S. G. Louie, "Excitonic effects and optical spectra of single-walled carbon nanotubes," *Phys. Rev. Lett.* **92**, 077402 (2004).
14. F. Wang, G. Dukovic, L. Brus, and T. F. Heinz, "The optical resonances in carbon nanotubes arise from excitons," *Science* **308**, 838–841 (2005).
15. A. Srivastava, H. Htoon, V. I. Klimov, and J. Kono, "Direct observation of dark excitons in individual carbon nanotubes: Inhomogeneity in the exchange splitting," *Phys. Rev. Lett.* **101**, 087402 (2008).
16. A. Gambetta, C. Manzoni, E. Menna, M. Meneghetti, G. Cerullo, G. Lanzani, S. Tretiak, A. Piryatinski, A. Saxena, R. L. Martin et al., "Real-time observation of nonlinear coherent phonon dynamics in single-walled carbon nanotubes," *Nat. Phys.* **2**, 515–520 (2006).
17. Y.-S. Lim, K.-J. Yee, J.-H. Kim, E. H. Hároz, J. Shaver, J. Kono, S. K. Doorn, R. H. Hauge, and R. E. Smalley, "Coherent lattice vibrations in single-walled carbon nanotubes," *Nano Lett.* **6**, 2696–2700 (2006).
18. J.-H. Kim, K.-J. Han, N.-J. Kim, K.-J. Yee, Y.-S. Lim, G. D. Sanders, C. J. Stanton, L. G. Booshehri, E. H. Hároz, and J. Kono, "Chirality-selective excitation of coherent phonons in carbon nanotubes by femtosecond optical pulses," *Phys. Rev. Lett.* **102**, 037402 (2009).
19. G. D. Sanders, C. J. Stanton, J.-H. Kim, K.-J. Yee, Y.-S. Lim, E. H. Hároz, L. G. Booshehri, J. Kono, and R. Saito, "Resonant coherent phonon spectroscopy of single-walled carbon nanotubes," *Phys. Rev. B* **79**, 205434 (2009).
20. Y.-S. Lim, J.-G. Ahn, J.-HG. Kim, K.-J. Yee, T. Joo, S.-H. Baik, E. H. Hároz, L. G. Booshehri, and J. Kono, "Resonant coherent phonon generation in single-walled carbon nanotubes through near-band-edge excitation," *ACS Nano* **4**, 3222–3226 (2010).
21. K. Kato, K. Ishioka, M. Kitajima, J. Tang, R. Saito, and H. Petek, "Coherent phonon anisotropy in aligned single-walled carbon nanotubes," *Nano Lett.* **8**, 3102–3108 (2008).

22. L. Lüer, C. Gadermaier, J. Crochet, T. Hertel, B. Brida, and G. Lanzani, "Coherent phonon dynamics in semiconducting carbon nanotubes: A quantitative study of electron–phonon coupling," *Phys. Rev. Lett.* **102**, 127401 (2009).

23. K. Makino, A. Hirano, K. Shiraki, Y. Maeda, and M. Hase, "Ultrafast vibrational motion of carbon nanotubes in different pH environments," *Phys. Rev. B* **80**, 245428 (2009).

24. J. Wang, M. W. Graham, Y. Ma, G. R. Fleming, and R. A. Kaindl, "Ultrafast spectroscopy of midinfrared internal exciton transitions in separated single-walled carbon nanotubes," *Phys. Rev. Lett.* **104**, 177401 (2010).

25. A. Jorio, G. Dresselhaus, and M. S. Dresselhaus, *Carbon Nanotubes: Topics Applied Physics*, Vol. **111**, Springer-Verlag, Berlin, Heidelberg, Germany, p. 371 (2008).

26. A. M. Rao, E. Richter, S. Bandow, B. Chase, P. C. Eklund, K. A. Williams, S. Fang, K. R. Subbaswamy, M. Menon, A. Thesis et al., "Diameter-selective Raman scattering from vibrational modes in carbon nanotubes," *Science* **275**, 187–191 (1997).

27. N. Ishii, E. Tokunaga, S. Adachi, T. Kimura, H. Matsuda, and T. Kobayashi, "Optical frequency- and vibrational time-resolved two-dimensional spectroscopy by real-time impulsive resonant coherent Raman scattering in polydiacetylene," *Phys. Rev. A* **70**, 023811 (2004).

28. T. Kobayashi, J. Du, W. Feng, and K. Yoshino, "Excited-state molecular vibration observed for a probe pulse preceding the pump pulse by real-time optical spectroscopy," *Phys. Rev. Lett.* **101**, 037402 (2008).

29. C. J. Bardeen, Q. Wang, and C. V. Shank, "Femtosecond chirped pulse excitation of vibrational wave packets in LD690 and bacteriorhodopsin," *J. Phys. Chem. A* **102**, 2759–2766 (1998).

30. T. Saito and T. Kobayashi, "Conformational change in azobenzene in photoisomerization process studied with chirp-controlled sub-10fs pulses," *J. Phys. Chem. A* **106**, 9436–9441 (2002).

31. R. Pausch, M. Heid, T. Chen, W. Kiefer, and H. Schwoerer, "Selective generation and control of excited vibrational wave packets in the electronic ground state of K2," *J. Chem. Phys.* **110**, 9560–9567 (1999).

32. C. H. Brito Cruz, J. P. Gordon, P. C. Becker, R. L. Fork, and C. V. Shank, "Dynamics of spectral hole burning," *IEEE J. Quant. Electron.* **24**, 261–269 (1988).

33. J. J. Baumberg, B. Huttner, R. A. Taylor, and J. F. Ryan, "Dynamic contributions to the optical stark effect in semiconductors," *Phys. Rev. B* **48**, 4695–4706 (1993).

34. B. Fluegel, N. Peyghambarian, G. Olbright, M. Lindberg, S. W. Koch, M. Joffre, D. Hulin, A. Migus, and A. Antonetti, "Femtosecond studies of coherent transients in semiconductors," *Phys. Rev. Lett.* **59**, 2588–2591 (1987).

35. M. Lindberg and S. W. Koch, "Transient oscillations and dynamic stark effect in semiconductors," *Phys. Rev. B* **38**, 7607–7614 (1988).

36. C. V. Shank, R. L. Fork, C. H. Brito Cruz, W. H. Knox, G. R. Fleming, and A. E. Siegman (Eds.), *Ultrafast Phenomena V*, Springer-Verlag, Berlin, Germany, p. 179 (1986).

37. R. A. Cheville, "Grischkowsky, far-infrared terahertz time domain spectroscopy of flames," *Opt. Lett.* **20**, 1646–1648 (1995).

38. J.-Y. Bigot, M. T. Portella, R. W. Schoenlein, C. J. Bardeen, A. Migus, and C. V. Shank, "Non-markovian dephasing of molecules in solution measured with three-pulse femtosecond photon echoes," *Phys. Rev. Lett.* **66**, 1138–1141 (1991).

39. C. H. Grossman and J. J. Schwendiman, "Ultrashort dephasing time measurements in nile blue polymer films," *Opt. Lett.* **23**, 624–626 (1998).

40. Y. Wang and T. Kobayashi, "Electronic and vibrational coherence dynamics in a cyanine dye studied using a few-cycle pulsed laser," *Chem. Phys. Chem.* **11**, 889–896 (2010).

41. T. Hattori and T. Kobayashi, "Femtosecond dephasing in a polydiacetylene film measured by degenerate four-wave mixing with an incoherent nanosecond laser," *Chem. Phys. Lett.* **133**, 230–234 (1987).

42. I. A. Walmsley and C. L. Tang, "The determination of electronic dephasing rates in time-resolved quantum-beat spectroscopy," *J. Chem. Phys.* **92**, 1568–1574 (1990).

43. M. Khalil, N. Demirdöven, and A. Tokmakoff, "Coherent 2D IR spectroscopy: Molecular structure and dynamics in solution," *J. Phys. Chem. A* **107**, 5258–5279 (2003).

44. E. C. Fulmer, P. Mukherjee, A. T. Krummell, and M. T. Zanni, "A pulse sequence for directly measuring the anharmonicities of coupled vibrations: Two-quantum two-dimensional infrared spectroscopy," *J. Chem. Phys.* **107**, 8067 (**2004**).

45. T. Kobayashi, Z. Nie, J. Du, K. Okamura, H. Kataura, Y. Sakakibara, and Y. Miyata, "Electronic relaxation and coherent phonon dynamics in semiconducting single-walled carbon nanotubes with several chiralities," *Phys. Rev. B* **88**, 035424 (2013).

46. A. Shirakawa, I. Sakane, M. Takasaka, and T. Kobayashi, "Sub-5-fs visible pulse generation by pulse-front-matched noncollinear optical parametric amplification," *Appl. Phys. Lett.* **74**, 2268–2270 (1999).

47. A. Baltuska, T. Fuji, and T. Kobayashi, "Visible pulse compression to 4 fs by optical parametric amplification and programmable dispersion control," *Opt. Lett.* **27**, 306–308 (2002).

48. M. S. Arnold, S. I. Stupp, and M. C. Hersam, "Enrichment of single walled carbon nanotubes by diameter in density gradients," *Nano Lett.* **5**, 713–718 (2005).

49. S. M. Bachilo, L. Balzano, J. E. Herrera, F. Pompeo, D. E. Resasco, and R. B. Weisman, "Narrow (n, m)-distribution of single-walled carbon nanotubes grown using a solid supported catalyst," *J. Am. Chem. Soc.* **125**, 11186–11187 (2003).

50. Y. Miyata, K. Yanagi, Y. Maniwa, T. Tanaka, and H. Kataura, "Diameter analysis of rebundled single-wall carbon nanotubes using X-ray diffraction: Verification of chirality assignment based on optical spectra," *J. Phys. Chem. C* **112**, 15997–16001 (2008).

51. S. B. Sinnott and R. Andreys, "Carbon nanotubes: Synthesis, properties, and applications," *Crit. Rev. Solid State Mater. Sci.* **26**, 145–249 (2001).

52. H. Kataura, Y. Kumazawa, Y. Maniwa, I. Umezu, S. Suzuki, Y. Ohtsuka, and Y. Achiba, "Optical properties of single-wall carbon nanotubes," *Synth. Met.* **103**, 2555–2558 (1999).

53. M. Joffre, C. B. A La. Guillaume, N. Peyghambarian, M. Lindberg, D. Hulin, A. Migus, S. W. Koch, and A. Antonetti, "Coherent effects in pump–probe spectroscopy of excitons," *Opt. Lett.* **13**, 276–278 (1988).

54. T. Kobayashi and W. Zhuan, "Spectral oscillation in optical frequency-resolved quantum-beat spectroscopy with a few-cycle pulse laser," *IEEE J. Quant. Electron.* **44**, 1232–1241 (2008).

55. T. Kobayashi, Z. Nie, B. Xue, H. Kataura, Y. Sakakibara, and Y. Miyata, "Real-time spectroscopy of single-walled carbon nanotubes for negative time delays by using a few-cycle pulse laser," *J. Phys. Chem. C* **118**, 3285–3294 (2014).

Section 9.3

1D Oligomers and Polymers

9.3.1 Fluorescence from Molecules and Aggregates in Polycrystalline Thin Films of α-Oligothiophenes

9.3.1.1 INTRODUCTION

Thiophene-based oligomers have been extensively studied theoretically and experimentally for a better understanding of the physical properties of polythiophenes [1–9]. They are also promising materials for various devices [10–15], such as light-emitting diodes [16–18]. However, the radiative relaxation kinetics of α-oligothiophene vacuum-evaporated thin films has not yet been fully studied, and the origin of their fluorescence is still unclear [19–23]. In order to improve the α-oligothiophene devices and to provide a basis for the understanding of the photophysical properties of more complex polymers, it is important to identify the electronic state from which the fluorescence originates and to understand the relaxation dynamics of the photoexcitations, which may be spatially extended and energetically broadened with a degree related intrinsically with the molecular chemical structure and extrinsically with its surrounding.

Previous studies have shown that the vacuum-evaporated films of α-oligothiophene Tn ~where n denotes the number of the thiophene rings) has a crystallographic structure [21,22,24–26]. The molecules are oriented parallel to each other with the long molecular axes nearly perpendicular to the substrate plane [21,22,25,26]. The close-packed arrangement gives rise to the formation of aggregates with dramatically different photophysical properties to those of isolated α-oligothiophenes in solution. For example, the absorption spectrum of the film sample reveals a dichroic and blue-shifted band, which was interpreted in terms of the formation of H-aggregates due to the strong intermolecular interactions by various groups [19,22,25,27].

Here I would like to briefly explain the H-aggregates and J-aggregates. When small molecular chromophores (not polymers) assemble in a solid state, they often form J-type or H-type aggregates, depending on the relative alignment of the transition dipole moments on adjacent molecules. In J-aggregates molecules stack in a head-to-tail configuration, while in H-aggregates, molecules stack predominantly in face-to-face configurations of transition dipole moments. The energy levels of the interaction dipoles form a band structure. In the case of head-to-tail stacking in the J-aggregates, the allowed transition becomes the lowest while in the case of card packing the allowed transition is concentrated at the band top resulting in the red shift of the absorption spectrum. In J-aggregates N molecules are coherently coupled and can emit intense fluorescence with N times transition dipole corresponding to N^2 times larger transition cross-section. Because of Kasha rule, the fluorescence takes place from the bottom of the band. Then intense fluorescence is emitted from the J-band bottom. While in the case of H-aggregates the molecular stacking is card-packing type resulting in the location of accumulated N transition moments at the highest band edge of the H-band. Then the fluorescence emission is suffered from the competition with radiationless relaxation to the lowest forbidden level of the H-band, from which relatively weak fluorescence is emitted.

DOI: 10.1201/9780429196577-53

Several stationary and time-resolved studies have been performed on the α-oligothiophene vacuum evaporated films; Chen et al. measured the nanosecond time-resolved transient absorption spectra of polycrystalline films of a-quaterthiophene $T4$, α-quinquethiophene $T5$, and α-sexithiophene $T6$ [28] Lane and co-workers studied the absorption spectra of charged excitations in α sexithiophene thin films induced by doping and photogeneration [29]. Periasamy et al. and Zamboni et al. performed one-photon and two photon fluorescence excitation spectroscopies on an asexithiophene $T6$ vacuum-evaporated film. They concluded that the 1^1Bu exciton band is located at 573 nm ~2.164 eV) and the 2^1A_g exciton band is at 544 nm ~2.279 eV) for the $T6$ film [20]. Furukawa et al. carried out detailed studies of infrared absorption and Raman spectra on several α-oligothiophene films [8]. Watanabe et al. performed the femtosecond transient spectroscopy and picosecond time-resolved fluorescence spectroscopy on a $T6$ film [30]. Studies of femtosecond transient spectra of $T6$ polycrystalline films were also presented by Klein et al. [23] and Lanzani et al. [31], respectively. Very recently, site selective and time-resolved spectroscopies have been used to resolve the aggregate emission from the excitonic emission both in $T6$ thin films and single crystal by Marks et al. [3], and the dipole sums with the Ewald method have been performed to propose the locations of the Davydov-splitting components in the $T6$ single crystal by Muccini et al. [4].

In this subsection, we report the experimental results of stationary fluorescence, fluorescence excitation, site-selective fluorescence, and femtosecond time-resolved fluorescence spectroscopies of vacuum-evaporated films of α-quaterthiophene ($T4$) and α-quinquethiophene ($T5$) at different temperatures. It is found that there are various energetically separated fluorescent species in the α-oligothiophene vacuum-evaporated films. In comparison with isolated molecules, they show changes in electronic energy levels and excited-state kinetics because of intermolecular interactions.

9.3.1.2   EXPERIMENT

Details of synthesis and purification methods of α-oligothiophenes are described in Refs. [32,33]. Thin films of $T4$ (375 nm thick) and $T5$ (600 nm thick) were prepared onto fused silica glasses by heading a small quantity of purified power in a boat under higher vacuum (background pressure 1×10^{-5} Pa). The pressure during the deposition was about 3×10^{-4} Pa. The deposition rates were 24 nm/min for the $T4$ film and 39 nm/min for the $T5$ film. The substrate was kept at room temperature during the deposition.

The absorption spectra were obtained with a Shimadzu UV-3101 PC spectrometer. The stationary fluorescence and fluorescence excitation spectra were measured with a Hitachi F-4500 fluorimeter, and corrections were made for the spectral sensitivity of the entire measuring system. For low-temperature measurements, a continuous liquid He-flow cryostat (Oxford, CF-1204) was utilized.

The experimental apparatus for the femtosecond time-resolved fluorescence measurements is described in detail elsewhere [9]. In brief, a mode-locked Ti sapphire laser (Clark-MAX, NJA-4) was utilized to produce pulses at 810 nm with a repetition rate of 100 MHz and a 600 mW average power. The second harmonic (1 mm β-BaB_2O_4 BBO crystal) of the Ti sapphire laser was focused into the film sample with an average excitation power of less than 3 mW. The fluorescence from the sample was collected with two lenses and subsequently focused into a 0.5 mm thick BBO crystal to be mixed with the fundamental beam for the type-I phasematching sum frequency generation. The time resolution of the system was estimated to be about 250 fs from the full width at half maximum of the cross-correlation trace between the excitation and the fundamental pulses.

9.3.1.3   RESULTS AND DISCUSSION

9.3.1.3.1   Absorption and Fluorescence Excitation Spectra

The absorption spectra of the $T4$ and $T5$ films at room temperature measured at normal incidence are shown in Figure 9.3.1.1a, together with the absorption spectrum of $T4$ in dichloromethane for comparison. The absorption spectrum of the film is dramatically different to that of the solution spectrum. The broad, blue-shifted bands at 3.65 eV for the $T4$ film and at 3.56 eV for the $T5$

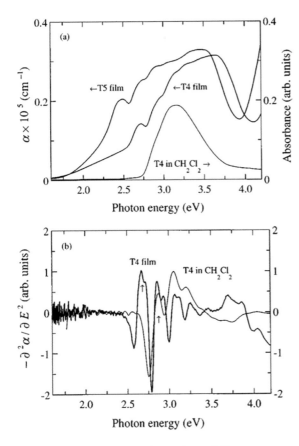

FIGURE 9.3.1.1 (a) Absorption spectra of the T4 and T5 films at room temperature measured at normal incidence (thick lines), and an absorption spectrum of the T4 in dichloromethane (thin line) at room temperature. (b) Negative second derivatives of the absorption spectra of the T4 film (thick line) and T4 in dichloromethane (thin line).

film correspond to the absorption of *H*-aggregates caused by strong intermolecular interactions [19,22,25,27]. The absorption spectrum of each film also reveals a structured band in the ultraviolet-visible region and a long-weak tail at the low-energy side.

The structured absorption bands in the region of 2.4–3.4 eV for the T4 and 2.35–3.35 eV for the T5 film are considered to be predominantly from disordered molecules lying at grain boundaries [19,22,34]. The structures observed are associated with the intramolecular electronic transition coupled to intramolecular vibrational modes ~electron-phonon coupling) [19,22,34]. In comparison with the absorption of isolated molecules, as shown in Figure 9.3.1.1a, the structured absorption band of the film sample is red-shifted. To determine the energy shift of the intramolecular electronic transition, $S_1 \leftarrow S_0 0$–0 (S_0 and S_1 are the ground and first excited singlet states, respectively), upon going from solution to film, we calculated second derivatives of the absorption spectra of the film and solution samples for each Tn; the $S_1 \leftarrow S_0 0$–0 transitions of the molecules in solution cannot be determined directly due to the broad absorption spectra [35].

As displayed in Figure 9.3.1.1b, the negative second derivative (NSD) spectrum of the T4 film exhibits oscillating structures. Four peaks at 2.67, 2.86, 3.07, and 3.25 eV are present. The average energy difference between adjacent peaks is about 0.19 eV, which is very close to the stretching vibrational energy of the C=C bond. The peak at 2.67 eV noted by an upward arrow in Figure 9.3.1.1b is assigned to the molecular $S_1 \leftarrow S_0 0$–0 transition of T4 in the evaporated film [34].

Three peaks and a small shoulder are also present in the NSD spectrum of T4 in dichloromethane at room temperature. The $S_1 \leftarrow S_0 0$–0 transition energy of T4 in dichloromethane is determined to be 2.87 eV [9,35]. Consequently, the energy shift of the $S_1 \leftarrow S_0 0$–0 transition of the T4 and T5 molecules upon going from solution to film is of the order of 0.20–0.21 eV. There are three possible mechanisms of the red-shift for the molecular transition in the solid. The first one is the larger stabilization of the excited state than in the ground state because of dielectric screening. The second one is the larger molecular interaction in film than in solution. The third one, which is associated

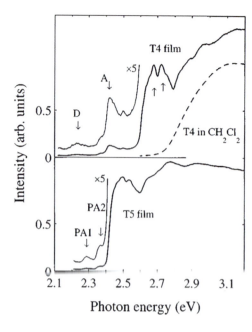

FIGURE 9.3.1.2 Fluorescence excitation spectra of the *T*4 film (solid line) at 4.2 K measured at an emission photon energy of 2.03 eV and the *T*5 film (solid line) at 4.2 K measured at an emission photon energy of 2.07 eV. The spectra in the low energy region (thin lines) are scaled with a factor of 5. The fluorescence excitation spectrum of *T*4 in dichloromethane (dashed line) at room temperature is shown for comparison.

with the molecular conformational change, is due to the extension of electronic wave functions of the oligothiophene molecule in the vacuum evaporated film compared to that in solution [9].

Figure 9.3.1.2 presents the fluorescence excitation (FE) spectra of the *T*4 and *T*5 films at 4.2 K and the FE spectrum of *T*4 in dichloromethane at room temperature. The lowest energy strong absorption band of the films shows fine structures; two peaks at 2.68 and 2.72 eV are observed for the *T*4 and two peaks at 2.50 and 2.54 eV for the *T*5 film. Previously, Fichoh et al. also observed similar fine structures in the absorption spectra of several α-oligothiophenes evaporated films at low temperature [34]. The fine structures in the FE spectrum are interpreted in terms of the coupling between the intramolecular π-π* transitions and various molecular vibrational modes, i.e., thiophene ring deformation modes [34,36]. In addition, we observed a very weak but structured tail band in the FE of each film at 4.2 K. In this band, two small peaks at 2.29 and 2.37 eV labeled as PA1 and PA2 are present in the excitation spectrum of the *T*5 film (see Figure 9.3.1.2). A small peak at 2.42 eV labeled as A and a bump at 2.24 eV labeled as D are also observed in the excitation spectrum of the *T*4 film. These observations show that in the evaporated film there are additional low energy, directly accessible excitations, which are absent for the molecule in solution.

9.3.1.3.2 FLUORESCENCE SPECTRA

The fluorescence spectra of the *T*4 and *T*5 films at 4.2 and 295 K are depicted in Figure 9.3.1.3. The fluorescence spectrum of the *T*4 film at 4.2 K consists of a weak and structured band in the region of 2.42–2.61 eV, a strong emission band at 2.35 eV, a distinct peak at 2.22 eV, and a small shoulder at 2.04 eV. The observation of the higher-energy peak at 2.61 eV shows the emission from high-energy species, which are assigned to the disordered molecules at the grain boundaries. From the inset of Figure 9.3.1.3, one can see that the spectral positions of the two peaks at 2.48 and 2.52 eV in the fluorescence spectrum are identical with those of two minima in the fluorescence excitation spectrum of the *T*4 film. These observations reflect reabsorption, i.e., the fluorescence from the high-energy emitting species is reabsorbed by some low-lying species. The reabsorption reduces the intensity of the fluorescence from the high-energy species. The above results indicate the inhomogenous distribution of energy levels in the evaporated films of α-oligothiophenes.

The fluorescence spectra of the *T*4 and *T*5 films at various temperatures are shown in Figure 9.3.1.4a and b, respectively. The fluorescence of the *T*4 film shows weak temperature dependence. In contrast, both the fluorescence intensity and spectrum of the *T*5 film are strongly dependent

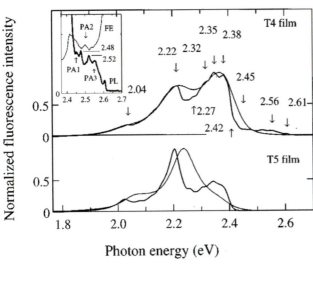

FIGURE 9.3.1.3 Normalized fluorescence spectra of the T4 and T5 films at 4.2 K (thick lines) and 295 K (thin lines). The excitation photon energies are 3.26 eV for the T4 film and 3.35 eV for the T5 film. The inset displays the fluorescence ~thick line) and fluorescence excitation (thin line) spectra in the spectral region of 2.35–2.7 eV for the T4 film at 4.2 K.

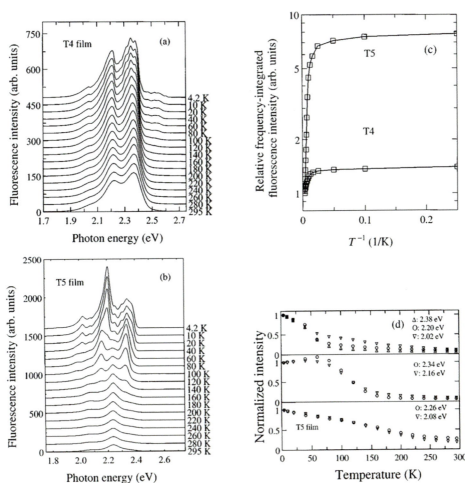

FIGURE 9.3.1.4 (a) Fluorescence spectra of the T4 film at various temperatures for an excitation photon energy of 3.26 eV. (b) Fluorescence spectra of the T5 film at various temperatures for an excitation photon energy of 3.35 eV. (c) Temperature dependence of frequency-integrated fluorescence intensities for the T4 and T5 films. (d) Fluorescence intensities of the T5 film vs temperature at various emission photon energies.

on the temperature, T. The frequency-integrated fluorescence intensity of each film is plotted as a function of T in Figure 9.3.1.4c. The fluorescence intensity of the $T5$ film is increased by eight times when the temperature is decreased from 295 to 4.2 K, while that of the $T4$ film is increased by only about 1.4. Such different temperature dependence between the $T4$ and $T5$ films indicates different emission mechanisms and fluorescence origin for the two films.

At room temperature, the fluorescence spectrum of the $T5$ film shows a peak at 2.24 eV and a shoulder at 2.08 eV. At 4.2 K, a strong emission band at 2.20 eV is present, and a high-energy band at 2.34 eV shows observable structures.

An additional peak observed at 4.2 K is located at 2.02 eV. To understand the dramatic change of the fluorescence line shape with decreasing temperature, we analyze the temperature-dependent intensity of the fluorescence at different emission photon energies. As can be seen from Figure 9.3.1.4d, there are three distinct features. (1) For the peaks at 2.38, 2.20, and 2.02 eV with an energy spacing of 0.18 eV, the fluorescence intensities show the same increasing tendency when the temperature is decreased. This observation suggests that these three peaks originate from the same emitting site. The 0.18 eV energy spacing is associated with the C=C vibrational progressions. (2) The fluorescence intensities at 2.34 and 2.16 eV, with an energy separation of 0.18 eV, show similar temperature dependence with each other, but different from that for case (1). This result indicates that these two peaks share one fluorescent origin which differs from the origin in case (1). (3) The fluorescence intensities at 2.26 and 2.08 eV, again with an energy separation of 0.18 eV, show nearly identical temperature-dependent features which are distinct from cases (1) and (2). This result suggests that the fluorescence at 2.26 and 2.08 eV is from an additional lowest-energy emitting site which differs from the origin in cases (1) and (2). Thus, the fluorescence spectrum of the $T5$ film at 4.2 K as shown in Figure 9.3.1.3 is a superposition of the emission spectra originated from three energetically different emitting sites. The fluorescence from the three different sites exhibits identical vibrational development with the 0.18 eV energy spacing. Similar phenomena have also been observed in a $T6$ film [37].

Based on the above observations, we suggest that the fluorescence peaks at 2.38, 2.20, and 2.02 eV correspond to the $S_1 \rightarrow S_0 0$–0, $S_1 \rightarrow S_0 0$–1, and $S_1 \rightarrow S_0 0$–2 transitions from one high-energy emitting site. The peaks at 2.34 and 2.16 eV are, respectively, attributed to the $S_1 \rightarrow S_0 0$–0 and $S_1 \rightarrow S_0 0$–1 transitions from site PA2, which shows a small absorption peak at 2.37 eV ~see Figure 9.3.1.2). The peaks at 2.26 and 2.08 eV are, respectively, ascribed to the $S_1 \rightarrow S_0 0$–0 and $S_1 \rightarrow S_0 0$–1 transitions from site PA1 which exhibits a small absorption peak at 2.29 eV (see Figure 9.3.1.2).

9.3.1.3.3 SITE-SELECTIVE FLUORESCENCE SPECTRA

We have performed site-selective fluorescence measurements on the $T4$ film at 4.2 K. The fluorescence spectra measured at excitation energies of 2.43, 2.53, 2.82, 3.02, 3.26, and 3.54 eV are shown in Figure 9.3.1.5a. For the excitation energy greater than 2.53 eV, the emission spectrum is almost independent of the excitation photon energy. The peaks of the fluorescence at 2.35, 2.22, and 2.04 eV are not shifted with changing excitation photon energy from 2.43 to 3.54 eV. Even at 2.43 eV excitation, the spectral positions of the peaks at 2.22 and 2.04 eV are still not shifted with respect to those at 3.02 eV excitation. However, changes in fluorescence intensity ratios of $I_{2.22}/I_{2.04}$, $I_{2.27}/I_{2.04}$, $I_{2.32}/I_{2.04}$, $I_{2.35}/I_{2.04}$, and $I_{2.38}/I_{2.04}$, are observed as the excitation photon energy is turned to 2.53 and 2.43 eV. For instance, at 3.02 eV excitation the ratios are 4.29, 3.30, 4.66, 5.51, and 5.25, respectively; but at 2.53 eV excitation, they are 4.46, 4.04, 5.83, 5.87, and 4.84, respectively. The intensity ratios are plotted against the excitation energy in Figure 9.3.1.5b.

The following three characteristic features, shown in Figure 9.3.1.5b, plotted for the spectral shape of $T4$ at 4.2 K.

1. At 2.53 eV excitation, the ratios of $I_{2.27}/I_{2.04}$ and $I_{2.32}/I_{2.04}$ are higher than those at 2.82 eV excitation. However, as the excitation energy decreased further to 2.43 eV, the two fluorescence intensity ratios decreased from those at 2.53 eV excitation. These results may reflect

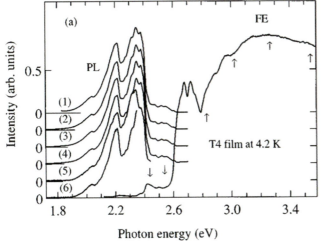

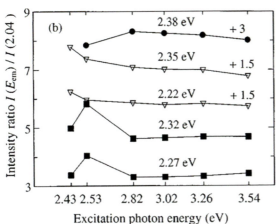

FIGURE 9.3.1.5 (a) Fluorescence (PL) and fluorescence excitation (FE) spectra of the *T*4 film at 4.2 K. The fluorescence spectra are measured at the excitation photon energies of 2.43 (1), 2.53 (2), 2.82 (3), 3.02 (4), 3.26 (5), and 3.54 eV (6), respectively. The excitation spectrum is measured at an emission photon energy of 2.03 eV. The arrows indicate the different excitation energies. (b) Intensity ratios of the fluorescence at 2.22, 2.27, 2.32, 2.35, and 2.38 eV to that at 2.04 eV vs excitation photon energy for the *T*4 film at 4.2 K. The intensity ratios of the fluorescence at 2.22, 2.35, and 2.38 eV are shifted upward by adding 1.5, 1.5, and 3, respectively, for a better view.

that the fluorescence at 2.27 and 2.32 eV are partly contributed by some species with original levels in the region of 2.43–2.53 eV.

2. At 2.53 eV excitation, the ratio $I_{2.38}/I_{2.04}$ is decreased from that at 2.82 eV excitation. This result may indicate that a part of the fluorescence at 2.38 eV stems from some species with original levels above 2.53 eV.

3. The increase of the ratios $I_{2.22}/I_{2.04}$ and $I_{2.35}/I_{2.04}$ as the excitation photon energy decreases from 2.82 to 2.53 and 2.43 eV, being different from cases (1) and (2), suggests that the fluorescence at 2.22 and 2.35 eV are mainly contributed by additional species with original energy levels below 2.43 eV. Therefore, the above results show that the fluorescence spectrum of the *T*4 film (see Figure 9.3.1.3) is a superposition of the emission spectra from various emitting species. The original levels of the predominant fluorescence in the *T*4 film are below 2.43 eV.

9.3.1.3.4 ASSIGNMENT OF FLUORESCENCE

In 1997, Bongiovanni et al. studied the conformations and optical properties of *T*5 molecules included in a perhydrotriphenylene matrix, both theoretically and experimentally [38]. They suggested that a weak absorption band observed at 2.43 is related to the 0–0 vibronic feature of a planar conformer. We have observed two fluorescence origin levels at 2.29 and 2.37 eV in the *T*5 vacuum evaporated film as shown in Figure 9.3.1.2. They are red-shifted by about 0.14 and 0.06 eV, respectively, in comparison with the lowest-energy band at 2.43 eV of the *T*5 molecule in the host matrix [38]. Therefore, the physics of the two origins at 2.29 and 2.37 eV is different from that of the isolated molecule. We name these energetically separated emitting sites in the region of 2.29–2.43 eV as pre-aggregate species. The pre-aggregate species are characterized by very low optical densities and red-shifted absorption maxima compared to those of the disordered molecules, implying that the pre-aggregate species are located in relatively ordered domains where the intermolecular interactions may be stronger than those in the disordered domains. The fluorescence spectrum from the different sites of the pre-aggregate species exhibits identical vibrational progression with an energy spacing of about 0.18 eV. The pre-aggregate species is the dominant fluorescence origin in the T5 film.

The pre-aggregate species may also be observed in the T4 film. On the basis of the fluorescence, fluorescence excitation, and site-selective fluorescence experimental results, the peaks at 2.27, 2.32, and 2.38 eV in the fluorescence spectrum may be assigned to the $S_1 \rightarrow S_0 0$–1 transitions with corresponding $S_1 \rightarrow S_0 0$–0 peaks at 2.45 (PA1), 2.50 (PA2), and 2.56 eV (PA3), respectively (see Figure 9.3.1.3). Therefore, below the first strong absorption band of the disordered molecules, there are pre-aggregate species which consist of several energetically separated sites with the fluorescence original levels in the region of 2.43–2.61 eV for the T4 film and 2.29–2.43 eV for the T5 film.

As mentioned in Sec. III B, the rather large difference in the fluorescence temperature dependence between the T4 and T5 films indicates that the electronic configurations of the predominant fluorescence species in the T4 film differ from those in the T5 film. The site-selective fluorescence measurements have demonstrated that there is additional fluorescence origin in the low-energy region for the T4 film. Previously, Birnbaum et al. performed measurements of fluorescence emission and fluorescence excitation spectra of T4 in tetradecane at 4.2 K [39]. They observed four independent, but nearly identical excitation/emission pairs with different dipole allowed origin and the same vibrational development. The lowest energy dipole allowed origin was observed to be at 2.75 eV. For the T4 vacuum-evaporated film, we have observed two small peaks at 2.24 and 2.42 eV in the excitation spectrum as shown in Figure 9.3.1.2. They are red-shifted at about 0.51 and 0.33 eV, respectively, in comparison with the lowest-energy band at 2.75 eV of the T4 molecule in the tetradecane lattice at 4.2 K. We propose that the strong emission band at 2.35 eV in the T4 film is mainly originated from an aggregate-like state at 2.42 eV labeled as A in the excitation spectrum (see Figure 9.3.1.2). Below the emission photon energy of 2.24 eV, a part of the fluorescence may stem from the T4 molecule trapped In physical defects in crystal, which shows a small absorption bump at 2.24 eV labeled as D in the excitation spectrum (Figure 9.3.1.2). Further evidence supporting the above assignments come from the time-resolved fluorescence spectra to be discussed in the following.

9.3.1.3.5 TIME-RESOLVED FLUORESCENCE SPECTRA

We have measured fluorescence decays in a wide emission spectral range for the T4 and T5 films. As with the stationary spectra, the fluorescence kinetics are quite different between the T4 and T5 films. The time-resolved fluorescence decay of the T4 film changes dramatically with the detected fluorescence wavelength, while no wavelength-dependent fluorescence decay is observed for the T5 film in the region of 2.03–2.38 eV within our experimental uncertainty ~see Figure 9.3.1.6). The fluorescence decays of the T5 film are better described by a stretched-exponential function $\exp[(-t/\tau)^\alpha]$. By fitting the stretched-exponential function to the experimental data measured at an emission photon energy of 2.21 eV for the T5 film at 5 K, the parameters τ and α are determined to be 3161 and 0.560.01 ps, respectively.

The fluorescence decays of the T4 film are shown in Figure 9.3.1.7a–c. In the higher energy region of 2.41–2.60 eV, the decays are nonexponential. The fluorescence decays at the emission photon energies of 2.45, 2.51, and 2.60 eV are well described by a sum of a single exponential and a stretched exponential functions,

$$I(t) = A_1 \exp(-t/\tau_1) + A_2 \exp\left[(-t/\tau_2)^\alpha\right] \qquad (9.3.1.1)$$

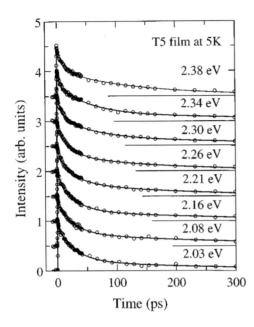

FIGURE 9.3.1.6 Fluorescence decay curves measured at different spectral positions for the T5 film at 5 K for an excitation photon energy of 3.02 eV.

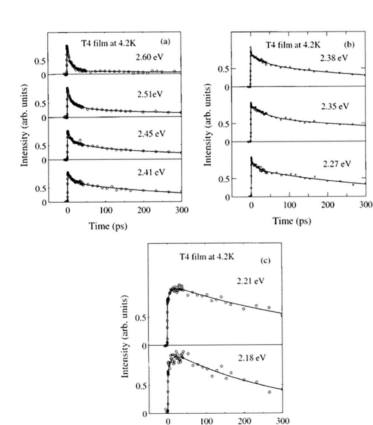

FIGURE 9.3.1.7 (a) Fluorescence decay curves of the T4 film at 4.2 K measured at different spectral positions for an excitation photon energy of 3.02 eV. (b) Fluorescence decay curves of the T4 film at 4.2 K measured at different spectral positions for an excitation photon energy of 3.02 eV. (c) Fluorescence decay curves of the T4 film at 4.2 K measured at different spectral positions for an excitation photon energy of 3.02 eV.

TABLE 9.3.1.1

Fluorescence Decay Parameters of the *T*4 Film at 4.2 K

E_{em} (eV)	τ_1 (ps)	a_1	τ_2 (ps)	α	a_2
2.60	340 ± 190	0.12	8 ± 1	0.75	0.88
2.51	375 ± 90	0.28	11 ± 1	0.64	0.72
2.45	320 ± 50	0.48	10 ± 2	0.61	0.52
2.27	390 ± 65	0.70	13 ± 5	0.68	0.30

Here $I(t)$ is the intensity at time t, A_i ($i=1$ and 2), τ_i ($i=1$ and 2), and α are fitting parameters. The parameters determined by fitting the experimental data to Eq. (9.3.1.1) are summarized in Table 9.3.1.1. The fact that the value of the time constant t_1 of the slow exponential decay component is very close to that of *T*4 in dichloromethane (390 ps) [35], supports the suggestion that the fluorescence spectrum contains a contribution from disordered molecules. From Table 9.3.1.1, one can also see that the values of t_2 and a of the fast stretched exponential decay component, similar to those of the *T*5 film, do not change much as a function of the detected spectral positions. This result may indicate again the contribution of the preaggregate species in the *T*4 film. Further, the temporal fluorescence at the emission photon energy of 2.27 eV is also well described by Eq. (9.3.1.1), and the values of t_2 and a as listed in Table 9.3.1.1 are very close to those at 2.45 eV. This fact supports our assignment that the fluorescence at 2.27 corresponds to the $S_1 \rightarrow S_0 0-1$ transition with associated $S_1 \rightarrow S_0 0-0$ peak at 2.45 eV ~PA1). The above results reinforce the conclusion that there are energetically different emitting species in the *T*4 film.

The stretched-exponential decay, Kohlrausch–Williams–Watts (KWW) relaxation, is usually observed in disordered systems [39]. Recently, the KWW law was used to describe the fluorescence kinetics of PPV [poly-(phenylene vinylene)] and LPPP [ladder-type poly-(para)-phenylene] films [40–42]. The fluorescence decay profiles in these polymers are dependent on the emission wavelength detected. These observations were interpreted consistently in terms of energetic relaxation of neutral excitations within an inhomogeneously broadened density of states ~DOS), i.e., photoexcitations undergo incoherent random walk among the chain segments until they reach a longer segment from which they cannot escape within their lifetime. In contrast, no obvious wavelength dependence of the fluorescence decay was observed in the *T*5 film. We may interpret these results in terms of that, below the pre-aggregate, there are further energetically low-lying states where the photoexcitations can migrate efficiently. One of the further low-lying states is proposed to be a dipole-forbidden state (*L*-band) of the *H* aggregate, because we have observed the dichroic and blue shifted *H*-band of the *H*-aggregate, both in the *T*4 and in the *T*5 films. Since the transition from the *L*-band to the ground state is dipole-forbidden, the nonradiative process from this state is expected to be highly dominant. Here we note that the above discussion is based on the very large Davydov–Splitting hypothesis proposed in Refs. [19,22,25,27]. This hypothesis has been questioned by Muccini et al. and Marks et al. recently; and they point out that the actual Davydov–Splitting value in *T*6 single crystal should be much lower [3,4]. Further experimental evidence is required to determine the locations of the Davydov–Splitting components in the *T*4 and *T*5 polycrystalline films.

For the *T*4 film, at the emission photon energy of 2.35 eV we observed a long-lived decay component with a time constant of about 670 ps, which is greater than that of the *T*4 molecule in solution ~about 390 ps) [35]. The decay time of the long-lived component obtained here is in agreement with a previous result [22]. This observation is suggestive of the presence of a new species, which is assigned to the dipole allowed aggregate with an associated absorption peak at 2.42 eV labeled as A in the excitation spectrum (see Figure 9.3.1.2).

At the emission energies of 2.18 and 2.21 eV for the *T*4 film (see Figure 9.3.1.7), the transient fluorescence exhibits a slow rise with a maximum at about 24 ps and subsequently reveals a slow single-exponential decay with a time constant of the order of 340–450 ps which is very close to the fluorescence decay time of the $T4$ molecule in solution. The fluorescence decay at the emission photon energy of 2.21 eV has been studied by changing temperature and polarization of the excitation light. It is found that the feature of the slow rise at 2.21 eV is almost independent of the temperature and polarization of the excitation light. Therefore, the slow rise observed may contain two messages. First, there is an additional energy level in the low-energy region. Second, the additional species can be excited through the migration of photoexcitations from energetically higher-lying species. Previously, Zamboni et al. performed site-selective fluorescence measurements on a $T6$ film. They observed that the fluorescence spectrum with two prominent bands shifted to lower energies as laser excitation energy decreases as much as 2000 cm^{-1} (~0.25 eV) below the 2.164 band. They assigned the states extending 2000 cm^{-1} below the 2.164 band to the $T6$ molecules located in physical defect sites in crystal [20]. Accordingly, it is suggested that the slow rise at 2.21 eV for the $T4$ film may be associated with energy transfer from the energetically high-lying species to deep-trap defects. The defect species show a very weak absorption band at 2.24 eV which is about 0.21 eV below the PA1 state at 2.45 eV as shown in Figure 9.3.1.2.

We have measured the fluorescence efficiency of our $T5$ film to be on the order of 0.023 at room temperature. The fluorescence intensity is increased by eight times as lowing the temperature from 295 to 4.2 K, indicating that a thermally activated nonradiative decay route is strongly annihilated at low temperature. At higher temperatures, the lattice thermal fluctuation may play a role in quenching the fluorescence of the $T5$ film. The fluorescence efficiency of our $T4$ film is about 0.1 at room temperature, which is higher than that of the $T5$ film. The fluorescence intensity of the $T4$ film is not changed much with the temperature, because the dominant fluorescence originated from the higher ordered system, the dipole-allowed aggregates. The formation of the dipole-allowed aggregates in the $T4$ film may open an efficient radiative decay channel competed with nonradiative decay route, resulting in higher fluorescence efficiency in comparison with that of the $T5$ film.

9.3.1.4 SUMMARY

We have studied the stationary and time-resolved fluorescence spectra of vacuum-evaporated polycrystalline films of a-quaterthiophene and a-quinquethiophene. The fluorescence spectrum of each film is found to be the superposition of emissions from various species and sites. In the $T4$ film, fluorescence is observed from disordered molecules that exhibit the solution-like decay features due to weak intermolecular interaction. The multiple emission spectra from preaggregate species which consist of several energetically separated emitting sites are observed in the $T5$ film. The interactions of each molecule with neighboring units in the pre-aggregate species may be greater than those of the disordered molecules lying at the grain boundary but weaker than those of the molecules in the aggregates. The fluorescence of the pre-aggregate species shows well-defined intramolecular vibrational progressions, and its time-resolved decay is a stretched exponential without emission wavelength dependence. The pre-aggregate is the major emitting species in the $T5$ film. In the $T4$ film, the fluorescence is predominantly from dipole-allowed aggregate and defect species. The formation of the fluorescent aggregate species in the $T4$ film opens an efficient radiative decay channel. The fluorescent aggregate species may also play a role in transferring the photoexcitations to the defects in crystal. The H-aggregate is considered to be a fluorescence quenching center. In addition, based on the results of the temperature-dependent fluorescence spectra, the lattice thermal fluctuation may reduce the fluorescence efficiency at higher temperatures but be frozen out at lower temperatures.

We present a diagram to illustrate schematically the various relaxation and recombination processes in Figure 9.3.1.8. On the basis of the above stationary and time-resolved results, it is suggested that the various hot excitations, excluding the defect species in the $T4$ film, can be generated directly by incident light (3.02 eV), which is expressed in Figure 9.3.1.8 as thick solid upward arrow (Ex). The hot excitons relax down to the red edges of the excited manifolds of the various species upon rapid "internal conversion" (IC), indicated by dashed downward arrows in Figure 9.3.1.8. From there, the excitations of the high-energy species can go to the ground state through both

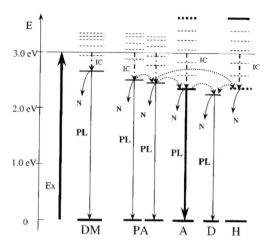

FIGURE 9.3.1.8 A scheme of the dynamics of the photoexcitations in the α-quaterthiophene vacuum-evaporated film. DM, Disordered molecule; PA, Pre-aggregate; A, Aggregate; D, Localized-defect; H, H-aggregate; Ex, excitation light; PL, Fluorescence; N, Nonradiative decay process; IC, Internal conversion.

radiative (thick downward arrow lines labeled as PL) and nonradiative decay routes (curve arrows noted by N). Additionally, the high-energy excitations can transfer the energy to the low-energy species (dashed curve arrows).

In this subsection, the fluorescence spectrum of the vacuum evaporated film of each α-oligothiophene studied is found to be the superposition of the emission spectra from various energetically different species. In vacuum-evaporated α-oligothiophene polycrystalline film, the molecules located at different domains in a grain experience nonequivalent intermolecular interactions, manifesting in different shifts in energy levels as well as various departure in fluorescence kinetics from those of isolated molecules. The efficiency of the formation of the fluorescent aggregates may relate to the chemical structure of the molecule [44]. In 1996, Bennati et al. suggested that the reduction of rotation barriers with increasing thiophene unites causes a distribution of twist angle rather than definite *cis* and *trans* configurations [43]. Kanemitsu et al. found that the molecular orientation of the oligomer with a longer chain length is very sensitive to the deposition rate [22]. It is very likely that the formation of the dipole allowed fluorescent aggregate species relates to the degree of the crystallinity of the film. The increase of the intramolecular disorder, as increasing the chain length, is considered to be the reason of difficulty of film formation with high quality crystallinity and of the formation of the dipole-allowed fluorescent aggregate species for longer oligomers. Furthermore, based on the above assignments, the Stokes shift is less than 0.07 eV for each fluorescent species and sites. From the discussion, it was considered that the elementary photoexcitations in the vacuum-evaporated films of the α-oligothiophenes are intra- and intermolecular Frenkel excitons. The contents of this subsection is based on the research activity of the following people: Aiping yang, Masashi Kuroda, Yotaro Shiraishi, Takayoshi Kobayashi [44].

REFERENCES

1. T. Kobayashi, M. Yoshizawa, U. Stamm, M. Taiji, and M. Hasegawa, *J. Opt. Soc. Am. B* **7**, 1558 (1990).
2. U. Stamm, M. Taiji, M. Yoshizawa, K. Yoshino, and T. Kobayashi, *Mol. Cryst. Liq. Cryst. A* **182**, 147 (1990).
3. R. N. Marks, M. Muccini, E. Lunedei, R. H. Michel, M. Murgia, R. Zamboni, C. Taliani, G. Horowitz, F. Garnier, M. Hopmeier, M. Oestreich, and R. F. Mahrt, *Chem. Phys.* **227**, 49 (1998).
4. M. Muccini, E. Lunedei, A. Bree, G. Horowitz, F. Garnier, and C. Taliani, *J. Chem. Phys.* **108**, 7327 (1998).
5. F. Garnier, R. Hajlaoui, A. Yassar, and P. Srivastava, *Science* **265**, 1684 (1994).
6. L. L. Miller, Y. Yu, E. Gunic, and R. Duan, *Adv. Mater.* **7**, 547 (1995).
7. D. Fichou, M.-P. Teulade-Fichou, G. Horowitz, and F. Demanze, *Adv. Mater.* **9**, 75 (1997).
8. Y. Furukawa, M. Akimoto, and I. Harada, *Synth. Met.* **18**, 151 (1988).
9. A. Yang, S. Hughes, M. Kuroda, Y. Shiraishi, and T. Kobayashi, *Chem. Phys. Lett.* **280**, 475 (1997).

10. G. Horowitz, D. Fichou, X. Z. Peng, Z. G. Xu, and F. Garnier, *Solid State Commun.* **72**, 381 (1997).
11. N. S. Sariciftci, U. Lemmer, D. Vacar, A. J. Heeger, and R. A. J. Janssen, *Adv. Mater.* **8**, 651 (1996).
12. A. Dodabalapur, L. Torsi, and H. E. Katz, *Science* **268**, 270 (1995).
13. N. Noma, T. Tsuzuki, and Y. Shirota, *Adv. Mater.* **7**, 647 (1995).
14. R. Hajlaoui, G. Horowitz, and F. Garnier, *Adv. Mater.* **9**, 389 (1997).
15. D. Fichou, J.-M. Nunzi, F. Charra, and N. Pfeffer, *Adv. Mater.* **6**, 64 (1994).
16. F. Geiger, M. Stoldt, H. Schweizer, P. Bauerle, and E. Umbach, *Adv. Mater.* **5**, 922 (1993).
17. T. Noda, H. Ogawa, N. Noma, and Y. Shirota, *Adv. Mater.* **9**, 720 (1997).
18. K. Uchiyama, K. Akimichi, S. Hotta, H. Noge, and H. Sakaki, *Synth. Met.* **63**, 57 (1994).
19. A. Yassar, G. Horowitz, P. Valat, V. Hmyene, F. Deloffre, P. Srivastava, P. Lang, and F. Garnier, *J. Phys. Chem.* **99**, 9155 (1995).
20. (a) N. Periasamy, R. Danieli, G. Ruani, R. Zamboni, and C. Taliani, *Phys. Rev. Lett.* **68**, 919 (1992); (b) R. Zamboni, N. Periasamy, G. Ruani, and C. Taliani, *Synth. Met.* **54**, 57 (1993).
21. P. Ostoja, S. Guerri, S. Rossini, M. Servidori, C. Taliani, and R. Zamboni, *Synth. Met.* **54**, 447 (1993).
22. Y. Kanemitsu, N. Shimizu, K. Suzuki, Y. Shiraishi, and M. Kuroda, *Phys. Rev. B* **54**, 2198 (1996).
23. G. Klein, C. Jundt, B. Sipp, A. A. Villaeys, A. Boeglin, A. Yassar, G. Horowitz, and F. Garnier, *Chem. Phys.* **215**, 131 (1997).
24. B. Servet, G. Horowitz, S. Ries, O. Lagorsse, P. Alnot, A. Yassar, F. Deloffre, P. Srivastava, R. Hajlaoui, P. Lang, and F. Garnier, *Chem. Mater.* **6**, 1809 (1994).
25. K. Hamano, T. Kurata, S. Kubota, and H. Koezuka, *Jpn. J. Appl. Phys., Part 2* **33**, L1031 (1994).
26. G. Barbarella, M. Zambianchi, A. Bongini, and L. Antolini, *Adv. Mater.* **4**, 282 (1992).
27. J. Egelhaaf, P. Bauerle, K. Rauer, V. Hoffmann, and D. Oelkrug, *Synth. Met.* **61**, 143 (1993).
28. X. Chen, K. Ichimura, D. Fichou, and T. Kobayashi, *Chem. Phys. Lett.* **185**, 286 (1991).
29. P. A. Lane, X. Wei, Z. V. Vardeny, J. Poplawski, E. Ehrenfreund, M. Ibrahim, and A. J. Frank, *Chem. Phys.* **210**, 229 (1996).
30. K. Watanabe, T. Asahi, H. Fukumura, H. Masuhara, K. Hamano, and T. Kurata, *J. Phys. Chem. B* **101**, 1510 (1997).
31. G. Lanzani, M. Nisoli, S. D. Silvestri, and F. Abbate, *Chem. Phys. Lett.* **264**, 667 (1997).
32. J. Nakahara, T. Konishi, S. Murabayashi, and M. Hoshino, *Heterocycles* **26**, 1793 (1987).
33. E. Schlte, G. Henke, G. Ruecker, and S. Foerster, *Tetrahedron* **24**, 1899 (1968).
34. D. Fichou, G. Horowitz, B. Xu, and F. Garnier, *Synth. Met.* **48**, 167 (1992).
35. A. Yang, M. Kuroda, Y. Shiraishi, and T. Kobayashi, *J. Phys. Chem. A* **102**, 3706 (1998).
36. (a) D. Birnbaum and B. E. Kohler, *J. Chem. Phys.* **95**, 4783 (1991); (b) **90**, 3506 (1989); (c) D. Birnbaum, D. Fichou, and B. E. Kohler, *ibid.* **96**, 165 (1991).
37. A. Yang, Doctoral Dissertation, Department of Physics, University of Tokyo, Japan (1997).
38. G. Bongiovanni, C. Botta, J. L. Bredas, J. Cornil, D. R. Ferro, A. Mura, A. Piaggi, and R. Tubino, *Chem. Phys. Lett.* **278**, 146 (1997).
39. R. A. Cheville and N. J. Halas, *Phys. Rev. B* **45**, 4548 (1992).
40. R. Kersting, B. Mollay, M. Rusch, J. Wenisch, G. Leising, and H. F. Kauffmann, *J. Chem. Phys.* **106**, 2850 (1997).
41. M. Scheidler, U. Lemmer, R. Kersting, S. Karg, W. Riess, B. Cleve, R. F. Mahrt, H. Kurz, H. Bassler, E. O. Gobel, and P. Thomas, *Phys. Rev. B* **54**, 5536 (1996).
42. R. F. Mahrt, T. Pauck, U. Lemmer, U. Siegner, M. Hopmeier, R. Hennig, H. Bassler, and E. O. Gobel, *Phys. Rev. B* **54**, 1759 (1996).
43. M. Bennati, A. Grupp, M. Mehring, and P. Bauerle, *J. Phys. Chem.* **100**, 2849 (1996).
44. A. Yang, M. Kuroda, Y. Shiraishi, T. Kobayashi, *J. Chem. Phys.* **109**, 8442 (1998).

9.3.2 Sequential Singlet Internal Conversion of $1B_u^- \rightarrow 3A_g^- \rightarrow 1B_u^- \rightarrow 2A_g^- \rightarrow (1A_g^-$ Ground) in All-*Trans*-Spirilloxanthin Revealed by Two-Dimensional Sub-5-fs Spectroscopy

9.3.2.1 INTRODUCTION

In bacterial photosynthetic systems, all-*trans*-carotenoids (Cars), generally having conjugated double bonds of numbers $n=9$–13, are specifically bound to the LH1 and LH2 antenna complexes and play an important function of light harvesting [1–4]. This function includes the transfer of energy to bacteriochlorophylls (BChls) after photo-absorption by Cars. Approximate C_{2h} symmetry of the all-*trans* conjugated chain gives rise to singlet-excited states that can be classified into kB_u^+, lB_u^-, mA_g^+, and nA_g^- groups; here, Pariser's signs, $+$ and $-$, indicate the symmetry of electronic configurations [5,6], and k, l, m, and n label the excited states in each symmetry group from the lowest to the higher energies. The Pariser-Parr-Pople calculations by the multi-reference method including singly- and doubly-excited configurational interactions (PPP-MR-SDCI) of shorter polyenes showed the presence of the low-lying $2A_g^-$, $1B_u^-$, $3A_g^-$, and $1B_u^+$ singlet states [7,8]. The selection rule shows that optical transitions are allowed (forbidden) between a pair of electronic states having different signs (the same sign) [5,6]. The energies of these excited states decrease as linear functions of $1/(2n+1)$ when n increases; the slope ratios were theoretically calculated to be $2A_g^- : 1B_u^- : 3A_g^- = 2 : 3.1 : 3.7$ [8] and experimentally found as 2:3.1:3.8 [9].

Time-resolved measurements by subpicosecond lasers [10–12] on a set of Cars with $n=9$–13 revealed the branched relaxation processes including triplet manifold [12]: $1B_u^+ \rightarrow 1B_u^- \rightarrow 2A_g^- \rightarrow 1A_g^-$ and $1B_u^+ \rightarrow T_2(1^3A_g) \rightarrow T_1(1^3B_u)$. However, this subpicosecond time-resolved absorption spectroscopy in the visible region could not identify the $3A_g^-$ state located between the $1B_u^+$ and $1B_u^-$ states for Cars with $n=11$–13 (Figure 9.3.2.1). This state was identified in the near-infrared region instead. Two different internal conversion pathways, one $1B_u^+ \rightarrow 1B_u^- \rightarrow$ for Cars with $n=9$ and 10 and the other $1B_u^+ \rightarrow 3A_g^- \rightarrow$ for Cars with $n=11$–13, which are consistent with the state ordering shown in Figure 9.3.2.1, were identified by two different types of transient absorptions originating from the second species, i.e., the $1B_u^-$ and $3A_g^-$ states [13].

Recently, the presence of the 'S_x' state between the $1B_u^+$ and $2A_g^-$ states was claimed for all-*trans*-β-carotene and lycopene (both $n=11$) by the use of 10–12 fs pulses, and the $1B_u^+$ and $1B_u^-$ ('S_x')

DOI: 10.1201/9780429196577-54

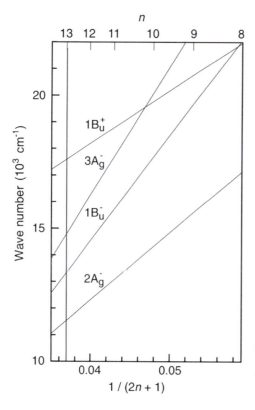

FIGURE 9.3.2.1 Energy diagram of the four excited states of Cars with conjugated double bonds n ¼9–13. The linear relations of the state energies as functions of $1/(2n+1)$ are based on Eqs. (9.3.2.1)–(9.3.2.4) of [9].

TABLE 9.3.2.1

Lifetime of the $1B_u^+$ and '$1B_u^-$' States of all-*trans*-β-Carotene and Lycopene[a]

Carotenoids	$1B_u^+$ (fs)	$1B_u^-$ ('S_x') (fs)
β-carotene	10±2	150
lycopene	9±2	90

[a] Cerullo et al. [14]

---

lifetimes were determined as shown in Table 9.3.2.1 [14]. However, no indication was found for the $3A_g^-$ state expected between these two states (see Figure 9.3.2.1).

In this chapter, we utilized much shorter pulses to identify the $3A_g^-$ state and to determine the intrinsic relaxation dynamics of spirilloxanthin in solution. We have been able to identify, by femtosecond absorption spectroscopy in the visible region, the $3A_g^-$ state in this particular Car ($n=13$), of which internal conversion processes are the fastest among the set of Cars with $n=11$–13 [13]. We have then analyzed the spectra and the lifetimes of the $1B_u^+$, $3A_g^-$, and $1B_u^-$ states by the singular-value decomposition (SVD) and global-fitting analysis (GFA) of a data matrix collected by sub-5-fs pump-probe pulses. High density of data with high S/N ratio taken with a multi-channel lock-in amplifier has enabled unambiguous determination of the decay dynamics and spectra by the SVD-GFA method. The sequential relaxation process has been clarified unambiguously including $1B_u^-$ and $3A_g^-$.

This letter is the first step of a systematic study of relaxation processes in the series of carotenoids to clarify the chain-length dependence of relaxation processes in polyenes with the different ordering of electronic states with different symmetry and various vibronic couplings in the states involved.

9.3.2.2 EXPERIMENTAL

All-*trans*-spirilloxanthin ($n=13$) was prepared as reported [9]. In the time-resolved difference transmission measurement, spirilloxanthin was dissolved in tetrahydrofuran and the concentration of spirilloxanthin solution was adjusted to OD$=1.5\,cm^{-1}$ at the absorption peak corresponding to the concentration of 1×10^{-5} M.

The setup for femtosecond time-resolved absorption spectroscopy was described elsewhere [15,16]. A sub-5-fs 1 kHz pulse train was generated from the noncollinear optical parametric amplifier (NOPA) [17–20] in the range of 500–750 nm. Energy of the pump pulse at the sample was ≈ 40 nJ and that of the probe pulse was about a quarter of the pump pulse. Weak pump–probe signals were measured with a multi-channel lock-in amplifier. The normalized transmittance changes were measured in the pump–probe delay-time ranging from −30 to 1000 fs with a 5- or 2-fs interval, and in the spectral region of 520–700 nm. The data shown in this chapter were taken with the 5-fs interval. The results obtained with 2- and 5-fs intervals were basically identical.

9.3.2.3 RESULTS AND DISCUSSION

9.3.2.3.1 CHARACTERIZATION OF FEMTOSECOND TIME-RESOLVED ABSORPTION SPECTRA: IDENTIFICATION OF SEQUENTIAL INTERNAL CONVERSION

9.3.2.3.1.1 Time-Resolved Absorption Spectra Near Zero Delay Time

Figure 9.3.2.2 shows time-resolved spectra in the region of −30 to 10 fs. It is well established that the optically active $1B_u^+$ state is the main excitation channel because of the strongly allowed transition [2].

The time-resolved spectra in this figure exhibit a strong negative peak in the region of the $1B_u^+(0) \leftarrow 1A_g^-(0)$ absorption with no corresponding signals in the longer-wavelength region of the ground-state absorption spectrum; this is probably due to the inhomogeneity in the ground-state absorption spectrum. In the region from 600 to 700 nm, another highly oscillating pattern appears between −30 and −15 fs. It can be attributed to the perturbed free induction decay [21]. This can provide information about the electronic dephasing time, which will be discussed in a separate paper.

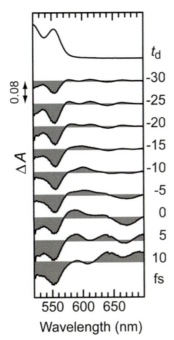

FIGURE 9.3.2.2 Femtosecond time-resolved absorption spectra of spirilloxanthin in the region of near zero delay times, showing the effects of interference between the 5 fs pump and probe pulses.

9.3.2.3.1.2   Time-Resolved Spectra with Positive Delay Times

Figure 9.3.2.3a shows time-resolved spectra in the 15–280 fs region. This region was chosen in the spectral analysis to avoid the effects of the above-mentioned interference pattern before and immediately after excitation. Time-resolved spectra after 300 fs, when a triplet state is generated [12], are not shown because our attention is focused on the *initial* internal conversion processes in the singlet manifold. Partially enlarged spectra are shown in Figure 9.3.2.3b to display the detailed time course of the transient spectrum between 35 and 65 fs. This set of time-resolved spectra shows *four* distinct difference-spectra, which we assign to the $1B_u^+$, $3A_g^-$ $1B_u^-$, and $2A_g^-$ states as follows.

1. *First* component corresponds to the initial difference spectrum (15 fs) in the figure. It is composed of a pair of negative ΔA peaks around 553 and 603 nm. The peak at 553 nm can be attributed to the stimulated emission (SE) associated with the $1B_u^+(0) \to 1A_g^-(0)$ and bleaching (BL) due to ground-state depletion with much longer lifetime. As clearly seen in Figure 9.3.2.3a, this negative peak at 553 nm decays rapidly with ≈ 10 fs decay time determined by the SVD method as discussed later. Thus, the peak intensity becomes nearly

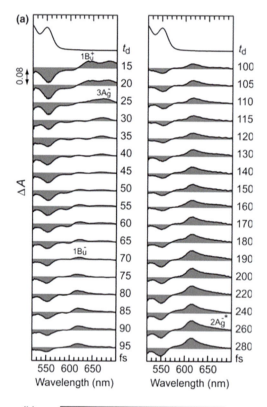

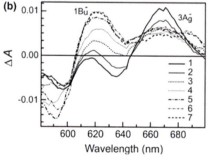

FIGURE 9.3.2.3 (a) Femtosecond time-resolved absorption spectra of spirilloxanthin in the positive delay times free from the interference effects seen in Figure 9.3.2.2. Typical spectra of the $1B_u^+$, $3A_g^-$, $1B_u^-$, and $2A_g^{-*}$ states appear in the time range indicated by the names in the figure. (b) Enlarged portions in the range of 580–700 nm between 35 fs (curve 1) and 65 fs (curve 7) with a 5 fs step.

constant and the peak wavelength is shifted to 551 nm, which is very close to the ground-state absorption peak (550 nm) indicating that the negative peak is mainly due to bleaching. The peak at 603 nm is associated with the $1B_u^+(0) \rightarrow 1A_g^-(1)$ transition. There is also a pair of transient absorption peaks around 640 and 685 nm. Note that the 603 nm peak is absent in the ground-state absorption spectrum shown on the top of Figure 9.3.2.3. These features were also clearly found in the 2-fs-interval measurement within the same probe-delay time range (data not shown). Therefore, the bleaching of the ground-state absorption is not expected to give rise to such a peak.

2. *Second* spectral component appearing at 35 fs in the time sequence in the distinct difference spectrum consists of a sequence of three negative peaks ascribable to the vibrational progression of the $1B_u^+$ stimulated emission, and a positive absorption peak around 670 nm. This spectral pattern can be attributed to the $3A_g^-$ state because the state in spirilloxanthin is energetically next to the $1B_u^+$ state, as can be seen in Figure 9.3.2.1. As will be discussed later, it has a bleaching ($\Delta A < 0$) peak at the same wavelength, 551 nm, which is again close to the ground-state absorption peak at 550 nm, indicating that the negative peak is due to ground state depletion.

3. *Third* component appearing in the sequence is the difference spectrum, which is found typically at 55 fs. It consists of a pair of positive ($\Delta A > 0$) signals to be attributable to the vibrational progression of the transient absorption centered around 630 and 686 nm. A shoulder around 580 nm also composes a progression as will be discussed later using the SVD-GFA data. This spectral pattern can be assigned to the $1B_u^-$ state based on a spectral comparison with the difference spectral pattern reported in detail [12]. The intensity of $3A_g^-$ and $1B_u^-$ state is relatively weak, but its spectral shapes can clearly be seen in Figure 9.3.2.4b (see Section 9.3.2.3.3).

4. *Fourth* component in the time sequence of difference spectrum (220 fs, for example) consists of the bleaching of the $1B_u^+(0) \leftarrow 1A_g^-(0)$ stationary-state absorption and a strong transient absorption peaked at 615 nm. Because of the close resemblance in the spectral shape, this can be definitely assigned to the $2A_g^-$ state as reported previously [12]. The transient absorption exhibits a tail within its lifetime on the longer-wavelength side, indicating that the vibrational relaxation is not completed within the probe delay time in the present experiment. Therefore, it should be more specifically assigned to a vibrationally-hot $2A_g^-(2A_g^{-*})$ state, as concluded in Refs. [22,23].

Thus, the internal conversion processes in the sequence of $1B_u^+ \rightarrow 3A_g^- \rightarrow 1B_u^- \rightarrow 2A_g^- \rightarrow$ have been clearly identified by the present two-dimensional time-resolved absorption spectroscopy using sub-5-fs pump-probe pulses. In particular, the $3A_g^-$ state, which were theoretically predicted and indirectly verified [9,13], is now detected directly in the visible region for the first time.

In the transient spectra after 220 fs, there is a strong peak of bleaching around 550 nm, while it is weaker between 60 and 220 fs. The value of ΔA (<0) between 35 and 50 fs with a peak at 552 nm is close to that after 220 fs indicating that there is an induced absorption contribution in the range of 530–570 nm, as clearly demonstrated by SVD-GFA in the next subsection.

9.3.2.3.2 ANALYSIS BY SVD AND GLOBAL-FITTING IN THE FRAMEWORK OF A SEQUENTIAL MODEL

The SVD method was applied to a time-resolved data matrix, in the 520–700 nm spectral region and 15–280 fs time range, consisting of 117×53 data points. Detailed procedures of the SVD-GFA were described elsewhere [24]. Signal intensities in the negative delay time region induced by the perturbed free induction decay [21] in the −30 to 10 fs were set to practically zero to form a baseline. The singular values obtained were as follows: $V_1 = 1.35$, $V_2 = 0.54$, $V_3 = 0.194$, $V_4 = 0.0853$, and

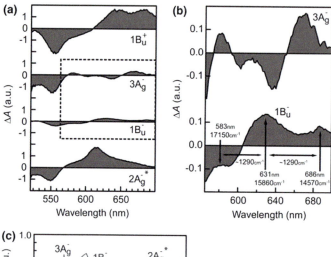

FIGURE 9.3.2.4 Results of SVD and GFA analysis using a sequential model. (a) SADS (species-associated difference spectra) and (b) Section of (a) indicated by a dashed box. (c) Time-dependent changes in population for the $1B_u^+$, $3A_g^-$, $1B_u^-$, and $2A_g^{-*}$ states. In (c), observed data points and fitting curves are shown in thinner and thicker curves respectively.

$V_5 = 0.066$. The first four major components exhibited basis spectra (s_i) and time profiles ($V_i t_i$) that were well-defined, but the fifth component could not be defined. A global fitting was performed by the use of a sequential model including the $1B_u^+$, $3A_g^-$, $1B_u^-$, and $2A_g^-$ states, the lifetime of the $2A_g^-$ state being fixed to be 1.4 ps, which was determined precisely in Ref. [12]. The lifetimes have been determined to be ~ 10 ± 2, 25 ± 2, and 140 ± 30 fs for the $1B_u^+$, $3A_g^-$, and $1B_u^-$ states, respectively (Table 9.3.2.2), and the results of this SVD-GFA are shown in Figure 9.3.2.4.

Figure 9.3.2.4a depicts the species-associated difference spectra (SADS) that are assigned to the $1B_u^+$, $3A_g^-$, $1B_u^-$, and $2A_g^{-*}$ states. The assignments described in the preceding section are basically the time evolution in accord with the state ordering in spirilloxanthin with $n=13$ (see the solid vertical line in Figure 9.3.2.1): The SADS of the $1B_u^+$ state reproduces the doubly-peaked transient absorption and a pair of $1B_u^+ \rightarrow 1A_g^-$ stimulated-emission peaks found in the measured time-resolved spectra at 15 and 20 fs. The SADS of the $3A_g^-$ state reproduces fairly well the vibrational progression of the $1B_u^+ \rightarrow 1A_g^-$ stimulated emission and the transient absorption around 670 nm that are seen in the observed spectrum at 35 fs. The difference-spectrum of the species in the time and spectral ranges has been determined for the first time. The SADS of the $1B_u^-$ state reproduces a pair of negative peaks ascribable to the vibrational progression of the $1B_u^+ \rightarrow 1A_g^-$ stimulated emission and the double-peaked transient absorption that are clearly seen in the time-resolved spectra in the 50–65 fs region. The SADS of the $2A_g^{-*}$ state exhibits bleaching of the $1B_u^+ \leftarrow 1A_g^-$ ground-state absorption and a strong peak at 615 nm with a shoulder on the longer-wavelength side due to the vibrational progression. (Note the difference in wavelength indicating that the stimulated emission peak is not seen here but the bleaching of the ground-state absorption.) The difference spectrum agrees well with the observed one in the time-resolved spectra around 220 fs; at this delay time the main contribution is due to the $2A_g^{-*}$ state.

Here again, we see some negative peaks in the SADS of the $1B_u^+$ state, which can be ascribed to the $1B_u^+(0) \rightarrow 1A_g^-(1)$ and $1B_u^+(0) \rightarrow 1A_g^-(2)$ stimulated emission. When the transient absorption

TABLE 9.3.2.2

Lifetimes of the Singlet Excited States of all-*trans*-Spirilloxanthin Determined by Time-Resolved Absorption Spectroscopy

State	Previous Works		Present Work
	Visible[a]	Near-infrared[b]	
$1B_u^+$	130 fs	≈10 fs	≈10 fs
$3A_g^-$	–	100 fs	25±2 fs
$1B_u^-$	180 fs		140±30 fs
$2A_g^-$	1.4 ps		1.4 ps

[a] Rondonuwu et al. [12]
[b] Fujii et al. [13]

peaks are taken into account, the series of vibrational progressions due to the $1B_u^+ \rightarrow 1A_g^-(v)$ transitions with $v=0-2$ appear with an interval of the 1100–1200 cm^{-1}, which is consistent with the progression in the fluorescence spectra of Cars with $n=9-13$ reported previously [25].

Figure 9.3.2.4b shows an enlarged portion of the $3A_g^-$ and $1B_u^-$ spectra. The spectral features of these two states can be well characterized from this figure. In the spectra of $1B_u^-$, there are positive ΔA peaks at 686 and 630 nm. There is another peak at 583 nm though the absolute value is negative. The energy difference between the 686 and 630 nm peaks and the 630 and 583 nm peaks are both 1290 cm^{-1}, corresponding to the C–C stretching of the molecule. Figure 9.3.2.4c shows the time-dependent population of the four singlet-excited states obtained by the SVD analysis and the observed data. The latter data points exhibit an oscillatory feature due to molecular vibration, but the observed curves are in general agreement with the fitting. Therefore, short-lived excited-state species including the $1B_u^+, 3A_g^-$, and $1B_u^-$ states are concluded to be time-resolved, and sequential formation and conversion process in the order, $1B_u^+ \rightarrow 3A_g^- \rightarrow 1B_u^- \rightarrow 2A_g^-$, are clearly resolved.

9.3.2.3.3 COMPARISON WITH THE PREVIOUS RESULTS OF SUBPICOSECOND TIME-RESOLVED ABSORPTION SPECTRA

The time-resolved spectra obtained by the use of 120 fs pulse of spirilloxanthin ($n=13$) (Figure 9.3.2.2e in [12]) can be contrasted to the present time-resolved spectra with a 5 fs interval as follows. In the spectral comparison, we must keep in mind that the apparent time-resolved spectra are the results of convolution of the excited-state dynamics with pump and probe pulses: (i) The spectrum immediately after excitation is close to the present spectrum at 15 fs, although the observed double-peaked transient absorption was more or less flat. (ii) The spectrum at 0.03 ps exhibited a broad transient absorption around 600 nm, which was assigned to the $1B_u^-$ state. Based on the present time-resolved spectra with much higher resolution and density of the two-dimensional data, it can be regarded as mixed contributions of transient species ($3A_g^-$, $1B_u^-$, and possibly $2A_g^-$ states). (iii) The spectrum at 0.06 ps reported in Ref. [12] shows a transient absorption in the same region, in which the contribution of the $2A_g^-$ transient absorption is observed much more clearly.

The above comparison demonstrates that the durations of the pump and probe pulses play an essential role in time-resolving the set of spectra originating from such short-lived species as the $1B_u^+$, $3A_g^-$, and $1B_u^-$ states.

In summary, we have clarified the serial relaxation process of all-*trans*-spirilloxanthin by unambiguous determination of the four difference absorption spectra of the excited states:

$1B_u^+ \rightarrow 3A_g^- \rightarrow 1B_u^- \rightarrow 2A_g^- \rightarrow 1A_g^-$ Especially, the $3A_g^-$ state, which had been observed only indirectly, has now been identified in real-time sequence, and the lifetime of the $1B_u^+$ state as short as $\approx 10\,\mathrm{fs}$ has also been evaluated. The contents of this subsection is mainly reproduction of the paper [26], which is the product of a collaborative work conducted by the following people: K. Nishimura, F. S. Rondonuwu, R. Fujii, J. Akahane, Y. Koyama, T. Kobayashi [26].

REFERENCES

1. H. A. Frank and R. J. Cogdell, in *Carotenoids in Photosynthesis*, edited by A. Young and G. Britton, Chapman & Hall, London, p. 252 (1993).
2. Y. Koyama, M. Kuki, P. O. Andersson, and T. Gillbro, *Photochem. Photobiol.* **63**, 243 (1996).
3. V. Sundström, T. Pullerits, and R. van Grondelle, *J. Phys. Chem. B* **103**, 2327 (1999).
4. T. Ritz, A. Damjanović, K. Schulten, J.-P. Zhang, and Y. Koyama, *Phostosynth. Res.* **66**, 125 (2000).
5. R. Pariser, *J. Chem. Phys.* **24**, 250 (1956).
6. P. R. Callis, T. W. Scott, and A. C. Albrecht, *J. Chem. Phys.* **78**, 16 (1983).
7. P. Tavan and K. Schulten, *J. Chem. Phys.* **85**, 6602 (1986).
8. P. Tavan and K. Schulten, *Phys. Rev. B* **36**, 4337 (1987).
9. K. Furuichi, T. Sashima, and Y. Koyama, *Chem. Phys. Lett.* **356**, 547 (2002).
10. J. Watanabe, J. Nakahara, and T. Kushida, *J. Lumin.* **58**, 194 (1994).
11. J. Watanabe, H. Takahashi, and J. Nakahara, *Chem. Phys. Lett.* **213**, 351 (1993).
12. F. S. Rondonuwu, Y. Watanabe, R. Fujii, and Y. Koyama, *Chem. Phys. Lett.* **376**, 292 (2003).
13. R. Fujii, T. Inaba, Y. Watanabe, Y. Koyama, and J.-P. Zhang, *Chem. Phys. Lett.* **369**, 165 (2003).
14. G. Cerullo, D. Polli, G. Lanzani, S. De Silvestri, H. Hashimoto, and R. J. Cogdell, *Science* **298**, 2395 (2002).
15. T. Kobayashi, A. Shirakawa, and T. Fuji, *IEEE J. Sel. Topics Quant. Electron.* **7**, 525 (2001).
16. T. Kobayashi, A. Shirakawa, H. Matsuzawa, and H. Nakanishi, *Chem. Phys. Lett.* **321**, 385 (2000).
17. A. Shirakawa and T. Kobayashi, *Appl. Phys. Lett.* **72**, 147 (1998).
18. A. Shirakawa, I. Sakane, M. Takasaka, and T. Kobayashi, *Appl. Phys. Lett.* **74**, 2268 (1999).
19. T. Kobayashi, T. Saito, and H. Ohtani, *Nature* **414**, 531 (2001).
20. A. Baltuska, T. Fuji, and T. Kobayashi, *Opt. Lett.* **27**, 306 (2002).
21. C. H. Brito Cruz, J. P. Gordon, P. C. Becker, R. L. Fork, and C. V. Shank, *IEEE J. Quant. Electron.* **24**, 261 (1988).
22. H. H. Billsten, D. Zigmantas, V. Sundström, and T. Polívka, *Chem. Phys. Lett.* **355**, 465 (2002).
23. F. L. de Weerd, I. H. M. van Stokkum, and R. van Grondelle, *Chem. Phys. Lett.* **354**, 38 (2002).
24. J.-P. Zhang, R. Fujii, P. Qian, T. Inaba, T. Mizoguchi, Y. Koyama, K. Onaka, Y. Watanabe, and H. Nagae, *J. Phys. Chem. B* **104**, 3683 (2000).
25. R. Fujii, T. Ishikawa, Y. Koyama, M. Taguchi, Y. Isobe, H. Nagae, and Y. Watanabe, *J. Phys. Chem. A* **105**, 5348 (2001).
26. K. Nishimura, F. S. Rondonuwu, R. Fujii, J. Akahane, Y. Koyama, and T. Kobayashi, *Chem. Phys. Lett.* **392**, 68–73 (2004).

9.3.3 Observation of Breather Exciton and Soliton in a Substituted Polythiophene with a Degenerate Ground State

9.3.3.1 INTRODUCTION

Soliton was first discovered in 1844 [1], which has been identified in many fields of nonlinear physics [2–7] including water waves, sound waves, matter waves, and electromagnetic waves [8]. According to simulations performed using the Su-Schrieffer-Heeger (SSH) Hamiltonian [9], a photogenerated electron-hole (e–h) pair evolves into a soliton-antisoliton pair $\left(S - \bar{S} \right)$ within 100 fs after photoexcitation because of barrier-free relaxation in a one-dimensional system. Matter-wave solitons have given rise to many interesting phenomena in the simplest conducting polymer, *trans*-polyacetylene (*trans*-PA) [10], including anomalous conductivity and huge optical nonlinearity [11]. The formation times of solitons in polyacetylene have been determined to be <150 fs [12]. Even though the existence of a soliton in *trans*-PA is well known and has been extensively studied, besides *trans*-PA, there has been no other systematic study of conjugated polymer systems. This is probably because of the scarcity of polymers with a degenerate ground state.

The soliton pair is spatially localized to form a dynamic bound state called a breather, which has also been theoretically predicted [9,13–16]. The excess energy of the photogenerated (*e–h*) pair over that of the soliton pair induces collective carbon-carbon (C–C) oscillations, namely the breather mode, due to electron-phonon coupling. Breathers predicted in Ref. [17] have been observed in *trans*-PA [18], which was found to have a period of 44 fs and an extremely short lifetime of ~50 fs. However, there is not yet currently a consensus among researchers as to whether breather is the primary photogenerated excitations and how they affect the ultrafast vibronic dynamics [18–23].

In the present work, the study of the dynamics of solitons and breather in a derivative of polythiophene which has a degenerate ground state (shown in Figure 9.3.3.1a) has been reported. Polythiophene is one of the most promising materials for various device applications, which makes a detailed understanding of the dynamics of electronic state and vibrational dynamics photoexcitations in them and their derivatives highly desirable. In addition, there has been no other spectroscopic study of solitons except for *trans*-PA before, it is of great interest to study other degenerate-ground-state polymers' dynamics and compare any differences with *trans*-PA. To the best of our knowledge, this is the first observation of the existence of a breather and solitons and their dynamics in a system other than *trans*-PA. To rationalize the experimental data, we performed a quantum-chemical excited-state molecular dynamics simulation, and its results are consistent with experimental data, enabling breather excitations to be analyzed and related coupled vibrational normal modes to be identified.

DOI: 10.1201/9780429196577-55

9.3.3.2 EXPERIMENTAL DESCRIPTIONS

In the present experiment, we utilized a nearly Fourier-transform limited visible-near-IR pulse generated from a noncollinear optical parametric amplifier (NOPA) seeded by a white-light continuum. The pump source of this NOPA system was a regenerative amplifier (Spectra Physics, Spitfire) with the following operating parameters: central wavelength, 800 nm; pulse duration, 50 fs; repetition rate, 5 kHz; average output power, 650 mW. We used a 1-mm-thick sapphire plate to generate the continuum spectrum. The NOPA output pulse was compressed with a pair of chirp mirrors and then with a prism pair, resulting in a nearly FT-limited pulse duration of 6.3 fs. Both the pump and probe pulses covered the spectral range extending from 515 to 716 nm [24], and the energies of them are about 50 and 6 nJ, respectively. The pump-probe signal at 128 different wavelengths was detected by a combined system of a polychromator and a multi-channel lock-in amplifier. Thanks to the extreme stability of the light source and noise reduction by the lock-in-detector; the time resolution is better than 1 fs, which was recognized by the difference between the time-resolved spectra at neighboring delay step obtained using the time-step of 0.2 fs.

The sample studied in this study was a thin film of PHTDMABQ, whose structure is shown in our previous paper [25] and in Figure 9.3.3.1a. It was dissolved in methanol and cast on a quartz substrate for stationary and time-resolved spectra measurements. All experiments were performed at room temperature (293 ± 1 K).

9.3.3.3 MOLECULE STRUCTURE

The structure of PHTDMABQ is depicted in Figure 9.3.3.1a, whose monomer is a derivative of thiophene. It appears ostensibly not to have a degenerate ground state; however, it has degeneracy due to the resonance of the inner structure of the polymer. In polythiophene, there is no degenerate ground state. After photoexcitation, bipolarons are generated. The two polarons in the bipolarons cannot be separated from each other because they are non-degenerate. However, in the case of PHTDMABQ, there are two mesomeric forms to form a repeat unit. In each repeat unit, there are two thiophene rings with both cis and trans configurations. They can exchange their mesomeric structures without energy requirement. Therefore, the ground state structure can be either structure A or structure B as shown in Figure 9.3.3.1a. Because of the resonance of the internal structure of PHTDMABQ, it can have a degenerate ground state and thus solitons can be generated in it.

9.3.3.4 QUANTUM-CHEMICAL METHODOLOGY

To analyze the experimental data, we used the Austin Model 1 (AM1) Hamiltonian and an excited-state molecular dynamics (ESMD) computational package, which is described in detail in [26] and [27], to follow photoexcitation adiabatic dynamics on a picosecond timescale for all calculations presented in this study. The ESMD approach calculates the excited-state potential energy as $E_e(q) = E_g(q) + \Omega(q)$ in the space of nuclear coordinates q that span the entire ($3N$–6) dimensional space, where N is the total number of atoms in the molecule. Here, $\Omega(q)$ is the electronic transition frequency to the lowest $1B_u$ (band-gap) state of the photoexcited molecule. The program efficiently calculates analytical derivatives of $E_e(q)$ with respect to each nuclear coordinate q_i to evaluate forces and to subsequently step along the excited-state hypersurface using these gradients. All computations start from vertical excitation at the optimal ground-state molecular geometry. The total molecular energy $E_e(q)$ is conserved if no dissipative processes are included. Subsequent analysis of the photoexcited trajectories of the excitation energy $\Omega(q, t)$ and oscillator strength $f(q, t)$ in Fourier space allows us to identify periods of participating vibrational motions. Alternatively, the minimum of the excited state potential energy surface can be calculated by including an artificial dissipative force in the equations of motion corresponding to the relaxed excited state geometry. To understand the formation of photoexcited breathers, we calculated the dynamics of the bandgap excited state in the 10-unit thiophene oligomer shown

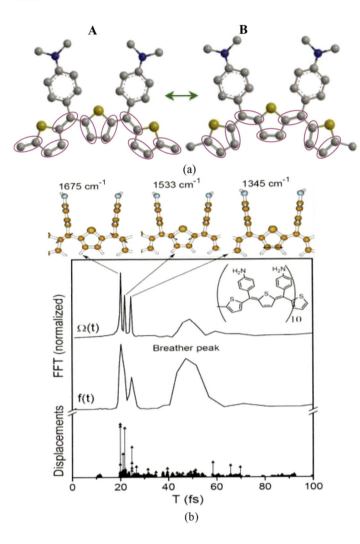

FIGURE 9.3.3.1 Two quantum mechanical resonant structures of (a) PHTDMABQ and (b) the excited state molecular dynamics simulations results. To make the degeneracy clearly, the pink elliptical circles denote the position of the C=C bond. (b) shows the normalized Fourier spectra of the lowest dipolar allowed excited state transition energy $\Omega(t)$ and its respective oscillator strength $f(t)$ trajectories (top two plots), and the amplitudes of dimensionless displacements Δ (stick spectrum, bottom panel) along normal modes calculated in the oligomer with 10 repeat units, as shown in the inset. The three molecular structures at the top schematically show vibrational normal modes with frequencies strongly coupled to the electronic system, which leads to the formation of the breather excitation. These correspond to the C=C vibration and C–C stretches as schematically shown in the middle panel.

in the inset of Figure 9.3.3.1b, where the alkyl side-chain has been replaced by hydrogen, effectively reducing the molecular size for calculations. This molecule is sufficiently long (10 nm) compared to the characteristic exciton size of about 2 nm for the infinite chain limit to be valid.

9.3.3.5 RESULTS AND DISCUSSION

9.3.3.5.1 ELECTRONIC RELAXATION AND MOLECULAR VIBRATION DYNAMICS

In this chapter, the decay dynamics of the electronic state and vibrational dynamics highly correlated through excitonic coupling (vibronic coupling) were observed under the same condition at the same time. The real-time absorbance change signal $\Delta A(\omega)$ (Figure 9.3.3.2a) shows the decay dynamics and spectral change associated with the change in the electronic state. On top of that signal, the modulation $\delta\Delta A(\omega)$ of the $\Delta A(\omega)$ due to molecular vibration can be used to study the vibrational dynamics of the system completely in the same condition as that electronic dynamics study. This situation is difficult to be realized in the experiment made by using conventional UV (VIS) pump-UV (VIS) probe experiment and time-resolved vibrational experiment.

The difference absorbance (ΔA) signals in Figure 9.3.3.2a exhibit oscillation due to molecular vibrations. As shown in the ΔA traces, the lifetimes of the electronic states consist of three

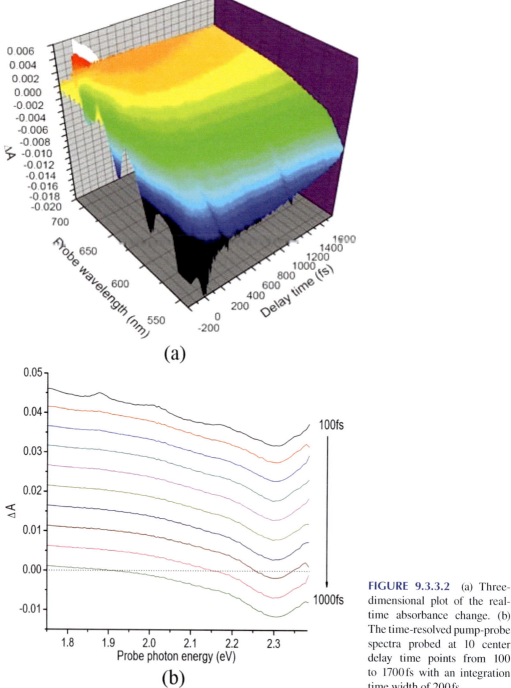

(a)

(b)

FIGURE 9.3.3.2 (a) Three-dimensional plot of the real-time absorbance change. (b) The time-resolved pump-probe spectra probed at 10 center delay time points from 100 to 1700 fs with an integration time width of 200 fs.

components: 62 ± 2 fs, 750 ± 20 fs, and >3 ps [27]. The Fourier power spectra in Figure 9.3.3.3 have peaks at 1111 ± 7, 1343 ± 7, and 1465 ± 7 cm^{-1} ($p1$, $p2$, and $p3$, respectively).

Theoretical calculations allow assigning these peaks to C–C stretching modes with different bond orders. Figure 9.3.3.1b shows the calculated dimensionless displacements Δ along the vibrational coordinates of the optimal geometries between the ground and excited states. This immediately enables us to identify the $p1$–$p3$ vibrational normal modes that are strongly coupled to the

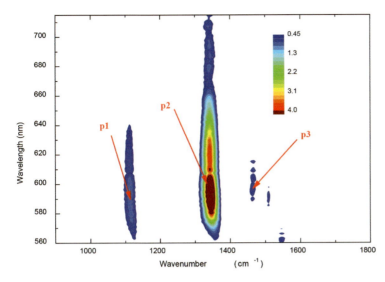

FIGURE 9.3.3.3 Fourier transform power spectra. The Fourier transform power spectrum at 615 nm is plotted as an example.

electronic excitation. These correspond to intra- and inter-thiophene ring C–C and C=C stretching motions (see Figure 9.3.3.1b, top structures). The highest, medium, and lowest frequencies are considered to correspond to C=C double bonds, a mixture of double and single bonds, and single bonds, respectively. The calculated vibrational frequencies (1345, 1533, and 1675 cm^{-1}) are overestimated by about 200 cm^{-1} compared to the experimental values, which is typical for semi-empirical calculations.

9.3.3.5.2 DYNAMICS OF BREATHER AND SOLITON

The requirement for the existence of solitons is a degenerate ground state structure. As mentioned before, PHTDMABQ has a degenerate ground state because the two quantum mechanical resonance structures have equivalent energies. In Figure 9.3.3.2b, the transient absorption spectra exhibit negative absorbance changes in the photon energy range 1.91–2.38 eV, while for photon energies smaller than 1.91 eV, the absorbance change is positive. This increase in the absorbance is attributed to the tail of the solitons not fully relaxing to the band-gap center, as is the case in trans-PA [18]. Therefore, the induced absorption observed in poly(substituted thiophene) is attributed to solitons. The positive value indicates the increased contribution of induced absorption due to a soliton with a peak near the mid gap, which is estimated to be around 1.4 eV.

Figure 9.3.3.4 shows the contour map obtained by the spectrogram analysis [28], which is suitable to study the dynamic process where the molecular geometrical relaxation or chemical reaction is accompanied by change in its vibrational frequency due to molecular structural change during the processes. As shown in Figure 9.3.3.4, in addition to the three prominent peaks $p1$, $p2$, and $p3$, there are five more peaks appearing as side bands of $p1$–$p3$ at 270, 500, 640, 1960, and 2200 cm^{-1}. These five modes are breather modes which are not visible in the two-dimensional Fourier power spectrum (Figure 9.3.3.3) or in the stationary resonance Raman spectrum because they have extremely broad widths due to their short lifetimes. We also note that there are no detectable normal modes with substantial intensity in this spectral region coupled to the electronic excitation as evidenced by lack of significant displacements calculated in Figure 9.3.3.1b. The amplitudes of the high-frequency modes in the 2000–2200 cm^{-1} range decrease rapidly due to their short vibrational periods, which cannot be properly resolved by the finite pulse widths of both pump and probe pulses. However, the frequency is not affected and the breather excitation is clearly visible for the three sidebands with the lower frequencies.

The lifetimes of the side bands were determined to be about 50, 32, and 40 fs for sidebands of $p1$, $p2$, and $p3$, respectively. They are corresponding to the lifetime of the breather mode as observed decay time of 50 fs in polyacetylene [18]. The theoretically predicted decay time is shorter than

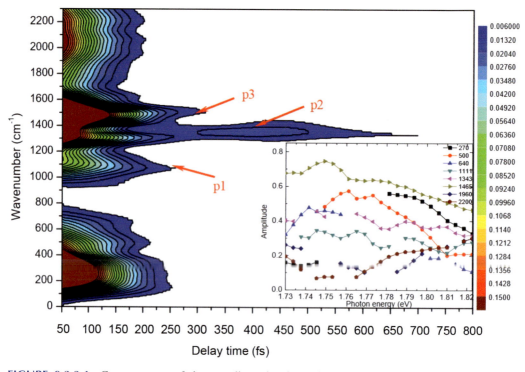

FIGURE 9.3.3.4 Contour maps of the two-dimensional Fourier power of the vibrational components obtained by spectrogram calculation for real-time data covering 680–690 nm (Inset is the probe photon energy dependence of the vibrational amplitude probed at a 110 fs gate delay time with a gate width of 120 fs (HWHM) in the spectrogram).

100 fs is again consistent with our observation. The result means the time for the breather mode to disappear followed by the separation into isolated solitons, and even shorter than that in polyacetylene. These short lifetimes are also corresponding to the shortest lifetime component of ~62 fs in the ΔA trace [25]. The ultrafast relaxation of ~50 fs of this nonlinear excitation ensures an ultrafast nonlinear response that can be used in all-optical switches.

The average frequency difference between the three main bands (1111, 1343, and 1465 cm^{-1}) and their corresponding side bands (270 and 1960 cm^{-1} for 1111 cm^{-1}, 500 and 2200 cm^{-1} for 1343 cm^{-1}, and 640 cm^{-1} for 1465 cm^{-1}) is calculated to be 843 ± 12 cm^{-1}, which is close to the experimentally observed value of about 760 cm^{-1} reported for polyacetylene [18], and theoretically expected values of 660–1000 cm^{-1} [14–17]. This frequency separation between the main bands and the satellite bands indicates that the breather modulates the C–C stretching modes with a period of ~40 fs generating the sidebands.

Consequently, these sideband peaks appear due to nonlinear electron-vibrational excited state dynamics. To analyze these processes from theoretical calculations, we compute the power spectra of 750 fs photoexcited trajectories of the excitation energy $\Omega(q, t)$ and oscillator strength $f(q, t)$ shown in Figure 9.3.1b [26,29]. Both plots show an additional broad peak centered around 50 fs, corresponding to a vibration in the range 550–800 cm^{-1}, which does not correspond to any of the vibrational normal modes that exhibit substantial coupling to the electronic degrees of freedom (compare the displacements peaks with the FFT trajectories in Figure 9.3.3.1b). Based on our previous computational studies, we assign this peak to a nonlinear breather excitation that occurs due to coupling of C–C vibrational motions. This vibrational excitation decays gradually over long time scales due to the dissipation of vibrational energy to internal vibrational degrees of freedom that are weakly coupled to the electronic system; this is similar to previous findings by us [26,29]. Power spectra of longer excited state trajectories (not shown) demonstrate diminishing breather peak.

Our calculations estimate the breather lifetime to be about 0.2 ps without accounting for intermolecular dissipation channels (baths). Such fast decay is considered to be reasonably in agreement with our experiment data of ~50 fs in case we take into account both intermolecular interactions and phonon energy leakage through the boundaries of conjugated segments (defects). These processes may reduce breather lifetime from 0.2 ps to ~50 fs [29]. As expected, the breather peak is more pronounced in the power spectrum of the oscillator strength (see the inset of Figure 9.3.3.4), since it is directly related to the modulation of the respective transition dipole moment [29].

Thanks to the new multi-channel detection system developed we could observe the molecular vibration-induced modulation $\delta\Delta A$ of the absorbance change ΔA at 128 different wavelengths simultaneously. In this way, the probe wavelength dependence can be used in detailed discussions on the photon energy dependence of modulation amplitudes of various modes. As shown in the inset of Figure 9.3.3.4, the breather strongly contributes to the variations in the oscillator strength. Except for the modes at 1960 and 2200 cm^{-1}, the signal sizes of all the modes exhibit an almost monotonic increase when the photon energy is reduced from 1.82 to 1.73 eV. This is consistent with the theoretical discussion in Ref. [26]. The frequencies of these two modes are nearly equal to the overtones of 1111 cm^{-1}, which may affect the dynamics of the breather. Because both amplitude modulation (AM) and frequency modulation (FM) can affect the vibrational amplitude of the single and double CC bonds [30], we also calculated the ratios of the amplitude of FM to that of AM; they were about 0.11 and 0.03 for the 1111 and 1343 cm^{-1} modes, respectively.

Similar feature is found for the positive absorbance change, as shown in Figure 9.3.3.2b. The magnitude of the positive absorbance change increases when the probe photon energy decreases. This provides evidence of the larger contribution of the breather to the modulation ($\delta\Delta A$) of the difference absorbance change (ΔA) due to soliton in the lower energy range. We can interpret this feature in terms of the modulation of the transient spectrum of soliton by the vibrational modes as follows.

Molecular vibration associated with the breather and soliton is expected to modulate the transition energy and transition probability, and the amplitudes are expected to be proportional to the zeroth, first, and second derivatives of the absorption and/or stimulated emission spectrum depending on the mechanism of induction of the wavepacket motions [31]. Since the absorption spectrum of soliton is expected to be close to the mid-gap, which is located at 1.4 eV in the case of polyacetylene, the spectral range of the present observation is higher energy tail of the soliton absorption as seen from the time-resolved spectrum as shown in Figure 9.3.3.2b. Therefore, all of the zeroth, first, and second derivatives of the absorption spectrum are expected to increase monotonically with decreasing probe photon energy.

Here we go back to the discussion of the above-mentioned exceptional behavior of the probe wavelength dependence of the amplitudes of the two modes at 1960 and 2200 cm^{-1}. It can be explained in the following way. The molecular vibration modulates the transition probability and transition energy because of the electronic distribution change nearly instantaneously following the motion of nuclei during the molecular vibration. The frequencies of the modes at 1960 and 2200 cm^{-1} are nearly equal to the overtones of 1111 cm^{-1}, which may affect the dynamics of the breather. The electronic transition is modulated periodically with a period of ~30, 17, and 15 fs corresponding to the frequencies of 1111, 1960, and 2200 cm^{-1}, respectively. Then at integer multiple of about 32 fs all of them contribute. In case the vibrations of the modes are in phase or out of phase then the amplitudes of them may be affected by either constructive or destructive interference.

We now discuss the effect of the differences in the sizes and structures of the repeat units. The repeat unit in the *trans*-PA is composed of one single bond and one double bond, while in PHTDMABQ, it is bulkier since it has two thiophene rings with both *cis* and *trans* configurations (see Figure 9.3.3.1). Since PHTDMABQ has a cis configuration in the thiophene ring, it is interesting to compare it with *cis*-PA. The lifetime of breather is expected to be longer than that in *trans*-PA. However, the decay time of the breather seems to be even shorter than that in *trans*-PA. That is because the lifetime is not determined by the separation of soliton pairs from the originally generated site in the polymer chain, but by the energy dissipation to internal vibrational freedom. Since PHTDMABQ has many more internal vibrational modes than *trans*-PA, its breather lifetime is even shorter than that of *trans*-PA.

9.3.3.6 CONCLUSIONS

We investigated the ultrafast dynamics that take place immediately after excitation in a polythiophene derivative with a degenerate ground state [32]. The simulation results of quantum-chemical excited-state molecular dynamics agree reasonably well with the experimental data by showing the formation of short-lived breather excitation. The breather lifetime was experimentally determined from the electronic spectral dynamics to be ~62 fs, which agrees with the time constants determined by the time-dependent signal intensity that appears as side peaks associated with the breather. Even though extensive theoretical studies have been conducted there was no experimental observation of the modulation of the C–C single and double stretching modes by the breather. In the experiment conducted in the work presented here, the modulation due to the coupling was observed for the first time in a system other than in *trans*-PA [32].

REFERENCES

1. J. S. Russell, *Report of the Fourteenth Meeting of the British Association for the Advancement of Science*, Murray, London, pp. 311–390 (1844),
2. N. S. Manton and P. Sutcliffe, *Topological Solitons*, Cambridge University Press, Cambridge, England (2004).
3. L. Khaykovich, F. Schreck, G. Ferrari, T. Bourdel, J. Cubizolles, L. D. Carr, Y. Castin, and C. Salomon, *Science* **296**, 1290 (2002).
4. K. E. Strecker, G. B. Partridge, A. G. Truscott, and R. G. Hulet, *Nature* **417**, 150 (2002).
5. M. Nakazawa, E. Yamada, and H. Kubota, *Phys. Rev. Lett.* **66**, 2625 (1991).
6. W. P. Su, J. R. Schrieffer, and A. J. Heeger, *Phys. Rev. Lett.* **42**, 1698 (1979).
7. A. M. Kosevich, B. A. Ivanov, and A. S. Kovalev, *Phys. Rep.* **194**, 117 (1990).
8. G. I. Stegeman and M. Segev, *Science* **286**, 1518 (1999).
9. W. P. Su and J. R. Schrieffer, *Proc. Natl. Acad. Sci. U.S.A.* **77**, 5626 (1980).
10. H. Shirakawa, *Rev. Mod. Phys.* **73**, 713 (2001).
11. H. Naarmann and N. Theophilou, *Synth. Met.* **22**, 1 (1987).
12. C. V. Shank, R. Yen, R. L. Fork, J. Orenstein, and G. L. Baker, *Phys. Rev. Lett.* **49**, 1660 (1982).
13. D. K. Campbell, *Nature* **432**, 455 (2004).
14. M. Sasai and H. Fukutome, *Prog. Theor. Phys.* **79**, 61 (1988).
15. S. R. Phillpot, A. R. Bishop, and B. Horovitz, *Phys. Rev. B* **40**, 1839 (1989).
16. S. Block and H. W. Streitwolf, *J. Phys.: Condens. Matter* **8**, 889 (1996).
17. A. R. Bishop, D. K. Campbell, P. S. Lomdahl, B. Horovitz, and S. R. Phillpot, *Phys. Rev. Lett.* **52**, 671 (1984).
18. S. Adachi, V. M. Kobryanskii, and T. Kobayashi, *Phys. Rev. Lett.* **89**, 027401 (2002).
19. G. S. Kanner, Z. V. Vardeny, G. Lanzani, and L. X. Zheng, *Synth. Met.* **116**, 71 (2001).
20. T. Kobayashi, J. Du, W. Feng, K. Yoshino, S. Tretiak, A. Saxena, and A. R. Bishop, *Phys. Rev. B* **81**, 075205 (2009).
21. K. M. Gaab and C. J. Bardeen, *J. Phys. Chem. B* **108**, 4619 (2004).
22. G. Lanzani, G. Cerullo, C. Brabec, and N. S. Sariciftci, *Phys. Rev. Lett.* **90**, 047402 (2003).
23. R. Lécuiller, J. Berréhar, C. Lapersonne-Meyer, and M. Schott, *Phys. Rev. Lett.* **80**, 4068 (1998).
24. A. Baltuška, T. Fuji, and T. Kobayashi, *Opt. Lett.* **27**, 306.
25. J. Du, Z. Wang, W. Feng, K. Yoshino, and T. Kobayashi, *Phys. Rev. B* **77**, 195205 (2008).
26. S. Tretiak, A. Saxena, R. L. Martin, and A. R. Bishop, *Proc. Natl. Acad. Sci. U.S.A.* **100**, 2185 (2003).
27. S. Tretiak, A. Saxena, R. L. Martin, and A. R. Bishop, *Phys. Rev. Lett.* **89**, 097402 (2002).
28. M. J. J. Vrakking, D. M. Villeneuve, and A. Stolow, *Phys. Rev. A* **54**, R37 (1996).
29. S. Tretiak, A. Piryatinski, A. Saxena, R. L. Martin, and A. R. Bishop, *Phys. Rev. B* **70**, 233203 (2004).
30. T. Teramoto, Z. Wang, V. M. Kobryanskii, T. Taneichi, and T. Kobayashi, *Phys. Rev. B* **79**, 033202 (2009).
31. T. Kobayashi, Z. Wang, and I. Iwakura, *New J. Phys.* **10**, 065009 (2008).
32. T. Kobayashi, J. Du, W. Feng, K. Yoshino, S. Tretiak, A. Saxena, and A. R. Bishop, *Phys. Status Solidi C* **8**, 74–79 (2011).

9.3.4 Ultrafast Electronic Relaxation and Vibrational Dynamics in a Polyacetylene Derivative

9.3.4.1 INTRODUCTION

One-dimensional conjugated polymers have been of interest for the last two decades because of their characteristic properties including the tailorability by chemical synthetic processes to change the side groups and physical or physicochemical modification processes of morphological, mechanical, electrical, and optical properties. The capability is useful for the preparation of materials with various electrooptical, optoelectrical, and photonic properties, which are of vital importance for varieties of applications, such as electroluminescent devices, nonlinear optical devices, and field-effect transistors [1–5]. Among them, the large ultrafast optical nonlinear property is based on the dimensionality and conjugation of the polymer [1,6]. Conjugation of p-electrons along the main chain of the polymer supports the correlated electrons to induce enhanced transition probability with a large transition dipole. This results in the large third-order nonlinearity because of the existence of the deviation from the bosonic properties of exciton due to the Pauli Exclusion Principle. Ultrafast relaxation is expected in such a one-dimensional system due to a barrierless potential between the free exciton and the self-trapped exciton [7].

The combination of the large third-order nonlinearities due to the electronic correlation and the ultrafast response is quite attractive for basic techniques such as the optoelectro-switching and optical information processing. Many experimental and theoretical studies have been made to clarify the mechanism which is responsible for the macroscopic nonlinear properties characteristic of this class of materials [8]. The optical nonlinearities of the one-dimensional conjugated polymers are closely related to geometrically relaxed excitations such as a pair of solitons, polarons, and a self-trapped exciton (STE). STE is equivalent to an exciton polaron and a neutral bipolaron and is formed via strong coupling between electronic excitations and lattice vibrations [2]. A free exciton formed in such a one-dimensional system spontaneously relaxes within 100 fs because of the absence of a barrier between the free exciton minimum and STE minimum of the potential curves [9] and changes the optical properties of the conjugated polymers, inducing the absorption coefficients and refractive indices [9,10]. Relaxation dynamics of photoexcitations in polydiacetylenes and polythiophene are thus related to the ultrafast nonlinear response dynamics. Their formation and relaxation processes are, therefore, quite essential and one of the most fundamental subjects to be investigated. The change is induced not only by the localized electronic excitation but also by vibrational excitation coupled to the electronic excitation through vibronic coupling start to modulate the molecular structure [11–14]. The structural modulation changes the energy-level scheme and the transition probability. The former changes the electronic spectrum and hence the intensity at some specific probe wavelength. The latter modulates the intensity in the way in the relevant homogeneous spectral range.

In this subsection, the delay time dependence of difference absorbance and time-resolved spectrum is shown in Section 9.3.4.3.1, and the effect of the electronic transition spectrum by molecular vibration is discussed in Section 9.3.4.3.2. Initial phases of the vibrational modes coupled to the electronic transition via impulsive excitation are used to identify the mode observed either to the ground state or to the excited state in Section 9.3.4.3.3. From the analysis of the dynamics of the mean distribution

DOI: 10.1201/9780429196577-56

energy of the vibration energy, the rate of the descending process of the vibrational ladder is calculated for several modes. Vibrational phase relaxation rate is also calculated from the FWHM of the Fourier spectrum in Section 9.3.4.3.4. Section 9.3.4.3.5 discusses the electronic phase relaxation obtained from the data in the negative time range when the probe pulse proceeds with the pump pulse.

9.3.4.2   EXPERIMENTAL

9.3.4.2.1   SAMPLE

The sample polymer studied is poly[o-TFMPA([o-(trifluoromethyl)phenyl]acetylene)] (hereafter abbreviated as PTTPA), whose molecular structure is shown in the inset of Figure 9.3.4.1. It was synthesized in the following way: The monomer, o-TFMPA ([o-(trifluoromethyl)phenyl]acetylene) was prepared according to the procedure by Okuhura [15]. Polymerization of the monomer was carried out under dry nitrogen. It was initiated by 1:l mixtures of WCl, or MoCIS with various organometallic cocatalyst (mixture of WCl, with Ph4Sn) and achieved molecular weights as high as 160 kDa (this is unit for molecular weight). Metal carbonyl-based catalysts were prepared by irradiation of carbon tetrachloride solution of a metal carbonyl with UV light (200-W high-pressure Hg lamp, distance 5 cm) at 30°C for 1 h. A mechanically strong film could be obtained by solution casting. The polymer was thermally fairly stable in the air. The high molecular weight, film formation, and fair thermal stability of the present polymer are notable characteristics, which are not seen in poly(phenylacetylene).

The polymerization using a WCl_6 or $MoCl_5$ catalyst was made to achieve a high (80–100%) yield. The polymers have molecular weights significantly larger than 10^6, and therefore, contain 10^4 repeat units in a chain. The synthesis and various chemical properties of the conjugated polymers are described and discussed in detail elsewhere [15,16].

9.3.4.2.2   ULTRAFAST SPECTROSCOPY

Using the 6.8 fs pulse, the pump-induced absorbance change (DA) in the PTTPA file sample was measured at 128 different wavelengths from 529 to 726 nm (2.34–1.71 eV) [17,18]. The pump and probe beams were both from the non-collinear parametric amplifier (NOPA), which is reported in detail in Refs. [19,20]. In a brief description, the pump source of the NOPA system is a commercially supplied regenerative amplifier (Spectra Physics, Spitfire). The central wavelength, pulse duration, power of the output, and repetition rate of this amplifier were 800 nm, 50 fs, 740 mW, and 5 kHz, respectively. The output pulse from the NOPA was compressed with a compressor composed of a pair of prisms and a set of two chirp mirrors. The polarizations of the pump and probe beams were parallel to each other.

The pump–probe experiment setup was described in detail in our Refs. [19–22]. The pump–probe signal was spectrally dispersed with a polychromator (JASCO, M25-TP) over 128 photon energies (wavelengths) from 1.65 to 2.23 eV (753–555 nm). It was detected by 128 sets of avalanche photodiodes and lock-in amplifiers with a reference from an optical chopper intersecting the pump pulse at the repetition rate of 2.5 kHz.

9.3.4.3   RESULTS AND DISCUSSION

9.3.4.3.1   DELAY TIME DEPENDENCE OF DIFFERENCE ABSORBANCE AND TIME-RESOLVED SPECTRUM

Figure 9.3.4.1 shows the absorption, spontaneous fluorescence, and stimulated emission spectra. The last one is calculated from the spontaneous emission spectrum of the polymer sample. The laser spectrum also shown in the figure is overlapping with the tail of the absorption.

Figure 9.3.4.2a shows the traces of difference absorbance $\Delta A(t)$ at 10 different probe photon energies. In the absorbance traces shown on the left-hand side of Figure 9.3.4.2a, the electronic and vibrational effects are both apparent. The former appears as slow changes, which decay slowly due to electronic relaxation. The latter effect appears as rapid modulation of the transition due to molecular vibration.

The time trace signals depicted in red color lines in Figure 9.3.4.2a are the multiplied traces by appropriate factors to show the modulation due to molecular vibration in the traces more clearly. Figure 9.3.4.2b shows the Fourier power spectra of DA(t) traces at the 10 different probe photon energies corresponding to the real-time traces in Figure 9.3.4.2a.

To investigate the mechanism of electronic relaxation, we first analyze the change of spectral shape during relaxation and then decay dynamics using analytic functions. Figure 9.3.4.3a shows the time-resolved spectrum integrated for 100 fs delay time duration with the center delay times between 100 and 1700 fs with a 100 fs step. For example, the spectrum at the center delay time of 400 fs is the time-resolved spectrum integrated over the 351–450 fs range. To see the change in the spectrum better, the time-resolved spectra normalized to the peak intensity are shown in Figure 9.3.4.3b. The spectral shit is to be discussed later in this Letter in terms of vibrational relaxation. Compared the time-resolved spectra with previous studies [23,24] of conjugated polymers, it can be concluded that dominant positive DA(t) in the whole spectral range of measurement is attributable to induced absorption from free excitons at short delay time and then to the self-trapped excitons at longer delay time except in the probe frequency range higher than 2.33 eV (18800 cm^{-1}).

The two-dimensional difference absorption spectrum in the pump–probe delay time range from 200 to 1800 fs is shown in Figure 9.3.4.4a. Figure 9.3.4.4b shows the two-dimensional Fourier power spectrum of $\Delta A(t)$, which is calculated for the time range of 50–1800 fs. The time range shorter than 50 fs is not included in the calculation to avoid the effect of coherent effect between the scattered light of pump pulse and the probe pulse. Since the spectra in Figure 9.3.4.3a and b are all integrated for 100-fs span, the effect of the spectral shift and the intensity due to molecular vibration is substantially reduced because the integration smears out the spectral shifting associated with the molecular vibration. The crossing points between the neighboring delay-time steps of 100 fs are shown in Figure 9.3.4.3c with the numbers of the participating delay times. The energies of the crossing points in 100–200, 200–300, 300–400, 400–500, and 500–600 fs descend rapidly from 2.18 to 2.31 eV (corresponding to 17600–18600 cm^{-1}). After 600 fs, the crossing points are heavily congested around 2.33 eV (18800 cm^{-1}), indicating that the spectral shift becomes much slower than the preceding delay time. The normalized spectrum in Figure 9.3.4.3b shows a rapid blue shift of the whole spectrum from the delay time just after excitation until the central delay time of 400 fs. The spectral shape is also changed rapidly in the same time range.

By the global fitting in the spectral range of 2.10–2.25 eV, two time constants were determined to be $\tau_1 = 20 \pm 2$ fs and $\tau_2 = 320 \pm 50$ fs. The former corresponds to the ultrafast geometrical relaxation from free excitons to self-trapped excitons in PTTPA in the absence of a barrier between them. The latter one, which is too short to be assigned to the population decay of STEs, namely the electronic population decay [23,24], is ascribed to the vibrational relaxation. Therefore, this blue-shift can

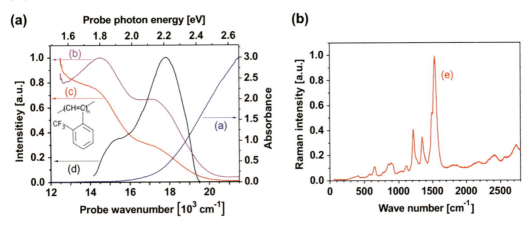

FIGURE 9.3.4.1 (a) Absorption, (b) spontaneous fluorescence, (c) stimulated emission spectra of the PTTPA film sample, and (d) the 6.8-fs laser spectrum. Inset is the molecular structure of PTTPA. Raman spectrum of PTTPA excited at 442 nm is depicted in (e).

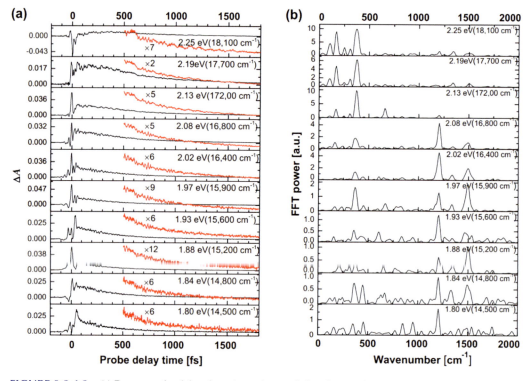

FIGURE 9.3.4.2 (a) Pump–probe delay time dependence of absorbance changes at 10 typical probe photon energies from 200 to 1800 fs. Magnified curves from 500 fs with appropriate magnification factors are also shown to clarify the molecular vibrations. (b) FFT amplitude spectra made by subtracting an exponential function from each real-time trace and taking FFT from 200 to 1800 fs.

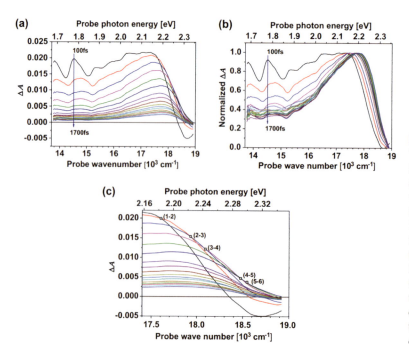

FIGURE 9.3.4.3 (a) Time-resolved spectrum integrated for 100 fs delay time duration with the center delay times between 100 and 1700 fs with a 100-fs step from the highest peak to the lowest peak. (b) Normalized spectra of (a). (c) Enlarged spectra from (a) to show the crossing points between the two neighboring delay times.

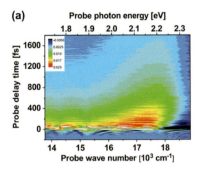
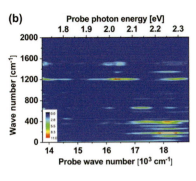

(a) Probe photon energy [eV]

(b) Probe photon energy [eV]

FIGURE 9.3.4.4 (a) Two-dimensional difference absorption spectrum. (b) Two-dimensional spectrum of FFT power of the difference absorption spectrum.

be explained in terms of the intra-chain thermalization process in which the vibrational quanta of modes with high vibrational frequencies are scattered, being converted to low-frequency modes via vibrational mode coupling. It results in the reduction of the mean vibrational energy of the population distributed over vibrational levels with many modes and with different quantum numbers in the electronic excited state, which is the initial state of the transition with positive DA. The dynamics of the process can be studied by using the average transition energy in the positive DA regions calculated by the first moment of the transition energy; this is discussed later in Section 9.3.4.3.4. The blue shift can be ascribed to the change in the induced absorption caused by that of the population distribution in the lowest excited state (i.e., the initial state of the observed transition in the induced absorption) without change in the energy position of the vacant higher excited state (the final state of the transition). See the discussion in later sections on the relaxation process from the viewpoint of the vibronic coupled signals.

9.3.4.3.2 THE EFFECT OF THE ELECTRONIC TRANSITION SPECTRUM BY MOLECULAR VIBRATION

As described in Section 9.3.4.3.1, Figure 9.3.4.2a shows the traces of difference absorbance $\Delta A(t)$ as a function of the probe delay time at 10 different probe photon energies. The traces show a highly modulating signal on top of the slowly varying absorbance change, $DA(t)$, due to electronic dynamics. The former rapid modulation in $\Delta A(t)$, represented by $\delta\Delta A(t)$, is due to a change in the transition spectra and/or those in the ground-state bleaching, stimulated emission, and excited-state absorption. This can be explained in terms of the stimulated Raman interaction described as K-type in the ground state or the Raman-like interaction described as V-type in the excited state of the spectral components corresponding to the pump and Stokes component [21]. The Fourier power spectra of the time-resolved modulation $\delta\Delta A(t)$ at the corresponding photon energies to the real-time traces in Figure 9.3.4.2a are shown in Figure 9.3.4.2b.

The signals of real-time vibration can be contributed from various mechanisms, as described below.

The first one arises from spectral changes including both intensity and shape due to the wavepacket motions in the ground and excited states. The intensity change at a specific probe wavelength can be due to the oscillation of the coefficients of the wavefunction of the wavepacket, which is a linear combination of vibrational eigenfunctions. Non-Condon effect can also be the origin of the intensity change. These correspond to the imaginary part of the nonlinear susceptibility of the pump–probe process.

Another contribution is induced by the molecular phase modulation, MPM, caused by a kind of third-order nonlinear effect described by the molecular vibration-induced Kerr effect [25]. This may be interpreted as a kind of cross-phase modulation (sometimes called XPM), where the refractive index is modulated in proportion to the vibrational amplitude, which in turn is proportional to the pump laser intensity. This effect corresponds to the real part of the nonlinear susceptibility. The observed spectral change due to this mechanism is simply described by

$$d\Delta A = (d\Delta A(\omega)/d\omega)d\omega \qquad (9.3.4.1)$$

The effects of Raman and Raman-like contributions with high frequencies are not included in this equation, since they are out of the probe-frequency range of the moment calculation in the case of high-frequency mode. As for the modes of lower frequencies, they are included because of the above type of contributions.

From Eq. (9.3.4.1), the dependence of the Fourier amplitude on the probe photon energy is expected to be given by the first derivative of the $\Delta A(\omega)$ spectrum if the wavepacket moves on the ground state potential modifying the stationary absorption spectrum of the ground state.

9.3.4.3.3 INITIAL PHASES OF THE VIBRATIONAL MODES COUPLED TO THE ELECTRONIC TRANSITION VIA IMPULSIVE EXCITATION

The initial phase of the molecular vibration is important for the assignment of the wavepacket responsible for the coherent modulation of the electronic transition intensity to either the excited state or the ground state. Figure 9.3.4.5a–h depicts the dependence of the initial phase and the Fourier power of molecular vibration on the probe photon energy obtained by the FT of the time-dependent difference absorbance from 50 to 1800 fs.

As for the 106 cm^{-1} data shown in Figure 9.3.4.5a, the phases are $(-1/2)\pi$ in the full range of observation except in the probe-photon energy range between 2.02 and 2.05 eV (16300 and 16500 cm^{-1}), where the Fourier power is relatively small. It can be safely concluded from this phase dependence that the mode with 106 cm^{-1} is mainly attributed to the ground-state wavepacket. The frequency of this mode is too low for easy detection by conventional Raman spectroscopy, but this mode can be observed very clearly. This is the advantage of real-time vibrational spectroscopy, which is not suffered from intense Rayleigh scattering appearing at the pumping time with no delay.

Rayleigh scattering results from the electric polarizability of the particles. The oscillating electric field of a light wave acts on the charges within a particle, causing them to move at the same

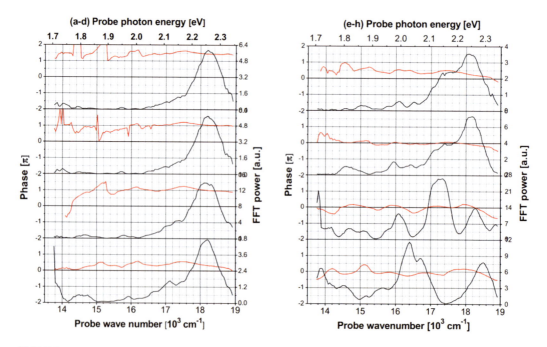

FIGURE 9.3.4.5 Probe photon energy dependencies of the initial phase (red) and FFT power (black) of molecular vibration for (a) 106, (b) 114, (c) 171, (d) 244, (e) 309, (f) 366, (g) 651, and (h) 1205 cm^{-1} obtained by the FFT of the time-dependent difference absorbance from 50 to 1800 fs. (For interpretation of the references to color in this figure legend, the reader is referred to the web version of this article.)

frequency. The particle, therefore, becomes a small radiating dipole whose radiation we see as scattered light. Because of the same frequency of the Rayleigh scattering as that of the pump, it is difficult to remove it by a spectroscopic filter in ultrafast spectroscopy.

In the same way, the mode with $114\,cm^{-1}$ shown in Figure 9.3.4.5b is well attributed to the excited state except in the spectral range around 2.00 and 2.35 eV (16100 and $18950\,cm^{-1}$). Figure 9.3.4.5c shows the third lowest frequency of $171\,cm^{-1}$. In this case, all the initial phases calculated are close to p ranging between 1.97 and 2.35 eV (15900 and $18950\,cm^{-1}$) except in the range of 1.75 and 1.85 eV. From these phases, the wavepacket generating modulation with $171\,cm^{-1}$ can be assigned to the excited state. The modulation in the range of 1.75 and 1.85 eV is considered to be mixed with the signal induced by the excited wavepacket. In the same way, the mode with $244\,cm^{-1}$ shown in Figure 9.3.4.5d is also due to the mixed contribution of the ground and excited states. The mode with $651\,cm^{-1}$ depicted in Figure 9.3.4.5g has a peculiar feature of oscillating spectral Fourier power in the full probe range. The phases are all close to 0 p around 1.77, 1.85, 2.05, and 2.26 eV (14300, 14900, 16500, and $18200\,cm^{-1}$), which are close to the peak positions of the FT power spectrum of this mode. Therefore, the signals corresponding to the peaks are dominantly due to the wavepacket in the excited state. The phase shown in Figure 9.3.4.5h is close to p/2 near their FT power peaks around 2.0 and 2.3 eV. Therefore, the mode having a large amplitude is concluded to be mainly due to the ground state. In the other probe range, the contribution from the excited state also has some sizable amount.

For discussion on the vibrational assignment of the peaks observed in the Fourier amplitude spectra, the frequencies of the Raman spectrum and those obtained by the FT of the real-time traces are listed together in Table 9.3.4.1. The Raman spectral pattern using the Ar laser, shown in Figure 9.3.4.1b, is quite different from that by the FT of real-time vibrational spectroscopy. This difference can be attributed to that of the resonance conditions in the excitation processes and the effect of the contribution of the wavepacket in the excited state. There are two more reasons for the

---

TABLE 9.3.4.1

Peaks in the Fourier Amplitude Spectra of Vibrational Modes Obtained in the Negative and Positive Ranges of ΔA Compared with Those Observed in the Raman Scattering Spectrum Using the Excitation Wavelength of 441.92 nm (w: Weak, m: Middle, s: Strong)

ΔA>0		ΔA<0		Raman (cm⁻¹)
Wavenumber (cm⁻¹)	Normalized FFT Power	Wavenumber (cm⁻¹)	Normalized FFT Power	Excited at 442 nm
106	0.34	114	0.38	
171	0.66	171	0.69	
244	0.23			
		260	0.50	
309	0.27			
366	1			404.64 (w)
		399	1	
448	0.34			576.0 (w)
		578	0.37	648.0 (w)
651	0.20	659	0.35	850.0 (w)
		757	0.18	878.9 (w)
				1108.4 (w)
1201	0.85	1203	0.67	1211.5 (m)
1335	0.20			1345.1 (m)
1465	0.24			
		1497	0.30	1489.5 (m)
		1530	0.25	1529.1 (s)

difference: For low-frequency modes, it is due to the difficulty in observation of Raman signals, while for high-frequency modes the FT amplitudes obtained from real-time traces are reduced by a factor of [22]

$$\left[\left(2\pi/t_p\right)^2 / \left(\left(\omega_m\right)^2 + \left(2\pi/t_p\right)^2\right)\right]$$

where x_m is the vibrational angular frequency of mode m and t_p is the pulse duration.

9.3.4.3.4 VIBRATIONAL-ENERGY LADDER DESCENDING PROCESS AND VIBRATIONAL PHASE RELAXATION

For a detailed study of the vibrational-energy relaxation process associated with the thermalization discussed in Section 9.3.4.3.1, the first moment of the induced absorption was calculated in the integration range of 529–726 nm (2.34–1.71 eV). As shown in Figure 9.3.4.6, the decay of the moment corresponding to the average transition energy reduces gradually with the delay time t_D. This trend can be reproduced by the following fit to the calculated first moment as follows:

$$\Delta E\left(t_D\right) = \left(\Delta E - \Delta E_0\right)\exp\left(-t_D/\tau_e\right) + \Delta E_0 \tag{9.3.4.2}$$

where the relaxation time, τ_e, the initial extra energy, ΔE, and the final energy after thermalization, ΔE_0, were taken as variable parameters. From the best fit to the observed curve, the parameters were found to be $\tau_e = 217 \pm 2$ fs, $\Delta E = 1.91 \pm 0.01$ eV, and $\Delta E_0 = 2.08 \pm 0.01$ eV.

Peaks in the Fourier amplitude spectra of vibrational modes were obtained in the negative and positive ranges of DA compared with those observed in the Raman scattering spectrum using the excitation wavelength of 441.92 nm. (w: weak, m: middle, s: strong).

The fitted curve was then subtracted from the observed curve of the first moment to calculate the FT after normalization of all the peak intensities in the FFT spectra by the maximum peak at 171 cm^{-1}. The corresponding vibrational dephasing time was calculated from the bandwidth of each mode shown in Figure 9.3.4.6b. Six most intense vibrational modes with wavenumbers of 103, 171, 258, 383, 1201,

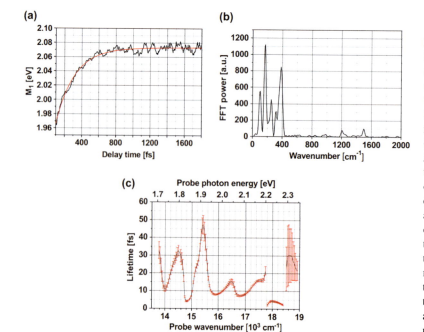

FIGURE 9.3.4.6 (a) Delay time dependence of the first moment (M_1) of the induced absorption spectra calculated by integrating in the range of 529–726 nm (2.34–1.71 eV). Curve M_1 is smoothed over the probe delay gate of 10 fs. Fitting to an exponential curve is also displayed. (b) FFT of the (a) curve after the fitting curve is subtracted from the experimental curve. (c) Decay times of the signal in the negative time range as a function of the probe photon energy.

and 1499 cm^{-1} can be observed in the normalized FFT spectrum in Figure 9.3.4.6b. The Fourier power associated with the mean energy (first moment) reducing with time provides the strength of the vibronic coupling which induces (mean) energy reduction. The modulations shown in the (mean) energy relaxation obtained by the first moment is the total effect contributed from each vibrational mode. Therefore, the first moment can be expressed by the individual contribution from each mode (ΔE_{vi}) as

$$M_1(t_D) = \Delta E(t_D) + \sum_i \Delta E_{vi} \exp\left(-t_D/T_{2i}^{vib}\right) \cos\left(\omega_{vi} t_D + \varphi_i\right),$$

(9.3.4.3)

$$(i = 1, 2, .., 6)$$

where ω_{vi} is the molecular vibrational frequency of each mode, T_{2i}^{vib} is the vibrational dephasing time, and the second term in the vibrational cos function is the initial phase. The individual contribution to the energy relaxation can be obtained from each of the six modes, out of all the modes including those not well observed under the noise level using their relative peak values in the FT power spectrum in Figure 9.3.4.6b, in combination with the total amount of the energy relaxation in the excited state, obtained from the first moment of induced absorption. Using the individual contribution divided by the vibrational frequency of itself, the Huang–Rhys factors corresponding to the transition from the lowest to the higher excited states for these six modes can be determined as listed in Table 9.3.4.2. This is the first determination of multi-dimensional Huang–Rhys factors; the data of these multi-dimensional values provide the structure of the multi-dimensional potential hypersurface of complex molecular and polymer systems.

9.3.4.3.5 ELECTRONIC PHASE RELAXATION OBTAINED FROM THE DATA IN THE NEGATIVE TIME RANGE

In pump–probe spectroscopy, the pump pulse perturbs the absorption spectrum of the medium, which is subsequently probed after a set time delay. This method implicitly assumes that the weak probe pulse does not induce any substantial excitation of the sample [26]. In the femtosecond regime, however, the difference absorption spectra cannot be directly interpreted as a change in the absorption spectrum

TABLE 9.3.4.2
Lifetimes and Huang–Rhys Factors of the Vibrational Modes Determined from the First Moment of ΔA

Peak Wave Number (cm^{-1})	Normalized FFT Power	Lifetime τ_{Mlvb} (ps)	Huang–Rhys Factor
*M*1			
49	0.026	1.67	
103	0.492	1.69	0.25
171	1	1.96	0.15
219	0.17	1.42	
258	0.415	1.78	0.10
313	0.273	2.21	
383	0.765	1.15	0.11
1201	0.071	1.80	0.01
1252	0.028	1.63	
1355	0.016	1.75	
1424	0.013	1.21	
1499	0.089	1.26	0.01
1550	0.011	2.01	

because of coherence effects. These effects fall into two categories: One is 'coherent coupling' due to the induced grating formed by the temporally coincident pump and the probe in the sample [26–29], and the other is the 'perturbed free polarization decay' generated by the probe and perturbed by the pump. In previous reports [26–33], the experimental results observed in the negative delay-time region were discussed in terms of perturbed free induction decay and coherent coupling. The difference absorption spectrum was calculated for a two-level system and applied to molecular systems [34,35]. We have here modified their treatment for the vibronic system in the following way: under the assumption that only one mode is coupled to the excitation to the exciton state but that it can readily be extended to a multi-mode system. The apparent absorbance difference, $\Delta A(x)$, observed in the negative time range in the rotating reference frame using the pump, $E_{pu}(t)$, and probe, $E_{pr}(t)$, fields is given by [36]:

$$\Delta A(\omega) \sim \text{Im}\left\{ \left| f_2(\omega) / e_{pr}(\omega) \right| F \left[E_{pu}(t) \left[F_1(t) \otimes \left(E_{pu}^*(t) P_{pr}(t) \right) \right] \right] \right\} \qquad (9.3.4.4)$$

Here $P_{pr}(t)$ is the macroscopic polarization in a molecular vibronic system propagating in the probe direction,

$f_2(\omega) = \Im|F_2(t)|$ is the FT of the following.

$$F_2(t) = (i\mu/h)\exp\left(-t/T_2^{el}\right)\exp\left(-i\Omega t\right)\exp\left(-t/T_2^{vib}\right)\exp\left(-i\left(\omega t + \phi\right)\right) \qquad (9.3.4.5)$$

$$F_1(t) = (2i\mu/h)\exp(-t/T_1)\exp\left(-i\left(\omega t + \phi\right)\right) \qquad (9.3.4.6)$$

$$\phi = \arctan\left[\left(\omega - \omega_{ba} + \omega_v\right)T_2^{vib}\right] \qquad (9.3.4.7)$$

Here, T_2^{el} and T_2^{vib} are the electronic and vibrational dephasing times, respectively, and x is the optical angular frequency corresponding to the 0–0 transition energy from the ground state to the electronic excited state. The symbol denotes convolution; l is the transition dipole moment; T_1 is the longitudinal electronic relaxation time; $X = x_{ba} x_1$ is the detuning between the pump field frequency x_1 and the transition frequency x_{ba}; x_m is the vibrational angular frequency.

This perturbed free polarization decay term represents the case when the probe pulse arrives earlier than the pump pulse and there is no temporal overlapping between them. The probe pulse generates electronic coherence in the sample with the duration of the electronic dephasing time. Then the intense pump field forms a grating, i.e., $E_{pu}(t)P_{pr}(t)$ term in Eq. (9.3.4.4), which interacts with another pump field to be diffracted into the probe direction, satisfying the causality. In the present case, the vibronic coupling expected to be strong in the conjugated electron system is the origin of the electronic spectrum of the ground state. Therefore, the polarization generated by the probe pulse that precedes the pump pulse is a vibronic transition, instead of pure electronic transition, which is associated with the transition between the ground vibrational level in the ground electronic state and the vibronically excited state. The wave-packet formation in the ground state requires two fields of the pump pulse. Therefore, this signal increases with the delay time and the time constant T_2 and disappears quickly at $t=0$ [29].

The decay times of the signal in the negative time range are functions of the probe photon energy, as shown in Figure 9.3.4.6c. The apparent lifetimes depend on the contribution of the coherent spike, which reduces the real dephasing time. The longest among the observed values is estimated to be the closest to the true value 47±5 fs around 1.91 eV in the figure. This is a reasonable duration of dephasing in condensed phase materials [36].

9.3.4.4   CONCLUSIONS

By utilizing the pump–probe data in the negative time range with sub-7 fs pulses, important information was obtained related to the electronic phase relaxation time and the frequencies of the vibrational modes due to the wavepacket motion in the electronic excited sate. The absorbance change observed in the 'negative' delay time range has been used for estimation of the electronic dephasing time to be 47 ± 5 fs. Coherent molecular vibration of a polymer in the excited state has been observed in the real-time trace without the effect of wave packet motion in the ground state, which usually hinders assignment of the signal to either the ground state or the excited state. This method would provide the novel method for the simultaneous measurement of the phase relaxation and population relaxation dynamics [37].

REFERENCES

1. T. Kobayashi (Ed.), *Nonlinear Optics of Organics and Semiconductors, Springer Proceeding Physics*, Vol. **36**, Springer, Berlin (1989).
2. A. J. Heeger, S. Kivelson, J. R. Schrieffer, and W.-P. Su, *Rev. Mod. Phys.* **60**, 781 (1988).
3. S. Etemad, Z. G. Soos, in *Spectroscopy of Advanced Materials*, edited by R. J. H. Clark and R. E. Hester, Wiley, New York, p. 87 (1991).
4. R. H. Friend, R. W. Gymer, A. B. Holmes, J. H. Burroughes, R. N. Marks, C. Taliani, D. D. C. Bradley, D. A. Dos Santos, J. L. Brédas, M. Lögdlund, and W. R. Salaneck, *Nature* **397**, 121 (1999).
5. F. Hide, M. A. Díaz-García, B. J. Schwartz, and A. J. Heeger, *J. Acc. Chem. Res.* **30**, 4301 (1997).
6. T. Kobayashi, *IEICE Trans. Fundam.* **E75**, 38 (1992).
7. E. I. Rashiba, in *Excitons, Selected Chapters*, edited by E. I. Rashiba and M. D. Sturge, Elsevier Science, Amsterdam, p. 273 (1987).
8. T. Kobayashi (Ed.), *Relaxation in Polymers*, World Scientific, Singapore (1993).
9. T. Kobayashi, M. Yoshizawa, U. Stamm, M. Taiji, and M. Hasegawa, *J. Opt. Soc. Am. B* **7**, 1558 (1990).
10. T. Kobayashi, M. Yoshizawa, T. Masuda, T. Higashimura, and T. Kobayashi, *IEEE J. Quant. Electron.* **28**, 2508 (1992).
11. H. Kano, T. Saito, A. Ueki, and T. Kobayashi, *Int. J. Mod. Phys. B* **15**, 3817 (2001).
12. H. Kano, T. Saito, and T. Kobayashi, *J. Phys. Chem. A* **106**, 3445 (2002).
13. T. Kobayashi, T. Teramoto, V. M. Kobryanskii, and T. Taneichi, *Synth. Met.* **159**, 1751 (2009).
14. J. Du and T. Kobayashi, *Chem. Phys. Lett.* **481**, 204 (2009).
15. K. Okuhara, *J. Org. Chem.* **41**, 1487 (1976).
16. T. Masuda, T. Hamano, K. Tsuchihara, and T. Higashimura, *Macromolecules* **23**, 1374 (1990).
17. A. Shirakawa, I. Sakane, and T. Kobayashi, *Opt. Lett.* **23**, 1292 (1998).
18. A. Baltuska, T. Fuji, and T. Kobayashi, *Opt. Lett.* **27**, 306 (2002).
19. T. Kobayashi, J. Zhang, and Z. Wang, *New J. Phys.* **11**, 013048 (2009).
20. T. Kobayashi and A. Yabushita, *Chem. Rec.* **11**, 99 (2011).
21. J. Du, T. Teramoto, K. Nakata, E. Tokunaga, and T. Kobayashi, *Biophys. J.* **101**, 995 (2011).
22. Y. Wang and T. Kobayashi, *Chem. Phys. Chem.* **11**, 889 (2011).
23. T. Kobayashi, A. Shirakawa, H. Matsuzawa, and H. Nakanishi, *Chem. Phys. Lett.* **321**, 385 (2000).
24. T. Kobayashi, M. Hirasawa, Y. Sakazaki, and H. Hane, *Chem. Phys. Lett.* **400**, 301 (2004).
25. N. Zhavoronkov and G. Korn, *Phys. Rev. Lett.* **88**, 203901 (2002).
26. C. H. Brito Cruz, R. L. Fork, W. H. Knox, and C. V. Shank, *Chem. Phys. Lett.* **132**, 341 (1986).
27. J. J. Baumberg, B. Huttner, R. A. Taylor, and J. F. Ryan, *Phys. Rev. B* **48**, 4695 (1993).
28. C. V. Shank, R. L. Fork, C. H. Brito Cruz, and W. Knox, in *Ultrafast Phenomena V*, edited by G. R. Fleming and A. E. Siegman, Springer, Berlin (1986).
29. C. H. Brito Cruz, J. P. Gordon, P. C. Becker, R. L. Fork, and C. V. Shank, *IEEE J. Quant. Electron.* **24**, 261 (1988).
30. B. Fluegel, N. Peyghambarian, G. Olbright, M. Lindberg, S. W. Koch, M. Joffre, D. Hulin, A. Migus, and A. Antonetti, *Phys. Rev. Lett.* **59**, 2588 (1987).
31. F. W. Wise, M. J. Rosker, G. L. Millhauser, and C. L. Tang, *IEEE J. Quant. Electron.* **23**, 1116 (1987).
32. M. Lindberg and S. W. Koch, *Phys. Rev. B* **38**, 7607 (1988).
33. J. P. Likforman, M. Joffre, G. Chériaux, and D. Hulin, *Opt. Lett.* **20**, 2006 (1995).
34. C. J. Bardeen and C. V. Shank, *Chem. Phys. Lett.* **203**, 535 (1993).
35. J. Y. Bigot, M. T. Portella, R. W. Schoenlein, C. J. Bardeen, A. Migus, and C. V. Shank, *Phys. Rev. Lett.* **66**, 1138 (1991).
36. T. Kobayashi, J. Du, W. Feng, and K. Yoshino, *Phys. Rev. Lett.* **101**, 037402 (2008).
37. T. Kobayashi, T. Iiyama, K. Okamura, J. Du, and T. Masuda, *Chem. Phys. Lett.* **567**, 6–13 (2013).

9.3.5 Ultrabroadband Time-Resolved Spectroscopy of Polymers

9.3.5.1 EFFECT OF ANNEALING ON THE PERFORMANCE OF P3HT: PCBM SOLAR CELLS

The demand for renewable energy sources has stimulated progress in the development of efficient photovoltaic devices, and organic solar cell research has achieved several critical milestones in recent decades. Replacing traditional inorganic semiconductor-based solar cells, organic solar cells have become established as a future photovoltaic technology because of their advantages of cost-effective production, large area, lightweight, and flexibility [1,2]. The highest reported power-conversion efficiency to date is 10.8% [3,4], whereas it barely reached 1% in the first reported polymer solar cell [5]. In the past couple of years, the polymer–fullerene heterojunction has dominated organic solar cell research [6,7]. For standard bulk polymer–fullerene heterojunction systems, the polymer poly(3-hexylthiophene) (P3HT) as the electron donor and the fullerene [6,6]-phenyl-C61-butyric acid methyl ester (PCBM) as the electron acceptor are typically blended to create a composite material that has been demonstrated to exhibit effective device performance [7].

In solar cell devices, an anode and cathode are necessary to collect separated charges. The anode is made of tin-doped indium oxide (ITO) coated with a layer of poly-ethylene-dioxythiophene: polystyrene-sulfonic acid (PEDOT:PSS). ITO is one of the most extensively used electrode materials because of its high electrical conductivity and optical transparency. The transparent, water-soluble PEDOT:PSS is used to smooth the rough ITO surface and further effectively collect the separated holes into the electrode because it has a higher work function. A metal layer (e.g., aluminum) serves as the cathode.

For a P3HT:PCBM device, as shown in the inset of Figure 9.3.5.1, ITO-coated glass substrates were used as the anode; they were modified by spin-coating with ~40 nm thick conductive PEDOT:PSS followed by baking at 150°C for 30 min. P3HT was blended with PCBM at a weight ratio of 2.5% and dissolved in 1,2-dichlorobenzene. The active layer was thermally annealed at 190°C for 10 min in a nitrogen-filled glove box before (preannealing) or after (post-annealing) deposition of the aluminum. The cathode, which had a 100 nm Al layer, was thermally evaporated onto the polymer film at a base pressure of 7.5×10^{-9} Pa to form an active area of $0.06\,cm^2$. The current density–voltage (J–V) characteristics for the devices were recorded under light illumination using standard solar irradiation of $100\,mW/cm^2$ with a xenon lamp as the light source and a computer-controlled voltage–current source meter.

Figure 9.3.5.1 shows the current density–voltage (J–V) characteristics of a device with the structure ITO/PEDOT:PSS/P3HT: PCBM/Al and different thermal annealing processes. As expected, the device fabricated using a preannealing process exhibits poor performance characteristics and its power conversion efficiency is only 1.63%. The other device, prepared with a post-annealing process, demonstrates better performance and its power conversion efficiency is 2.88%. In terms of device performance, both the open-circuit voltage (V_{OC}) and short-circuit current (J_{SC}) are improved by the post-annealing process.

The open-circuit voltage, V_{OC}, is the maximum voltage available from a solar cell, and this occurs at zero current. The open current voltage, J_{SC}, corresponds to the amount of forward bias on the solar

DOI: 10.1201/9780429196577-57

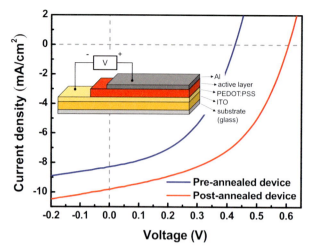

FIGURE 9.3.5.1 Current density–voltage (J–V) characteristics of solar cells of ITO/PEDOT:PSS/P3HT:PCBM/Al with pre-(top curve) and post-annealing (bottom curve) processes. Inset: device architecture of a bulk heterojunction solar cell device. The Al layer is the cathode. The active layer is a semiconducting polymer/fullerene blend. ITO coated with PEDOT:PSS serves as the anode. These layers are deposited on a glass substrate.

cell due to the bias of the solar cell junction with the light-generated current. The short-circuit current is the maximum current from a solar cell and occurs when the voltage across the device is zero.

Some crucial studies [8–11] have pointed out the reasons for the higher performance in post-annealed devices. However, the microscopic viewpoint of high-performance devices has yet to be considered. Here, the fundamental carrier dynamics directly correlated to the efficiency of charge transport in solar cell devices are studied by ultrafast spectroscopy [12]. The stationary absorbance spectra of pre- and postannealed devices in the visible range show that the absorbance increases rapidly at photon energies greater than ~1.9 eV, which demonstrates that the polymer has a wide absorption band [6]. The spectra of both pre- and post-annealed devices exhibit the π–π transition of P3HT at 2.05 and 2.23 eV [13,14].

Time-resolved spectroscopy using sub-10-fs visible pulses from a broadband OPA (as demonstrated in Section 9.3.5.2.B.2) was performed at room temperature. The output pulses of a broadband OPA were separated into pump and probe pulses. The fluences of the pump and probe pulses at the sample were 2.7 and 0.3 mJ/cm², respectively. The changes in the sample induced by a pump pulse were obtained by detecting the change in absorption (ΔA) of probe pulses as a function of probe delay time. The femtosecond time evolutions were derived by delaying the relative arrival times of the pump and probe pulses rapidly by a fast-scan stage (Section 9.3.5.2.C.1). The probe pulse was dispersed using a polychromator into a 96 branch fiber bundle, the other end of which was separated into 96 fiber branches and connected to APDs with a spectral resolution of 2.56 nm, i.e., ~10 meV (Section 9.3.5.2.C.2). Therefore, the time-resolved absorption differences at 96 probe wavelengths were simultaneously detected at the photodiodes. The detected signals were sent to a multichannel lock-in amplifier to be spectrally resolved for simultaneous detection of low-intensity signals over the entire spectral region.

The time- and photon-energy-resolved transient absorption difference, $\Delta A(\omega, t)$ of the pre- and post-annealed devices was measured using the pump–probe technique. Figure 9.3.5.2a and c show 2D plots of the ΔA spectra as functions of time and photon energy. The ΔA spectrum is positive at photon energies of less than ~1.98 eV, which is attributed to the induced absorption for transitions from the first excited state to higher states. The negative ΔA at photon energies greater than ~1.98 eV is due to stimulated emission from the excited state and photobleaching due to ground state depletion. The two peaks at 2.05 and 2.23 eV represent the π–π transition in P3HT. Figure 9.3.5.3 illustrates the relaxation processes of the P3HT:PCBM blend, which are excited by pump pulses with a photon energy of >1.9 eV (the absorption gap energy) [15,16]. In the composite samples, it is estimated that more than 60% of the incident photons are absorbed by the polymer [17]. Therefore, the sample excited by the pump pulses generates excited electron–hole pairs, primarily in P3HT molecules. The excited electrons at the lowest unoccupied molecular orbital (LUMO) of P3HT are

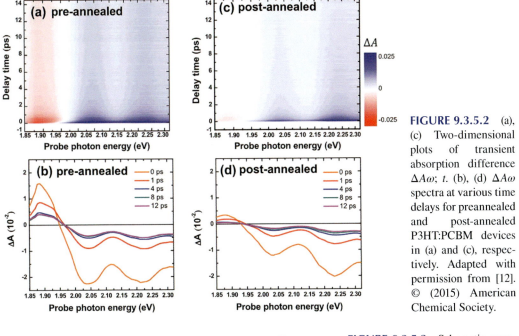

FIGURE 9.3.5.2 (a), (c) Two-dimensional plots of transient absorption difference $\Delta A \omega$; t. (b), (d) $\Delta A \omega$ spectra at various time delays for preannealed and post-annealed P3HT:PCBM devices in (a) and (c), respectively. Adapted with permission from [12]. © (2015) American Chemical Society.

FIGURE 9.3.5.3 Schematic representation of ultrafast carrier dynamics after photoexcitation. E^D_{LUMO}, the LUMO of the electron donor; E^D_{HOMO}, the highest occupied molecular orbital (HOMO) of the electron donor; E^A_{LUMO}, the LUMO of the electron acceptor; E^A_{HOMO}, the HOMO of the electron acceptor. In this study, the electron donor and electron acceptor are P3HT and PCBM, respectively. τ is the time constant for the relaxation processes. Adapted with permission from [12]. © (2015) American Chemical Society.

transferred to the LUMO of PCBM, and the holes remain in the P3HT to form a bounded polaron pair (BPP) with the excited electrons. The time constant for this interfacial charge transfer is measured as ~90 fs [16,18]. The generated BPP then relaxes to the ground state via the parallel processes of dissociation into separated polarons, trapping by defect states, and recombination. The time constants for dissociation into the separated polarons and defect trapping are reported to be ~0.95 and ~2.8 ps [16], respectively.

According to the scenario just described the real-time traces for $\Delta A \omega$; t are expressed by the equation

$$\Delta A(t) = A_{\mathrm{CT}} e^{\frac{t}{\tau_{\mathrm{CT}}}} + A_{\mathrm{SP}} \left(-e^{\frac{t}{\tau_{\mathrm{CT}}}} + e^{-\frac{t}{\tau_{\mathrm{SP}}}} \right) + A_{\mathrm{trap}} \left(-e^{-\frac{t}{\tau_{\mathrm{CT}}}} + e^{-\frac{t}{\tau_{\mathrm{trap}}}} \right) + A_{\mathrm{Recomb}} \qquad (9.3.5.1)$$

where the suffixes CT, SP, trap, and Recomb correspond to charge transfer, separated polarons (dissociated BPP), trapped BPP, and carrier recombination, respectively. The time constant for carrier recombination is beyond the measurement range in this study. For the post-annealed device, the time constants τ_{CT}, τ_{SP}, and τ_{trap} are ~0.13, ~0.68, and ~8.48 ps, respectively. For the preannealed device, the time constants τ_{CT}, τ_{SP}, and τ_{trap} are ~0.13, ~0.54, and ~2.6 ps, respectively. The relaxation processes for the BPP, especially for trapping by defect states (τ_{trap}), apparently have a longer lifetime in the post-annealed device.

This implies that the excited carriers in the E_{LUMO}^{A} state have a longer lifetime and so are more likely to be dissociated into photocarriers which further produce a photocurrent. Accordingly, this longer lifetime of the excited carriers in the post-annealed device may explain the increase in the J_{SC} value. However, in reality, there are several relaxation channels for excited carriers in the E_{LUMO}^{A} state, such as dissociation into separate polarons, trapping by defect states, and recombination. Most excited carriers in the E_{LUMO}^{A} state are trapped by defect states or recombine with opposite charges without contributing to the photocurrent, and then these excited carriers certainly do not increase the value of J_{SC} even though they have a longer lifetime in the E_{LUMO}^{A} state. Consequently, it is necessary to determine how many excited carriers in the E_{LUMO}^{A} state relax through each channel, an issue that is still unresolved.

It should be emphasized that pump–probe spectroscopy with high time and photon energy resolution can be further used to show the relative amount of photoexcited carriers relaxed through each of the relaxation processes in the E_{LUMO}^{D} state (see Figure 9.3.5.3), which is the key to understanding any improvement in device performance. The percentage of carriers relaxed through every channel is calculated using the coefficients A_{CT}, A_{SP}, A_{trap}, and A_{Recomb}. In addition, the percentage of each component in the region of 1.98–2.13 eV for stimulated emission can be further estimated (Figure 9.3.5.4). The percentage of charge transfer increases by 4.5% for the post-annealed devices. This demonstrates that interfacial charge transfer from an electron donor (P3HT) to an electron acceptor (PCBM) in the postannealed devices is more efficient than that in the preannealed devices. There are 1.8% more separated polarons in the post-annealed devices than in the preannealed devices, but there is 6.4% less recombination in the post-annealed devices. Consequently, more charges are transferred from the electron donor (P3HT) to the electron acceptor (PCBM).

9.3.5.2 CONCLUSION AND PERSPECTIVES

In conclusion, in this short chapter, we describe the analyses of each relaxation process in P3HT:PCBM solar cells show that there are increases in the charge transfer and the number of separated polarons and a decrease in the amount of recombination between excited carriers, which is one of the physical mechanisms responsible for enhanced performance after a post-annealing process. These findings are consistent with observations of the annealing-dependent surface morphology and vertical distribution of P3HT:PCBM blends, which provides key information for the design of high-performance solar cells.

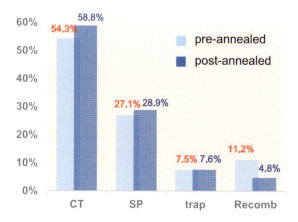

FIGURE 9.3.5.4 Average percentages of each of the relaxation processes shown in Figure 9.3.5.3 at 1.98–2.13 eV. Adapted with permission from [86]. © (2015) American Chemical Society.

These important results and conclusions indicate that ultrabroadband time-resolved spectroscopy provides a powerful means of studying the interactions between quasi-particles, which are the basis for composing a material. Using broadband OPA, we can observe several energy levels simultaneously with extremely high time resolution and study the correlations among them. This provides much clearer physical insight into the interactions of quasi-particles in several novel types of condensed matter. The development of ultrabroadband light sources continues. For example, the generation of ultrabroadband MIR coherent light using four-wave difference-frequency generation from two-color femtosecond pulses in gases has been demonstrated [18]. This type of ultrabroadband light source extending to the MIR region as well as the terahertz region is desirable and extremely important for investigating the detailed ultrafast dynamics in solids, as the bandgap energy or a number of optical transitions have resonance energies in this frequency region. The study described here in this subsection shows an example of ultrashort pulse laser for the characterization of organic solar cells from the viewpoint of basic physics in the system showing how useful the basic study is to obtain the essential point of improvement of the specification of organic cells [19]. The research work described in this subsection was conducted by the following people in collaboration: C.-W. Luo, Y.-T. Wang, A. Yabushita, and T. Kobayashi [19].

REFERENCES

1. T. Stubhan, M. Salinas, A. Ebel, F. C. Krebs, A. Hirsch, M. Halik, and C. J. Brabec, "Increasing the fill factor of inverted P3HT:PCBM solar cells through surface modification of Al-doped ZnO via phosphonic acid anchored C60 SAMs," *Adv. Energy Mater.* **2**, 532–535 (2012).
2. F. C. Krebs, R. Søndergaard, and M. Jørgensen, "Printed metal back electrodes for R2R fabricated polymer solar cells studied using the LBIC technique," *Sol. Energy Mater. Sol. Cells* **95**, 1348–1353 (2011).
3. Y. Liu, J. Zhao, Z. Li, C. Mu, W. Ma, H. Hu, K. Jiang, H. Lin, H. Ade, and H. Yan, "Aggregation and morphology control enables multiple cases of high-efficiency polymer solar cells," *Nat. Commun.* **5**, 5293 (2014).
4. C. W. Tang, "Two-layer organic photovoltaic cell," *Appl. Phys. Lett.* **48**, 183–185 (1986).
5. C. Winder and N. S. Sariciftci, "Low bandgap polymers for photon harvesting in bulk heterojunction solar cells," *J. Mater. Chem.* **14**, 1077–1086 (2004).
6. G. Dennler, M. C. Scharber, and C. J. Brabec, "Polymer-fullerene bulk heterojunction solar cells," *Adv. Mater.* **21**, 1323–1338 (2009).
7. H. J. Kim, J. H. Park, H. H. Lee, D. R. Lee, and J.-J. Kim, "The effect of Al electrodes on the nanostructure of poly(3-hexylthiophene): Fullerene solar cell blends during thermal annealing," *Org. Electron.* **10**, 1505–1510 (2009).
8. J. J. Fang, H. W. Tsai, I. C. Ni, S. D. Tzeng, and M. H. Chen, "The formation of interfacial wrinkles at the metal contacts on organic thin films," *Thin Solid Films* **556**, 294–299 (2014).
9. A. Orimo, K. Masuda, S. Honda, H. Benten, S. Ito, H. Ohkita, and H. Tsuji, "Surface segregation at the aluminum interface of poly(3 hexylthiophene)/fullerene solar cells," *Appl. Phys. Lett.* **96**, 043305 (2010).
10. W. H. Tseng, H. Lo, J. K. Chang, I. H. Liu, M. H. Chen, and C. I. Wu, "Metal-induced molecular diffusion in [6,6]-phenyl-C61-butyric acid methyl ester poly(3-hexylthiophene) based bulk-heterojunction solar cells," *Appl. Phys. Lett.* **103**, 183506 (2013).
11. Y.-T. Wang, M.-H. Chen, C.-T. Lin, J.-J. Fang, C.-J. Chang, C.-W. Luo, A. Yabushita, K.-H. Wu, and T. Kobayashi, "Use of ultrafast time-resolved spectroscopy to demonstrate the effect of annealing on the performance of P3HT:PCBM solar cells," *ACS Appl. Mater. Interfaces* **7**, 4457–4462 (2015).
12. V. Shrotriya, J. Ouyang, R. J. Tseng, G. Li, and Y. Yang, "Absorption spectra modification in poly(3-hexylthiophene): Methanofullerene blend thin films," *Chem. Phys. Lett.* **411**, 138–143 (2005).
13. Y. Zhao, Z. Xie, Y. Qu, Y. Geng, and L. Wang, "Solvent-vapor treatment induced performance enhancement of poly(3-hexylthiophene): Methanofullerene bulk-heterojunction photovoltaic cells," *Appl. Phys. Lett.* **90**, 043504 (2007).
14. X. Ai, M. C. Beard, K. P. Knutsen, S. E. Shaheen, G. Rumbles, and R. J. Ellingson, "Photoinduced charge carrier generation in a poly(3 hexylthiophene) and methanofullerene bulk heterojunction investigated by time-resolved terahertz spectroscopy," *J. Phys. Chem. B* **110**, 25462–25471 (2006).
15. Y. H. Lee, A. Yabushita, C. S. Hsu, S. H. Yang, I. Iwakura, C. W. Luo, K. H. Wu, and T. Kobayashi, "Ultrafast relaxation dynamics of photoexcitations in poly(3-hexylthiophene) for the determination of the defect concentration," *Chem. Phys. Lett.* **498**, 71–76 (2010).

16. T. J. Savenije, J. E. Kroeze, M. M. Wienk, J. M. Kroon, and J. W. Warman, "Mobility and decay kinetics of charge carriers in photoexcited PCBM/PPV blends," *Phys. Rev. B* **69**, 155205 (2004).

17. C. J. Brabec, G. Zerza, G. Cerullo, S. De Silvestri, S. Luzzati, J. C. Hummelen, and S. Sariciftci, "Tracing photoinduced electron transfer process in conjugated polymer/fullerene bulk heterojunctions in real time," *Chem. Phys. Lett.* **340**, 232–236 (2001).

18. Y. Nomura, H. Shirai, K. Ishii, N. Tsurumachi, A. A. Voronin, A. M. Zheltikov, and T. Fuji, "Phase-stable subcycle mid-infrared conical emission from filamentation in gases," *Opt. Express* **20**, 24741–24747 (2012).

19. C.-W. Luo, Y.-T. Wang, A. Yabushita, and. T. Kobayashi, "Ultra-broadband time-resolved spectroscopy in novel types of condensed matter," *Optica* **3**, 82–92 (2016).

Section 9.4

2D Topological Materials

9.4.1 Ultrabroadband Time-Resolved Spectroscopy of Topological Insulators

9.4.1.1 INTRODUCTION

To understand condensed materials as demonstrated by band theory, one could imagine that electrons behave as an extended plane wave. This theory derives an energy band structure for electrons in a periodic lattice of atoms, and electrons in the material may be described as having or not having a bandgap [1,2]. In this description, electrons in the material can be considered as being in a "sea" of the averaged motion of the other quasi-particles. This approach of nearly free particles is valid in some well-understood materials, such as most metals, as the interaction strength between quasi-particles is negligible compared with their kinetic energy. However, strong correlation between the quasi-particles leads to a new type of behavior in some important materials. Such materials are difficult to describe theoretically because strong interactions between quasi-particles cause phenomena that cannot be predicted by studying the behavior of individual particles alone, and these interactions play a major role in determining the properties of such systems. The seemingly simple material NiO, as the typical example of metal–insulator transitions, would be expected to be a good conductor with a partially filled 3D band [3]. However, the strong Coulomb repulsion between electrons makes NiO an insulator. Therefore, this type of strongly correlated material cannot be understood using a free-electron-like scenario. In addition to the metal–insulator transitions just mentioned [3,4], there are numerous physical properties arising from the effects of strong correlations, e.g., high-T_c superconductivity [5], colossal magnetoresistance [6], heavy fermions [7], multiferroics [8], and low-dimensional phenomena [9]. Accordingly, the crucial correlations between quasi-particles can be responsible for significant characteristics of some materials, and so it is extremely important to discover the underlying interactions among quasiparticles in these materials.

Because interactions among quasi-particles are known to play an important role in understanding condensed matter, experimental techniques that can unambiguously clarify these interactions are needed. Studies have demonstrated the ability of numerous methods to indirectly estimate correlation among quasi-particles by measuring certain related physical characteristics, such as the carrier mobility [10], Shubnikov–de Haas oscillations [11], the thermoelectric power [12], the Burstein-Moss shift [13], the Raman shift [14], and the Faraday rotation [15].

Here the Shubnikov–de Haas oscillations, the Burstein-Moss shift, and the Faraday rotation are briefly explained below.

The Shubnikov-de Haas oscillations are oscillations of the resistivity parallel to the current ow in the edge states of a 2D electron gas in an applied magnetic field (B) Therefore they are related to the Quantum–Hall effect. The Shubnikov-de Haas oscillations have a 1/B-periodicity.

The Burstein–Moss shift, is the phenomenon in which the apparent band gap of a semiconductor is increased as the absorption edge is pushed to higher energies as a result of some states close to the conduction band being populated. This is observed for a degenerate electron distribution such as that found in some degenerate semiconductors.

The Faraday rotation is a physical magneto-optical phenomenon. The Faraday effect causes a polarization rotation which is proportional to the projection of the magnetic field along the direction of the light propagation. Formally, it is a special case of gyroelectromagnetism obtained when the dielectric permittivity tensor is diagonal. This effect occurs in most optically transparent dielectric

DOI: 10.1201/9780429196577-59

materials (including liquids) under the influence of magnetic fields. The Faraday effect has applications in measuring instruments. For instance, it has been used to measure optical rotatory power and for remote sensing of magnetic fields such as fiber optic current sensors. The Faraday effect is used in spintronics research to study the polarization of electron spins in semiconductors. Faraday rotators can be used for amplitude modulation of light, and are the basis of optical isolators and optical circulators; such components are required in optical telecommunications and other laser applications. A related magneto-optical effect is an optical Kerr effect.

As an example, electron–electron interaction is usually studied by transport measurements. However, the contribution of electron–electron interaction to the resistivity can be observed only at low temperatures because the electrical resistance is primarily dominated by electron–phonon scattering above the Debye temperature. On the other hand, the electron–phonon interaction strength can be determined from the phonon linewidths obtained through Raman or neutron scattering, which are easily influenced by selection rules and inhomogeneous broadening. Moreover, the quasi-particle interactions obtained by the techniques just mentioned are derived mainly from stationary experiments.

Ultrafast spectroscopy is one of the desired techniques that enable direct observation of transient interactions among quasiparticles. With transient spectroscopy, the so-called pump–probe measurement, we can monitor energy transfer among quasiparticles and even specify the interaction strength, as shown in Figure 9.4.1.1. Taking advantage of developments in the area of ultrashort pulses in recent decades [16–20], ultrafast optical spectroscopy can provide the required time resolution for studying ultrafast primary phenomena on the characteristic time scales of electron, phonon, and spin dynamics [21–33], that is, in the range of femtoseconds (10^{-15}s), picoseconds (10^{-12}s), and nanoseconds (10^{-9}s). Advanced progress in pulsed lasers has also extended ultrafast spectroscopy from the visible region to the mid-infrared (MIR) and ultraviolet (UV) regions using nonlinear techniques such as optical parametric amplification, sum and difference frequency generation, and four-wave mixing [34–36].

9.4.1.2 BROADBAND TIME-RESOLVED SPECTROSCOPY

9.4.1.2.1 DEVELOPMENT

Compared with monochromatic detection, the use of spectral broadband probes for optical measurements provides the ability to record responses at various wavelengths simultaneously and obtain much broader insight into the underlying physics. In 1964, optical broadband detection was first used by Jones and Stoicheff [37]. In that influential work, a continuum light source was generated by incident maser radiation on liquid toluene to study the induced Raman absorption of liquid benzene.

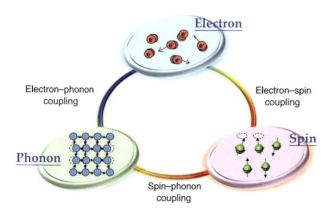

FIGURE 9.4.1.1 Schematic representation of a system including electron, phonon, and spin degrees of freedom.

The sample was irradiated simultaneously with a monochromatic excitation light of frequency v_0 and a continuum probe light. The excited atoms and molecules change their energy states by hv_M and absorb frequencies at the Stokes frequency, $v_0 - v_M$, and anti-Stokes frequency, $v_0 + v_M$, from the continuum probe light. This breakthrough in broad spectrum generation enabled the first Raman absorption spectrum measurement.

Seven years later, a broadband pulsed light source was applied to transient absorption spectroscopy for studying nonradiative relaxation processes in excited molecules [38,39]. Photoisomerization on a time scale of picoseconds was observed in 3-3′diethyloxadicarbocyanine iodide by broadband detection in the visible region [39]. In 1979, pioneering work by Shank et al., who used a time-resolved white-light continuum pulse probe to study GaAs thin films, opened a new era for dynamic studies in solid-state physics [40]. In this period, broadband probe pulses were generated by focusing the 750 nm pulses from a Nd:YAG amplifier into a cell containing water. The spectral range of interest, 785–835 nm, was selected using filters. By taking advantage of broadband detection, the entire relaxation process of band filling and bandgap renormalization was clearly observed within 0.5 ps. In addition, the relaxation processes of excited carriers in the heavy, light, and split-off hole bands were simultaneously observed by broadband detection [41,42]. Furthermore, the dynamics of magnetoexcitons in GaAs quantum wells were studied using a spin-resolved broadband probe [43]. Circularly polarized pump and probe beams were used to resolve excitonic interactions with regard to angular momentum states. By using the broadband probe, interactions between magnetoexcitons generated by the pump pulses at various angular states were unambiguously revealed. Magnetoexcitons with identical spins repel each other and cause a blueshift, whereas those with opposite spins attract each other, causing a redshift.

This brief review shows how the evolution of broadband time-resolved spectroscopy was driven by the development of light sources. In the 1970s, most continuum light sources were generated by self-phase modulation (SPM) and stimulated Raman scattering by focusing a colliding pulse mode-locked dye laser on a medium [44–46] or fiber [47], or generated by the fluorescence from a scintillator dye [48]. In the 1990s, femtosecond light sources were significantly improved with the development of solid-state laser materials [49], e.g., a sapphire crystal (Al_2O_3) doped with titanium ions (Ti:sapphire) and the chirp-pulse amplification technique [50]. Solid-state lasers have allowed optical parametric conversion to extend the spectral range of femtosecond pulses to the UV [35], visible [34], and IR [51,52] regions. A novel method, the noncollinear optical parametric amplifier (NOPA), was proposed to provide a broad spectrum with a sub-10-fs pulse width [53,54] or even a sub-5-fs pulse width [19,20,55], and the pulse width has recently even reached 2.4 fs [56].

These advanced broadband light sources with ultrashort pulse duration have been applied to various research areas, including ultrafast chemical reactions, photoisomerization, biophysics, and solid-state materials. By using their extremely high time resolution, the relaxation processes during trans–cis isomerization in the retinal chromophore of bacteriorhodopsin have been revealed by studying the real-time vibrational dynamics [57]. The environmentally affected vibrational photoisomerization processes of push–pull substituted azobenzene dye have been disclosed [58]. Broadband detection has enabled the demonstration of the energy transfer channels and efficiencies in photosynthetic light harvesting [59]. The pathways for exciton fine structure relaxation in CdSe nanorods have also been revealed [60]. Furthermore, the electron–phonon interaction strength in high-T_c superconductors was unambiguously determined recently [61].

Time-resolved spectroscopy provides a versatile and effective tool for studying the dynamics of photoexcited carriers in real time. Moreover, light sources with a broad spectrum and high time resolution enable the unambiguous discovery of the dynamics of energy transfer among quasi-particles both extensively and correctly. According to the measured energy transfer rate, the interplay among electron, phonon, spin, and orbital measurements in various materials or at the interfaces between dissimilar functional materials can be clearly revealed. Therefore, we can further understand some previously baffling issues in crystalline and dissimilar functional materials.

9.4.1.2.2 FEMTOSECOND LIGHT SOURCES

9.4.1.2.2.1 Narrowband Optical Parametric Amplifier

A regenerative amplifier (RGA) seeded with a Ti:sapphire laser oscillator served as a light source for a homemade optical parametric amplifier (OPA). The type-I ($e \rightarrow o+o$) β-BBO (barium borate, BaB_2O_4)-based OPA was pumped by the second harmonic of the RGA (wavelength, 400 nm; repetition rate, 5 kHz) to obtain output pulses in the visible range of 500–700 nm.

The second harmonic of the RGA was generated by 800 nm pulses focused on a BBO crystal. The 400 nm beam was separated into two copies; one served as the pump beam of the parametric interaction process and the other was used to obtain a white-light continuum as the signal beam. The white-light continuum was produced by focusing the 400 nm pulses on a sapphire disk to induce SPM. Then the white-light continuum passed through a prism pair to remove the fundamental light (400 nm). Both the signal (the white-light continuum) and pump (400 nm) beams were focused on a BBO crystal to realize a NOPA. The phase-matching condition is satisfied in the visible broadband region from 500 to 700 nm, so the wavelength of the amplified signal beam can be selected by adjusting the delay between the pump pulse and the linearly chirped visible seed pulse. As a result, the output wavelength of the signal beam can be tuned from 500 to 700 nm continuously (Figure 9.4.1.2).

It is noteworthy that the constructed narrowband OPA system is based on information on the carrier envelope phase structure in the NOPA. The noncollinear geometry in the OPA process is used to amplify the broad visible tuning range. For the present purposes of the narrowband OPA, the seed pulse is positively chirped with further insertion of the prism. Therefore, the OPA spectrum can be linearly adjusted to have a single color by changing the delay between the pump pulse and the seed pulse. Additionally, the gain of the OPA was saturated to avoid variation in its output caused by the fluence of the RGA pulses.

9.4.1.2.2.2 Broadband Optical Parametric Amplifier

For the broadband OPA, we also used the noncollinear configuration, which is the same as that used in the narrowband OPA. However, the signal beam, namely, the femtosecond continuum, is generated by the SPM of an 800 nm pulse focused on a 2 mm sapphire plate. To amplify the femtosecond visible broadband continuum in the full bandwidth, the signal pulses are precompressed using a pair of ultrabroadband chirped mirrors. The signal beam is then noncollinearly overlapped with the pump pulse and focused on a BBO crystal with the noncollinear angle of α 3.7° to fulfill the broadband phase-matching condition [19,55].

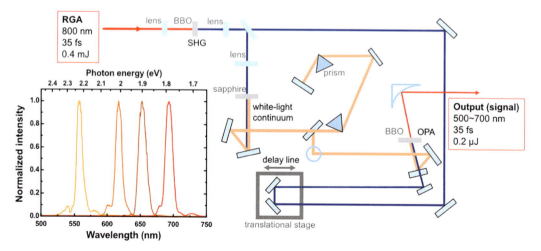

FIGURE 9.4.1.2   Schematic diagram of BBO-based OPA. The OPA was pumped by 400 nm pulses, and its tunable range is 480–700 nm with a pulse duration of 35 fs. Inset: normalized output spectra of the wavelength-tunable OPA, ranging from 500 to 700 nm.

Finally, both the pulse front-tilted pump beam and the noncollinearly incident signal beam are focused on a BBO crystal. The signal beam is then amplified through the OPA process. The pulse energy is 86 nJ after the first-stage amplification. Both pump and signal beams are reflected back to the BBO crystal by concave mirrors in the confocal configuration for the second stage amplification. The resulting amplified output pulse energy is 144 nJ, which is not affected by the fluence of the RGA pulses owing to the gain saturation of the OPA.

By using a total pulse compressor, the pulse width was compressed to ~9 fs. The output pulse energy after compression is 40 nJ. Amplitude and phase characterization of the compressed broadband OPA pulses was verified by a second-harmonic generation frequency-resolved optical gating (SHG FROG) in a very thin BBO crystal (5 µm). The corresponding spectrum and the measured SHG FROG are shown in Figure 9.4.1.3; it exhibits a wide visible spectrum with a nearly constant phase.

9.4.1.2.3 PUMP–PROBE SPECTROSCOPY

As shown in Figure 9.4.1.4, the general idea of the pump–probe technique is that responses from a sample induced by a pump pulse are investigated by detecting the changes in the reflectivity (ΔR) or transmissivity (ΔT) of probe pulses as a function of the delay time between pump and probe pulses. However, the difference in absorbance should be obtained indirectly from ΔR and ΔT. The absorbance of the target material without excitation, i.e., before a pump pulse arrives, is where I_o is the intensity of the probe pulse. T and R are transmitted and reflected intensities, respectively, of the probe pulse from the sample. Therefore, the difference in absorbance, ΔA $A0-A$, can be derived as follows:

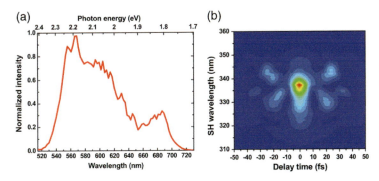

FIGURE 9.4.1.3 (a) Broadband OPA output spectrum, which covers almost the entire visible range. (b) Measured second-harmonic generation frequency-resolved optical gating (SHG FROG) trace of the output broadband OPA pulses.

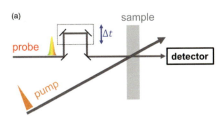

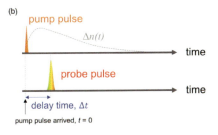

FIGURE 9.4.1.4 (a) Schematic diagram of the pump–probe technique. Time delay (Δt) between pump and probe pulses can be controlled by a mechanical delay line. (b) Fundamental principle of pump–probe spectroscopy. Time-dependent refractive index changes $n(t)$ of the sample induced by pump pulses can be observed by detecting the intensity variations of probe pulses.

The femtosecond time evolutions are derived by delaying the relative arrival times of the pump and probe pulses using a mechanical delay line. The fluence of the probe beam is usually much weaker than that of the pump beam to avoid a second excitation in the samples. Further, the polarizations of the pump and probe pulses are set perpendicular to each other to avoid interference between the pump and probe beams [62].

9.4.1.2.3.1 Fast-Scan Techniques

In pump–probe measurements, traditional scanning methods collect data step by step, which is time-consuming and easily influenced by the ultrashort pulse laser's instability. The instability of the light sources thus hinders the precise determination of electronic decay dynamics and may introduce systematic errors. This makes it difficult to obtain reproducible and reliable experimental data. However, a fast-scan pump–probe spectroscopic system that can complete a single scan in 5 s has been developed [63]. The rapid scan system is described in detail in [63].

In the fast-scan method, the signal (ΔR or ΔT) is collected while the delay time is scanned rapidly in 500 steps across the scanning range. A fast-scan stage with a total accessible pulse delay range of 15 ps is controlled by an external voltage generated by a digital/analog converter. At each delay point, the signal is obtained in 10 ms and stored in the memory of the lock-in amplifier. Averaged values of the data are collected for several 100 scans. This method provides a good signal-to-noise ratio and avoids the effects of laser fluctuations, including instability of the laser output power and pulse width.

9.4.1.2.3.2 Broadband Detection Techniques

To detect the relatively weak signal (on the order of 10^{-4} or less) in pump–probe measurements at multiple probe wavelengths, a multichannel lock-in amplifier developed by our group was used for time-resolved spectroscopy. The avalanche photodiodes (APDs) were commercial products optimized for UV to visible light detection.

As shown in Figure 9.4.1.5, the probe pulse was dispersed by a polychromator into a 96-branch fiber bundle whose other end was separated into 96 fiber branches and connected to APDs. Therefore, the time-resolved absorption differences at 96 probe wavelengths were simultaneously detected at the photodiodes. The detected signals were sent to a multichannel lock-in amplifier to be spectrally resolved for simultaneous detection of tiny changes in the probe intensity over the entire spectral region.

9.4.1.3   ULTRAFAST DYNAMICS IN NOVEL CONDENSED MATTER

9.4.1.3.1   Spin-Valley Coupled Polarization in Monolayer MoS₂

The discovery of graphene initiated a new era for two-dimensional (2D) materials in condensed matter physics. In particular, much attention has been focused on single-layer semiconducting materials. For example, the transition metal dichalcogenide MoS_2 exhibits unique physical, optical, and

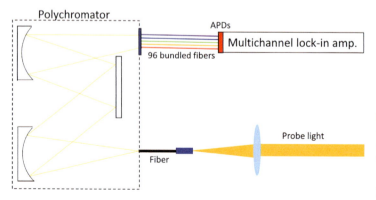

FIGURE 9.4.1.5 Schematic diagram of the multichannel lock-in amplifier for the broadband pump–probe measurement system. APDs, avalanche photodiodes.

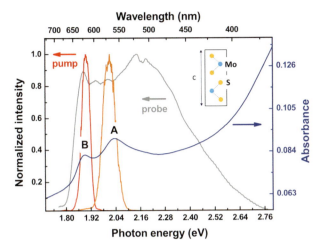

FIGURE 9.4.1.6   Spectra of 1.89 eV pump pulse (red), 2.01 eV pump pulse (orange), broadband visible probe pulse (gray), and the stationary absorption of monolayer MoS₂ at room temperature (blue). Inset: lattice structure of MoS₂ in the out-of-plane direction. Adapted from [72].

electrical properties correlated with its atomic layered structure. MoS₂ is a 2D material consisting of a horizontal single layer of molybdenum stacked vertically between two single layers of sulfur, as shown in the inset of Figure 9.4.1.6. The sulfur layers are held together by weak van der Waals forces, allowing MoS₂ sheets to be easily separated. Unlike pristine graphene, which does not have a bandgap for applications, MoS₂ possesses an indirect bandgap of 1.2 eV in bulk form, similar to that of silicon, and a direct bandgap of 1.8 eV as an atomically thin monolayer [64,65]. Moreover, the inherent coupling between the valley and spin in monolayer MoS₂ provides a noteworthy characteristic for spintronics [66], valleytronics [67], and semiconductor devices [9,68]. For example, monolayer MoS₂ exhibits a high channel mobility $(200 \, cm^2 \, V^{-1} \, s^{-1})$ and current ON/OFF ratio (1×10^8) when it is used as the channel material in a field-effect transistor [68].

Highly polarized luminescence in monolayer MoS₂ has been observed with resonant excitation. Furthermore, the valley–spin lifetime was previously predicted to be larger than 1 ns [69]. However, a time-resolved study of the polarized photoluminescence (PL) demonstrated that the carrier spin flip has a time scale of several picoseconds, which is limited by the time resolution of the time-resolved PL measurement system [70]. Mai et al. further observed that the polarized exciton A decays within only several 100 femtoseconds according to optical pump–probe measurements [71]. This controversial situation was resolved by a conclusive study of the full dynamics and physical nature of polarized excitons in monolayer MoS₂, including the spin–valley coupling [72].

The absorption spectrum of monolayer MoS₂ in Figure 9.4.1.6 clearly shows A (1.89 eV) and B (2.04 eV) excitonic transitions, which indicate the splitting of the valence band at the K valley due to spin–orbit coupling [64,65]. The pump pulse was generated by an OPA (as demonstrated in Section 9.4.1.2.B.1), and the excitation energy was set to be resonant with either exciton A or B. A probe pulse with a visible broadband spectrum was produced by SPM of an RGA pulse in a sapphire plate. To distinguish the nonequivalent K and K^0 valleys, the polarizations of the pump and probe beams were adjusted to be circular by broadband quarter-wave plates. The pump (probe) beam was focused on the sample in a spot with an area of $1.3 \times 10^{-4} \, cm^2$ $(0.7 \times 10^{-4} \, cm^2)$ and a pulse energy of 40 μJ (3 μJ). The time resolution of the measuring system was estimated to be 30 fs. The transient absorbance changes of the probe pulses induced by the pump pulses were detected by a CCD camera at all probe wavelengths simultaneously. The sample was mounted inside a cryostat to control the environmental temperature of the samples.

Figure 9.4.1.7 shows a 2D display of the photon-energy and time-resolved transient absorbance difference $\Delta A\omega$; t at 78 K. For the measurements, the polarizations of the broadband probe pulses were adjusted to be σ and σ^-, whereas the pump pulses were set to σ circular polarization and 1.89 eV to resonate with exciton A. For both the σ^+ and σ^- probes, the time-resolved spectra exhibited

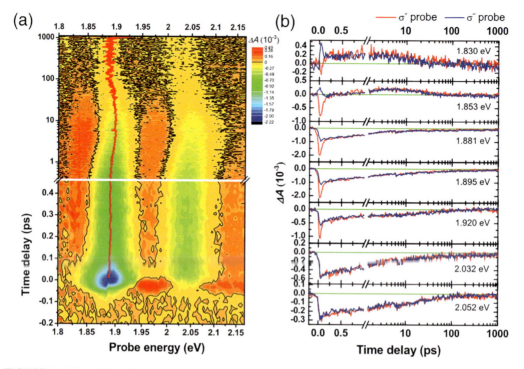

FIGURE 9.4.1.7 (a) Transient absorbance difference (ΔA) induced by excitation using σ circularly polarized pump pulses with a photon energy of 1.89 eV and probed by σ circularly polarized pulses at 78 K. Black curves are contours for ΔA zero. Red line indicates expected values of transition A as a function of the time delays and probe photon energies. (b) Probe delay time traces of ΔA at various probe photon energies. Red and blue lines represent σ and σ^- probes, respectively, and horizontal green lines show ΔA 0. Adapted from [72].

negative ΔA in the spectral band of excitons A and B. The negative ΔA signal could be caused by stimulated emission from the excited state and/or photobleaching due to depletion of the ground state and population of the excited state. The lifetime of photobleaching is usually much longer than that of stimulated emission because stimulated emission occurs only within the lifetime of the excited state whereas photobleaching remains until the ground state is fully repopulated.

Obviously, ΔA is significantly probe polarization dependent within 100 fs, as shown in Figure 9.4.1.7b. The nonlinear optical response after 100 fs is independent of the angular momentum of the initially excited distribution, which indicates that the initial polarization distribution relaxes to some quasi-equilibrium states at 100 fs. Following fast spin-polarization relaxation, the peaks in the ΔA spectrum show a blueshift before 10 ps and a redshift after 10 ps, as shown by the red line in Figure 9.4.1.7a. We note that the pump-induced response at the $K^{(t)}$ valley is observed even when exciton A at the K valley is excited by the σ pump in contrast to cases of excitons coupled to pump and probe pulses, which do not share common states. This unexpected phenomenon could be explained by various possible mechanisms, e.g., dark excitons generated by pump pulses [71], weakening of the excitonic binding energy [43], or dielectric screening from the excited excitons [43].

The measured time-resolved traces of $\Delta A(\omega, t)$ were fitted using the sum of three exponential functions and a constant term, as follows:

$$\Delta A(\omega,t) = \Delta A_{\text{spin}}(\omega)e^{-\frac{t}{\tau_{\text{spin}}}} + \Delta A_{\text{exciton}}(\omega)e^{-\frac{t}{\tau_{\text{exciton}}}} + \Delta A_{\text{carrier}}(\omega)e^{-\frac{t}{\tau_{\text{carrier}}}} + \Delta A_{e-b}(\omega) \quad (9.4.1.1)$$

The fitting results are shown in Figure 9.4.1.8. For the σ^+ probe, the time constant and τ_{spin}, τ_{exciton}, and τ_{carrier} are 55 ± 7 fs, 1.02 ± 0.22, and probe 26.32 ± 5.41 ps, respectively. For the σ^- probe, they

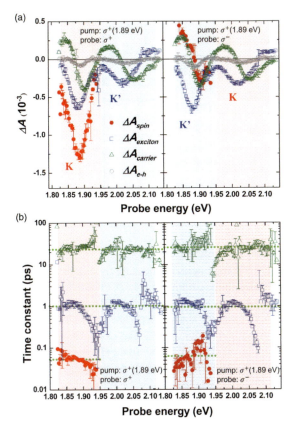

FIGURE 9.4.1.8 Triple exponential fitting results of time-resolved ΔA data excited by 2.01 eV and σ pump pulse at 78 K. Left column, σ probe; right column, σ^- probe. (a) ΔA spectra, (b) time constant of each component. Dotted lines indicate estimated values. Adapted from [72].

are 63 ± 42 fs, 0.96 ± 0.49, and 25.72 ± 8.61 ps, respectively. Because of the small signal amplitudes at photon energies of ~1.93 and ~2.08 eV, the fitting error is large. A comparison of the spectra of the σ and σ^- probes in Figure 9.4.1.8a reveals that $\Delta A_{\text{exciton}}$, $\Delta A_{\text{carrier}}$, and $\Delta A_{\text{e-h}}$—but not Δ_{Aspin}— exhibit similar dependences on the probe photon energy. Thus, these three relaxation processes occur regardless of the initial polarization distribution within 100 fs. However, ΔA_{spin} is completely different for the σ and σ^- probes in that it depends on the relative polarizations between the probe and pump beams. For the σ pump and σ probe, ΔA_{spin} is negative and possesses larger amplitude than for the σ pump and σ^- probe. This implies that the polarized exciton A at the K valley excited by the σ pump pulses leads to intense photobleaching and stimulated emission only when the probe pulses have the same circular polarization as the pump pulses and the probe photon energy overlaps the band of exciton A. Moreover, the spectral shape of ΔA fits well with excitonic transition A, which indicates that the valley polarization is efficiently excited at the high-symmetry K point [73]. On the other hand, the σ^- probe pulses generate exciton A with opposite spin polarization at the K^0 valley. The presence of excitons with opposite polarization leads to generation of biexcitons, which are the origin of induced absorption ($\Delta A > 0$) at ~1.87 eV [71,74], as shown in the right panel of Figure 9.4.1.8a.

After the excitons are dissociated to become free carriers in highly excited states, the exciton peak exhibits a redshift, as shown by the red line in Figure 9.4.1.7a. This shift is attributed to intravalley scattering of free carriers, in which electrons relax to the bottom of the conduction band and holes relax to the top of the valence band. The $\Delta A_{\text{carrier}}$ spectra show the sum of bleaching at the transition energy peaks and the induced absorption of a broad conduction band. Thus, the intermediate relaxation time $\tau_{\text{carrier}} \sim 25$ ps was assigned to the intraband transition of free carriers. The decay time of $\Delta A_{\text{e-h}}$ is found to be too long to be determined in the present work. The constant term $\Delta A_{\text{e-h}}$ represents only bleaching behavior, which can be attributed to electron–hole recombination in the direct band. The recombination time was estimated to be ~300 ps in a previous study [75].

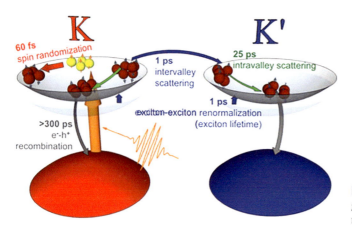

FIGURE 9.4.1.9 Schematic diagram of the relaxation processes in monolayer MoS_2. Adapted from [72].

The present study completely elucidates the fairly comprehensive ultrafast dynamics of spin-polarized excitons in monolayer MoS_2, as schematically shown in Figure 9.4.1.9. Owing to the high temporal resolution and visible broadband detection, the time constants for the 60 fs spin-polarized exciton decay, 1 ps exciton dissociation (intervalley scattering), and 25 ps hot carrier relaxation (intravalley scattering) are clearly identified. Moreover, substantial intervalley scattering strongly diminished the spin–valley coupled polarization under off-resonant excitation. These results provide a complete understanding of spin–valley coupled polarization anisotropy and carrier dynamics of atomic-layer MoS_2, which can further help us to develop ultrafast multilevel logic gates.

9.4.1.4 CONCLUSION AND PERSPECTIVES

In conclusion, in this short review chapter, we describe the fascinating relationships between the energy, spin, and valley in monolayer MoS_2. By using these degrees of freedom, optically driven logic gates can be realized. A two-level logic gate can be operated by sequential excitation with circularly polarized 2.01 eV (resonant with exciton B) and 1.98 eV (resonant with exciton A) pulses at room temperature. According to our time-resolved studies, the nonequilibrium population between the K and K^0 valley lasts for ~1 ps in monolayer MoS_2, making it an excellent candidate material for ultrafast optical control. For application to high-rate optical pulse control, the problem of accumulation of the remnant coherence after the control pulse always exists. Thus, the following pulse to control the succeeding step must wait for decoherence of the target, which limits the bandwidth of optical spin control devices.

These important results and conclusions indicate that ultrabroadband time-resolved spectroscopy provides a powerful means of studying the interactions between quasi-particles, which are the basis for composing a material. Using broadband OPA, we can observe several energy levels simultaneously with extremely high time resolution and study the correlations among them. This provides much clearer physical insight into the interactions of quasi-particles in several novel types of condensed matter. The development of ultrabroadband light sources continues. This type of ultrabroadband light source extending to the MIR region as well as the terahertz region is desirable and extremely important for investigating the detailed ultrafast dynamics in solids, as the bandgap energy or number of optical transitions have resonance energies in this frequency region. The research described in this subsection is based on cooperative activity among the following people: C.-W. Luo, Y.-T. Wang, A. Yabushita, T. Kobayashi [76].

REFERENCES

1. N. W. Ashcroft and N. D. Mermin, *Solid State Physics*, Harcourt, Fort Worth (1976).
2. C. Kittel, *Introduction to Solid State Physics*, 7th ed., Wiley, New York (1996).

3. D. Adler, "Mechanisms for metal-nonmetal transitions in transition-metal oxides and sulfides," *Rev. Mod. Phys.* **40**, 714–736 (1968).

4. M. Imada, A. Fujimori, and Y. Tokura, "Metal-insulator transitions," *Rev. Mod. Phys.* **70**, 1039–1263 (1998).

5. P. A. Lee, N. Nagaosa, and X.-G. Wen, "Doping a Mott insulator: Physics of high-temperature superconductivity," *Rev. Mod. Phys.* **78**, 17–85 (2006).

6. A. P. Ramirez, "Colossal magnetoresistance," *J. Phys. Condens. Matter* **9**, 8171–8199 (1997).

7. H. V. Löhneysen, T. Pietrus, G. Portisch, H. G. Schlager, A. Schröder, M. Sieck, and T. Trappmann, "Non-Fermi-liquid behavior in a heavy fermion alloy at a magnetic instability," *Phys. Rev. Lett.* **72**, 3262–3265 (1994).

8. W. Eerenstein, N. D. Mathur, and J. F. Scott, "Multiferroic and magnetoelectric materials," *Nature* **442**, 759–765 (2006).

9. Q. H. Wang, K. Kalantar-Zadeh, A. Kis, J. N. Coleman, and M. S. Strano, "Electronics and optoelectronics of two-dimensional transition metal dichalcogenides," *Nat. Nanotechnol.* **7**, 699–712 (2012).

10. J. R. Haynes and W. Shockley, "The mobility and life of injected holes and electrons in germanium," *Phys. Rev.* **81**, 835–843 (1951).

11. K. F. Komatsubara, "Effect of electric field on the transverse magnetoresistance in n-indium antimonide at 1.5 K," *Phys. Rev. Lett.* **16**, 1044–1047 (1966).

12. J. Zucker, "Thermoelectric power of hot carriers," *J. Appl. Phys.* **35**, 618–621 (1964).

13. H. Heinrich and W. Jantsch, "Experimental determination of the electron temperature from Burstein-shift experiments in gallium antimonide," *Phys. Rev. B* **4**, 2504–2508 (1971).

14. T. R. Hart, R. L. Aggarwal, and B. Lax, "Temperature dependence of Raman scattering in silicon," *Phys. Rev. B* **1**, 638–642 (1970).

15. H. Heinrich, "Infrared Faraday effect and electron transfer in many valley semiconductors: N-GaSb," *Phys. Lett. A* **32**, 331–332 (1970).

16. G. M. Gale, M. Cavallari, T. J. Driscoll, and F. Hache, "Sub-20-fs tunable pulses in the visible from an 82-MHz optical parametric oscillator," *Opt. Lett.* **20**, 1562–1564 (1995).

17. M. K. Reed, M. S. Armas, M. K. Steiner-Shepard, and D. K. Negus, "30-fs pulses tunable across the visible with a 100-kHz Ti:sapphire regenerative amplifier," *Opt. Lett.* **20**, 605–607 (1995).

18. G. Cerullo, M. Nisoli, S. Stagira, and S. De Silvestri, "Sub-8-fs pulses from an ultrabroadband optical parametric amplifier in the visible," *Opt. Lett.* **23**, 1283–1285 (1998).

19. A. Shirakawa, I. Sakane, M. Takasaka, and T. Kobayashi, "Sub-5-fs visible pulse generation by pulse-front-matched noncollinear optical parametric amplification," *Appl. Phys. Lett.* **74**, 2268–2270 (1999).

20. T. Kobayashi and A. Baltuška, "Sub-5 fs pulse generation from a noncollinear optical parametric amplifier," *Meas. Sci. Technol.* **13**, 1671–1682 (2002).

21. T. Kobayashi and A. Yabushita, "Transition-state spectroscopy using ultrashort laser pulses," *Chem. Rec.* **11**, 99–116 (2011).

22. C. W. Luo, P. S. Tseng, H.-J. Chen, K. H. Wu, and L. J. Li, "Dirac fermion relaxation and energy loss rate near the Fermi surface in monolayer and multilayer graphene," *Nanoscale* **6**, 8575–8578 (2014).

23. C. W. Luo, H. J. Wang, S. A. Ku, H.-J. Chen, T. T. Yeh, J.-Y. Lin, K. H. Wu, J. Y. Juang, B. L. Young, T. Kobayashi, C.-M. Cheng, C.-H. Chen, K.-D. Tsuei, R. Sankar, F. C. Chou, K. A. Kokh, O. E. Tereshchenko, E. V. Chulkov, Y. M. Andreev, and G. D. Gu, "Snapshots of Dirac fermions near the Dirac point in topological insulators," *Nano Lett.* **13**, 5797–5802 (2013).

24. C. W. Luo, H.-J. Chen, C. M. Tu, C. C. Lee, S. A. Ku, W. Y. Tzeng, T. T. Yeh, M. C. Chiang, H. J. Wang, W. C. Chu, J.-Y. Lin, K. H. Wu, J. Y. Juang, T. Kobayashi, C.-M. Cheng, C.-H. Chen, K.-D. Tsuei, H. Berger, R. Sankar, F. C. Chou, and H. D. Yang, "THz generation and detection on Dirac fermions in topological insulators," *Adv. Opt. Mater.* **1**, 804–808 (2013).

25. H.-J. Chen, K. H. Wu, C. W. Luo, T. M. Uen, J. Y. Juang, J.-Y. Lin, T. Kobayashi, H. D. Yang, R. Shankar, F. C. Chou, H. Berger, and J. M. Liu, "Phonon dynamics in $Cu_xBi_2Se_3$ (x = 0, 0.1, 0.125) and Bi_2Se_2 crystals studied using femtosecond spectroscopy," *Appl. Phys. Lett.* **101**, 121912 (2012).

26. L. Y. Chen, J. C. Yang, C. W. Luo, C. W. Liang, K. H. Wu, J.-Y. Lin, T. M. Uen, J. Y. Juang, Y. H. Chu, and T. Kobayashi, "Ultrafast photoinduced mechanical strain in epitaxial $BiFeO_3$ thin films," *Appl. Phys. Lett.* **101**, 041902 (2012).

27. C. W. Luo, I. H. Wu, P. C. Cheng, J.-Y. Lin, K. H. Wu, T. M. Uen, J. Y. Juang, T. Kobayashi, D. A. Chareev, O. S. Volkova, and A. N. Vasiliev, "Quasiparticle dynamics and phonon softening in FeSe superconductors," *Phys. Rev. Lett.* **108**, 257006 (2012).

28. H. S. Shih, L. Y. Chen, C. W. Luo, K. H. Wu, J.-Y. Lin, J. Y. Juang, T. M. Uen, J. M. Lee, J. M. Chen, and T. Kobayashi, "Ultrafast thermoelastic dynamics of $HoMnO_3$ single crystals derived from femtosecond optical pump-probe spectroscopy," *New J. Phys.* **13**, 053003 (2011).

29. H. C. Shih, T. H. Lin, C. W. Luo, J.-Y. Lin, T. M. Uen, J. Y. Juang, K. H. Wu, J. M. Lee, J. M. Chen, and T. Kobayashi, "Magnetization dynamics and Mn^{3+} d–d excitation in hexagonal $HoMnO_3$ revealed by wavelength tunable time-resolved femtosecond spectroscopy," *Phys. Rev. B* **80**, 024427 (2009).

30. C. W. Luo, C. C. Hsieh, Y.-J. Chen, P. T. Shih, M. H. Chen, K. H. Wu, J. Y. Juang, J.-Y. Lin, T. M. Uen, and Y. S. Gou, "Spatial dichotomy of quasiparticle dynamics in underdoped thin-film $YBa_2Cu_3O_{7-\delta}$ superconductors," *Phys. Rev. B* **74**, 184525 (2006).

31. C. W. Luo, P. T. Shih, Y.-J. Chen, M. H. Chen, K. H. Wu, J. Y. Juang, J.-Y. Lin, T. M. Uen, and Y. S. Gou, "Spatially resolved relaxation dynamics of photoinduced quasiparticle in underdoped $YBa_2Cu_3O_{7-\delta}$," *Phys. Rev. B* **72**, 092506 (2005).

32. C. W. Luo, K. Reimann, M. Woerner, T. Elsaesser, R. Hey, and K. H. Ploog, "Phase-resolved nonlinear response of a two-dimensional electron gas under femtosecond intersubband excitation," *Phys. Rev. Lett.* **92**, 047402 (2004).

33. C. W. Luo, M. H. Chen, S. P. Chen, K. H. Wu, J. Y. Juang, J.-Y. Lin, T. M. Uen, and Y. S. Gou, "Spatial symmetry of superconducting gap in $YBa_2Cu_3O_{7-\delta}$ obtained from femtosecond spectroscopy," *Phys. Rev. B* **68**, 220508 (2003).

34. K. R. Wilson and V. V. Yakovlev, "Ultrafast rainbow: Tunable ultrashort pulses from a solid-state kilohertz system," *J. Opt. Soc. Am. B* **14**, 444–448 (1997).

35. V. Petrov, F. Seifert, O. Kittelmann, J. Ringling, and F. Noack, "Extension of the tuning range of a femtosecond Ti:sapphire laser amplifier through cascaded second-order nonlinear frequency conversion processes," *J. Appl. Phys.* **76**, 7704–7712 (1994).

36. T. Fuji and T. Suzuki, "Generation of sub-two-cycle mid-infrared pulses by four-wave mixing through filamentation in air," *Opt. Lett.* **32**, 3330–3332 (2007).

37. W. J. Jones and B. P. Stoicheff, "Inverse Raman spectra: Induced absorption at optical frequencies," *Phys. Rev. Lett.* **13**, 657–659 (1964).

38. M. R. Topp, P. M. Rentzepis, and R. P. Jones, "Time resolved picosecond emission spectroscopy of organic dye lasers," *Chem. Phys. Lett.* **9**, 1–5 (1971).

39. D. Magde and M. W. Windsor, "Picosecond flash photolysis and spectroscopy: 3, 3'-diethyloxadicarbocyanine iodide (DODCI)," *Chem. Phys. Lett.* **27**, 31–36 (1974).

40. C. V. Shank, R. L. Fork, R. F. Leheny, and J. Shah, "Dynamics of photoexcited GaAs band-edge absorption with subpicosecond resolution," *Phys. Rev. Lett.* **42**, 112–115 (1979).

41. R. W. Schoenlein, W. Z. Lin, E. P. Ippen, and J. G. Fujimoto, "Femtosecond hot-carrier energy relaxation in GaAs," *Appl. Phys. Lett.* **51**, 1442–1445 (1987).

42. C. J. Stanton, D. W. Bailey, and K. Hess, "Femtosecond-pump, continuum-probe nonlinear absorption in GaAs," *Phys. Rev. Lett.* **65**, 231–234 (1990).

43. J. B. Stark, W. H. Knox, and D. S. Chemla, "Spin-resolved femtosecond magnetoexciton interactions in GaAs quantum wells," *Phys. Rev. B* **46**, 7919–7922 (1992).

44. R. R. Alfano and S. L. Shapiro, "Observation of self-modulation and small-scale filaments in crystals and glasses," *Phys. Rev. Lett.* **24**, 592–594 (1970).

45. R. R. Alfano and S. L. Shapiro, "Emission in the region 4000 to 7000 Å via four-photon coupling in glass," *Phys. Rev. Lett.* **24**, 584–587 (1970).

46. R. R. Alfano and S. L. Shapiro, "Direct distortion of electronic clouds of rare-gas atoms in intense electric fields," *Phys. Rev. Lett.* **24**, 1217–1220 (1970).

47. C. Lin, V. T. Nguyen, and W. G. French, "Wideband near-I. R. Continuum (0.7–2.1 μm) generated in low-loss optical fibres," *Electron. Lett.* **14**, 822–823 (1978).

48. C. Lin and R. H. Stolen, "New nanosecond continuum for excited-state spectroscopy," *Appl. Phys. Lett.* **28**, 216–218 (1976).

49. D. E. Spence, P. N. Kean, and W. Sibbett, "60-fsec pulse generation from a self-mode-locked Ti:sapphire laser," *Opt. Lett.* **16**, 42–44 (1991).

50. J. Squier, F. Salin, G. Mourou, and D. Harter, "100-fs pulse generation and amplification in $Ti:Al_2O_3$," *Opt. Lett.* **16**, 324–326 (1991).

51. P. E. Powers, R. J. Ellingson, W. S. Pelouch, and C. L. Tang, "Recent advances of the Ti:sapphire-pumped high-repetition-rate femtosecond optical parametric oscillator," *J. Opt. Soc. Am. B* **10**, 2162–2167 (1993).

52. M. K. Reed and M. K. S. Shepard, "Tunable infrared generation using a femtosecond 250 kHz Ti:sapphire regenerative amplifier," *IEEE J. Quant. Electron.* **32**, 1273–1277 (1996).

53. A. Shirakawa, I. Sakane, and T. Kobayashi, "Pulse-front-matched optical parametric amplification for sub-10-fs pulse generation tunable in the visible and near infrared," *Opt. Lett.* **23**, 1292–1294 (1998).

54. Y. Kida and T. Kobayashi, "Generation of sub-10 fs ultraviolet Gaussian pulses," *J. Opt. Soc. Am. B* **28**, 139–148 (2011).

55. T. Kobayashi and A. Shirakawa, "Tunable visible and near-infrared pulse generator in a 5 fs regime," *Appl. Phys. B* **70**, S239–S246 (2000).

56. K. Okamura and T. Kobayashi, "Octave-spanning carrier-envelope phase stabilized visible pulse with sub-3-fs pulse duration," *Opt. Lett.* **36**, 226–228 (2011).

57. T. Kobayashi, A. Yabushita, T. Saito, H. Ohtani, and M. Tsuda, "Sub-5-fs real-time spectroscopy of transition states in bacteriorhodopsin during retinal isomerization," *Photochem. Photobiol.* **83**, 363–369 (2007).

58. C. C. Hsu, Y. T. Wang, A. Yabushita, C. W. Luo, Y. N. Hsiao, S. H. Lin, and T. Kobayashi, "Environment-dependent ultrafast photoisomerization dynamics in azo dye," *J. Phys. Chem. A* **115**, 11508–11514 (2011).

59. G. Cerullo, C. Manzoni, L. Lüer, and D. Polli, "Time-resolved methods in biophysics. 4. Broadband pump-probe spectroscopy system with sub 20 fs temporal resolution for the study of energy transfer processes in photosynthesis," *Photochem. Photobiol. Sci.* **6**, 135–144 (2007).

60. C. Y. Wong, J. Kim, P. S. Nair, M. C. Nagy, and G. D. Scholes, "Relaxation in the exciton fine structure of semiconductor nanocrystals," *J. Phys. Chem. C* **113**, 795–811 (2009).

61. C. Gadermaier, A. S. Alexandrov, V. V. Kabanov, P. Kusar, T. Mertelj, X. Yao, C. Manzoni, D. Brida, G. Cerullo, and D. Mihailovic, "Electron phonon coupling in high-temperature cuprate superconductors determined from electron relaxation rates," *Phys. Rev. Lett.* **105**, 257001 (2010).

62. C. W. Luo, Y. T. Wang, F. W. Chen, and H. C. Shih, "Eliminate coherence spike in reflection-type pump-probe measurements," *Opt. Express* **17**, 11321–11327 (2009).

63. A. Yabushita, Y. H. Lee, and T. Kobayashi, "Development of a multiplex fast-scan system for ultrafast time-resolved spectroscopy," *Rev. Sci. Instrum.* **81**, 063110 (2010).

64. K. F. Mak, C. Lee, J. Hone, J. Shan, and T. F. Heinz, "Atomically thin MoS_2: A new direct-gap semiconductor," *Phys. Rev. Lett.* **105**, 136805 (2010).

65. A. Splendiani, L. Sun, Y. Zhang, T. Li, J. Kim, C.-Y. Chim, G. Galli, and F. Wang, "Emerging photoluminescence in monolayer MoS_2," *Nano Lett.* **10**, 1271–1275 (2010).

66. I. Zutic, J. Fabian, and S. Das Sarma, "Spintronics: Fundamentals and applications," *Rev. Mod. Phys.* **76**, 323–410 (2004).

67. A. Rycerz, J. Tworzydlo, and C. W. J. Beenakker, "Valley filter and valley valve in graphene," *Nat. Phys.* **3**, 172–175 (2007).

68. B. Radisavljevic, A. Radenovic, J. Brivio, V. Giacometti, and A. Kis, "Single-layer MoS_2 transistors," *Nat. Nanotechnol.* **6**, 147–150 (2011).

69. K. F. Mak, K. He, J. Shan, and T. F. Heinz, "Control of valley polarization in monolayer MoS_2 by optical helicity," *Nat. Nanotechnol.* **7**, 494–498 (2012).

70. D. Lagarde, L. Bouet, X. Marie, C. R. Zhu, B. L. Liu, T. Amand, P. H. Tan, and B. Urbaszek, "Carrier and polarization dynamics in monolayer MoS_2," *Phys. Rev. Lett.* **112**, 047401 (2014).

71. C. Mai, A. Barrette, Y. Yu, Y. G. Semenov, K. W. Kim, L. Cao, and K. Gundogdu, "Many-body effects in valleytronics: Direct measurement of valley lifetimes in single layer MoS_2," *Nano Lett.* **14**, 202–206 (2014).

72. Y.-T. Wang, C.-W. Luo, A. Yabushita, K.-H. Wu, T. Kobayashi, C.-H. Chen, and L.-J. Li, "Ultrafast multi-level logic gates with spin-valley coupled polarization anisotropy in monolayer MoS_2," *Sci. Rep.* **5**, 8289 (2015).

73. Q. Wang, S. Ge, X. Li, J. Qiu, Y. Ji, J. Feng, and D. Sun, "Valley carrier dynamics in monolayer molybdenum disulphide from helicity resolved ultrafast pump-probe spectroscopy," *ACS Nano* **7**, 11087–11093 (2013).

74. E. J. Sie, Y.-H. Lee, A. J. Frenzel, J. Kong, and N. Gedik, "Biexciton formation in monolayer MoS_2 observed by transient absorption spectroscopy," *arXiv:1312.2918* (2013).

75. H. Shi, R. Yan, S. Bertolazzi, J. Brivio, B. Gao, A. Kis, D. Jena, H. G. Xing, and L. Huang, "Exciton dynamics in suspended monolayer and few-layer MoS_2 2D crystals," *ACS Nano* **7**, 1072–1080 (2013).

76. C.-W. Luo, Y.-T. Wang, A. Yabushita, T. Kobayashi, "Ultra-broadband time-resolved spectroscopy in novel types of condensed matter," *Optica* **3**, 82–92 (2016).

9.4.2 Phonon Dynamics in Cu$_x$Bi$_2$(x50, 0.1, and 0.125) and Bi$_2$Se$_2$ Crystals Studied Using Ultrafast Spectroscopy

9.4.2.1 INTRODUCTION

Topological insulators (TIs) are recently discovered states of quantum matter characterized by a full gap with degenerated spins in the bulk while exhibiting polarized spins and no gap on the surface [1,2]. The salient electronic, optical, and magnetic properties of these materials provide innovative opportunities not only for research in fundamental physics but also for potentially revolutionary applications such as quantum computers and spintronic devices [1–3]. Bi-based compounds such as Bi$_2$Se$_3$ and Bi$_2$Te$_3$ have been extensively investigated due to their intriguing thermoelectric properties and their uniqueness of being three-dimensional TIs.

Here in the following, the above-mentioned spintronics is explained very briefly. Spintronics also known as spin electronics, is the study of the intrinsic spin of the electron and its associated magnetic moment, in addition to its fundamental electronic charge, in solid-state devices. The field of spintronics concerns spin-charge coupling in metallic systems; the analogous effects in insulators fall into the field of multiferroics. Spintronics fundamentally differs from traditional electronics in that, in addition to charge state, electron spins are exploited as a further degree of freedom, with implications in the efficiency of data storage and transfer.

Recently, the effect of doping atoms is one of the most important issues for Bi-based TIs. Previous research showed that the "robust" conducting edge states did not disappear even when some impurity atoms are doped into TIs [4–7]. The doping atoms can also induce significant modification on surface electronic states, such as opening up a gap at the Dirac point [4] and shifting the Fermi level [5]. Moreover, Bi-based TIs doped with Fe, Mn, and Cr can exhibit other interesting physical properties, such as ferromagnetism [8,9] and anomalous quantized Hall effect [10], while superconductivity has been observed in Bi-based TIs doped with Cu [6,7]. Recent investigations on superconductivity exhibited in Cu$_x$Bi$_2$Se$_3$ provide the platforms for studying the interplay between topological and broken-symmetry order [6]. However, the locations of the doped Cu atoms still remain a subject of debate. The Cu atoms can either be intercalated between van der Waals-like bonded Se planes or substitute Bi atoms in the lattice [7]; these two different possibilities cause different structural deformations in the Bi$_2$Se$_3$ matrix. Previous studies on Bi$_2$Te$_3$ also indicate that the application of extra pressure on the sample can induce superconductivity while the topological surface states remain well-defined [11,12]. Furthermore, it has been demonstrated that small changes in lattice parameters and atomic positions in Bi$_2$Te$_3$ can result in surface states that are slightly different from those in the original Bi$_2$Te$_3$ [11]. This calculation result implies that the structural strain might be relevant to the superconductivity observed in some TIs. Thus, to understand the underlying mechanism that gives rise to superconductivity in Cu$_x$Bi$_2$Se$_3$, it is essential to investigate the changes in the crystal structure that are caused by the doped Cu atoms. Such changes in the crystal structure can be studied by probing the changes in phonon dynamics using ultrafast spectroscopy.

DOI: 10.1201/9780429196577-60

Bi_2Se_3 is a narrow-gap semiconductor with a rhombohedral crystal structure belonging to the $D_{3d}^5 \left(R\bar{3}m \right)$ space group. The Bi_2Se_3 crystal structure is constructed by repeated quintuple layers (QLs) arranged along the c-axis. Each QL is stacked in a sequence of Se(1)-Bi-Se(2)-Bi-Se(1) atomic layers and is weakly bonded to its neighboring QLs by van der Waals interaction. This property makes the native Bi_2Se_3 crystal a layered compound such that both substitution and intercalation can take place for a small percentage of foreign atoms added into this material. For example, the crystal structure of a Bi-rich Bi-Se compound, Bi_2Se_2, is similar to that of Bi_2Se_3, but it has a bi-layer Bi-Bi slab inserted between two QLs [13,14]. That is, the additional Bi atoms are intercalated into the Bi_2Se_3 matrix to form the Bi_2 layers. In this case, each QL will feel an extra Coulomb force resulting from the layer of doped atoms. This Coulomb interaction distorts the structure of the QL by changing its bond length and bond angle. Moreover, because this interaction also suppresses the original vibration of the QL, the corresponding phonon dynamics of the QL is changed slightly. As we will show later, it causes a red shift in the phonon frequency for the intercalation case. In addition to intercalation, the doped atoms can also substitute the Bi atom inside the QL. Previous Raman scattering studies on rhombo-hedral V_2–VI_3 compounds reveal a blue shift in the phonon frequency when the lighter Sb atom substitutes the Bi atom in the QL of $Bi_{2-y}Sb_yTe_3$ [10]. Clearly, composition change in the QL also affects the phonon dynamics. Because the phonon frequency shift caused by substitution and that caused by intercalation have opposite signs, the structural deformation and the locations of doped atoms in a TI can be resolved by measuring the sign and magnitude of the phonon frequency shift.

Ultrafast time-resolved pump-probe spectroscopy has proven to be a powerful tool in unravel-ing the carrier, lattice, and spin dynamics in various materials. An ultrashort laser pulse that is sufficiently shorter than the period of a particular vibrational mode can excite the mode with a high degree of temporal and spatial coherence in a material. Therefore, propagation and localiza-tion of coherent phonons can be observed directly in the time domain [15,16]. Recently, coherent phonons generated by ultrashort laser pulses have been observed in a wide variety of materials, such as semiconductors, metals, superconductors, colossal magnetoresistive manganites, and mul-tiferroics [17–21]. Both coherent optical phonons (COPs) and coherent acoustic phonons (CAPs) have been measured in thermoelectric materials and topological insulators, such as Sb2Te3 [22,23], Bi_2Te_3 [24], and Bi_2Se_3 [25,26]. Because the COPs measured in femtosecond pump-probe spectros-copy are Raman active, the corresponding COP frequency should show up in conventional Raman spectra. Indeed, in previous studies performed on rhombohedra 1 V_2–VI_3 compounds [22–26] using ultrafast pump-probe experiments, a damped oscillation of a subpicosecond period that is observed in measured transient variations of reflectivity or transmission ($\Delta R/R$ or $\Delta T/T$) is attrib-uted to the displacive excitation of COPs associated with the Raman active modes. Another damped oscillation of a longer period of tens of picoseconds, originally identified by Thomsen et al. [15], is attributed to the CAP oscillations generated by the interference of the probe beams reflected from the sample surface and the strain pulse that propagates longitudinally at sound velocity. The relationship between the period τ_f of the slow oscillation and the longitudinal sound velocity v_s is

$$\tau_f = \lambda / \left[2 v_s \left(\sqrt{n^2 - \sin^2 \theta} \right) \right],$$ where k is the probe wavelength, n is the refractive index at λ, and θ is the incident angle of the probe beam. Consequently, one can obtain the sound velocity by measuring the CAP oscillations provided that the refractive index of the material is known.

In the present study, we use optical-pump optical-probe femtosecond time-domain spectroscopy to investigate the dynamics of coherent phonons in Bi, Bi_2Se_2, and $Cu_xBi_2Se_3$ ($x = 0, 0.1, 0.125$) single crystals. Two damped oscillations, namely the fast and slow oscillation components, in the measured $\Delta R/R$ curves are observed in these samples and are attributed to COP and CAP, respec-tively. Analyses carried out on the COP oscillation (the fast component) in the Bi_2Se_2 crystal reveal that the shift of the A_{1g} phonon mode frequency of the Bi layer and that of the QL have opposite signs and are respectively associated with the shortening and stretching of the bond/chain length. The deformation of the QL in $Cu_xBi_2Se_3$ crystals is also clearly observed. The structural deforma-tion and the shift of the A_{1g}^1 phonon frequency associated with the effects resulting from the doped

Cu atoms are discussed. From the phonon frequency shifts and the lifetime of the COP, the locations of the Cu atoms in Cu$_x$Bi$_2$Se$_3$ can be determined, and the influence of the Cu atoms in QL is subsequently resolved.

9.4.2.2 EXPERIMENTAL

Single crystal samples of Cu$_x$Bi$_2$Se$_3$ ($x = 0$, 0.1, 0.125), Bi$_2$Se$_2$, and Bi have been prepared using stoichiometric mixtures of 5N purity of Bi, Se, and Cu in sealed evacuated quartz tubes. Bi crystals were grown by the vertical Bridgman method between 450°C and 270°C of the thermal gradient of 1 C/cm near the solidification point with a pulling rate of 0.1 mm/h. Cu$_x$Bi$_2$Se$_3$ crystals were grown with slow-cooling method from 850°C to 650°C at a rate of 2°C/h and then quenched in cold water. Single crystal Bi$_2$Se$_2$ has been prepared using the vertical Bridgman method. Preliminary homogenization was carried out in a horizontal tube furnace at 35°C for 75 h. The sealed ampoules were then passed through a vertical Bridgman furnace between 650°C and 600°C of thermal gradient 1 C/cm near the solidification point. The pulling rate was kept at 0.2 mm/h. The resulting crystals could be cleaved easily, and the freshly cleaved plane showed a silvery shining mirror-like surface. All of the samples were kept in a vacuum tank to avoid surface oxidation. Before each experiment, the sample was cleaved using a scotch tape to ensure that a flat and reflective surface was obtained. Figure 9.4.2.1 shows the x-ray diffraction patterns of Bi, Bi$_2$Se$_2$, and Cu$_x$Bi$_2$Se$_3$ (x¼ 0, 0.1, 0.125) single crystals. The diffraction peaks reveal pure (001)-oriented reflections without any discernible impurity phase, indicating that all the single crystal samples have a pure c-axis oriented normal to the cleaved surface.

For standard pump-probe measurements, a commercial mode-locked Ti:sapphire laser system providing ultrashort pulses of 100 fs-pulse width at a repetition rate of 5 MHz and a center wavelength of 800 nm was used. The fluences of the pump beam and the probe beam were 1 and 0.067 mJ/cm², respectively. The pump beam was focused on the Cu$_x$Bi$_2$Se$_3$ single crystals to a diameter of 35 lm, and the probe beam was focused to a diameter of 25 lm at the center of the pump beam spot. The polarizations of the pump and probe beams were orthogonal to each other and were perpendicular to the c-axis of the Cu$_x$Bi$_2$Se$_3$ single crystals. All the samples were stuck on a copper sample holder in a cryostat and kept at a constant temperature of 293 K before laser illumination. A linear motor stage was used to vary the delay time between the pump and probe pulses. The small reflected signals were detected by using a photodiode and a lock-in amplifier.

9.4.2.3 RESULTS AND DISCUSSION

Figure 9.4.2.2a–e displays the typical $\Delta R/R$ signals as a function of delay time for all samples measured at room temperature.

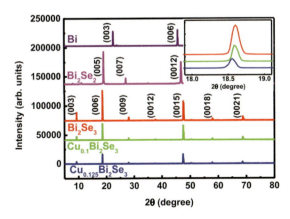

FIGURE 9.4.2.1 X-ray diffraction patterns of Bi, Bi$_2$Se$_2$, and Cu$_x$Bi$_2$Se$_3$ ($x = 0$, 0.1, 0.125) crystals. Inset: Magnified patterns around the (006) peak of Cu$_x$Bi$_2$Se$_3$.

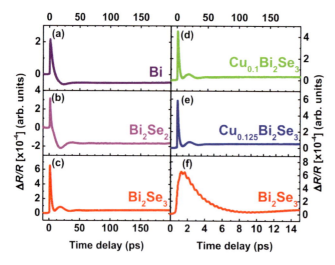

FIGURE 9.4.2.2 Temporal variations of $\Delta R/R$ signals for (a) Bi, (b) Bi_2Se_2, (c) Bi_2Se_3, (d) $Cu_{0.1}Bi_2Se_3$, and (e) $Cu_{0.125}Bi_2Se_3$ crystals at room temperature for the pump–probe energy 1.55 eV. (f) $\Delta R/R$ signal of Bi_2Se_3 crystal observed on a short timescale. Note that the horizontal time scale for panel (f) is different from the other panels.

In general, the time evolution of each $\Delta R/R$ curve can be separated into several components corresponding to different energy-transfer processes: A fast component of a sub-picosecond time scale characterizes the thermalization between electrons and optical phonons, while a subsequent slow component of a time scale of several picoseconds characterizes the thermalization between electrons and acoustic phonons [26]. After the electron-lattice relaxation through these processes, a quasi-constant component is observed, which might represent heat diffusion out of the illuminated area on the sample [25]. In addition, two damped oscillation components of different periods are superimposed on the $\Delta R/R$ curves. Only the slow oscillations can be seen in Figure 9.4.2.2a–e because the $\Delta R/R$ signals in these figures are plotted on long timescales. The fast oscillation component of a Bi_2Se_3 crystal is shown in Figure 9.4.2.2f. This component can be extracted by removing the adjacent average background from the measured $\Delta R/R$ signal; the result is displayed in Figure 9.4.2.3c. The frequency of this component is centered at 2.148 THz; it can be assigned as the A_{1g}^1 COP mode of Bi_2Se_3, based on comparison with continuous-wave (CW) Raman spectroscopy [27]. The slow oscillation components, as shown in Figure 9.4.2.2a–e, are attributed to the CAPs generated by ultrafast laser pulses. The frequency of the CAP for the Bi_2Se_3 crystal is centered at 0.033 THz (the corresponding period is 30 ps), and the oscillation decays completely within 60 ps. According to the pulse strain model, the disappearance of the slow oscillation around 60 ps is determined by the penetration depth of the probe beam at 800 nm. Taking the refractive index of Bi_2Se_3 crystal reported in Ref. [28], the sound velocity is estimated to be 1996.33 m/s at room temperature.

Figure 9.4.2.2c–e also reveals that the periods of the slow oscillations of $Cu_xBi_2Se_3$ ($x \frac{1}{4}$ 0, 0.1, 0.125) crystals vary slightly (from 29.9 to 30.2 ps). By contrast, the frequencies of the fast oscillations display a significant distribution among the samples studied and are associated with the changes in the chain length of the QL and in the lattice constant c. Thus, in the following, we concentrate on the COP behaviors exhibited by the Bi, Bi_2Se_2, and $Cu_xBi_2Se_3$ ($x = 0$, 0.1, 0.125) single crystals and attempt to clarify the changes in the crystal structure of Cu-doped Bi_2Se_3.

Figure 9.4.2.3a–e shows the fast oscillation component for every single crystal. For the Bi_2Se_2 crystal, shown in Figure 9.4.2.3b, the strength of the oscillatory signal is significantly smaller and its relaxation time much shorter than those for the other crystals. Besides, this oscillatory signal also exhibits a complex behavior in the first few picoseconds. To understand this complex oscillation, we use the methods of short-time Fourier transformation (STFT) [29–31] and fast Fourier transformation (FFT) to analyze the oscillatory signals. A Blackman window with a full width at half-maximum (FWHM) of 2400 fs was used as a gate function in the STFT calculation [30].

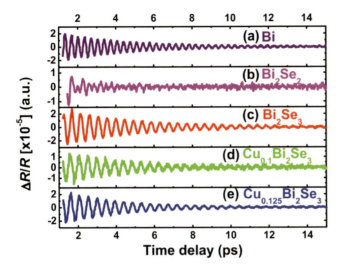

FIGURE 9.4.2.3 High-frequency oscillatory temporal variations of $\Delta R/R$ signals for all samples: (a) Bi crystal, (b) Bi$_2$Se$_2$ crystal, (c) Bi$_2$Se$_3$ crystal, (d) Cu$_{0.1}$Bi$_2$Se$_3$ crystal, (e) Cu$_{0.125}$Bi$_2$Se$_3$ crystal.

Figure 9.4.2.4a–e displays the STFT spectra of the oscillatory signals shown in Figure 9.4.2.3a–e, respectively. As is evident from Figure 9.4.2.4b, two characteristic peaks centered at 2.025 THz and 3.319 THz are observed for the Bi$_2$Se$_2$ crystal, which is respectively very close to the A_{1g}^1 mode frequencies of the Bi (Figure 9.4.2.4a) and Bi$_2$Se$_3$ (Figure 9.4.2.4c) crystals.

However, both peaks are slightly shifted away from the characteristic frequencies of the A_{1g}^1 mode in pure Bi and Bi$_2$Se$_3$ crystals. These shifts of the spectral peaks can be explained by the changes in bond lengths due to the involvement of a covalent contribution in the bonding between the Se-Bi-Se-Bi-Se five-layer slab and the Bi-Bi two-layer slab in the Bi$_2$Se$_2$ crystal [13,14]. By comparison to pure Bi and Bi$_2$Se$_3$ crystals, this additional covalent contribution rebalances the distance and angle of the Bi-Bi bonds and the Se-Bi-Se-Bi-Se chains. The chain length of Se-Bi-Se-Bi-Se QL stretches from 11.752 Å in Bi$_2$Se$_3$ to 11.797 Å in Bi$_2$Se$_2$ (Ref. [14]) such that the A_{1g}^1 mode frequency of the QL layer structure decreases from 2.148 THz in a Bi$_2$Se$_3$ crystal to 2.025 THz in a Bi$_2$Se$_2$ crystal. By contrast, the bond length of the Bi-Bi bond in Bi$_2$Se$_2$ shortens from 3.056 Å in Bi to 2.987 Å in Bi$_2$Se$_2$ (Ref. [14]) such that the A_{1g}^1 mode frequency of the Bi$_2$ layer structure increases from 2.928 THz in a Bi crystal to 3.319 THz in a Bi$_2$Se$_2$ crystal. For the Se-Bi-Se-Bi-Se chain, the frequency shift is 0.123 THz for a 0.38% change in the chain length, while for the Bi-Bi bond, the frequency shift is 0.391 THz for a 2.26% change in the bond length. These results indicate that in Bi$_2$Se$_2$, where extra Bi atoms are intercalated between QLs, the phonon frequency of the A_{1g}^1 mode of the QL exhibits a red shift compared to that in Bi$_2$Se$_3$. Therefore, it appears that the microstructural deformation in the Bi$_2$Se$_2$ crystals can be accurately resolved by measuring the magnitude and sign of the phonon frequency shift of the QL.

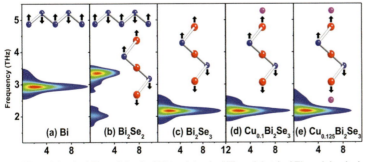

FIGURE 9.4.2.4 STFT spectrograms of high-frequency oscillatory signals for all samples: (a) Bi crystal, (b) Bi$_2$Se$_2$ crystal, (c) Bi$_2$Se$_3$ crystal, (d) Cu$_{0.1}$Bi$_2$Se$_3$ crystal, (e) Cu$_{0.125}$Bi$_2$Se$_3$ crystal. Schematic diagrams of the Raman A_{1g}^1 modes for the samples are also shown (Bi atoms are shown in blue, Se in red and Cu in pink).

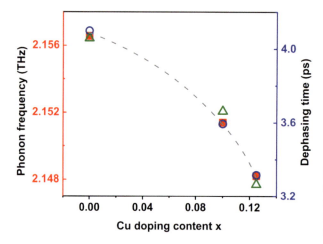

FIGURE 9.4.2.5 Phonon frequency (red square) and dephasing time of A_{1g}^1 phonon mode (blue circle) and change in lattice constant c (green triangle) for $Cu_xBi_2Se_3$ crystal as a function of Cu doping concentration.

Figure 9.4.2.4d and e are the STFT spectra for the $Cu_{0.1}Bi_2Se_3$ and $Cu_{0.125}Bi_2Se_3$ crystals, respectively. The STFT spectra reveal that the lifetime of the A_{1g}^1 phonon mode decreases with increasing Cu concentration. To quantitatively analyze this behavior, a damped oscillation waveform $A_{OSC}\cos\left(2\pi f_{OSC}t+\phi\right)e^{-t/\tau_{dephasing}}$ was used to fit the original data shown in Figure 9.4.2.3 to get the dephasing time ($\tau_{dephasing}$) and the phonon frequency (f_{osc}) for samples of various Cu concentrations. Figure 9.4.2.5 shows the fitting results obtained from samples of various Cu concentrations. These results show that both the phonon frequency and the dephasing time decrease with increasing Cu concentration, indicating that the addition of Cu atoms does indeed deform the Se-Bi-Se-Bi-Se chain in $Cu_xBi_2Se_3$ crystals, even though this deformation is much smaller than that in Bi_2Se_2 crystals. The inverse relation between the phonon frequency and the Cu concentration provides us with information about the locations of the Cu atoms. As mentioned above, for the substitution case where a Bi atom is replaced by a doping atom, the Raman spectrum reveals that the phonon frequency of the A_{1g}^1 mode increases with increasing doping concentration, while for the intercalation case where a doping atom is intercalated between two QLs, the phonon frequency of the A_{1g}^1 mode decreases with increasing doping concentration. Therefore, the photon frequency shifts seen in Figure 9.4.2.5 for $Cu_xBi_2Se_3$ crystals hint at intercalation for the Cu atoms. Additionally, as also shown in Figure 9.4.2.5, the lattice constant c increases slightly with increasing Cu concentration, implying that the QL chain in $Cu_xBi_2Se_3$ is stretched by the doping Cu atoms. The stretch of the QL chain length can be interpreted as follows: When the doping Cu atoms are intercalated between QLs, they form a mediated layer to strengthen the interaction between QLs instead of the weak van der Waals interaction, thus stretching the chain length of the QL. It is reasonable to conclude from our experimental results that the Cu atoms are intercalated between every pair of QLs, and these intercalated Cu atoms cause an effective interaction to slightly deform the QLs.

9.4.2.4   CONCLUSION

In conclusion, we have systematically studied the A_{1g}^1 COP dynamics in Bi, Bi_2Se_2, and $Cu_xBi_2Se_3$ ($x=0$, 0.1, 0.125) single crystals using femtosecond pump-probe reflectivity spectroscopy [32]. Because the phonon frequency shift caused by substitution and that caused by intercalation have opposite signs, the structural deformation and the locations of the doping atoms in a TI can be resolved by measuring the sign and magnitude of the phonon frequency shift. The frequency shifts of the phonon modes in Bi-rich Bi_2Se_2 crystals indicate that the extra Bi atoms are intercalated into the Bi_2Se_3 matrix and form a Bi_2 layer between the QLs. From the red shift of the A_{1g}^1 phonon frequency associated with the doping of Cu atoms, we also conclude that the additional Cu atoms are predominantly intercalated between every pair of QLs in $Cu_xBi_2Se_3$ crystals. The relationship between structural deformation and superconductivity, however, needs further investigations; in particular, the temperature-dependent phonon dynamics in $Cu_xBi_2Se_3$ crystals are crucial for understanding this relationship. Experimental work along these directions is in progress. The research described in this subsection was conducted in collaboration with the following people: H.-J. Chen, 1 K. H. Wu, C. W. Luo, T. M. Uen, J. Y. Juang, J.-Y. Lin, T. Kobayashi, H.-D. Yang, R. Sankar, F. C. Chou, H. Berger, and J. M. Liu [32].

REFERENCES

1. M. Z. Hasan and C. L. Kane, *Rev. Mod. Phys.* **82**, 3045 (2010).
2. X.-L. Qi and S.-C. Zhang, *Rev. Mod. Phys.* **83**, 1057 (2011).
3. Q.-K. Xue, *Nat. Nanotechnol.* **6**, 197 (2011).
4. L. A. Wray, S.-Y. Xu, Y. Xia, D. Hsieh, A. V. Fedorov, Y. S. Hor, R. J. Cava, A. Bansil, H. Lin, and M. Z. Hasan, *Nat. Phys.* **7**, 32 (2011).
5. D. Hsieh, Y. Xia, D. Qian, L. Wray, J. H. Dil, F. Meier, J. Osterwalder, L. Patthey, J. G. Checkelsky, N. P. Ong, A. V. Fedorov, H. Lin, A. Bansil, D. Grauer, Y. S. Hor, R. J. Cava, and M. Z. Hasan, *Nature (London)* **460**, 1101 (2009).
6. L. A. Wray, S.-Y. Xu, Y. Xia, Y. S. Hor, D. Qian, A. V. Fedorov, H. Lin, A. Bansil, and M. Z. Hasan, *Nat. Phys.* **6**, 855 (2010).
7. Y. S. Hor, A. J. Williams, J. G. Checkelsky, P. Roushan, J. Seo, Q. Xu, H. W. Zandbergen, A. Yazdani, N. P. Ong, and R. J. Cava, *Phys. Rev. Lett.* **104**, 057001 (2010).
8. Y. S. Hor, P. Roushan, H. Beidenkopf, J. Seo, D. Qu, J. G. Checkelsky, L. A. Wray, D. Hsieh, Y. Xia, S.-Y. Xu, D. Qian, M. Z. Hasan, N. P. Ong, A. Yazdani, and R. J. Cava, *Phys. Rev. B* **81**, 195203 (2010).
9. T. M. Schmidt, R. H. Miwa, and A. Fazzio, *Phys. Rev. B* **84**, 245418 (2011).
10. R. Yu, W. Zhang, H.-J. Zhang, S. C. Zhang, X. Dai, and Z. Fang, *Science* **329**, 61 (2010).
11. J. L. Zhang, S. J. Zhang, H. M. Weng, W. Zhang, L. X. Yang, Q. Q. Liu, S. M. Feng, X. C. Wang, R. C. Yu, L. Z. Cao, L. Wang, W. G. Yang, H. Z. Liu, W. Y. Zhao, S. C. Zhang, X. Dai, Z. Fang, and C. Q. Jin, *Proc. Natl. Acad. Sci. U.S.A.* **108**, 24 (2011).
12. C. Zhang, L. Sun, Z. Chen, X. Zhou, Q. Wu, W. Yi, J. Guo, X. Dong, and Z. Zhao, *Phys. Rev. B* **83**, 140504(R) (2011).
13. H. Lind and S. Lidin, *Solid State Sci.* **5**, 47 (2003).
14. H. Lind, S. Lidin, and U. H€aussermann, *Phys. Rev. B* **72**, 184101 (2005).
15. C. Thomsen, H. T. Grahn, H. J. Maris, and J. Tauc, *Phys. Rev. B* **34**, 4129 (1986).
16. H. J. Zeiger, J. Vidal, T. K. Cheng, E. P. Ippen, G. Dresselhaus, and M. S. Dresselhaus, *Phys Rev. B* **45**, 768 (1992).
17. A. V. Bragas, C. Aku-Leh, S. Costantino, A. Ingale, J. Zhao, and R. Merli, *Phys. Rev. B* **69**, 205306 (2004).
18. R. N. Kini, A. J. Kent, N. M. Stanton, and M. Henini, *Appl. Phys. Lett.* **88**, 134112 (2006).
19. I. Matsubara, S. Ebihara, T. Mishina, and J. Nakahara, *Phys. Rev. B* **79**, 054110 (2009).
20. Y. H. Ren, M. Trigo, R. Merlin, V. Adyam, and Q. Li, *Appl. Phys. Lett.* **90**, 251918 (2007).
21. H. C. Shih, L. Y. Chen, C. W. Luo, K. H. Wu, J.-Y. Lin, J. Y. Juang, T. M. Uen, J. M. Lee, J. M. Chen, and T. Kobayashi, *New J. Phys.* **13**, 053003 (2011).
22. Y. Wang, X. Xu, and R. Venkatasubramanian, *Appl. Phys. Lett.* **93**, 113114 (2008).
23. Y. Li, V. A. Stoica, L. Endicott, G. Wang, C. Uher, and R. Clarke, *Appl. Phys. Lett.* **97**, 171908 (2010).
24. N. Kamaraju, S. Kumar, and A. K. Sood, *Europhys. Lett.* **92**, 47007 (2010).
25. N. Kumar, B. A. Ruzicka, N. P. Butch, P. Syers, K. Kirshenbaum, J. Paglione, and H. Zhao, *Phys. Rev. B* **83**, 235306 (2011).
26. J. Qi, X. Chen, W. Yu, P. Cadden-Zimansky, D. Smirnov, N. H. Tolk, I. Miotkowski, H. Cao, Y. P. Chen, Y. Wu, S. Qiao, and Z. Jiang, *Appl. Phys. Lett.* **97**, 182102 (2010).
27. W. Richter, H. Kohler, and C. R. Becker, *Phys. Stat. Sol. (B)* **84**, 619 (1977).
28. J. Zhang, Z. Peng, A. Soni, Y. Zhao, Y. Xiong, B. Peng, J. Wang, M. S. Dresselhaus, and Q. Xiong, *Nano Lett.* **11**, 2407 (2011).
29. M. R. Portnoff, *IEEE Trans. Acoust., Speech, Signal Process.* **28**, 55 (1980).
30. A. H. Nuttall, *IEEE Trans. Acoust., Speech, Signal Process.* **29**, 84 (1981).
31. A. Yabushita, T. Kobayashi, and M. Tsuda, *J. Phys. Chem. B* **116**, 1920 (2012).
32. H.-J. Chen, K. H. Wu, C. W. Luo, T. M. Uen, J. Y. Juang, J.-Y. Lin, T. Kobayashi, H.-D. Yang, R. Sankar, F. C. Chou, H. Berger, and J. M. Liu, *Appl. Phys. Lett.* **101**, 12912 (2012).

9.4.3 Ultrafast Multi-Level Logic Gates with Spin-Valley Coupled Polarization Anisotropy in Monolayer MoS$_2$

9.4.3.1 INTRODUCTION

Structural inversion symmetry together with time reversal symmetry allows monolayer MoS$_2$ to possess the same magnitude of magnetic moments but the opposite signs at the K and K' valleys [1,2]. Furthermore, spin-orbit coupling separates the spin-up and spin-down states of the valence band [3–5], which plays a crucial role in spintronics [6], valleytronics [7], and semiconductor devices [8,9]. As shown in Figure 9.4.3.1a, there is fascinating coupling between energy, spin, and valley.

Valleytronics is a word from the combination of "valley" and "electronics." It is an area of physics especially semiconductors in which local extrema called "valleys" in the electronic band structure

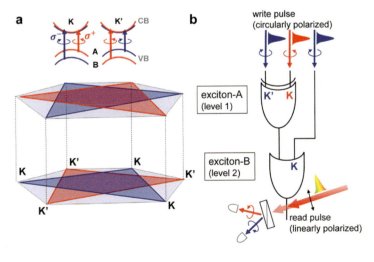

FIGURE 9.4.3.1 Schematics of an optically driven ultrafast room-temperature and multi-level logic gate with monolayer MoS$_2$. (a) The band diagram of monolayer MoS$_2$ at the K and K' valleys. The blue and red colors represent spin-up and spin-down states, respectively. (b) A two-level MoS$_2$ was produced by self-phase modulation in a sapphire plate (see page 147 in Subsection 4.2, page 154 in Subsection 4.3.2, and page 171 in Subsection 5.1.4). As shown in Figure 9.4.3.3, the absorption spectrum of a monolayer MoS$_2$ clearly presents A (1.89 eV) and B (2.04 eV) excitonic transitions, which indicates the splitting of the valence band at the K valley due to spin-orbit coupling [14,15]. In order to distinguish the non-equivalent K and K' valleys, polarizations of pump and probe pulses were adjusted to be circularly polarized by quarter wave plates.

DOI: 10.1201/9780429196577-61

are the main workhorses. Some semiconductors have multiple "valleys" in the electronic band structure of the first Brillouin zone and are known as multivalley semiconductors. Valleytronics is the technology of control over the valley degree of freedom, a local maximum/minimum on the valence/conduction band, of such multivalley semiconductors. The term is in analogy to spintronics. While in spintronics the internal degree of freedom of spin instead of charge in electronics, is utilized to store, manipulate and read out bits of information in electronics, the valleytronics is an extension of the idea to perform similar tasks using the multiple extrema of the band structure, so that the information of 0s and 1s would be stored as different discrete values of the crystal momentum.

Utilizing the degrees of freedom of monolayer; valleytronics, MoS_2, and optically driven logic gates can be realized. Figure 9.4.3.1b illustrates that a two-level logic gate can be operated by being sequentially excited with circularly polarized 2.01 eV (resonant with exciton B) and 1.98 eV (resonant with exciton A) pulses at room temperature. According to our time-resolved studies, the nonequilibrium population between the K and K9 valley lasts for, 1 ps in monolayer MoS_2, which is an excellent candidate material for ultrafast optical control. For the application to a high-rate optical pulse control, the problem of the accumulation of the remnant coherence after the control pulse always exists. Thus, the following pulse to control the succeeding step must wait for the decoherence of the target. This limits the bandwidth of optical spin-controlling devices.

Circularly polarized luminescence from monolayer MoS_2 has been demonstrated to have the same helicity as the circular polarization of an excitation laser [3–5,10,11]. This highly polarized luminescence has only been observed with resonant excitation. Furthermore, the valley-spin lifetime was previously predicted to be >1 ns[3]. However, a time-resolved study of polarized photoluminescence (PL) exhibited that the carrier spin flip time is in the time scale of several picoseconds, which is limited by the time resolution of the PL measurement system [12]. Mai et al. [13] further observed that the polarized exciton A decays only within several 100 femtoseconds, according to optical pump-probe measurements. Under the controversial situation, the full dynamics and physical insight of the polarized excitons in monolayer MoS_2, including the spin-valley coupling, have not yet been conclusively studied.

In this work, we propose optically driven ultrafast two-level MoS_2-logic gates through a systematic study of the ultrafast dynamics of monolayer MoS_2 with 30-fs time resolution. The identification of a single atomic layer of MoS_2 is confirmed by photoluminescence spectrum, Raman spectrum and AFM measurement, as shown in Figure 9.4.3.2. Through the resonant and off-resonant excitations in the direct bandgap, the valley polarization dynamics in monolayer MoS_2 were clearly observed. The pump pulse used in this study was generated by a wavelength-tunable optical parametric amplifier. Meanwhile, a probe pulse with a visible broadband spectrum gate can be written

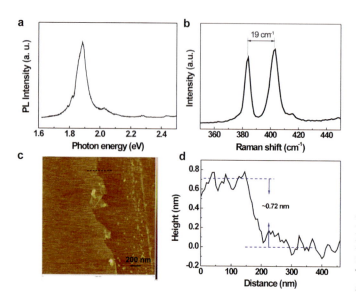

FIGURE 9.4.3.2 Characterization of monolayer MoS_2. (a) Photoluminescence spectrum. (b) Raman spectrum. (c) AFM measurement of monolayer MoS_2. (d) The height profile of MoS_2 gives an average thickness of ~0.72 nm.

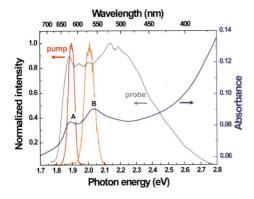

FIGURE 9.4.3.3 Spectra of pump–probe pulses and the stationary absorbance of monolayer MoS₂. Spectra of 1.89 eV pump pulse, 2.01 eV pump pulse, broadband visible probe pulse and the stationary absorption of monolayer MoS₂ at room temperature.

by circularly polarized 2.01 eV (resonant with exciton B) and 1.98 eV (resonant with exciton A) pulses and read by a linearly polarized pulse with a visible broadband spectrum.

Figure 9.4.3.4 shows a 2D display of photon energy- and time-resolved transient difference absorbance $\Delta A(v, t)$ at 78 K. In the measurements, the polarizations of broadband probe pulses were adjusted to be σ^+ and σ^- while the pump pulses were set as σ^+ circular polarization and 1.89 eV to resonate with exciton A. For both σ^+ and σ^- probes, the time-resolved spectra exhibited negative A in the spectral band of exciton A and exciton B. The negative ΔA signal could be caused by stimulated emission from the excited state and/or photobleaching due to the depletion of the ground state and the population of the excited state. The lifetime of photobleaching is usually much longer than that of stimulated emission because a stimulated emission occurs only within the lifetime of the excited state, whereas photobleaching occurs until the ground state is fully repopulated. Obviously, ΔA is significantly probe polarization-dependent within 100 fs as shown in Figure 9.4.3.4c and d. The nonlinear optical response after 100 fs is independent of the angular momentum of the initially excited distribution, which indicates that the initial polarization distribution relaxes to some quasi-equilibrium states in 100 fs. Following fast spin-polarization relaxation, the peaks in the ΔA spectrum show a blue-shift before 10 ps and a red-shift after 10 ps as shown in Figure 9.4.3.4a and b. We note that the pump-induced response at the K' valley is observed even when exciton A at the K valley is excited by the s^1 pump, in contrast to cases of excitons coupled to pump and probe pulses, which do not share common states. This unexpected phenomenon could be explained through various possible mechanisms, e.g., the dark excitons generated by pump pulses [13], the weakening of the excitonic binding energy [16] or the dielectric screening from the excited excitons [16]. Here general on the dark exciton is briefly mentioned below.

The fundamental optical excitation in semiconductors is an electron–hole pair with antiparallel spins: the 'bright' exciton. Bright excitons in optically active, direct-bandgap semiconductors and their nanostructures have been thoroughly studied. In quantum dots, bright excitons provide an essential interface between light and the spins of interacting confined charge carriers. Recently, complete control of the spin state of single electrons and holes in these nanostructures has been demonstrated, a necessary step towards quantum information processing with these two-level systems. In principle, the bright exciton's spin could also be used directly as a two-level system. However, because of its short radiative lifetime, its usefulness is limited. An electron–hole pair with parallel spins forms a long-lived, optically inactive 'dark exciton', and has received less attention as it is mostly regarded as an inaccessible excitation.

The measured time-resolved traces of $\Delta A(v, t)$ were fitted using the sum of three exponential functions and a constant term, as in equation.

$$\Delta A(\omega, t) = \Delta A_{\text{spin}}(\omega)e^{-\frac{t}{\tau_{\text{spin}}}} + \Delta A_{\text{exciton}}(\omega)e^{-\frac{t}{\tau_{\text{excition}}}} + \Delta A_{\text{carrier}}(\omega)e^{-\frac{t}{\tau_{\text{carrier}}}} + \Delta A_{e-h}(\omega) \qquad (1).$$

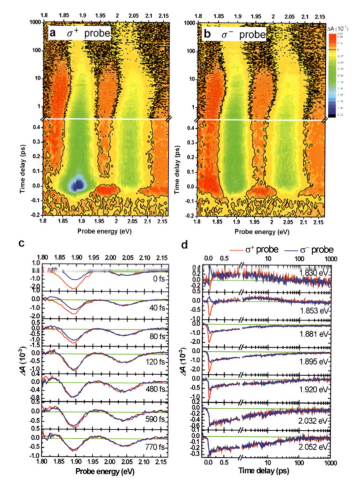

FIGURE 9.4.3.4 Transient difference absorbance (ΔA). ΔA induced by excitation using σ^+ circularly polarized pump pulse with the photon energy of 1.89 eV and probed by (a) σ^+ and (b) σ^- circularly polarized pulse at 78 K. The black curves are contours of ΔA being zero. (c) Time-resolved DA spectra at various time delays between pump and probe pulses. (d) Probe delay time traces of ΔA at various probe photon energies. In (c) and (d), the red and blue lines represent σ^+ and σ^- probe, respectively. The horizontal lines show $\Delta A = 0$.

The fitting results are shown in Figure 9.4.3.5. For the σ^+ probe, the time constants, ΔA_{spin}, $\Delta A_{exciton}$, and $\Delta A_{carrier}$ are 55 ± 7 fs, 1.0 ± 0.2, and 26.3 ± 5.4 ps, respectively. For the σ^- probe, they are 60 ± 40 fs, 0.96 ± 0.49, and 25.7 ± 8.6 ps, respectively. Due to the small signal amplitudes at the photon energies of ~1.93 and ~2.08 eV, the fitting error is large. Comparing the spectra of the σ^+ and σ^- probe in Figure 9.4.3.5a $\Delta A_{exciton}$, $\Delta A_{carrier}$ and ΔA_{e-h} exhibit similar dependence on the probe photon energy with the exception of ΔA_{spin}. Thus, these three relaxation processes take place independent of the initial polarization distribution within 100 fs. However, ΔA_{spin} is completely different for the cases of σ^+ and σ^- probes in that it is dependent on the relative polarization between the probe and pump beams. In the case of the σ^+ pump and σ^+ probe, ΔA_{spin} is negative and possesses a larger amplitude than in the case of the σ^+ pump and σ^- probe. This implies that the polarized exciton A at the K valley excited by the σ^+ pump pulses leads to intense photobleaching and stimulated emission only when the probe pulses have the common circular polarization to the pump pulses and the probe photon energy overlaps the band of exciton A. Moreover, the spectral shape of ΔA fits well with excitonic transition A, which indicates that the valley polarization is efficiently excited at the high symmetry K point [17]. On the other hand, the σ^- probe pulses generate exciton A with opposite spin polarization at the K' valley. Excitons with opposite polarization lead to the generation of biexcitons, which is the origin of induced absorption ($\Delta A > 0$) at ~1.87 eV [13,18], as shown in the right panel of Figure 9.4.3.5a.

The mean transition energy of excitonic band A is further calculated at every time delay, i.e., the time-dependent energy gap between excited electrons and holes, is plotted in the inset of Figure 9.4.3.5a. Since the photobleaching (caused by the same polarizations s¹ for the pump and probe) and

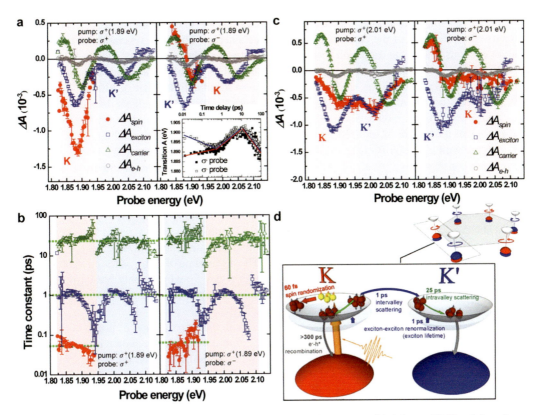

FIGURE 9.4.3.5 Triple exponential fitting results of the delay time traces of ΔA data at 78 K and the scheme of relaxation processes. (a), (b) Excited by 1.89 eV and σ^+ pump pulse. (c) Excited by 2.01 eV and σ^+ pump pulse. Left column: σ^+ probe. Right column: σ^- probe. Solid circles (red), open squares (blue), open triangles (green), and open circles (gray) represent the components for spin randomization, exciton dissociation, hot carrier relaxation and electron-hole recombination, respectively. (a) and (c) ΔA spectra, (b) Time constant of each component. Horizontal dotted lines indicate the estimated values. The three horizontal dotted lines clearly show the existing three time constants. Inset of (a): time-dependent (in log scale) mean energy of transition band A excited by 1.89 eV and σ^+ pump pulse at 78 K. The solid squares are the σ^+ probe and the open squares are the σ^- probe. (d) Schematics of the relaxation processes in monolayer MoS_2.

biexciton formation (caused by opposite polarizations σ^+ for pump and σ^- for probe), the results of the σ^+ and σ^- probes before 100 fs show distinct energy differences in transition A. After randomization of the polarized excitons generated by a pump pulse, the energy difference disappears. Blue shift and red shift take place when the polarizations of the pump and probe are the same and opposite, respectively. Similar energy difference (~13 meV at t, 0 ps in this study) is also observed on the CdSe nanocrystals with the splitting of bright-dark exciton states [19,20]. It is found that the flipping transition time scales with the energy split of bright-dark exciton states. Once the energy split increases to, ~14 meV for the small size of nanocrystals, the flipping transition time is in the range of tens of femtoseconds. This implies that the energy difference between the K and K9 valleys caused by the inhomogeneity of initially excited population in the K and K9 valleys leads to the fast spin randomization time. Moreover, the exciton size (diameter ~ 1.86 nm [21]) and spin randomization time (~60 fs) of monolayer MoS_2 also satisfy the size dependence of spin flip rate in semiconductor nanocrystals [19,20]. The behaviors can be rephrased as follows. The time constant $\Delta A_{spin} \sim 60$ fs reflects the lifetime of spin-polarized exciton A. It has relevant decay time but different behaviors between the co-circular polarizations and anticircular polarizations of the pump and probe beams. Since the hole spin states are non-degenerated at the K and K' valleys, the relaxation of hole spin is blocked by spin-valley coupling. On the contrary, the spin states of electrons can be easily destroyed because the spin states of the conduction band are degenerate. Thus, the relaxation of the electron spin of excitons causes the transition from optically active excitons to dark excitons.

The difference absorbance, $\Delta A_{\mathrm{exciton}}$, probed by σ^+ and σ^- circular polarizations have the same photobleaching phenomena, i.e., the blue shift of exciton A in the time range of ~10 ps. This implies that the excess population of excitons is created in both the K and K' valleys through intervalley scattering, which further reveals the relaxation of hole spin polarization. Renormalization of the self-energy of the exciton is induced by mutual exciton-exciton interaction leading to photobleaching together with a blue shift of the exciton band within the exciton lifetime [22]. Therefore, A_{exciton} with a time constant of τ_{exciton} ~1 ps represents the exciton intervalley transition and dissociation rate.

After the excitons are dissociated to become free carriers in highly excited states, the exciton peak exhibits a red shift. This shift is attributed to the intravalley scattering of free carriers, in which electrons relax to the bottom of the conduction band and holes relax to the top of the valence band. The $\Delta A_{\mathrm{carrier}}$ spectra show the sum of bleaching at transition energy peaks and the induced absorption of the broad conduction band. Thus, the intermediate relaxation time τ_{carrier} ~ 25 ps was assigned to the intraband transition of free carriers. The decay time of ΔA_{e-h} is found to be too long to be determined in the present work. The constant term ΔA_{e-h}, exhibits only bleaching behavior which can be attributed to the electron-hole recombination in the direct band. The recombination time was estimated to be ~300 ps in a previous study [23].

Time-resolved measurements were also performed with the σ^+ pump and σ^+ probe at room temperature (293 K). The fitting results of the measured pump-probe data are shown in Figure 9.4.3.6. The amplitude of ΔA becomes lower, and the peak positions shift to lower energy at 293 K compared with the result at 78 K. The red shift of ΔA implies the band gap is reduced with increasing temperatures [24]. The amplitude ratio of ΔA_{spin} to $\Delta A_{\mathrm{exciton}}$ at different temperatures are more or less similar, i.e., 2.22±0.27 for 78 K and 1.81±0.49 for 293 K. This indicates that the spin-valley coupled polarizations are almost the same at various temperatures [12].

The same experiment was also performed by changing the excitation energy from 1.89 to 2.01 eV to be resonant to the band of exciton B instead of exciton A (see Figure 9.4.3.3). Figure 9.4.3.5c

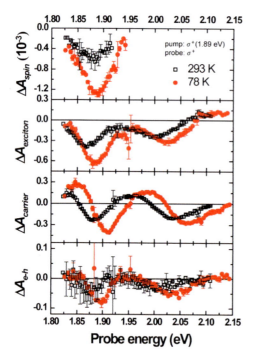

FIGURE 9.4.3.6 Comparison of the triple exponential fitting results at 78 and 293 K. Triple exponential fitting results of the delay time traces of DA data excited by 1.89 eV with σ^+ pump and σ^- probe pulse at 293 K (black open squares) and 78 K (red solid circles).

shows the spectra of three relaxation components obtained from the fitting using Eq. (9.4.3.1). ΔA_{spin} has a finite size not only in the energy range of exciton B but also the energy range slightly higher than that of exciton A. This is because the excitation at 2.01 eV with a σ^+ circular polarization simultaneously populates both exciton B at the K9 valley, resulting from the resonance excitation caused by the σ^+ pump, and the higher energy side of exciton A at the K valley. Therefore, the calculated difference absorbance ΔA_{spin} covers the band of transition B and the region of the high energy side of transition A. Additionally, $\Delta A_{\text{exciton}}$ is comparable with ΔA_{spin} in the energy range of exciton B and even larger than ΔA_{spin} in the energy range of exciton A. Consequently, the signal of spin-valley coupled polarization can be sizable in the energy range of exciton B, which is the resonance excitation energy of the pump, as in the case shown in Figure 9.4.3.4a, but its polarization anisotropy becomes much smaller. On the other hand, the spin-valley coupled polarization in the energy region of exciton A, which is off-resonant, is strongly diminished. This kind of reduction in spin-valley coupled polarization anisotropy, as observed in the PL experiments [11,12], could be attributed to efficient intervalley scattering.

In conclusion, the present study completely elucidates the fairly comprehensive ultrafast dynamics of spin-polarized excitons in monolayer MoS$_2$ [25]. Owing to the high temporal resolution and visible broadband detection, the time constants for the 60-fs spin-polarized exciton decay, 1-ps exciton dissociation (intervalley scattering), and 25-ps hot carriers relaxation (intravalley scattering) have been clearly identified. Temperature-dependent measurements further disclose the transition energy shifting and conservation of spin-valley coupled polarizations at various temperatures. Moreover, substantial intervalley scattering strongly diminished the spin-valley coupled polarizations under off-resonant excitation condition. These results provide a complete understanding of spin-valley coupled polarization anisotropy and the carrier dynamics of atomic layer MoS$_2$, which can further help us to develop ultrafast multi-level logic gates at room temperature.

The research presented in this subsection was performed cooperatively by the following people: Yu-Ting Wang, Chih-Wei Luo, Atsushi Yabushita, Kaung-Hsiung Wu, Takayoshi Kobayashi, Chang-Hsiao Chen and Lain-Jong Li [25].

REFERENCES

1. W. Yao, D. Xiao, and Q. Niu, "Valley-dependent optoelectronics from inversion symmetry breaking," *Phys. Rev. B* **77**, 235406 (2008).
2. D. Xiao, G.-B. Liu, W. Feng, X. Xu, and W. Yao, "Coupled spin and valley physics in monolayers of MoS$_2$ and other group-VI dichalcogenides," *Phys. Rev. Lett.* **108**, 196802 (2012).
3. K. F. Mak, K. He, J. Shan, and T. F. Heinz, "Control of valley polarization in monolayer MoS$_2$ by optical helicity," *Nat. Nanotechnol.* **7**, 494–498 (2012).
4. G. Sallen, et al. "Robust optical emission polarization in MoS$_2$ monolayers through selective valley excitation," *Phys. Rev. B* **86**, 081301(R) (2012).
5. H. Zeng, J. Dai, W. Yao, D. Xiao, and X. Cui, "Valley polarization in MoS$_2$ monolayers by optical pumping," *Nat. Nanotechnol.* **7**, 490–493 (2012).
6. I. Z̆utić, J. Fabian, and S. D. Sarma, "Spintronics: Fundamentals and applications," *Rev. Mod. Phys.* **76**, 323–410 (2004).
7. A. Rycerz, J. Tworzydlo, and C. W. J. Beenakker, "Valley filter and valley valve in graphene," *Nat. Phys.* **3**, 172–175 (2007).
8. B. Radisavljevic, A. Radenovic, J. Brivio, V. Giacometti, and A. Kis, "Single-layer MoS$_2$ transistors," *Nat. Nanotechnol.* **6**, 147–150 (2011).
9. Q. H. Wang, K. Kalantar-Zadeh, A. Kis, J. N. Coleman, and M. S. Strano, "Electronics and optoelectronics of two-dimensional transition metal dichalcogenides," *Nat. Nanotechnol.* **7**, 699–712 (2012).
10. T. Cao, et al. "Valley-selective circular dichroism of monolayer molybdenum disulphide," *Nat. Commun.* **3**, 887 (2012).
11. G. Kioseoglou, et al. "Valley polarization and intervalley scattering in monolayer MoS$_2$," *Appl. Phys. Lett.* **101**, 221907 (2012).
12. D. Lagarde, et al. "Carrier and polarization dynamics in monolayer MoS$_2$," *Phys. Rev. Lett.* **112**, 047401 (2014).

13. C. Mai, et al. "Many-body effects in valleytronics: Direct measurement of valley lifetimes in single-layer MoS_2," *Nano Lett.* **14**, 202–206 (2014).

14. K. F. Mak, C. Lee, J. Hone, J. Shan, and T. F. Heinz, "Atomically thin MoS_2: A new direct-gap semiconductor," *Phys. Rev. Lett.* **105**, 136805 (2010).

15. A. Splendiani, et al. "Emerging photoluminescence in monolayer MoS_2," *Nano Lett.* **10**, 1271–1275 (2010).

16. J. B. Stark, W. H. Knox, and D. S. Chemla, "Spin-resolved femtosecond magnetoexciton interactions in GaAs quantum wells," *Phys. Rev. B* **46**, 7919–7922 (1992).

17. Q. Wang, et al. "Valley carrier dynamics in monolayer molybdenum disulphide from helicity-resolved ultrafast pump-probe spectroscopy," *ACS Nano* **7**, 11087–11093 (2013).

18. E. J. Sie, Y.-H. Lee, A. J. Frenzel, J. Kong, and N. Gedik, "Biexciton formation in monolayer MoS2 observed by transient absorption spectroscopy," *arXiv Prepr:1312.2918* (2013). Available online: http://arxiv.org/abs/1312.2918 (accessed on 1 October 2014).

19. C. Y. Wong, J. Kim, P. S. Nair, M. C. Nagy, and G. D. Scholes, "Relaxation in the exciton fine structure of semiconductor nanocrystals," *J. Phys. Chem. C* **113**, 795–811 (2009).

20. J. Kim, C. Y. Wong, and G. D. Scholes, "Exciton fine structure and spin relaxation in semiconductor colloidal quantum dots," *Acc. Chem. Res.* **42**, 1037–1046 (2009).

21. T. Cheiwchanchamnangij and W. R. L. Lambrecht, "Quasiparticle band structure calculation of monolayer, bilayer, and bulk MoS_2," *Phys. Rev. B* **85**, 205302 (2012).

22. N. Peyghambarian, et al. "Blue shift of the exciton resonance due to exciton–exciton interactions in a multiple-quantum-well structure," *Phys. Rev. Lett.* **53**, 2433–2436 (1984).

23. H. Shi, et al. "Exciton dynamics in suspended monolayer and few-layer MoS_2 2D crystals," *ACS Nano* **7**, 1072–1080 (2013).

24. T. Korn, S. Heydrich, M. Hirmer, J. Schmutzler, and C. Schüller, "Low-temperature photocarrier dynamics in monolayer MoS_2," *Appl. Phys. Lett.* **99**, 102109 (2011).

25. Y.-T. Wang, C.-W. Luo, A. Yabushita, K.-H. Wu, T. Kobayashi, C.-H. Chen, and L.-J. Li, "Ultrafast multi-level logic gates with spin-valley coupled polarization anisotropy in monolayer MoS_2," *Sci. Rep.* **55**, 8289 (2015).

9.4.4 Femtosecond Time-Evolution of Mid-Infrared Spectral Line Shapes of Dirac Fermions in Topological Insulators

9.4.4.1 INTRODUCTION

Time-resolved spectroscopy is important in various fields, such as determining the exotic carrier dynamics of Tis [1–7]. The photon energy (~100 meV) of a MIR is less than the bulk band gap of TIs and has a quite different energy to the resonance energy of phonon absorptions. Therefore, MIR light sources are eminently suited to the study of SSTs in topological surface states (TSSs). The existing literature [8–22] reports the existence of a spectral line shape in the MIR region but there is no clear consensus. The explanation for FCA based on the Drude model has been adapted [11,12,17], but some studies give conflicting results [14,18] with considering more resonance factors. SSTs have also been reported [8–22] but these studies do not clarify the absorption mechanisms for SSTs and FCA using static MIR spectroscopic techniques.

This study unambiguously demonstrates the time evolution of MIR spectral line shapes in TIs using an optical pump and ultra-broadband MIR probe spectroscopy [23]. The MIR probe-pulses with a supercontinuum of 200–5000 cm^{-1} (or 25–620 meV) and a pulse width of 8.2 fs are generated using four-wave different-frequency generation (DFG) in nitrogen gas. This novel spectroscopy technique has the advantages of a wide bandwidth for standard Fourier-transform-infrared spectroscopy (FTIR) [24] and it allows femtosecond time-resolution by generating ultrashort pulses from nonlinear crystals using DFG. Two types of TI crystals are used for the experiments in this study. One is n-type Bi_2Te_2Se with a bulk/surface carrier concentration of 12.5×10^{18} cm^{-3}/5.5×10^{12} cm^{-2}, which is a bulk-conduction-electron-rich crystal. The other is p-type information), which features a higher ratio of surface to bulk carrier concentration. Figure 9.14.1 shows the clear presence of a bulk-conduction band (BCB) in Bi_2Te_2Se, but not in Sb_2TeSe_2.

9.4.4.2 EXPERIMENTAL

Optical pump and ultra-broadband MIR probe spectroscopy [23] consist of three stages: (i) 800-nm optical pulses with a duration of 30 fs were generated, (ii) ultra-broadband MIR probe pulses were generated in nitrogen, and (iii) chirped pulses were generated for detection. The fundamental pulses (800 nm) and the second harmonic pulses (400 nm, which were generated by a type I β-BaB_2O_4 crystal with a thickness of 0.1 mm) from a Ti:sapphire amplifier (790 nm, 30 fs, 0.85 mJ at 1 kHz, Femtopower compactPro, FEMTOLASERS) were focused into nitrogen gas to generate MIR pulses. The filamentation occurred *via* four-wave DFG when the pulse was focused using a concave mirror ($r = 1$ m). The length of the filament was ~3 cm. The bandwidth and the duration of the generated MIR pulses were 200–5000 cm^{-1} and 8.2 fs, respectively. When the MIR pulses were reflected from the sample with an incident angle of 45°, they were converted to ~400-nm pulses for the detection

DOI: 10.1201/9780429196577-62

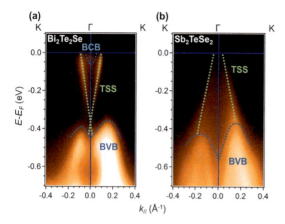

FIGURE 9.4.4.1 The angle-resolved photoemission spectroscopy (ARPES) images of Bi_2Te_2Se and Sb_2TeSe_2 single crystals. (a) The ARPES image of a Bi_2Te_2Se single crystal measured with 22 eV photon energy. (b) The ARPES image of a Sb_2TeSe_2 single crystal measured with 24 eV photon energy. All single crystals were the same pieces as those used in ultrafast experiments for the consistency of all measurements. The single crystals were in-situ cleaved under a base pressure of 5.1×10^{-11} torr at 85 K just before measurements. ARPES experiment was conducted National Synchrotron Radiation Research Center in Taiwan using a BL21B1 beamline. The photoemission spectra were recorded with a Scienta R4000 hemispherical analyzer. The polarization vector was always in the angular dispersion plane. The overall energy resolution is about 12 meV. The green dash lines represent as the TSS of crystals, and the blue dash lines show the bulk-conduction-band (BCB) and bulk valance band (BVB). The Dirac point in Sb_2TeSe_2 was estimated at 189 meV above the Fermi level (see S1 in Supplementary Information). A notable difference of band structure exists between Bi_2Te_2Se and Sb_2TeSe_2, the Dirac point of Bi_2Te_2Se is embedded in the BVB. In contrast to Bi_2Te_2Se, Sb_2TeSe_2 has an isolated Dirac cone and surface carriers cannot be scattered easily by bulk carriers. This difference in their band structure makes a significant difference in optical measurement results.

using a chirped-pulse up conversion (CPU) in nitrogen gas. A third 800-nm beam was transmitted through dispersive materials, including four BK7 glass plates (thickness: 10 mm) and one ZnSe plate (thickness: 5 mm), to produce chirped pulses. The converted visible (VIS) spectrum was measured by a spectrometer with an electron-multiplying CCD camera (SP-2358 and ProEM+1600, Princeton Instruments). The time resolution was estimated to be ~60 fs. To prevent significant absorption from vapor, the system was placed in boxes whose interior was purged with nitrogen.

9.4.4.3 RESULTS

9.4.4.3.1 ULTRA-BROADBAND MIR $\Delta R/R$ SPECTRA OF FCA AND SSTS IN TOPOLOGICAL INSULATORS

The typical ultra-broadband MIR $\Delta R/R$ spectra for Bi_2Te_2Se and Sb_2TeSe_2 are respectively shown in Figure 9.4.4.2a and b. These two spectra are significantly different. Along the wavenumber axis, there is a positive change in the lower frequency region and a negative change in the high-frequency region, which indicates a blueshift in the plasma edge for Bi_2Te_2Se after pumping (see Figure 9.14.2c). The zero-crossing line, $L_{0,X}$ (dashed line), in Figure 9.14.2a also shows a rapid blue-shift at the beginning of the delay time and then slowly (>50 ps) returns to the original position. However, the value of Figure 9.4.1. The angle-resolved photoemission spectroscopy (ARPES) images of Bi_2Te_2Se and Sb_2TeSe_2 single crystals: (i) The ARPES image of a Bi_2Te_2Se single crystal measured with 22 eV photon energy. (ii) The ARPES image of a Sb_2TeSe_2 single crystal measured with 24 eV photon energy. All single crystals were the same pieces as those used in ultrafast

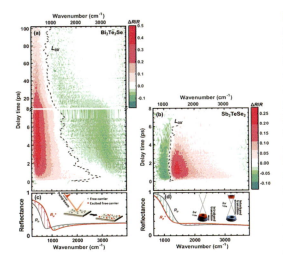

FIGURE 9.4.4.2 The time-resolved ultra-broadband MIR $\Delta R/R$ spectra for Bi2Te2Se and Sb2TeSe2 single crystals and the schematics of the theoretical model. (a) and (b) the 2D plots of wavenumber- and time-resolved reflectance change ($\Delta R/R$) spectra with an optical pump fluence of 101 μJ/cm^2 for Bi$_2$Te$_2$Se (a) and Sb$_2$TeSe$_2$ (b) single crystals. The red and green colors respectively represent the parts with a positive change and a negative change. The zero-crossing line is marked $L_{0,X}$ as a black dashed line. (c) Shows the p-polarized reflectivity before pumping (R_p, gray solid-line. Assume N is 12.5×10^{18} cm^{-1}, so $\omega_p = 1880$ cm^{-1} with $m^* = 0.32$ and $\varepsilon_0 = 23.7$) and after pumping (R_p^*, red solid-line. Assume N is 25×10^{18} cm^{-1} so $\omega_p = 2630$ cm^{-1} with $m^* = 0.32$ and $\varepsilon_0 = 23.7$) for the Drude model and (d) shows the p-polarized reflectivity before pumping (R_p, gray solid-line. Assuming $\mu = 50$ meV at room temperature) and after pumping (R_p^*, red solid-line. Assuming $\mu = 40$ meV at room temperature) for the SST-Kubo model.

experiments for the consistency of all measurements. The single crystals were *in-situ* cleaved under a base pressure of 5.1×10^{-11} torr at 85 K just before measurements. ARPES experiment was conducted National Synchrotron Radiation Research Center in Taiwan using BL21B1 beamline. The photoemission spectra were recorded with a Scienta R4000 hemispherical analyzer. The polarization vector was always in the angular dispersion plane. The overall energy resolution is about 12 meV. The green dash lines represent as the TSS of crystals, and the blue dash lines show the bulk-conduction-band (BCB) and bulk valance-band (BVB). The Dirac point in Sb$_2$TeSe$_2$ was estimated at 189 meV above the Fermi level (see S1 in Supplementary Information). A notable difference of band structure exists between Bi$_2$Te$_2$Se and Sb$_2$TeSe$_2$, the Dirac point of Bi$_2$Te$_2$Se is embedded in the BVB. In contrast to Bi$_2$Te$_2$Se, Sb$_2$TeSe$_2$ has an isolated Dirac cone and surface carriers cannot be scattered easily by bulk carriers. This difference in their band structure makes a significant difference in optical measurement results.

$\Delta R/R$ for Sb$_2$TeSe$_2$ shows a red shift in the plasma edge after pumping. It is worthy of note that the zero-crossing line, $L_{0,X}$ (dashed line) in Figure 9.4.4.2b is red-shifted until ~2 ps and then returns to the original position at ~6 ps, which is much faster than the change for Bi$_2$Te$_2$Se. Generally, there is a blue shift in the plasma edge because there is an increase in the carrier concentration [25], which is explained by the Drude model. The red shift in the $\Delta R/R$ spectrum of Sb$_2$TeSe$_2$ until ~2 ps is not explained by the Drude model because there is a decrease in the carrier concentration after pumping. It is found that the SST model using the Kubo formula [20] (SST-Kubo model), which has been successfully used to explain the transitions of Dirac cone in grapheme [20], explains the novel phenomena that are observed in p-type Sb$_2$TeSe$_2$.

By comparing the band mapping results of Bi$_2$Te$_2$Se and Sb$_2$TeSe$_2$ in Figure 9.4.4.1 a notable difference between Bi$_2$Te$_2$Se and Sb$_2$TeSe$_2$ can be found that the Dirac point of Bi$_2$Te$_2$Se is embedded in the BVB. The surface carriers cannot avoid scattering from bulk carriers, and the major change of optical property might be dominated by bulk carrier. In contrast to Bi$_2$Te$_2$Se, Sb$_2$TeSe$_2$ has an isolated Dirac cone and thus the surface carriers cannot be scattered easily by bulk carriers, that is why the SST is a major factor in Sb$_2$TeSe$_2$. Besides, the difference between bulk FCA of the Bi$_2$Te$_2$Se and SST of Sb$_2$TeSe$_2$ could be attributed to the intrinsic responses with a 1.55-eV excitation. As the schematics of Figure 9.14.4e and j, the final and initial states of the excitation process are different. The photoexcited carriers of the former are excited from the valence band maximum to the second conduction band [2,4,26], which is far from the Fermi level. For the latter case, the photoexcited carriers are excited from a deep valance band to the states near the Fermi level consisting of an isolated Dirac cone [5]. Therefore, the MIR probe beam tends to detect the free carriers of the conduction band in Bi_2Te_2Se, and the SST near the Fermi level in Sb_2TeSe_2.

9.4.4.3.2 Quantitative Analysis of the Ultra-Broadband MIR ΔR/R Spectra

To quantitatively reveal the hidden mechanism, the Drude model and the SST-Kubo model are used to fit the ultra-broadband MIR $\Delta R/R$ spectra for n-type Bi_2Te_2Se and p-type Sb_2TeSe_2 TIs. It is initially assumed that before and after pumping, all reflectivity R_p (gray solid-line, before pumping) and R_p^* (red solid-line, after pumping) have similarly shaped spectra for both the Drude model (Figure 9.14.2c) and the SST-Kubo model (Figure 9.4.4.2d). After pumping, the reflection spectrum shifts because there is an increase in the free carrier concentration. In terms of the Drude model, the dielectric function ε_D is:

$$\varepsilon_D(\omega) = \varepsilon_\infty - \frac{\omega_p^2}{\omega^2 + i\Gamma\omega} \tag{9.4.4.1}$$

where ε_∞ is the permittivity at an infinite frequency, ω is the frequency, ω_p is the plasma frequency and $\wp$ is the plasma scattering rate. The carrier concentration N is related to the effective mass m^* by the equation, $N = m^* w_p^2/4pe^2$. However, Falkovsky et al. estimated the reflectivity by considering the SSTs20. The dielectric function using the Kubo formula is:

$$\varepsilon_F(\omega) = -\frac{8T}{(\omega^2 + i\tau^{-1}\omega d_{TSS})}\left(\frac{e^2}{\hbar}\right)\ln\left[2\cosh\left(\frac{\mu}{2T}\right)\right]$$

$$+ \frac{1}{d_{TSS}}\left(\frac{e^2}{\hbar}\right)\left[\frac{i\pi}{\omega}G\left(\frac{\omega}{2}\right) - 4\int_0^\infty d\zeta G(\xi) - \frac{\left[G(\zeta) - G\left(\frac{\omega}{2}\right)\right]}{(\omega^2 - 4\zeta^2)}\right] \tag{9.4.4.2}$$

where μ is the chemical potential, T is the carrier temperature, G is the Fermi-Dirac distribution function, τ^{-1} is the collision rate for TSSs, which depends on the density of impurities, and d_{TSS} is the optical penetration depth of the TSSs. The first and second terms respectively represent the intraband transitions and the inter-band transitions in Dirac cone. Both models are applied under the "quasi-equilibrium" state in a view of sub-10 fs probe pulse (see S2 of Supplementary Information). The penetration depth of ultra-broadband MIR in TIs is few μm (see S3 of Supplementary Information).

As previously mentioned, the increase of N in the Drude model represents the change in the electronic population after pumping. In Figure 9.14.2c, the estimated value of N for R_p^* is larger than that for R_p, which results in a blue shift in the plasma edge. In the SST-Kubo model, the photoexcitation has a significant impact on μ and T and induces changes in the reflection spectrum. In terms of the ground state of p-type Sb_2TeSe_2, both the smaller number of carriers in the vicinity of the Dirac point and the higher electron temperature result in a reduction in Γ^{20}. Therefore, after pumping, the reduction in the chemical potential μ causes a change in the reflection spectrum from R_p to R_p^*, as shown in Figure 9.14.2d. This result is in good qualitative agreement with the $\Delta R/R$ spectrum in Figure 9.14.2b.

9.4.4.3.2.1 Ultrafast Time-Evolution of the Ultra-Broadband MIR $\Delta R/R$ Spectra

Figure 9.14.3 shows the typical time evolution of the MIR $\Delta R/R$ spectrum and the fitted curves. As mentioned previously, the photoexcited carrier dynamics in n-type Bi_2Te_2Se is dominated by FCA and can be fitted well with the Drude model, as shown in Figure 9.14.3a. For Sb_2TeSe_2, the contribution of FCA to the photoexcited carrier dynamics cannot be neglected.

Therefore, the $\Delta R/R$ spectra are fitted with the modified dielectric function of $\varepsilon_{DF}(\omega + \delta\omega) = \varepsilon_D(\omega + \delta\omega) + \varepsilon_F(\omega + \delta\omega)$, where δ_ω is a shifted frequency in fitting. This is called the Drude-SST-Kubo model. Figure 9.14.3b shows that this model fits the MIR $\Delta R/R$ spectrum at various delay times quite well. The details of the fitting are presented in the Method section.

9.4.4.4 DISCUSSION

The fitting results in Figure 9.14.4a and b are of interest, in particular the time evolutions of ω_p, Γ, N, μ, and T in TIs. During the pumping process, the 1.55-eV pump photons excite the electrons to a higher BCB from the occupied states [1]. For Bi_2Te_2Se, both ω_p and Γ respectively exhibit growth and relaxation dynamics. Although it is difficult to obtain the real value of N because there is no m^*, it is still possible to obtain the temporal evolution of N through $N\omega_p^2$, as shown in Figure 9.14.4c and d. The serious shift of ω_p (~3.7 times after photoexcitation) equivalents to the dramatic enhancement of photo-excited concentration (see S4 in Supplementary Information). This photoexcited carrier mainly experiences FCA in bulk states (BSs), as shown by the notation of probe(1) and probe(2) in Figure 9.14.4e, or in TSSs, as shown by the notation of probe(3). A bi-exponential decay function is further used to obtain the reduction times for the concentration of photoexcited carriers. This has a maximum within ~2.2 ps and then undergoes two relaxation processes for 1.5 ps and 8.4 ps. The fast relaxation process is caused by the thermal diffusion in BCB and TSS [27,28], or the acoustic-phono assistant process [3]. The slow

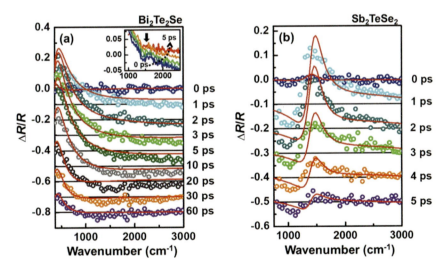

FIGURE 9.4.4.3 Typical ultra-broadband MIR $\Delta R/R$ spectra with fitting curves, taking account of free carrier absorption (FCA) and surface state transition (SST). $\Delta R/R$ as a function of the wavenumber for different delay times for (a) Bi_2Te_2Se and (b) Sb_2TeSe_2, using a pump fluence of 101 μJ/cm². The open circles represent the experimental data and each $\Delta R/R$ spectrum is shifted for clarity. The red lines are the fitting curves with the Drude model and b the Drude-SST-Kubo model ($\delta_\omega = -690\,cm^{-1}$). The insert in a shows the Moss-Burstein shift, as indicated by the arrow. This is the phenomenon in which the apparent band gap of a semiconductor is increased as the absorption edge is pushed to higher energies as a result of some states close to the conduction band being populated. Between the top solid line and the bottom solid line are the spectra that are respectively obtained by averaging the data from 0–1 (red), 1–2 (orange), 2–3 (bright green), 3–4 (green), and 4–5 (blue) ps.

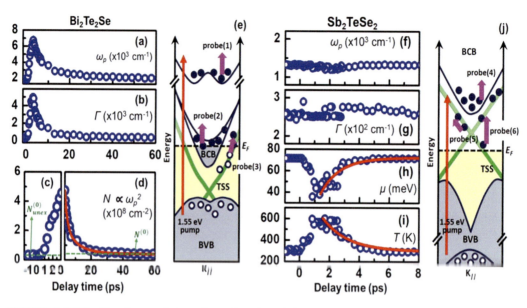

FIGURE 9.4.4.4 The time-evolution of ω_P, Γ, μ, T and the schematic energy band structure of TIs for pump/probe processes. The fitting results (a–d) and the pump-probe scheme (e) for Bi_2Te_2Se and (f–i) and for Sb_2TeSe_2 (j). (a,b) respectively show the time-evolution of the fitting parameter ω_P and Γ for the Drude model of a Bi_2Te_2Se crystal. (c,d) show the partial trace of the squared values of ω_P before and after 3 ps (c) and (d), respectively. The red line in d shows the bi-exponential fitting that is described in the Method section. The green dashed lines in c marked by $N_{unex}^{(0)}$ ($3.42 \times 10^7 \mathrm{cm}^{-2}$) relate to the concentration of unexcited carriers and (d) marked by $N^{(0)}$ ($4.16 \times 10^7 \mathrm{cm}^{-2}$) represent the height of the constant term from the fitting curve. The time-domain traces for $(f)\omega_P(g)\varepsilon_{F,\,intra} = -2e^2|\mu|/h(h)$, and (i) carrier temperature T are obtained using the Drude-SST-Kubo model. The red lines in (h), (i) show the single-exponential fitting that is mentioned in the Method section. The red arrows and pink arrows respectively show the 1.55 eV (800 nm) pumping and the MIR probing in (e), (j). The notation E_F in (e), (j) is the Fermi energy.

one is consistent with the results of time-resolved ARPES [1,2]. Additionally, the appearance of a Moss-Burstein shift (until ~6 ps) near the bulk band gap (see the inset in Figure 9.14.3a) also indicates the recombination in BSs [15]. However, the value of N (i.e., $N^{(0)}$ in Figure 9.14.4d) does not recover to its original value (i.e., $N_{unex}^{(0)}$ in Figure 9.14.4c) within the limited delay time (~ 100 ps). This inconsistency between $N^{(0)}$ and $N_{unex}^{(0)}$ is explained by the long-lived recombination process. There are several scenarios proposed for this long relaxation process. First, it is generally assigned to the photo-voltage effect [29]. Moreover, huge Rashba-splitting effect has been clearly observed in BCB [30,31], which might cause the long-time relaxation processes like indirect bandgap semiconductors [25].

For p-type Sb_2TeSe_2, the Drude-SST-Kubo model is used to fit the results in Figure 9.14.4f–i. It is worth emphasizing that the relative changes in μ and T are more distinct than the changes in ω_p and Γ. Even though the MIR probe-pulses can also detect FCA (even though it originates from TSSs or BSs as shown by the notation of probe(4) in Figure 9.14.4j), the $\Delta R/R$ spectra are significantly dominated by SST's contribution in the Dirac cone (see the notations of probe(5) and probe(6) in Figure 9.14.4j). Figure 9.14.4h shows that after the deep valance electrons are excited to the upper Dirac cone [5], μ reaches a minimum at ~1 ps and it takes ~1.28 ps for the recombination process according to the fitting for a single exponential decay function. Besides, the hot-carrier temperature reaches ~600 K and recovers to room temperature after 1.68 ps, which results are consistent with the time-resolved ARPES results for Sb_2Te_3 [5–7]. Therefore, this hot-carrier temperature decay would be resulted from the thermal diffusion between BVB and TSS [27].

By taking account of the difference of the number of states between bulk and surface, when the electrons are photo-excited, the chemical potential should shift towards the higher energy direction. In Drude-SST-Kubo model, the chemical potential (μ) and the carrier temperature (T) are associated with the surface state, and the SST-Kubo term is consisted by inter-transition and intra-transition of Dirac cone. For $\mu \gg T$, the intra-transition term could be derived to the form [20] $\varepsilon_{F,\,intra} = -e^2|\mu|\pi\hbar(\omega^2 + i\omega\Gamma_{im})$ which coincides with the Drude expression, and the effective plasma frequency $\omega_{p,\,s}$ could be further expressed as $(e^2|\mu|\pi\hbar)^{1/2}$. More precisely, the contribution of excited charges to "ω_p" in Dirac cone is considered by the intra-transition term. In other words, the Drude term in the Drude-SST-Kubo model represents the excited carriers which are out of the surface state. For p-type Sb_2TeSe_2 with the Fermi level located at the lower energy part of the Dirac cone, after photoexcitation, the chemical potential shifts to the higher energy direction, and further indicates the red-shift of the plasma edge and decreasing of the density of states. From the fitting result of smaller ω_p (~$1.3 \times 10^3 cm^{-1}$) on Sb_2TeSe_2, it shows the lower contribution from the excited bulk carriers, which is consistent with the results in Figure 9.14.2b.

Summary. The ultrafast dynamics of Dirac fermions and bulk free carriers in the TIs, n-type Bi_2Te_2Se and p-type Sb_2TeSe_2 single crystals, are studied using time-resolved ultra-broadband MIR spectroscopy.

Methods. In the n-type Bi_2Te_2Se is dominated by bulk carriers because the $\Delta R/R$ spectra show a blue-shift in the plasma edge due to FCA. For p-type Sb_2TeSe_2, the dynamics is dominated by the Dirac fermion from the red-shift of the plasma edge in the $\Delta R/R$ spectra. This study shows that the MIR absorption peaks for FCA and SST in TIs can be distinguished and demonstrates the importance of time-resolved ultra-broadband MIR spectroscopy for gapless or small band gap exotic materials.

Experimental Setup. Optical pump and ultra-broadband MIR probe spectroscopy 23 consists of three stages: (i) 800-nm optical pulses with a duration of 30 fs were generated, (ii) ultra-broadband MIR probe pulses were generated in nitrogen and (iii) chirped pulses were generated for detection. The fundamental pulses (800nm) and the second harmonic pulses (400 nm, which were generated by a type I β-BaB2O4 crystal with a thickness of 0.1 mm) from a Ti:sapphire amplifier (790 nm, 30 fs, 0.85 mJ at 1 kHz, Femtopower compactPro, FEMTOLASERS) were focused into nitrogen gas to generate MIR pulses. The filamentation occurred via four-wave DFG when the pulse was focused using a concave mirror ($r = 1$ m). The length of the filament was ~3 cm. The bandwidth and the duration of the generated MIR pulses were 200–5000 cm^{-1} and 8.2 fs, respectively. When the MIR pulses were reflected from the sample with an incident angle of 45°, they were converted to ~400-nm pulses for the detection using a chirped-pulse up conversion (CPU) in nitrogen gas. A third 800-nm beam was transmitted through dispersive materials, including four BK7 glass plates (thickness: 10 mm) and one ZnSe plate (thickness: 5 mm), to produce chirped pulses. The converted visible (VIS) spectrum was measured by a spectrometer with an electron-multiplying CCD camera (SP-2358 and ProEM+1600, Princeton Instruments). The time resolution was estimated to be ~60 fs. To prevent significant absorption from vapor, the system was placed in boxes whose interior was purged with nitrogen.

Retrieving the MIR spectra from an up-converted spectra and calibrating the spectra of the VIS pulse to MIR region. The MIR spectrum from, especially the sharp absorption peaks, can be seriously distorted after CPU measurements. That is to say, the dispersion of chirped pulses causes additional oscillations in the spectrum [32,33]. The CPU signal $\left(E_{CP}^2(t-\tau)E_{MIR}^*(t)\right)$ was obtained by performing four-wave DFG(FWDFG $E_{VWM}(t)$) between the chirped pulse $\left(E_{CP}^2(t-\tau)\right)$ and the MIR pulse $\left(E_{MIR}(t)\right)$ The chirped pulse is written as:

$$E_{CP}(t) = \varepsilon_{CP}(t)e^{iw_{r+i\frac{1}{2}w^{m}]2}} \tag{9.4.4.3}$$

where $\varepsilon_{CP}(t)$ represents the envelope, $\omega^{(0)}$ is the central angular frequency, and $\omega^{(1)}$ is a parameter. The MIR pulse can be divided into a main part $E_{MIR}^{(0)}(t)$ and a free induction decay part $E_{MIR}^{(1)}(t)$. Substituting $E_{MIR}(t) = E_{MIR}^{(0)}(t) + E_{MIR}^{(0)}(t)$ yields:

$$E_{\text{FWM}}(t) = E_{\text{CP}}^2(t)E_{\text{MIR}}^{(0)\bullet}(t) + E_{\text{CP}}^2(t)E_{\text{MIR}}^{(1)\bullet}(t) = E_{\text{FWM}}^{(0)}(t) + E_{\text{FWM}}^{(1)}(t) \qquad (9.4.4.4)$$

where $E_{\text{FWM}}^{(0)}(t)$ can be assumed to be the Dirac delta function $\delta(t)$ due to the short duration of MIR pulse. Using the Wiener-Khinchin theorem and these assumptions, the autocorrelation $C_A(t)$ of $E_{\text{FWM}}(t)$ is formed by [33]

$$C_A(t) = \int dt' E_{\text{FWM}}^{\bullet}(t') E_{\text{FWM}}(t'+t)$$

$$= \delta(t) + E_{\text{MIR}}^{(1)\bullet}(t)\varepsilon_{\text{CP}}^2(t)e^{12\omega^{(0)}t + L^{(2)}t^2} \qquad (9.4.4.5)$$

$$+ E_{\text{MIR}}^{(1)}(-t)\varepsilon_{\text{CP}}^{\bullet2}(-t)e^{i2\omega^{(0)}t - i\omega^{(1)}t^2}$$

A similar autocorrelation form $C'_A(t)$ is obtained for a pulse that is up-converted using a monochromatic pulse by multiplying $e^{-i\omega^{(1)}t^2 \sin(t)}$, so that Eq. (9.4.4.5) becomes [33]

$$C'_A(t) = \delta(t) + E_{\text{MIR}}^{(1)\bullet}(t)\varepsilon_{\text{CP}}^2(t)e^{i2\omega^{(0)}t} + E_{\text{MIR}}^{(1)}(-t)\varepsilon_{\text{CP}}^{\bullet2}(-t)e^{i2\omega^{(0)}t} \qquad (9.4.4.6)$$

Therefore, the original MIR spectrum with shift $2\omega^{(0)}t$ is acquired using the measured up-converted power spectrum and the known value of $\omega^{(1)}$ for the chirped pulse. Finally, the wavenumber is calibrated using a binomial fitting of the three absorption peaks, including carbon dioxide($\sim2300\,\text{cm}^{-1}$) and water vapor (~1600 and $\sim3700\,\text{cm}^{-1}$).

Analyses using the Drude, SST-Kubo and Drude-SST-Kubo models. In this study, the dielectric function ε in the Drude model, the SST-Kubo model and the Drude-SST-Kubo model is used to calculate the p-polarized reflectivity R_p using the Fresnel equation (with an incident angle of 45°)as:

$$R_p = \frac{\varepsilon \cos\theta - \sqrt{\varepsilon - \sin^2\theta}}{\varepsilon \cos\theta + \sqrt{\varepsilon - \sin^2\theta}} \qquad (9.4.4.7)$$

The transient $\Delta R/R$ is obtained by:

$$\frac{\Delta R}{R} = \frac{R_p^{\bullet} - R_p^0}{R_p^0} \qquad (9.4.4.8)$$

where the superscripts "•" and "0" of respectively represent the reflectivity with and without optical pumping The fitting with the Drude model is performed using the software, RefFIT, The fitting with the Drude-SST-Kubo model uses four parameters: ω_p, Γ, μ and T. To limit the computational load without losing the accuracy, the grid search method and an interval search algorithm with few iterations are used. After obtaining all possible values for the 4 parameters, the most appropriate parameter set P^i is selected by calculating the minimum root-mean-square deviation between the data and calculated results at the jth iteration. More specifically, using the grid search method, the value of P^i at the jth iteration can be obtained. The best interval is decided using the neighboring points of P^i. In this analysis, four parameters produce the 8 neighboring points. Using this interval, the next iteration $j+1$ of the grid search is undertaken. Therefore, the accuracy is exponentially increased.

The conditions, R_p^0 are determined using the ARPES results and the FTIR spectra. For $\text{Bi}_2\text{Te}_2\text{Se}_2$, R_2^0 is calculated using the Drude model with $\varepsilon_\infty = 23.7$, $\omega_p = 1880\,\text{cm}^{-1}$, and $\Gamma = 272\,\text{cm}^{-1}$, which values are obtained by fitting the FTIR spectra using the RefFIT program [34]. For Sb_2TeSe_2, R_p^Φ is determined using the Drude-SST-Kubo model with $\varepsilon_\infty = 19.4$, $\omega_p = 1320^{-1}$, $\Gamma = 253\,\text{cm}^{-1}$, $d_{\text{rSS}} = 1.4\,\text{nm}$, $\mu = 72\,\text{meV}$, and $T = 297\,\text{K}$, The former 4 parameters are obtained by fitting with fixed values of μ and T using the grid search method and an interval search algorithm, as described previously. If μ is sufficiently large. It can be estimated as:

$$\mu = \sqrt{\pi N_{rSS} h \nu_{rSS}} \tag{9.4.4.9}$$

where N_{rSS} is the surface carrier concentration ($\sim 2.2 \times 10^{12}\,\text{cm}^{-2}$). The parameter N_{rSS} is expressed as:

$$N_{rSS} = \frac{A_{Fs}}{A_{BZ}A_{\nu C}} = \frac{\pi K_F^2}{(4\pi^2/a^2)a^2} = \frac{K_F^2}{4\pi} \tag{9.4.4.10}$$

where A_{FS} is the area of the Fermi surface, A_{FZ} is the area per Brillouin zone, $A_{\nu C}$ is the area per unit cell and K_p is the Fermi-wavenumber ($\sim 5.2 \times 10^6\,\text{cm}^{-1}$ from ARPES). The parameter $\nu_{rss} = 4.12 \times 10^7\,\text{cm/s}$ is estimated from the gradient of the Dirac cone from ARPES. More ARPES information of TIs is shown in S1 of Supplementary Information.

Exponential fitting in Figure 9.4.4.4. The red line in Figure 9.4.4.4d shows the bi-exponential fitting for $N^{(0)} + N^{(1)} \exp\left[-t/\tau_{N,1}\right] + N^{(2)} \exp\left[-t/\tau_{N,2}\right]$ for ω^2 (proportional to the time evolution of N) with a delay time t, where the parameters $N^{(0)} = 4.16 \times 10^9\,\text{cm}^{-2}$, $N^{(1)} = 1.79 \times 10^9\,\text{cm}^{-2}$, $N^{(2)} = 2.6 \times 10^8\,\text{cm}^{-2}$, $\tau_{N,1} = 1.5$ ps and $\tau_{N,2} = 8.4$ ps. The red line in (h) shows the single-exponential fitting for $\mu^{(0)} + \mu^{(1)} \exp\left[t/\tau_\mu\right]$ for the transient chemical potential $\mu(t)$, where $\mu^{(0)} = 72$ meV is static chemical potential, $\mu^{(1)}$ is 99meV and $\tau_\mu = 1.28$ ps. The red curve in Figure 9.14.4i is fitted using a single-exponential function of $T^{(0)} + T^{(1)} \exp\left[-t/\tau_r\right]$ and the time evolution of the temperature, where $T^{(0)}$ represents the room temperature, $T^{(1)}$ is 770K and τ_T is 1.68 ps

The collaborative work presented in this subsection was performed by the following people: Tien-Tien Yeh, Chien-Ming Tu, Wen-Hao Lin, Cheng-Maw Cheng, Wen-Yen Tzeng, Chen-Yu Chang, Hideto Shirai, Takao Fuji, Raman Sankar, Fang-Cheng Chou, Marin M. Gospodinov, Takayoshi Kobayashi, Chih-Wei Luo [35].

REFERENCES

1. M. Hajlaoui et al., "Tuning a Schottky barrier in a photoexcited topological insulator with transient Dirac cone electron-hole asymmetry," *Nat. Commun.*, 3003 (2014).
2. M. Neupane et al., "Gigantic surface lifetime of an intrinsic topological insulator," *Phys. Rev. Lett.* **115**, 116801 (2015).
3. J. Qi et al., "Ultrafast carrier and phonon dynamics in Bi_2Se_3 crystals," *Appl. Phys. Lett.* **97**, 182102 (2010).
4. M. C. Wang, S. Qiao, Z. Jiang, S. N. Luo, and J. Qi, "Unraveling photoinduced spin dynamics in the topological insulator Bi_2Se_3," *Phys. Rev. Lett.* **116**, 036601 (2016).
5. J. Sánchez-Barriga et al., "Ultrafast spin-polarization control of Dirac fermions in topological insulators," *Phys. Rev. B* **93**, 155426 (2016).
6. S. Zhu et al., "Ultrafast electron dynamics at the Dirac node of the topological insulator Sb_2Te_3," *Sci. Rep.* **5**, 13213 (2015).
7. J. Reimann, J. Guddle, K. Kuroda, E. V. Chulkov, and U. Hofer, "Spectroscopy and dynamics of unoccupied electronic states of the topological insulators Sb_2Te_3 and Sb_2Te_2S," *Phys. Rev. B.* **90**, 081106(R) (2014).
8. W. S. Whitney et al., "Gate-variable mid-infrared optical transitions in a $(Bi_{1-x}Sb_x)_2Te_3$ topological insulator," *Nano Lett.* **17**, 255–260 (2017).
9. C. W. Luo, P. S. Tseng, H.-J. Chen, K. H. Wu, and L.J. Li, "Dirac fermion relaxation and energy loss rate near the Fermi surface in monolayer and multilayer graphene," *Nanoscale.* **6**, 8575 (2014).
10. C. W. Luo et al., "Snapshots of Dirac fermions near the Dirac point in topological insulators," *Nano Lett.* **13**, 5797–5802 (2013).
11. T. Dong, R.-H. Yuan, Y.-G. Shi, and N.-L. Wang, "Temperature-induced plasma frequency shift in Bi_2Te_3 and $Cu_xBi_2Se_3$," *Chin. Phys. Lett.* **30**, 127801 (2013).
12. S. V. Dordevic, M. S. Wolf, N. Stojilovic, H. Lei, and C. Petrovic, "Signatures of charge inhomogeneities in the infrared spectra of topological insulators Bi_2Se_3, Bi_2Te_3 and Sb_2Te_3," *J. Phys.: Condens. Matter.* **25**, 075501 (2013).

13. S. V. Dordevic et al., "Fano q-reversal in topological insulator Bi_2Se_3," *J. Phys.: Condens. Matter.* **28**, 165602 (2016).

14. S. V. Dordevic et al., "Magneto-optical effects in $Bi_{1-x}As_x$ with $x = 0.01$: Comparison with topological insulator $Bi_{1-x}Sb_x$ with $x = 0.20$," *Phys. Status Solidi B.* **251**, 1510–1514 (2014).

15. A. D. LaForge et al., "Optical characterization of Bi_2Se_3 in a magnetic field: Infrared evidence for magnetoelectric coupling in a topological insulator material," *Phys. Rev. B.* **81**, 125120 (2010).

16. P. D. Pietro et al., "Optical conductivity of bismuth-based topological insulators," *Phys. Rev. B.* **86**, 045439 (2012).

17. C. Martin et al., "Bulk Fermi surface and electronic properties of $Cu_{0.07}Bi_2Se_3$," *Phys. Rev. B.* **87**, 201201(R) (2013).

18. A. A. Reijnders et al., "Optical evidence of surface state suppression in Bi-based topological insulators," *Phys. Rev. B.* **89**, 075138 (2014).

19. K. F. Mak, L. Ju, F. Wang, and T. F. Heinza, "Optical spectroscopy of graphene: From the far infrared to the ultraviolet," *Solid State Commun.* **152**, 1341–1349 (2012).

20. L. A. Falkovsky, "Optical properties of graphene," *J. Phys.: Conf. Ser.* **129**, 012004 (2008).

21. Y. Yao et al., "Electrically tunable metasurface perfect absorbers for ultrathin midinfrared optical modulators," *Nano Lett.* **14**, 6526–6532 (2014).

22. Y. Wang et al., "Observation of ultrahigh mobility surface states in a topological crystalline insulator by infrared spectroscopy," *Nat. Commun.* **8**, 366 (2017).

23. H. Shirai, T.-T. Yeh, Y. Nomura, C.-W. Luo, and T. Fuji, "Ultrabroadband midinfrared pump-probe spectroscopy using chirped-pulse up-conversion in gases," *Phys. Rev. Appl.* **3**, 051002 (2015).

24. G. D. Smith and R. A. Palmer, in *Handbook of Vibrational Spectroscopy*, edited by J. M. Chalmers and P. R. Griffiths, Vol. 1, J. Wiley and Sons, New York (2006).

25. T.-T. Yeh et al., "Ultrafast carrier dynamics in Ge by ultra-broadband mid-infrared probe spectroscopy," *Sci. Rep.* **7**, 40492 (2017).

26. C.-M. Tu et al., "Manifestation of a Second Dirac Surface State and Bulk Bands in THz Radiation from Topological Insulators," *Sci. Rep.* **5**, 14128 (2015).

27. J. Sánchez-Barriga et al., "Laser-induced persistent photovoltage on the surface of a ternary topological insulator at room temperature," *Appl. Phys. Lett.* **110**, 141605 (2017).

28. A. Sterzi et al., "Bulk diffusive relaxation mechanisms in optically excited topological insulators," *Phys. Rev. B.* **95**, 115431 (2017).

29. T. Yoshikawa et al., "Photovoltage on the surface of topological insulator via optical aging," *Appl. Phys. Lett.* **112**, 192104 (2018).

30. Z.-H. Zhu et al., "Rashba spin-splitting control at the surface of the topological insulator Bi_2Se_3," *Phys. Rev. Lett.* **107**, 186405 (2011).

31. B. Zhou et al., "Controlling the carriers of topological insulators by bulk and surface doping," *Semicond. Sci. Technol.* **27**, 12 (2012).

32. Y. Nomura et al., "Single-shot detection of mid-infrared spectra by chirped-pulse upconversion with four-wave difference frequency generation in gases," *Opt. Express.* **21**, 18249–18254 (2013).

33. T. Fuji, H. Shirai, and Y. Nomura, "Ultrabroadband mid-infrared spectroscopy with four-wave difference frequency generation," *J. Opt.* **17**, 094004 (2015).

34. A. Kuzmenko, "RefFIT." http://optics.unige.ch/alexey/reffit.html (accessed 23 May 2012) (2016).

35. T.-T. Yeh, C.-M. Tu, W.-H. Lin, C.-M. Cheng, W.-Y. Tzeng, C.-Y. Chang, H. Shirai, T. Fuji, R. Sankar, F.-C. Chou, M. M. Gospodinov, T. Kobayashi, and C.-W. Luo, "Femtosecond time-evolution of mid-infrared spectral line shapes of Dirac fermions in topological insulators," *Sci. Rep.* **10**, 9803 (2020). https://doi.org/10.1038/s41598-020-66720-4.0.9803.

Section 10

Conductors and Superconductors

Section 10.1

Super Conductors

10.1.1 Dichotomy of Photoinduced Quasiparticle on CuO_2 Planes of $YB_2Cu_3O_7$ Directly Revealed by Femtosecond Polarization Spectroscopy

10.1.1.1 INTRODUCTION

Optical characterizations have become an important tool in modern materials research. For instance, polarized light may be used to investigate the anisotropic optical responses in anisotropic materials [1]. Recently, a great deal of research related to the dichotomy phenomena of quasiparticles (QP) has been reported and discussed in the community of strongly correlated electrons. By the standard time-resolved measurements, the dichotomy between coherent nodal QP excitations and incoherent antinodal excitations has been associated with the abrupt change in the sign of transient reflectivity R and the kinetics of QP decay in $Bi_2Sr_2Ca_{1-y}Dy_yCu_2O_{8+\delta}$ crystal [2]. However, the conclusions of dichotomy in $Bi_2Sr_2Ca_{1-y}Dy_yCu_2O_{8+\delta}$ crystal are not directly from femtosecond spectroscopy. In addition, it is extremely difficult to measure the nodal characteristics in a small size single crystal by optical light. A natural alternative is to use thin film for such investigations. Here we demonstrate a method that combines both polarized femtosecond spectroscopy and thin films with specific orientations to reveal the dichotomy of photoinduced QP dynamics directly.

10.1.1.2 EXPERIMENT

To study the dichotomy of QP dynamics in the strongly correlated electron system, the polarization of pulses in the nearly collinear polarized pump-probe scheme (as shown in Figure 10.1.1.1) should be controlled to probe the optical responses along each axis individually. The femtosecond pulses from a mode-locked Ti:sapphire laser, which produced a 75 MHz train of 20 fs pulses with a central wavelength of 800 nm, were prechirped via two prisms and split into two parts i.e., pump and probe beams by a beam splitter. One of both was modulated at 87 kHz by an acousto-optic modulator AOM and served as a pump beam. The intensity and polarization electric field, **E** of pulses can be adjusted by a $\lambda/2$ plate and a polarizer. The reflective intensity changes ΔR and the reflective intensity (R) of the probe beam were, respectively, measured via a lock-in amplifier and a multimeter as a function of delay time t. Moreover, the $\Delta R/R(t, \phi_{pump}, \phi_{probe}, \theta)$ curves along various directions on the surface of a sample can be obtained by rotating the polarization of pulses at nearly normal incidence $0°$. If the polarization of pulses is perpendicular to the c axis of 110 films see the inset of Figure 10.1.1.1, one is able to measure the responses $\Delta R/R(t, 90, 90, \theta)$ along various directions on the ab plane by changing

DOI: 10.1201/9780429196577-65

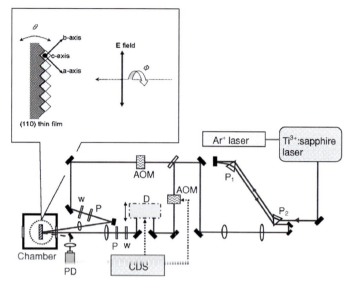

FIGURE 10.1.1.1 The experimental setup for pump-probe spectroscopy. Code: AOM: acousto-optic modulator. P: polarizer. w: $\lambda/2$ plate. PD: photodiode. D: delay stage. CDS: control and detection system. P_1, P_2: prism pair for pulse compression. Both solid and dashed lines represent the laser beam paths and the dotted lines stand for the electrical signal connection. Inset: ϕ is the angle between the c-axis of the samples and the polarization of the pump (or probe) pulses. θ is the angle between the surface of samples and the polarization of the pump (or probe) pulses.

the angle. Thus, the three-dimensional polarization-dependent (or orientation-resolved femtosecond) time-resolved spectroscopy in the layered-structure materials could be doubtless carried out. Under this idea, the typical layered-structure material, $YBa_2Cu_3O_7$ (YBCO), was the testing sample in this study. In addition, three types of oriented YBCO thin films, i.e., (001), (100), and (110), were prepared by pulsed laser deposition. For (001), (100), and (110) YBCO films, the zero resistance transition temperature (T_c) are 90.2, 89.7, and 88.2 K, respectively. All of them are well-characterized oriented thin films with >97% in-plane alignment, of which detail could be found in Refs. [3–5].

10.1.1.3   RESULTS AND DISCUSSION

The method combining both polarized femtosecond spectroscopy and specific thin films with various orientations could be simply illustrated in Figure 10.1.1.2. Here we consider some charges C_a, C_b, and C_{ab} located along the a-axis, b-axis, and ab-diagonal in a cubic lattice structure, respectively. Under a proper driving force (e.g., the electric field of light), each charge could only move along one specific direction noted by its suffix. For the electric field $\mathbf{E}_a$ which is parallel with the a-axis and incident along the c axis [i.e., $\mathbf{k}$ in Figure 10.1.1.2a], the charge C_a will be driven and the charge C_{ab} will also be driven by the decomposed part of E_a which is parallel with the ab diagonal. The responses from other polarizations in Figure 10.1.1.2 are summarized in Table 10.1.1.1.

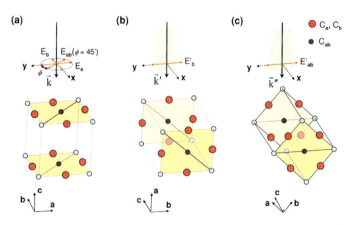

FIGURE 10.1.1.2 (a–c) Schematic illustrations of the geometric relation between the polarized light and the two-dimensional 2D charges in an anisotropic structure. C_a, C_b, and C_{ab} are the charges along the a-axis, b-axis, and ab-diagonal, respectively. $\mathbf{k}$, $\mathbf{k}'$, and $\mathbf{k}''$ are the wave vectors of the polarized light. ϕ is the angle between the y-axis and $\mathbf{E}$ field.

TABLE 10.1.1.1

Driven Charges under Various Polarization Configurations in Figure 10.1.1.2a

Polarization of Incident Light	Driven Charges
E_a	C_a, C_{ab}
E_b	C_b, C_{ab}
E_{ab}	C_a, C_b, C_{ab}
E'_b	C_b
E'_{ab}	C_{ab}

According to the results in Table 10.1.1.1, there is no way to get the pure response of each type of charge via the simple configuration in Figure 10.1.1.2a, i.e., the propagating direction of incident light is perpendicular to the plane to be investigated. For example, the inset of Figure 10.1.1.3 clearly demonstrates that the responses on the twin-free (001) YBCO thin films are polarization-independent. The $\Delta R/R$ measured at various polarization angles ($\phi = \phi_{pump} = \phi_{probe} = 20°$, $50°$, and $90°$) are essentially identical. In this measuring configuration, the **E** field on ab-plane can be decomposed into two parts that are along the a-axis and b-axis, respectively. Therefore, the average responses of a-axis, b-axis, and ab-diagonal are always observed in (001) YBCO thin films, whether the polarization direction is parallel to the $a(b)$-axis or not. Furthermore, the usual (001) YBCO thin films often contain significant twins, making it difficult to resolve the intrinsic properties along respective crystalline orientation. Thus, there is only one way to obtain the pure responses of charge C_b by the configuration in Figure 10.1.1.2b or charge C_{ab} by the configuration in Figure 10.1.1.2c. For the direction of incident light **k'** or **k''**, the propagating direction of the EM fields must lie along either the a/b axis or the ab diagonal so that the **E** field, which is perpendicular to the propagating direction, cannot be decomposed into the components along any directions on the ab plane [i.e., **E** field cannot be decomposed on the

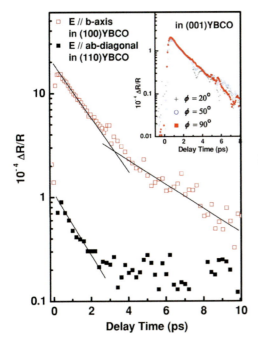

FIGURE 10.1.1.3 $\Delta R/R$ semilogarithmic plots of vs pump-probe delay time at 60 K on various crystalline orientations. The $\Delta R/R$ signal in the $E\|b$-axis configuration was measured for (100) YBCO films by the polarized light with the wave vector k shown in Figure 10.1.1.2b. The $\Delta R/R$ signal in the $E\|ab$-diagonal configuration was measured for (110) YBCO films by the polarized light with the wave vector **k''** shown in Figure 10.1.1.2c. The solid lines are guides to the eye emphasizing the relaxation behavior of the photoinduced carriers along various crystalline orientations. Inset: $\Delta R/R$ signal measured at 70 K for a (001)YBCO thin film in several configurations with different angles ϕ between the y axis and **E** field [Figure 10.1.1.2a] presented on a semilogarithmic scale.

shaded plane in Figure 10.1.1.2b and c]. Therefore, the E_b field of incident light with propagating direction $\mathbf{k}'$ could only drive the charge C_b. Similarly, the E_{ab}' field of incident light with propagating direction $\mathbf{k}''$ could only drive the charge C_{ab}.

As shown in Figure 10.1.1.3, the photoinduced $\Delta R/R$ responses along the b axis (or antinodal direction) and ab diagonal (or nodal direction) have been directly measured in the (100) and (110) YBCO thin films, respectively. It is extremely difficult to perform the same measurements in (001) thin films or single crystals. The $\Delta R/R$ along the b axis is dramatically distinct from the other one along the ab diagonal not only in the amplitude but also in the relaxation dynamics. Two relaxation processes can be definitely observed along the b axis. Conversely, the slower relaxation process which could be associated with a generic manifestation of the superconducting gap opening [6] is absent along the ab diagonal (i.e., the nodal direction) within our experimental resolution. This disappearance of a QP relaxation bottleneck may be due to the complete shrinking of superconducting gap along the nodal direction. Furthermore, the amplitude difference of $\Delta R/R$ is about 5–10 times larger in the b axis than that in the ab-diagonal direction. These results evidently demonstrate that the dichotomy of QP relaxation between the b axis and ab diagonal.

The data displayed in Figure 10.1.1.4 depict the amplitude evolution of the normalized $\Delta R(t-0)/R$ as a function of the reduced temperature (T/T_c) along various crystalline orientations on the ab plane of YBCO. For the b and a axis, the amplitude of $\Delta R(t=0)/R$ dramatically changes near T_c (dashed circle in Figure 10.1.1.4), which suggests the opening of the superconducting gap. However, there is no such kind of sharp boundary along the ab diagonal. The monotonic temperature evolution of $\Delta R(t=0)/R$ along the nodal direction may be dominated by the dynamics of QP thermalization [7] or recombination [8]. This result strongly suggests the behaviors of photoinduced QPs between nodal and antinodal directions are markedly different and obviously indicate that the symmetry of superconducting gap in the YBCO superconductors is d-wave symmetry. Moreover, this dichotomy phenomenon is consistent with the observations by other experimental methods. For example, the angle-resolved photoemission spectroscopy evidence that a nodal-antinodal dichotomous character does not only exist in the cuprate superconductors, e.g., underdoped $(La_{2-x}Sr_x)CuO_4$ [9] and lightly doped $Ca_{2-x}Na_xCuO_2Cl_2$, [10] but also in the colossal magnetoresistive bilayer manganite $La_{1.2}Sr_{1.8}Mn_2O_7$ [11].

In summary, we have demonstrated a method which combines both polarized femtosecond spectroscopy and thin film preparation with specific orientations to directly reveal the dichotomy of photoinduced QPs on the CuO_2 planes of YBCO. The combined method will be a promising way to resolve the photoinduced structural phonon structures [12].

The research described in the present section is performed by the following people: C. W. Luo, L. Y. Chen, Y. H. Lee, K. H. Wu, J. Y. Juang, T. M. Uen, J.-Y. Lin, Y. S. Gou, and T. Kobayashi [12].

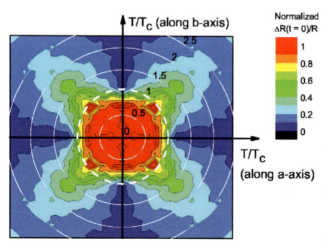

FIGURE 10.1.1.4 Color online 2D distribution of the amplitude of $\Delta R(t=0)/R$ as a function of the reduced temperature T/T_c on the ab-plane of YBCO. Inside the dashed circle is the superconducting zone where the temperature T is below T_c.

REFERENCES

1. S. Tajima, G. D. Gu, S. Miyamoto, A. Odagawa, and N. Koshizuka, *Phys. Rev. B* **48**, 16164 (1993).
2. N. Gedik, M. Langner, J. Orenstein, S. Ono, Y. Abe, and Y. Ando, *Phys. Rev. Lett.* **95**, 117005 (2005).
3. C. W. Luo, M. H. Chen, S. J. Liu, K. H. Wu, J. Y. Juang, T. M. Uen, J.-Y. Lin, J.-M. Chen, and Y. S. Gou, *J. Appl. Phys.* **94**, 3648 (2003).
4. S. J. Liu, J. Y. Juang, K. H. Wu, T. M. Uen, Y. S. Gou, J. M. Chen, and J.-Y. Lin, *J. Appl. Phys.* **93**, 2834 (2003).
5. C. W. Luo, M. H. Chen, C. C. Chiu, K. H. Wu, J. Y. Juang, T. M. Uen, J.-Y. Lin, and Y. S. Gou, *J. Low Temp. Phys.* **131**, 545 (2003).
6. J. Demsar, B. Podobnik, V. V. Kabanov, Th. Wolf, and D. Mihailovic, *Phys. Rev. Lett.* **82**, 4918 (1999).
7. G. P. Segre, N. Gedik, J. Orenstein, D. A. Bonn, R. Liang, and W. N. Hardy, *Phys. Rev. Lett.* **88**, 137001 (2002).
8. J. Demsar, R. D. Averitt, V. V. Kabanov, and D. Mihailovic, *Phys. Rev. Lett.* **91**, 169701 (2003).
9. X. J. Zhou, T. Yoshida, D.-H. Lee, W. L. Yang, V. Brouet, F. Zhou, W. X. Ti, J. W. Xiong, Z. X. Zhao, T. Sasagawa, T. Kakeshita, H. Eisaki, S. Uchida, A. Fujimori, Z. Hussain, and Z.-X. Shen, *Phys. Rev. Lett.* **92**, 187001 (2004).
10. M. K. Shen, F. Ronning, D. H. Lu, F. Baumberger, N. J. C. Ingle, W. S. Lee, W. Meevasana, Y. Kohsaka, M. Azuma, M. Takano, H. Takagi, and Z.-X. Shen, *Science* **307**, 901 (2005).
11. N. Mannella, W. L. Yang, X. J. Zhou, H. Zheng, J. F. Mitchell, J. Zaanen, T. P. Devereaux, N. Nagaosa, Z. Hussain, and Z.-X. Shen, *Nature* **438**, 474 (2005).
12. C. W. Luo, L. Y. Chen, Y. H. Lee, K. H. Wu, J. Y. Juang, and T. M. Uen, J.-Y. Lin, Y. S. Gou, and T. Kobayashi, *J. Appl. Phys.* **102**, 033909 (2007).

10.1.2 Ultrafast Dynamics and Phonon Softening in $Fe_{1+y}Se_{1-x}Te_x$ Single Crystals

10.1.2.1 INTRODUCTION

In 2008, the discovery of the superconductors $LaFeAsO_{1-x}F_x$ with a T_c around 26K at ambient pressure [1] initiated investigations on the diversified family of Fe-based pnictides. Shortly afterwards, the T_c was further increased to 56K by substituting La with other rare earths [2]. Since then, other Fe-based superconductors (FeSCs) have been successively found, including $Ba_{1-x}K_xAs_2Fe_2$ (122-type) with $T_c \leqq 38K$ [3], LiFeAs (111-type) with T_c $6 \leqq 18K$ [4] and FeSe (11-type) with $T_c \leqq$ 10K [5]. In these new FeSCs, the interplay between electronic structure, phonons, magnetism and superconductivity is very rich and would help us understand the origin of high-T_c superconductivity. Among various FeSCs, however, the iron chalcogenide FeSe [5] stands out because of its structural simplicity, which consists of iron-chalcogenide layers stacking one another with the same Fe^{2+} charge state as the iron pnictides. Additionally, the T_c of FeSe has been increased further to 37K at 7GPa [6]; meanwhile, the partial replacement of Se with Te in FeSe also yields $T_c \sim 14K$ [7], which stimulated much interest in the properties of $FeSe_{1-x}Te_x$.

Recently, the existence of precursor superconductivity above T_c that competes with the spin-density wave order [8] and a pseudogap-like feature with onset around 200K [9] were, respectively, observed on underdoped (Ba, K)Fe_2As_2 and nearly optimally doped $SmFeAsO_{0.8}F_{0.2}$ by optical pump-probe studies. Moreover, a coherent lattice oscillation was also found in Co-doped $BaFe_2As_2$ using time-resolved pump-probe reflectivity with 40fs time resolution [10]. These results have unambiguously shown that femtosecond pump-probe spectroscopy is a protocol to study the simultaneous presence of electrons, phonons, and magnons and the interactions between them.

Therefore, further studies of the quasiparticle (QP) dynamics in FeSCs and its evolution with time and temperature are indispensable for understanding the mechanism of high-T_c superconductivity in FeSCs. In this paper, we report the time-resolved femtosecond spectroscopy study of the $Fe_{1+y}Se_{1-x}Te_x$ single crystals to elucidate the electronic structure and the QP dynamics.

10.1.2.2 EXPERIMENTS

In this study, $Fe_{1.14}Te$ and $Fe_{1.05}Se_{0.2}Te_{0.8}$ single crystals were grown with an optical zone melting technique [11]. The FeSe single crystals were grown in evacuated quartz ampoules using a KCl/$AlCl_3$ flux [12]. The crystalline structure of the samples was examined by x-ray diffraction. The magnetic properties were obtained by temperature dependence of the magnetic susceptibility $\chi(T)$ as shown in the insets of Figure 10.1.2.1. The superconducting transition temperatures of FeSe and $Fe_{1.05}Se_{0.2}Te_{0.8}$ are 8.8 and 10K, respectively. For non-superconductive $Fe_{1.14}Te$, pronounced anomalies can be seen at 125K in the inset of Figure 10.1.2.1c. Below this temperature, $\chi(T)$ exhibits clear irreversibility between zero-field cooling (ZFC) and field cooling (FC) magnetization data. This may be due to the magnetite (Fe_3O_4) impurities and related to the Verwey transition, which is observed in magnetite at 120–125K [13]. According to recent neutron-diffraction experiments in

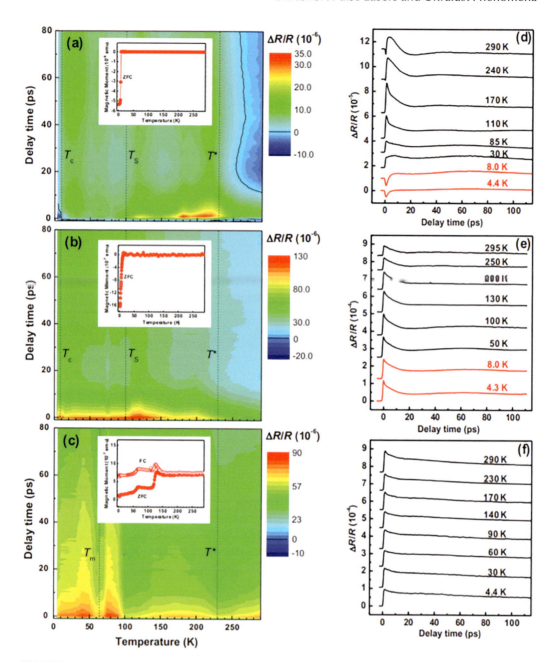

FIGURE 10.1.2.1 Temperature and delay time dependence of the two-dimensional (2D) $\Delta R/R$ in (a) FeSe, (b) $Fe_{1.05}Se_{0.2}Te_{0.8}$ and (c) $Fe_{1.14}Te$ single crystals. Inset show the temperature dependence of the magnetic susceptibility for (a) FeSe in $H = 20$ Oe, (b) $Fe_{1.05}Se_{0.2}Te_{0.8}$ in $H = 80$ Oe and (c) $Fe_{1.14}Te$ in $H = 80$ Oe. Panels (d)–(f) are the selected $\Delta R/R$ at some typical temperatures in panels (a), (b) and (c), respectively.

$Fe_{1+y}Te$ [14], the significant drop at 65K (T_m) corresponds to an antiferromagnetic (AFM) ordering with a rather complex magnetic structure and to a simultaneous structural transition from tetragonal $P4/nmm$ symmetry to either monoclinic $P2_1/m$ or orthorhombic $Pmmn$ symmetry.

The femtosecond spectroscopy measurement was carried out using a dual-color pump-probe system (for the laser light source, the repetition rate is 5.2 MHz, the wavelength is 800 nm and the pulse duration is 100 fs) and an avalanche photodetector with the standard lock-in technique. The fluences of the pump beam and the probe beam are 2.48 and 0.35 μJ/cm², respectively. The pump pulses have the corresponding photon energy (3.1 eV) where the higher absorption occurred in the absorption spectrum of FeSe [15] and hence can generate electronic excitations. The photo-induced QP dynamics is studied by measuring the photoinduced transient reflectivity changes ($\Delta R/R$) of a probe beam with a photon energy of 1.55 eV.

10.1.2.3 TEMPERATURE-DEPENDENT $\Delta R/R$

Figure 10.1.2.1 shows the 2D $\Delta R/R$ taken in $Fe_{1+y}Se_{1-x}Te_x$ single crystals. For the case of FeSe, there appear four temperature regions. Above 230K, T^* (region I), there is a fast negative response with a relaxation time of about 1.5 ps together with a periodic oscillation in which the minima occur at $\sim$21 and $\sim$125 ps, respectively. When the temperature decreases below 230K (region II), a positive and slow response appears and $\Delta R/R$ gradually becomes smaller until $T = 90K$ (T_s). Below 90K (region III), the slow positive response disappears and is replaced by a complicated mixture of the positive and negative components as discussed later. For $T < T_c$ (region IV), a long-lived negative response appears like the one in region I.

Qualitatively similar features were also observed in a $Fe_{1.05}Se_{0.2}Te_{0.8}$ single crystal as shown in Figure 10.1.2.1b. However, the negative oscillations above T^* were smeared due to the doping of Te and completely disappear on a fully Te-doped sample of $Fe_{1.14}Te$ as shown in Figure 10.1.2.1c. Additionally, in an $Fe_{1.05}Se_{0.2}Te_{0.8}$ single crystal (Figure 10.1.2.1b) the positive $\Delta R/R$ becomes larger in amplitude with decreasing temperature. This temperature-dependent positive $\Delta R/R$ also markedly shows anomalies at 125 and 65K in Figure 10.1.2.1c of an $Fe_{1.14}Te$ single crystal, which are associated with the appearance of Fe_3O_4 impurities [13] and the magnetic phase transition [14,16] as shown in the inset of Figure 10.1.2.1c, respectively.

In the pump-probe measurements, the electronic excitations generated by the pump pulses result in a swift rise of $\Delta R/R$ at zero time delay as shown in Figure 10.1.2.2. The observed excitation was triggered by transferring the electrons from d valence band of Fe to d conduction band of Fe [17]. At zero time delay, the number of excited electrons generated by this non-thermal process is related to the amplitude of $\Delta R/R$. These high-energy electrons accumulated in the d conduction band of Fe release their energy through the emission of longitudinal-optical (LO) phonons within several picoseconds [18]. The LO phonons further decay into longitudinal acoustic (LA) phonons via anharmonic interactions, i.e. transferring energy to the lattice. These relaxation processes can be detected using a probe beam as shown in Figures 10.1.2.1 and 10.1.2.2. In the two-temperature model, the electrons and phonons (or lattice) are in thermal quasi-equilibrium with two different time-dependent temperatures T_e and T_l. After the excitation of pump pulses, the increase of electron temperature is dramatically larger than that of phonons (T_e can reach several thousands of Kelvin above T_l) because of the much smaller heat capacity in the electron subsystem. Then, both subsystem temperatures of electrons and phonons will become equal through electron-phonon coupling. Namely, the T_e decreases with a timescale of sub-ps to ps by transferring energy to phonons [19]. Following that, the T_l will decrease with a time scale of several ps to several hundreds of ps by phonon population decay (inelastic scattering) or dephasing (elastic scattering) [19]. According to the two-temperature model, relaxation processes ($t>0$) of $\Delta R/R$ in $Fe_{1+y}Se_{1-x}Te_x$ single crystals can be phenomenologically described by lines are the fitting curves using Eq. (10.1.2.1). Solid lines are the fitting curves without the oscillation component in Eq. (10.1.2.1). Insets show the oscillation component (subtract the solid line from the open circles) and their Fourier transformation.

$$\frac{\Delta R}{R} = A_e e^{-t/\tau_e} + A_{LO}e^{-t/\tau_{LO}} + A_0 + A_{LA}e^{-t/\tau_{LA}} \sin\left(\frac{2\pi t}{T(t)} + \phi\right) \quad (10.1.2.1)$$

The first term on the right-hand side of Eq. (10.1.2.1) is the decay of A_e with a relaxation time τ_e, which is proportional to the initial excited electron (photoexcited QP) population number per unit cell [20]. In the second term, A_{LO} is proportional to the high-energy phonon population number per unit cell and decay with the relaxation time τ_{LO}. The third term describes energy loss from the

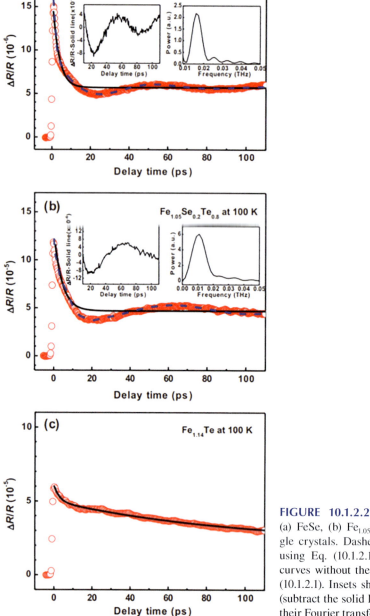

FIGURE 10.1.2.2 Selected $\Delta R/R$ curves for (a) FeSe, (b) $Fe_{1.05}Se_{0.2}Te_{0.8}$ and (c) $Fe_{1.14}Te$ single crystals. Dashed lines are the fitting curves using Eq. (10.1.2.1). Solid lines are the fitting curves without the oscillation component in Eq. (10.1.2.1). Insets show the oscillation component (subtract the solid line from the open circles) and their Fourier transformation.

hot spot to the ambient environment within the time scale of a microsecond, which is far longer than the period of the measurement (~150 ps) and hence is taken as a constant. The last term is the chirped oscillation component associated with strain pulse propagation[1]: A_{LA} is the amplitude of the

---

[1] In the displacive excitation of coherent phonon (DECP for absorbing media) mechanism [21], photoexcitation induces changes in the electronic energy distribution function, and consequently, the crystal lattice starts to oscillate around the new equilibrium position $A_0(t)$, which is proportional to the photoexcited carrier density $n(t)$. In the first order, only the A_{1g} totally symmetric modes are currently excited by the DECP mechanism. In 2002, Stevens et al. [22] further solved the equation of motion for an LO vibrational mode to obtain the coherent phonon amplitude (or population), $A_{LO} \propto Im (\varepsilon)$. Namely, the changes in phonon population A_{LO} cause a variation of the imaginary part of the dielectric constant ε. Then, changes in the imaginary part of ε vary the refractive index and cause further changes in reflectivity (R) in materials. Therefore, the $\Delta R/R$ in FeSe is proportional to the population of LO phonons, $\Delta R/R \propto A_{LO}$.

oscillation; τ_{LA} is the damping time; $T(t)$ is the time-dependent period; and ϕ is the initial phase of the oscillation.

Like the dashed lines in Figures 10.1.2.2a and b, Eq. (10.1.2.1) can fit the $\Delta R/R$ data very well in FeSe and $Fe_{1.05}Se_{0.2}Te_{0.8}$ single crystals. However, the fitting for $\Delta R/R$ in $Fe_{1.14}Te$ has almost no need for the oscillation component in Eq. (10.1.2.1), as shown in Figure 10.1.2.2c. Consequently, each component in $\Delta R/R$ described above can be extracted using Eq. (10.1.2.1). The results of the extraction are shown in Figure 10.1.2.3. For the negative and fast component (A_e) of $\Delta R/R$ only observed at $T > 230$K and $T < 90$K, it gradually increases as T decreases from 90K; meanwhile, it also suppresses the positive and fast component that appeared at 100–200K in Figure 10.1.2.1d. This trend is closely related to the strong AFM spin fluctuations below $T = T_s$ as revealed by ^{77}Se NMR measurements [23]. The relaxation of QP associated with the spin fluctuations between T_c and 90K (in Figure 10.1.2.3b) is ~1.5–2 ps, which is almost temperature-independent. Intriguingly, as $T < T_c$, A_e dramatically shrinks as shown in the inset of Figure 10.1.2.3a; meanwhile, the QP relaxation time increases rapidly as shown in the inset of Figure 10.1.2.3b. Correspondingly, the spin–lattice relaxation rate $1/T_1$ also decreases rapidly due to the onset of superconductivity [23]. These certainly indicate that the growth of A_e associated with spin fluctuations at low temperatures is suppressed by the appearance of superconductivity. Thus, spin fluctuations and superconductivity are competing factors in the FeSe system. The above results provide strong experimental evidence of competing orders in FeSe, which are consistent with the theoretical calculations [24]. It is noted that experimental evidence on the competing orders was also reported in the underdoped (Ba, K)Fe$_2$As$_2$ system [8]. The presence of a gap in the QP density of states gives rise to a bottleneck in carrier relaxation, which is clearly observed in the relaxation time τ_e close to T_c (see the inset of Figure 10.1.2.3b). For the slower component (solid circles in

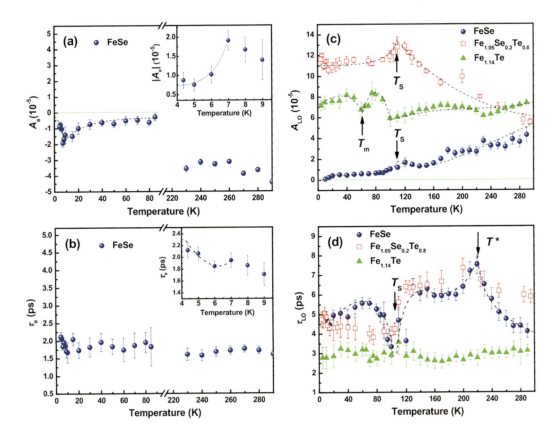

FIGURE 10.1.2.3 Temperature dependence of the amplitude (a) A_e, (c) A_{LO} and the relaxation time (b) τ_e, (d) τ_{LO} by fitting Eq. (10.1.2.1). Insets of (a) and (b) show a part of the temperature-dependent A_e and τ_{LO} on an enlarged scale. Dashed lines are a guide to the eyes.

Figure 10.1.2.3c), the amplitude A_{LO} monotonically decreases as T decreases except for a small kink at T_s and then completely disappears in the superconducting state. In contrast, the relaxation time τ_{LO} (solid circles in Figure 10.1.2.3d) exhibits two marked anomalies at both $T^* \sim 230K$ and $T_s \sim 90K$. According to the scenario of relaxation processes described in Ref. [19] (see footnote 1), τ_{LO} is the relaxation time of LO phonon population via anharmonic decay into LA phonons.

The divergences of temperature-dependent τ_{LO} at $T_s \sim 90K$ and $T^* \sim 230K$ imply a possibly efficient way or a bottleneck for the energy relaxation from LO phonons to LA phonons. If we take a look at the temperature-dependent LA phonon energy in Figure 10.1.2.5, we only find one significant drop around 90K, where the structural transition from the tetragonal phase to the orthorhombic phase occurs [25]. The smaller LA phonon energy causes the energy of LO phonons to release to LA phonons more efficiently and results in a significantly shorter relaxation time τ_{LO} around 90K. Moreover, the structural transition also leads to the more fluctuant τ_{LO} around 90K. For an abnormally long relaxation time τ_{LO} of around 230K, however, we cannot find any significant corresponding changes in the temperature-dependent LA phonon energy of Figure 10.1.2.5. Thus, this bottleneck in the LO phonon energy transfer is possibly due to the photo-induced QPs rather than the LA phonons. As shown in Figure 10.1.2.3a, the sudden disappearance of A_r below 230K may undermine the energy release efficiency of LO phonons. The sign change of the Seebeck coefficient of FeSe possibly due to an elusive higher-temperature phase transition was found to be also at T^* [26].

All the above anomalies of A_{LO} and τ_{LO} in FeSe were also found in superconductive $Fe_{1.05}Se_{0.2}Te_{0.8}$, but almost disappear in non-superconductive $Fe_{1.14}Te$. For the case of $Fe_{1.14}Te$, we only observed the abnormal changes of A_{LO} at $T_m \sim 65K$ and near 125K, which were caused by the magnetic phase transition and the Verwey transition as shown in the inset of Figure 10.1.2.1c, respectively. This implies that the anomalies of A_{LO} and τ_{LO} at $T_s \sim 90K$ and $T^* \sim 230K$ may be associated with the superconductivity in FeSCs. Namely, both phase transition at $T^* \sim 230K$ and $T_s \sim 90K$ would be the key effect to cause superconductivity at low temperature [27].

10.1.2.4   ELECTRON–OPTICAL PHONON COUPLING STRENGTH

By fitting the $1R/R$ curves with Eq. (10.1.2.1), dynamic information on QPs and phonons is available, which includes the number of QPs, the relaxation time of QPs and the energy of phonons. In a metal, the photo-induced QPs relaxation time is governed by transfer of energy from electrons to phonons with electron–phonon coupling strength λ [28]:

$$\frac{1}{\tau_e} = \frac{3\hbar\lambda\langle\omega^2\rangle}{\pi k_B T_e} \qquad (10.1.2.2)$$

where $\lambda\langle\omega^2\rangle$ is the second moment of the Eliashberg function and T_e can be further described by [29]

$$T_e = \left\langle \sqrt{T_i^2 + \frac{2(1-R)F}{l_s\gamma}e^{-z/l_s}} \right\rangle \qquad (10.1.2.3)$$

where T_i is the initial temperature of electrons, R is the unperturbed reflectivity at 400 nm, F is the pumping fluence and γ is the linear coefficient of heat capacity due to the electronic subsystem. The mean value is taken for the depth z going from the crystal surface down to the skin depth $l_s \sim 24$ nm (it was estimated from the skin depth of an electromagnetic wave in a metal, $\lambda/4\pi k$). All the parameters for the calculation of electron–phonon coupling strength are listed in Table 10.1.2.1. For the estimation of $\langle\omega^2\rangle$, some vibrational modes are more efficiently coupled to QPs than others. In the case of Co-doped $BaFe_2As_2$, the symmetric A_{1g} mode is coherently excited by photoexcitation and efficiently coupled [10]. Consequently, we take the A_{1g} mode into account in the present case of $Fe_{1+y}Se_{1-x}Te_x$, which is the strongest phonon mode in the electron–phonon spectral function, $\alpha^2F(\omega)$

TABLE 10.1.2.1

The Parameters for Estimating at $T = 20$ K of the $Fe_{1+y}Se_{1-x}Te_x$ Single Crystals

	T_c (K)	R (400 nm)	F (μJ/cm^{-2})	γ (mJ/(mol K^2))	τ_e^a (ps)	A_{1g}^b (meV)	$\lambda\langle\omega^2\rangle^c$ (meV^2)	λ
FeSe	8.8	0.25	9.92	5.73	1.75	19.9	61.3	0.16
$Fe_{1.05}Se_{0.2}Te_{0.8}$	10	0.20	15.76	57.5	4.32	20.0	4.5	0.01
$Fe_{1.14}Te$	–	0.11	17.36	32.0	2.86	19.7	4.0	0.01

^{a} From Refs. [12,33].
^{b} From Refs. [34-36].
^{c} Obtained from Eq. (10.1.2.2).

[17]. By Eq. (10.1.2.2), the consequent electron–phonon (A_{1g} mode) coupling strength, $\lambda = 0.16$, in FeSe. This value is consistent with the theoretical results of $\lambda = 0.17$ [17] obtained by using a linear response within the generalized gradient approximation. For the case of $Fe_{1.05}Se_{0.2}Te_{0.8}$, we obtained $\lambda = 0.01$, which is smaller than the value of $\lambda = 0.16$ in FeSe even if it possesses a higher $T_c \sim 10$K. Furthermore, we can use the McMillan formula [30], $T_c = (\langle\omega\rangle/1.2)\exp\{-[1.40(1+\lambda)]/[\lambda-\mu^*(1+0.62\lambda)]\}$, to evaluate the critical temperature T_c. Taking $\langle\hbar\omega\rangle = 19.9$ meV and $\mu^* = 0$ [31], we obtain $T_c \sim 0.08$K for FeSe and ~ 0K for $Fe_{1.05}Se_{0.2}Te_{0.8}$, which are far below the actual T_c of about 8.8 and 10K, respectively. Therefore, the electron pairing mechanism in $Fe_{1+y}Se_{1-x}Te_x$ cannot be explained only by the electron–phonon interactions. Recently, the electron–phonon coupling strength of $\lambda \sim 0.12$ was measured in the $Ba(Fe_{0.92}Co_{0.08})_2As_2$ system, which is also too small to sustain its T_c of 24K [32]. These results strongly imply that a phonon-mediated process cannot be the only mechanism leading to the formation of superconducting pairs in FeSCs.

10.1.2.5   ACOUSTIC PHONON SOFTENING

Further insight into the phase transition observed at $\sim$90 and $\sim$230K in $Fe_{1+y}Se_{1-x}Te_x$ is provided by the study of the oscillation component of $\Delta R/R$. The temperature dependence of a strain pulse (LA phonons) propagation was clearly observed in the oscillation feature of $\Delta R/R$ after subtracting the decay background (i.e. the first, second and third terms in Eq. (10.1.2.1) and the solid lines in Figure 10.1.2.2), as shown in Figures 10.1.2.2a and b and 10.1.2.4. This oscillation is caused by the propagation of strain pulses inside $Fe_{1+y}Se_{1-x}Te_x$ single crystals, namely the interference between the probe beams reflected from the crystal surface and the wave front of the propagating strain pulse [37]. At high temperatures, the damping time is very short and the oscillation is sustained only for one period. However, the number of oscillation periods markedly increases around 100K in FeSe (see Figure 10.1.2.4a); hence, the damping time becomes much longer. Besides, the oscillation period significantly increases below 90K. This means that the LA phonons can propagate further into the interior of FeSe crystals with an orthorhombic structure. Similar features were also observed in $Fe_{1.05}Se_{0.2}Te_{0.8}$. Nevertheless, the characteristics of the temperature-dependent oscillation component in superconductive FeSe and $Fe_{1.05}Se_{0.2}Te_{0.8}$ are almost obscured in the non-superconductive $Fe_{1.14}Te$.

By Fourier transformation of the oscillation component at $T = 100$K in the right inset of Figure 10.1.2.2a of FeSe, the phonon frequency is found to be 16 GHz. The phonon energy is estimated to be $\sim$0.07 meV. The coherent acoustic phonon detected by a pump–probe reflectivity measurement can be described as a Brillouin scattering [38] phenomenon occurring in the materials after excitation of pump pulses. The scattering condition is $q_{phonon} = 2nk_{probe}\cos(\theta_i)$, where q_{phonon} is the phonon wave vector, n is the real part of the refractive index, and the probe photon has a wave vector k_{probe} arriving at an incident angle θ_i (inside crystals) with respect to the surface normal.

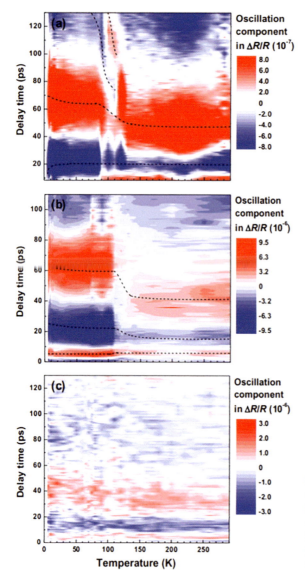

FIGURE 10.1.2.4 Temperature-dependent oscillation component of $\Delta R/R$ in (a) FeSe, (b) $Fe_{1.05}Se_{0.2}Te_{0.8}$ and (c) $Fe_{1.14}Te$ single crystals, which were obtained by subtracting the decay background (the first, second and third terms in equation (10.1.2.1)) from $\Delta R/R$ of Figure 10.1.2.1. Dashed lines are a guide to the eyes.

Here the basic textbook-like information of the Brillouin scattering and Raman scattering is described as follows.

Brillouin scattering is caused by the interaction of light with the material waves in a medium. It is associated with the dependence of refractive index on the material properties of the medium. The index of refraction of a transparent material changes under deformation such as compression-distension or shear-skewing. If the medium is a solid crystal, a lattice chain or a viscous liquid or gas, then the low-frequency atomic-chain-deformation waves within the transmitting medium could be, for example, mass oscillation (acoustic) modes (phonons), charge displacement modes in dielectrics, (polaritons), and magnetic spin oscillation modes in magnetic materials (magnons).

Raman scattering is another phenomenon that involves inelastic scattering of light caused by the vibrational properties of matter. The detected range of frequency shifts and other effects are different from Brillouin scattering. In Raman case, photons are scattered by vibrational and rotational transitions associated with the bonds between nearest neighbor atoms, while Brillouin scattering results from the scattering of photons caused by large scale naturally relevant to low-frequency phonons. The effects of the two phenomena provide very different information about the sample materials: Raman spectroscopy in most cases can provide the information of the chemical composition of transmitting medium and/or molecular structure, while Brillouin scattering can provide the material's properties on a larger scale – such as its elastic or mechanical behavior. The frequency shifts in Brillouin scattering obtained in Brillouin spectroscopy are detected with an interferometer while Raman scattering uses either an interferometer or a dispersive (grating) spectrometer. This is because of the difficulty of construction of dispersive spectrometer in the former spectroscopy due to too small shift in the Brillouin scattering.

The result of the interaction between the light-wave and the carrier-deformation wave is that a fraction of the transmitted light-wave changes its momentum (thus its frequency and energy) in preferential directions, as if by diffraction caused by an oscillating 3-dimensional diffraction grating.

Following this scattering condition, the probe beam acts as a filter to select the acoustic wave propagating along the scattering plane symmetry axis, i.e. the normal to the crystal surface, and traveling with the wave vector q_{phonon}. The energy of the acoustic wave is

$$E_{\text{phonom}} = \hbar\omega_{\text{phonon}} = \hbar q_{\text{phonon}} \upsilon_s = \hbar 2 n \upsilon_s k_{\text{probe}} \cos(\theta_i) \qquad (10.1.2.4)$$

where υ_s is the sound velocity along the normal direction of the crystal surface. Using $\lambda_{\text{probe}} = 800\,\text{nm}$, $n_{\text{probe}} = 2$ [15], $\theta_i = 2.5°$ (estimated from the incident angle (5°) of the probe beam by Snell's law) and $\upsilon_s = 3.58\,\text{km/s}$ [39], the phonon energy, E_{phonon}, is calculated to be 0.077 meV, which is very close to the result, 0.07 meV, directly obtained from the above $\Delta R/R$ measurements.

Moreover, we further investigate the temperature dependence of the LA phonon energy as shown in Figure 10.1.2.5. For the case of FeSe, the phonon energy dramatically drops by 60%[2] around 90K where a structural phase transition occurs and then remains constant at low temperatures. Additionally, we found that the LA phonons also soften by 6% in the superconducting state as shown in the inset of Figure 10.1.2.5, which is consistent with the larger distance between the first depth and the first peak in Figure 10.1.2.4a. Very recently, the phonon softening near the structure transition in $BaFe_2As_2$ and Co-doped $BaFe_2As_2$ was observed by inelastic x-ray scattering [40] and resonant ultrasound spectroscopy [27,41], respectively. Fernandes et al. [27] found 16% softening of shear modulus in $BaFe_{1.84}Co_{0.16}As_2$ at $T_c = 22\text{K}$. For the non-superconducting case of $BaFe_2As_2$, however, a rather large softening of 90% was observed around 130K, where the structural and AFM phase transition temperature occurs. Similarly, a large phonon softening due to structural phase transition and a

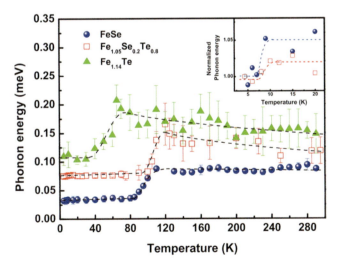

FIGURE 10.1.2.5 Temperature dependence of the phonon energy derived from the oscillation component in Figure 10.1.2.4. The inset shows a part of the temperature-dependent phonon energy on an enlarged scale. Dashed lines are a guide to the eyes.

---

[2] Here we assume that the refractive index n of FeSe is temperature-independent.

rather small phonon softening due to the superconducting phase transition were also unambiguously observed in superconductive FeSe and $Fe_{1.05}Se_{0.2}Te_{0.8}$. These results suggest that the reduction of phonon energy at both the structural and the superconducting phase transition is a general feature of FeSCs. The above phonon softening may participate in superconductive pairing, albeit not the mechanism responsible for high T_c in FeSCs. Additionally, in the non-superconductive $Fe_{1.14}Te$ a marked anomaly was clearly observed at around 65K, which is just the temperature of magnetic phase transition as shown in the inset of Figure 10.1.2.1c. Namely, the LA phonons in $Fe_{1.14}Te$ softened with the appearance of AFM ordering through the magnetoelastic effect. In Kulic and Haghighirad's´ theoretical calculations [42], they also predicted the existence of giant magnetoelastic effects at the transition from the magnetic state to the non-magnetic state in Fe-pnictides. Consequently, the phonon softening in FeSe and $Fe_{1.05}Se_{0.2}Te_{0.8}$ accompanying the simultaneous appearance of spin fluctuations shown in Figure 10.1.2.3a may also be caused by the magnetoelastic effect.

10.1.2.6   SUMMARY

The ultrafast QP dynamics and phonon softening in $Fe_{1+y}Se_{1-x}Te_x$ single crystals studied by dual-color femtosecond spectroscopy are described in this subsection. From the relaxation time τ_e of $\Delta R/R$, the electron–phonon coupling strength was obtained to be $\lambda = 0.16$ for FeSe and $\lambda = 0.01$ for $Fe_{1.05}Se_{0.2}Te_{0.8}$. The anomalous changes of amplitude (A_e, A_{LO}) and relaxation time (τ_e, τ_{LO}) in the temperature-dependent $\Delta R/R$ are clearly observed at 90K (T_s) and 230K (T^*) and show the existence of phase transition in FeSe and $Fe_{1.05}Se_{0.2}Te_{0.8}$. Moreover, the energy of LA phonons as a function of temperature estimated from the oscillation component of $\Delta R/R$ markedly softens at T_c and the temperatures of the structural and magnetic phase transitions through the magnetoelastic effects. The results and discussion in this subsection provide a vital understanding of the competing picture between the spin fluctuations and the superconductivity, and the role of phonons in Fe-based superconductors [43]. The research described here in this subsection was performed by the following people in collaboration: C.- W. Luo, I.- H.Wu, P.- C. Cheng, J.-Y. Lin, K.- H. Wu, T.- M. Uen, J. -Y Juang, T. Kobayashi, Y.-C. Wen, T.-W. Huang, K.-W. Yeh, M.-K. Wu, D. A. Chareev, O. S.Volkova and A. N. Vasiliev [43].

REFERENCES

1. Y. Kamihara, T. Watanabe, M. Hirano, and H. Hosono, *J. Am. Chem. Soc.* **130**, 3296–3297 (2008).
2. C. Wang, et al., *Europhys. Lett.* **83**, 67006 (2008).
3. M. Rotter, M. Tegel, and D. Johrendt, *Phys. Rev. Lett.* **101**, 107006 (2008).
4. X. C. Wang, Q. Q. Liu, Y. X. Lv, W. B. Gao, L. X. Yang, R. C. Yu, F. Y. Li and C. Q. Jin, *Solid State Commun.* **148**, 538 (2008).
5. F. C. Hsu, et al., *Proc. Natl Acad. Sci. USA* **105**, 14262 (2008).
6. S. Margadonna, Y. Takabayashi, Y. Ohishi, Y. Mizuguchi, Y. Takano, T. Kagayama, T. Nakagawa, M. Takata, and K. Prassides, *Phys. Rev. B* **80**, 064506 (2009).
7. B. C. Sales, A. S. Sefat, M. A. McGuire, R. Y. Jin, and D. Mandrus, *Phys. Rev. B* **79**, 094521 (2009).
8. E. E. M. Chia, et al., *Phys. Rev. Lett.* **104**, 027003 (2010).
9. T. Mertelj, V. V. Kabanov, C. Gadermaier, N. D. Zhigadlo, S. Katrych, J. Karpinski, and D. Mihailovic, *Phys. Rev. Lett.* **102**, 117002 (2009).
10. B. Mansart, D. Boschetto, A. Savoia, F. Rullier-Albenque, A. Forget, D. Colson, A. Rousse, and M. Marsi, *Phys. Rev. B* **80**, 172504 (2009).
11. K. W. Yeh, C. T. Ke, T. W. Huang, T. K. Chen, Y. L. Huang, P. M. Wu, and M. K. Wu, *Cryst. Growth Des.* **9**, 4847 (2009).
12. C. W. Luo, et al., *Phys. Rev. Lett.* **108**, 257006 (2012).
13. F. Walz, *J. Phys.: Condens. Matter* **14**, R285 (2002).
14. E. E. Rodriguez, C. Stock, P. Zajdel, K. L. Krycka, C. F. Majkrzak, P. Zavalij, and M. A. Green, *Phys. Rev. B* **84**, 064403 (2011).
15. X. J. Wu, et al., *Appl. Phys. Lett.* **90**, 112105 (2007).

16. V. Gnezdilov, et al., *Phys. Rev. B* **83**, 245127 (2011).
17. A. Subedi, L. Zhang, D. J. Singh, and M. H. Du, *Phys. Rev. B* **78**, 134514 (2008).
18. F. S. Krasniqi, S. L. Johnson, P. Beaud, M. Kaiser, D. Grolimund, and G. Ingold, *Phys. Rev. B* **78**, 174302 (2008).
19. M. Hase, K. Ishioka, J. Demsar, K. Ushida, and M. Kitajima, *Phys. Rev. B* **71**, 184301 (2005).
20. V. V. Kabanov, J. Demsar, B. Podobnik, and D. Mihailovic, *Phys. Rev. B* **59**, 1497 (1999).
21. H. J. Zeiger, J. Vidal, T. K. Cheng, E. P. Ippen, G. Dresselhaus, and M. S. Dresselhaus, *Phys. Rev. B* **45**, 768 (1992).
22. T. E. Stevens, J. Kuhl, and R. Merlin, *Phys. Rev. B* **65**, 144304 (2002).
23. T. Imai, K. Ahilan, F. L. Ning, T. M. McQueen, and R. J. Cava, *Phys. Rev. Lett.* **102** 177005 (2009).
24. H. Shi H, Z. B. Huang, J. S. Tse, and H. Q. Lin, *J. Appl. Phys.* **110**, 043917 (2011)
25. T. M. McQueen, A. Williams, P. W. Stephens, J. Tao, Y. Zhu, V. Ksenofontov, F. Casper, C. Felser, and R. J. Cava, *Phys. Rev. Lett.* **103**, 057002 (2009).
26. T. M. McQueen, et al., *Phys. Rev. B* **79**, 014522 (2009).
27. R. M. Fernandes, L. H. VanBebber, S. Bhattacharya, P. Chandra, V. Keppens, D. Mandrus, M. A. McGuire, B. C. Sales, A. S. Sefat, and J. Schmalian, *Phys. Rev. Lett.* **105**, 157003 (2010).
28. P. B. Allen, *Phys. Rev. Lett.* **59**, 1460 (1987).
29. D. Boschetto, E. G. Gamaly, A. V. Rode, B. Luther-Davies, D. Glijer, T. Garl, O. Albert, A. Rousse, and J. Etchepare, *Phys. Rev. Lett.* **100**, 027404 (2008).
30. W. L. McMillan, *Phys. Rev.* **167**, 331 (1968).
31. L. Boeri, O. V. Dolgov, and A. A. Golubov, *Phys. Rev. Lett.* **101**, 026403 (2008).
32. B. Mansart, D. Boschetto, A. Savoia, F. Rullier-Albenque, F. Bouquet, E. Papalazarou, A. Forget, D. Colson, A. Rousse, and M. Marsi, *Phys. Rev. B* **82**, 024513 (2010).
33. T. J. Liu et al, *Nature Mater.* **9**, 718 (2010).
34. P. Kumar, A. Kumar, S. Saha, D. V. S. Muthu, J. Prakash, S. Patnaik, U. V. Waghmare, A. K. Ganguli, and A. K. Sood, *Solid State Commun.* **150**, 557 (2010).
35. K. Okazaki, S. Sugai, S. Niitaka, and H. Takagi, *Phys. Rev. B* **83**, 035103 (2011).
36. T.-L. Xia, D. Hou, S. C. Zhao, A. M. Zhang, G. F. Chen, J. L. Luo, N. L. Wang, J. H. Wei, Z.-Y. Lu, and Q. M. Zhang, *Phys. Rev. B* **79**, 140510 (2009).
37. C. Thomsen, H. T. Grahn, H. J. Maris, and J. Tauc, *Phys. Rev. B* **34**, 4129 (1986).
38. L. Brillouin, *Ann. Phys.* **17**, 88 (1922).
39. S. Chandra, and A. K. M. A. Islam, *Physica C* **470**, 2072 (2010).
40. J. L. Niedziela, D. Parshall, K. A. Lokshin, A. S. Sefat, A. Alatas, and T. Egami, *Phys. Rev. B* **84**, 224305 (2011).
41. T. Goto, R. Kurihara, K. Araki, K. Mitsumoto, M. Akatsu, Y. Nemoto, S. Tatematsu, and M. Sato, *J. Phys. Soc. Japan* **80**, 073702 (2011).
42. M. L. Kulic, and A. A. Haghighirad, *Europhys. Lett.* **87**, 17007 (2009).
43. C. W. Luo, I. H. Wu, P. C. Cheng, J.-Y. Lin, K. H. Wu, T. M. Uen, J. Y. Juang, T. Kobayashi, Y. C. Wen, T. W. Huang, K. W. Yeh, M. K. Wu, D. A. Chareev, O. S. Volkova, and A. N. Vasiliev, *New J. Phys.* **14**, 103053 (2012).

10.1.3 Quasiparticle Dynamics in FeSe Superconductors Studied by Femtosecond Spectroscopy

10.1.3.1 INTRODUCTION

The discovery of $LaFeAsO_{1-x}F_x$ with $T_c \sim 26$ K [1] initiated the investigations of the diverse family of Fe-based superconductors (FeSC), e.g., $Ba_{1-x}K_xAs_2Fe_2$ (122-type) with $T_c \leq 38$ K [2], LiFeAs (111-type) with $T_c \leq 18$ K [3], and FeSe (11-type) with $T_c \leq 10$ K [4]. Among various FeSCs, the iron chalcogenide FeSe [4] stands out due to its structure simplicity, which consists of iron-chalcogenide layers stacking one by another with the same Fe^{+2} charge state as the iron pnictides. This so-called "11" system is so simple that it could be the key structure to understanding the origin of high-T_c superconductivity [5]. There has been considerable concern over the interplay between electronic structure, phonons, magnetism and superconductivity in 11-type FeSe. Therefore, further studies of their quasiparticle dynamics are indispensable to understanding the high T_c mechanism in FeSCs. Here we report the time-resolved femtosecond spectroscopy of FeSe single crystals to elucidate the electronic structure and the quasiparticle (QP) dynamics.

10.1.3.2 EXPERIMENTS

In this study, the FeSe single crystals were grown in evacuated quartz ampoules using a $KCl/AlCl_3$ flux [6]. The crystalline structure of the samples was examined by X-ray diffraction. The low-temperature feature related to superconducting transition is at $T_c = 8.8$ K.

The femtosecond spectroscopy measurement was performed by using a dual-color pump-probe system (for the laser light source, the repetition rate is 5.2 MHz, the wavelength is 800 nm, and the pulse duration is 100 fs) and an avalanche photodetector with the standard lock-in technique. The fluences of the pump beam and the probe beam are 9.92 and 1.40 $\mu J/cm^2$, respectively. The pump pulses have corresponding photon energy (3.1 eV) where the higher absorption occurred in the absorption spectrum of FeSe [7] and hence can generate electronic excitations. The photoinduced QP dynamics is studied by measuring the photoinduced transient reflectivity changes ($\Delta R/R$) of a probe beam with photon energy of 1.55 eV.

10.1.3.3 RESULTS AND DISCUSSION

Figure 10.1.3.1 shows the typical transient reflectivity changes ($\Delta R/R$) taken at various temperatures on a FeSe single crystal. Above 230 K, there is a fast negative response with a relaxation time of about 1.5 ps together with a long period oscillation. When the temperature decreases below 230 K, a positive and slow response appears and $\Delta R/R$ gradually becomes smaller until $T = 90$ K. Below 90 K, the slow positive response disappears and is replaced by a complicated mixture of the positive and negative components. For $T < T_c$ (8.8 K), a negative response appears as that one in the region above 230 K. To figure out the QP relaxation processes after excitation, we try to fit the $\Delta R/R$ curves as shown in Figure 10.1.3.1 and obtain the relaxation time of QPs. In the case of Co-doped

DOI: 10.1201/9780429196577-67

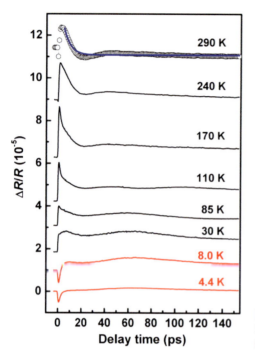

FIGURE 10.1.3.1 Temperature-dependent $\Delta R/R$ curves in a FeSe single crystal. The *solid line* at 290 K is the fitting curve using an exponential decay function.

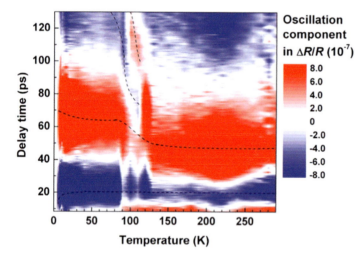

FIGURE 10.1.3.2 Temperature-dependent oscillation component of $\Delta R/R$ in a FeSe single crystal, which was obtained by subtracting the decay background (solid line in Figure 10.1.3.1) from $\Delta R/R$ in Figure 10.1.3.1. Dashed lines are a guide to the eyes.

$BaFe_2As_2$, the symmetric A_{1g} mode is coherently excited by photoexcitation and efficiently coupled [8]. Consequently, we take the A_{1g} mode into account in the present case of FeSe, which is the strongest phonon mode in electron-phonon spectral function, $\alpha^2F(\omega)$ [9]. Then, we can obtain the electron–phonon coupling strength, $\lambda=0.16$, in FeSe from Allen's model [10]. This value is consistent with the theoretical results of $\lambda=0.17$ [9] obtained by using linear response within the generalized gradient approximation (GGA).

However, the temperature-dependent $\Delta R/R$ in FeSe cannot be solely fitted by an exponential decay function as shown in the case of 290 K in Figure 10.1.3.1. By subtracting the decay background (e.g., the solid line in Figure 10.1.3.1), a significant oscillation component is clearly observed as shown in Figure 10.1.3.2. This oscillation is caused by the propagation of strain pulses inside a FeSe single crystal, namely the interference between the probe beams reflected from the crystal surface and the wavefront of the propagating strain pulse [11]. At high temperatures, the damping time is very short and the oscillation sustains only for one period. However, the number of oscillation period markedly increase around 100 K in FeSe; hence the damping time becomes much longer. Besides, the oscillation period significantly increases below 90 K. This means that the longitudinal-acoustic (LA) phonons can propagate further into the interior of FeSe crystals with the orthorhombic structure. According to the difference between the first trough (at 23.52 ps) and the second trough (at 72.78 ps) at $T = 110$ K in Figure 10.1.3.2, the phonon frequency is found to be 20.3 GHz. The phonon energy is estimated to be ~0.087 meV. It is worth to note that the phonon energy drops by 60% around 90 K where a structural phase transition occurs and by 6% at superconducting transition temperature, which is consistent with the larger distance between the first trough and first peak in Figure 10.1.3.2.

Very recently, the phonon softening near the structure transition in $BaFe_2As_2$ and Co-doped $BaFe_2As_2$ was observed by inelastic X-ray scattering [12] and resonant ultrasound spectroscopy [13,14], respectively. Fernandes et al. [13] found the 16% softening of shear modulus in $BaFe_{1.84}Co_{0.16}As_2$ at $T_c = 22$ K. For the non-superconducting case of $BaFe_2As_2$, however, the rather large softening of 90% was observed around 130 K where is the structural and AFM phase transition temperature. Here, the larger phonon softening due to structural phase transition and rather small phonon softening due to the superconducting phase transition is also observed in 11-type FeSe. These results suggest that the reduction of phonon energy at both the structural and the superconducting phase transitions is a general feature in FeSCs and may participate in the superconductive pairing, albeit not the mechanism responsible for high T_c in FeSCs.

10.1.3.4 SUMMARY

We have studied the ultrafast quasiparticle dynamics and phonon softening in FeSe single crystals by dual-color femtosecond spectroscopy [15]. The relaxation time of quasiparticles reveals an electron–phonon coupling strength $\lambda = 0.16$. Moreover, the energy of LA phonons at 110 K was estimated to be 0.087 meV from the oscillation component of $\Delta R/R$, which markedly softens around both the structural phase transition and superconducting transition. Our results provide the vital understanding of the role of phonons in Fe-based superconductors. The study was summarized and was expected to be a milestone of understanding the role of phonons in Fe-based superconductors. The research described in this subsection was performed by the following people in collaboration: C.-W. Luo, I-H. Wu, P.-C. Cheng, J.-Y Lin, K.-H. Wu, T.-M. Uen, J.-Y. Juang, T. Kobayashi, D.-A. Chareev, O.-S. Volkova, and A.- N. Vasiliev [15].

REFERENCES

1. Y. Kamihara, T. Watanabe, M. Hirano, and H. Hosono, *J. Am. Chem. Soc.* **130**, 3296 (2008).
2. M. Rotter, M. Tegel, and D. Johrendt, *Phys. Rev. Lett.* **101**, 107006 (2008).
3. X. C. Wang, Q. Q. Liu, Y. X. Lv, W. B. Gao, L. X. Yang, R. C. Yu, F. Y. Li, and C. Q. Jin, *Solid State Commun.* **148**, 538 (2008).
4. F. C. Hsu, J. Y. Luo, K. W. Yeh, T. K. Chen, T. W. Huang, P. M. Wu, Y. C. Lee, Y. L. Huang, Y. Y. Chu, D. C. Yan, and M. K. Wu, *Proc. Natl. Acad. Sci. USA* **105**, 14262 (2008).
5. D. C. Johnston, *Adv. Phys.* **59**, 803 (2010).
6. C. W. Luo, I. H. Wu, P. C. Cheng, J.-Y. Lin, K. H. Wu, T. M. Uen, J. Y. Juang, T. Kobayashi, D. A. Chareev, O. S. Volkova, and A. N. Vasiliev, *Phys. Rev. Lett.* **108**, 257006 (2012).
7. X. J. Wu, D. Z. Shen, Z. Z. Zhang, J. Y. Zhang, K. W. Liu, B. H. Li, Y. M. Lu, B. Yao, D. X. Zhao, B. S. Li, C. X. Shan, X. W. Fan, H. J. Liu, and C. L. Yang, *Appl. Phys. Lett.* **90**, 112105 (2007).
8. B. Mansart, D. Boschetto, A. Savoia, F. Rullier-Albenque, A. Forget, D, Colson, A. Rousse, and M. Marsi, *Phys. Rev. B* **80**, 172504 (2009).
9. A. Subedi, L. Zhang, D. J. Singh, and M. H. Du, *Phys. Rev. B* **78**, 134514 (2008).
10. P. B. Allen, *Phys. Rev. Lett.* **59**, 1460 (1987).

11. C. Thomsen, H. T. Grahn, H. J. Maris, and J. Tauc, *Phys. Rev. B* **34**, 4129 (1986).

12. J. L. Niedziela, D. Parshall, K. A. Lokshin, A. S. Sefat, A. Alatas, and T. Egami, *Phys. Rev. B* **84**, 224305 (2011).

13. R. M. Fernandes, L. H. VanBebber, S. Bhattacharya, P. Chandra, V. Keppens, D. Mandrus, M. A. McGuire, B. C. Sales, A. S. Sefat, and J. Schmalian, *Phys. Rev. Lett.* **105**, 157003 (2010).

14. T. Goto, R. Kurihara, K. Araki, K. Mitsumoto, M. Akatsu, Y. Nemoto, S. Tatematsu, and M. Sato, *J. Phys. Soc. Jpn.* **80**, 073702 (2011).

15. C.-W. Luo, I.-H. Wu, P.-C, Cheng, J. Y. Lin, K.-H. Wu, T.-M. Uen, J.-Y. Juang, T. Kobayashi, D.-A. Chareev, O.-S. Volkova, and A.-N. Vasiliev, *J. Supercond Nov Magn.* **26**, 1213–1215 (2013).

Section 10.2

THz, MIR Spectroscopy of Materials

10.2.1 Dirac Fermions Near the Dirac Point in Topological Insulators

10.2.1.1 INTRODUCTION

The discovery of 3D topological insulators (TIs) [1] initiated a new era of condensed matter physics [2,3]. As Dirac fermions play a crucial role in determining the performances of any real TI devices, a better understanding of the bulk state and the surface state, [4–14] and the coupling mechanisms between them, is imperative. From a practical point of view, contact-free optical techniques, such as second harmonic generation [11], terahertz time-domain spectroscopy [15], UV–visible–IR reflectance and transmission spectroscopy [16], and optical pump-probe spectroscopy [17–21], would be the most feasible schemes to investigate the characteristics of TIs. However, the surface signatures are easily overwhelmed by the bulk contributions. Recent TrARPES studies have shown the surface carrier-population in TIs can be induced by photoexcitation [12,13] and can separately obtain the temperature and chemical potential relaxation of both the surface and the bulk [14]. Nevertheless, the ultrafast behavior of Dirac fermions near the Dirac point and their detailed energy-dependent coupling with phonons remain elusive for the lack of probes with the appropriate energy range (~100 meV) specific to the Dirac cone. We further take the advantage of the appropriate probe photon energies in the optical pump mid-infrared probe (OPMP) spectroscopy to explore the nonequilibrium dynamics of TIs. The mid-infrared photon energy range (87–153 meV < bandgap energy of 300 meV in Bi_2Se_3) naturally selects the transitions limited within the Dirac cone, and the femtosecond-time and millielectronvolt-energy resolution allow us to distinguish the individual dynamics of both the surface and the bulk. Such an ultrafast midinfrared approach has potentially provided significant insights to other correlation physics in strongly correlated materials, for example, electronic phase transition in $BaFe_2As_2$ superconductors [22] and phonon resonances in optimally doped YBa2Cu3O7$_{-\delta}$ [23].

Figure 10.2.1.1 provides a synopsis of the OPMP spectra for all samples investigated. The doping levels of samples span a wide range, as listed in Table 10.2.1.1, (#1: $n = 51.5 \times 10^{18}$ cm^{-3}, #2: $n = 13.9 \times 10^{18}$ cm^{-3}, #3: $n = 5.58 \times 10^{18}$ cm^{-3}, #4: $n = 0.25 \times 10^{18}$ cm^{-3}) and show corresponding ARPES images.[24] For the case of Bi_2Se_3 #1 with a high carrier concentration ($n = 51.5 \times 10^{18}$ cm^{-3}), a positive peak is clearly observed in $\Delta R/R$. This positive peak gradually diminishes as n decreases (#1→#4 in Figure 10.2.1.1a). Noticeably, an additional negative peak appears for the cases of $n = 5.58 \times 10^{18}$ cm^{-3} and $n = 0.25 \times 10^{18}$ cm^{-3}, and its amplitude is inversely proportional to n.

10.2.1.2 RESULTS AND DISCUSSION

To elucidate the origins of both the positive and negative signals, a model is shown in Figure 10.2.1.2a for the optical pumping (1.55 eV) and mid-infrared probing processes in the schematic energy band structure of the TIs based on the ARPES image in Figure 10.2.1.1b. Because the used probe photon energy (87–153 meV) of the mid-infrared (mid-IR) is much smaller than the band gap of ~300 meV in Bi_2Se_3 (as shown in the ARPES images of Figure 10.2.1.1b), the interband transitions between the valence band (VB) and the conduction band (CB) of the bulk are not allowed to occur. Thus, the free carrier absorption in the CB (mid-IR probe (1) in Figure 10.2.1.2a) and Dirac cone surface state (mid-IR probe (2) in Figure 10.2.1.2a) will dominate the probe processes, which are responsible for the positive and negative peaks in $\Delta R/R$, respectively. To confirm this assignment and reveal

DOI: 10.1201/9780429196577-69

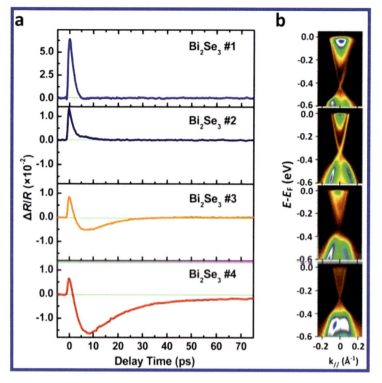

FIGURE 10.2.1.1 Carrier concentration (n) dependence of the transient change in reflectivity $\Delta R/R$ in Bi_2Se_3 single crystals. (a) $\Delta R/R$ of samples #1 ($n = 51.5 \times 10^{18}$ cm^{-3}), #2 ($n = 13.9 \times 10^{18}$ cm^{-3}), #3 ($n = 5.58 \times 10^{18}$ cm^{-3}), and #4 ($n = 0.25 \times 10^{18}$ cm^{-3}) with a pumping fluence of 34 μJ/cm^2 and probing photon energy of 141 meV. (b) ARPES band dispersion images on samples of (a) [24].

TABLE 10.2.1.1

Fermi Energy and Carrier Concentration of Bulk and Surface States for Various Samples Grown by Different Methods (Vertical Bridgman, Modified Floating Zone)

Code	$E_F - E_{Dirac\ point}$ (meV)	Carrier Concentraion		$n_{surface}/$ $(n_{surface} + n_{bulk}d)$
		n_{bulk} (10^{18} cm^{-3})	$n_{surface}$ (10^{13} cm^{-2})	
#1	422	$^-$51.5 0.84	$^-$1.45	0.11
#2	325	$^-$13.9 0.26	$^-$0.83	0.20
#3	284	$^-$5.58 0.25	$^-$0.72	0.35
#4	260	$^-$0.25 0.01	$^-$0.47	0.89

All samples are n-type. "$d = 23.5$ nm" is the penetration depth of 800-nm pumping light

the physical meanings of the positive peak in $\Delta R/R$, the photon energy dependence of $\Delta R/R$ for #1 sample is shown in Figure 10.2.1.2b. By decreasing the photon energy, $\Delta R/R$ gradually changes from positive to negative. Around 136 meV (1100 cm^{-1}), there are some intermediate signals mixed with both positive and negative peaks, corresponding to deep in the Fourier transform infrared (FTIR) reflectance spectrum (the inset of Figure 10.2.1.2b). After pumping, the excited carriers suffer the so-called intervalley scattering (see Supporting Information), leading to the redshift of the reflectance spectra. Therefore, the reflectivity increases as a function of time with a large probing photon energy, which is higher than the position of 136 meV deep in the reflectance spectra due to plasma edge. On the contrary, the reflectivity decreases as a function of time with a small probing photon energy, which is lower than the position of 136 meV deep in the reflectance spectra. Similar results were also observed in a typical semiconductor n-type GaAs [25].

Compared to the $\Delta R/R$ curves and ARPES images in Figure 10.2.1.1, the amplitude of the positive peak in $\Delta R/R$ gradually shrinks as the bulk carrier concentration decreases (see Table

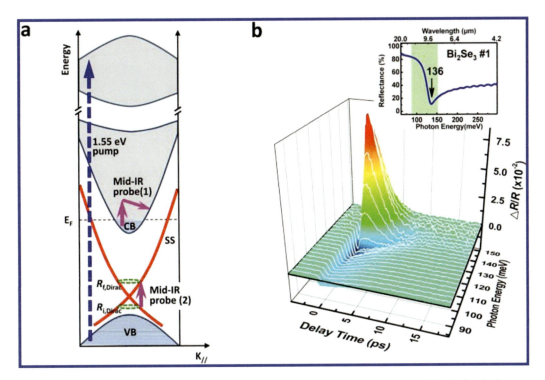

FIGURE 10.2.1.2   Schematic energy band structure and photon energy dependence of $\Delta R/R$ in a bulk state. (a) Schematic energy band structure of TIs according to the ARPES images in Figure 1b and the optical pump mid-IR probe processes. CB: conduction band. VB: valence band. SS: surface state. $R_{i,\mathrm{Dirac}}$ and $R_{f,\mathrm{Dirac}}$: the circumferences of initial and final states in Dirac cone for mid-IR probing. (b) With a pumping fluence of 38 μJ/cm², the $\Delta R/R$ of Bi_2Se_3 #1 at various photon energies (wavenumber) from 87 to 153 meV (700 to 1234 cm⁻¹). Inset: the Fourier transform infrared (FTIR) reflectance spectrum of Bi_2Se_3 #1. The gray area indicates the range of the mid-IR photon energy used in this study.

10.2.1.1). On the other hand, the negative peak in $\Delta R/R$ increases as the bulk and surface carrier concentrations decrease. However, the negative peak of $\Delta R/R$ increases dramatically with an increasing ratio of the surface carrier concentration to the total carrier concentration $[n_{\mathrm{surface}}/(n_{\mathrm{surface}}+n_{\mathrm{bulk}}d)$ in Table 10.2.1.1], implying an intimate relation between the negative peak of $\Delta R/R$ and Dirac fermions. Besides, the $\Delta R/R$ signal significantly depends on the pumping fluences shown in Figure 10.2.1.3a. Interestingly, the positive peak of $\Delta R/R$ has a stronger dependence on the pumping fluences than the negative peak does. Therefore, the negative peak still subsists at the low pumping fluence of 3.3 μJ/cm² [26], while the positive peak almost vanishes. This means the mid-IR probe process (1) in the bulk state (see Figure 10.2.1.2a) can be suppressed by reducing the pumping fluences; meanwhile, the mid-IR probe process (2) (see Figure 10.2.1.2a) associated with the negative peak can be preserved at the low pumping fluence (see Supporting Information). Here, we can conclude that the positive (or negative) signal within a several picoseconds time scale in $\Delta R/R$ is due to the process (1) of the mid-IR probe in the bulk state of Bi_2Se_3.

The relation between the negative peak and Dirac fermions can be certified in a quantitative way. According to the Fermi Golden rule, the amplitude of the negative peak should be proportional to the transition probability $(T_{i\rightarrow f})$ between the initial and final density of states in the Dirac cone. With an increase in the probing photon energy, the amplitude of the negative peak increases. Owing to the large positive signal before 5 ps in samples #1 and #2, this probing photon energy dependence of the negative peak amplitude cannot be easily disclosed. However, this dependence was clearly observed in both samples #3 and #4. The experimental data are fitted well by the R_i, Dirac×R_f, Dirac (dashed

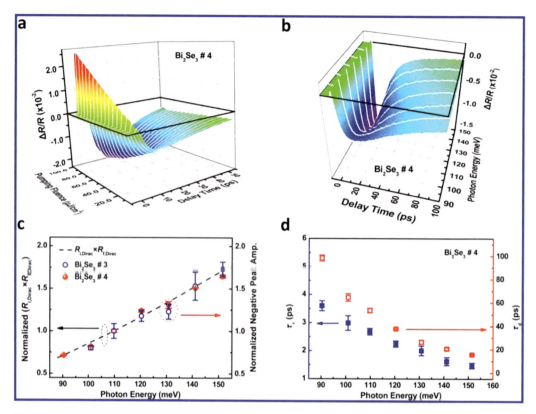

FIGURE 10.2.1.3 Pumping fluence and photon energy dependence of $\Delta R/R$ and its amplitude and rising (decay) time in the surface state. (a) With probing photon energy of 141 meV, the $\Delta R/R$ of Bi_2Se_3 #4 at various pumping fluences from 3.3–105 μJ/cm^2. (b) With a pumping fluence of 3.3 μJ/cm^2, the $\Delta R/R$ of Bi_2Se_3 #4 at various photon energies from 90 to 152 meV. (c) The photon energy-dependent negative peak amplitude of $\Delta R/R$ in panel b. The photon energy dependence of the normalized absorption probability (dashed line, that is, $R_{i,\text{Dirac}} \times R_{f,\text{Dirac}}$ in Figure 10.2.1.2a) of the mid-IR probe beam in the Dirac cone surface state. (d) The photon energy-dependent rising time (τ_r) and decay time (τ_d) of $\Delta R/R$ in (b).

line in Figure 10.2.1.3c, $R_{i,\text{Dirac}}$ and $R_{f,\text{Dirac}}$ are the circumferences of rings in Figure 10.2.1.2a), which is proportional to the transition rate between the initial and final density of states for the mid-IR probe process (2) in the Dirac cone (see Supporting Information). This strongly indicates the negative peak of $\Delta R/R$ is dominated by the mid-IR probe process (2) in the Dirac cone (see Figure 10.2.1.2a). Consequently, the ultrafast dynamics of the Dirac fermions can be clearly disclosed by the negative peak of $\Delta R/R$. The above experiments were carried out at the low pumping fluence of 3.3 μJ/cm^2 to avoid disturbance of the positive peak from the bulk state, as shown in Figure 10.2.1.3b.

The rise time (τ_r) and decay time (τ_d) of the negative peak of $\Delta R/R$ significantly depends on the probing photon energy, as in Figure 10.2.1.3d. The rising time of the negative peak of $\Delta R/R$ also becomes longer when the probed regime is closer to the Dirac point. On the basis of the above observations, we can further establish the ultrafast relaxation picture for Dirac fermions in TIs. Immediately following the 1.55 eV pumping, the major process is the carriers in the bulk valence band (BVB) are excited to the bulk conduction band (BCB). The carrier recombination between the BCB and BVB can be ignored in this study due to the time scale for such a process is typically ≫1 ns [27]. Consequently, the unoccupied states in BVB caused by pumping would mainly be refilled through the bottom part of the upper Dirac cone that almost overlaps with the top of BVB at the same momentum space, as shown in the ARPES images of Figure 10.2.1.1b. This implies the carriers in this part of the Dirac cone can be easily transferred into the unoccupied states in BVB and increasing the number of the unoccupied states near the Dirac point enhances the absorption channel for the mid-IR process (2) in the Dirac cone (Figure 10.2.1.2a). Therefore, the reflectivity of the mid-IR probing light decreases within 1.47–3.60 ps, that is, the rising time of the negative peak in Figure 10.2.1.3b and d. Once the carriers in the Dirac cone relax into BVB, the BCB (like a carrier reservoir) subsequently injects the excited carriers into the unoccupied states in the Dirac cone to diminish the absorption channel for the mid-IR process (2) (Figure 10.2.1.2a). This leads to the increased mid-IR reflectivity within 14.8–87.2 ps, consistent with the ARPES results [12,13] of a nonequilibrium population of the surface state persisting for >10 ps. The several tens of picoseconds in decay time, which is much longer than the rising time of several picoseconds, is because the carriers in BCB cannot directly transfer into the top of the Dirac cone without overlaps occurring between them and other auxiliaries, for example, phonons. A movie showing the relaxation processes of Dirac fermions in the Dirac cone after pumping is included in the Supporting Information.

A simple description of Dirac fermion is described below.

In particle physics fermion is a spin-1/2 particle which is different from its antiparticle (perhaps except neutrinos) and therefore are Dirac fermions. They are modeled by Dirac equation. A Dirac fermion Is equivalent to two Weyl fermions. The counterpart to a Dirac fermion is Majorana fermion, a particle that must be its own antiparticle.

Phonons have been considered as the main medium in the relaxation of Dirac fermions [14,28–31]. Here, we follow this approach. The photon energy dependence of the rising time implies the coupling strength (λ) between Dirac fermions and phonons varies at different positions of the Dirac cone. According to the second moment of the Eliashberg function [32], the λ is inversely proportional to the relaxation time (τ_e) of excited electrons as shown below,

$$\lambda\langle\omega^2\rangle \propto \frac{1}{\tau_e}, \qquad (10.2.1.1)$$

where ω is the phonon energy that couples with the electrons. For the estimate of $\langle\omega^2\rangle$, some vibrational modes are more efficiently coupled to Dirac fermions than others are. In the case of Bi_2Se_3, the symmetric A_{1g} [1] mode of ~8.9 meV is coherently excited by photoexcitation and efficiently coupled [21,33]. Taking $\tau_e = \tau$ in Figure 10.2.1.3d and $T_e=370$ K (obtained from Ref. [14] at the low pumping fluence as mentioned above) to estimate the coefficient of $(\pi k_B T_e/3\hbar)$ in Eq. (10.2.1.1), the photon energy dependence of the Dirac fermion-phonon coupling strength is $\lambda=0.08$–0.19, as shown in Figure 10.2.1.4a. Recently, the ARPES measurements have reported inconsistent electron–phonon coupling strength in Bi_2Se_3 varying from a rather small $\lambda\sim0.08$ [30] to a larger $\lambda\sim0.25$ [31]. The Dirac fermion– phonon coupling strength measured by the present OPMP becomes significantly

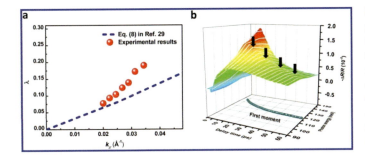

FIGURE 10.2.1.4 Dirac fermion–phonon coupling strength as a function of momentum and photon energy dependence of the first moment. (a) Momentum-dependent electron–phonon coupling strength (λ) in Dirac cone surface state and compared with the theoretical results (dashed line) from Ref. [29]. (b) Three-dimensional plot of the $-\Delta R/R$ in Figure 10.2.1.3b as a function of photon energy at various time delays and the time-dependent first moment (solid dots). The arrows mark the position of absorption peaks at different delay time.

smaller near the Dirac point (the point of $K_{//}=0$ in Figure 10.2.1.4a), which has a qualitatively similar tendency to the estimate of equation 8 (dashed line in Figure 4a) in Ref. [29]. The present results suggest that the variation of λ from ARPES may be due to the different explored regimes (i.e., the different chemical potentials) in the Dirac cone. Besides, the time-resolved ARPES experiments also showed similar results. Wang et al. [14] reported that the surface cooling rate decrease with the Fermi level (i.e., closing to the Dirac point). Because the cooling time is inversely proportional to the surface cooling rate, these time-resolved ARPES results are consistent with those in Figure 10.2.1.3d. If the Dirac fermions are closer to the Dirac point, they will have weaker coupling with the phonons to suppress the scattering with phonons. This also implies the effective mass of Dirac fermions in the surface state gradually decreases as the Dirac fermions approach the Dirac point, in agreement with the results in graphene [34]. Consequently, this study further provides a possibility to control the characteristics of Dirac fermions for various applications in TIs such as terahertz optoelectronics, spintronics, quantum computation, and magnetic memories.

Finally, a closer look at the 3D plot of $-\Delta R/R$ as a function of photon energy at various delays in Figure 10.2.1.4b reveals the absorption peak (marked by arrows in Figure 10.2.1.4b) suffers a red shift with the time delay. This implies the unoccupied density of states in the Dirac cone shift as a function of time, that is, the energy of carrier loss as a function of time. According to the first moment, $(\int (\Delta R/R)E_{photon}\,dE_{photon})/(\int (\Delta R/R)dE_{photon})$, we estimate the energy loss rate of carriers in the Dirac cone. As shown in Figure 10.2.1.4b, the solid dots represent the first moment at different time, which is associated with the red shift of the absorption peak in Figure 10.2.1.4b. An exponential fit to the time-dependent first moment in Figure 10.2.1.4b gives a relaxation time of 14.8 ps within the range of 15 meV. Therefore, the energy loss rate of Dirac fermions in the Dirac cone is ~1 meV/ps, which is larger than that of ~0.64 meV/ps in GaAs estimated from Ref. [25] but smaller than that of ~17.7 meV/ps in graphene with Dirac cone [35]. This parameter measured by OPMP would be extremely important for designing optoelectronics, especially in the terahertz range [36].

10.2.1.3 CONCLUSION

We have studied Dirac fermions near the Dirac point in topological Insulators and clarified dynamics by determining the loss rate of carriers in the cone as a function of delay time. The results can provide important information for optoelectronics.

The research presented n this subsection was conducted by the following people in collaboration: C.-W. Luo, H.-J. Wang, S.-A. Ku, H.-J. Chen, T.-T. Yeh, J.-Y. Lin, K.-H. Wu, J.-Y. Juang, B.-L. Young, T. Kobayashi, C.-M. Cheng, C.-H. Chen, K.-D. Tsuei, R. Sankar, F. Chou, K. Kokh, O. E. Tereshchenko, E. V. Chulkov, Yu. M. Andreev, and Genda Gu [36].

REFERENCES

1. D. Hsieh, D. Qian, L. Wray, Y. Xia, Y. Y. S. Hor, R. J. Cava, and M. Z. Hasan, "A topological Dirac insulator in a quantum spin Hall phase," *Nature* 452, 970–974 (2008).
2. M. Z. Hasan, and C. J. Kane, "Colloquium: Topological insulators," *Rev. Mod. Phys.* 82, 3045–3067 (2010).
3. X.-L. Qi, and S.-C. Zhang, "Topological insulators and superconductors," *Rev. Mod. Phys.* 83, 1057–1110 (2011).
4. Y. Xia, D. Qian, D. Hsieh, L. Wray, A. Pal, H. Lin, A. Bansil, D. Grauer, Y. S. Hor, R. J. Cava, and M. Z. Hasan, "Observation of a large gap topological-insulator class with a single Dirac cone on the surface," *Nat. Phys.* 5, 398–402 (2009).
5. D. Hsieh, et al, "A tunable topological insulator in the spin helical Dirac transport regime," *Nature* 460, 1101–1106 (2009).
6. Y. L. Chen, et al., "Experimental realization of a three-dimensional topological insulator, Bi_2Te_3," *Science* 325, 178–181 (2009).
7. T. Zhang, et al., "Experimental demonstration of topological surface states protected by time-reversal symmetry," *Phys. Rev. Lett.* 103, 266803 (2009).

8. P. Roushan, J. Seo, C. V. Parker, Y. S. Hor, D. Hsieh, D. Qian, A. Richardella, M. Z. Hasan, R. J. Cava, and A. Yazdani, "Topological surface states protected from backscattering by chiral spin texture," *Nature* 460, 1106–1110 (2009).

9. Z. Alpichshev, J. G. Analytis, J.-H. Chu, I. R. Fisher, Y. L. Chen, Z. X. Shen, A. Fang, and A. Kapitulnik, "STM imaging of electronic waves on the surface of Bi_2Te_3: topologically protected surface states and hexagonal warping effects," *Phys. Rev. Lett.* 104, 016401 (2010).

10. S. Kim, et al., "Surface scattering via bulk continuum states in the 3D topological insulator Bi_2Se_3," *Phys. Rev. Lett.* 107, 056803 (2011).

11. D. Hsieh, J. W. McIver, D. H. Torchinsky, D. R. Gardner, Y. S. Lee, and N. Gedik, "Nonlinear optical probe of tunable surface electrons on a topological insulator," *Phys. Rev. Lett.* 106, 057401 (2010).

12. J. A. Sobota, S. Yang, J. G. Analytis, Y. L. Chen, I. R. Fisher, P. S. Kirchmann, and Z.-X. Shen, "Ultrafast optical excitation of a persistent surface-state population in the topological insulator Bi_2Se_3," *Phys. Rev. Lett.* 108, 117403 (2012).

13. M. Hajlaoui, et al.," Ultrafast surface carrier dynamics in the topological insulator Bi_2Te_3," *Nano Lett.* 12, 3532–3536 (2012).

14. Y. H. Wang, D. Hsieh, E. Sie, H. Steinberg, D. Gardner, Y. Lee, P. Jarillo-Herrero, and N. Gedik, "Measurement of intrinsic Dirac fermion cooling on the surface of the topological insulator Bi_2Se_3 using time-resolved and angle-resolved photoemission spectroscopy," *Phys. Rev. Lett.* 109, 127401 (2012).

15. R. V. Aguilar, et al., "Terahertz response and colossal Kerr rotation from the surface states of the topological insulator Bi_2Se_3," *Phys. Rev. Lett.* 108, 087403 (2012).

16. A. D. LaForge, A. Frenzel, B. C. Pursley, T. Lin, X. Liu, X. J. Shi, and D. N. Basov, "Optical characterization of Bi_2Se_3 in a magnetic field: Infrared evidence for magnetoelectric coupling in a topological insulator material," *Phys. Rev. B* 81, 125120 (2010).

17. N. Kamaraju, S. Kumar, and A. K. Sood, "Temperature-dependent chirped coherent phonon dynamics in Bi_2Te_3 using high-intensity femtosecond laser pulses," *Europhys. Lett.* 92, 47007 (2010).

18. J. Qi, et al., "Ultrafast carrier and phonon dynamics in Bi_2Se_3 crystals," *Appl. Phys. Lett.* 97, 182102 (2010).

19. Kumar, N., B. A. Ruzicka, N. P. Butch, P. Syers, K. Kirshenbaum, J. Paglione, and H. Zhao, "Spatially resolved femtosecond pump-probe study of topological insulator Bi_2Se_3," *Phys. Rev. B* 83, 235306 (2011).

20. D. Hsieh, F. Mahmood, J. W. McIver, D. R. Gardner, Y. S. Lee, and N. Gedik, "Selective probing of photoinduced charge and spin dynamics in the bulk and surface of a topological insulator," *Phys. Rev. Lett.* 2011, 107, 077401.

21. H.-J. Chen, et al., "Phonon dynamics in $Cu_xBi_2Se_3$ (x=0, 0.1, 0.125) and Bi_2Se_2 crystals studied using femtosecond spectroscopy," *Appl. Phys. Lett.* 101, 121912 (2012).

22. K. W. Kim, A. Pashkin, H. Schafer, M. Beyer, M. Porer, T. Wolf, C. Bernhard, J. Demsar, R. Huber, and A. Leitenstorfer, "Ultrafast transient generation of spin-density-wave order in the normal state of $BaFe_2As_2$ driven by coherent lattice vibration," *Nat. Mater.* 11, 497–501 (2012).

23. Pashkin, et al., "Femtosecond response of quasiparticles and phonons in superconducting $YBa_2Cu_3O_{7-\delta}$ studied by wideband terahertz spectroscopy," *Phys. Rev. Lett.* 105, 067001 (2010).

24. C. W. Luo, "THz generation and detection on Dirac fermions in topological insulators," *Adv. Opt. Mater.* 1, 886 (2013).

25. N. A. van Dantzig, and P. C. M. Planken, "Time-resolved far-infrared reflectance of n-type GaAs," *Phys. Rev. B* 59, 1586 (1999).

26. If one absorbed photon generates one photoinduced carrier, the maximum photoinduced carrier density can be estimated by $\Delta n = (1-R) \times F/(E \times \delta)$, where R=0.55 is the reflectance, F=3.3 $\mu J/cm^2$ is pumping fluence, E=2.48×10^{-19}J (= 1.55 eV) is the pumping photon energy, δ=23.5 nm is the penetration depth. For the pumping fluence of 3.3 $\mu J/cm^2$, the photoinduced carrier density Δn is around $2.54 \times 10^{18} cm^{-3}$. Figure S12 in Supporting Information further shows that the pump–probe experiments were performed at the weak perturbation limit and the linear response. Additionally, the interband transitions between BVB and BCB dominate the excitation process as discussed in the section S9 of Supporting Information. This indicates that most carriers in BCB and surface states still keep "cold" during pumping.

27. The electrons and holes in Dirac cone also possibly recombine cross the Dirac point. However, the asymmetric decay of carrier population in Dirac cone observed by time-resolved ARPES experiment[13] has demonstrated that the phonon-assisted cooling of hot carriers is the main relaxation process, which is the focus point in this study.

28. X. Zhu, L. Santos, R. Sankar, S. Chikara, C. Howard, F. C. Chou, C. Chamon, and M. El-Batanouny, "Interaction of phonons and Dirac fermions on the surface of Bi_2Se_3: a strong Kohn anomaly," *Phys. Rev. Lett.* 107, 186102 (2011).

29. X. Zhu, L. Santos, C. Howard, R. Sankar, F. C. Chou, C.; Chamon, and M. El-Batanouny, "Electron-phonon coupling on the surface of the topological insulator Bi_2Se_3 determined from surface phonon dispersion measurements," *Phys. Rev. Lett.* 108, 185501 (2012).

30. Z.-H. Pan, A. V. Fedorov, D. Gardner, Y. S. Lee, S. Chu, and T. Valla, "Measurement of an exceptionally weak electron-phonon coupling on the surface of the topological insulator Bi_2Se_3 using angle resolved photoemission spectroscopy," *Phys. Rev. Lett.* 108, 187001 (2012).

31. R. C. Hatch, M. Bianchi, D. Guan, S. Bao, J. Mi, B. B. Iversen, L. Nilsson, L. Hornekær, and P. Hofmann, "Stability of the Bi_2Se_3(111) topological state: Electron-phonon and electron-defect scattering," *Phys. Rev. B* 83, 241303 (2011).

32. P. B. Allen, "Theory of thermal relaxation of electrons in metals," *Phys. Rev. Lett.* 59, 1460 (1987).

33. V. Chis, I. Y. Sklyadneva, K. A. Kokh, V. A.; Volodin, O. E. Tereshchenko, and E. V. Chulkov, "Vibrations in binary and ternary topological insulators: first-principles calculations and Raman spectroscopy measurements," *Phys. Rev. B* 86, 174304 (2012).

34. K. S. Novoselov, A. K. Geim, S. V. Morozov, D. Jiang, M. I. Katsnelson, I. V. Grigorieva, S. V. Dubonos, and A. A. Firsov, "Two-dimensional gas of massless Dirac fermions in graphene. *Nature* 438, 197–200 (2005).

35. C.-W. Luo, H.-J. Wang, S.-A. Ku, H.-J. Chen, T.-T. Yeh, J.-Y. Lin, K.-H. Wu, J.-Y. Juang, B.- L. Young, T. Kobayashi, C.-M. Cheng, C.-H. Chen, K.-D. Tsuei, R. Sankar, F. Chou, K. Kokh, O. E. Tereshchenko, E. V. Chulkov, Yu. M. Andreev, and G. Gu, "Snapshots of Dirac fermions near the Dirac point in topological insulators," *Nano Lett.* 13, 5792–5802 (2013).

10.2.2 Helicity-Dependent Terahertz Emission Spectroscopy of Topological Insulator

10.2.2.1 INTRODUCTION

Recently, the exotic properties of topological surface states (TSSs) in three-dimensional topological insulators (TIs) have attracted much attention due to their potentials in the applications of spintronics [1–5]. TSSs have been confirmed using several techniques, such as angle-resolved photoemission spectroscopy (ARPES) [4–6] and scanning tunneling microscopy [7–10], etc. In terms of the optical coupling of TSSs, it has been demonstrated that the helicity-dependent photocurrent originating from TSSs can be manipulated by the optical helicity and these results demonstrate the potential for optoelectronic devices based on the TSSs of TIs [11–16]. However, these results alone are inadequate to fully unveil the intriguing characteristics of TSS-photon coupling because of the limitations of indispensable electrodes in transport measurements and, in particular, the effects of TIs' crystalline orientation on photocurrents and the large signal background from thermoelectric currents. To address these problems, contact-free techniques have been proposed as viable alternatives. Terahertz emission spectroscopy is a useful contact-free technique for spintronics. For instance, transient spin currents [17] and photocurrents [18] that are generated by optical pulses on magnetic heterostructures have been recently studied by using terahertz emission spectroscopy.

Here, we demonstrate the observation of helicity-dependent terahertz emissions that originate from helicity-dependent photocurrents in TI Sb_2Te_3 thin films. Using the time-domain decomposition and recombination of the terahertz signals, the time-domain traces of the circular photogalvanic effect (CPGE), the linear photogalvanic effect (LPGE), and the photon drag effect (PDE) coefficients are extracted individually. It is worth noting that both the CPGE and the LPGE show similar characteristics in both the time and frequency domains and their polarities are coincident with the rotational symmetry of the Dirac cone of TIs. Furthermore, the anisotropic PDE is also clearly identified by the unique analysis as well as direct measurements.

10.2.2.2 EXPERIMENTS

Optical pulses with a central wavelength of 800 nm and a pulse duration of ~75 fs were focused on the (111) surface of a Sb_2Te_3 thin film to generate terahertz radiation, as shown schematically in Figure 10.2.2.1a. The Sb_2Te_3 samples with 45-nm thickness were grown on ($1\bar{1}02$) sapphire substrates by using molecular beam epitaxy, and the TSSs are clearly observed by using ARPES, as shown in Figure 10.2.2.1c. The details of sample preparation and the terahertz emission measurement are discussed in Appendix A. When circularly polarized optical pulses excite the Sb_2Te_3 thin films at an incident angle θ, an asymmetric distribution is generated in a helical Dirac cone because of the selection rules for Dirac fermions. Hot carriers are also generated and annihilate via relaxation processes on a picosecond time scale [19–23]. For TSSs, in-plane helicity-dependent photocurrents are generated in the direction (along the x-axis) perpendicular to the plane of incidence [11–13] and radiate helicity-dependent terahertz emissions. Upon reversing the helicity of incident optical

DOI: 10.1201/9780429196577-70

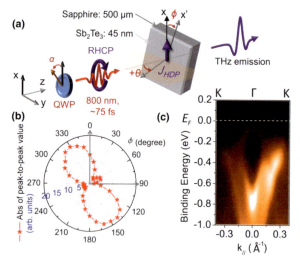

FIGURE 10.2.2.1 (a) Right-hand circularly polarized (RHCP) optical pulses illuminate a topological insulator Sb_2Te_3 thin film at an incident angle $+\theta$ and generate a helicity-dependent photocurrent in the direction perpendicular to the incident plane (y-z plane). The polarization of optical pulses is controlled by rotating a quarter-wave plate (QWP) with an angle α. (b) The crystal-orientation-dependent (ϕ-dependent) absolute peak-to-peak amplitude of terahertz emission waveforms with linear polarization of optical pulses (along the x axis, $\alpha = 0°$) at nearly normal incidence ($\theta \sim +1°$). (c) The ARPES image of a used sample in this study. The Fermi level is denoted as E_F.

pulses, the direction of the helicity-dependent photocurrents is also reversed. Therefore, the polarity reversal of the emitted terahertz pulses can be observed. The helicity-dependent effect scales with the incident angle θ, and it can be confirmed by changing the incident angle from $+\theta$ to $-\theta$ and vice versa.

The out-of-plane (i.e., perpendicular to the surface of TIs) bulk transient currents near the surface of TIs might contribute to terahertz emissions [24,25], and these can be avoided by polarization settings of terahertz detection (see Appendix A). In TIs, terahertz emissions originating from optical rectification (OR) have been observed recently [24–26]. In this study, we used linearly polarized (along the x-axis) optical pulses to generate terahertz radiation from the Sb_2Te_3 thin films. As shown in Figure 10.2.2.1b, the azimuthal-scan (ϕ scan) results for the peak-to-peak terahertz amplitudes of the terahertz emissions show twofold symmetry.

10.2.2.3 RESULTS AND DISCUSSION

The helicity-dependent terahertz emissions were characterized by rotating a quarter-wave plate (QWP) with an angle α at different incident angles θ, as shown in Figure 10.2.2.2. The α-scan measurements along different crystalline orientations of TIs give more profound insights into the origins of terahertz emissions. According to the results for OR that are shown in Figure 10.2.2.1b, two orientations are selected: $\phi=0°$ and $90°$ (see Appendix B). Figures 10.2.2.2a and b show the polar plots of terahertz waveforms (0–5 ps) for different-helicity optical excitations (α scan) at $\theta=-40°$ and $+40°$, respectively for $\phi=0°$. Twofold symmetries are clearly observed in the α-scan patterns. This phenomenon becomes more pronounced as the incident angle θ increases.

Figures 10.2.2.2c and d show that the time-domain terahertz waveforms from the Sb_2Te_3 thin film are generated by linearly polarized (LP, $\alpha=0°$), right-hand circularly polarized (RHCP, $\alpha=45°$), and left-hand circularly polarized (LHCP, $\alpha=135°$) optical excitations. The polarity of the emitted terahertz radiation is reversed because the photon helicity is reversed. Furthermore, for helicity-fixed optical excitation (both RHCP and LHCP), the polarity reversal also occurs when the incident angle alternates from $+40°$ to $-40°$. These results are consistent with the scenario for the helicity-dependent terahertz emission: The spin-polarized current that is generated by incident photon spin is the main contributor to the process.

In terms of the optical control of TSSs, helicity-dependent photocurrents that are generated by optical pulses in TIs have been reported [11–13] and these are described as $J_{HDP}(\alpha)=C \sin(2\alpha)+L_1 \sin(4\alpha)+L_2 \cos(4\alpha)+D$, where C is the coefficient of the helicity-dependent CPGE and L_1 describes the helicity-independent LPGE. Both C and L_1 are related to the TSSs, and it has been theoretically

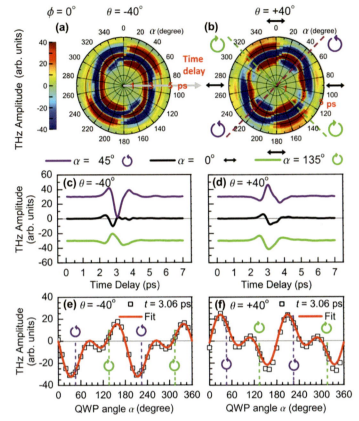

FIGURE 10.2.2.2 (a and b) Polar plots of terahertz waveforms (0–5 ps) as a function of α at $\theta = -40°$ and $+40°$, respectively, for $\phi = 0°$. The colors represent the amplitude of the terahertz emissions, (c and d) Terahertz waveforms for excitation with linearly polarized, righthand circularly polarized, and left-hand circularly polarized optical pulses at $\theta = -40°$ and $+40°$, respectively. (e), (f) The α-dependent terahertz amplitude at $t = 3.06$ ps. The red solid lines are the best fits with Eq. (10.2.2.1). The symbol $\leftrightarrow$, the counterclockwise arrow, and the clockwise arrow denote linearly polarized (black: $\alpha = 0°$, along the x axis), right-hand circularly polarized (purple: $\alpha = 45°$), and left-hand circularly polarized (green: $\alpha = 135°$) incident photons, respectively.

predicted that both the CPGE and LPGE are linked by the Berry phase in spin-orbit coupled quantum well structures [27]. Macroscopically, both the CPGE and LPGE can be described by third-rank tensors [11,14,28–30]. Both require the electric-field component of incident light in the direction of the sample's surface normal and are odd in the incident angle θ [14,28,30]. L_2 describes the PDE, and the bulk thermoelectric current contributes to D. The PDE can be described by a fourth-rank tensor and is associated with the linear momentum transfer between incident photons and electrons. In terms of the transient current, the terahertz electric field is described as $E_{THz}(\alpha, t) \propto \partial J_{HDP}(\alpha, t)/\partial t$. OR also contributes to the terahertz emissions. Therefore, the helicity-dependent terahertz emission from TIs is described as follows:

$$E_{THz}(\alpha, t) \propto C'(t)\sin(2\alpha) + L_1'(t)\sin(4\alpha) + L_2^1(t)\cos(4\alpha) + O(t) \qquad (10.2.2.1)$$

The coefficient $C'(t)$ describes the terahertz radiation that originates from the helicity-dependent CPGE. $L_1'(t)$ and $L_2'(t)$ are the coefficients for the LPGE and the PDE, respectively. OR mainly contributes to $O(t)$. OR is a second-order nonlinear optical process, and the nonlinear polarization of OR follows the time dependents of the incident light intensity [31]. Full descriptions about the dependences of CPGE, LPGE, PDE, and OR on ϕ, θ, and α are shown in Appendix C. To test this model, we use Eq. (10.2.2.1) to fit the experimental data by choosing the specific moment at which the α-dependent terahertz amplitude shows the largest variation. Figures 10.2.2.2e and f show the time-domain fits for the α scans [Figures 10.2.2.2a and b] with a time delay $t = 3.06$ ps at $\theta = -40°$ and $+40°$, for $\phi = 0°$.

Obviously, Eq. (10.2.2.1) fits the experimental results very well, and the polarity reversals are significant at the chosen moment. The time-domain traces are also produced for all coefficients and the results are discussed in detail later.

The separation of the signals due to the Dirac fermions from the massive bulk contributions in topological insulators is an important concern in modern condensed matter physics. ARPES measurements have successfully achieved this goal and the dynamics of Dirac fermions are also obtained by using a time-resolved ARPES technique [20,21,32]. To achieve a similar goal, we develop a method that is suitable for extracting all coefficients based on Eq. (10.2.2.1) for every time delay and separating the signals of the Dirac fermions from the other bulk contributions. Figures 10.2.2.3a–d show the time-domain traces of the coefficients $C'(t)$, $L_1'(t)$, $L_2'(t)$, and $O(t)$, which are extracted individually from different α scans, for $1\phi=0°$, and all data are offset to ensure greater clarity. Comparing Figure 10.2.2.3a with Figure 10.2.2.3b, all the extracted time-domain traces for the coefficients) show similar characteristics to those for the coefficients $C'(t)$. When the incident angle θ changes its sign, the polarities of the time-domain traces for both $C'(t)$ and $L_1'(t)$ change correspondingly. In the fast-Fourier-transform (FFT) spectra of $C'(t)$ and $L_1'(t)$ [Figures 10.2.2.3f and g], the peaks for all the spectra are located at ~0.61 THz, which is in particularly good agreement with the results for the time-domain analysis. These results strongly indicate that both $C'(t)$ and $L_1'(t)$ share the same physical origin.

It is also worth noting that the amplitude of the time-domain traces evolves with the incident angle θ. Considering the Fresnel coefficient and the procedure for optical-fluence normalization, Figure 10.2.2.3e shows the calibrated peak-to-peak terahertz amplitudes for all traces in Figures 10.2.2.3a–d as a function of θ. Obviously, the sign for the peak-to-peak terahertz amplitudes of both $C'(t)$ and $L_1'(t)$ changes when the sign of θ changes. The best fit for $C'(t)$ (red line) almost intersects the origin and this result is consistent with the characteristics of CPGE: The photon spin is orthogonal to the two-dimensional spin texture of the Dirac cone at normal incidence. For $L_1'(t)$ (black line), an offset is seen at $\theta=+1°$, which may be caused by OR. Although the sign change can also be observed for $L_2'(t)$ in Figure 10.2.2.3e, the polarity reversal seems to be difficult to identify in Figure 10.2.2.3c. This might be due to the interferences from OR. On the contrary, at $\phi=90°$, the polarity reversal features of the PDE are not only clearly observed in time-domain analysis (see Appendix C) but are also revealed by the direct measurements.

The photon drag effect has been observed in graphene by using terahertz emission spectroscopy [33]. In previous reports of photocurrent measurements [11–13], the coefficient $L_2'(t)$ associated with the PDE describes the experimental results well but PDE has only been confirmed by a recent terahertz laser-driven experiment [34]. In this study, the anisotropic PDE is not only directly measured

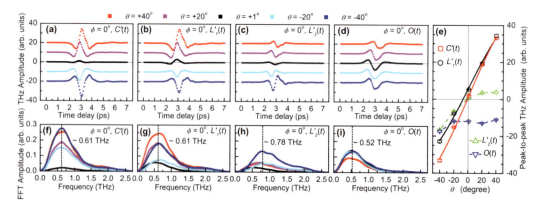

FIGURE 10.2.2.3   (a)–(d) The time-domain traces for the coefficients $C'(t)$, $L_1'(t)$, $L_2'(t)$, and $O(t)$ are extracted [using Eq. (10.2.2.1)] individually from the α-dependent terahertz amplitudes with various time delays t at $\phi = 0°$. In (a)–(d), all data are offset to ensure greater clarity. (f)–(i) The corresponding spectra of the time-domain traces in (a)–(d) by FFT. (e) The peak-to-peak terahertz amplitude of the time-domain traces in (a–d) as a function of the incident angle θ. (e) plot of peak-to-peak THz amplitude plotted against the incident angle θ. The red and black lines are the best sinusoidal fits for $C'(t)$, and $L_1'(t)$, respectively.

by linear-polarized optical excitation but also demonstrated by the time-domain recombination of $L_2'(t) + O(t) = 90°$. As shown in Figure 10.2.2.4a, the polarity of the time-domain terahertz waveforms that are generated by linear-polarized ($\alpha = 0°$) optical pulses changes when the sign of the incident angle θ changes from $+40°$ to $-40°$, which is direct evidence of PDE. Figure 10.2.2.4d shows the FFT spectra for the directly measured terahertz waveforms that are shown in Figure 10.2.2.4a. The peaks for all the spectra are located at a frequency of ~ 0.73 THz.

We also observe similar evidence in time-domain decomposition and recombination of the α scans at $\phi = 90°$ (see Appendix D). As mentioned previously OR can affect other helicity-independent terms and can have some residual contribution to the $L_2'(t)$ and $O(t)$ terms in the normal incidence case ($\theta = +1°$). It is worth emphasizing that these residuals are opposite in sign. Therefore, the time-domain traces of $L_2'(t)$ and $O(t)$ terms are combined, as shown in Figure 10.2.2.4b. A comparison of Figures 10.2.2.4a and b shows that the combined traces of $L_2'(t) + O(t)$ are almost the same as the directly measured terahertz waveforms at $\alpha = 0°$.

The FFT spectra of $L_2'(t) + O(t)$ [in Figure 10.2.2.4e], which has a peak at ~ 0.73 THz, also agree with that of the directly measured results in Figure 10.2.2.4d. In Figure 10.2.2.4c, all the peak-to-peak terahertz amplitudes for the direct measurements (direct PDE), $L_2'(t)$ and $L_2'(t) + O(t)$, change their signs when the sign of θ changes. These features are well simulated with the sinusoidal function. Obviously, the term $L_2'(t) + O(t)$ is much closer to the directly measured value than the $L_2'(t)$ term alone. These results not only verify the reliability of this time-domain analysis but also prove that $L_2'(t)$ clearly represents the PDE.

A comparison of the time-domain decomposition and recombination for $\phi = 0°$ and $90°$ shows that all the extracted time-domain traces of the coefficient $L_1'(t)$ behave similarly to that of the coefficient $C'(t)$. The polarities of the time-domain traces $C'(t)$ and $L_1'(t)$ at $\phi = 90°$ are the same as those

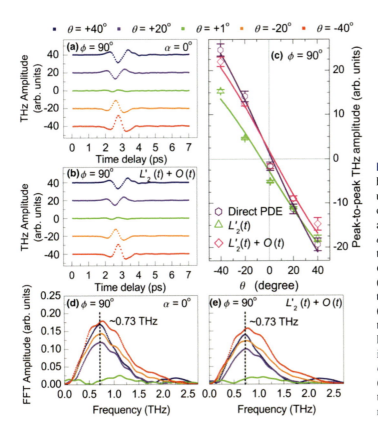

FIGURE 10.2.2.4 (a) The terahertz emissions are generated using linearly polarized ($\alpha = 0°$) optical pulses at various incident angles from $\theta = -40°$ to $+40°$, at $\phi = 90°$. (b) The combination of the time domain traces $L_2'(t)$ and $O(t)$ at $\phi = 90°$ (see Appendix D). (d and e) The corresponding spectra of the time-domain traces in (a), (b) by FFT. (c) The peak-to-peak terahertz amplitudes of the terahertz emissions from direct PDE in (a) (purple hexagons), $L_2'(t) + O(t)$ (pink diamonds), and $L_2'(t)$ (green triangles), as a function of the incident angle θ. The solid lines represent the best sinusoidal fits.

at $\phi = 0°$, and their FFT spectra also have the same shape with a peak at ~0.59 THz (see Appendix D). All these results are in good agreement with the rotational symmetry of the Dirac cone in TIs. Furthermore, in spin-orbit coupled quantum well structures, both CPGE and LPGE photocurrents associated with the Berry phase have been theoretically predicted to have equal magnitude [27]. In this study, the experimental results of the time-domain traces and the peak-to-peak analysis of both $C'(t)$ and $L_1'(t)$ are close at both $\phi = 0°$ and $90°$ [Figures 10.2.2.3e, and Figure 10.2.2.8e (in Appendix D)]. The tendency in the experimental observations agrees with this theoretical prediction.

Recently, time-resolved ARPES measurement of Sb_2Te_3 single crystals has been reported [22]. An asymmetric distribution in the Dirac cone due to circularly polarized optical excitation has been confirmed and these results are coincident with the results for helicity-dependent photocurrents in the transport experiments [11,13]. Based on the time-resolved ARPES results and considering transient-current radiation and far-field diffraction [35,36], we simulate the terahertz emission waveforms and spectra from different positions in the TSSs (see Appendix E). The peaks of terahertz spectra are located at around 0.56–0.60 THz for the positions near the Dirac point. ($E - E_F < 0.3$ eV, Figure 10.2.2.9d in Appendix E). These peak positions are coincident well with the peak positions of the $C'(t)$ and $L_1'(t)$ spectra for both $\phi = 0°$ and $90°$ [see Figure 10.2.2.3, and Figure 10.2.2.8 (in Appendix D)]. These results give strong evidence that the $C''(t)$ and $L_2'(t)$ originate from TSSs. Therefore, it can be expected that the photoexcited carriers near the Dirac point in the TSSs mainly contribute to the helicity-dependent terahertz emission from a TI Sb_2Te_3 under circularly polarized optical excitations.

Recently, some TI-terahertz-emission works based on circularly polarized excitation have been reported, and these works either use circular dichroism or single delay-time analysis [37,38]. Helicity-dependent terahertz emission measurements of Bi_2Se_3 thin films have been demonstrated, and a threefold periodicity with a constant offset in the azimuthal scan has been observed [38]. This observation has been ascribed to the circular photon drag effect, and the origin of the constant offset part is still unclear due to the limited information from experimental results and model fitting [38]. Regarding transient current radiation, the frequency of few-cycle terahertz pulses is inversely proportional to the carrier relaxation time. Longer carrier relaxation times result in lower-frequency terahertz emissions. It is well known that the carrier relaxation times in TSSs of the typical TIs, e.g., Bi_2Se_3 and Bi_2Te_3, are 10ps and are longer than those in the bulk states [21,23,32,39,40]. In the case of Bi_2Se_3, the carrier relaxation time in TSSs is ~10ps at 70 K [32], and the estimated frequency of the terahertz emissions from Bi_2Se_3 would be ~0.1 THz, which is close to the results (~0.23 THz) of Ref. [38]. For Sb_2Te_3 at room temperature, the carrier relaxation time in the TSSs is ~1.2 ps from time-resolved ARPES results [22]. Therefore, the estimated frequency of the terahertze missions from Sb_2Te_3 is ~0.8 THz, which is consistent with our results of 0.61 THz for CPGE and LPGE in this study. The time-domain decomposition-recombination method developed in this study could be further applied to other TIs.

10.2.2.4 SUMMARY AND CONCLUSIONS

In summary, we have demonstrated that the helicity-dependent terahertz emissions from topological insulator Sb_2Te_3 thin films can be manipulated by using ultrafast optical pulses. Using the time-domain decomposition and recombination, the terahertz waveforms that originate from the CPGE, the LPGE, and the PDE are extracted individually. Both the CPGE and the LPGE results agree with the rotational symmetry of the Dirac cone, as verified by different crystalline orientation measurements. Anisotropic PDE is also observed by both direct measurements using linearly polarized light and the time-domain decomposition-recombination analysis. Furthermore, the spectra of time-domain traces for the CPGE and the LPGE coefficients agree with the simulated terahertz spectra of the transient photoexcited carriers near the Dirac point, which are observed by using time-resolved ARPES. The observations of this study not only demonstrate the importance of field-resolved terahertz emissions but also pave the way toward applications of helicity-dependent terahertz emission spectroscopy in spintronics.

The contents in the present subsection including the following appendices are the products of the tight collaboration among the following people: Chien-Ming Tu, Yi-Cheng Chen, Ping Huang, Pei-Yu Chuang, Ming-Yu Lin, Cheng-Maw Cheng, Jiunn-Yuan Lin, Jenh-Yih Juang, Kaung-Hsiung Wu, Jung-Chun A. Huang, Way-Faung Pong, Takayoshi Kobayashi, and Chih-Wei Luo [41].

APPENDIX A: SAMPLE PREPARATION AND TERAHERTZ EMISSION MEASUREMENT

In this study, Sb_2Te_3 thin films were grown by using molecular beam epitaxy (MBE) on (1102)-R-plane sapphire substrates; in-plane mismatch is around 12%. The sapphire substrate was heated to 1000°C for1h to remove contaminants. High-purity Sb (99.999%) and Te (99.999%) were evaporated by Knudsen cells and the fluxes were calibrated *in situ* by using a quartz crystal microbalance. The base pressure for the MBE system was less than 1×10^{-10} Torr and the growth pressure for the Sb_2Te_3 thin films was maintained at less than 1×10^{-9} Torr. The Sb and Te effusion cell temperatures were selected so that the fluxing ratio Te/Sb was 12. The Sb deposition rate was 1A° /min and that for Te was 12A° /min. The substrate temperature was maintained at 230°C throughout the growth. The single-crystal structure of the film was obtained by using *in situ* reflection high-energy electron diffraction (RHEED), as shown in Figure 10.2.2.5a. The Sb_2Te_3 samples with 45-nm thickness were used for terahertz emission experiments, and a capping layer (~10nm) of Se was deposited on the surface of Sb_2Te_3 thin films to prevent oxidation or reaction with the TI thin films.

The crystal structure of R-plane sapphires is rectangle, and it shows a twofold symmetry. Although the symmetry of Sb_2Te_3 (111) is hexagonal, the lattice mismatch between Sb_2Te_3 thin films and R-plane sapphire substrates may be induced during the thin-film-growth process. Optical rectification (OR) is a second-order nonlinear optical process, and the φ-scan patterns of second-order nonlinear-optical effects strongly depend on the crystal structure of samples. We performed x-ray diffraction measurements on the samples. As shown in Figure 10.2.2.5b, the x-ray diffraction ϕ-scan (ϕ_{XRD}) pattern of a Sb_2Te_3 (111) thin film shows a twofold symmetry. This result is coincident with the OR-ϕ-scan pattern as shown in Figure 10.2.2.1b in the main text.

In the terahertz emission experiments, a mode-locked Ti:sapphire oscillator was used to generate an optical pulse train with a central wavelength of 800nm, a pulse duration of ~75fs, and a repetition rate of 5.1MHz. Optical pump pulses illuminated the (111) surface of the TI thin films to

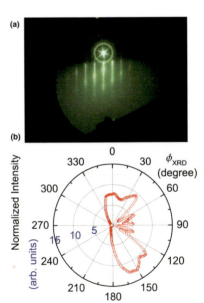

FIGURE 10.2.2.5 (a) The *in-situ* reflection high-energy electron diffraction (RHEED) pattern of the used Sb_2Te_3 thin film. (b) The x-ray diffraction ϕ-scan (ϕ_{XRD}) pattern of the Sb_2Te_3 thin film on a *R*-plane sapphire substrate.

generate terahertz radiation. The pulse energy of the pump beam was around 31.4 nJ and the spot size (diameter) on the surface of samples was about 450 μm at normal incidence. Therefore, the pump fluence was about 19.4 μJ/cm². The terahertz radiation emitted from the sample was collimated by two off-axis parabolic mirrors and focused on a 1-mm-thick <110> ZnTe slab for electro-optic sampling. A wire-grid polarizer was used to purify the polarization of the terahertz radiation along the x-axis (in-plane) and electro-optic sampling was also used to detect terahertz polarization along the x-axis. All the experiments were performed at room temperature and in a chamber purged by dry nitrogen gas, to avoid the absorption of water vapor. The helicity-dependent terahertz emissions were characterized by rotating a quarter-wave plate with an angle α (α scan). The rotation of the quarter-wave plate changes the photon polarization from linearly polarized ($\alpha=0°$), to right-hand circularly polarized ($\alpha=45°$), to linearly polarized ($\alpha=90°$), to left-hand circularly polarized ($\alpha=135°$), and to linearly polarized ($\alpha=180°$). This type of polarization change shows a period of 180°.

Firstly, we performed normal-incident excitation ($\theta=+1°$ and linearly polarized pump beam) and rotated the samples over 360° (OR ϕ-scan) to determine the contributions from OR. At $\phi=90°$, the contributions of OR almost reach to zero, and this means that the terahertz emission generated by other mechanisms can be unambiguously identified along this crystalline orientation. To double-check our idea, the other ϕ angle of 0° [30° offset from the OR maximum in Figure 10.2.2.1b of the main text] with a portion of the OR signal was selected for the same analyses as performed at $\phi=90°$. This offset of 30° between the OR maximum and $\phi=0°$ is just to avoid the large OR component and keep a reasonable amplitude of terahertz signal for the subsequent analyses. As shown in the main text, the time-domain traces, and spectra of both $C'(t)$ and $L_1'(t)$ at $\phi=0°$ are really coincident with those at $\phi=90°$.

APPENDIX B: TIME-DOMAIN FITS FOR THE HELICITY-DEPENDENT TERAHERTZ RADIATION AT $\Phi=0°$ AND 90°

Crystalline-orientation-dependent measurements can give a clearer insight into the origins of terahertz emissions. The α-scan patterns and the time-domain fits at different incident angles θ for $\phi=0°$ and 90° are shown in Figures 10.2.2.6 and 10.2.2.7, respectively. At nearly normal incidence [$\theta=+1°$, Figures 10.2.2.6(g) and 10.2.2.7(g)] in both orientations, obviously, the α-scan patterns show clear fourfold symmetry, and the main contributor is optical rectification (OR). By contrary, at oblique incidences ($\theta=-40°$ and $+40°$), for both $\phi=0°$ and 90°, the α-scan patterns show clear twofold symmetry, and it means that the helicity-dependent circular photogalvanic effect (CPGE) affects the results. For the same incident angle, for example, $\theta=+40°$, the α-scan patterns [Figures 10.2.2.6m and 10.2.2.7m] are similar but not identical. It indicates that the anisotropic photon drag effect (PDE) contributes to the results and it is discussed later.

According to Eq. (10.2.2.1) in the main text,

$E_{THz}(\alpha,t) \propto C'(t)\sin(2\alpha) + L_1'(t)\sin(4\alpha) + L_2'(t)\cos(4\alpha) + O(t)$ we use this equation to fit the experimental data at different time delay. Apparently, Eq. (10.2.2.1) fits the experimental results very well at all incident angles for both $\phi=0°$ and 90°, as shown in Figures 10.2.2.6 and 10.2.2.7.

In Figures 10.2.2.2d–f (see main text), the polarity reversals of the terahertz waveforms by right-hand circularly polarized (RHCP) and left-hand circularly polarized (LHCP) optical excitations are clearly demonstrated. However, the terahertz amplitudes for RHCP and LHCP optical excitations are not equal, as shown in Figures 10.2.2.6 and 10.2.2.7. This means that other helicity-independent effects are also involved in the terahertz emissions, and only circularly polarized optical excitation would not be sufficient to disentangle the intrinsic underlying physical mechanisms.

APPENDIX C: DEPENDENCE OF CIRCULAR

Photogalvanic Effect, Linear Photogalvanic Effect, Photon Drag Effect, and Optical Rectification on ϕ, θ, and α

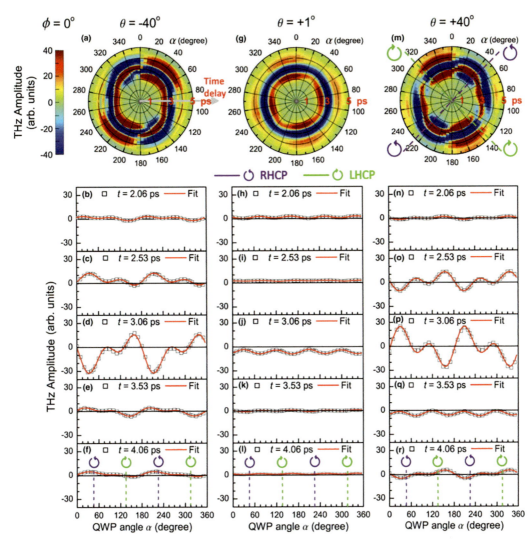

FIGURE 10.2.2.6 $\phi = 0°$: (a) $\theta = -40°$, (g) $\theta = +1°$, and (m) $\theta = +40°$ as defined in the top panels. The top panels show the polar plots for the terahertz waveforms (0–5 ps). The colors represent the amplitude of the terahertz radiation. (b)–(f) For $\theta = -40°$, the time-domain amplitudes of the terahertz waveforms at $t = 2.06, 2.53, 3.06, 3.53$, and 4.06 ps are shown. The red solid lines are the fits to the data. The results for $\theta = +1°$ ($\theta = +40°$) are shown in (h)–(l) [(n)–(r)].

Macroscopically, the photocurrent J_λ generated by the electric field $\mathbf{E}$ of incident light is shown as follows [11,14,28–30]:

$$J_\lambda = \sum_\mu \gamma_{\lambda\mu} i(\mathbf{E} \times \mathbf{E})_\mu + \frac{1}{2} \sum_{\mu\nu} \chi_{\lambda\mu\nu} \left(E_\mu E_\nu^* + E_\nu E_\mu^* \right) + \sum_{\delta\mu\nu} T_{\lambda\delta\mu\nu} q_\delta E_\mu E_\nu^*,$$

where $\gamma_{\lambda\mu}$ is a third-rank pseudotensor for the CPGE. $\chi_{\lambda\mu\nu}$ is a fourth-rank tensor for the LPGE. $T_{\lambda\delta\mu\nu}$ is a fourth-rank tensor for the PDE. q_δ is the photon linear momentum. The indices run through the spatial coordinate.

The CPGE photocurrents change sign as the polarization of the incident light changes from the right-hand circular polarization to the left-hand circular polarization [29,30]. Therefore, the CPGE

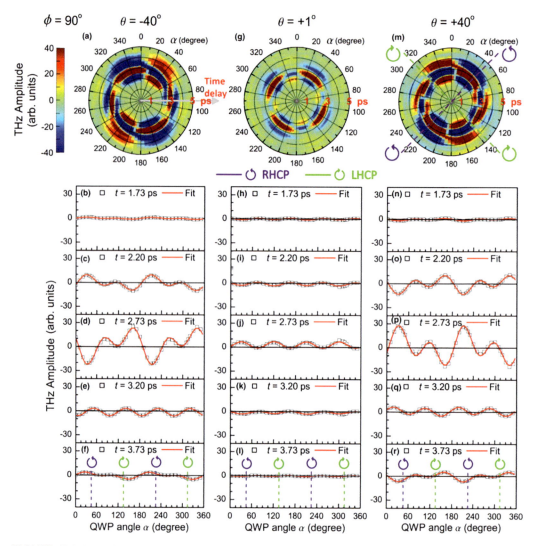

FIGURE 10.2.2.7 $\phi = 90°$ (a) $\theta = -40°$, (g) $\theta = +1°$, and (m) $\theta = +40°$ as defined in the top panels. The top panels show the polar plots for the terahertz waveforms (0–5 ps). The colors represent the amplitude of the terahertz radiation. (b)–(f) For $\theta = -40°$, the time-domain amplitudes for the terahertz waveforms at $t = 1.73$, 2.20, 2.73, 3.20, and 3.73 ps are shown. The red solid lines are the fits to the data. The results for $\theta = +1°$ ($\theta = +40°$) are shown in (h)–(l) [(n)–(r)].

shows 2α periodicity. On the other hand, because the LPGE has no response to the changes in the helicity of the incident light as well as the second-order dependence on the electric field of the incident light, the LPGE shows 4α periodicity [14,30]. Macroscopically, both the CPGE and LPGE require the electric-field component in the direction of the sample's surface normal, and both are odd in the incident angle θ [14,28]. As the incident angle of light changes sign, e.g., from $+\theta$ to $-\theta$, with respect to the surface normal of a sample, the incident electrical-field component (in the direction of the surface normal) changes sign as well, and the direction of the LPGE photocurrent also reverses. Therefore, the time-domain traces of the LPGE flip.

In quantum well structures, it has been theoretically predicted that both CPGE and LPGE are linked by the Berry phase, and the Berry phase generates both PGEs with equal magnitude [27]. Microscopically, for TIs, the CPGE originates from the TSSs and satisfies the rotational symmetry of the Dirac cone [29]. Therefore, the CPGE should show the same characteristics at different crystal orientations. Nevertheless, a fully microscopic theory of the LPGE in TIs is still absent.

TABLE 10.2.2.1

Dependence of CPGE, LPGE, PDE and OR on ϕ, θ, and α

	ϕ: Crystalline Orientation	θ: Incident Angle ($\theta \to -\phi$)	α: QWP Angle
CPGE	Macroscopic: $\gamma_{\lambda\mu}$	Polarity: sign change	2α-Periodicity
	Macroscopic:TSS[a]		
	ϕ-Independent		Sin2α
LPGE	Macroscopic: $\chi_{\lambda\mu\nu}$	Polarity: sign change	4α-Periodicity
	Macroscopic:TSS[a]		
	ϕ-Independent		Sin4α[b]
PDE	$T_{\lambda\delta\mu\nu}$	Polarity: sign change	4α-periodicity
	ϕ-Dependent		Cos4α[c]
OR	$\chi^{(2)}$[d]	Polarity: no sign change	4α-periodicity
	ϕ-Dependent		Cos4α[e]

[a] Both CPGE and LPGE are linked by Berry phase in quantum well structures [27].
[b] The experimental results (time-domain traces and spectra)of $L_1'(t)$ show that the Sin4α term is dominant and are coincident with those of $C'(t)$.
[c] The experimental results (time-domain traces and spectra)of $L_2'(t)$ show that the cos4α term is dominant and are coincident with those of the direct PDE.
[d] $\chi^{(2)}$: The second-order nonlinear-optical tensor [31].
[e] OR contributes mainly to the cos4α term, and it is confirmed by using a <110> ZnTe single crystal as a terahertz emitter through OR effect (see Figure 10.2.2.10 in Appendix F)

The PDE is described by a fourth-rank tensor $T_{\lambda\delta\mu\nu}$. It depends on not only photon linear momentum q_δ but also the polarization of incident light [14,28]. The main feature of the PDE is that its sign changes as the incident angle of the linearly polarized incident light reverse from $+\theta$ to $-\theta$. It may also show the anisotropic behaviors (i.e., ϕ dependent) because of the fourth-rank tensor.

OR is one of the second-order nonlinear-optical processes, and basically, it is described by the third-rank nonlinear-optical tensor $\chi^{(2)}$ which is the same as the well-known second-harmonic generation in non-centrosymmetric materials (i.e., P OR$=\varepsilon_0\chi^{(2)}|E|^2$, P OR is the nonlinear polarization of OR, and ε_0 is the electric permittivity of free space). This indicates that the nonlinear polarization P OR of OR depends on the crystal orientation and the intensity profile of incident light [31]. Besides, both right-hand circular polarization and left-hand circular polarization of the incident light are identical in the OR process. Therefore, it shows 4α periodicity, and no polarity reversal for OR as the sign of the incident angle θ changes. The details of the dependences of CPGE, LPGE, PDE, and OR on ϕ, θ, and α are shown in Table 10.2.2.1.

APPENDIX D: TIME-DOMAIN DECOMPOSITION AND RECOMBINATION OF THE α-DEPENDENT TERAHERTZ WAVEFORMS AT $\Phi = 90°$

The time-domain decomposition-recombination procedure is also applied to the terahertz waveforms measured at $\phi=90°$. At $\phi=90°$, OR reaches almost zero, and this means that the terahertz emission generated by other mechanisms can be unambiguously identified along this crystalline orientation. In Figures 10.2.2.8a and b, all the extracted time-domain traces of the coefficients $L_1'(t)$ show characteristics similar to that of the coefficients $C'(t)$. The fast-Fourier-transform (FFT) spectra for both $C'(t)$ and $L_1'(t)$ have the same shape with a peak at ~0.59 THz and this value is remarkably close to that at $\phi=0°$ (0.61 THz). As shown in Figure 10.2.2.8c, the polarity of the time domain traces $L_2'(t)$ is reversed when the sign of the incident angle θ changes. This is the main feature for photon drag effect.

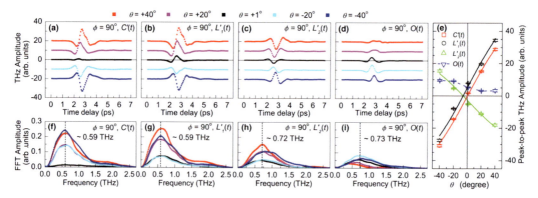

FIGURE 10.2.2.8 (a)–(d) The time-domain traces for the coefficients $C'(t)$, $L_1'(t)$, $L_2'(t)$, and $O(t)$ are extracted individually from different α scans [Figures 10.2.2.7a–c, $\theta = -40°$, $+1°$, and $+40°$] at $\phi = 90°$. In (b), all extracted time-domain traces for the coefficients $L_1'(t)$ show characteristics similar to that of the coefficients $C'(t)$ in (a). In (c), the polarity of the time-domain traces $L_2'(t)$ is reversed when the sign of the incident angle θ changes. (a)–(d) All of the data are offset to ensure greater clarity. (f)–(i) The corresponding FFT spectra for the time-domain traces in (a)–(d). (f) All of the spectral centers for the coefficients $C'(t)$ are located at 0.59 THz, which coincides with those for the coefficients $L_1'(t)$ in (g). (e) The peak-to-peak values for the time-domain traces shown in (a–d) as a function of the incident angle θ. The red, black, and green lines are the best sinusoidal fits for $C'(t)$, $L_1'(t)$, and $L_2'(t)$, respectively.

APPENDIX E: ESTIMATION OF THE TERAHERTZ-EMISSION SPECTRA FOR DIRAC FERMIONS BY USING PHOTOEMISSION DYNAMICS FROM TIME-RESOLVED ARPES MEASUREMENTS

Time-resolved ARPES measurements on TIs provide valuable insights into the carrier dynamics of TIs under circularly polarized optical excitation. Recently, direct optical transitions from deeper-lying bulk states to TSSs in Sb_2Te_3 single crystals at room temperature have been observed [22]. Using circularly polarized optical pulses (800 nm, 1.55 eV) causes an asymmetric distribution in the TSSs and these results are in agreement with the results for helicity-dependent photocurrent in the transport experiments. We extracted the ARPES image and the photoemission dynamics of the TSSs from Ref. [22] and the digitalized figures are shown in Figures 10.2.2.9a and b. Figure 10.2.2.9a shows the direct optical transitions from deeper lying bulk states to the TSSs and the circularly polarized optical pulses result in an asymmetric distribution in the TSSs.

Figure 10.2.2.9b shows the photoemission dynamics of different positions (1–5) in the TSSs and the relaxation time of the photoemission intensity increases as the position in the TSSs becomes closer to the Dirac point. In Ref. [22], the positions in which energy is less than ~0.3 eV above the Fermi level show relaxation properties similar to that of position 5. Therefore, we choose positions 1–5 for discussion.

The photocurrent density is described by the standard expression, $J = -e \sum_k v_g(k) n(k)$, where k is the momentum state in the band, $v_g(k)$ is the group velocity, and $n(k)$ is the distribution function [28,42,43]. In terms of transient current radiation, the relaxation time of photoexcited carriers dominates the terahertz emission. Thus, $\partial n(k)/\partial t$ is mainly responsible for the terahertz emission process. Finally, we can simulate the terahertz emission waveforms and spectra from the distribution variation of different states in the TSSs. By taking the far-field diffraction [35,36] into account, the terahertz waveforms from the contributions of positions 1–5 in the TSSs are obtained and shown in Figure 10.2.2.9c. The corresponding FFT spectra are shown in Figure 10.2.2.9d. The spectral centers tend toward the low-frequency region as the position becomes closer to the Dirac point. This result agrees with the characteristics of

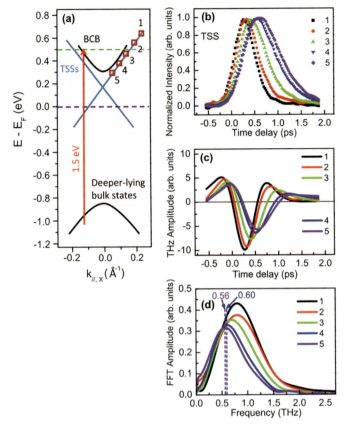

FIGURE 10.2.2.9 (a) The schematics for direct optical transitions from deeper-lying bulk states to TSSs in Sb_2Te_3. (b) The normalized photoemission intensities at the different positions (energy-momentum windows described in Ref. [22]) of the TSSs as a function of the pump probe time delay in the time-resolved ARPES measurements. Both (a) and (b) are digitalized figures that are extracted from Ref. [22]. In (a) the numbers 1-5 seem to indicate the time course after pulse excitaion. In (b) and (c), gradual growth delay ca be clearly observed. In (d), spectral peak shit with time in the direction of lower frequency (red shift) can be observed. Considering transient-current radiation with far-field diffraction, the terahertz waveforms and FFT spectra for the contributions from different positions in the TSS are shown in (c) and (d), respectively.

the time-domain photoemission intensity traces. Surprisingly, the spectral centers (0.56–0.60 THz) of positions 4 and 5 ($E - E_F < 0.3eV$) are coincident with that of both $C'(t)$ and $L'_1(t)$ for $\phi = 0°$ and $90°$ in this study.

These results provide strong evidence that $C'(t)$ and $L'_1(t)$ originate from TSSs. Nevertheless, all of the positions in the TSSs must be considered (sum over states in the TSSs) for the final terahertz emission. It is noted that most electrons are concentrated in the region near the Dirac point [22] due to the electron bottleneck effect near the Dirac point [32]. Therefore, it is to be expected that the photoexcited carriers near the Dirac point in the TSSs mainly contribute to helicity-dependent terahertz emission from a TI Sb_2Te_3 under circularly polarized optical excitations. In general, the filter function of electro-optic sampling (EOS) is necessary to be considered for final terahertz waveforms and spectra. However, the effects of EOS filter function can be neglected due to the flat spectrum in the low-frequency region (<2 THz) [44–46].

Rashba spin-split bulk states would result in spin-polarized photocurrent, and spin-polarized photocurrent may contribute to helicity-dependent terahertz emissions. In general, surface heavy doping and gas absorptions would result in Rashba splitting for the bulk bands. It has been shown that the contribution from Rashba spin-split bulk states in Bi_2Se_3 can be neglected due to the cancellation effect of the two oppositely spin-polarized Fermi surfaces of Rashba bulk bands as well as difficulties for strong band bending without surface doping [11]. For MBE-grown Sb_2Te_3 thin films, it has been shown that the robustness of the TSSs in Sb_2Te_3 is more superior to those in Bi_2Se_3 and Bi_2Te_3 by electrical transport measurements [47]. Furthermore, it has been shown that the aging effect (surface band bending) in Sb_2Te_3 thin films is not obvious, and even the Rashba effect has not been observed by ARPES measurements [48]. This means that strong surface band bending has not been observed in MBE-grown Sb_2Te_3 thin films. Therefore, we believe that the

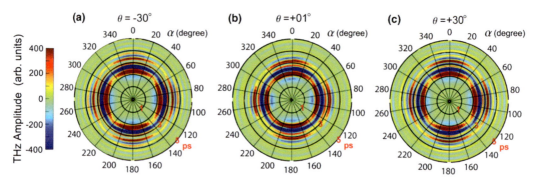

FIGURE 10.2.2.10 (a)–(c) α-scan patterns for terahertz waveforms (0–5 ps) from a 110 ZnTe single crystal at $\theta = -30°$, + 1°, and +30°.

contributions from Rashba spin-split bulk states to CPGE and LPGE can be neglected in our experimental results.

Furthermore, the spectra centers (~0.73THz) of PDE (dominated by bulk states) all show blue-shift with respect to the spectra (~0.61 THz) of CPGE and LPGE. This indicates that terahertz emission spectra from the bulk states are higher than those from TSSs.

APPENDIX F: HELICITY-INDEPENDENT TERAHERTZ RADIATION FROM A <110>ZnTe SINGLE CRYSTAL

To confirm our model, we also use a popular terahertz emitter, a <110> ZnTe single crystal of 1-mm thickness, to perform an α scan for $\theta=-30°$, +1°, and +30° under the same condition. The mechanism of the terahertz emission from ZnTe is OR. As shown in Figure 10.2.2.10, apparently, all α-scan patterns show the same 4α periodicity (cos4α), and no polarity reversal for terahertz waveforms as the incident angle of optical pulses changes from $\theta=-30°$ to +30°.

REFERENCES

1. J. E. Moore, *Nature* **464**, 194 (2010).
2. M. Hasan, and C. Kane, *Rev. Mod. Phys.* **82**, 3045 (2010).
3. X.-L. Qi, and S.-C. Zhang, *Phys. Today* **63**, 33 (2010).
4. Y. L. Chen, J. G. Analytis, J.-H. Chu, Z. K. Liu, S.-K. Mo, X. L. Qi, H. J. Zhang, D. H. Lu, X. Dai, Z. Fang, S. C. Zhang, I. R. Fisher, Z. Hussain, and Z.-X. Shen, *Science* **325**, 178 (2009).
5. Y. Xia, D. Qian, D. Hsieh, L. Wray, A. Pal, H. Lin, A. Bansil, D. Grauer, Y. S. Hor, R. J. Cava, and M. Z. Hasan, *Nat. Phys.* **5**, 398 (2009).
6. D. Hsieh, D. Qian, L. Wray, Y. Xia, Y. S. Hor, R. J. Cava, and M. Z. Hasan, *Nature* **452**, 970 (2009).
7. P. Roushan, J. Seo, C. V. Parker, Y. S. Hor, D. Hsieh, D. Qian, A. Richardella, M. Z. Hasan, R. J. Cava, and A. Yazdani, *Nature* **460**, 1106 (2009).
8. T. Zhang, P. Cheng, X. Chen, J.-F. Jia, X. Ma, K. He, L. Wang, H. Zhang, X. Dai, Z. Fang, X. Xie, and Q.-K. Xue, *Phys. Rev. Lett.* **103**, 266803 (2009).
9. Z. Alpichshev, J. G. Analytis, J.-H. Chu, I. R. Fisher, Y. L. Chen, Z. X. Shen, A. Fang, and A. Kapitulnik, *Phys. Rev. Lett.* **104**, 016401 (2010).
10. S. Kim, M. Ye, K. Kuroda, Y. Yamada, E. E. Krasovskii, E. V. Chulkov, K. Miyamoto, M. Nakatake, T. Okuda, Y. Ueda, K. Shimada, H. Namatame, M. Taniguchi, and A. Kimura, *Phys. Rev. Lett.* **107**, 056803 (2011).
11. J. W. McIver, D. Hsieh, H. Steinberg, P. Jarillo-Herrero, and N. Gedik, *Nat. Nanotechnol.* **7**, 96 (2012).
12. J. Duan, N. Tang, X. He, Y. Yan, S. Zhang, X. Qin, X. Wang, X. Yang, F. Xu, Y. Chen, W. Ge, and B. Shen, *Sci. Rep.* **4**, 4889 (2014).
13. C. Kastl, C. Karnetzky, H. Karl, and A. W. Holleitner, *Nat. Commun.* **6**, 6617 (2014).

14. P. Olbrich, L. E. Golub, T. Herrmann, S. N. Danilov, H. Plank, V. V. Bel'kov, G. Mussler, Ch. Weyrich, C. M. Schneider, J. Kampmeier, D. Grützmacher, L. Plucinski, M. Eschbach, and S. D. Ganichev, *Phys. Rev. Lett.* **113**, 096601 (2014).

15. J. Tang, L.-T. Chang, X. Kou, K. Murata, E. S. Choi, M. Lang, Y. Fan, Y. Jiang, M. Montazeri, W. Jiang, Y. Wang, L. He, and K. L. Wang, *Nano Lett.* **14**, 5423 (2014).

16. K. N. Okada, N. Ogawa, R. Yoshimi, A. Tsukazaki, K. S. Takahashi, M. Kawasaki, and Y. Tokura, *Phys. Rev. B* **93**, 081403 (2016).

17. T. Kampfrath, M. Battiato, P. Maldonado, G. Eilers, J. Nötzold, S. Mährlein, V. Zbarsky, F. Freimuth, Y. Mokrousov, S. Blügel, M. Wolf, I. Radu, P. M. Oppeneer, and M. Münzenberg, *Nat. Nanotechnol.* **8**, 256 (2013).

18. T. J. Huisman, R. V. Mikhaylovskiy, J. D. Costa, F. Freimuth, E. Paz, J. Ventura, P. P. Freitas, S. Blügel, Y. Mokrousov, Th. Rasing, and A. V. Kimel, *Nat. Nanotechnol.* **11**, 455 (2016).

19. D. Hsieh, F. Mahmood, J. W. McIver, D. R. Gardner, Y. S. Lee, and N. Gedik, *Phys. Rev. Lett.* **107**, 077401 (2011).

20. Y. H. Wang, D. Hsieh, E. J. Sie, H. Steinberg, D. R. Gardner, Y. S. Lee, P. Jarillo-Herrero, and N. Gedik, *Phys. Rev. Lett.* **109**, 127401 (2012).

21. M. Hajlaoui, E. Papalazarou, J. Mauchain, G. Lantz, N. Moisan, D. Boschetto, Z. Jiang, I. Miotkowski, Y. P. Chen, A. Taleb Ibrahimi, L. Perfetti, and M. Marsi, *Nano Lett.* **12**, 3532 (2012).

22. J. Sánchez-Barriga, E. Golias, A. Varykhalov, J. Braun, L. V. Yashina, R. Schumann, J. Minár, H. Ebert, O. Kornilov, and O. Rader, *Phys. Rev. B* **93**, 155426 (2016).

23. C. W. Luo, H. J. Wang, S. A. Ku, H.-J. Chen, T. T. Yeh, J.-Y. Lin, K. H. Wu, J. Y. Juang, B. L. Young, T. Kobayashi, C.-M. Cheng, C.-H. Chen, K.-D. Tsuei, R. Sankar, F. C. Chou, K. A. Kokh, O. E. Tereshchenko, E. V. Chulkov, Yu. M. Andreev, and G. D. Gu, *Nano Lett.* **13**, 5797 (2013).

24. C. W. Luo, H.-J. Chen, C. M. Tu, C. C. Lee, S. A. Ku, W. Y. Tzeng, T. T. Yeh, M. C. Chiang, H. J. Wang, W. C. Chu, J.-Y. Lin, K. H. Wu, J. Y. Juang, T. Kobayashi, C.-M. Cheng, C.-H. Chen, K.-D. Tsuei, H. Berger, R. Sankar, F. C. Chou, et al., *Adv. Opt. Mater.* **1**, 804 (2013).

25. C.-M. Tu, T.-T. Yeh, W.-Y. Tzeng, Y.-R. Chen, H.-J. Chen, S.-A. Ku, C.-W. Luo, J.-Y. Lin, K.-H. Wu, J.-Y. Juang, T. Kobayashi, C.-M. Cheng, K.-D. Tsuei, H. Berger, R. Sankar, and F. -C. Chou, *Sci. Rep.* **5**, 14128 (2015).

26. L. G. Zhu, B. Kubera, K. F. Mak, and J. Shan, *Sci. Rep.* **5**, 10308 (2015).

27. J. E. Moore and J. Orenstein, *Phys. Rev. Lett.* **105**, 026805 (2010).

28. S. D. Ganichev, and W. Prettl, *J. Phys. Condens. Matter* **15**, R935 (2003).

29. P. Hosur, *Phys. Rev. B* **83**, 035309 (2011).

30. A. Junck, Ph.D. thesis, Freie Universität Berlin, 2015.

31. R. W. Boyd, *Nonlinear Optics*, 3rd ed., Academic Press, London, p. 7, (2008)

32. J. A. Sobota, S. Yang, J. G. Analytis, Y. L. Chen, I. R. Fisher, P. S. Kirchmann, and Z.-X. Shen, *Phys. Rev. Lett.* **108**, 117403 (2012).

33. J. Maysonnave, S. Huppert, F. Wang, S. Maero, C. Berger, W. de Heer, T. B. Norris, L. A. De Vaulchier, S. Dhillon, J. Tignon, R. Ferreira, and J. Mangeney, *Nano Lett.* **14**, 5797 (2014).

34. H. Plank, L. E. Golub, S. Bauer, V. V. Bel'kov, T. Herrmann, P. Olbrich, M. Eschbach, L. Plucinski, C. M. Schneider, J. Kampmeier, M. Lanius, G. Mussler, D. Grützmacher, and S. D. Ganichev, *Phys. Rev. B* **93**, 125434 (2016).

35. J. W. Goodman, *Introduction to Fourier Optics*, McGraw-Hill, New York, pp. 53–54 (1988).

36. P. Kužel, M. A. Khazan, and J. Kroupa, *J. Opt. Soc. Am. B* **16**, 1795 (1999).

37. L. Braun, G. Mussler, A. Hruban, M. Konczykowski, T. Schumann, M. Wolf, M. Münzenberg, L. Perfetti, and T. Kampfrath, *Nat. Commun.* **7**, 13259 (2016).

38. S. Y. Hamh, S.-H. Park, S.-K. Jerng, J. H. Jeon, S.-H. Chun, and J. S. Lee, *Phys. Rev. B* **94**, 161405 (2016).

39. M. Hajlaoui, E. Papalazarou, J. Mauchain, L. Perfetti, A. Taleb Ibrahimi, F. Navarin, F. Monteverde, P. Auban-Senzier, C. R. Pasquier, N. Moisan, D. Boschetto, M. Neupane, M. Z. Hasan, T. Durakiewicz, Z. Jiang, Y. Xu, I. Miotkowski, Y. P. Chen, S. Jia, H. W. Ji, et al., *Nat. Commun.* **5**, 3003 (2014).

40. R. Valdés Aguilar, J. Qi, M. Brahlek, N. Bansal, A. Azad, J. Bowlan, S. Oh, A. J. Taylor, R. P. Prasankumar, and D. A. Yarotski, *Appl. Phys. Lett.* **106**, 011901 (2015).

41. C. –M. Tu, Y. –C. Chen, P. Huang, P. –Y. Chuang, M.-Y. Lin, C. –M. Cheng, J. –Y. Lin, J. –Y. Juang, K. –H. Wu, J. –C. A. Huang, W. –F. Pong, T. Kobayashi, and C.-W. Luo, *Phys. Rev. B* **96**, 195407 (2017).

42. S. D. Ganichev and W. Prettel, *Phys. E (Amsterdam, Neth.)* **14**, 166 (2002).

43. A. Junck, G. Refael, and F. von Oppen, *Phys. Rev. B* **88**, 075144 (2013).

44. G. Gallot, J. Zhang, R. W. McGowan, T.-I. Jeon, and D. Grischkowsky, *Appl. Phys. Lett.* **74**, 3450 (1999).

45. G. Gallot, and D. Grischkowsky, *J. Opt. Soc. Am. B* **16**, 1204 (1999).
46. C. M. Tu, S. A. Ku, W. C. Chu, C. W. Luo, J. C. Chen, and C. C. Chi, *J. Appl. Phys.* **112**, 093110 (2012).
47. Y. Takagaki, A. Giussani, K. Perumal, R. Calarco, and K.-J. Friedland, *Phys. Rev. B* **86**, 125137 (2012).
48. G. Wang, X. Zhu, J. Wen, X. Chen, K. He, L. Wang, X. Ma, Y. Liu, X. Dai, Z. Fang, J.-F. Jia, and Q.-K Xue, *Nano Res.* **3**, 874 (2010).

10.2.3 Femtosecond Time-Evolution of Mid-Infrared Spectral Line Shapes of Dirac Fermions in Topological Insulators

Time-resolved spectroscopy is important in various fields, such as determining the exotic carrier dynamics of Tis [1–7]. The photon energy (~100 meV) of a MIR is less than the bulk band gap of TIs and has a quite different energy to the resonance energy of phonon absorptions. Therefore, MIR light sources are eminently suited to the study of SSTs in topological surface states (TSSs). The existing literature [8–22] reports the existence of a spectral line shape in the MIR region but there is no clear consensus. The explanation for FCA based on the Drude model has been adapted [11,12,17], but some studies give conflicting results [14,18] with considering more resonance factors. SSTs have also been reported [8–22] but these studies do not clarify the absorption mechanisms for SSTs and FCA using static MIR spectroscopic techniques.

This study unambiguously demonstrates the time evolution of MIR spectral line shapes in TIs using an optical pump and ultra-broadband MIR probe spectroscopy [23]. The MIR probe-pulses with a supercontinuum of 200–5000 cm^{-1} (or 25–620 meV) and a pulse width of 8.2 fs are generated using four-wave different-frequency generation (DFG) in nitrogen gas. *This novel spectroscopy technique has the advantages of a wide bandwidth for standard Fourier-transform-infrared spectroscopy (FTIR) [24] and it allows femtosecond time-resolution by generating ultrashort pulses from nonlinear crystals using DFG.* Two types of TI crystals are used for the experiments in this study. One is n-type Bi_2Te_2Se with a bulk/surface carrier concentration of $12.5 \times 10^{18} cm^3 / 5.5 \times 10^{12} cm^2$, which is a bulk-conduction-electron-rich crystal. The other is p-type information), which features a higher ratio of surface to bulk carrier concentration. Figure 10.2.3.1 shows the clear presence of a bulk-conduction band (BCB) in Bi_2Te_2Se, but not in Sb_2TeSe_2.

10.2.3.1 RESULTS

Ultra-broadband MIR $\Delta R/R$ spectra of FCA and SSTs in topological insulators. The typical ultra-broadband MIR $\Delta R/R$ spectra for Bi_2Te_2Se and Sb_2TeSe_2 are respectively shown in Figure 10.2.3.2a and b. These two spectra are significantly different. Along the wavenumber axis, there is a positive change in the lower frequency region and a negative change in the high-frequency region, which indicates a blue shift in the plasma edge for Bi_2Te_2Se after pumping (see Figure 10.2.3.2c). The zero-crossing line, $L_{0,X}$ (dashed line), in Figure 10.2.3.2a also shows a rapid blue shift at the beginning of the delay time and then slowly (>50 ps) returns to the original position. However, the value of $\Delta R/R$ for Sb_2TeSe_2 shows a red shift in the plasma edge after pumping. It is worthy of note that the zero-crossing line, $L_{0,X}$ (dashed line) in Figure 10.2.3.2b is red-shifted until ~2 ps and then returns to the original position at ~6 ps, which is much faster than the

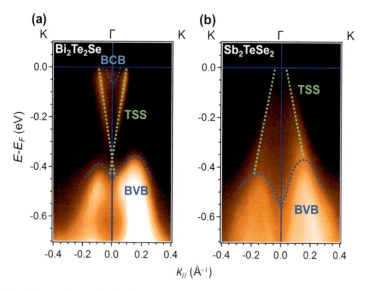

FIGURE 10.2.3.1 The angle-resolved photoemission spectroscopy (ARPES) images of Bi_2Te_2Se and Sb_2TeSe_2 single crystals: (a) The ARPES image of a Bi_2Te_2Se single crystal measured with 22 eV photon energy. (b) The ARPES image of a Sb_2TeSe_2 single crystal was measured with 24 eV photon energy. All single crystals were the same pieces as those used in ultrafast experiments for the consistency of all measurements. The single crystals were *in-situ* cleaved under a base pressure 5.1×10^{11} torr at 85 K just before measurements. ARPES experiment was conducted National Synchrotron Radiation Research Center in Taiwan using a BL21B1 beamline. The photoemission spectra were recorded with a Scienta R4000 hemispherical analyzer. The polarization vector was always in the angular dispersion plane. The overall energy resolution is about 12 meV. The green dash lines represent as the TSS of crystals, and the blue dash lines show the bulk-conduction-band (BCB) and bulk valance-band (BVB). The Dirac point in Sb_2TeSe_2 was estimated at 189 meV above the Fermi level. A notable difference of band structure exists between Bi_2Te_2Se and Sb_2TeSe_2, the Dirac point of Bi_2Te_2Se is embedded in the BVB. In contrast to Bi_2Te_2Se, Sb_2TeSe_2 has an isolated Dirac cone and surface carriers cannot be scattered easily by bulk carriers. This difference in their band structure makes a significant difference in optical measurement results.

change for Bi_2Te_2Se. Generally, there is a blue shift in the plasma edge because there is an increase in the carrier concentration [25], which is explained by the Drude model. The red shift in the $\Delta R/R$ spectrum of Sb_2TeSe_2 until ~2 ps is not explained by the Drude model because there is a decrease in the carrier concentration after pumping. It is found that the SST model using the Kubo formula [20] (SST-Kubo model), which has been successfully used to explain the transitions of Dirac cone in graphene [20], explains the novel phenomena that are observed in p-type Sb_2TeSe_2.

By comparing the band mapping results of Bi_2Te_2Se and Sb_2TeSe_2 in Figure 10.2.3.1, a notable difference between Bi_2Te_2Se and Sb_2TeSe_2 can be found that the Dirac point of Bi_2Te_2Se is embedded in the BVB. The surface carriers cannot avoid scattering from bulk carriers, and the major change of optical property might be dominated by bulk carrier. In contrast to Bi_2Te_2Se, Sb_2TeSe_2 has an isolated Dirac cone and thus the surface carriers cannot be scattered easily by bulk carriers, that is why the SST is a major factor in Sb_2TeSe_2. Besides, the difference between bulk FCA of the Bi_2Te_2Se and SST of Sb_2TeSe_2 could be attributed to the intrinsic responses with a 1.55-eV excitation. As the schematics of Figure 10.2.3.4e and j, the final and initial states of excitation process are different. The photoexcited carriers of the former are excited from the valence band maximum to the second conduction band [2,4,26], which is far from the Fermi level. For the latter case, the photoexcited carriers are excited from a deep valance band to the states near Fermi level consisting of an isolated Dirac cone[5]. Therefore, the MIR probe beam tends to detect the free carriers of conduction band in Bi_2Te_2Se, and the SST near Fermi level in Sb_2TeSe_2.

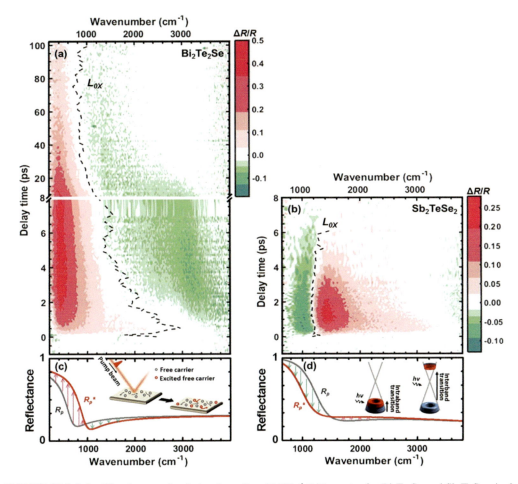

FIGURE 10.2.3.2 The time-resolved ultra-broadband MIR $\Delta R/R$ spectra for Bi_2Te_2Se and Sb_2TeSe_2 single crystals and the schematics of the theoretical model: (a) and (b) the 2D plots of wavenumber- and time-resolved reflectance change ($\Delta R/R$) spectra with an optical pump fluence of 101 μJ/cm^2 for Bi_2Te_2Se (a) and Sb_2TeSe_2 (b) single crystals. The red and green colors (low frequency range and high frequency range correspondingly in case (a) and vice versa in case (b)). respectively represent the parts with a positive change and a negative change. The zero-crossing line is marked L_0. as a black dashed line. (c) shows the p-polarized reflectivity before pumping (R_p, gray solid-line. Assume N is 12.5×10^{18} cm^{-1}, so $\omega_p = 1880$ cm^{-1} with $m^* = 0.32$ and $\varepsilon_0 = 23.7$) and after pumping (R_p^*, red solid-line. Assume N is 12.5×10^{18} cm^{-1} so $\omega_p = 2630$ cm^{-1} with $m^* = 0.32$ and $\varepsilon_0 = 23.7$) for the Drude model and (d) shows the p-polarized reflectivity before pumping (R_p, gray solid-line. Assuming $\mu = 50$ meV at room temperature) and after pumping (R_p^*, red solid-line. Assuming $\mu = 40$ meV at room temperature) for the SST-Kubo model.

Quantitative analysis of the ultra-broadband MIR $\Delta R/R$ spectra. To quantitatively reveal the hidden mechanism, the Drude model and the SST-Kubo model are used to fit the ultra-broadband MIR $\Delta R/R$ spectra for n-type Bi_2Te_2Se and p-type Sb_2TeSe_2 TIs. It is initially assumed that before and after pumping, all reflectivity R_p (gray solid-line, before pumping) and R_p^* (red solid-line, after pumping) have similarly shaped spectra for both the Drude model (Figure 10.2.3.2c) and the SST-Kubo model (Figure 10.2.3.2d). After pumping, the reflection spectrum shifts because there is an increase in the free carrier concentration. In terms of the Drude model, the dielectric function ε_D is:

$$\varepsilon_D(\omega) = \varepsilon_\infty - \frac{\omega_p^2}{\omega^2 + i\Gamma\omega} \qquad (10.2.3.1)$$

where ε_∞ is the permittivity at an infinite frequency, ω is the frequency, ω_p is the plasma frequency and Γ is the plasma scattering rate. The carrier concentration N is related to the effective mass m^* by the equation, $N = m^* \omega_p^2 / 4\pi e^2$. However, Falkovsky et al. estimated the reflectivity by considering the SSTs [20]. The dielectric function using the Kubo formula is:

$$\varepsilon_F(\omega) = -\frac{8T}{\left(\omega^2 + i\tau^{-1}\omega\right)d_{TSS}} \left(\frac{2\pi e^2}{h}\right) \ln\left[2\cosh\left(\frac{\mu}{2T}\right)\right] + \frac{1}{d_{TSS}}\left(\frac{2\pi e^2}{h}\right)$$

$$\times \left[\frac{i\pi}{\omega}G\left(\frac{\omega}{2}\right) - 4\int_0^\infty d\xi G(\xi) - \frac{\left[G(\xi) - G\left(\frac{\omega}{2}\right)\right]}{\left\{\omega^2 - 4\xi^2\right\}}\right] \tag{10.2.3.2}$$

where μ is the chemical potential, T is the carrier temperature, G is the Fermi-Dirac distribution function, τ^{-1} is the collision rate for TSSs, which depends on the density of impurities, and d_{TSS} is the optical penetration depth of the TSSs. The first and second terms respectively represent the intra-band transitions and the inter-band transitions in the Dirac cone. Both models are applied under the "quasi-equilibrium" state in a view of sub-10 fs probe pulse. The penetration depth of ultra-broadband MIR in TIs is few μm.

As previously mentioned, the increase of N in the Drude model represents the change in the electronic population after pumping. In Figure 10.2.3.2c, the estimated value of N for R_p^* is larger than that for R_p, which results in a blue shift in the plasma edge. In the SST-Kubo model, the photoexcitation has a significant impact on μ and T and induces changes in the reflection spectrum. In terms of the ground state of p-type Sb_2TeSe_2, both the smaller number of carriers in the vicinity of the Dirac point and the higher electron temperature results in a reduction in μ [20]. Therefore, after pumping, the reduction in the chemical potential μ causes a change in the reflection spectrum from R_p to R_p^*, as shown in Figure 10.2.3.2d. This result is in good qualitative agreement with the $\Delta R/R$ spectrum in Figure 10.2.3.2b.

Ultrafast time-evolution of the ultra-broadband MIR $\Delta R/R$ spectra. Figure 10.2.3.3 shows the typical time-evolution of the MIR $\Delta R/R$ spectrum and the fitted curves. As mentioned previously, the photoexcited carrier dynamics in n-type Bi_2Te_2Se is dominated by FCA and can be fitted well with the Drude model, as shown in Figure 10.2.3.3a. For Sb_2TeSe_2, the contribution of FCA to the photoexcited carrier dynamics cannot be neglected. Therefore, the $\Delta R/R$ spectra are fitted with the modified dielectric function of $\varepsilon_{DF}(\omega + \delta\omega) = \varepsilon_D(\omega + \delta\omega) + \varepsilon_F(\omega + \delta\omega)$, where $\delta\omega$ is a shifted frequency in fitting (This is called the Drude-SST-Kubo model). Figure 10.2.3.3b shows that this model fits the MIR $\Delta R/R$ spectrum at various delay times quite well. The details of the fitting are presented in the Method section.

10.2.3.2   DISCUSSION

The fitting results in Figure 10.2.3.4a and b are of interest, in particular the time evolutions of ω_p, Γ, N, μ, and T in TIs. During the pumping process, the 1.55-eV pump photons excite the electrons to a higher BCB from the occupied states [1]. For Bi_2Te_2Se, both ω_p and Γ respectively exhibit growth and relaxation dynamics. Although it is difficult to obtain the real value of N because there is no m^*, it is still possible to obtain the temporal evolution of N through $N \propto \omega_p^2$, as shown in Figure 10.2.3.4c and d. The huge shift of ω_p (~3.7 times after photo-excitation) is equivalent to the dramatic enhancement of photo-excited concentration. This photoexcited carrier mainly experiences FCA in bulk states (BSs), as shown by the notation of probe(1) and probe(2) in Figure 10.2.3.4e, or in TSSs, as shown by the notation of probe(3). A bi-exponential decay function is further used to obtain the reduction times for the

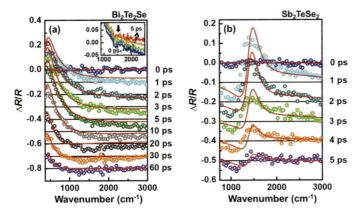

FIGURE 10.2.3.3 Typical ultra-broadband MIR $\Delta R/R$ spectra with fitting curves, taking account of free carrier absorption (FCA) and surface state transition (SST). $\Delta R/R$ as a function of the wavenumber for different delay times for (a) Bi_2Te_2Se and (b) Sb_2TeSe_2, using a pump fluence of 101 µJ/cm². The open circles represent the experimental data and each $\Delta R/R$ spectrum is shifted for clarity. The red lines are the fitting curves with the Drude model and b the Drude-SST-Kubo model ($\delta\omega = -690$ cm^{-1}). The insert in (a) shows the Moss-Burstein shift, as indicated by the arrow. Between the top solid line and the bottom solid line are the spectra that are respectively obtained by averaging the data from 0–1 ps (red), 1–2 ps (orange), 2–3 ps (bright green), 3–4 ps (green) and 4–5 ps (blue) ps.

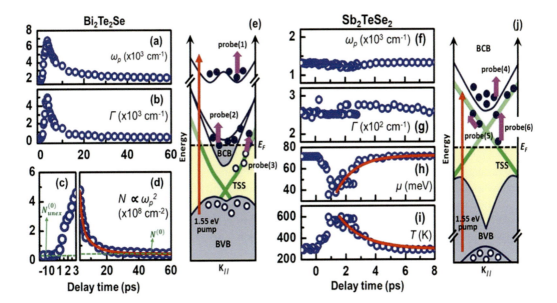

FIGURE 10.2.3.4 (a and b) The time-evolution of ω_p, Γ, μ, T and the schematic for a Bi_2Te_2Se crystal. (c and d) show the partial trace of the squared values of ω_p before c and after 3 ps (d) The red line in d shows the bi-exponential fitting that is described in the Method section. The green dashed lines in (c) marked by $N_{unex}^{(0)}$ (3.42×10^7 cm^{-2}) relate to the concentration of unexcited carriers and (d) marked by $N^{(0)}$ (4.16×10^7 cm^{-2}) represent the height of the constant term from the fitting curve. The time-domain traces for (f) ω_p, (g) Γ, (h) μ and (i) carrier temperature T are obtained using the Drude-SST-Kubo model. The red lines in (h), (i) show the single-exponential fitting that is mentioned in the Method section. The red arrows and pink arrows respectively show the 1.55 eV (800 nm) pumping and the MIR probing in (e), (j). The notation E_F in (e), (j) is the Fermi energy.

concentration of photoexcited carriers. This has a maximum within ~2.2 ps and then undergoes two relaxation processes for 1.5 and 8.4 ps. The fast relaxation process is caused by the thermal diffusion in BCB and TSS [27,28], or the acoustic-phono assistant process [3]. The slow one is consistent with the results of time-resolved ARPES [1,2]. Additionally, the appearance of a Moss-Burstein shift (until ~6 ps) near the bulk band gap (see the inset in Figure 10.2.3.3a) also indicates the recombination in BSs [15]. However, the value of N (i.e., $N^{(0)}$ in Figure 10.2.3.4d) does not recover to its original value (i.e., $N_{unex}^{(0)}$ in Figure 10.2.3.4c) within the limited delay time (~ 100 ps). This inconsistency between $N^{(0)}$ and $N_{unex}^{(0)}$ is explained by the long-lived recombination process. There are several scenarios proposed for this long relaxation process. First, it is generally assigned to the photo-voltage effect [29]. Moreover, a huge Rashba-splitting effect has been clearly observed in BCB [30,31], which might cause long-time relaxation processes like indirect band-gap semiconductors [25].

Here in the following, Rashba-splitting is briefly explained.

The Rashba effect is a momentum-dependent splitting of spin bands in bulk crystals and low-dimensional condensed matter systems (such as heterostructures and surface states) similar to the splitting of particles and anti-particles in the Dirac Hamiltonian. The splitting is a combined effect of spin–orbit interaction and asymmetry of the crystal potential, in particular in the direction perpendicular to the two-dimensional plane as applied to surfaces and heterostructures. The Rashba spin-orbit coupling is typical for systems with uniaxial symmetry, e.g., for hexagonal crystals of CdS and CdSe for which it was originally found and perovskites, and also for heterostructures where it develops as a result of a symmetry breaking field in the direction perpendicular to the 2D surface.

For p-type Sb_2TeSe_2, the Drude-SST-Kubo model is used to fit the results in Figure 10.2.3.4f–i. It is worth emphasizing that the relative changes in μ and T are more distinct than the changes in ω_p and Γ. Even though the MIR probe-pulses can also detect FCA (even though it originates from TSSs or BSs as shown by the notation of probe(4) in Figure 10.2.3.4j), the $\Delta R/R$ spectra are significantly dominated by SST's in the Dirac cone (see the notations of probe(5) and probe(6)). Figure 10.2.3.4h shows that after the deep valance electrons are excited to the upper Dirac cone [5], μ reaches a minimum at ~1 ps and it takes ~1.28 ps for the recombination process according to the fitting for a single exponential decay function. Besides, the hot-carrier temperature reaches ~600 K and recovers to room temperature after 1.68 ps, which results are consistent with the time-resolved ARPES results for Sb_2Te_3 [5–7]. Therefore, this hot-carrier temperature decay would have resulted from the thermal diffusion between BVB and TSS [27].

By taking account of the difference in the number of states between bulk and surface, when the electrons are photo-excited, the chemical potential should shift towards the higher energy direction. In the Drude-SST-Kubo model, the chemical potential (μ) and the carrier temperature (T) are associated with the surface state, and the SST-Kubo term is consisted of inter-transition and intra-transition of the Dirac cone. For $\mu \gg T$, the intra-transition term could be derived from the form [20]

$$\varepsilon_{F,\text{intra}} = -2e^2 |\mu|/h\left(\omega^2 + i\omega\Gamma_{im}\right)$$

which coincides with the Drude expression, and the effective plasma frequency $\omega_{p,s}$ could be further expressed as $\sqrt{2e^2|\mu|/he^2}$. More precisely, the contribution of excited charges to "ω_p" in Dirac cone is considered by the intra-transition term. In other words, the Drude term in the Drude-SST-Kubo model represents the excited carriers which are out of the surface state. For p-type Sb_2TeSe_2 with the Fermi level located at the lower energy part of the Dirac cone, after photoexcitation, the chemical potential shifts to the higher energy direction, and further indicates the redshift of the plasma edge and decreasing of the density of states. From the fitting result of smaller ω_p (~1.3×10^3 cm^{-1}) on Sb_2TeSe_2, it shows the lower contribution from the excited bulk carriers, which is consistent with the results in Figure 10.2.3.2b.

Summary. The ultrafast dynamics of Dirac fermions and bulk-free carriers in the TIs, n-type Bi_2Te_2Se and p-type Sb_2TeSe_2 single crystals, are studied using time-resolved ultra-broadband

MIR spectroscopy. The dynamics in the n-type Bi_2Te_2Se is dominated by bulk carriers because the $\Delta R/R$ spectra show a blue shift in the plasma edge due to FCA. For p-type Sb_2TeSe_2, the dynamics is dominated by the Dirac fermion from the red shift of the plasma edge in the $\Delta R/R$ spectra. This study shows that the MIR absorption peaks for FCA and SST in TIs can be distinguished and demonstrates the importance of time-resolved ultra-broadband MIR spectroscopy for gapless or small band gap exotic materials. The research described in this subsection includes the following appendices was performed by the following people in collaboration [32]: Tien-Tien Yeh, Chien-Ming Tu, Wen-Hao Lin, Cheng-Maw Cheng, Wen-Yen Tzeng, Chen-Yu Chang, Hideto Shirai, Takao Fuji, Raman Sankar, Fang-Cheng Chou, Marin M. Gospodinov, Takayoshi Kobayashi, Chih-Wei Luo.

10.2.3.3   METHODS

Experimental setup. Optical pump and ultra-broadband MIR probe spectroscopy [23] consists of three stages: (i) 800-nm optical pulses with a duration of 30 fs were generated, (ii) ultra-broadband MIR probe pulses were generated in nitrogen and (iii) chirped pulses were generated for detection. The fundamental pulses (800 nm) and the second harmonic pulses (400 nm, which were generated by a type I ®-BaB_2O_4 crystal with a thickness of 0.1 mm) from a Ti:sapphire amplifier (790 nm, 30 fs, 0.85 mJ at 1 kHz, Femtopower compactPro, FEMTOLASERS) were focused into nitrogen gas to generate MIR pulses. The filamentation occurred *via* four-wave DFG when the pulse was focused using a concave mirror ($r = 1$ m). The length of the filament was ~3 cm. The bandwidth and the duration of the generated MIR pulses were 200–5000 cm^{-1} and 8.2 fs, respectively. When the MIR pulses were reflected from the sample with an incident angle of 45°, they were converted to ~400-nm pulses for detection using a chirped-pulse up conversion (CPU) in nitrogen gas. A third 800-nm beam was transmitted through dispersive materials, including four BK7 glass plates (thickness: 10 mm) and one ZnSe plate (thickness: 5 mm), to produce chirped pulses. The converted visible (VIS) spectrum was measured by a spectrometer with an electron-multiplying CCD camera (SP-2358 and ProEM+1600, Princeton Instruments). The time resolution was estimated to be ~60 fs. To prevent significant absorption from vapor, the system was placed in boxes whose interior was purged with nitrogen.

     Retrieving the MIR spectra from an up-converted spectra and calibrating the spectra of the VIS pulse to MIR region. The MIR spectrum form, especially the sharp absorption peaks, can be seriously distorted after CPU measurements. That is to say, the dispersion of chirped pulses causes additional oscillations in the spectrum [33,34]. The CPU signal ($E_{CP}^2 (t-\tau) E_{MIR}^*(t)$) was obtained by performing four-wave DFG (FWDFG, $E_{FWM}(t)$) between the chirped pulse ($E_{CP}^2 (t-\tau)$) and the MIR pulse ($E_{MIR}(t)$). The chirped pulse is written as:

$$E_{CP}(t) = \varepsilon_{CP}(t) \exp\left(i\left(\omega^{(0)}t + \omega^{(1)}t^2(1/2)\right)\right) \tag{10.2.3.3}$$

where $\varepsilon_{CP}(t)$ represents the envelope, $\omega^{(0)}$ is the central angular frequency, and $\omega^{(1)}$ is a chirp parameter. The MIR pulse can be divided into the main part $E_{MIR}^{(0)}(t)$ and a free induction decay part $E_{MIR}^{(1)}(t)$. Substituting $E_{MIR}(t) = E_{MIR}^{(0)}(t) + E_{MIR}^{(1)}(t)$ yields:

$$E_{FWM}(t) = E_{CP}^2(t)E_{MIR}^{(0)}{}^*(t) + E_{CP}^2(t)E_{MIR}^{(1)}{}^*(t) = E_{FWM}^{(0)}(t) + E_{FWM}^{(1)}(t) \tag{10.2.3.4}$$

where $E_{FWM}^{(0)}(t)$ can be assumed to be the Dirac delta function $\delta(t)$ due to the short duration of MIR pulse. Using the Wiener–Khinchin theorem and these assumptions, the autocorrelation $C_A(t)$ of $E_{FWM}(t)$ is formed by the following equation. [33]

$$C_A(t) = \int dt' E_{FWM}^*(t') E_{FWM}(t'+t)$$

$$= \delta(t) + E_{MIR}^{(1)*}(t)\varepsilon_{CP}^2(t)e^{i2\omega^{(0)}t + i\omega^{(1)}t^2} + E_{MIR}^{(1)}(-t)\varepsilon_{CP}^{*2}(-t)e^{i2\omega^{(0)}t - i\omega^{(1)}t^2} \tag{10.2.3.5}$$

A similar autocorrelation form e (1) $2t$ sign$(\cdot)t$, so that Eq. $(C_A{}^2(\cdot)t$ is obtained for a pulse that is up-converted using a monochromatic pulse [5]) becomes

$$C_A'(t) = \delta(t) + E_{\mathrm{MIR}}^{(1)*}(t)\varepsilon_{\mathrm{CP}}^2(t)e^{i2\omega^{(0)}t} + E_{\mathrm{MIR}}^{(1)}(-t)\varepsilon_{\mathrm{CP}}^{*2}(-t)e^{i2\omega^{(0)}t} \tag{10.2.3.6}$$

Therefore, the original MIR spectrum with shift $2\omega^{(0)}t$ is acquired using the measured up-converted power spectrum and the known value of $\omega^{(1)}$ for the chirped pulse. Finally, the wavenumber is calibrated using a binomial fitting of the three absorption peaks, including carbon dioxide ($\sim$2300 cm^{-1}) and water vapor ($\sim$1600 and $\sim$3700 cm^{-1}).

Analyses using the Drude, SST-Kubo and Drude-SST-Kubo models. In this study, the dielectric function Σ in the Drude model, the SST-Kubo model and the Drude-SST-Kubo model are used to calculate the p-polarized reflectivity R_p using the Fresnel equation (with an incident angle of 45°) as:

$$R_p = \frac{\varepsilon\cos\theta - \sqrt{\varepsilon - \sin^2\theta}}{\varepsilon\cos\theta + \sqrt{\varepsilon - \sin^2\theta}} \tag{10.2.3.7}$$

The transient $\Delta R/R$ is obtained by:

$$\frac{\Delta R}{R} = \frac{R_P^* - R_p^0}{R_p^0} \tag{10.2.3.8}$$

where the superscripts "*" and "0" of R_p respectively represent the reflectivity with and without optical pumping. The fitting with the Drude model is performed using the software, RefFIT [34]. The fitting with the Drude-SST-Kubo model uses 4 parameters: ω_p, Γ, μ, and T. To limit the computational load without losing accuracy, the grid search method and an interval search algorithm with few iterations are used. After obtaining all possible values for these 4 parameters, the most appropriate parameter set P^j is selected by calculating the minimum root-mean-square deviation between the data and the calculated results at the jth iteration. More specifically, using the grid search method, the value of P^s at the jth iteration can be obtained. The best interval is decided using the neighboring points of P^j. In this analysis, 4 parameters produce the 8 neighboring points. Using this interval, the next iteration $j+1$ of the grid search is undertaken. Therefore, the accuracy is exponentially increased.

The conditions, R_p^0, are determined using the ARPES results and the FTIR spectra. For Bi$_2$Te$_2$Se, R_p^0 is calculated using the Drude model with $\varepsilon = 23.7$, $\omega_p = 1880$ cm^{-1} and $\Gamma = 272$ cm^{-1}, which values are obtained by fitting the FTIR spectra using the RefFIT program [35]. For Sb$_2$TeSe$_2$, R_p^0 is determined using the Drude-SST-Kubo model with $\varepsilon_\infty = 19.4$, $\Gamma_p = 1320$ cm^{-1}, $\Gamma = 253$ cm^{-1}, $d_{\mathrm{TSS}} = 1.4$ nm, $\mu = 72$ meV and $T = 297$ K. The former 4 parameters are obtained by fitting with fixed values of μ and T using the grid search method and an interval search algorithm, as described previously. If μ is sufficiently large, it can be estimated as:

$$\mu = \sqrt{\pi N_{\mathrm{TSS}}}\,\hbar\upsilon_{\mathrm{TSS}} \tag{10.2.3.9}$$

where N_{TSS} is the surface carrier concentration ($\sim$2.2$\times$10^{12} cm^{-2}). The parameter N_{TSS} is expressed as:

$$N_{\mathrm{TSS}} = \frac{A_{\mathrm{FS}}}{A_{\mathrm{BZ}}A_{\mathrm{UC}}} = \frac{\pi K_F^2}{\left(\dfrac{4\pi^2}{a^2}\right)(a^2)} = \frac{K_F^2}{4\pi} \tag{10.2.3.10}$$

where A_{FS} is the area of the Fermi surface, A_{BZ} is the area per Brillouin zone, A_{UC} is the area per unit cell and K_F is the Fermi-wavenumber ($\sim$5.2$\times$10^6 cm^{-1} from ARPES). The parameter $\upsilon_{\mathrm{TSS}} = 4.12\times10^7$ cm/s is estimated from the gradient of the Dirac cone from ARPES.

Exponential fitting in Figure 10.2.3.4. The red line in Figure 10.2.3.4d shows the bi-exponential fitting for $N^{(0)}+N^{(1)}\exp[-t/\tau_{N,1}]+N^{(2)}\exp[-t/\tau_{N,2}]$ for ω_p^2 (proportional to the time evolution of N) with a delay time of t, where the parameters $N^{(0)}=4.16\times10\,\text{cm}^{-2}$, $N^{(1)}=1.79\times10^9\,\text{cm}^{-2}$, $N^{(2)}=2.6\times10^8\,\text{cm}^{-2}$, $\tau_{N,1}=1.5$ ps, and $\tau_{N,2}=8.4$ ps. The red line in (h) shows the single-exponential fitting for $\mu^{(0)}+\mu^{(1)}\exp[-t/\tau_\mu]$ for the transient chemical potential $\mu(t)$, where $\mu^{(0)}=72$ meV is static chemical potential, $\mu^{(1)}$ is 99 meV and $\tau_\mu = 1.28$ ps. The red curve in Figure 10.2.3.4i is fitted using a single-exponential function of $T^{(0)}+T^{(1)}\exp[-t/\tau_T]$ and the time evolution of the temperature, where $T^{(0)}$ represents the room temperature, $T^{(1)}$ is 770 K and τ_T is 1.68 ps.

REFERENCES

1. M. Hajlaoui, et al., "Tuning a Schottky barrier in a photoexcited topological insulator with transient Dirac cone electron-hole asymmetry," *Nat. Commun.* **5**, 3003 (2014).
2. M. Neupane, et al., "Gigantic surface lifetime of an intrinsic topological insulator," *Phys. Rev. Lett.* **115**, 116801 (2015).
3. J. Qi, et al., "Ultrafast carrier and phonon dynamics in Bi_2Se_3 crystals," *Appl. Phys. Lett.* **97**, 182102 (2010).
4. M. C. Wang, S. Qiao, Z. Jiang, S. N. Luo, and J. Qi, "Unraveling photoinduced spin dynamics in the topological insulator Bi_2Se_3," *Phys. Rev. Lett.* **116**, 036601 (2016).
5. J. Sánchez-Barriga, et al., "Ultrafast spin-polarization control of Dirac fermions in topological insulators," *Phys. Rev. B* **93**, 155426 (2016).
6. S. Zhu, et al., "Ultrafast electron dynamics at the Dirac node of the topological insulator Sb_2Te_3," *Sci. Rep.* **5**, 13213 (2015).
7. J. Reimann, J. Gudde, K. Kuroda, E. V. Chulkov, and U. Hofer, "Spectroscopy and dynamics of unoccupied electronic states of the topological insulators Sb_2Te_3 and Sb_2Te_2S," *Phys. Rev. B* **90**, 081106 (2014).
8. W. S. Whitney, et al., "Gate-variable mid-infrared optical transitions in a $(Bi_{1-x}Sb_x)_2Te_3$ topological insulator," *Nano Lett.* **17**, 255–260 (2017).
9. C. W. Luo, P. S. Tseng, H.-J. Chen, K. H. Wu, and L. J. Li, "Dirac fermion relaxation and energy loss rate near the Fermi surface in monolayer and multilayer graphene," *Nanoscale.* **6**, 8575 (2014).
10. C. W. Luo, et al., "Snapshots of Dirac fermions near the Dirac point in topological insulators. *Nano Lett.* **13**, 5797–5802 (2013).
11. T. Dong, R.-H. Yuan, Y.-G. Shi, and N.-L. Wang, "Temperature-induced plasma frequency shift in Bi_2Te_3 and $Cu_xBi_2Se_3$," *Chin. Phys. Lett.* **30**, 127801 (2013).
12. S. V. Dordevic, M. S. Wolf, N. Stojilovic, H. Lei, and C. Petrovic, "Signatures of charge inhomogeneities in the infrared spectra of topological insulators Bi_2Se_3, Bi_2Te_3 and Sb_2Te_3," *J. Phys.: Condens. Matter.* **25**, 075501 (2013).
13. S. V. Dordevic, et al., "Fano q-reversal in topological insulator Bi_2Se_3," *J. Phys.: Condens. Matter.* **28**, 165602 (2016).
14. S. V. Dordevic, et al., "Magneto-optical effects in $Bi_{1-x}As_x$ with $x=0.01$: Comparison with topological insulator $Bi_{1-x}Sb_x$ with $x=0.20$," *Phys. Status Solidi B* **251**, 1510–1514 (2014).
15. A. D. LaForge, et al., "Optical characterization of Bi_2Se_3 in a magnetic field: Infrared evidence for magnetoelectric coupling in a topological insulator material," *Phys. Rev. B* **81**, 125120 (2010).
16. P. D. Pietro, et al., "Optical conductivity of bismuth-based topological insulators," *Phys. Rev. B* **86**, 045439 (2012).
17. C. Martin, et al., "Bulk Fermi surface and electronic properties of $Cu_{0.07}Bi_2Se_3$," *Phys. Rev. B.* **87**, 201201 (2013).
18. A. A. Reijnders, et al., "Optical evidence of surface state suppression in Bi-based topological insulators," *Phys. Rev. B.* **89**, 075138 (2014).
19. K. F. Mak, L. Ju, F. Wang, and T. F. Heinza, "Optical spectroscopy of graphene: From the far infrared to the ultraviolet," *Solid State Commun.* **152**, 1341–1349 (2012).
20. L. A. Falkovsky, "Optical properties of graphene," *J. Phys.: Conf. Ser.* **129**, 012004 (2008).
21. Y. Yao, et al., "Electrically tunable metasurface perfect absorbers for ultrathin midInfrared optical modulators," *Nano Lett.* **14**, 6526–6532 (2014).
22. Y. Wang, et al., "Observation of ultrahigh mobility surface states in a topological crystalline insulator by infrared spectroscopy," *Nat. Commun.* **8**, 366 (2017).
23. H. Shirai, T.-T. Yeh, Y. Nomura, C.-W. Luo, and, T. Fuji, "Ultrabroadband midinfrared pump-probe spectroscopy using chirped-pulse up-conversion in gases," *Phys. Rev. Appl.* **3**, 051002 (2015).

24. G. D. Smith, and R. A. Palmer, in *Handbook of Vibrational Spectroscopy*, edited by J. M. Chalmers, and P. R. Griffiths, Vol. 1, J. Wiley and Sons, New York, (2006).

25. T.-T. Yeh, et al., "Ultrafast carrier dynamics in Ge by ultra-broadband mid-infrared probe spectroscopy," *Sci. Rep.* **7**, 40492 (2017).

26. C.-M. Tu, et al., "Manifestation of a second Dirac surface state and bulk bands in THz radiation from topological insulators," *Sci. Rep.* **5**, 14128 (2015).

27. J. Sánchez-Barriga, et al., "Laser-induced persistent photovoltage on the surface of a ternary topological insulator at room temperature," *Appl. Phys. Lett.* **110**, 141605 (2017).

28. A. Sterzi, et al., "Bulk diffusive relaxation mechanisms in optically excited topological insulators," *Phys. Rev. B* **95**, 115431 (2017).

29. T. Yoshikawa, et al., "Photovoltage on the surface of topological insulator via optical aging," *Appl. Phys. Lett.* **112**, 192104 (2018).

30. Z.-H. Zhu, et al., "Rashba spin-splitting control at the surface of the topological insulator Bi_2Se_3," *Phys. Rev. Lett.* **107**, 186405 (2011).

31. B. Zhou, et al., "Controlling the carriers of topological insulators by bulk and surface doping," *Semicond. Sci. Technol.* **27**, 12 (2012).

32. T. T. Yeh, C. M. Tu, W. H. Lin, C. M. Cheng, W. Y. Tzeng, C. Y. Chang, H. Shirai, T. Fuji, R. Sankar, F. C. Chou, M. M. Gospodinov, T. Kobayashi, and C.-W. Luo "Femtosecond time-evolution of mid-infrared spectral line shapes of Dirac fermions in topological insulators,"*Sci Rep.* **10**, 1–8 (2020).

33. Y. Nomura, et al., "Single-shot detection of mid-infrared spectra by chirped-pulse upconversion with four-wave difference frequency generation in gases," *Opt. Express.* **21**, 18249–18254 (2013).

34. T. Fuji, H. Shirai, and Y. Nomura, "Ultrabroadband mid-infrared spectroscopy with four-wave difference frequency generation," *J. Opt.* **17**, 094004 (2015).

35. A. Kuzmenko, RefFIT. http://optics.unige.ch/alexey/reffit.html, (Date of access:23/05/2012) (2016).

10.2.4 Ultrafast Carrier Dynamics in Ge by Ultra-Broadband Mid-Infrared Probe Spectroscopy

For semiconductors, the physical parameters, e.g., carrier scattering rate, mobility, and concentration are important for applications in electronics and opt-electronics, especially for high-speed devices such as photodetectors. The infrared (IR) absorption spectroscopy has been demonstrated to be a convincing method for investigating the optical properties of materials in the IR region and some other physical parameters explicitly relevant to the IR spectra [1]. Generally, the whole absorption feature in common materials typically extends a rather broad spectral range. Consequently, the broadband spectrum can capture the absorption feature even without studying the dependence of carrier concentration or effective mass. However, the conventional infrared (IR) absorption spectroscopy can provide only stationary information without dynamic behavior. More than a decade ago, by the intensity modulation of the IR light source, the nanosecond (ns) time resolution was achieved [2]. Higher time resolution experiment has been desired. Recently, A group in Japan [3–5] generated sub-10 fs ultra-broadband IR pulses in air plasma with much broader width over $5000\,cm^{-1}$ than that generated by different frequency generation (DFG) in several nonlinear crystals [6,7]. By utilizing such pulsed source in the optical pump-probe experiments, it can immediately provide the time-dependent physical parameters for dynamic investigations and applications.

Based on this novel ultrafast light source, Shirai et al. [8] performed the transient pump-probe spectroscopy for Ge bulk crystal with 70-fs-time-resolution. However, the transient spectra in Ge obtained by the optical pump mid-IR probe spectroscopy have not been discussed in detail yet. In this paper, we present more analyses and discussions for the difference reflection spectra $[\Delta R(\omega)/R(\omega)]$ in Ge because of the difficulty in transmission spectra due to the opaque property in the range below $2\,\mu m$. Besides, the difference transmission signal $(\Delta T/T)$ is heavily suppressed by the absorption of surface excited carriers causing difficulties for the analyses. Compared with the transmission configuration for practical applications, the measurements of $\Delta R(\omega)/R(\omega)$ in the reflection configuration are more widely applied to various types of materials, including opaque materials, transparent materials [9], bulk [6,10], thin films [11], and hetero-structures [12,13]. Moreover, the ultra-broadband and 70-fs time-resolved spectra developed in this study are wide enough to provide more reliable fitting results and able to fully reveal the evolution of most of the features in spectra. For example, we have obtained the time-dependent carrier plasma frequency, concentration, scattering rate, and mobility by using the free carrier absorption model. Additionally, we discuss the mechanism of photoexcited carrier relaxation processes through numerical analyses. Last but not least, we have found that a novel oscillation feature in time-resolved difference reflection spectra around $2000\,cm^{-1}$ prominently appears in the case of high pumping fluence, which is concluded to be due to the Lorentz oscillation with the Coulomb force within 20 ps.

DOI: 10.1201/9780429196577-72

10.2.4.1 EXPERIMENTS

An intrinsic (100) Ge crystal wafer of 0.5-mm thick was used as a sample. We use a Ti:sapphire multipass amplifier system (800 nm, 30 fs, 0.85 mJ at 1 kHz, Femtopower compactPro, FEMTOLASERS) as a light source. The output pulse is split into three with two beam splitters. The first pulse is used to generate an ultra-broadband mid-infrared (MIR) probe pulse, the second pulse is used as an optical pump pulse, and the third pulse is used for a chirped pulse. The MIR probe pulse (ω_0) with 8.2-fs-pulse duration is generated by combining the fundamental (800 nm, ω_1) and second harmonic (SH, 400 nm, ω_2) pulses with the four-wave difference frequency generation (FWDFG, $\omega_1 + \omega_1 - \omega_2 \rightarrow \omega_0$) through filamentation in air. By using the optical pump (800 nm) and ultra-broadband MIR probe spectroscopy, the reflectivity change ($\Delta R/R$) transients of Ge in the region from 200 to 5000 cm^{-1} can be obtained. For detection as shown in Figure 10.2.4.1, the MIR pulses reflected from the sample are converted to visible pulses (ω_2, 400– 500 nm) for detection through chirped-pulse up conversion (CPU, $\omega_1 + \omega_1 - \omega_0 \rightarrow \omega_2$). The chirped pulse is obtained from the 800 nm pulse through four BK7 (thickness: $t = 10$ mm) substrates and a ZnSe ($t = 5$ mm) substrate at the Brewster angles. The up-converted spectrum is measured by electron-multiplying charge-coupled device camera (EMCCD, SP-2358 and ProEM + 1600, Princeton Instruments). The $\Delta R/R$ spectrum is obtained by up-converted probe beam spectrum for each delay with or without pump. To prevent the absorption of carbon dioxide and water vapor, the whole system is purged with nitrogen. The details of experiments have been reported in our previous work [8].

10.2.4.2 RESULTS AND DISCUSSION

Photoexcited carrier dynamics. Figure 10.2.4.2 shows an example of $\Delta R/R$ spectrum of Ge in the MIR region. The feature of plasma edge with positive $\Delta R/R$ (red color) below 1000 cm^{-1} and negative $\Delta R/R$ (blue color) above 1000 cm^{-1} can be clearly observed from zero delay time up to 400 ps. As shown in Figure 10.2.4.3a, the $\Delta R/R$ dramatically shrinks with increasing wavenumber and it crosses zero to negative in the range of 750–2000 cm^{-1}. Additionally, the position of minimum $\Delta R/R$ (or plasma edge) gradually shifts toward a low-wavenumber region as the delay time increases; meanwhile, the negative hump (yellow area) also gradually narrow down. Similar phenomena were observed also by Carroll et al. [14] in bulk Ge with 100-ps resolution. These features can be qualitatively described by the Drude model, which treats the free carriers in a solid as the point charges with random collisions. Using the stationary reflectance $R = 0.24$ for Ge [8], the dynamic reflectance $R(t) = R + \Delta R(t) = R \times \{1 + [\Delta R(t)/R]\}$ is used to fit the experimental data in Figure 10.2.4.2.

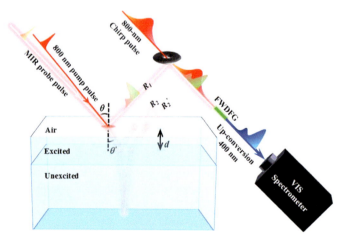

FIGURE 10.2.4.1 Schematics of the 800-nm pump and ultra-broadband mid-infrared (MIR) probe spectroscopy and the detection scheme with chirped-pulse upconversion. R_1: the 1st reflection of probe beam. R_2: the 2nd reflection of probe beam from the interface between excited and unexcited regions. R_2': the 2nd reflection of probe beam from the backside of a Ge sample. d: the depth of excited region. FWDFG: four-wave difference frequency generation. θ: incident angle of probe beam. θ': refraction angle of probe beam.

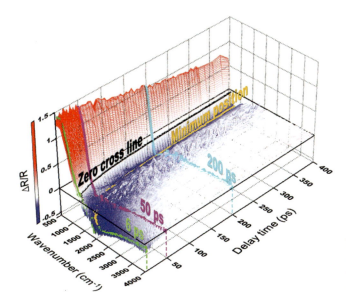

FIGURE 10.2.4.2 The reflectivity change ($\Delta R/R$) transients as a function of wavenumbers in Ge after exciting with the pump fluence of 135 µJ/cm².

By using the software of RefFIT [15], the $R(t)$ of p-wave IR probe can be fitted with [see the red lines in Figure 2.10.4.3a]

$$R_p = \left| \frac{n^2 \cos\theta - \sqrt{n^2 - \sin^2\theta}}{n^2 \cos\theta + \sqrt{n^2 - \sin^2\theta}} \right|^2$$

$$= \left| \frac{\varepsilon \cos\theta - \sqrt{\varepsilon - \sin^2\theta}}{\varepsilon \cos\theta + \sqrt{\varepsilon - \sin^2\theta}} \right|^2$$

(10.2.4.1)

where n: complex refractive index, $\theta = 45°$: incident angle, and $\varepsilon(\omega)$ complex dielectric constant given by

$$\varepsilon(\omega) = \varepsilon_\infty - \frac{\omega_p^2}{\omega^2 + i\Gamma\omega}$$

(10.2.4.2)

(ω: angular frequency, ω_p: plasma frequency, Γ: scattering rate, and ε_∞: permittivity at infinite frequency). Experimental incident angle is set at $\theta=45°$. For the best fit, the fitting parameter of ε_∞ is between 15 and 18, which closes to the theoretical estimation of $\varepsilon_\infty = 16$ [16]. Besides, the time evolution of ω_p and Γ can be obtained as in Figure 10.2.4.3b. Both ω_p and Γ significantly decrease with increasing the delay time and then remain constant for the longer delay time. On the contrary, the carrier mobility $\mu(=e\tau/m^*)$, where is the electron charge, m^* is the carrier effective mass, τ is the average scattering time, which is equal to $1/\Gamma$) rises with increasing the delay time due to the reduction of carrier concentration [17–19]. After 200 ps, the carrier mobility μ maintains to be ~350 cm²V⁻¹s⁻¹.

Moreover, the photoexcited carriers are only generated near the surface of the Ge sample due to the small penetration depth l_{800} of 0.2 µm for the 800-nm pump beam (defined as the inverse of absorption coefficient, where $\alpha=49322.85$ cm⁻¹ at 800 nm [20]). For the ultra-broadband MIR probe beam, the penetration depth is wavelength- and time-dependent. According to $l_{MIR}(t)=c/[2n_2(t)\omega]$, where c is the vacuum light speed, $n_2(t)$ is the imaginary part of the time-dependent refractive index and ω is the MIR angular frequency, the penetration depth $l_{MIR}(t)$ of MIR probe beam is estimated

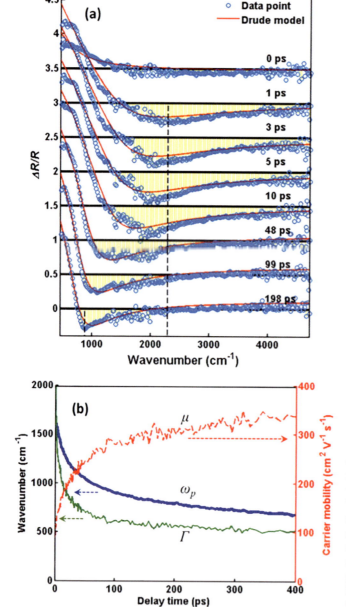

FIGURE 10.2.4.3 (a) The time-dependent $\Delta R/R$ as a function of wavenumbers in Ge at different delay time, which obtained from Figure 10.2.4.2. The blue-opened circles are experimental data. The red-dashed lines are the fitting curves with the Drude model of Eq. (10.2.4.1). (b) The time evolution of carrier mobility (μ), plasma frequency (ω_p), and scattering rate (Γ) obtained from the fitting in (a).

at different delay time. Prior to the pump pulse excitation, Ge is partially transparent (~30%) in the MIR range. The penetration depth of the MIR probe beam is only a few μm at 3 ps after pump pulse excitation. This is much smaller than the sample thickness of 500 μm. However, after 200 ps, it becomes larger than the sample thickness reaching a few hundred μm resulting in the appearance of the backside reflection feature (R_2' in Figure 10.2.4.1) of the sample. Additionally, the detection depth l_d of CPU system can be estimated by

$$l_d = T_{ch} \cdot \frac{c}{\dfrac{2 \cdot n_1}{\cos \theta'} - 2 \cdot \tan \theta' \cdot \sin \theta} \qquad (10.2.4.3)$$

where $T_{ch}=400$ fs is the duration of the chirped pulse, θ' is the refraction angle in Ge (see Figure 10.2.4.1), n_1 is the real part of refractive index of Ge. Taking $n_1=4$ [21], $\theta=45°$, and $\theta'=10.2°$, the detecting depth l_d is 15.2 μm, which is much smaller than the sample thickness. Therefore, our measurements are free from signal contamination by the backside reflection in the sample. However, $l_d=15.2$ μm is longer than the 0.2-μm penetration depth of pump beam. Therefore, interesting to say that the probe MIR beam can monitor both excited and unexcited regions simultaneously under the present study condition. The detailed analyses of the carrier relaxation processes is discussed in the following sections.

Transient carrier diffusion effect. As mentioned in the last section, the photoexcited carriers are generated nearby the surface of Ge (within 0.2 μm) by pump beam. Besides the short-range collisions among photoexcited carriers, the photoexcited carriers also diffuse from the excited region to the unexcited part due to the spatial gradient of photoexcited carrier concentration. The time evolution of carrier concentration N can be obtained by Eq. (10.2.4.4) [22]

$$N = \frac{\varepsilon_0 m^*}{e^2} \omega_p^2 \qquad (10.2.4.4)$$

where m^* denotes the carrier effective mass, ε_0 is the vacuum permittivity, e is the electron charge. The effective mass is $0.34m_e$ for a split-off hole [23], where m_e is the electron mass. Because of the time-dependent ω_p as shown in Figure 10.2.4.3b, we can further obtain the time evolution of carrier concentration N, which decreases gradually after pumping [see Figure 10.2.4.4a]. Moreover, the decrease of carrier concentration further causes the red shift of the position of spectrum minimum.

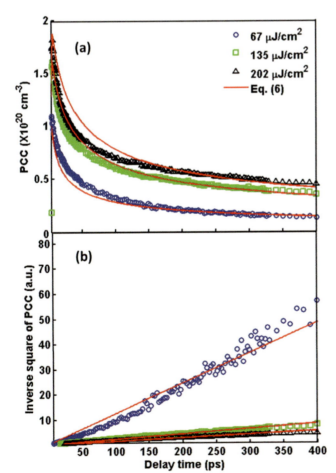

FIGURE 10.2.4.4 (a) Time evolution of photoexcited carrier concentration (PCC) obtained from Figure 10.2.4.3b with Eq. (10.2.4.4) at various pump fluences. The red solid lines show the fitting with Eq. (10.2.4.6). (b) Inverse square of photoexcited carrier concentration (PCC) as a function of delay time. The red solid lines show the fitting of Eq. (10.2.4.6) with $x\sim0$.

By changing the pump fluence (F) from 67 to 135 μJ/cm^2, the photoexcited carrier concentration increases significantly. However, the photoexcited carrier concentration shows saturation when the pump fluence is further increased from 135 to 202 μJ/cm^2. To explain the reduction of photoexcited carrier concentration quantitatively, the following differential equation with diffusion term is invoked [24],

$$\frac{dN}{dt} = D \cdot \frac{d^2 N}{dx^2} + G \tag{10.2.4.5}$$

where D is the diffusion coefficient, G is the carrier generation rate assuming to be much faster than the diffusion rate. By solving Eq. (10.2.4.5), we can obtain the analytic solution as follows,

$$N_d(x,t) = \frac{N_0}{\sqrt{\pi D t}} e^{-\frac{x^2}{4Dt}} \tag{10.2.5.6}$$

where N_0 is the total number of photoexcited carriers, which can be determined by integrating N_d along 1-dimension depth direction x perpendicular to the sample surface. In the range close to the surface of Ge (i.e., $x \sim 0$), $N_d^2 \propto t^{-1}$ is expected in case the dynamics is only due to the diffusion process. By fitting the data in Figure 10.2.4.4b via Eq. (10.2.4.6) with $x \sim 0$, the diffusion coefficient can be obtained from the slope, e.g., $D = 66$ cm^2/s for the pump fluence of 67 μJ/cm^2, which is consistent with the theoretical calculation of 65 cm^2/s [25]. Moreover, the diffusion coefficient D significantly decreases to 20 and 18 cm^2/s when the pump fluence increases to 135 and 202 μJ/cm^2, respectively, as listed in Table 10.2.4.1.

TABLE 10.2.4.1
List of Fitting Parameters in Figures 10.2.4.4 and 10.2.4.5 and Previous Works

Type of PCC	F (μJ/cm^2)	Experimental Method	N ($\times 10^{20}$) 1/cm^3)	D (cm^2/s)	γ_A ($\times 10^{-30}$ cm^6/s)	τ_A (ps)	α ($\times 10^{-5}$ 1/cm)
N_d	67	Current work	1.3	66	—	—	—
	135		1.6	20	—	—	—
	202		1.8	18	—	—	—
N_S	67	Current work	1.3	20	1.2	49.3	—
	135		1.6	8	0.5	78.1	—
	202		1.8	8	0.5	61.7	—
N_B	67	Current work	1.3	20	2.0	29.6	1/1.2
	135		1.6	20	3.0	13.0	1/8
	202		1.8	20	2.0	15.4	1/8
—	—	Calculation [25]		65	—	—	—
—	—	1.06-μm pump MIR probe [26]	0.7	—	3.2	31.9	—
—	7400	1.06-μm pump 1.55-μm probe [27]	3.4	—	0.11	78.6	—
—	—	Transient gratings [28]	0.17	53	—	—	—

Type of PCC: type of photoexcited carrier concentration. N: photoexcited carrier concentration. N_d: photoexcited carrier concentration with diffusion effect [via Eq. (10.2.4.6)]. N_S: photoexcited carrier concentration on surface [via Eq. (10.2.4.8)]. N_B: photoexcited carrier concentration in bulk [via Eq. (10.2.4.9)]. F: pump fluence. D: diffusion coefficient. γ_A: Auger coefficient. $\tau_A = 1/(\gamma_A N^2)$. α: absorption coefficient.

High-order transient effects. Even though the diffusion model qualitatively reproduces the dynamics of photoexcited carrier concentration, especially in the long delay-time range, the difference between experimental data and diffusion model is substantial at shorter delay than 150 ps as shown in Figures 10.2.4.4a and b. This implies that other mechanisms might involve in the relaxation processes of photoexcited carriers of Ge in short delay time, such as bandgap renormalization, recombination effect, and intervalley scattering. The bandgap renormalization usually happens after short pulse excitation because of the intimate relation between the gap size and the carrier concentration. However, to observe the bandgap renormalization effect, the measurements of transmittance [29,30] or photoluminescence [31] are indispensable. In Hamberg's works [29], moreover, they propose a clear picture for the roles of reflectance and transmittance, which can provide information of plasma oscillation and band absorption, respectively. Therefore, the difference reflection spectra in this study would primarily represent the signals of plasma oscillation rather than the bandgap renormalization, which can be furthermore definitively neglected in our fittings.

Ge is an indirect-bandgap semiconductor material. After photoexcitation, the intervalley scattering from the Γ valley to a side valley dominates the carrier transformation in hundreds of fs [32,33], and take few μs for recombination at the Γ point [34,35]. This is the main process for changing the photoexcited carrier concentration in Ge. Especially for the pump in p-type Ge, the relaxation processes from split-off hole band to upper hole band and scattering between the heavy hole and light hole bands could be observed [36,37]. However, the carrier relaxation processes inside of the split-off band, heavy-hole band, and light hole band do not induce the changes of photoexcited carrier concentration. Actually, we do observe the reduction of photoexcited carrier concentration in the short delay time region, which cannot be simply explained by the diffusion mechanism. Therefore, several other relaxation processes, e.g. the recombination, surface recombination, radiative recombination, and Auger process [38], should be involved in the analysis particularly for high pump fluence as in the following equation,

$$\frac{dN}{dt} = D \cdot \frac{d^2 N}{dx^2} - \gamma_r \cdot N - \gamma_S \cdot N - \gamma_R \cdot N^2 - \gamma_A \cdot N^3 + G \qquad (10.2.4.7)$$

where N is the carrier concentration, D is the diffusion coefficient, γ_r is the recombination rate, γ_S is the surface recombination coefficient, γ_R is the radiative recombination coefficient, γ_A is the Auger coefficient, and G is the Gaussian-type generation function for a laser pulse. In order to solve the nonlinear Eq. (10.2.4.7), it is rewritten by the Crank-Nicolson form [39] as described in Supplementary Information. If we simply consider that the photoexcited carriers are just generated or only can be detected on the surface, the photoexcited carrier concentration on the surface can be expressed as

$$N_S(t) = \int_0^{d_s} N(x,t) \cdot \delta(x = 0) dx \qquad (10.2.4.8)$$

where $N(x, t)$ is the solution of Eq. (10.2.4.7), d_s is the sample thickness and $\delta(x=0)$ is the Dirac delta function. As shown by the green lines in Figure 10.2.4.5, the experimental data are fitted well with Eq. (10.2.4.8) for the case of low pump fluence 67 μJ/cm². However, it cannot be applied to the cases of high pump fluence, especially below 100 ps.

Additionally, the penetration depth l_{800} of 800-nm pump beam is about 0.2 μm. As mentioned above, the detection depth l_d is around 15.2 μm. This indicates that it is necessary to consider all photoexcited carriers in bulk rather than only on the surface. Therefore, the photoexcited carrier concentration in bulk is expressed as

$$N_B(t) = \frac{\int N(x,t) \cdot e^{-\alpha x} dx}{\int e^{-\alpha x} dx} \qquad (10.2.4.9)$$

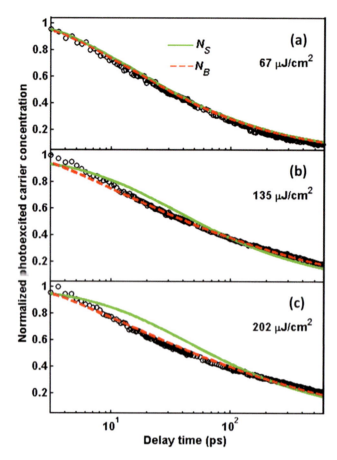

FIGURE 10.2.4.5 The experimental data in Figure 10.2.4.4a are presented lines composed of small open black circles in a normalized semi-logarithmic scale at various pump fluences of (a) 67 μJ/cm², (b) 135 μJ/cm², (c) 202 μJ/cm². The light-solid and dashed lines are fitted by the Eqs. (10.2.4.8) and (10.2.4.9), respectively.

where $N(x, t)$ is the solution of Eq. (7) and α is the absorption coefficient ($= 1/l_{MIR}$). The experimental data in Figure 10.2.4.5 can fit well with Eq. (10.2.4.9) for different pump fluence. When pump fluence increases from 65 to 135 μJ/cm², the substantial decrease in α is clearly shown in Table 10.2.4.1 indicating that a longer penetration depth ($l_{MIR} \sim 0.8\,\mu m$) of MIR probe beam for higher pump fluence. Moreover, for the same l_{MIR}, the saturation effect is also found for further increase in the pump fluence to 202 μJ/cm².

Based on the well-fit red-dashed lines in Figure 10.2.4.5, we can further discuss the importance of each term in Eq. (10.2.4.7). For the second term of Eq. (10.2.4.7), the time scale of recombination is in the order of μs, which is much longer than the measuring range of 400 ps in this study. In the third term of Eq. (10.2.4.7), the surface band bending causes surface recombination. Without the special surface treatment, the surface recombination velocity is about 1300 cm/s [40], and its time scale is still in μs. For the radiative recombination [the fourth term of Eq. (10.2.4.7)], the recombination rate in the bulk Ge with indirect band gap is $\sim 10^{-10}\,cm^3/s$ [26], which is smaller than the commonly found value of the order of $10^{-8}\,cm^3/s$ for direct band gap. Thus, the relaxation process of radiative recombination is also negligible in the present experimental condition (the critical value of γ_R for this study is $10^{-9}\,cm^3/s$).

As discussed above, the Auger effect dominates the relaxation within 100 ps. The fitting results in Table 10.2.4.1 show that the Auger coefficient γ_A ($2–3 \times 10^{-30}\,cm^6/s$) is independent of pump fluence (F), i.e., the photoexcited carrier concentration (N). According to the relation of $1/\tau_A = \gamma_A N^2$ [26], we further estimate the recombination time τ_A of Auger process, which is in the range of 13–30 ps and dependent on pump fluence. For high pump fluence, e.g., $F = 135$ and 202 μJ/cm², the τ_A becomes small to imply that the efficiency of Auger process would be dramatically enhanced

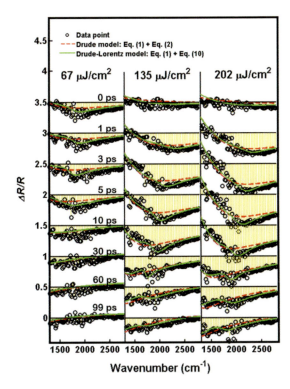

FIGURE 10.2.4.6 The time-dependent $\Delta R/R$ as a function of wavenumbers in Ge at different delay time, which obtained from Figure 10.2.4.2 at various pump fluences (lines composed of small open black circles). The origin of abscissa is shifted by 0.5 for each column from the bottom to the top. The black-opened circles are experimental data. The red-dashed lines are the fitting curves with the Drude model of Eqs. (10.2.4.1) and (10.2.4.2). The green-solid lines are the fitting curves with the DrudeLorentz model of Eqs. (10.2.4.1) and (10.2.4.10).

by high photoexcited carrier concentration. On the other hand, the diffusion coefficient decreases down to 20 cm²/s with including the Auger process. By the Einstein relation, the value of D/μ at high carrier concentration is ~0.07 [41]. Taking $\mu = 350$ cm² V⁻¹s⁻¹ obtained in Figure 10.2.4.3, thus, the D becomes 24.5 cm²/s which is consistent with the fitting results listed in Table 10.2.4.1.

Lorentz force for the photoexcited carriers. A closer look at the wavenumber dependence of $\Delta R/R$ at several delay times in Figure 10.2.4.6 reveals the fitting of the Drude model suffers a significant deviation around 2000 cm⁻¹, especially for high pump fluence. This implies that some driving forces exist among the photoexcited carriers, which we ascribe to the Lorentz force. Here, we further modified the Drude model with including the Lorentz force, i.e., the so-called Drude-Lorentz model [42]. In Eq. (10.2.4.1), thus, the angular frequency-dependent permittivity is given by

$$\varepsilon(\omega) = \varepsilon_\infty - \frac{\omega_p^2}{\omega^2 + i\Gamma\omega} + \frac{G_s\omega_p^2}{\omega_0 - \omega^2 - i\Gamma_L\omega} \tag{10.2.4.10}$$

where ε_∞ is the permittivity at an infinite frequency, ω is the frequency, ω_p is the plasma frequency, Γ is the scattering rate, G_s is related to the oscillator strengths, ω_0 is the resonance frequency, and Γ_L is the damping coefficient.

As shown in Figure 10.2.4.6, the green-solid lines of the Drude-Lorentz model can fit the $\Delta R/R$ rather well at different delay time. Interestingly, the difference between the Drude model and the Drude-Lorentz model, i.e. the Lorentz term, is strongly dependent on the pump fluence and delay time. In the cases of high pump fluence, the Lorentz term becomes more dominant and survives for longer time. From the fitting in Figure 10.2.4.6, the time-dependent resonance frequency ω_0 can be obtained as shown in Figure 10.2.4.7. For all pump fluence, ω_0 shows a remarkable red shift below 20 ps.

These results indicate that the photoexcited carriers are bound by a kind of spring force $F_s = m^*\omega_0^2 r$ with distance r. If the Coulomb collision could serve as the spring force, the carrier would be pulled back by the Coulomb force. Even though the paths and directions of collision are

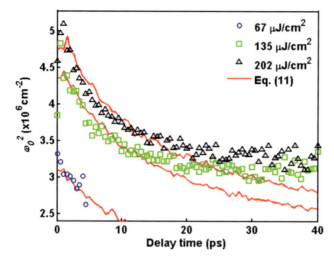

FIGURE 10.2.4.7 The square of resonance frequency (ω_0^2) of Lorentz term in Eq. (10.2.4.10) as a function of delay time for various pump fluences. The solid lines are obtained using Eq. (10.2.4.11).

random, the motion of carriers can be considered as a simple harmonic oscillation along a specific direction within short delay time. Here, we simply adopted the Coulomb force F_C to be the spring force F_s, which is just the binding force in Lorentz term. Thus, we have $F_C = F_s + c$ (where c is a phenomenological proportionality constant), and then the ω_0 can be expressed as

$$\omega_0^2 = \frac{ke^2}{r^3 m^*} + \frac{c}{rm^*} \qquad (10.2.4.11)$$

where k is the Coulomb's constant, r is the effective distance between the neighboring carriers (which is estimated by $(1/n^3)^{1/2}$ and n is the time-dependent carrier concentration), m^* is the effective mass, and c is 6×10^{-11} N. As shown in Figure 10.2.4.7, Eq. (10.2.4.11) can fit the resonance frequency ω_0 quite well below 20 ps. These results indicate that the oscillating feature of $\Delta R/R$ around 2000 cm^{-1} comes from the Lorentz oscillation. Moreover, this Lorentz oscillation is driven by the Coulomb force during the collision among the photoexcited carriers.

10.2.4.3 SUMMARY

We have studied the photoexcited carrier dynamics in Ge using 800-nm pump and ultra-broadband MIR probe spectroscopy [28]. The time evolutions of carrier mobility, plasma frequency, scattering rate, and carrier concentration have been extracted through the wavelength- (from 200 to 5000 cm^{-1}) and time-dependent (below 400 ps) $\Delta R/R$ by fitting with the Drude model. For the reduction of photoexcited carrier concentration, the Auger recombination with the Auger coefficient of $2-3 \times 10^{-30}$ cm^6/s dominates the relaxation processes of photoexcited carriers within 100 ps. On the other hand, the long-timescale relaxation process is dominated by the diffusion effect with diffusion coefficient of about 20 cm^2/s. Moreover, a novel oscillation feature is clearly observed in time-dependent trace of $\Delta R/R$ around 2000 cm^{-1}, especially in the cases of high pump fluence, which is considered to be due to the Lorentz oscillation raised by the Coulomb force exerted just after excitation. From the study, it can be expanded photocarrier dynamics with an MIR probe of picosecond resolution. The method is very useful to be applied to other systems than semiconductors.

The research represented here is a cooperative activity of the following people [43]: Tien-TienYeh, Hideto Shirai, Chien-Ming Tu, Takao Fuji, Takayoshi Kobayashi, and Chih-Wei Luo.

REFERENCES

1. M. A. Ordal, et al., "Optical properties of the metals Al, Co, Cu, Au, Fe, Pb, Ni, Pd, Pt, Ag, Ti, and W in the infrared and far infrared," *Appl. Opt.* **22**, 1099–1119 (1983).
2. J. M. Chalmers, and P. R. Griffiths, *Handbook of Vibrational Spectroscopy*, John Wiley & Sons, New York, p. 625 (2002).
3. T. Fuji, and T. Suzuki, "Generation of sub-two-cycle mid-infrared pulses by four-wave mixing through filamentation in air," *Opt. Lett.* **32**, 3330–3332 (2007).
4. Y. Nomura, et al., "Phase-stable sub-cycle mid-infrared conical emission from filamentation in gases," *Opt. Express* **20**, 24741–24747 (2012).
5. T. Fuji, and Y. Nomura, "Generation of phase-stable sub-cycle mid-infrared pulses from filamentation in nitrogen," *Appl. Sci.* **3**, 122–138 (2013).
6. C. W. Luo, et al., Snapshots of Dirac fermions near the Dirac point in topological insulators," *Nano Lett.* **13**, 5797–5802 (2013).
7. F. Seifert, V. Petrov, and M. Woerner, "Solid-state laser system for the generation of midinfrared femtosecond pulses tunable from 3.3 to 10 μm," *Opt. Lett.* **19**, 2009–2011 (1994).
8. H. Shirai, T.-T. Yeh, Y. Nomura, C.-W. Luo, and T. Fuji, Ultrabroadband midinfrared pump-probe spectroscopy using chirped pulse up-conversion in gases," *Phys. Rev. Appl.* **3**, 051002 (2015).
9. C. W. Luo, P. S. Tseng, H.-J. Chen, K. H. Wu, and L. J. Li, "Dirac fermion relaxation and energy loss rate near the Fermi surface in monolayer and multilayer graphene," *Nanoscale* **6**, 8575–8578 (2014).
10. C. W. Luo, et al., "Quasiparticle dynamics and phonon softening in FeSe superconductors," *Phys. Rev. Lett.* **108**, 257006 (2012).
11. Y. Chen, et al., "Ultrafast photoinduced mechanical strain in epitaxial BiFeO$_3$ thin films," *Appl. Phys. Lett.* **101**, 041902 (2012).
12. H. Park, M. Gutierrez, X. Wu, W. Kim, and X.-Y. Zhu, "Optical probe of charge separation at organic/inorganic semiconductor interfaces," *J. Phys. Chem. C* **117**, 10974–10979 (2013).
13. M. Panahandeh-Fard, et al., "Ambipolar charge photogeneration and transfer at GaAs/P3HT heterointerfaces," *J. Phys. Chem. Lett.* **5**, 1144–1150 (2014).
14. L. Carroll, et al., "Ultra-broadband infrared pump-probe spectroscopy using synchrotron radiation and a tunable pump," *Rev. Sci. Instrum.* **82**, 063101 (2011).
15. A. Kuzmenko, RefFIT. http://optics.unige.ch/alexey/reffit.html, (Date of access: 23/05/2012) (2016).
16. M. I. Gallant, and H. M. van Driel, "Infrared reflectivity probing of thermal and spatial properties of laser-generated carriers in germanium," *Phys. Rev. B* **26**, 2133 (1982).
17. M. B. Prince, "Drift mobilities in semiconductors. I. Germanium," *Phys. Rev.* **92**, 681 (1953).
18. G. Kaiblinger-Grujin, H. Kosina, and S. Selberherr, "Influence of the doping element on the electron mobility in *n*-silicon," *J. Appl. Phys.* **83**, 3096–3101 (1998).
19. G. Masetti, M. Severi, and S. Solmi, "Modeling of carrier mobility against carrier concentration in arsenic-, phosphorus-, and boron doped silicon," *IEEE Trans. Electron Dev* **30**, 764–769 (1983).
20. E. D. Palik, *Handbook of Optical Constants of Solids Set: Handbook of Thermo-Optic Coefficients of Optical Materials with Applications*, Academic Press, Cambridge, MA, p. 474 (1997).
21. H. H. Li, "Refractive index of silicon and germanium and its wavelength and temperature derivatives," *J. Phys. Chem. Ref. Data* **9**, 561–658 (1980).
22. M. Van Exter, and D. Grischkowsky, "Carrier dynamics of electrons and holes in moderately doped silicon," *Phys. Rev. B* **41**, 12140 (1990).
23. H. Y. Fan, "Infra-red absorption in semiconductors," *Rep. Prog. Phys.* **19**, 107 (1956).
24. W. C. Dunlap, "Diffusion of impurities in germanium," *Jr., Phys. Rev.* **94**, 1531 (1954).
25. W. Van Roosbroech, "The transport of added current carriers in a homogeneous semiconductor," *Phys. Rev.* **91**, 282 (1953).
26. L. Carroll, et al., "Direct-gap gain and optical absorption in germanium correlated to the density of photoexcited carriers, doping, and strain," *Phys. Rev. Lett.* **109**, 057402 (2012).
27. D. H. Auston, C. V. Shank, and P. LeFur, "Picosecond optical measurements of band-to-band auger recombination of high-density plasmas in germanium," *Phys. Rev. Lett.* **35**, 1022 (1975).
28. A. L. Smirl, S. C. Moss, and J. R. Lindle, "Picosecond dynamics of high-density laser-induced transient plasma gratings in germanium. *Phys. Rev. B* **25**, 2645 (1982).
29. I. Hamberg, C. G. Granqvist, K.-F. Berggren, B. E. Sernelius, and L. Engström, Band-gap widening in heavily Sn-doped In$_2$O$_3$. *Phys. Rev. B* **30**, 3240 (1984).
30. Y.-T. Wang, et al., "Ultrafast multi-level logic gates with spin-valley coupled polarization anisotropy in monolayer MoS$_2$. *Sci. Rep.* **5**, 8289 (2015).

31. A. Steinhoff, et al., "Efficient excitonic photoluminescence in direct and indirect band gap monolayer MoS$_2$," *Nano Lett.* **15**, 6841–6847 (2015).

32. G. Mak and H. M. van Driel, "Femtosecond transmission spectroscopy at the direct band edge of germanium," *Phys. Rev. B* **49**, 16817 (1994).

33. X. Q. Zhou, H. M. Van Driel, and G. Mak, Femtosecond kinetics of photoexcited carriers in germanium. *Phys. Rev. B* **50**, 5226 (1994).

34. R. N. Hall, "Electron-hole recombination in germanium," *Phys. Rev.* **87**, 387 (1952).

35. T. Timusk, "Far-infrared absorption study of exciton ionization in germanium," *Phys. Rev. B* **13**, 3511 (1976).

36. M. T. P. Oberli, et al., "Time resolved dynamics of holes in p-type germanium photoexcited by femtosecond infrared pulses," *Braz. J. Phys.* **26**, 520–524 (1996).

37. M. Woerner, T. Elsaesser, and W. Kaiser, "Inter-valence-band scattering and cooling of hot holes in p-type germanium studied by picosecond infrared pulses," *Phys. Rev. B* **41**, 5463 (1990).

38. N. G. Nilsson, "Band-to-band Auger recombination in silicon and germanium," *Phys. Scr.* **8**, 165 (1973).

39. J. Crank, and P. Nicolson, A practical method for numerical evaluation of solutions of partial differential equations of the heat conduction type. *Proc. Camb. Phil. Soc.* 43(1), 50–67 (1947).

40. N. Derhacobian, et al., "Determination of surface recombination velocity and bulk lifetime in detector grade silicon and germanium crystals," *IEEE Trans. Nucl. Sci.* **41**, 1026–1030 (1994).

41. P. T. Landsberg, *Recombination in Semiconductors*, Cambridge University Press, Cambridge, p. 83, (1991).

42. T. Nishio, and H. Uwe, Optical-phonon precursory behavior towards semiconductor–metal transition in Ba$_{1-x}$K$_x$BiO$_3$," *J. Phys. Soc. Jpn.* **72**, 1274–1278 (2003).

43. T.-T. Yeh, H. Shirai, C.-M. Tu, T. Fuji, T. Kobayashi, and C.-W. Luo, "Ultrafast carrier dynamics in Ge by ultra-broad band mid infrared probe spectroscopy," *Sci. Rep.* 7, 40492 (2017). DOI: 10.1038/srep40492.

Section 11

Chemical Reactions and Material Processing

Section 11.1

Chemical Reactions

11.1.1 Transition State in a Prevented Proton Transfer Observed in Real Time

11.1.1.1 INTRODUCTION

The identification of transition states (TSs) provides detailed information about reaction mechanisms supplementing information obtained by other means. In the 1990s, several theoretical methods were proposed for inferring the structures of TSs. They were the only methods then available, but their reliability was often questionable. This lack of a robust method prompted chemists to develop an experimental method for visualizing ultrafast changes in molecular structure that proceed via TSs. The experimental realization of femtosecond dynamic studies was developed through the pioneering work of Zewail [1]. More recently, ultrashort pulses [2] whose durations are much shorter than typical molecular vibrational periods have been used to observe structural changes during chemical reactions including TSs [3]. This novel visualization method observes the frequency shifts of the relevant molecular vibration modes. For instance, the photoisomerization dynamics of azobenzene [3c–3e], bacteriorhodopsin [3a,3b], and the proton-transfer (PT) reaction [3f–3j] have been elucidated using ultrafast spectroscopy.

Indigo and indigodisulfonate salt are commonly used for dying due to their outstanding photostabilities, which enable them to exist for a long time without undergoing decoloration.

Another reason for their common usage is that these dyes are chemically absorbed strongly on cellulose and other fibers. By contrast, several other indigo derivatives, such as perinaphthothioindigo [4], have relatively high photoisomerization efficiencies (quantum yield: µ0.25). The photostability of indigo and indigodisulfonate salt were studied in previous works [5], suggesting the existence of an ultrafast PT reaction in the excited state, which is much faster than photoisomerization. In the present work, the TS in the PT of indigodisulfonate salt was identified by real-time observation of frequency changes and the obtained result was also supported by theoretical analysis. This study experimentally clarifies the riddle of the ultrahigh photostability of indigo and indigodisulfonate salt.

11.1.1.2 EXPERIMENTAL

Pump-Probe Experiment. A noncollinear optical parametric amplifier (NOPA) was used to obtain an ultra-broadband visible pulse, which was compressed to 5fs for the ultrafast pump-probe measurement [2f] A Ti:sapphire regenerative amplifier (Spectra-Physics, model Spitfire, 150mJ, 100fs, 5kHz at 805nm) was used as a laser source to generate pump and seed pulses of the NOPA. The amplified signal pulse after the double-pass NOPA with a spectrum extending from 525 to 725nm was compressed with the main compressor, resulting in a pulse duration of 5fs which is nearly Fourier transform limited (Figure 11.1.1.2b). The experiments were performed at pump and probe pulse intensities of 2580 and 480 GW/cm², respectively. The focal areas of the pump and probe pulse laser were 100 and 75 mm², respectively. The polarizations of the pump and the probe pulses are parallel to each other. Sodium indigodisulfonate saturated in anhydrous methanol and methanol-d, sodium indigodisulfonate saturated in anhydrous DMSO, and potassium indigodisulfonate satu-rated in anhydrous methanol in a 1-mm cell were used as samples at 295 ± 1 K. The time-resolved difference transmittance DT from 525 to 725 nm was measured simultaneously with a delay time step of 1 fs in the time range of ~100 to 800 fs by a 128channel lock-in amplifier coupled to a poly-chromator. The spectral resolution of the total system was approximately 1.6 nm.

Computational Methods. The Gaussian 03 program was used for the calculations [6]. Geometric optimizations were performed using the CASSCF/6-31G*//B3LYP/6-311++G**, TD-B3LYP/6-311++G**//B3LYP/6311++G**, TD-BLYP/6-311++G**//BLYP/6-311++G**, TD-BP86/6-311++G**//BP86/6-311++G**, CIS/631++G**//CIS/6-31++G**, and CIS/6-31G*//CIS/631G* methods and basis sets. Calculations were performed without assuming symmetry. 5d functions were used for the d orbital. Frequency calculations were performed for all of the obtained structures at the same level, excluding CIS/631++G**//CIS/6-31++G**. It was confirmed that all the frequencies were real for the ground states and one imaginary frequency existed for the TS. Vectors of the imaginary frequencies directed the reaction mode and intrinsic reaction coordinate calculations were further performed to confirm that the obtained TSs were on the saddle points of the energy surface between the reactant and the product. However, a concerted pathway could be obtained by molecular structure optimiza-tion only under the condition that the molecule was symmetric with respect to its symmetric center. Theoretical results for the concerted pathway TSs had two imaginary frequencies that were ascribed to n_{sN-H-O} and $n_{asN-H-O}$. When the calculation was performed starting from the calculated results of concerted pathway TSs, which have two imaginary frequencies as initial structures, without sym-metry, a more stable TS of the stepwise pathway was obtained.

Spectroscopy. The Ultraviolet/Visible (UV-vis) absorption spectrum of sodium indigodisulfo-nate methanol solution (3×10^{-6} M) was recorded on an absorption spectrometer (CARY 50, Varian, JAPAN). Emission spectrum of the sample solution (3×10^{-6} M) was recorded on a fluorescence spectrophotometer (FP-6500, JASCO Corp., JAPAN). Both of the measurements were performed using the sample solutions in 1 cm² quartz cells at room temperature 293 ± 1 K.

11.1.1.3   RESULTS AND DISCUSSION

11.1.1.3.1   INVESTIGATION OF REACTION MECHANISMS OF PROTON TRANSFER (THEORY)

As the PT mechanism of indigo and indigodisulfonate salt after photoexcitation, two possible mechanisms can be considered, i.e., a concerted pathway (two PTs at the same time) or a step-wise pathway (one PT after the other). To determine the potential landscape in the two pathways, theoretical analyses were performed using several methods. Figure 11.1.1.1 shows the results cal-culated by CASSCF/6-31G*// B3LYP/6–311++G**. It represents the calculated results obtained by other methods, which gave essentially the same results (Figure 1, Figures S1-S5 and Table S1 in the Supporting Information (SI)). The calculated activation energy of the concerted pathway in the excited state was larger than 40 kcal/mol (Figure 11.1.1.1a), ruling out the possibility of a concerted PT. A microscopic mechanism of the concerted pathway can be given as follows. The $C^{1'}=C^1-C^2$ (a_1) and $C^{1'}=C^1-N$ (a_2) bond angles decrease (a_1: 125.8° → 118.7°, a_2: 126.0° → 123.7°) in the TS of the concerted PT (TS$_c$), as the carbonyl group changes into alcohol. In the case of the concerted PT mechanism, the change in the bond angle induces deformation of the molecular framework out of the original molecular plane, leading to a bent structure, which dramatically increases the activation energy.

The stepwise pathway (Figure 11.1.1.1b) is favorable because the activation energy of the TS in the first PT (TS$_1$) was calculated to be 5.3 kcal/mol, which is sufficiently low to form the monoal-cohol intermediate. However, the second PT is not favorable because the energy surface from the monoalcohol to the product increases gradually.

Therefore, theoretical analysis suggests that the stepwise pathway is energetically more favorable than the concerted pathway. We next investigated these pathways experimentally.

(a)

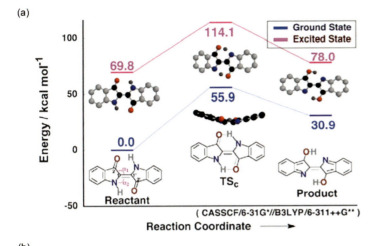

(b)

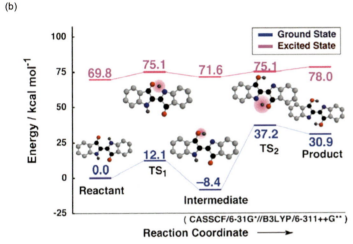

FIGURE 11.1.1.1 Reaction pathways of the proton transfer in indigo from CASSCF/6-31G*// B3LYP/6-311++G**. (a) Concerted pathway. (b) Stepwise pathway.

11.1.1.4 DIRECT OBSERVATION OF TRANSITION STATE (METHANOL SOLUTION OF INDIGODISULFONATE SALT)

For the direct observation of the dynamics after photoexcitation, 5-fs pump-probe measurement was performed to identify the reaction pathway including the TS. A methanol solution of sodium indigo-disulfonate, in which photoisomerization does not occur, was used as the sample (Figure 11.1.1.2a). Methanol was used as a solvent because it has no molecular vibration signal in the frequency range examined. The absorption spectrum of the sample has a peak around 598 nm and the fluorescence spectrum of the sample excited at 600 nm has a peak around 650 nm (Figure 11.1.1.2b). The structure in the excited state is not thought to be rigid because the fluorescence quantum yield is very low (0.0015) [3h] and the fluorescence spectral peak is red-shifted from the UV absorption peak (dotted line in Figure 11.1.1.2b).

Figure 11.1.1.3 shows the real-time traces of difference absorbance DA (calculated as $DA = -\log_{10}(1 + DT/T)$, where T and DT are transmittance and transmittance change induced by the pump, respectively) at 128 probe wavelengths following the 5-fs pulse excitation of the methanol solution of sodium indigodisulfonate. Using a multichannel lock-in amplifier of 128 channels, we observed a time-resolved signal over a broad spectral range. The sign for DA is positive in the probe wavelength of 715–725 nm due to mainly the induced absorption in the transition from S_1 state to S_n state(s) ($S_n \leftarrow S_1$ transition). Bleaching and stimulated emission still can be observed within the range of

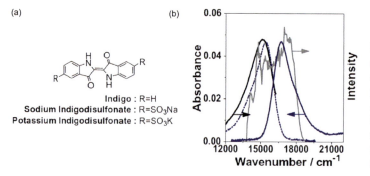

FIGURE 11.1.1.2 (a) Molecular structures of indigo, sodium indigodisulfonate, and potassium indigodisulfonate. (b) Absorption (indigo), its mirror image (indigo dots), fluorescence spectrum at 600 nm excitation (black) of sodium indigodisulfonate [51].

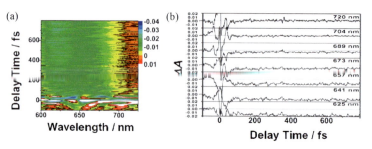

FIGURE 11.1.1.3 (a) A two-dimensional display of the absorbance change on the probe delay time and the wavelength following the 5-fs pulse excitation of the methanol solution of sodium indigodisulfonate [51]. (b) Real-time traces of absorbance change.

715–725 nm where DA is positive, even though their influence becomes smaller at longer wavelengths. The DA in the spectral range from 705 to 715 nm oscillates around zero, reflecting the signals of positive DA (induced absorption) and negative DA (bleaching and/or stimulated emission). The DA in the spectral range <705 nm has a negative value due to bleaching of the ground-state absorption and stimulated emission of sodium indigodisulfonate. Therefore, dynamics of molecular vibration of excited states can be most sensitively and selectively observed in the spectral range of 715–725 nm. To study the chemical reaction in the excited state, the data in this probe spectral range, 715–725 nm, was used for discussion in this work.

From the refractive index of methanol of $n = 1.3195 + 3053.64/l^2 - 3.41636 \times 10^7/l^4 + 2.62128 \times 10^{12}/-l^6$ given in Ref. [7], the pulse width after transmission through a 1-mm cell was calculated to be about 55 fs (Figure 11.1.1.4).

The pulse width broadening in transmission through the solution cell attenuates the amplitude of molecular vibration observed in the signal as shown in Figure 11.1.1.5a. This paper discusses vibration modes between 1000 and 3750 cm^{-1}. Figure 11.1.1.5b shows the transmission efficiency of photons passing through a glass cell filled with indigo-saturated methanol solution. Figure 11.1.1.5c shows the attenuation ratio of the vibration amplitude for vibration frequencies between 1000 and 3750 cm^{-1} in transmission through the 1-mm glass cell, calculated from Figure 11.1.1.5a and b. Their vibration amplitudes cannot be properly resolved by the stretched pulse whose duration is longer than their periods. However, the vibration frequency to be observed is not affected by the stretched pulse width. Hence, their frequency shifts can still be discussed correctly.

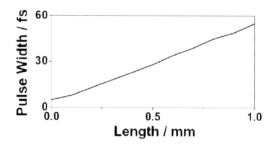

FIGURE 11.1.1.4 Pulse width after passing through 1 mm optical cell filled with methanol.

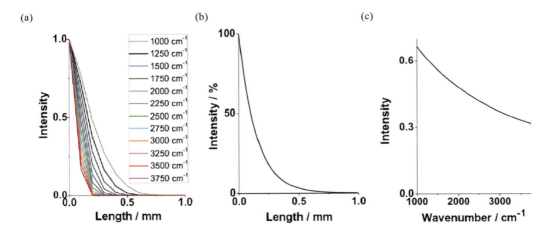

FIGURE 11.1.1.5 (a) Simulation results showing FFT power attenuation for vibration frequencies between 1000 and 3750 cm^{-1} after transmission through a glass cell filled with methanol solution. Different colored curves show the calculation results for 1000 (gray), 1250 (black), 1500 (indigo), 1750 (olive), 2000 (violet), 2250 (blue), 2500 (green), 2750 (cyan), 3000 (orange), 3250 (pink), 3500 (red), and 3750 cm^{-1} (wine). (b) Transmission efficiency of photons passing through a glass cell filled with indigo-saturated methanol solution. (c) Attenuation ratio of the vibration amplitude whose frequencies are from 1000 to 3750 cm^{-1} after transmission through 1mm of the indigo-saturated methanol solution.

Figure 11.1.1.6 shows spectrograms [8] obtained by applying a sliding-window Fourier transform (Eq 11.1.1.1).

$$S(\omega,\tau) = \int_0^\infty S(t)g(t-\tau)\exp(-i\omega t)\,dt, g(t)$$

$$= 0.42 - 0.5\cos(2\pi t/T) + 0.08\cos(2\pi t/T)$$

(11.1.1.1)

Using a Blackman window function with a full width at half maximum (FWHM) of 120 fs, the spectrogram was calculated from the real-time traces. The frequency resolution of the spectrogram is <30 cm^{-1}. The data in the region of 0 fs delay time could not be analyzed in terms of incoherent electronic transition probabilities proportional to the population modulated by vibration because of the strong interference between the scattered pump and probe pulses.

The spectrogram measured at 720 nm (Figure 11.1.1.6a) shows the smallest influence of bleaching and stimulated emission since the signal at 720 nm has the largest DA compared to the data in the positive DA region. Therefore, the spectrogram measured at 720 nm was used as a representative example of excited state dynamics. Similar spectrogram patterns were observed also at probe wavelengths of 715 nm as shown in Figure 11.1.1.6b. Bleaching and stimulated emission still can be observed within the range of 705–725 nm where DA is positive, even though their influence becomes smaller at longer wavelengths. Therefore, the spectrogram of 715 nm shows the influence of bleaching and stimulated emission more effectively than that of 720 nm, which gives the spectrogram of 715 nm an osculant shape between that of 720 nm and that of 700 nm.

In the other probe spectral range (<705 nm), both the bleaching and stimulated emission can take part. There is a peak in the absorption spectrum of sodium indigodisulfonate at 598 nm (Figure 11.1.1.2b). Therefore, the ground-state depletion is effective in the spectral range as is also confirmed by the negative sign of DA observed. The observed signal in the spectral range shows a C=C stretching vibration mode of the ground state of sodium indigodisulfonate (Figure 11.1.1.6c).

A peak around 1700 cm^{-1}, corresponding to a C=O stretching mode ($n_{C=O}$) of reactant (R) [9], appeared just after the excitation. This frequency gradually red-shifted and another peak around 1250 cm^{-1} appeared at 270 fs delay time, which can be attributed to the C–O single bond stretching mode (n_{C-O}), the frequency of which was reported to be about 1250 cm^{-1} [5c,10]. The presence of

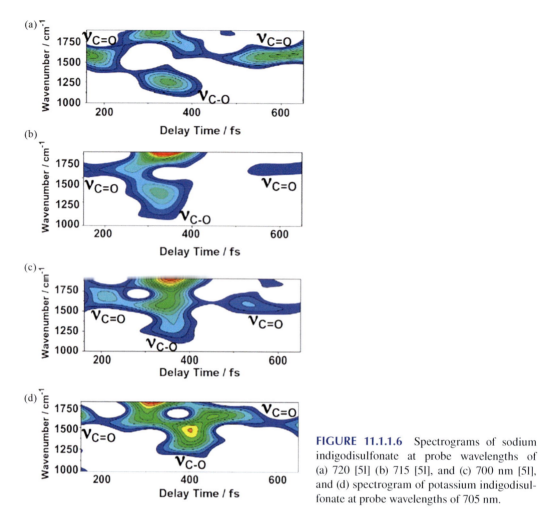

FIGURE 11.1.1.6 Spectrograms of sodium indigodisulfonate at probe wavelengths of (a) 720 [5I] (b) 715 [5I], and (c) 700 nm [5I], and (d) spectrogram of potassium indigodisulfonate at probe wavelengths of 705 nm.

this peak indicates the formation of the C–OH group in the intermediate (I). In addition, a spectrogram obtained by methanol solution of potassium indigodisulfonate is similar in the distribution of the Fourier power (Figure 11.1.1.6d).

11.1.1.5 COMPARISON BETWEEN EXPERIMENTAL RESULTS AND THEORETICAL RESULTS TD-B3LYP/6-311++G**//B3LYP/6311++G**.

The observed ultrafast dynamics of the indigodisulfonate salts can be explained by the theoretical results shown in Figure 11.1.1.1 and Table 11.1.1.1. The results of TDB3LYP/6-311++G**//B3LYP/6-311++G** were used as a representative example. Just after the excitation, two identical C=O groups (R) were observed to give rise to the peak centered around $1700 \, cm^{-1}$ (Figure 11.1.1.6), in good agreement with the calculated frequency of $1696 \, cm^{-1}$ (Table 11.1.1.1). This peak splits into two red- and blue-shifted peaks in the delay time from 200 to 270 fs. The reason for this frequency shift can be explained as follows. The electron density in one of the two C=O bonds, which is the acceptor of the transferred proton in the first PT, decreases after photoexcitation by p-electron delocalization extending to the transferred proton. This leads to the reduction of $n_{C=O}$, which is consistent with the monotonic red shift from $1696 \, cm^{-1}$ (R) to $1588 \, cm^{-1}$ (TS$_1$) found in the calculated frequencies. On the other hand, the electron density in the C=O bond that does not participate in the first PT is increased after the photoexcitation, which leads to a blue shift of the peak. This is also consistent with the monotonic blue shift of $n_{C=O}$ from 1696 ® to $1751 \, cm^{-1}$ (TS$_1$) in the calculation result. Another possible effect of the shifts of

TABLE 11.1.1.1

Theoretical Vibrational Frequencies and NBO Result

	Method: Basis Set	B3LYP 6-311++G**	BLYP 6-311++G**	BP86 6-311++G**	CIS 6-31++G**	CIS 6-31G*	Exp./cm^{-1}
		\multicolumn{6}{c}{Theoretical Results of Frequency/cm^{-1}}					
Reactant	C=O(1)	1696	1604	1626	-	1855	1700[a]
	C=O(2)	1696	1604	1626	-	1604	1700[a]
TS	C=O(1)	1588	1513	1544	-	1513	Red shift
	C=O(2)	1751	1651	1675	-	1651	Blue shift
Intermediate	C–O(1)	1276	1234	1241	-	1234	1250[b]
	C=O(2)	1763	1656	1678	-	1656	1770[b]
		\multicolumn{6}{c}{Bond Order}					
Reactant	C=O(1)	1.83	1.81	1.81	1.88	1.88	
	C=O(2)	1.83	1.81	1.81	1.88	1.88	
Intermediate	C–O(1)	0.99	0.99	0.99	0.99	0.99	
	C=O(2)	1.89	1.87	1.86	1.91	1.91	
		\multicolumn{6}{c}{Second-Order Perturbative Estimates of "Donar(C=O)-Acceptor(H)" Interaction in NBO Basis/kcal mol^{-1}}					
Reactant		8.34	36.79	11.31	7.65	-[c]	
Intermediate		<0.5	<0.5	<0.5	<0.5	-[c]	
		\multicolumn{6}{c}{Distance between C=O and N-H/Å}					
Reactant		2.281	2.294	2.264	2.275	2.267	
Intermediate		2.416	2.456	2.483	2.319	2.317	

[a] Frequency calculated in the spectrogram at reaction time 100 fs.
[b] Frequency calculated in the spectrogram at reaction time 360 fs.
[c] It is thought that the hydrogen bond was not calculated using CIS/6-31G* because there is no diffuse function.

$n_{C=O}$ are explained as follows. The indigodisulfonate has a symmetric structure. Therefore, just after the photoexcitation of the indigosulfonate, its structure still maintains symmetry resulting in anharmonic coupling of symmetric and asymmetric $\eta_{C=O}$ modes. The subsequent proton transfer breaks the symmetry reducing the anharmonic coupling between the two C=O bonds, which is also thought to contribute to the frequency shift of $n_{C=O}$.

The presence of a new peak around 1250 cm^{-1} at 270 fs delay time indicates that the C–OH is formed by the generation of the monoalcohol intermediate, which agrees well with the theoretical results that suggest the appearance of n_{C-O} (I) at 1276 cm^{-1}. According to the theory, the remaining C=O of the generated monoalcohol is expected to be blue shifted by ca. 70 cm^{-1} [from 1696 ⓡ to 1763 cm^{-1} (I)], which is in good agreement with the experimentally observed blue shift from 1700 ⓡ to 1770 cm^{-1} (I). After the generation of the monoalcohol (I), the peak around 1700 cm^{-1} ($n_{C=O}$) is reproduced in the time range >500 fs. This is in agreement with the theoretical result that the monoalcohol (I) is energetically unstable compared with the reactant, so that the monoalcohol returns to the reactant rather than forming a product.

The experiments show that the change from C–O to C=O associated with the change from the alcohol to the carbonyl cannot be observed in the back reaction (400–500 fs). One reason for the absence of spectral change is that the vibration dephases due to inhomogeneous and homogeneous broadening. Another reason is that the reaction time of the return path from the alcohol to the carbonyl is too fast to show the change compared with that of the forward path from the carbonyl to the alcohol.

The natural bond orbital (NBO) analysis [11] also supports the experimental observations by showing a second PT does not occur. At first brief description of NBO is described below.

The Natural Atomic Orbitals (NAOs) incorporate two important physical effects that distinguish them from isolated-atom natural orbitals as well as from standard basis orbitals:

i. The spatial diffuseness of NAOs is optimized for the effective atomic charge in the molecular environment (i.e., more contracted if A is somewhat cationic; more diffuse if A is somewhat anionic). NAOs therefore automatically incorporate the important "breathing" responses to local charge shifts that usually require variational contributions from multiple basis functions of variable range (double zeta, triple zeta, or higher) to describe accurately.

ii. The outer fringes of NAOs incorporate the important nodal features due to steric (Pauli) confinement in the molecular environment (i.e., increasing oscillatory features and higher kinetic energy as neighboring NAOs begin to interpenetrate, preserving the interatomic orthogonality required by the Pauli exclusion principle). The valence NAOs of atom A therefore properly incorporate both the inner nodes that preserve orthogonality to its own atomic core as well as the outer nodes that preserve orthogonality to filled orbitals on other atoms B. Both features are necessary for realistic steric properties in the molecular environment (i.e., proper Fermi-Dirac anticommutators of the associated second-quantized NAO field operators), but both are commonly ignored in standard basis orbitals.

The bond order of C=O and the distance between H–N and O=C, which is not responsible for the PT, increase from 1.83 (R) to 1.89 (I) and from 2.28 (R) to 2.42Å (I), respectively, and the degree of delocalization of π-electrons in C=O bond(s) to the proton is reduced. The NBO analysis also suggests that electron delocalization stabilizes the reactant molecule, but not in the monoalcohol intermediate. In this way, the increase in distance between the H and O atoms supports the experimental results that indicate the first PT takes place, but the second does not.

The transfer of such vibrational coherence or even the creation of coherence by a chemical reaction has been discussed by Jean and Fleming [12], which explains the weak oscillations clearly observed in the electronic curve crossing. The vibration frequency is too high to be explained by resonant electronic couplings. The high-frequency mode is thought to be a result of coherences between eigenstates with large projections onto diabatic states with different numbers of vibrational quanta. The bare electronic coupling and diabatic vibrational frequency have comparable magnitudes causing considerable interfusion of nonresonant diabatic states. Therefore, the vibrational coherence can be transferred into the product via the electronic curve crossing under sufficiently strong electronic coupling [13].

These results clearly provide direct evidence that PT takes place after photoexcitation in indigodisulfonate salts, and they show that the PT mechanism is a stepwise pathway. The first PT causes the reaction from reactant to intermediate with a high efficiency of >90% estimated from the level of error and noise in the spectrogram, but the generated intermediate also returns to the reactant with high efficiency (>90%). As a result, the total reactivity from reactant to intermediate is negligibly small. Therefore, indigo is very stable and resists discoloration, providing a final answer as to whether or not PT occurs in indigo [3]. Direct evidence was obtained indicating that even though PT occurs, the system returns to the original reactant within 0.5 ps. It was concluded that the PT does not take place at the S_1 state on a picosecond scale, and hence indigo is unusually photostable.

Kinetic Isotope Effect (Experiment). For confirmation of the stepwise mechanism in indigodisulfonate salts, the kinetic isotope effect (KIE) was studied by observing the real-time molecular vibration frequency of a deuterated sodium indigodisulfonate (Figure 11.1.1.7a). A peak around 1700 cm^{-1}, corresponding to a C=O stretching mode Ⓡ, appeared just after the excitation. This frequency is red-shifted and another peak appeared around 1250 cm^{-1} at ~450 fs after the photoexcitation in deuterated sodium indigodisulfonate, which is in good agreement with the reported value of C_6H_5OD ($n_{C-O} = 1250$ cm^{-1}) [10b].

The reaction rate in deuterated sodium indigodisulfonate was slowed down to $(2.0 \pm 0.1) \times 10^{12}$ s^{-1} from $(3.0 \pm 0.1) \times 10^{12}$ s^{-1} in the non-deuterated system. The primary KIE obtained in the experimental

(a)

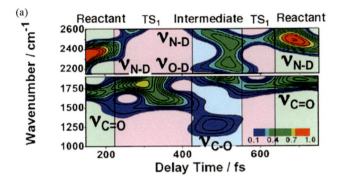

(b)

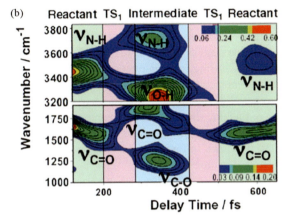

FIGURE 11.1.1.7 (a) Spectrograms of deuterated methanol solution of deuterated sodium indigodisulfonate [5m]. (b) Spectrograms of methanol solution of sodium indigodisulfonate [5l].

results predicts that the PT rate in the deuteride system was 44% slower than that in the nondeuteride system. The corresponding rate ratio (k_a^H/k_a^D) between H and D ranges widely from 1.15 (in the early TS) to several hundred (in the late TS).

The excited state of sodium indigodisulfonate is expected to have an early TS due to strong hydrogen bonding, and the observed k_a^H/k_a^D ratio was 1.7, which is larger than the expected value (1.15) in the early TS. This is caused by the decrease in k_a^H, indicating that the tunnel effect takes place in the PT of the non-deuterated system. Figure 11.1.1.7a shows that the frequency of $n_{C=O}$ is red-shifted from 1700 to about 1500 cm^{-1}, followed by blue shift to about 1600 cm^{-1}, and red shift to about 1250 cm^{-1}. On the other hand, Figure 11.1.1.7b shows that $n_{C=O}$ is gradually red-shifted from 1700 to 1250 cm^{-1} without the slow modulation observed in Figure 11.1.1.7a. This slow modulation of the $n_{C=O}$ frequency found for the deuterated system is considered to be induced by the coupling of $n_{C=O}$ mode with low-frequency modes. The low-frequency modes can be considered to be scissoring modes of $C^{1'}=C^1-C^2$ (a_1) and $C^{1'}=C^1-N$ (a_2) associated with the deuteron transfer. The mechanism of the mode coupling inducing the modulation can be explained as follows.

In the first deuteron transfer, the distance between N–D and C=O decreases for the deuteron to be transferred to the acceptor. Associated with these processes, the $C^{1'}=C^1-C^2$ bond angle and $C^{1'}=C^1-N$ bond angle are expected to decrease and increase, respectively, with a decrease in the carbon-carbon bond order changing from carbon-carbon double bond to carbon-carbon single bond. These bond angle changes in turn trigger the scissoring modes of $C^{1'}=C^1-C^2$ and $C^{1'}=C^1-N$. On the other hand, the oscillatory feature of $n_{C=O}$ was not found in the non-deuterated sample. It is thought to be due to much faster proton-transfer rate than the scissoring period.

In addition, in the spectrogram of sodium indigodisulfonate, a peak around 3350 cm^{-1} appeared just after the excitation and was assigned to the N–H stretching mode (n_{N-H}) (Figure 11.1.1.7b). The frequency of the peak displayed a gradual red shift and another peak around 3185 cm^{-1} appeared at ca. 270 fs after the photoexcitation. The peak located at 3185 cm^{-1} was attributed to the O–H stretching mode (n_{O-H}), because the new peak appeared at almost the same time as that for n_{C-O}. The red shift was thought to be due to PT-induced generation of the monoalcohol (I), which has an O–H instead of N–H. The vibrational frequency of the latter (3185 cm^{-1}) is lower than the former (3350 cm^{-1}) by about $(14/16)^{1/2}$. On the other hand, the electron density in the other N–H bond, which does not participate in the PT, is increased during the PT, leading to the blue-shift of the peak, because hydrogen bonding is weakened during the PT. When the generation of the monoalcohol (I) is complete, the peak around 3350 cm^{-1} (n_{N-H}) is reproduced, as shown at the later delay time (>500 fs) in Figure 11.1.1.7b.

In the spectrogram of deuterated sodium indigodisulfonate, a peak appeared around 2380 cm^{-1} just after the photoexcitation and was assigned to a N–D stretching mode (n_{N-D}) (Figure 11.1.1.7a). This frequency is lower than that of non-deuterated sodium indigodisulfonate by a factor of $(2)^{-1/2}$. This mode frequency also showed a gradual red-shift and another peak appeared around 2200 cm^{-1} at ca. 500 fs delay time. The peak located at 2200 cm^{-1} is thought to be due to the O–D stretching mode (n_{O-D}) because the new peak appeared at the same time as n_{C-O}. The reason for the lowering of the frequency from 2380 to 2200 cm^{-1} is that the PT during the generation of the monoalcohol (I) changes N–D to O–D, causing the red shift of the molecular vibration frequency by a factor of $(14/16)^{1/2}$. Meanwhile, the electron density in the other N–H bond, which does not participate in the PT, is increased after the photoexcitation, leading to the blue shift of the peak due to the weakened hydrogen bonding. After the generation of monoalcohol (I), the peak around 2380 cm^{-1} (n_{N-D}) was restored in the longer delay time region of >700 fs, as shown in Figure 11.1.1.7a. These results are completely consistent with that for the carbonyl group dynamics.

The small delay between the appearance of the n_{O-H} and n_{C-O} seen in Figure 11.1.1.7 can be explained as follows. The theoretical calculation shows that the distance between C=O and N–H decreases considerably in generating the O–H bond, from 2.279Å in the reactant to almost half of that in the TS (1.197 Å), and even shorter in the intermediate (0.986 Å). However, the C=O bond length changes only slightly from the reactant (1.236 Å) to the TS (1.296 Å) and the intermediate (1.334 Å), to generate the C–O bond. Therefore, the appearance of n_{C-O} is delayed slightly compared with that of n_{O-H}.

Solvent Effect. As mentioned above, in the case of methanol solution of indigodisulfonate salts, ultrafast PT in the excited state takes place by the stepwise mechanism. However, whether the PT takes place intramolecularly or intermolecularly is not understood, because the methanol is a protic solvent. Therefore, the DMSO solution, which is aprotic solvent, of sodium indigodisulfonate was used as a sample. The obtained spectrogram is shown in Figure 11.1.1.8. In the spectrogram, a peak around 1700 cm^{-1}, due to a C=O stretching mode of reactant, appeared just after the excitation. This frequency gradually red-shift in time toward 200 fs, and other peaks around 1250 and 1750 cm^{-1} started to appear. The peak around 1250 cm^{-1} can be attributed to the C–O single bond stretching mode (n_{C-O}). The presence of this peak indicates the formation of the C–OH group in the intermediate. After the generation of the monoalcohol (I), the peak around 1700 cm^{-1} ($n_{C=O}$) is reproduced in the time range >350 fs. If the PT takes place by intramolecular reaction in the methanol solution, the reaction time of the PT in methanol solution is close to that in the DMSO solution. These results show that the reaction time scales are almost the same between the DMSO solution and the methanol solution, but the reaction time of the PT in the DMSO solution is a little faster

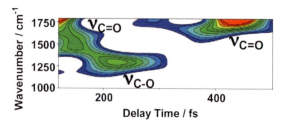

FIGURE 11.1.1.8 Spectrogram of DMSO solution of sodium indigodisulfonate.

than that in the methanol solution. Therefore, the possibility of intramolecular PT reaction in the methanol solution is high. However, the possibility of intermolecular PT reaction in the methanol solution cannot be denied. Anyway, these results confirm that PT takes place after photoexcitation in indigodisulfonate salts in both protic and aprotic solvents, and they show that the PT mechanism is a stepwise pathway. Moreover, the unstable monoalcohol (I) generated by PT then reverts to the reactant indigo compound.

11.1.1.6 CONCLUSION

In summary, the structural change during ultrafast proton transfer via a transition state was observed using an ultrashort pulse laser. It was concluded that a photoexcited proton transfer takes place in indigodisulfonate salts by the stepwise mechanism. To date, the reason why indigo is photostable could not be elucidated for over 100 years. The answer was obtained by the direct experimental observation in this work. The efficiency of photodiscoloration of indigodisulfonate salts caused by photoisomerization is suppressed by a single proton transfer after photoexcitation, because the molecular structure was fixed in the planar form by the formation of a monoalcohol intermediate generated by the single proton transfer. The unstable monoalcohol intermediate generated by the proton transfer then reverts to the reactant indigo compound. Thus, time-resolved spectroscopy with time resolution of a few femtoseconds provides a new way to clarify mechanisms and stimulate novel ideas for the development of new chemical reactions. Clarification of detailed reaction mechanisms is enabled by the application of this technique to the study of ultrafast dynamics in various fields, such as photochemistry, photophysics, and photobiology. The contents of this section are based on the research products of collaborative works by the following people: Izumi Iwakura, Atsushi Yabushita, and Takayoshi Kobayashi [14].

SUPPORTING INFORMATION

The energies of the most stable conformers and energy profiles of various methods. This material is available free of charge on the Web at: http://www.csj.jp/journals/bcsj/.

REFERENCES

1. A. H. Zewail, *J. Phys. Chem. A* **104**, 5660 (2000).
2. a) P. C. Becker, R. L. Fork, C. H. B. Cruz, J. P. Gordon, and C. V. Shank, *Phys. Rev. Lett.* **60**, 2462 (1988). b) A. Stingl, Ch. Spielmann, F. Krausz, and R. Szipöcs, *Opt. Lett.* **19**, 204 (1994). c) J. Zhou, C.-P. Huang, C. Shi, M. M. Murnane, and H. C. Kapteyn, *Opt. Lett.* **19**, 126 (1994). d) A. Baltuska, Z. Wei, M. S. Pshenichnikov, and D. A. Wiersma, *Opt. Lett.* **22**, 102 (1997). e) M. Nisoli, S. De Silvestri, O. Svelto, R. Szipöcs, K. Ferencz, Ch. Spielmann, S. Sartania, and F. Krausz, *Opt. Lett.* **22**, 522 (1997). f) A. Baltuška, T. Fuji, and T. Kobayashi, *Opt. Lett.* **27**, 306 (2002).
3. a) R. A. Mathies, C. H. B. Cruz, W. T. Pollard, and C. V. Shank, *Science* **240**, 777 (1988). b) T. Kobayashi, T. Saito, and H. Ohtani, *Nature* **414**, 531 (2001). c) T. Saito, and T. Kobayashi, *J. Phys. Chem. A* **106**, 9436 (2002). d) T. Fujino, S. Y. Arzhantsev, T. Tahara, *Bull. Chem. Soc. Jpn.* **75**, 1031 (2002). e) J. Bredenbeck, J. Helbing, A. Sieg, T. Schrader, W. Zinth, C. Renner, R. Behrendt, L. Moroder, J. Wachtveitl, and P. Hamm, *Proc. Natl. Acad. Sci. U.S.A.* **100**, 6452 (2003). f) A. Mühlpfordt, U. Even, and N. P. Ernsting, *Chem. Phys. Lett.* **263**, 178 (1996). g) C. Chudoba, E. Riedle, M. Pfeiffer, and T. Elsaesser, *Chem. Phys. Lett.* **263**, 622 (1996). h) S. Lochbrunner, A. J. Wurzer, E. Riedle, and *J. Chem. Phys.* **112**, 10699 (2000). i) M. Rini, A. Kummrow, J. Dreyer, E. T. J. Nibbering, and T. Elsaesser, *Faraday Discuss.* **122**, 27 (2003). j) M. Rini, J. Dreyer, E. T. J. Nibbering, and T. Elsaesser, *Chem. Phys. Lett.* **374**, 13 (2003). k) S. J. Schmidtke, D. F. Underwood, and D. A. Blank, *J. Am. Chem. Soc.* **126**, 8620 (2004).
4. C. P. Klages, K. Kobs, and R. Memming, *Chem. Phys. Lett.* **90**, 51 (1982).
5. a) W. R. Brode, E. G. Pearson, and G. M. Wyman, *J. Am. Chem. Soc.* **76**, 1034 (1954). b) J. Weinstein, and G. M. Wyman, *J. Am. Chem. Soc.* **78**, 2387 (1956). c) W. Lüttke, and M. Klessinger, *Chem. Ber.* **97**, 2342 (1964). d) M. Klessinger, and W. Lüttke, *Chem. Ber.* **99**, 2136 (1966). e) G. M. Wyman, *J. Chem. Soc. D*,

24, 1332 (1971). f) T. Kobayashi, and P. M. Rentzepis, *J. Chem. Phys.* **70**, 886 (1979). g) T. Elsaesser, W. Kaiser, and W. Lüttke, *J. Phys. Chem.* **90**, 2901 (1986). h) J. Sérgio Seixas de Melo, A. P. Moura, and M. J. Melo, *J. Phys. Chem. A* **108**, 6975 (2004). i) Y. Nagasawa, R. Taguri, H. Matsuda, M. Murakami, M. Ohama, T. Okada, and H. Miyasaka, *Phys. Chem. Chem. Phys.* **6**, 5370 (2004). j) J. Sérgio Seixas de Melo, R. Rondão, H. D. Burrows, M. J. Melo, S. Navaratnam, R. Edge, and G. Voss, *ChemPhysChem* **7**, 2303 (2006). k) A. Doménech, M. T. Doménech-Carbó, and M. L. Vázquez de Agredos Pascual, *J. Solid State Electrochem.* **11**, 1335 (2007). l) I. Iwakura, A. Yabushita, and T. Kobayashi, *Chem. Lett.* **38**, 1020 (2009). m) I. Iwakura, A. Yabushita, and T. Kobayashi, *Chem. Phys. Lett.* **484**, 354 (2010).

6. M. J. Frisch, G. W. Trucks, H. B. Schlegel, G. E. Scuseria, M. A. Robb, J. R. Cheeseman, J. A. Montgomery, Jr., T. Vreven, K. N. Kudin, J. C. Burant, J. M. Millam, S. S. Iyengar, J. Tomasi, V. Barone, B. Mennucci, M. Cossi, G. Scalmani, N. Rega, G. A. Petersson, H. Nakatsuji, M. Hada, M. Ehara, K. Toyota, R. Fukuda, J. Hasegawa, M. Ishida, T. Nakajima, Y. Honda, O. Kitao, H. Nakai, M. Klene, X. Li, J. E. Knox, H. P. Hratchian, J. B. Cross, V. Bakken, C. Adamo, J. Jaramillo, R. Gomperts, R. E. Stratmann, O. Yazyev, A. J. Austin, R. Cammi, C. Pomelli, J. W. Ochterski, P. Y. Ayala, K. Morokuma, G. A. Voth, P. Salvador, J. J. Dannenberg, V. G. Zakrzewski, S. Dapprich, A. D. Daniels, M. C. Strain, O. Farkas, D. K. Malick, A. D. Rabuck, K. Raghavachari, J. B. Foresman, J. V. Ortiz, Q. Cui, A. G. Baboul, S. Clifford, J. Cioslowski, B. B. Stefanov, G. Liu, A. Liashenko, P. Piskorz, I. Komaromi, R. L. Martin, D. J. Fox, T. Keith, M. A. Al-Laham, C. Y. Peng, A. Nanayakkara, M. Challacombe, P. M. W. Gill, B. Johnson, W. Chen, M. W. Wong, C. Gonzalez, and J. A. Pople, *Gaussian 03, Revision D.02*, Gaussian, Inc., Wallingford CT, 2004.

7. I. Z. Kozma, P. Krok, and E. Riedle, *J. Opt. Soc. Am. B* **22**, 1479 (2005).

8. M. J. J. Vrakking, D. M. Villeneuve, and A. Stolow, *Phys. Rev. A* **54**, R37 (1996).

9. a) I. T. Shadi, B. Z. Chowdhry, M. J. Snowden, and R. Withnall, *Chem. Commun.* **40**, 1436 (2004). b) R. Giustetto, F. X. Llabrés i Xamena, G. Ricchiardi, S. Bordiga, A. Damin, R. Gobetto, and M. R. Chierotti, *J. Phys. Chem. B* **109**, 19360 (2005).

10. a) R. L. Hinman, *J. Org. Chem.* **29**, 1449 (1964). b) G. Keresztury, F. Billes, M. Kubinyi, and T. Sundius, *J. Phys. Chem. A* **102**, 1371 (1998).

11. E. D. Glendening, A. E. Reed, J. E. Carpenter, and F. Weinhold, Gaussian NBO, Version 3.1.

12. J. M. Jean, and G. R. Fleming, *J. Chem. Phys.* **103**, 2092 (1995).

13. F. Rosca, A. T. N. Kumar, X. Ye, T. Sjodin, A. A. Demidov, and P. M. Champion, *J. Phys. Chem. A* **104**, 4280 (2000). b) A. T. N. Kumar, F. Rosca, A. Widom, and P. M. Champion, *J. Chem. Phys.* **114**, 701 (2001).

14. I. Iwakura, A. Yabushita, and T. Kobayashi, *Bull. Chem. Soc. Jpn*, **84**, 164–171 (2011).

11.1.2 Environment-Dependent Ultrafast Photoisomerization Dynamics in Azo Dye

11.1.2.1 INTRODUCTION

Over the past decades, trans-cis photoisomerization has been one of the most widely investigated photochromic reactions. Azoaromatic dyes [1] are typical systems that exhibit trans-cis photoisomerization. They have been extensively investigated and various experimental techniques have been used to clarify the trans-cis photoisomerization mechanism [2–4]. Irradiation by linearly polarized light results in reversible trans-cis-trans isomerization via a hole-burning mechanism. This isomerization reorients the molecules, causing the dye to become birefringent by making the azobenzene groups sensitive to optical polarization [5]. Thus, azoaromatic dyes can be used in various applications such as dynamic volume holograms [6,7] waveguide media [8], and light-driven molecular scissors [9]. To improve their performance, it is essential to elucidate the ultrafast dynamics of the photoisomerization process. There have been several recent studies of the ultrafast dynamics of azobenzene dyes; for example, Koller et al. [10] used infrared spectroscopy to investigate 4-nitro40-(dimethylamino)azobenzene and Poprawa-Smoluch et al. [11] used absorption spectroscopy to study Disperse Red 1. Disperse Red 19 (DR19) is a commercial azobenzene nonlinear optical chromophore with a large ground-state dipole moment of 8 D [12,13]. In the present report, we investigate the ultrafast dynamics of solution and film samples of DR19 using absorbance difference spectroscopy with a femtosecond time resolution.

11.1.2.2 EXPERIMENTAL SECTION

The azobenzene dye DR19 (dye content 95%) was purchased from Sigma-Aldrich Co. and used after recrystallization. A solution sample was prepared by dissolving 0.1 wt % DR19 in trimethylolpropane triglycidyl ether (TMTE; technical grade; Sigma-Aldrich). A polymer film sample of DR19 was prepared as follows: 0.1 wt % DR19 was dissolved in TMTE and a harden agent (1,2-diaminopropane). The solution was cast on a flat glass (doctor Blade method) [14] to form a polymer film. The polymer film was estimated to have an average thickness of 0.5 mm by a slide calliper.

The solution and film samples were respectively stored in a sealed vial and a drybox in a dark-adaptive ambient until immediately prior to being used. Figure 11.1.2.1 shows stationary absorption spectrum of the solution and film samples. The absorption peaks at 494 and 500 nm for the solution and film samples, respectively, are due to the strongly allowed $\pi\pi^*$ transition [15].

Time-resolved spectroscopy of the samples was performed using ultrashort visible pulses generated by a noncollinear optical parametric amplifier [16,17] seeded by a 5 kHz regenerative Ti: sapphire amplifier (Coherent Inc. Legend-USP-HE). These ultrashort visible pulses had a broad spectrum that extended from 514 to 758 nm (see Figure 11.1.2.1). The instrument response time was adjusted to be 9 fs by monitoring the pulse width using second-harmonic-generation frequency-resolved optical gating (SHG-FROG) measurements [18] (see below).

DOI: 10.1201/9780429196577-76

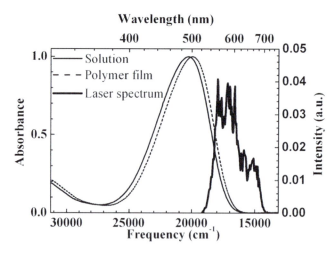

FIGURE 11.1.2.1 Normalized absorption spectra of DR19 in solution (solid line) and a polymer film (dashed line). The spectrum of NOPA output (thick solid line) is used for both pump and probe pulses.

In the pump-probe measurements of the solution sample, the pulse width was compressed to 9 fs in a 1-mm glass cell as follows. We first inserted a 1 mm thick glass plate in front of the entrance of the SHG-FROG system and adjusted pulse compressor to obtain 9 fs pulses. It set the pulse width to be 9 fs after passing through the 1-mm thick glass. The glass walls of the glass cell used for the solution measurement have the same thickness of 1 mm. If the 1 mm thick glass is removed and the pulse comes into the glass cell, the pulse width becomes 9 fs inside the glass cell passing through one side of the 1 mm thick glass wall.

In the pump-probe measurements of the film sample, the pulse width was also adjusted to 9 fs without inserting any glass plate. Therefore, the instrument response time was 9 fs in both cases. The pump beam was focused to a 1.4×10^5 cm^2 spot. The pump and probe pulse energies were 8 and 0.8 nJ, respectively.

The probe pulse intensity at each probe wavelength was simultaneously detected by Si avalanche photodiodes attached to a monochromator (Princeton Instruments, SpectraPro 2300i). A mechanical chopper was inserted in the optical path of the pump pulse and was operated at a frequency of 2.5 kHz so that it was in synchronization with the laser pulses. Changes in the transmitted intensity of the probe pulse induced by the pump pulse were processed by a 96-channel lock-in amplifier referenced to the frequency and phase of the chopper. Preamplifiers were used to amplify the electric current measured by the avalanche photodiodes prior to lock-in detection. Because the preamplifier had a cutoff frequency of 5 kHz, harmonic noise of the 2.5 kHz reference frequency was not detected in the observed signals.

For time-resolved spectroscopy of the solution sample, the sample was placed in a 1 mm thick quartz flow cell (Starna Cells Inc., 48-Q-1; flow rate: 50 mL/min) and was recirculated by a peristaltic pump to ensure that each laser pulse irradiated a fresh sample. To prevent the photogenerated cis-isomer from accumulating during measurements, the peristaltic pump was connected to a reservoir containing a large volume (60 mL) of the solution sample. During time-resolved measurements of the absorbance difference of the polymer film sample, the position of the sample was changed after each scan to ensure that photodamage did not accumulate during the measurement. Each scan of the time-resolved absorption difference spectra was performed in 5 s using a multiplex fast-scan system [19]. In the data analysis, we used data averaged over 40 scans. When the sample was irradiated at the same position for half an hour, the sample was damaged and caused strong scattering light. Meanwhile, the photodamage was negligible in the scan time of 5 s.

In the present measurement, the beam diameter on the sample is about 20 μm, which is about 10 larger than the period of the optically inscribed grating reported by Barrett et al. [20] Therefore, the sample still exists inside the beam spot even after the irradiation.

All experiments were performed at room temperature.

11.1.2.3   RESULTS AND DISCUSSION

Ultrafast Dynamics of the Solution Sample. Figure 11.1.2.2a shows a two-dimensional plot of absorbance difference (ΔA) spectrum of the DR19 solution sample measured for delays from 0.1 to 1.4 ps (short delay region) and probe wavelength from 514 to 758 nm. Figure 11.1.2.2b shows the ΔA spectrum measured between 1 and 14 ps (long delay region) in the same probe wavelength region. The black curves in Figure 11.1.2.2a and b are contours for which ΔA is zero. Figure 11.1.2.2c and d show ΔA traces for five different probe wavelengths in Figure 11.1.2.2a and b.

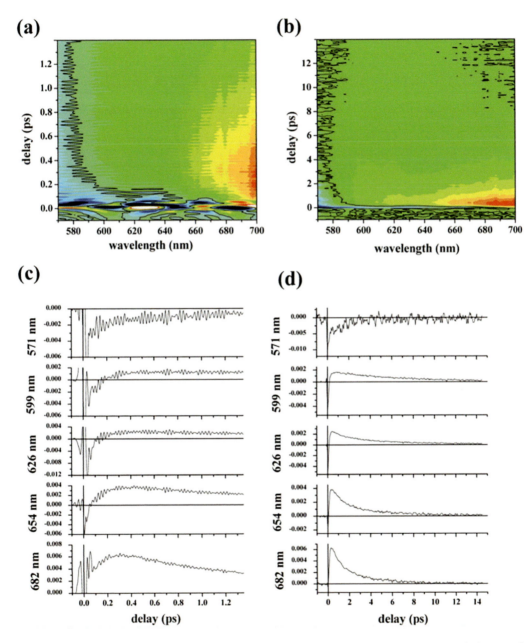

FIGURE 11.1.2.2   Measured 2D time-resolved absorbance difference spectra of DR19 in a solution sample (a) scanned from 0.1 to 1.4 ps and (b) that scanned from 1 to 14 ps. Black curves represent contours where the absorbance difference is zero ($\Delta A = 0$). (c and d) Time-resolved ΔA traces at five different probe wavelengths picked up from (a) and (b), respectively.

ΔA is generally negative at a probe wavelength below 580 nm. At these wavelengths, the ground state has an absorption band. Thus, ΔA is thought to be negative due to absorption bleaching, which implies that the electron population is depleted when DR19 is photoexcited by pump pulses. The positive ΔA observed at probe wavelengths longer than 580 nm was attributed to induced absorption of the excited state.

At first, we analyzed the exponential decay of the time traces to study the ultrafast dynamics of the electronic states during and after photoexcitation of DR19.

The ΔA traces in Figure 11.1.2.2c and d contain three decay components with lifetimes of ~0.1, ~1, and ~10 ps. Therefore, the ΔA traces should be fitted by a triple exponential function given by

$$f(t,\lambda) = \Delta A_1(\lambda)e^{-t/\tau_1} + \Delta A_2(\lambda)e^{-t/\tau_2} + \Delta A_0(\lambda)e^{-t/\tau_2} \quad (\tau_1 < \tau_2 < \tau_3) \qquad (11.1.2.1)$$

The τ_1 (~0.1 ps) lifetime component decays, becoming negligible at delay longer than 0.4 ps. Therefore, we determined the other two time constants (τ_2 and τ_3) by fitting the ΔA trace in the long delay region between 0.4 and 14 ps using the following double exponential function

$$f(t,\lambda) = \Delta A_2(\lambda)e^{-t/\tau_2} + \Delta A_3(\lambda)e^{-t/\tau_3} \quad (\tau_2 < \tau_3) \qquad (11.1.2.2)$$

Global fitting analysis estimated the time constants τ_2 and τ_3 to be 1.11 (0.13 and 4.65 (0.56 ps, respectively. Using the ΔA traces and the estimated time constants, the spectra of these lifetime components were calculated by the least-squares method (see Figure 11.1.2.3).

The instrument response time of the present measurements was estimated to be 9 fs, which is considered negligible for delays longer than 50 fs. Therefore, we determined the shortest time constant (τ_1) and its spectral component (ΔA_1) by fitting the ΔA trace in the short delay region from 50 fs to 1.4 ps. In the fitting analysis, we applied the global fitting using the following double exponential function by fixing the parameter τ_2 to 1.11 ps (as estimated above). Note that the τ_3 decay component can be considered constant for delays up to 1.4 ps. The function $f(t, \lambda)$ is given by

$$f(t,\lambda) = \Delta A_1(\lambda)e^{-t/\tau_1} + \Delta A_2(\lambda)e^{-t/\tau_2} + \Delta A_0(\lambda) \quad (\tau_1 < \tau_2) \qquad (11.1.2.3)$$

where $\Delta A_0(\lambda)$ indicates a spectral component with an infinite lifetime. Global fitting analysis estimated the time constant τ_1 to be 104 (12 fs and its spectral component $\Delta A_1(\lambda)$ was obtained by the least-squares method (see Figure 11.1.2.3).

The shortest time constant, τ_1, had been only roughly estimated in previous studies because of their limited time resolutions. In the present study using a 9 fs pulse, τ_1 was accurately estimated to be 104 ± 12 fs. Its spectral component, $\Delta A_1(\lambda)$, does not reflect the spectral profile of the stationary absorption spectrum shown in Figure 11.1.2.1. Therefore, the shortest decay component does not

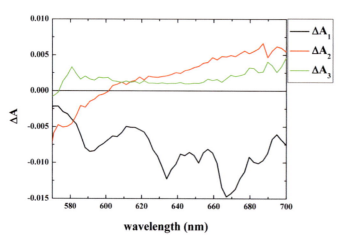

FIGURE 11.1.2.3 ΔA_1, ΔA_2, and ΔA_3 are spectral components of the DR19 solution sample, which decay with the lifetimes of τ_1, τ_2, and τ_3, respectively. The curves are of ΔA_1, ΔA_2, ΔA_3 from bottom to top at wavelength of 600 nm.

include decay to the electronic ground state, and it was attributed to large amplitude wave packet motion on the excited-state potential surface out of the Franck Condon (FC) region [21]. The negative spectrum of ΔA_1 is considered to be the stimulated emission spectrum of the FC state.

The time constant τ_2 was estimated to be 1.11 ± 0.13 ps. Its spectral component, $\Delta A_2(\lambda)$, is negative, reflecting the spectral profile of the stationary absorption spectrum at probe wavelengths shorter than 580 nm. Therefore, the wave packet generated by the photoexcitation is considered to be spread over the potential energy surface of the electronic excited state and to locate the conical intersection to the ground state via photoisomerization around the N=N double bond; this process is responsible for the time constant of τ_2 [22]. When probe wavelength is longer than 620 nm, $\Delta A_2(\lambda)$ is positive and induced absorption is the dominant process.

The time constant τ_3 was determined to be 5.86 ± 2.13 ps. Its spectral component, $\Delta A_3(\lambda)$, has a flat profile with a positive value reflecting induced absorption. Koller et al. [10] elucidated that vibrational cooling in the ground state occurs in 4-nitro-40(dimethylamino)azobenzene with a time constant of 5.5 ps. Therefore, τ_3 is assigned to the vibrational cooling time in the ground state of DR19.

Next, we analyzed the oscillation observed in the time traces shown in Figure 11.1.2.2c and d. The periodic oscillations reflect molecular vibrations that occur during the reaction after photoexcitation; these oscillations have been theoretically studied [23] and observed in various experiments [19,24,25]. Fourier transform of the time trace measured at 626 nm (see Figure 11.1.2.4a) was

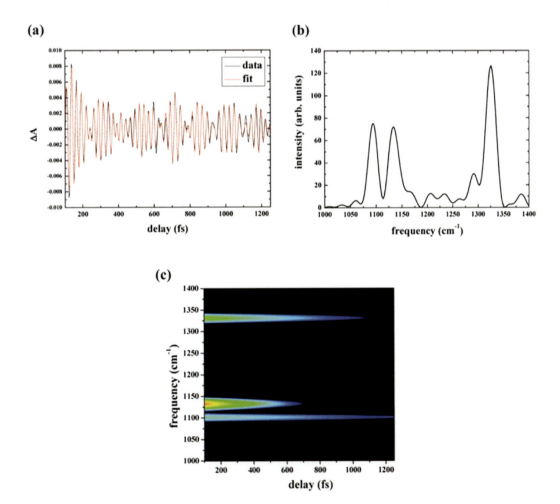

FIGURE 11.1.2.4 (a) Time trace of DR19 in solution measured at 626 nm (black curves) and the time trace reconstructed in LP-SVD analysis (red curves). (b) Fourier power spectrum of the measured time trace shown in (a). (c) Two-dimensional view of the intensity of the vibrational modes obtained in the LPSVD analysis.

TABLE 11.1.2.1
Results Obtained in the LP-SVD Analysis for DR19 in Solution

Frequency (cm⁻¹)	Amplitude	Decay (fs)
1100	107	2510
1131	285	458
1330	141	1411

calculated as shown in Figure 11.1.2.4b. It agrees with the Raman spectrum [26], confirming that the oscillations in the time-resolved trace are caused by molecular vibrations.

The vibrational dynamics was analyzed using linear prediction and singular value decomposition (LP-SVD) [27–29]. Table 11.1.2.1 shows the frequencies, amplitudes, and lifetimes of the vibrational modes obtained in the LP-SVD analysis. The obtained parameters are also shown as a two-dimensional view in Figure 11.1.2.4c. When the obtained parameters were used, the time trace was reconstructed, as shown in Figure 11.1.2.4a, agreeing with the measured time trace. The vibration modes of 1100, 1131, and 1330 cm⁻¹ are assigned to ΦN stretching, CH deformation, and NO_2 symmetric stretching modes, respectively, referring to the Raman study [26]. The result of the LP-SVD analysis shows that the CH deformation has distinctly high amplitude and fast lifetime among these three vibration modes. It reflects that photoisomerization of the DR19 in solution causes large deformation around the CH bond in its primary process.

Comparison with Ultrafast Dynamics of the Film Sample. Figure 11.1.2.5a shows a two-dimensional plot of the ΔA spectrum of the DR19 film sample measured for delays from 0.1 to 1.4 ps (short delay region) and probe wavelengths from 514 to 758 nm.

Figure 11.1.2.5b shows the ΔA spectrum measured between 1 and 14 ps (long delay region) in the same probe wavelength region. The black curves in Figure 11.1.2.5a and b are contour lines on which ΔA is zero. Figure 11.1.2.5c and d show ΔA traces at five different probe wavelengths in Figure 11.1.2.5a and b.

Similar to the solution sample, the negative ΔA at delays shorter than 580 nm is assigned to absorption bleaching caused by depletion of the ground state and the positive ΔA observed at probe wavelengths longer than 580 nm reflects induced absorption of the excited state.

The ΔA traces of the film sample in Figure 11.1.2.5c and d also include three decay components with lifetimes of ~0.1, ~1, and ~10 ps. Therefore, the ΔA traces were fitted in the same manner as the traces of the solution sample.

We first fitted the ΔA trace in the long delay region using Eq (11.1.2.2) for delays from 0.4 to 14 ps. Global fitting analysis estimated the time constants of τ_2 and τ_3 to be 1.35 (0.19 and 5.77 (0.81 ps, respectively. The spectral components of the obtained time constants were obtained by the least-squares method (see Figure 11.1.2.6).

Compared with this data for DR19 in the film sample, ΔA_2 has a large positive amplitude in the wavelength region reflecting the existence of induced absorption longer than 620 nm in solution (see Figure 11.1.2.3). The reason why the induced absorption appears in the solution sample of DR19 is thought to be explained as follows. The DR19 has a push-pull substituted structure with a large dipole moment of 8 D, and the solvent of TMTE is a polar solvent with a dipole moment of 3 D. Therefore, there is a strong dipole interaction between the DR19 and the solvent in the solution sample. The strong interaction modifies the electronic states of the higher excited state, which opens a way for the induced absorption via transition from the first excited state to the higher excited state.

The shortest time constant (τ_1) and its spectral component (ΔA_1) were then estimated by fitting the ΔA trace in the short delay region from 50 fs to 1.4 ps. Global fitting of the ΔA traces using Eq (11.1.2.3) estimated the time constant τ_1 to be 74 (10 fs). The spectral component of τ_1 ($\Delta A_1(\lambda)$) was obtained by the least squares-method and is plotted in Figure 11.1.2.6.

The shortest time constant, τ_1, was estimated to be 74 (10 fs). For the same reason, given above for the solution sample, the shortest decay component was attributed to a large amplitude wave

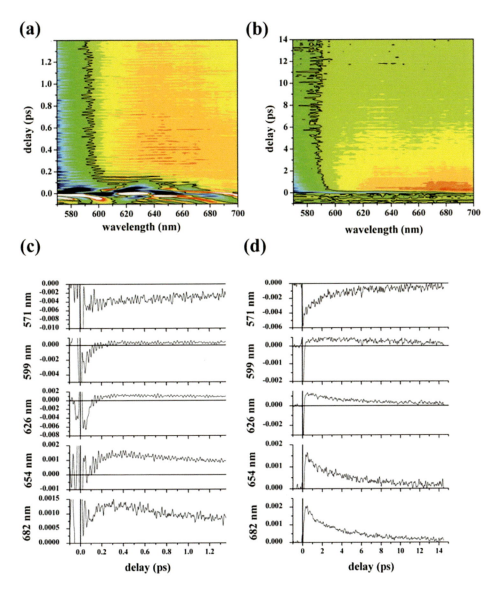

FIGURE 11.1.2.5 Measured 2D time-resolved absorbance difference spectra of DR19 in a film sample (a) scanned from −0.1 to 1.4 ps and (b) that scanned from −1 to 14 ps. Black curves represent contours where the absorbance change is zero ($\Delta A = 0$). (c and d) Time-resolved ΔA traces at five typical several probe wavelengths picked up from (a) and (b), respectively.

packet motion on the excited-state potential surface out of the FC region. The negative spectrum of ΔA_1 was considered to be the stimulated emission spectrum of the FC state. DR19 is known to have a large dipole moment of up to 8 D [12,13] because of its pushpull substituted structure. Therefore, photoexcitation of DR19 generates a charge-transfer (CT) state via electron transfer. Inter- and intramolecular modes adapt within τ_1 to the new charge distribution generated by CT excitation. The solution and film samples had similar shortest time constant τ_1, which implies that solutions and films of DR19 have comparable adaptation speeds.

The time constant τ_2 is estimated to be 1.35 (0.19 ps). $\Delta A_2(\lambda)$ is negative at wavelength below 610 nm reflecting the spectral profile of the stationary absorption spectrum. Therefore, τ_2 in the film sample was also assigned to the time for the photoexcited wave packet in the excited state to find the conical intersection to the ground state via photoisomerization around the NN double bond; this

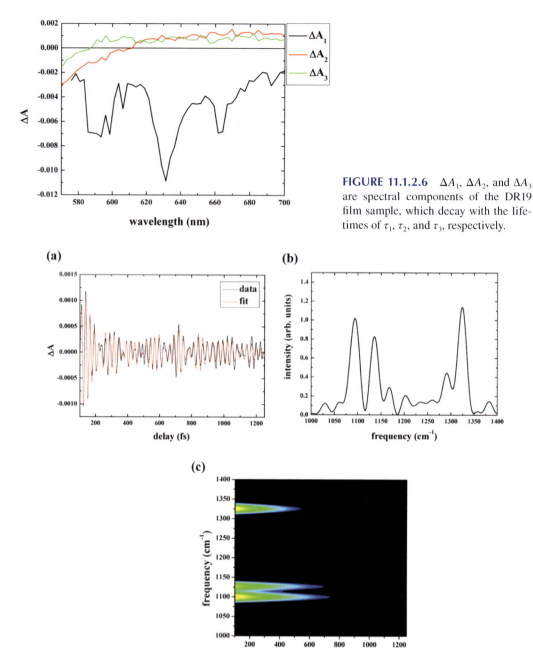

FIGURE 11.1.2.6 ΔA_1, ΔA_2, and ΔA_3 are spectral components of the DR19 film sample, which decay with the lifetimes of τ_1, τ_2, and τ_3, respectively.

(a)

(b)

(c)

FIGURE 11.1.2.7   (a) Time trace of DR19 in a film sample measured at 626 nm (black curves) and the time trace reconstructed in LP-SVD analysis (red curves). (b) Fourier power spectrum of the measured time trace shown in (a). (c) Two-dimensional view of the intensity of the vibrational modes obtained in the LP-SVD analysis.

process is responsible for the time constant τ_2. When the probe wavelength is longer than 610 nm, $\Delta A_2(\lambda)$ is positive and induced absorption is the dominated process.

The time constant of τ_3 was determined to be 5.77 (0.81 ps). Its spectral component, $\Delta A_3(\lambda)$, has a flat profile with a positive induced absorption. Just as for the solution sample, τ_3 reflects the vibrational cooling time in the ground electronic state of DR19.

Fourier transform of the time trace measured at 626 nm (see wa) was also calculated for DR19 in a film sample as shown in Figure 11.1.2.7b. It agrees with the result obtained for the

TABLE 11.1.2.2

Results Obtained in the LP-SVD Analysis for DR19 in a Film Sample

Frequency (cm^{-1})	Amplitude	Decay (fs)
1099	35	518
1126	27	588
1324	36	369

solution sample (see Figure 11.1.2.4b) and the Raman spectrum [26]. The LP-SVD analysis was performed for the time trace of the film sample. Table 11.1.2.2 shows the frequencies, amplitudes, and lifetimes of the vibrational modes obtained in the LP-SVD analysis. The obtained parameters are also shown as a two-dimensional view in Figure 11.1.2.7c. When the obtained parameters were used, the time trace was reconstructed as shown in Figure 11.1.2.7a, agreeing with the measured time trace. The assignments of 1099, 1126, and 1324 cm^{-1} are the same as those of the solution sample assigned to ΦN stretching, CH deformation, and NO$_2$ symmetric stretching modes, respectively, referring the Raman study [26]. The result of the LP-SVD analysis shows that all of those three modes have similar amplitude and fast lifetime of ~500 fs. It reflects that photoisomerization of the DR19 in the film sample causes structural deformation equally among the ΦN bond, CH bond, and NO$_2$ group. The result is significantly different from that of the DR19 in solution, in which only the CH deformation was dominantly excited. It is thought to be that the strong dipole interaction between TMTE and DR19 in the solution sample suppresses deformations along the dipole moment of the DR19 only allowing the CH deformation.

11.1.2.4 SUMMARY

The photoisomerization of DR19 in solution and a polymer film were investigated in absorbance difference measurements over a broadband visible spectral range and with a time resolution of 9 fs. The DR19 has a push-pull substituted structure, which forms the CT state via electron transfer after photoexcitation.

Therefore, the film sample may have different dynamics from the solution sample because significant intermolecular CT in the film modifies the relaxation pathway of the CT state. The observed time traces contain three decay components with lifetimes of ~0.1, ~1, and ~10 ps. We estimated the three time constants (τ_1, τ_2, and τ_3) and their spectral components (ΔA_1, ΔA_2, and ΔA_3) by fitting the time-resolved ΔA trace with a triple exponential function.

The shortest time constant, τ_1, had been roughly estimated in previous studies. In this study, τ_1 was precisely estimated for both the solution sample (104±12 fs) and film (74±10 fs) samples. Its spectral component, ΔA_1, is negative, reflecting stimulated emission from the FC state. The τ_1 decay component was attributed to a large amplitude wave packet motion on the excited-state potential surface out of the FC region.

The spectral component, $\Delta A_2(\lambda)$, is negative reflecting the spectral profile of the stationary absorption spectrum for probe wavelengths shorter than 580 nm. This implies that the wave packet generated by the photoexcitation is spread over the potential energy surface of the electronic excited state and finds the conical intersection to the ground state via photoisomerization around the NN double bond; this process is responsible for the time constant of τ_2.

The longest spectral component, $\Delta A_3(\lambda)$, has a flat profile and is positive due to induced absorption. Its time constant, τ_3, is much longer than τ_2, reflecting the vibrational cooling time in the ground electronic state of DR19.

In the present study, we clarified the ultrafast dynamics of a push-pull substituted azobenzene dye, DR19, which is known to have a large dipole moment and is promising for various applications [30]. Ultrafast time-resolved absorbance difference spectroscopy is expected to provide essential information for photonic applications and is a promising alternative method for evaluating switching performance. The contents of this section are cooperative research among the following people: Chun-Chih Hsu, Yu-Ting Wang, Atsushi Yabushita, Chih-Wei Luo, Takayoshi Kobayashi [30].

REFERENCES

1. G. S. Kumar, and D. C. Neckers, *Chem. Rev.* 89, 1915–1925 (1989).
2. H. Rau, and E. J. L€uddecke, *J. Am. Chem. Soc.* 104, 1616–1620 (1982).
3. Stuart, C. M., R. R. Frontiera, and R. A. Mathies, *J. Phys. Chem. A* 111, 12072–12080 (2007).
4. R. D. Curtis, J. W. Hilborn, G. Wu, M. D. Lumsden, R. E. Wasylishen, and J. A. Pincock, *J. Phys. Chem.* 97, 1856–1861 (1993).
5. A. Natansohn, P. Rochon, J. Gosselin, and S. Xie, *Macromolecules* 25, 2268–2273 (1992).
6. S. Bian, and M. G. Kuzyk, *Opt. Lett.* 27, 1761–1763 (2002).
7. W. Zhang, S. Bian, S. Kim, and M. G. Kuzyk, *Opt. Lett.* 27, 1105–1107 (2002).
8. C. C. Jung, M. Rutloh, and J. Stumpe, *J. Phys. Chem. B* 109, 7865–7871 (2005).
9. T. Muraoka, K. Kinbara, and T. Aida, *Nature* 440, R512–515 (2006).
10. F. O. Koller, C. Sobotta, T. E. Schrader, T. Cordes, W. J. Schreier, A. Sieg, and P. Gilch, *Chem. Phys.* 341, 258–266 (2007).
11. M. Poprawa-Smoluch, J. Baggerman, H. Zhang, H. P. A. Maas, L. De Cola, and A. M. J. Brouwer, *Phys. Chem. A* 110, 11926–11937 (2006).
12. B. Park, P. Paoprasert, I. In, J. Zwickey, P. E. Colavita, R. J. Hamers, P. Gopalan, and P. G. Evans, *Adv. Mater.* 19, 4353–4357 (2007).
13. J. M. Choi, J. Lee, D. K. Hwang, J. H. Kim, J. H.S. Im, and E. Kim, *Appl. Phys. Lett.* 88, 043508 (2006).
14. R. E. Mistler, *Am. Ceram. Soc. Bull.* 77, 82–86 (1998).
15. D€urr, H., and Bouas-Laurent, H. *Photochromism: Molecules and Systems*, Elsevier, Amsterdam; Boston, MA, pp. 165192 (2003).
16. A. Shirakawa, I. Sakane, and T. Kobayashi, *Opt. Lett.* 23, 1292–1294 (1998).
17. A. Shirakawa, I. Sakane, M. Takasaka, and T. Kobayashi, *Appl. Phys. Lett.* 74, 2268–2270 (1999).
18. T. Kobayashi, and A. Baltuska, *Meas. Sci. Technol.* 13, 1671–1682 (2002).
19. A. Yabushita, Y. H. Lee, T. Kobayashi, *Rev. Sci. Instrum.* 81, 063110 (2010).
20. C. J. Barrett, A.L. Natansohn, P. L. and Rochon, *J. Phys. Chem.* 100, 8836–8842 (1996).
21. B. Schmidt, C. Sobotta, S. Malkmus, S. Laimgruber, M. Braun, W. Zinth, and P. Gilch, *J. Phys. Chem. A* 108, 4399–4404 (2004).
22. C. W. Chang, Y. C. Lu, T. T. Wang, and E. Diau, *J. Am. Chem. Soc.* 126, 10109–10118 (2004).
23. W. T. Pollard, S.-Y. Lee, and A. Mathies, *Chem. Phys.* 92, 4012–4029 (1990).
24. D. Polli, L. Lüer, and G. Cerullo, *Rev. Sci. Instrum.* 78, 103108 (2007).
25. A. Yabushita, and T. Kobayashi, *J. Phys. Chem. B* 114, 4632–4636 (2010).
26. I. G. Marino, D. Bersami, P. P. Lottici, L. Tosini, and A. Montero, *J. Raman Spectrosc.* 31, 555–558 (2000).
27. H. Barhuijsen, R. De Beer, W. M. M. J. Bovee, J. H. N. Creyhgton, and D. Van Ormondt, *Magn. Reson. Med.* 2, 86–89 (1985).
28. H. Barhuijsen, R. De Beer, and D. J. Van Ormondt, *Magn. Reson.* 1985, 64, 343–346.
29. A. E. Johnson, and A. B. Myers, *J. Chem. Phys.* 104, 2497–2507 (1996).
30. C.-C. Hsu, Y.-T. Wang, A. Yabushita, C.-W. Luo, and T. Kobayashi, *J. Phys. Chem. A* 115, 11508–11514 (2011).

11.1.3 Direct Observation of Denitrogenation Process of 2,3-diazabicyclo [2.2.1] hept-2-ene (DBH) Derivatives, Using a Visible 5-fs Pulse Laser

11.1.3.1 INTRODUCTION

The denitrogenation of azoalkanes with a functional group of -N=N-, such as 2,3-diazabicyclo[2.2.1] hept-2-ene derivatives (DBHs) and azobisisobutyronitrile (AIBN), is an important reaction that generates cleanly radical intermediates [1]. The reaction mechanism of the denitrogenation of DBH derivatives has been studied and discussed since 1967, when a double inversion process [2–4] was observed in the formation of 2,3-dideuterobicyclo[2.1.0]pentane during the thermal denitrogenation of exo-5,6dideutero-2,3-diaza-bicyclo[2.2.1]hept-2-ene (DBH) [5–12].

Recent computational studies predicted that the denitrogenation mechanism of DBH derivatives, concerted versus stepwise, is largely dependent on the substituents X at C(7) (Scheme 11.1.3.1). Thus, the concerted denitrogenation process via the transition state TS_C is energetically favored for the parent DBH (X=H). Meanwhile, the stepwise denitrogenation via the intermediate DZ and two transition states TS_{S1} and TS_{S2} was found to be favored for the denitrogenation of 7,7-dialkoxy-substituted DBH derivatives (X=OR) [13,14]. The substituent effect on the denitrogenation can be rationalized by the most stable electronic configuration of the resulting biradicals BR [13]. Thus, the symmetric non-bonded molecular orbital (ψ_S) is the HOMO for the dialkoxy-substituted biradical, for which the concerted denitrogenation is a symmetry forbidden process. The concerted denitrogenation is a symmetry-allowed process for the electron-donating substituted azoalkane such as DBH, in which the two electrons are selectively occupied in the antisymmetric non-bonded molecular orbital (w_A). Experimentally, the preferred mechanism of stepwise denitrogenation was suggested by the aryl-group effect on the rate constant of the denitrogenation reaction of 1,4-diaryl-7,7-dialkoxy-substituted DBH derivatives [15].

In the present study, the denitrogenation processes of 7,7-diethoxy-substituted DBH (X=OEt) and DBH (X=H) were studied using an ultrafast spectroscopy system with a visible ultra-short pulse laser developed by Kobayashi and coworkers [16,17]. Typical vibrational modes of the DBH derivatives have vibrational periods of 60–80 fs (butterfly flap mode or ring bending mode) and 20 fs (CH_2 scissoring mode). Since the pulse width of the visible pulse used in the present work was much shorter than the vibrational periods, the observed signal was modulated by the molecular vibrations reflecting the real-time amplitude of the modes. During chemical reactions, molecular structural changes including transition states can be traced by the time-dependent wavenumber shifts of the relevant molecular vibrational modes [18]. The direct observation of the molecular vibrational changes induced by the impulsive excitation with a visible 5-fs pulse clarified the denitrogenation processes.

DOI: 10.1201/9780429196577-77

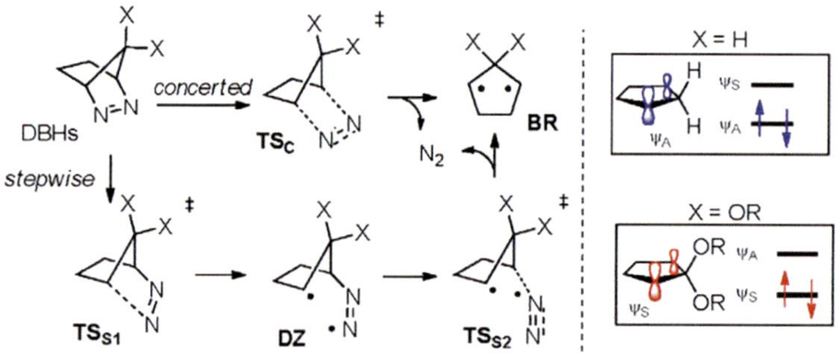

SCHEME 11.1.3.1 Substituent (X) effect on denitrogenation mechanism of DBH derivatives.

11.1.3.2 EXPERIMENTAL

To obtain the visible ultrashort pulses, we have generated broadband visible light with high intensity in a non-collinear optical parametric amplifier (NOPA), and its pulse width was compressed by using a chirp mirror pair and a prism compressor, as described below.

The light source of the NOPA was a regenerative amplifier (Spectra-Physics, model Spitfire) with 150 μJ of power, a central wavelength of 790 nm, pulse duration of 100 fs, and a repetition rate of 5 kHz. The NOPA generated visible broadband extending from 525 to 725 nm. A beam splitter separated the visible broadband light pulse into two copies of the pulse to be used as a pump pulse and a probe pulse, respectively. The chirp-mirror pair and the prism compressor were adjusted to compress their pulse widths as 5-fs. The polarizations of the pump and the probe pulses were parallel to each other. The focal spot areas of the pump and the probe pulses were 100 and 75 μm^2, respectively. The probe pulse was dispersed using a polychromator (300 grooves/mm, 500 nm blazed). A 128-channel fiber bundle sends each spectral component of the dispersed probe pulse to each piece of 128 avalanche photodiodes simultaneously. The time-resolved difference transmittance, ΔT, was measured simultaneously by avalanche photodiodes in the range of 525–725 nm. The signal-to-noise ratio was improved by coupling the signals of avalanche photo-diodes to a 128-channel lock-in amplifier.

A neat liquid sample of 7,7-diethoxy-substituted DBH derivative [19] and a CH_2Cl_2 solution sample of DBH [20] (100 mg/100 μl) were prepared for a pump–probe measurement. Because DBH is a solid sample, the solvent is needed to make a solution. To avoid the solvent effect on changing the denitrogenation mechanism, aprotic and non-polar solvent, i.e., CH_2Cl_2, was used for the experiment. A liquid cell with an optical path length of 1 mm was used to contain the sample. All measurements were performed at room temperature.

11.1.3.3 RESULTS AND DISCUSSION

11.1.3.3.1 PUMP–PROBE EXPERIMENTAL RESULTS

The azo-chromophore (-N=N-) of DBH derivatives has an absorption band at around 350 nm (Figure 11.1.3.1a). This clearly indicates that the photodenitrogenation of the DBH derivative does not occur under the conditions of one-photon absorption of the visible 5-fs pulse laser (Figure 11.1.3.1b). The ultrashort visible pulse excites vibration modes coherently in the electronic ground state through the stimulated Raman process, which triggers a reaction in the electronic ground state like thermal excitation by heating [21–23].

Pump–probe measurements were performed for the 7,7diethoxy DBH derivative and the parent DBH. Figure 11.1.3.2 shows two-dimensional displays of absorbance changes (ΔA) in the probe delay time from 200 to 2900 fs following the visible 5-fs pulse excitation of the two DBH derivatives. The signal of DA was calculated as $\Delta A = \log_{10}(1 + \Delta T/T)$, where T and ΔT are the transmittance and the transmittance change, respectively, induced by the pump pulse. The absorbance change was oscillating around zero, which indicates that the signal modulation is not due to the population

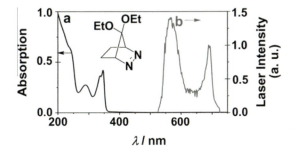

FIGURE 11.1.3.1 (a) Absorption spectrum of 7,7-diethoxy-substituted DBH derivative (black curves) and (b) the visible 5-fs pulse laser (gray curves) spectra.

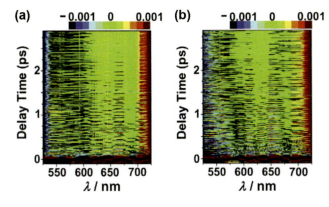

FIGURE 11.1.3.2 Two-dimensional displays of the absorbance changes plotted two dimensionally against the probe delay time and the wavelength following the visible 5-fs pulse excitation of (a) the 7,7-diethoxy-substituted DBH derivative and (b) DBH.

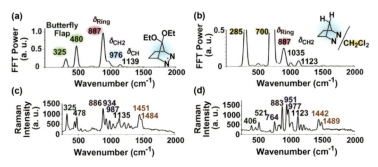

FIGURE 11.1.3.3 Fourier power spectra of the real-time trace from 200 to 800 fs of (a) the 7,7-diethoxy-substituted DBH and (b) DBH. Raman spectra of (c) the 7,7-diethoxy-substituted DBH and (d) DBH.

dynamics of the electronic excited states. The Fourier power spectra were obtained from the real-time traces from 200 to 800 fs (Figure 11.1.3.3a and b). The wavenumber resolution of Fourier power spectrum was estimated to be 16 cm^{-1}.

The wavenumber modes observed in Figure 11.1.3.3a reflect the vibrational modes of the 7,7-diethoxy DBH derivative. The wavenumber modes were assigned to the butterfly flap modes (325 and 480 cm^{-1}), the ring bending mode (δ_{Ring}: 887 cm^{-1}), the CH$_2$ wagging mode (δ_{CH2}: 976 cm^{-1}), and the CH deformation mode (δ_{CH}: 1139 cm^{-1}), whose assignments were done based on the reported assignments for DBH [24] and frequency calculations [25]. The wavenumber modes around 285 and 700 cm^{-1} in Figure 11.1.3.3b, for DBH solution in CH$_2$Cl$_2$, were assigned to the C–Cl scissoring mode (δ_{C-Cl}) and to the C–Cl stretching mode (ν_{C-Cl}) in the CH$_2$Cl$_2$ solvent molecule. The other observed frequencies shown in Figure 11.1.3.3b can be assigned to the ring-bending mode (δ_{Ring}: 887 cm^{-1}), to the CH$_2$ wagging mode (δ_{CH2}: 1035 cm^{-1}), and to the CH deformation mode (δ_{CH}: 1123 cm^{-1}) [24]. The frequencies observed by the pump–probe measurements (Figure 11.1.3.3a and b) agreed well with the respective ground-state Raman frequencies (Figure 11.1.3.3c and d). Since the pump laser is not resonant with the electronic transition in the DBH molecule, the observed modulation of ΔT is due to the wavepacket in the ground state. Hence, the frequencies of the FT of the real-time traces are expected to agree with the Raman wavenumber in the electronic ground state.

The pump–probe spectroscopy and Raman measurement are members of a broad class of nonlinear optical techniques related to the third-order optical polarization and the corresponding susceptibility. Therefore, the pump–probe signal shows Raman active modes. However, the two methods differ by their measurement mechanism, which results in a difference in the signal intensities.

The Fourier phases of the real-time traces showed that the observed molecular vibrations follow a sine-like oscillation, which confirmed that the observed signals reflect the wavepacket dynamics in the electronic ground state (Figure 11.1.3.4a and b). The pump intensity dependence of the vibrational amplitudes also supports assignment of the observed oscillating signal to the ground state, as described below. Three sets of measurements were performed maintaining the probe intensity at 20 GW/cm^2 (Figure 11.1.3.4c and d). In the case of the 7,7-diethoxy DBH derivative (X=OEt) with pump intensities of 120, 135, and 160 GW/cm^2, the powers $p(i)$ of the pump intensity dependence $I^{p(i)}$ for three modes (325, 887, and 976 cm^{-1}) were determined to be $p(325\,cm^{-1})=1.1\pm0.1$, $p(887\,cm^{-1})=0.9\pm0.1$, and $p(976\,cm^{-1})=1.2\pm0.7$ (Figure 11.1.3.4c). In the case of DBH with pump intensities of 105, 123, and 135 GW/cm^2, the power of the pump intensity dependence of the vibrational amplitude was determined to be $p(887\,cm^{-1})=0.70\pm0.01$ (Figure 11.1.3.4d). The linear pump intensity dependence suggests that the observed oscillation of DA is not due to the dynamics of the electronic excited state generated by multi-photon absorption but is predominantly due to the wavepacket generated in the ground state through the stimulated Raman process.

11.1.3.3.2 SPECTROGRAM

The time-dependent wavenumber shifts of the relevant molecular vibrational modes were analyzed using Spectrograms, which have been used for time-resolved analysis of Fourier power spectra. Spectrograms [26] shown in Figure 11.1.3.5 were calculated by applying a sliding-window Fourier transform to the DA traces averaged over 10 probe wavelengths. A Blackman window function with a full width half maximum of 240 fs was used for the spectrograms, as shown in Eq. (11.1.3.1). The spectrogram trace was calculated by shifting the window at 10-fs step in the delay time.

$$S(\omega,\tau) = \int_0^\infty S(t)g(t-\tau)\exp(-i\omega t)\,dt, \; g(t)$$

$$= 0.42 - 0.5\cos 2\pi t + 0.08\cos 4\pi t$$

(11.1.3.1)

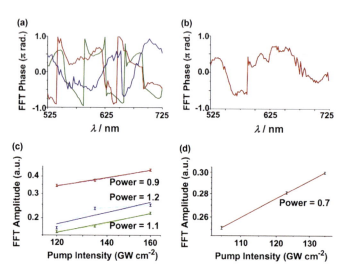

(a) FFT Phase (π rad.) — λ / nm — 525, 625, 725

(b) FFT Phase (π rad.) — λ / nm — 525, 625, 725

(c) FFT Amplitude (a.u.) — Power = 0.9, Power = 1.2, Power = 1.1 — Pump Intensity (GW cm^{-2}) — 120, 140, 160

(d) FFT Amplitude (a.u.) — Power = 0.7 — Pump Intensity (GW cm^{-2}) — 110, 120, 130

FIGURE 11.1.3.4 Fourier phase spectra of the observed molecular vibrations of (a) the 7,7-diethoxy-substituted DBH with frequencies of 325 cm^{-1} (green curve), 887 cm^{-1} (red curve), and 976 cm^{-1} (blue curves) and (b) DBH with wavenumber of 887 cm^{-1} (red curve). Pump intensity dependencies of vibrational amplitudes the same condition of the probe intensity of 20 GW/cm^2 are shown for (c) the 7,7-diethoxy-substituted DBH with frequencies of 325 cm^{-1} (green curve and green crosses), 887 cm^{-1} (red curve and red crosses), and 976 cm^{-1} (blue curve and blue crosses) and (d) DBH with wavenumber of 887 cm^{-1} (red curve and red crosses).

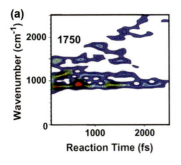
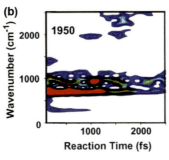

FIGURE 11.1.3.5 Spectrograms of (a) the 7,7-diethoxy-substituted DBH derivative and (b) DBH.

The wavenumber resolution of the spectrogram is 30 cm^{-1}. The data in the vicinity of a zero delay was disturbed by the interference between the scattered pump and the probe pulses. Molecular structure deformation during the reaction process along the reaction coordinate changes the oscillation wavenumber of the wave packet oscillating along the coordinate perpendicular to the reaction coordinate. The transfer of vibrational coherence, or even the generation of coherence, in the chemical reaction was previously discussed and demonstrated [27–31]. When the molecule stays either in the reactant, the intermediate, or the product, the vibration frequencies are thought to be constant during the lifetime of the relevant state although the vibration frequencies might be different between those electronic states. The vibrational frequencies are considered to change gradually when the molecule undergoes the structural change from the reactant, through the intermediate, and finally to the product.

In the reaction of 7,7-diethoxy DBH derivative (Figure 11.1.3.5a), vibrational bands appeared around 900, 1050, and 1450 cm^{-1} just after the photo-excitation. These modes can be assigned to the ring bending modes, the CH$_2$ wagging mode, and the CH$_2$ scissoring mode, respectively. The CH$_2$ wagging mode started to blue-shift after photo-excitation, and two new bands appeared at around 1300 and 1750 cm^{-1} at ca. 1 ps after the photo-excitation, which are assigned to the CH bending mode and NN stretching mode, respectively (see Figure 11.1.3.5 and the Scheme 11.1.3.2).

Meanwhile, just after photo-excitation of DBH (Figure 11.1.3.5b), the CH$_2$ bending modes appeared at around 1000 and 1450 cm^{-1} and vibrational modes of the CH$_2$Cl$_2$ solvent appeared at around 300 and 700 cm^{-1}. At ca. 1 ps after the photo-excitation, a vibrational band appeared at around 1950 cm^{-1} being assigned to the NN stretching mode, which has a higher wavenumber than that observed for the 7,7-diethoxy-substituted DBH (1750 cm^{-1}).

11.1.3.3.3 Denitrogenation Mechanism

As mentioned in the Introduction (Scheme 11.1.3.1), two possible mechanisms, either the concerted pathway or the stepwise pathway [1–14], can be considered to understand the dynamics observed in the denitrogenation of DBH derivatives. The electron density of the NN bond is supposed to be higher in the transition state (TSc) of the concerted process than those in the intermediate (DZ) and the transitions states (TS$_{S1}$ and TS$_{S2}$) for the stepwise process (Scheme 11.1.3.1). Thus, the NN stretching vibrational mode of TSc should have higher wavenumber than those in DZ, TS$_{S1,}$ and TS$_{S2}$. Indeed, it was also predicted by a theoretical calculation performed at the (U)B3LYP/6–31G(d) level showing that the NN stretching modes (v_{NN}, scaled by 0.961) of the model intermediary species TS$^0_{S1}$ (v_{NN}=1718 cm^{-1}, imaginary wavenumber (v^i)=240 cm^{-1}), DZ^0 (v_{NN}=1777 cm^{-1}), and TS$^0_{S2}$ (v_{NN}=1869 cm^{-1}, imaginary wavenumber (v^i)=405 cm^{-1}) have lower wavenumber than that of the transition state TS^0_C (v_{NN}=1917 cm^{-1}, imaginary wavenumber (v^i)=492 cm^{-1}) (see Schemes 11.1.3.2 and 11.1.3.3) [13,25].

Spectrogram analysis of the measured time-resolved traces found a large wavenumber difference in the NN stretching mode at a 1 ps delay between the 7,7-diethoxy DBH derivative and the parent DBH, i.e. 1750 cm^{-1} in Figure 11.1.3.5a and 1950 cm^{-1} in Figure 11.1.3.5b. Considering the above

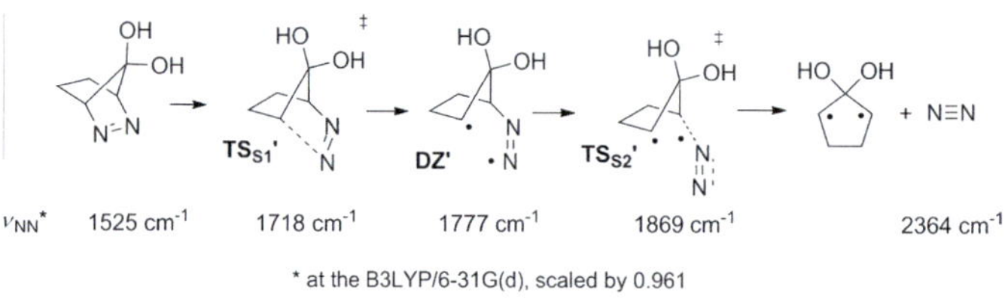

SCHEME 11.1.3.2 Wavenumber change of DBH derivative (X = OH) during the stepwise denitrogenation, see Ref. [25]. By comparing the Spectrograms and the above scheme, the stepwise denitrogenation is considered to be involved in the reaction [25].

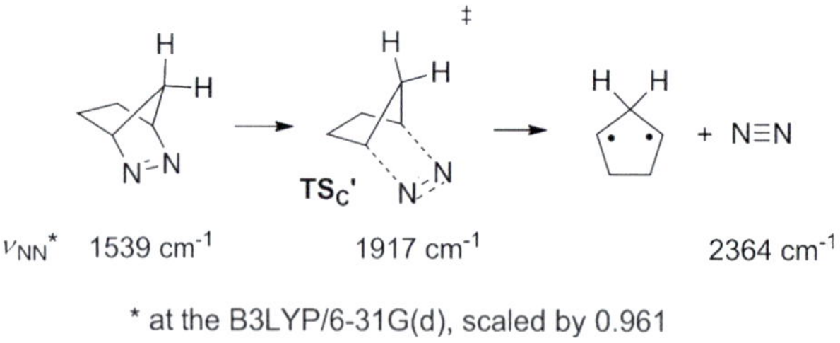

ν_{NN}^* 1525 cm^{-1} 1718 cm^{-1} 1777 cm^{-1} 1869 cm^{-1} 2364 cm^{-1}

* at the B3LYP/6-31G(d), scaled by 0.961

ν_{NN}^* 1539 cm^{-1} 1917 cm^{-1} 2364 cm^{-1}

* at the B3LYP/6-31G(d), scaled by 0.961

SCHEME 11.1.3.3 Wavenumber change of DBH (X = H) during the concerted denitrogenation, see Ref. [25].

prediction given by the theoretical calculation, the observed wavenumber difference elucidates that the denitrogenation of the 7,7-diethoxy DBH derivative proceeds through the stepwise pathway, and that of the parent DBH proceeds through the concerted pathway. After a 1 ps delay, the NN stretching frequencies increased with a delay in both of the compounds, reflecting the dissociation of nitrogen. The observed blue shift of the NN stretching wavenumber also agrees with the computational prediction.

11.1.3.4   CONCLUSION

In summary, the reaction mechanism of the thermal denitrogenation of DBH derivatives was investigated using a visible 5-fs pulse laser, by which the time-dependent wavenumber shift of molecular vibrations is directly observed [32]. The concerted denitrogenation process was found in the thermolysis of the parent DBH. However, the stepwise nitrogen dissociation process was observed for the denitrogenation of the 7,7-diethoxy DBH derivative. These results agree well with the computational predictions. The ultrashort visible pulse excited vibrational modes coherently in the electronic ground state through the stimulated Raman processes, which precedes the reaction in the electronic ground state like the thermal excitation under heating. The research presented in this subsection was conducted by collaboration among the following people: M. Abe, I. Iwakura, A. Yabushita, S. Yagi, J. Liu, K. Okumura, and T. Kobayashi [32].

REFERENCES

1. P.S. Engel, *Chem. Rev.* **80**, 99 (1980).
2. W.R. Roth, and M. Martin, *Ann. Chem.* **702**, 1 (1967).
3. W.R. Roth, and M. Martin, *Tetrahedron Lett.* **8**, 4695 (1967).
4. E. L. Allred, and R. L. Smith, *J. Am. Chem. Soc.* **91**, 6766 (1969).
5. W. Adam, H. Garca, V. Marti, and J. N. Moorthy, *J. Am. Chem. Soc.* **121**, 9475 (1999).
6. W. Adam, M. Diedering, and A. Trofimov, *J. Phys. Org. Chem.* **17**, 643 (2004).
7. W. Adam, T. Oppenlaunder, and G. Zang, *J. Org. Chem.* **50**, 3303 (1985).
8. N. Yamamoto, M. Olivucci, P. Celani, F. Bernardi, and M. A. Robb, *J. Am. Chem. Soc.* **120**, 2391 (1998).
9. A. Sinicropi, C. S. Page, W. Adam, and M. Olivucci, *J. Am. Chem. Soc.* **125**, 10947 (2003).
10. B. A. Lyons, J. Pfeifer, T. H. Peterson, and B. K. Carpenter, *J. Am. Chem. Soc.* **115**, 2427 (1993).
11. M. B. Reyes, and B. K. Carpenter, *J. Am. Chem. Soc.* **120**, 1641 (1998).

12. M. B. Reyes, and B. K. Carpenter, *J. Am. Chem. Soc.* **122**, 10163 (2000).
13. M. Abe, C. Ishihara, S. Kwanami, and A. Masuyama, *J. Am. Chem. Soc.* **127**, 10 (2005).
14. M. Hamaguchi, M. Nakaishi, T. Nagai, T. Nakamura, and M. Abe, *J. Am. Chem. Soc.* **129**, 12981 (2007).
15. C. Ishihara, and M. Abe, *Aust. J. Chem.* **63**, 1615 (2010).
16. A. Baltuska, T. Fuji, and T. Kobayashi, *Opt. Lett.* **27**, 306 (2002).
17. T. Kobayashi, A. Shirakawa, and T. Fuji, *IEEE J. Quant. Electron.* **7**, 525 (2001).
18. T. Kobayashi, T. Saito, and H. Ohtani, *Nature* **414**, 531 (2001).
19. S. L. Buchwalter, and G. L. Closs, *J. Am. Chem. Soc.* **131**, 4688 (1979).
20. P. G. Gassman, and K. T. Mansfield, *Org. Synth.* **49**, 1 (1969).
21. I. Iwakura, A. Yabushita, and T. Kobayashi, *J. Am. Chem. Soc.* **131**, 688 (2009).
22. I. Iwakura, A. Yabushita, and T. Kobayashi, *Chem. Lett.* **39**, 374 (2010).
23. I. Iwakura, A. Yabushita, and T. Kobayashi, *Chem. Phys. Lett.* **501**, 567 (2011).
24. D. Gernet, and W. Kiefer, *J. Anal. Chem.* **362**, 84 (1998).
25. S. Yagi, Y. Hiraga, R. Takagi, and M. Abe, *J. Phys. Org. Chem.* **24**, 894 (2011).
26. M. J. J. Vrakking, D. M. Villeneuve, and A. Stolow, *Phys. Rev. A* **54**, R37 (1996).
27. J. M. Jean, and G. R. Fleming, *J. Chem. Phys.* **103**, 2092 (1995).
28. F. Rosca, A. T. N. Kumar, X. Ye, T. Sjodin, A. A. Demidov, and P. M. Champion, *J. Phys. Chem. A* **104**, 4280 (2000).
29. M. H. Vos, F. Rappaport, J.-C. Lambry, J. Breton, and J.-L. Martin, *Nature* **363**, 320 (1993).
30. R. J. Stsnley, and S. G. Boxer, *J. Phys. Chem.* **99**, 859 (1995).
31. Q. Wang, R. W. Schoenlein, L. A. Peteanu, R. A. Mathies, and C. V. Shank, *Science* **266**, 422 (1994).
32. M. Abe, I. Iwakura, A. Yabushita, S. Yagi, J. Liu, K. Okumura, and T. Kobayashi, *Chem. Phys. Lett.* **527**, 79–83 (2012).

11.1.4 Photo-Impulsive Reactions in the Electronic Ground State without Electronic Excitation
Non-Photo, Non-Thermal Chemical Reactions

11.1.4.1 INTRODUCTION

Chemical reactions can be generally classified into two groups: photoreactions and thermal reactions. When a chemical compound absorbs light of which the photon energy is equal to the electronic transition energy, the compound is excited into an electronic excited (EE) state. The excitation of the electronic state increases lability, which triggers a photoreaction in the EE state. In contrast, thermal reactions proceed in the electronic ground (EG) state due to thermal excitation of molecular vibrational modes of the EG state. Therefore, the frontier orbital that participates in thermal reactions differs from that for photoreactions, which frequently causes differences in reactivity.

For example, irradiation of allyl phenyl ether (APE) with ultraviolet (UV) light at its absorption band ($\lambda < 400$ nm) causes a photo-Claisen rearrangement via a radical intermediate and produces not only o-allyl phenol but also p-allyl phenol and phenol (Scheme 11.1.4.1(1)) [1]. In contrast, thermal excitation of APE proceeds with a [3,3]-sigmatropic rearrangement, in which a six-membered transition state appears under a supra– supra facial reaction and produces 6-allyl-cyclohexa-2,4-dienone. The 6-allyl-cyclohexa-2,4-dienone converts itself into the more stable o-allyl phenol product under keto–enol tautomerism (Scheme 11.1.4.1(2)) [2].

We have previously reported [3,4] that when the pump photon energy is lower than the minimum electronic transition energy, molecular vibrational modes of the EG state are excited via an induced Raman process, which triggers the reaction in the EG state without converting photon energy to thermal energy [5–7]. In the present work, we have applied selective triggers for the Claisen rearrangements of APE. As a result, the Claisen rearrangement in the EE state (photoreaction) and that in the EG state (photo-impulsive reaction in the EG state) can be selectively induced using few-optical-cycle pulses of UV and visible radiation, respectively (Figure 11.1.4.1). It was confirmed that the photo-impulsive reaction pathway in the EG state (as for thermal reactions) is different from that in the EE state (as for photoreactions).

11.1.4.2 EXPERIMENTAL

11.1.4.2.1 FEW-OPTICAL-CYCLE ULTRAVIOLET PULSES

A hollow-fiber compression system was used to obtain a few optical-cycle broadband UV pulses [8]. A Ti: sapphire regenerative amplifier (Coherent, Legend Elite-USP; 2.5 mJ, 35 fs, 1 kHz at

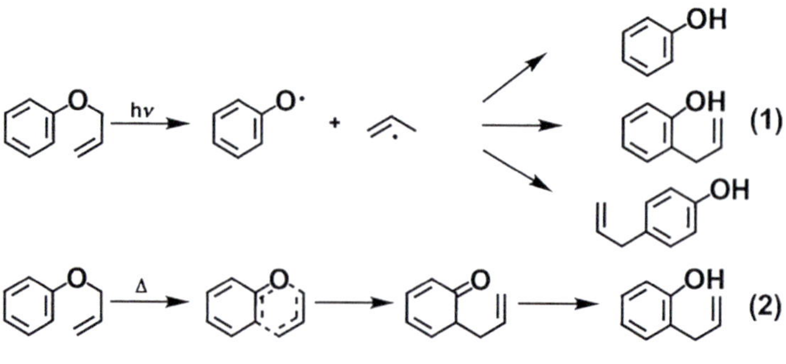

Scheme 1

SCHEME 11.1.4.1 Photoche-mically-allowed Claisen rear-rangement of allyl phenyl ether by few-optical-cycle ultraviolet pulse irradiation.

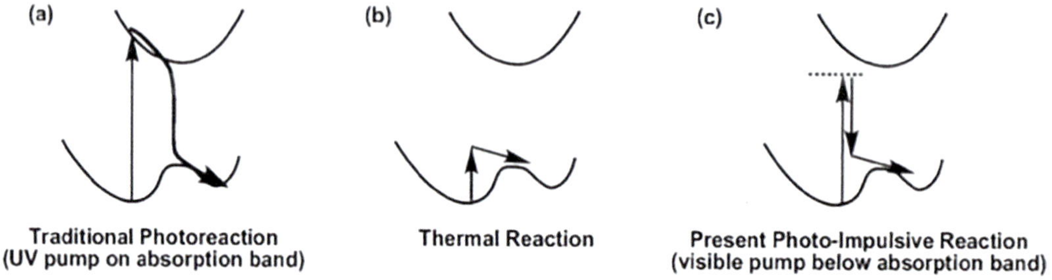

FIGURE 11.1.4.1 The three reaction mechanisms under investigation: (a) traditional photoreaction in the electronic excited state; (b) thermal reaction in the electronic ground state; (c) the present photo-impulsive reaction in the electronic ground state.

800 nm) was wavenumber-doubled in a beta barium borate (BBO) crystal to generate second-harmonic laser pulses at 400 nm, which were focused into a hollow fiber. The pulses were spectrally broadened in the fiber to 360–440 nm and the time duration was compressed to as short as 8 fs using a pulse compressor. The focal areas of the pump pulse and the probe pulse were 100 and 75 mm², respectively.

11.1.4.2.2 Few-Optical-Cycle Visible Pulses

A Ti: sapphire regenerative amplifier (Spectra Physics, Spitfire; 150 mJ, 100 fs, 5 kHz at 805 nm) was used to pump a noncollinear optical parametric amplifier (NOPA) to obtain few optical-cycle broadband visible pulses [9]. The signal pulse was amplified by the double-pass NOPA in the broadband spectral region from 525 to 725 nm and the pulse compressor compressed the pulse duration to 5 fs. The focal areas of the pump pulse and the probe pulse were 100 and 75 mm², respectively.

11.1.4.2.3 Sample Cell

A liquid cell with an optical path length of 1 mm was used to contain the sample during measurement at 295 ± 1 K. Neat liquid of APE (CAS: 1746-13-0, Tokyo Chemical Industry) was stored in the cell for use as a sample. Under irradiation by the UV pulses, light absorption by the sample causes thermal accumulation and bubbling if the pulses irradiate a fixed point. Therefore, during UV pulse irradiation, the sample cell was continuously translated in the plane perpendicular to the probe beam. The shape of the translation trajectory was near-circular and octagon-like, and the speed was ca. 5 mm/s, which corresponds to a sample displacement of ca. 10 mm during the pulse period of 1 ms. This procedure reduced the probability of sample photodamage to a minimum.

11.1.4.2.4 PUMP–PROBE MEASUREMENT

The probe pulse was dispersed using a polychromator (300 grooves per mm, 500 nm blazed for visible pulses and 300 nm blazed for UV pulses). The dispersed probe wavelength components were guided to 128 avalanche photodiodes via a 128-channel bundled fiber. The signal-to-noise ratio was improved by coupling the signals of the avalanche photodiodes to a 128-channel lock-in amplifier.

11.1.4.2.5 QUANTUM CHEMICAL CALCULATION

The Gaussian 03 program [10] was used for the calculations without assuming symmetry. Geometric optimization was performed using the B3LYP/6–311+G(d, p) method and basis set (5d functions were used for the d orbital). Wavenumber calculations were performed for all of the obtained structures at the same level. All the frequencies were confirmed as real for the ground states and one imaginary wavenumber for the transition state (TS). Vectors of the imaginary frequencies directed the reaction mode and intrinsic reaction coordinate calculations were further performed to confirm that the obtained TSs were on the saddle points of the energy surface between the reactant and product.

11.1.4.3 RESULTS

11.1.4.3.1 VIBRATIONAL DYNAMICS IN THE REACTION UNDER FEW-OPTICAL CYCLE ULTRAVIOLET PULSE IRRADIATION (SEE FIGURE 11.1.4.1A)

APE has absorption bands in the UV region ($\lambda < 400$ nm); therefore, irradiation with few-optical-cycle UV pulses ($\lambda = 360$–440 nm) caused photo-excitation at the absorption band, which excited the electronic state of APE and photo-Claisen rearrangement (radical reaction) proceeded in the EE state (Figure 11.1.4.2).

A pump–probe measurement of APE was performed to investigate the photo-Claisen rearrangement using a few optical-cycle UV pulses (see Section 11.1.4.1.2.1). Figure 11.1.4.3a shows a two-dimensional view of the observed time-resolved transmission change ($\Delta T/T$), which indicates the appearance of an induced absorption at 1.8 ps after photo-excitation.

A fast Fourier transform (FFT) of $\Delta T/T$ between 200 and 1200 fs was calculated using the Hanning window function, as shown in Figure 11.1.4.3b. The calculated FFT power spectrum has peaks at 814, 993, 1278, and 1644 cm^{-1}. Spectrogram analysis [11] was performed using a Blackmann window function with a 400-fs full width at half maximum (FWHM) to elucidate the dynamics of the vibrational modes (see Figure 11.1.4.3c). Vibrational modes at 1007, 1080, 1186, 1320, 1487, and 1670 cm^{-1} were evident immediately after the photo-excitation. The highest peak at 1007 cm^{-1} is separated into two components; one exhibited a red shift and the other did not shift until 2.5 ps after photo-excitation. At a delay time of 1.7 ps, the peak at 1670 cm^{-1} disappeared and a new peak appeared at 1533 cm^{-1}. The peak at 1080 cm^{-1} underwent a gradual blueshift with the generation of a new peak at 1425 cm^{-1}.

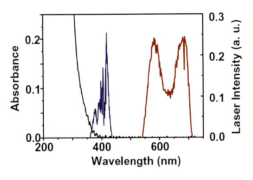

FIGURE 11.1.4.2 An absorption spectrum of allyl phenyl ether (black curve located at the leftmost place), and laser spectra of the few-optical-cycle UV (blue curve located at the central place) and visible (red curve located at the rightmost place) pulses.

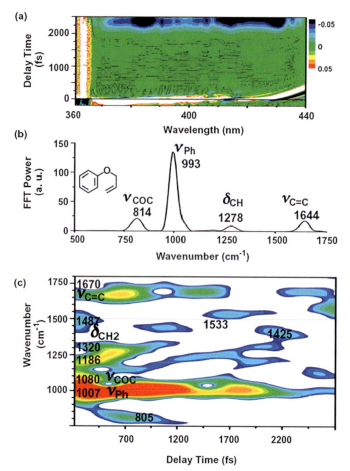

FIGURE 11.1.4.3 Experimental results observed using UV pulses. (a) A two-dimensional display of the change in absorbance of allyl phenyl ether as a function of probe delay time and wavelength. (b) A FFT power spectrum of the oscillating components of the time-resolved absorbance change. (c) The observed time dependence of the vibrational spectra.

11.1.4.3.2 VIBRATIONAL DYNAMICS IN THE REACTION UNDER FEW-OPTICAL CYCLE VISIBLE PULSE IRRADIATION (SEE FIGURE 11.1.4.1c)

The pump–probe measurement was also performed using a few optical-cycle visible pulses (see Section 11.1.4.2.2). The photon energy of the visible pulses ($\lambda = 525$–$725\,\text{nm}$) is lower than the minimum electronic transition energy of APE ($\lambda < 400\,\text{nm}$) (Figure 11.1.4.2).

A two-dimensional view of the measured $\Delta T/T$ is shown in Figure 11.1.4.4a. The standard deviation of the induced absorbance difference ΔA, oscillating around the estimated zero absorbance change, was less than 5×10^{-5}, which is negligible when compared with the oscillation amplitude ($\Delta \delta A$) of 3×10^{-4}. The deviation from zero is due to a very small accumulated steady-state population of EGs.

The FFT of $\Delta T/T$ between 200 and 1200 fs was calculated using a Hanning window function, as shown in Figure 11.1.4.4b. The calculated FFT power spectrum has peaks at 814, 993, 1286, and $1644\,\text{cm}^{-1}$. Spectrogram analysis was performed using a Blackmann window function with 400 fs-FWHM to elucidate the dynamics of the vibrational modes (see Figure 11.1.4.4c). Vibrational modes at 820, 1000, 1230, 1300, 1420, 1470, 1595, and $1650\,\text{cm}^{-1}$ were evident immediately after photoexcitation. The peak at $1000\,\text{cm}^{-1}$ disappeared at a delay time of 700 fs without any shift in wavenumber. The mode of $1650\,\text{cm}^{-1}$ underwent a gradual red shift with the generation of a new peak at $1580\,\text{cm}^{-1}$ at a delay time of 750 fs. A new peak appeared at $1750\,\text{cm}^{-1}$ after a delay time of 1 ps and disappeared at a delay time of 2 ps.

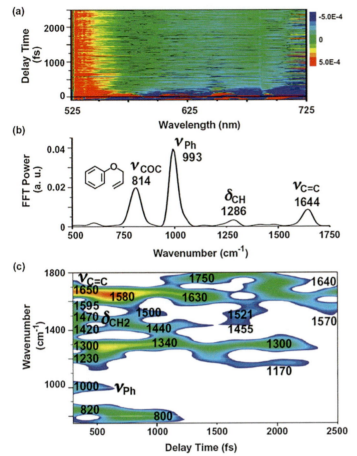

FIGURE 11.1.4.4 Experimental results observed using visible pulses [4]. (a) A two-dimensional display of the change in absorbance of allyl phenyl ether as a function of the probe delay time and wavelength. (b) A FFT power spectrum of the oscillating components of the time-resolved absorbance change. (c) The observed time dependence of the vibrational spectra.

11.1.4.3.3 THEORETICAL VIBRATIONAL DYNAMICS OF THE PHOTO- AND THERMAL REACTIONS

A density functional formalism was used to calculate molecular vibrational changes in the photo-Claisen rearrangement and the thermal Claisen rearrangement. Figure 11.1.4.5 shows a measured Raman spectrum of APE and the calculated wavenumbers of the molecular vibrational modes. The measured results show that symmetric stretching of the phenyl group, C–O–C symmetric stretching of the ether group, C–H deformation of the phenyl group, C–O–C stretching of the ether group and C–H$_2$ twisting of the methylene group, C–H deformation of the allyl group, C–H$_2$ wagging of the methylene group, C–H$_2$ scissoring of the allyl group, C=C stretching of the phenyl group, and C=C stretching of the allyl group appear at 997, 1032, 1173, 1244, 1291, 1419, 1458, 1595, and 1648 cm^{-1}, respectively. The values in Figure 11.1.4.5 (the measured Raman wavenumbers and the corresponding calculated wavenumbers) were used to estimate scaling factors (see Table 11.1.4.1).

In the photo-Claisen rearrangement (see Table 11.1.4.2), an allyl radical appears with a C–C* stretching mode at 1510 cm^{-1}, and a phenoxy radical appears with ring stretching mode of the phenyl group (805 and 1008 cm^{-1}) and C–O stretching (1480 cm^{-1}).

In the thermal Claisen rearrangement (see Table 11.1.4.3), a six-membered structure appears with C–O–C asymmetric stretching of the ether group (815 cm^{-1}), C–H deformation of the allyl group (1319 cm^{-1}), and CQC stretching in the allyl and phenyl groups (1542 cm^{-1}). The subsequent intermediate, 6-allyl-cyclohexa-2,4-dienone, has a C–H$_2$ twisting of the methylene group, C–H deformation of the allyl group, C–H$_2$ scissoring deformation of the allyl group, CQC stretching of the allyl group, and CQO stretching at 1196, 1300, 1448, 1637, and 1720 cm^{-1}, respectively. The final product, o-allyl phenol, appears with CQC stretching of the phenyl group (1584 cm^{-1}) and CQC stretching of the allyl group (1635 cm^{-1}).

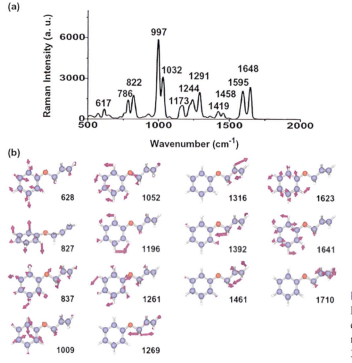

FIGURE 11.1.4.5 (a) Measured Raman spectrum of allyl phenyl ether. (b) Molecular vibrational modes calculated using B3LYP/6-311+G8(d, p).

TABLE 11.1.4.1

Wavenumber Scaling Factors of the Vibrational Modes

	Molecular Vibrational Mode cm⁻¹												
	δ_{ph}	δ_{CH}	δ_{ph}	δ_{ph}	ν_{COC}	δ_{CH}	ν_{COC}	δ_{CH_2}	δ_{CH}	δ_{CH_2}	δ_{CH_2}	$\nu_{C=C}$	$\nu_{C=C}$	$\nu_{C=C}$
	Ph	Ph	Ph	Ph	Ether	Ph	Ether	Methylene	Allyl	Methylene	Allyl	Ph	Ph	Allyl
Exp.	617	786	822	997	1032	1173	1244	1244	1291	1419	1458	1595	1595	1648
Calc.	628	827	837	1009	1052	1196	1261	1269	1316	1392	1461	1623	1641	1710
S.F.[a]	0.98	0.95	0.98	0.99	0.98	0.98	0.99	0.98	0.98	1.02	1.00	0.98	0.97	0.96

[a] Scaling Factor (S.F.) = Experimental wavenumber (Exp.)/Calculated wavenumber (Calc.).

TABLE 11.1.4.2

Calculated Vibrational Wavenumbers of Chemical Species Involved in the Photo-Claisen Rearrangement of Allyl Phenyl Ether

	Calculated Molecular Vibration Wavenumber[a] cm⁻¹										
	ν_{ph}	ν_{CCC}	ν_{COC}	δ_{CH}	δ_{CH_2}	δ_{CH}	δ_{CH_2}	δ_{CH_2}	ν_{CCC}	$\nu_{C=C}$	$\nu_{C=O}$
Compound	Ph	Allyl	Ether	Ph	Methylene	Allyl	Methylene	Allyl	Allyl	Ph	Allyl
Reactant		1009	1052	1196	1269	1316	1392	1461		1641	1710
Allyl radical		1036					793		1269		1510
Phenoxy radical	805	1008		1480	1166					1585	

[a] Not scaled by any scaling factor.

TABLE 11.1.4.3

Calculated Vibrational Wavenumbers of Chemical Species Involved in the Thermal Claisen rearrangement of Allyl Phenyl Ether

	Calculated Molecular Vibration Wavenumber^a cm^{-1}									
	δ_{ph}	ν_{ph}	ν_{COC}	δ_{CH_2}	δ_{CH}	δ_{CH_2}	δ_{CH_2}	$\nu_{C=C}$	$\nu_{C=C}$	$\nu_{C=O}$
Compound	Ph	Ph	Ether	Methylene	Allyl	Methylene	Allyl	Ph	Allyl	
Reactant	822	997	1032	1244	1291	1419	1458	1595	1648	
TS	815				1319		1434	1507	1542	
Keto-intermediate				1196	1300		1448	1555	1637	1720
Phenol product								1584	1635	

^{a} Scaled using scaling factors shown in Table 11.1.4.1.

11.1.4.4 DISCUSSION

11.1.4.4.1 PHOTOCHEMICALLY-ALLOWED CLAISEN REARRANGEMENT OF ALLYL PHENYL ETHER BY FEW-OPTICAL-CYCLE ULTRAVIOLET PULSE IRRADIATION

APE has an absorption band at wavelengths shorter than 400 nm. Therefore, irradiation with few-optical-cycle UV pulses having a broadband spectrum of 360–440 nm excites the electronic state of APE to the EE state. Molecular vibrations

are coherently excited via vibronic interaction, which enables observation of the EE state dynamics via molecular vibrations. Figure 11.1.4.3c shows a calculated spectrogram that reflects the vibrational dynamics. The observed vibrational dynamics of the photoreaction during reaction under few-optical-cycle UV pulses were in agreement with the theoretical vibrational dynamics of the photoreaction, which confirms that the vibrational dynamics follow those for the photoreaction, as explained in the following.

The CQC stretching of the allyl group was observed at 1670 cm^{-1} immediately after the photo-excitation and then disappeared at a delay time of 1.7 ps when the allyl radical was produced, and a new peak of C–C stretching of the allyl radical appeared at 1533 cm^{-1}. The peak of symmetric stretching in the phenyl group observed at 1000 cm^{-1} just after photo-excitation was separated into two modes as the phenoxy radical was generated; one of the two modes was redshifted to 805 cm^{-1} and the other experienced no change in wavenumber. The peak observed at 1080 cm^{-1} just after photo-excitation was gradually blue-shifted with the generation of the C–O group and a new peak at 1425 cm^{-1} appeared at a delay time of ca. 2 ps.

The vibrational mode dynamics indicate that active radical species are generated at 1.8–2.0 ps after photo-excitation. Phenoxy radicals have an absorption band at 360–420 nm [12] and the transitional absorption appeared at a delay time of 1.8 ps (see Figure 11.1.4.3a). Thus, irradiation with few-optical-cycle pulses induces photoreaction by excitation of the electronic state into the EE state if the laser spectrum overlaps with the absorption band of the sample that corresponds to its electronic transition.

11.1.4.4.2 THERMALLY-ALLOWED CLAISEN REARRANGEMENT OF ALLYL PHENYL ETHER BY FEW-OPTICAL-CYCLE VISIBLE PULSE IRRADIATION

The reaction induced by irradiation with few-optical-cycle visible pulses (525–725 nm) (Figure 11.1.4.4c) was completely different from that for the few-optical-cycle UV pulse irradiation (Figure 11.1.4.3c). The observed vibrational dynamics were in agreement with the theoretical vibrational dynamics of the thermal Claisen reaction (Table 11.1.4.3), which indicates that the reaction is the thermally-allowed Claisen rearrangement in the EG state, as explained in the following.

The symmetric stretching of the phenyl group observed at $1000\,cm^{-1}$ immediately after the photo-excitation disappeared at a delay time of ca. 700 fs. The $C–H_2$ deformation of the methylene group observed at 1230 and $1420\,cm^{-1}$ also disappeared at the same delay time of 700 fs. The CQC stretching of the allyl group observed at $1650\,cm^{-1}$ was red-shifted to $1580\,cm^{-1}$ near a delay time of 750 fs.

The C–O bond was firstly weakened before generation of the six-membered structure. When the six-membered intermediate was generated, the C=C bond of the allyl group became an aromatic-like C=C bond observed at $1580\,cm^{-1}$ (the C=C stretching mode of benzene is $1585\,cm^{-1}$). At a delay time of 1 ps, 6-allyl-cyclohexa-2,4-dienone was generated, as evidenced by a new peak due to carbonyl stretching at $1750\,cm^{-1}$. The peak at $1750\,cm^{-1}$ disappeared at 2 ps because the unstable 6-allyl-cyclohexa-2,4-dienone converted to the stable enol form. Thus, when the photon energy of the excitation light is lower than the minimum electronic transition energy, irradiation with few-optical-cycle pulses induces a photo-impulsive reaction in the EG state without converting photon energy to thermal energy, which is different from a traditional photoreaction

11.1.4.4.3 Non-Photo, Non-Thermal Chemical Reaction

In the case of the thermal Claisen rearrangement of APE, the activation energy was calculated to be about 33.9 kcal/mol using B3LYP/6–311+G(d, p). This value agrees with earlier reports [13]. Meanwhile, the few-optical-cycle visible pulses ($\lambda = 525$–725 nm) have a spectrum bandwidth of $5200\,cm^{-1}$. The bandwidth corresponds to ~14.5 kcal/mol, which is much lower than the activation energy of the thermal Claisen rearrangement. The reason why the Claisen rearrangement proceeds by the few-optical-cycle visible pulse is explained in the following.

The photo-impulsive reaction in the EG state excites some vibrational modes, which include modes related with the reaction. Therefore, the photo-impulsive reaction can proceed effectively with excitation energy lower than that of the thermal reaction exciting all of the vibrational modes. The photo-impulsive reaction in the EG state proceeds the reaction keeping coherence of the molecular vibrations, and it is no wonder that the activation barrier of the photo-impulsive reaction is different from that of the incoherent thermal reaction.

Meanwhile, the frontier orbitals of the photo-impulsive reaction in the EG state are thought to be the same as those of the thermal reaction. Both of the two reactions have the same reaction pathway because the reaction mechanism is dominated by orbital symmetry.

The photo-impulsive reaction is thought to require the following two key factors to trigger the reaction.

1. The width of the excitation pulse is much shorter than the molecular vibrational periods and should be as short as a few oscillation periods of the optical electric field (few-optical cycle pulse).
2. The photon energy of the excitation pulse should be lower than the minimum electronic transition energy.

Future challenges will focus on obtaining a definitive understanding of the driving mechanism for the reaction.

11.1.4.5 CONCLUSIONS

Traditional thermal reactions excite all vibrational modes of molecules in the reaction system. The photo-impulsive reaction induced with the excitation of vibrational modes in the EG state with the few-optical-cycle pulse is coherently triggered. Induced Raman processes excite only a fraction of molecular vibrational modes to high-level vibrational excited states. This results in only a fraction of molecules becoming "hot molecules" with highly excited vibrational states that trigger the reactions in the EG state. Hot molecules generally lose energy during the collision with surrounding solvent molecules. In the reaction studied in the present paper [14], the chemical reaction of hot molecules proceeds within approximately 2 ps, which is much faster than collisions with surrounding molecules. The reaction presented in Figure 11.1.4.4c follows the same reaction pathway as that of the symmetry allowed thermal Claisen rearrangement in the EG state. Therefore, although the possibility of thermal reaction may not be completely ruled out, the photo-impulsively reaction in the EG state with the few-optical-cycle pulse, which is neither a photoreaction nor a thermal reaction, is highly possible as a novel reaction scheme [14]. The cooperative research presented in this subsection was conducted by the following people: I. Iwakura, A. Yabushita, J. Liu, K. Okamura, and T. Kobayashi [14].

REFERENCES

1. M. S. Kharasch, G. Stampa, and W. Nudenberg, *Science*, **116**, 309 (1952); F. Galindo, *J. Photochem. Photobiol., C*, **6**, 123–138 (2005).
2. L. Claisen, *Ber. Dtsch. Chem. Ges.*, **45**, 3157–3166 (1912); A. M. M. Castro, *Chem. Rev.*, **104**, 2939–300 (2004).
3. I. Iwakura, A. Yabushita, and T. Kobayashi, *Chem. Lett.*, **39**, 374–375 (2010).
4. I. Iwakura, A. Yabushita, and T. Kobayashi, *Chem. Phys. Lett.*, **501**, 567–571 (2011).
5. The laser temperature-jump technique: E. Bemberg, and P. Läuger, *J. Membr. Biol.*, **11**, 177–194 (1973); J. T. Knudtson, and E. M. Eyring, *Annu. Rev. Phys. Chem.*, **25**, 255–274 (1974); W. Brock, G. Stark, and P. S. Jordan, *Biophys. Chem.*, **13**, 329–348 (1981); G. Stark, M. Strässle and Z. Takácz, *J. Membr. Biol.*, **89**, 23–37 (1986); C. M. Phillips, Y. Mizutani and R. M. Hochstrasser, *Proc. Natl. Acad. Sci. U. S. A.*, **92**, 7292–7296 (1995); K. Yamamoto, Y. Mizutani and T. Kitagawa, *Biophys. J.*, **79**, 485–495 (2000); T. Yatsuhash and N. Nakashima, *Bull. Chem. Soc. Jpn.*, **74**, 579–593 (2001); J. Kubelka, *Photochem. Photobiol. Sci.*, 2009, **8**, 499–512.
6. The pressure-jump technique: M. Schiewek, and A. Blume, *Eur. Biophys. J.*, **38**, 219–228 (2009); A. Blume, and M. Hillmann, *Eur. Biophys. J.*, **13**, 343–353 (1986); K. Elamrani, and A. Blume, *Biochemistry*, **22**, 3305–3311 (1983); J. Erbes, A. Gabke, G. Rapp, and R. Winter, *Phys. Chem. Chem. Phys.*, **2**, 151–162 (2000).
7. The PH-jump technique: J. H. Clark, S. L. Shapiro, A. J. Campillo, and K. R. Winn, *J. Am. Chem. Soc.*, **101**, 746–748 (1979); K. K. Smith, K. J. Kaufmann, D. Huppert, and M. Gutman, *Chem. Phys. Lett.*, **64**, 522–527 (1979).
8. J. Liu, Y. Kida, T. Teramoto, and T. Kobayashi, *Opt. Express*, **18**, 4664–4672 (2010).
9. A. Baltuska, T. Fuji, and T. Kobayashi, *Opt. Lett.*, **27**, 306–308 (2002).
10. M. J. Frisch, G. W. Trucks, H. B. Schlegel, G. E. Scuseria, M. A. Robb, J. R. Cheeseman, J. A. Montgomery, Jr., T. Vreven, K. N. Kudin, J. C. Burant, J. M. Millam, S. S. Iyengar, J. Tomasi, V. Barone, B. Mennucci, M. Cossi, G. Scalmani, N. Rega, G. A. Petersson, H. Nakatsuji, M. Hada, M. Ehara, K. Toyota, R. Fukuda, J. Hasegawa, M. Ishida, T. Nakajima, Y. Honda, O. Kitao, H. Nakai, M. Klene, X. Li, J. E. Knox, H. P. Hratchian, J. B. Cross, V. Bakken, C. Adamo, J. Jaramillo, R. Gomperts, R. E. Stratmann, O. Yazyev, A. J. Austin, R. Cammi, C. Pomelli, J. W. Ochterski, P. Y. Ayala, K. Morokuma, G. A. Voth, P. Salvador, J. J. Dannenberg, V. G. Zakrzewski, S. Dapprich, A. D. Daniels, M. C. Strain, O. Farkas, D. K. Malick, A. D. Rabuck, K. Raghavachari, J. B. Foresman, J. V. Ortiz, Q. Cui, A. G. Baboul, S. Clifford, J. Cioslowski, B. B. Stefanov, G. Liu, A. Liashenko, P. Piskorz, I. Komaromi, R. L. Martin, D. J. Fox, T. Keith, M. A. Al-Laham, C. Y. Peng, A. Nanayakkara, M. Challacombe, P. M. W. Gill, B. Johnson, W. Chen, M. W. Wong, C. Gonzalez and J. A. Pople, *Gaussian 03, (Revision D.02)*, Gaussian, Inc., Wallingford CT, 2004.
11. M. J. J. Vrakking, D. M. Villeneuve, and A. Stolow, *Phys. Rev. A: At., Mol., Opt. Phys.*, **54**, R37–R40 (1996).
12. K. Heimi, *Chem. Pharm. Bull.*, **22**, 718 (1974).
13. S. Yamabe, S. Okumoto, and T. Hayashi, *J. Org. Chem.*, **61**, 6218 (1996); B. Gmez, P. K. Chattaraj, E. Chamorro, R. Contreras, and P. Fuentealba, *J. Phys. Chem. A*, **106**, 11227 (2002).
14. I. Iwakura, A. Yabushita, J. Liu, K. Okamura, and T. Kobayashi, *Phys. Chem. Chem. Phys.*, **14**, 9696–9701 (2012).

11.1.5 The Reaction Mechanism of Claisen Rearrangement Obtained by Transition State Spectroscopy and Single Direct-Dynamics Trajectory

11.1.5.1 INTRODUCTION

Phenomena which are too fast to be directly observed by our eye can be visualized by observing them using strobe lights. The use of such a stroboscopic method to observe ultrafast chemical bond breaking and formation in chemical reactions has been a long-awaited dream for chemists. Lord G. Porter was awarded the Nobel Prize in Chemistry for his contribution to the technique of flash photolysis [1].

Since the first report of laser oscillation in 1960 [2], laser pulses have been used as strobe lights and their time duration has been kept trying to be shortened as short as attosecond order [3–5]. In the field of ultrafast optical measurement, the shorted pulse of femtosecond strobe light enabled us to observe electronic and vibration spectra in transition states of photoreactions. Zewail, who was awarded the Nobel Prize in Chemistry for his pioneering work on femtosecond time-resolved spectroscopy [6], has proposed "transition state spectroscopy" as a study of transition state realizing the chemists dream to observe chemical bond breaking and formation. Generally, heavy atom—hydrogen stretching vibrational modes (3000–3800 cm^{-1}) have a period of 11–9 fs, and carbonyl stretching vibrational mode and C=C bond stretching vibrational mode (1600–1750 cm^{-1}) show the vibration in a period of 21–19 fs. Therefore, using laser pulses whose duration is much shorter than the vibration periods, molecular motion in those vibrational modes can be time-resolved observing the modulation of transition probability of the corresponding wavelength in real-time. It means that molecular structure changes in photoreactions can be observed by measuring the real-time amplitudes of molecular vibrations from which time-dependent frequencies are calculated [7–9]. In addition, we have previously reported [8–11] that when the pump photon energy is lower than the minimum electronic transition energy, molecular vibrational modes of the electronic ground state are excited via an induced Raman process, which triggers the reaction in the electronic ground state without converting photon energy to thermal energy. As a result, thermally allowed reactions also can be observed by measuring the real-time amplitudes of molecular vibrations, from which time-dependent frequencies are calculated. In this work, a visible 5-fs laser pulse, which is much shorter than those vibration periods, was used to observe molecular structure changes in thermally allowed Claisen rearrangements, including their transition states.

DOI: 10.1201/9780429196577-79

11.1.5.2 RESULTS AND DISCUSSIONS

11.1.5.2.1 TRANSITION STATE SPECTROSCOPY OF THE CLAISEN REARRANGEMENT OF ALLYL VINYL ETHER

We have performed pump-probe measurement of neat liquid of allyl vinyl ether (AVE). Measured time-resolved absorption change traces were analyzed by time-frequency analysis [12] using a Blackmann window function of 400 fs FWHM. The result is shown in Figure 11.1.5.1a [10], whose x-axis and y-axis correspond to reaction time after 5 fs pulse irradiation and time-dependent molecular vibration frequency, respectively. In the spectrogram, the molecular vibrational modes appear immediately after the 5 fs pulses irradiation being assigned to those of AVE; C–O–C symmetric stretching vibrational mode ($v_{s\,C-O-C}$) of the ether group (900 cm^{-1}), C–H deformation vibrational modes (δ_{C-H}) of the allyl and the vinyl groups (1290 cm^{-1} and 1320 cm^{-1}, respectively), C–H$_2$ deformation vibrational mode (δ_{C-H2}) of the methylene group (1500 cm^{-1}) and C=C stretching vibrational modes ($v_{C=C}$) of the allyl and the vinyl groups (1650 cm^{-1}). These molecular vibrational modes agree well with the Raman data of AVE (Figure 11.1.5.1b).

The CH$_2$ deformation vibrational mode of the methylene group and the C–O–C symmetric stretching vibrational mode of the ether group disappear at about 800 fs delay. It implies that the C^4–O bond is weakened or broken in the first step of the reaction. The wavenumber shifts of the C=C bonds stretching modes also suggest that the C^4–O bond is weakened. Just after the 5 fs pulses irradiation, the C=C bond stretching vibrational mode of the vinyl and that of the allyl groups appear around 1650 cm^{-1}. Then, the C^4–O bond weakening causes the electronic density of the allyl and the vinyl groups to decrease and increase, respectively. Therefore, the C=C bond stretching vibrational mode observed at 1650 cm^{-1} was separated into a red-shifted mode toward 1570 cm^{-1} and in a blue-shifted mode toward 1690 cm^{-1}.

(a)

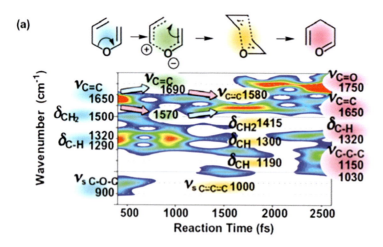

(b)   (c)

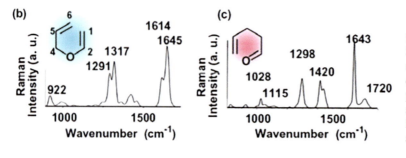

FIGURE 11.1.5.1 (a) Spectrogram of a Claisen rearrangement induced by visible 5 fs pulses [10]; (b) Raman spectrum of AVE; (c) Raman spectrum of allyl acetaldehyde.

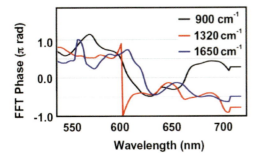

FIGURE 11.1.5.2 Fourier initial phase spectra of 900 cm^{-1} (black line), 1320 cm^{-1} (red line), and 1650 cm^{-1} (blue line) modes. Even by colorless curves, these three phases corresponding to 900, 1320, and 1650 cm^{-1} are located from top to bottom at the wavelength of 570 nm .

After the C^4–O bond weakening, electrons transfer from the vinyl group to the allyl group to form a weak C^1–C^6 bond, which causes an increase and decrease of the electronic density of the allyl and the vinyl groups, respectively. Thus, the C^5=C^6 bonds stretching the vibrational mode of the allyl group are blue-shifted from 1570 to 1580 cm^{-1}, and the C^1=C^2 bonds stretching the vibrational mode of the vinyl group is red-shifted from 1690 to 1580 cm^{-1}. In addition, the electron transfer from the vinyl group to the allyl group makes the C=C bonds of both the allyl and the vinyl groups equivalent having the same wavenumber of 1580 cm^{-1} around 1500 fs delay, which implies that aromatic-like C=C bonds are formed. This result shows that the generated intermediate has an aromatic-like six-membered structure. Finally, C^4–O bond breaking and C^1–C^6 bond formation proceed simultaneously to generate allyl acetaldehyde being observed in the appearance of molecular vibrational modes around 2,000 fs delay. The frequencies of the new modes at 1030, 1150 and 1750 cm^{-1} agree well with the Raman data of allyl acetaldehyde (Figure 11.1.5.1c), which can be assigned to the C–C–C symmetric stretching vibrational mode ($v_{s\ C-C-C}$), the C–C–C asymmetric stretching vibrational mode ($v_{as\ C-C-C}$), and the C=O stretching vibrational mode ($v_{C=O}$), respectively.

In addition, the initial phases of the observed molecular vibrational modes appeared immediately after the 5 fs pulses irradiation (900, 1320, and 1650 cm^{-1}) are close to sine-like within ±0.21 radian (Figure 11.1.5.2). Therefore, this result conforms that the observed molecular vibrational modes are associated with the wavepacket in the ground state.

[3,3]-Sigmatropic rearrangements of allyl aryl ethers were reported by Claisen in 1912 [13], which has been followed by broad variations of the Claisen rearrangements as below. After the first report of [3,3]-sigmatropic rearrangements of allyl vinyl ether in 1938 by Schuler and Murphy [14] and its first kinetic study [15], the [3,3]-sigmatropic rearrangement of allyl vinyl ether has been widely studied as a most simple model of the Claisen rearrangement in kinetic isotope effect measurements in experiments and reaction mechanism analysis in theoretical calculations [16–31]. In the transition state of the Claisen rearrangement, substitution and solvent were reported to affect the competing processes of C^4–O bond breaking and C^1–C^6 bond formation, resulting in changes in the detailed structure of the transition states [32].

In the following basic textbook-like description of rearrangement reaction is described.

Three key rearrangement reactions are 1,2-rearrangements, pericyclic reactions and olefin metathesis. A 1,2-rearrangement is an organic reaction where a substituent moves from one atom to another atom in a chemical compound. In a 1,2 shift, the movement involves two adjacent atoms but moves over larger distances are possible. Skeletal isomerization is not normally encountered in the laboratory, but is the basis of large applications in oil refineries. In general, straight-chain alkanes are converted to branched isomers by heating in the presence of a catalyst. Examples include the isomerization of n-butane to isobutane and pentane to isopentane. Highly branched alkanes have favorable combustion characteristics for internal combustion engines.

In general, three possible mechanisms have been suggested for the Claisen rearrangement mechanism of allyl vinyl ether [16–31].

1. In the first mechanism, the reaction proceeds in a synchronous concerted pathway *via* an aromatic-like transition state.

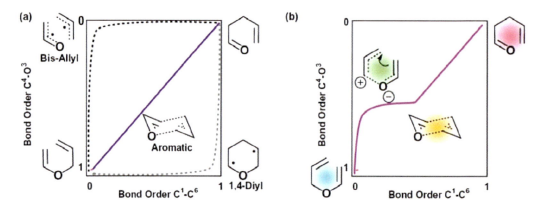

FIGURE 11.1.5.3 Transition-state profile for the Claisen rearrangement. (a) Purple curve: a synchronous concerted pathway reaction *via* an aromatic-like TS. Black curve: a stepwise pathway reaction *via* a bis-allyl-like TS. Gray curve: a stepwise pathway reaction *via* a 1-4-diyl-like TS; (b) Red curve: this work.

2. The second possible mechanism proposes an asynchronous stepwise pathway *via* a bis-allyl-like transition state, in which C^4–O bond breaking takes place in the first step of the reaction.
3. The third possible mechanism indicates an asynchronous stepwise pathway *via* a 1–4-diyl-like transition state, in which C^1–C^6 bond formation takes place in the first step of the reaction. In this work, we have observed intermediates and transition states which indicate a new possible mechanism for the Claisen rearrangement. The mechanism is described by a three-step pathway. At first, the C^4–O bond is weakened to generate a bis-allyl-like intermediate. Next, the formation of a weak C^1–C^6 bond results in the generation of an aromatic-like six-membered intermediate. Finally, C^4–O bond breaking and C^1–C^6 bond formation occurs simultaneously to generate allyl acetaldehyde (Figure 11.1.5.3). This resembles the transition state reported in the Claisen rearrangement of alkoxy allyl enol ether [33].

In any general thermally activated reaction, the reactant spends a substantial part of the total reaction time waiting for the very rare circumstance to get sufficient energy in the reaction coordinate. Once it does, the rest of the reaction is very fast. In other words, the reaction does not progress along the reaction coordinate at a constant speed. However, the calculated spectrogram in Figure 11.1.5.1 showed that the reaction progressed in a different way as follows. The first step of the reaction (generation of a bis-allyl-like intermediate) proceeds in 800–1000 fs. The second step generates an aromatic-like six-membered intermediate in 300–500 fs. The final step of the three-step pathway finishes in several tens to several hundred femtoseconds. The observed reaction timescale was confirmed by theoretical calculation of single direct-dynamics trajectory.

11.1.5.2.2 SINGLE DIRECT-DYNAMICS TRAJECTORY

Molecular dynamics of the Claisen rearrangement of AVE was simulated by dynamic reaction coordinate (DRC) calculations with large kinetic energy equally assigned to all degrees of freedom of the molecule in an isolated condition. The calculations were performed with GAMESS 2009 program package [34]. Trajectories were computed at the B3LYP/6–311G+(d, p) level of theory. Excess kinetic energy of 8–12 kcal/mol was provided for each freedom degree in the calculation. Several trajectory calculations with different directions of initial velocities generated randomly were carried out. A bow-like structured AVE obtained in the intrinsic reaction coordinate (IRC) calculation was used as an initial structure of the DRC calculation (Figure 11.1.5.4).

Figure 11.1.5.5 summarizes typical time evolutions of length of bonds that dissociate (C^4–O) and form (C^1–C^6) in the Claisen rearrangement. A successful trajectory which underwent the

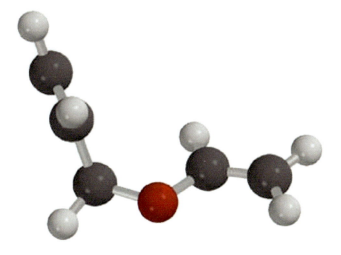

FIGURE 11.1.5.4   Bow-like structured AVE obtained in the IRC calculation.

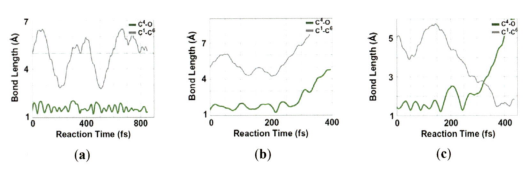

FIGURE 11.1.5.5   Bond length changes of C^4–O bond (green curve) and C^1–C^6 bond (gray curve) observed in the DRC trajectories with initial kinetic energies of (a) 8 kcal/mol, (b) 10 kcal/mol, and (c) 12 kcal/mol.

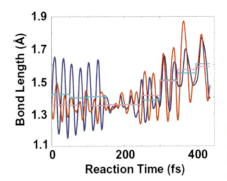

FIGURE 11.1.5.6   Time evolutions of bond lengths of C^1=C^2 (blue curve) and C^5=C^6 (red curve) in the trajectory with initial kinetic energy of 10 kcal/mol. Means in 50 fs periods are also indicated (light blue and pink lines, respectively).

Claisen rearrangement was observed in a calculation with the initial kinetic energy of 10 kcal/mol (Figure 11.1.5.5b). Dissociation of the C^4–O bond and decrease of distance between C^1 and C^6 lead to the aromatic-like six-membered structure around 320 fs after large elongation and recovery of the dissociating bond length at 200 fs.

Time evolutions of bond lengths of C^1=C^2 and C^5=C^6 in the reactive trajectory (Figure 11.1.5.6) indicate that the first structure of the trajectory calculation shown in Figure 11.1.5.4 corresponds to the structure in the 800–1,000 fs time region of the spectrogram. Those bond lengths are observed to be distinctly different, ~1.35 and ~1.42 Å, respectively, from 0 to 200 fs, which agree with C=C stretching modes appearing at 1690 and 1570 cm^{-1} around 1,000 fs after photo-excitation in the spectrogram.

After the initial phase, those bond lengths in the trajectory become almost identical when C^4–O bond starts to break and the aromatic-like six-membered structure forms from 200–to 320 fs. This behavior is consistent with a merge of the two C=C stretching bands to a single band of the aromatic-like six-membered structure appearing at $1580 \, cm^{-1}$ around 1,500 fs delay in the spectrogram.

In the reactive trajectory, the product is generated within ~150 fs after the formation of the aromatic-like six-membered structure. The time constant is half of that observed in the spectrogram. The discrepancy arises presumably due to the isolated condition of the trajectory calculation which lacks interaction with surrounding molecules. Molecular friction caused by the interaction with surrounding molecules in the neat solvent used in the experiment could slow the product formation.

In the case of the initial kinetic energy of 8 kcal/mol, the C^4–O bond did not dissociate and in turn, the Claisen rearrangement was not observed (Figure 11.1.5.5a). Although C^1 and C^6 atoms approached closely during the trajectory, thermal elongation of the C^4–O bond was not enough to initiate the rearrangement. On the other hand, when the initial kinetic energy of 12 kcal/mol was supplied, the larger excess energy led to fragmentation, *i.e.*, C^4–O bond dissociation without bond formation between C^1 and C^6 atoms (Figure 11.1.5.5c). However, the fragmentation would be strongly suppressed in a neat solvent due to caging effect. Furthermore, an NMR measurement confirmed that the product was pure allyl acetaldehyde, whereas such fragmentations are expected to generate various product species. We therefore conclude that the rearrangement proceeds through the aromatic-like six-membered structure as observed in the trajectory with the initial kinetic energy of 10 kcal/mol.

11.1.5.3   EXPERIMENTAL

11.1.5.3.1   VISIBLE 5-FS LASER SYSTEM

The ultrashort pulse laser [35] and ultrafast spectroscopy system used in the measurement are described elsewhere [36] and it is briefly summarized in the following. The output pulse from a Ti:sapphire regenerative amplifier (Spectra Physics Spitfire) with 100 fs duration, centered at 790 nm, and 5 kHz repetition rate was separated by a beam splitter in two pulses. One of the two pulses was focused into a β-BaB$_2$O$_4$ (BBO) crystal (0.4 mm-thick, θ=29°) to generate second harmonic (SH) pulse, which was used as a pump pulse in the following optical parametric amplification. The other pulse of the separated two pulses was focused in a 2 mm-thick sapphire plate to generate femtosecond white light broadening spectral bandwidth by third-order nonlinear effect of self-phase modulation. The white light pulse was amplified in a non-collinear optical parametric amplifier (NOPA) pumped by the SH pulse. In the NOPA, the angle between the SH pump pulse and the white light seed pulse was set to be 3.7° in the non-linear crystal (type-I BBO crystal, 1 mm-thick, θ=31.5°) to satisfy the phase matching condition in broad visible spectral region, which results in broadband amplification of the white light seed pulse. A prism pair and a chirped mirror pair were used to compensate material dispersion compressing the pulse duration as short as 5 fs to be used for pump-probe measurement. The pulse duration can be compressed more as short as 3.9 fs by inserting an additional chirp compressor made of a diffraction grating and a deformable mirror.

11.1.5.3.2   "THE REACTION IN THE ELECTRONIC GROUND STATE",
TRIGGERED BY THE VISIBLE 5-FS PULSE

Most organic compounds have absorption bands in the ultraviolet region, therefore the visible 5 fs pulse with a broad bandwidth of 525–725 nm does not excite their electronic states by single-photon excitation but induces their molecular vibration via stimulated Raman process with Λ-type interaction in the ground state or V-type interaction in the excited state [37]. The 5 fs pulse has a bandwidth of $5200 \, cm^{-1}$, which can excite high energy vibration bands as high as in the case of thermal excitation at 7500 K. Meanwhile, being different from the standard thermal reaction, certain vibration modes are selectively excited by the selection rule and cross-section of the stimulated Raman process. Activated vibration modes have high vibration quantum number compared with that of thermal excitation at 7500 K, while other vibration modes keep a low vibration quantum number of that at room temperature. This direct excitation of vibration modes and its relaxation and transfer to other modes in several hundred femtoseconds [38] will excite vibration modes on the reaction pathway. Thus, this impulsive excitation by the visible 5 fs pulse triggers "the reaction in the electronic ground state", which is different from that induced by electronic state excitation under photo irradiation conditions.

11.1.5.4 CONCLUSIONS

In conclusion, the Claisen rearrangement of allyl vinyl ether was triggered by a new scheme exciting the sample by visible 5 fs pulses whose photon energy is much lower than the absorption band of the sample. Observing the molecular vibration frequency changes in the reaction, including its transition states, elucidated the reaction mechanism of the Claisen rearrangement in the new scheme [39]. The time constants of transformation from straight-chain structure to aromatic-like six-membered ring structure forming the C^1–C^6 bond were estimated from the observed dynamics of the molecular vibration modes. It was compared with the molecular dynamics simulated by dynamic reaction coordinate calculations with large kinetic energy. The result clarifies that the reaction proceeds via three steps showing agreement between the observed molecular vibration frequency change and that predicted in the dynamic reaction coordinate calculations. This finding provides a new hypothesis and discussion, helping the development of the field of reaction mechanism analysis.

The research described in this section was conducted by the following people in collaboration: Izumi Iwakura, Yu Kaneko, Shigehiko Hayashi, Atsushi Yabushita and Takayoshi Kobayashi [39].

REFERENCES

1. R. G. W. Norish, and G. Porter, "Chemical reactions produced by very high light intensities," *Nature* **164**, 658 (1949).
2. T. H. Maiman, "Stimulated optical radiation in ruby," *Nature* **187**, 493–494 (1960).
3. M. Hentschel, R. Kienberger, Ch. Spielmann, G. A. Reider, N. Milosevic, T. Brabec, P. Corkum, U. Heinzmann, M. Drescher, and F. Krausz, "Attosecond metrology," *Nature* **414**, 509–513 (2001).
4. J. Itatani, J. Levesque, D. Zeidler, H. Niikura, H. Pépin, J. C. Kieffer, P. B. Corkum, and D. M. Villeneuve, "Tomographic imaging of molecular orbitals," *Nature* **432**, 867–871 (2004).
5. F. Krausz, and M. A. Ivanov, "Attosecond physics," *Rev. Mod. Phys.* **81**, 163–234 (2009).
6. A. H. Zewail, "Femtochemistry: Atomic-scale dynamics of the chemical bond," *J. Phys. Chem. A* **104**, 5660–5694 (2000).
7. T. Kobayashi, T. Saito, and H. Ohtani, "Real-time spectroscopy transition state in bacteriorhodopsin during retinal isomerization," *Nature* **414**, 531–534 (2001).
8. I. Iwakura, "The experimental visualization of molecular structural changes during both photochemical and thermal reactions by real-time vibrational spectroscopy," *Phys. Chem. Chem. Phys.* **13**, 5546–5555 (2011).
9. T. Kobayashi, and A. Yabushita, "Transition-state spectroscopy using ultrashort laser pulses," *Chem. Rec.* **11**, 99–116 (2011).
10. I. Iwakura, A. Yabushita, and T. Kobayashi, "Direct observation of the molecular structural changes during the Claisen rearrangement including the transition state," *Chem. Lett.* **39**, 374–375 (2010).
11. I. Iwakura, A. Yabushita, J. Liu, K. Okamura, and T. Kobayashi, "Photo-impulsive reactions in the electronic ground state without electronic excitation: Non-photo, Non-thermal chemical reactions," *Phys. Chem. Chem. Phys.* **14**, 9696–9701 (2012).
12. M. J. J. Vrakking, D. M. Villeneuve, and A. Stolow, "Observation of fractional revivals in molecular wavepackets," *Phys. Rev. A* **54**, R37–R40 (1996).
13. L. Claisen, "Rearrangement of phenol allyl ethers into C-allylphenols," *Chem. Ber.* **45**, 3157–3166 (1912).
14. C. D. Hurd, and M. A. Pollack, "The rearrangement of vinyl allyl ethers," *J. Am. Chem. Soc.* **60**, 1905–1911 (1938).

15. F. W. Schuler, and G. W. Murphy, "The kinetics of the rearrangement of vinyl allyl ether," *J. Am. Chem. Soc.* **72**, 3155–3159 (1950).

16. A. M. M. Castro, "Claisen rearrangement over the past nine decades," *Chem. Rev.* **104**, 2939–3002 (2004).

17. J. J. Gajewski, J. Jurayj, D. R. Kimbrough, M. E. Gande, B. Ganem, and B. K. Carpenter, "The mechanism of rearrangement of chorismic acid and related compounds," *J. Am. Chem. Soc.* **109**, 1170–1186 (1987).

18. R. L. Vance, N. G. Rondan, K. N. Houk, F. Jensen, W. T. Borden, A. Komornicki, and E. Wimmer, "Transition structures for the Claisen rearrangement," *J. Am. Chem. Soc.* **110**, 2314–2315 (1988).

19. O. Wiest, K. A. Black, and K. N. Houk, "Density functional theory isotope effects and activation energies for the cope and Claisen rearrangements," *J. Am. Chem. Soc.* **116**, 10336–10337 (1994).

20. H. Hu, M. N. Kobrak, C. Xu, and S. Hammes-Schiffer, "Reaction path Hamiltonian analysis of dynamical solvent effects for a Claisen rearrangement and a Diels Alder reaction," *J. Phys. Chem. A* **104**, 8058–8066 (2000).

21. J. G. Hill, P. B. Karadakov, and D. L. Cooper, "A spin-coupled study of the Claisen rearrangement of allyl vinyl ether," *Theor. Chem. Acc.* **115**, 212–220 (2006).

22. M. J. S. Dewar, and E. F. Healy, "Ground states of molecules. 68. MNDO study of the Claisen rearrangement," *J. Am. Chem. Soc.* **106**, 7127–7131 (1984).

23. M. J. S. Dewar, and C. Jie, "Mechanism of the Claisen rearrangement of allyl vinyl ethers," *J. Am. Chem. Soc.* **111**, 511–519 (1989).

24. J. J. Gajewski, and N. D. Conrad, "Aliphatic Claisen rearrangement transition state structure from secondary .alpha.-deuterium isotope effects. *J. Am. Chem. Soc.* **101**, 2747–2748 (1979).

25. J. J. Gajewski, K. R. Gee, and J. Jurayj, "Energetic and rate effects of the trifluoromethyl group at C-2 and C-4 on the aliphatic Claisen rearrangement," *J. Org. Chem.* **55**, 1813–1822 (1990).

26. L. Kupczyk-Subotkowska, W. H. Saunders Jr., H. J. Shine, and W. Subotkowski, "Carbon kinetic isotope effects and transition structures in the rearrangements of allyl vinyl ethers. 2-(Trimethylsilyloxy)- and 2-(Methoxycarbonyl)-3-oxa-1,5-hexadiene," *J. Am. Chem. Soc.* **116**, 7088–7093 (1994).

27. M. M. Davidson, and I. H. Hillier, "Aqueous acceleration of the Claisen rearrangement of allyl vinyl ether: A hybrid, explicit solvent, and continuum model," *J. Phys. Chem.* **99**, 6748–6751 (1995).

28. J. J. Gajewski, "The Claisen rearrangement. Response to solvents and substituents: The case for both hydrophobic and hydrogen bond acceleration in water and for a variable transition state," *Acc. Chem. Res.* **30**, 219–225 (1997).

29. V. Aviyente, H. Y. Yoo, and K. N. Houk, "Analysis of substituent effects on the Claisen rearrangement with Ab Initio and density functional theory," *J. Org. Chem.* **62**, 6121–6128 (1997).

30. H. Y. Yoo, and K. N. Houk, "Theory of substituent effects on pericyclic reaction rates: Alkoxy substituents in the Claisen rearrangement," *J. Am. Chem. Soc.* **119**, 2877–2884 (1997).

31. M. P. Meyer, A. J. DelMonte, and D. A. Singleton, "Reinvestigation of the isotope effects for the Claisen and aromatic Claisen rearrangements: The nature of the Claisen transition states," *J. Am. Chem. Soc.* **121**, 10865–10874 (1999).

32. J. Rehbein, and M. Hiersemann, Mechanistic Aspects of the Aliphatic Claisen Rearrangement, in *The Claisen Rearrangement*, 1st ed., edited by M. Hiersemann, U. Nubbemeyer, Wiley-VCH, Winheim, Germany, pp. 525–557 (2007).

33. R. M. Coates, B. D. Rogers, S. J. Hobbs, D. P. Curran, and D. R. Peck, "Synthesis and Claisen rearrangement of alkoxyallyl enol ethers. Evidence for a dipolar transition state," *J. Am. Chem. Soc.* 1987, **109**, 1160–1170.

34. M. W. Schmidt, K. K. Baldridge, J. A. Boatz, S. T. Elbert, M. S. Gordon, J. H. Jensen, S. Koseki, N. Matsunaga, K. A. Nguyen, S. J. Su, et al., "General atomic and molecular electronic structure system," *J. Comput. Chem.* 1993, **14**, 1347–1363.

25. A. Shirakawa, I. Sakane, M. Takasaka, T. Kobayashi, "Sub-5-fs visible pulse generation by pulse-front-matched noncollinear optical parametric amplifier," *Appl. Phys. Lett.* 1999, **74**, 2268–2270.

36. T. Kobayashi, I. Iwakura, A. Yabushita, "Excitonic and vibrational nonlinear processes in a polydiacetylene studied by a few-cycle pulse laser," *New J. Phys.* 2008, **10**, 065016.

37. J. Du, T. Teramoto, K. Nakata, E. Tokunaga, T. Kobayashi, "Real-time vibrational dynamics in chlorophyll a studied with a few-cycle pulse laser," *Biophys. J.* 2001, **101**, 995–1003.

38. T. Yagasaki, S. Saito, "Ultrafast intermolecular dynamics of liquid water: A theoretical study on two-dimensional infrared spectroscopy," *J. Chem. Phys.* 2008, **128**, 154521.

39. I. Iwakura, Y. Kaneko, S. Hayashi, A. Yabushita, and T. Kobayashi, "Ultrafast electronic relaxation and vibrational dynamics in a polyacetylene derivative," *Molecules* **18**, 1995–2004 (2013).

11.1.6 A New Reaction Mechanism of Claisen Rearrangement Induced by Few-Optical-Cycle Pulses

Demonstration of Nonthermal Chemistry by Femtosecond Vibrational Spectroscopy

11.1.6.1 INTRODUCTION

The Claisen rearrangement is one of the most popular sigmatropic rearrangements in organic chemistry. Along with the Cope rearrangement, it is known for its high stereoselectivity and is extremely useful in organic synthesis. The Claisen rearrangement was first reported by Claisen [1] for allyl aryl (or vinyl) ethers, spurring the development of various other reactions [2–4]. In 1938, it was found that allyl vinyl ether generates allyl acetaldehyde upon heating at 255°C [5], which was later found to be generated via [3,3]-sigmatropic rearrangement (Figure 11.1.6.1). Following the first kinetic study of the Claisen rearrangement in 1950 [6], various investigations of its reaction mechanism have been reported. The Claisen rearrangement is thought to proceed through a six-membered transition state (TS) by a supra-supra facial reaction following Woodward–Hoffmann rules [7] and frontier orbital theory [8]. Experimental stereochemical outcomes [9–12] and theoretical calculations [13–15] implicated the six-membered chair-form TS. However, the understanding of the mechanism of the Claisen rearrangement in greater detail has remained elusive (Figure 11.1.6.1) [4,14–25]. In one proposed mechanism, the reaction progresses along a synchronous concerted pathway via an aromatic-like TS. Another possible mechanism involves the reaction progressing by an asynchronous concerted pathway in which either C^1–C^6 bond formation or C^4–O bond cleavage occurs in advance of the other. An asynchronous reaction in which the C^1–C^6 bond forms first may proceed through a 1,4-diyl-like TS, whereas that in which the C^4–O bond breaks first may proceed through a bis-allyl-like TS. As the simplest example of the Claisen rearrangement, allyl vinyl ether was studied by quantum chemical calculations and the kinetic isotope effect.

In this work, the Claisen rearrangement was studied directly by observing the reaction process through time-resolved vibration spectroscopy using a pulsed laser with few optical cycles [26,27]. Direct observation of the Claisen rearrangement has not been performed by any other methods to date. Here, we observed molecular structural changes during the "nonthermal Claisen

DOI: 10.1201/9780429196577-80

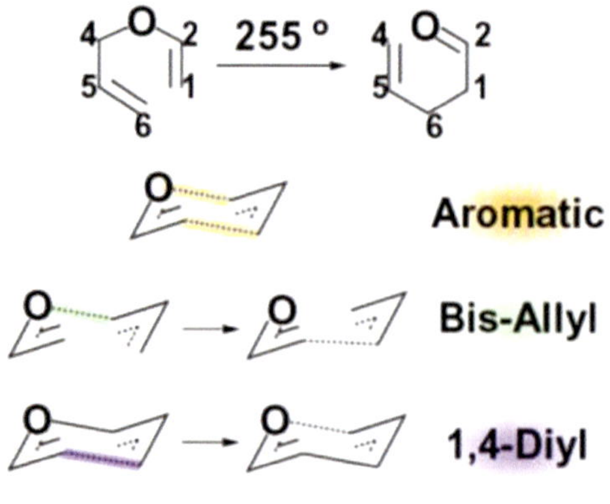

FIGURE 11.1.6.1   Claisen rearrangement of allyl vinyl ether and three proposed reaction mechanisms.

rearrangement" process in the electronic ground state, including the TS, as changes in the instantaneous wavenumber of molecular vibrations. The reaction mechanism of the nonthermal Claisen rearrangement in the electronic ground state is elucidated. The observed dynamics of the reaction process will provide important information to clarify the reaction mechanism of the thermal Claisen rearrangement process, even though the reaction processes of the two kinds of Claisen rearrangement may differ.

11.1.6.2   EXPERIMENTAL

11.1.6.2.1   VISIBLE FEW-OPTICAL-CYCLE PULSES

To generate a broadband intense laser pulse for a visible few-optical-cycle broadband pulse, we have used a non-collinear optical parametric amplifier (NOPA) described elsewhere [26]. In short, the pump source of the NOPA is a Ti:sapphire regenerative amplifier (Spectra-Physics, Spitfire; 150 µJ, 100 fs, 5 kHz at 805 nm). Visible white light generated by self-phase modulation in a sapphire plate was amplified in the double-pass NOPA having a broadband spectrum extending from 525 to 725 nm. Its time duration was compressed with a main compressor, resulting in the pulse duration of 5 fs, which is nearly Fourier-transform limited. The polarizations of the pump pulse and the probe pulse were parallel to each other. The focal areas of the pump pulse and the probe pulse were 100 and 75 µm^2, respectively.

11.1.6.2.2   ULTRAVIOLET FEW-OPTICAL-CYCLE PULSES

Second harmonic generation (SHG) of a Ti:sapphire regenerative amplifier (Coherent, Legend EliteUSP; 2.5 mJ, 35 fs, 1 kHz at 800 nm) was focused into a hollow-core fiber. Argon gas filled in the hollow-core fiber has broadened spectral width to generate a few-optical-cycle broadband ultraviolet (UV) pulse [27]. The broadened spectrum coming out from the hollow-core fiber was extending from 360 to 440 nm. The main compressor compressed the pulse duration as 8 fs. The polarizations of the pump pulse and the probe pulse were parallel to each other. The focal areas of the pump pulse and the probe pulse were 100 and 75 µm^2, respectively.

11.1.6.2.3   SAMPLE CELL

Neat liquid of allyl vinyl ether and allyl phenyl ether were used as samples. The time-resolved signal was measured by storing the sample in a liquid cell with a 1-mm optical path length. When the UV pulse irradiates at a fixed point, UV light absorption by the sample causes thermal accumulation and bubbling. Therefore, we have translated the sample cell continuously in the plane perpendicular to the probe beam to avoid thermal accumulation and bubbling. The cell was translated in a near-circular and octagon-like trajectory with the speed of ca. 5 mm/s, thus, the sample was displaced ca. 10 μm during the pulse period of 1 ms suppressing the sample photodamage to minimum. All of the measurements were performed at a room temperature of 295 ± 1 K.

11.1.6.2.4 PUMP–PROBE MEASUREMENT

A polychromator (300 grooves/mm, 500 nm blazed for visible pulses, and 300 nm blazed for UV pulses) has spectrally dispersed the probe spectrum coupling probe wavelength components to 128 avalanche photodiodes via a 128-channel bundled fiber. Lock-in amplifiers of the same number were utilized at all avalanche photodiodes to improve the signal-to-noise ratio.

Here a simple explanation of blazed grafting for monochromator is described.

A blazed grating has a constant line spacing determining the magnitude of the wavelength splitting caused by the grating. The grating lines possess a triangular, sawtooth-shaped cross section, forming a step structure. The steps are tilted at the so-called blaze angle with respect to the grating surface.

The blaze angle is optimized to maximize efficiency for the wavelength of the used light. Descriptively, this means is chosen such that the beam diffracted at the grating and the beam reflected at the steps are both deflected in the same direction. Commonly blazed gratings are manufactured in the so-called Littrow configuration.

11.1.6.2.5 THEORETICAL CALCULATION

The theoretical calculation was performed by the Gaussian 03 program [28] optimizing geometry by the B3LYP/6–311+G** method and basis set. Calculations were performed without assuming symmetry using $5d$ functions for the d orbital. For all of the obtained structures, frequencies were calculated at the same level. Calculation results have confirmed that all the frequencies were real for the ground states and one imaginary frequency existed for the TS. Vectors of the imaginary frequencies directed the reaction mode and intrinsic reaction coordinate (IRC) calculations have confirmed that the obtained TSs were on the saddle points of the energy surface between the reactant and the product.

11.1.6.3 RESULTS AND DISCUSSION

11.1.6.3.1 CLAISEN REARRANGEMENT OF ALLYL VINYL ETHER

Allyl vinyl ether has an absorption band located at shorter than 220 nm (Figure 11.1.6.2a) that cannot be excited by either one- or two-photon absorption of few-optical-cycle pulses in the visible range from 525 to 725 nm (Figure 11.1.6.2b). Therefore, a visible few-optical-cycle pulse excites molecular vibrations in allyl vinyl ether only in the electronic ground state [29,30].

Figure 11.1.6.2c shows an averaged real-time trace of ΔA obtained by pump–probe measurement over 16 probe channels with a delay time range of –100 to 3000 fs. In the pump–probe measurement using visible few-optical-cycle pulses, the observed signal reflects the wave packet motion; i.e., fine oscillation in the real-time trace of ΔA reflects the transformation in periodic molecular structure caused by molecular vibrations. Therefore, spectrogram analysis [31] of the real-time traces was performed using a Blackman window function with a full width at half-maximum (FWHM) of 400 fs. Spectrogram analysis shows the time evolution of the vibration modes after the laser pulse irradiation.

Figure 11.1.6.3b shows the calculated spectrogram, where x- and y-axes show reaction time and instantaneous vibrational wavenumber, respectively, and pseudocolor reflects the power of fast Fourier transform (FFT). In the spectrogram, the molecular vibrations observed just after visible-pulse excitation were caused by the reactant allyl vinyl ether, as shown in the following assignment. The observed wavenumbers can be assigned to C=C stretching vibrations ($\nu_{C=C}$) of the allyl and

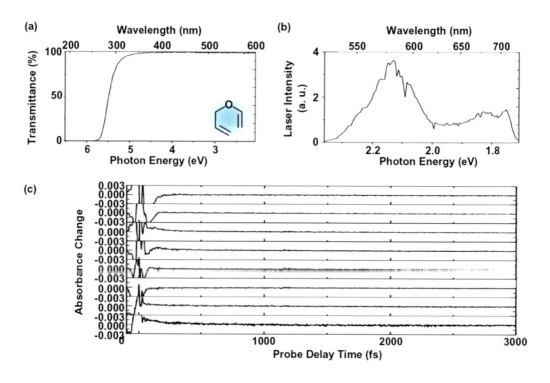

FIGURE 11.1.6.2   (a) Transmittance spectrum of allyl vinyl ether. (b) Spectrum of visible few-optical-cycle pulses. (c) Realtime traces of absorbance changes (ΔA) of allyl vinyl ether averaged over 16 probe channels in the delay time range between –100 and 3000 fs [30]. The probe wavelength regions of the eight traces are 700–725, 675–700, 650–675, 625–650, 600–625, 575–600, 550–575, and 525–550 nm, respectively, from top to bottom.

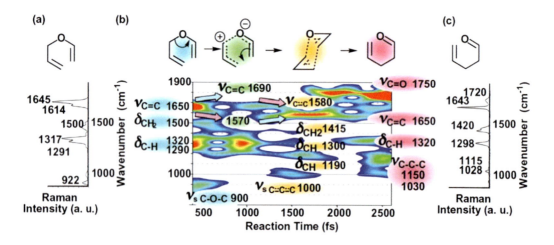

FIGURE 11.1.6.3   (a) Raman spectrum of allyl vinyl ether. (b) Spectrogram of nonthermal Claisen rearrangement of allyl vinyl ether induced by visible few-optical-cycle pulses [30]. (c) Raman spectrum of allyl acetaldehyde.

vinyl groups (1650 cm^{-1}), C–H$_2$ deformation vibration (δ_{C-H2}) of the methylene group (1500 cm^{-1}), C–H deformation vibrations (δ_{C-H}) of the vinyl and allyl groups (1320 and 1290 cm^{-1}, respectively), and C–O–C symmetric stretching vibration ($\nu_{s\,C-O-C}$) of the ether group (900 cm^{-1}). These observed wavenumbers agree well with the Raman spectrum of the electronic ground state of allyl vinyl ether

(Figure 11.1.6.3a), which confirms that the pump–probe observations closely reflect the molecular vibration dynamics of the electronic ground state of allyl vinyl ether.

Molecular structure deformation during the reaction process along the reaction coordinate changes the vibrational wavenumber of the wave packet vibrating along the coordinate perpendicular to the reaction one. There are 3N-7 coordinates perpendicular to the reaction coordinate in a molecule composed of N atoms with 3N-6 normal modes. Therefore, when the molecule stays between the two species of reactant, the intermediates, and the product, the vibrational wavenumbers in a spectrogram should change. If the molecule stays as the reactant, intermediate, or product, the vibrational wave numbers should not shift during the lifetime of the relevant state. In the following, we discuss the real-time frequency shifts of the molecular vibration wavenumbers observed in the spectrogram of allyl vinyl ether.

The disappearance of the two modes of the δ_{CH2} (1500 cm^{-1}) and $\nu_{s\ C-O-C}$ (900 cm^{-1}) bands at a delay of around 800 fs shows that the C^4–O bond is weakened or broken in the first stage of the reaction. The signal from $\nu_{C=C}$ of the vinyl and allyl groups, observed at 1650 cm^{-1} just after visible pulse irradiation, was separated into a blue-shifted mode moving toward 1690 cm^{-1} and a red-shifted mode heading toward 1570 cm^{-1} in the probe-delay region from 500 to 800 fs. These wavenumber shifts also suggest that the C^4–O bond is weakened, because electron transfer from the allyl group to the vinyl group causes the electronic density in the vinyl and allyl groups to increase and decrease, respectively.

After the C^4–O bond weakens, electrons are transferred in the opposite direction from the vinyl group to the allyl group to form a weak C^1–C^6 bond. This transfer decreases the electronic density along the C^1=C^2 bond in the vinyl group, resulting in a red shift of $\nu_{C=C}$ from 1690 to 1580 cm^{-1}. In the allyl group, the electronic density along the C^5=C^6 bond increased, inducing the observed blue shift of $\nu_{C=C}$ from 1570 to 1580 cm^{-1}, and the C^4–C^5 single bond changed into a C^4=C^5 double bond. Thus, the three C=C bonds in the vinyl and allyl groups (C^1=C^2, C^4=C^5, and C^5=C^6) become equivalent at a delay of about 1500 fs, all exhibiting a band at 1580 cm^{-1}. The appearance of a band at around 1580 cm^{-1} indicates that aromatic-like C=C bonds were formed because the aromatic $\nu_{C=C}$ in benzene appears at 1585 cm^{-1} [32]. This implies that the generated intermediate does not have the perfect C6 symmetry of a benzene ring, but instead has a six-membered structure with aromatic C=C bonds, which is also supported by the appearance of δ_{C-H} and $\nu_{s\ C-C-C}$ at 1190 and 1000 cm^{-1}, respectively (δ_{C-H} of benzene is observed at 1180 cm^{-1} [32]). Finally, C^4–O bond cleavage and C^1–C^6 bond formation are thought to proceed simultaneously (i.e., in a synchronous concerted process) to generate the product, allyl acetaldehyde. Therefore, new bands that appeared at a delay of 2000 fs reflect the formation of the product. These new bands at 1750, 1150, and 1030 cm^{-1} can be assigned to the C=O stretching vibration ($\nu_{C=O}$), C–C–C asymmetric stretching vibration ($\nu_{as\ C-C-C}$), and C–C–C symmetric stretching vibration ($\nu_{s\ C-C-C}$), respectively, of allyl acetaldehyde. The wavenumbers of these new modes correlate well with the Raman spectrum of allyl acetaldehyde (Figure 11.1.6.3c) that was synthesized by the oxidation of 4-penten-1-ol.

We also performed a pump–probe experiment using the synthesized product (allyl acetaldehyde) to compare with the measurements obtained for the reactant. No wavenumber shift was observed in the pump–probe measurements of the product, indicating that no reaction occurred.

NMR spectra of allyl vinyl ether before and after the pump–probe experiment further confirmed the formation of allyl acetaldehyde. We performed a pump–probe experiment with a small amount of allyl vinyl ether in a glass cell (10 mm^3). The NMR spectrum of allyl vinyl ether after the measurement indicated the presence of allyl acetaldehyde with a fraction of 1% w/w (Figure 11.1.6.4). The quantum yield of the photoinduced process was estimated to be about 0.01. As described above, time-resolved observation of the conformational changes during the [3,3]-sigmatropic rearrangement of allyl vinyl ether was performed by observing shifts in molecular vibrations. The [3,3]-sigmatropic rearrangement of allyl vinyl ether in the electronic ground state was triggered by irradiation with visible few-optical-cycle pulses, and here this reaction in the electronic ground state is called nonthermal Claisen rearrangement.

In general, three possible mechanisms are suggested to explain the Claisen rearrangement. One of the proposed mechanisms is a synchronous concerted pathway reaction via an aromatic-like TS

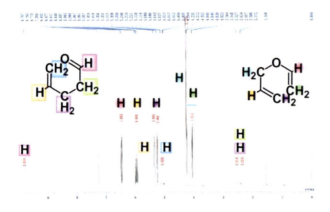

FIGURE 11.1.6.4 NMR spectrum of a sample after neat allyl vinyl ether was irradiated with visible few-optical-cycle pulses during time-resolved measurement [30].

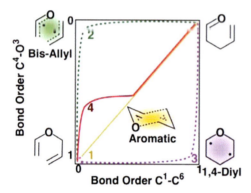

FIGURE 11.1.6.5 Map of alternative routes for the Claisen rearrangement [34]. 1: Synchronous concerted reaction pathway via an aromatic-like TS. 2: Stepwise reaction pathway via a bis-allyl-like TS. 3: A stepwise reaction pathway via a 1-4-diyl-like TS. 4: Mechanism proposed in this work [33].

(yellow line 1 in Figure 11.1.6.5). Two other possible mechanisms are stepwise pathway reactions via a bis-allyl-like

TS in which C^4–O bond cleavage takes place in the first stage of the reaction (green dotted line 2 in Figure 11.1.6.5) or a 1–4-diyl-like TS in which C^1–C^6 bond formation takes place in the first stage of the reaction (purple dotted line 3 in Figure11.1.6.5). In this work, a new possible mechanism for the following three-stage pathway is proposed for the Claisen rearrangement using the data obtained by exciting allyl vinyl ether using visible few-optical-cycle pulses [30,33]. First, the C^4–O bond is weakened to generate a bisallyl-like intermediate. Formation of a weak C^1–C^6 bond then generates an aromatic-like intermediate. Finally, C^4–O bond cleavage and C^1–C^6 bond formation occur simultaneously to generate allyl acetaldehyde (red line 4 in Figure 11.1.6.5).

We then compared the observed nonthermal Claisen rearrangement process in the electronic ground state with the standard Claisen rearrangement process. Geometric optimization of TSs and IRCs was calculated using B3LYP/6–311+G** (Figure 11.1.6.6). In the reactant, the C^1=C^2 bond in the vinyl group is longer than the C^5=C^6 bond in the allyl group, so $\nu_{C=C}$ of the vinyl group appears at lower wave number than that of the allyl. In the first stage of the reaction, the molecular structure of allyl vinyl ether changes from its normal chain structure to a ring structure. Along with this structural change, the C^4–O bond length increases (green line 1 in Figure 11.1.6.6) causing a blue shift of $\nu_{s\,C-O-C}$ followed by its disappearance. Associated with this change, the C^1=C^2 bond length decreases (blue line 3 in Figure 11.1.6.6) and $\nu_{C=C}$ of the vinyl group exhibits a blue shift. In contrast, $\nu_{C=C}$ of the allyl group exhibits a red shift as the length of the C^5=C^6 bond increases (pink line 4 in Figure 11.1.6.6).

In the next stage of the reaction, along with the formation of a weak C^1–C^6 bond, the C^1=C^2 bond length increases, which induces a red shift of $\nu_{C=C}$ of the vinyl group. The bond orders of C^1=C^2 and C^5=C^6 bonds become similar because the wavenumbers of C^1=C^2 and C^5=C^6 stretching modes are

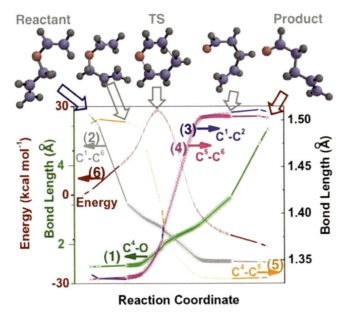

FIGURE 11.1.6.6 IRC calculation at the B3LYP/6-311+G** level. The six curves show bond lengths of (1) C^4–O, (2) C^1–C^6, (3) C^1=C^2 of the vinyl group, (4) C^5=C^6 of the allyl group, (5) C^4–C^5 of the allyl group, and (6) calculated IRC energy of the reaction pathway. In each molecular model, C, O, and H are dark, light, and small spheres, respectively.

almost equal. In addition, the C^4–C^5 bond of the allyl group (orange line 5 in Figure 11.1.6.6) also becomes equivalent to the C^1=C^2 and C^5=C^6 bonds. The lengths of all three C=C bonds were 1.39 Å, showing that they can be assigned as aromatic C=C bonds.

After this process, the C^1=C^2 and C^5=C^6 bonds of the vinyl and allyl groups, respectively, become longer and their signals disappear because they become Raman inactive. In contrast, the C^4–C^5 bond of the allyl group shortens and forms a new C=C bond.

The change in bond length calculated by IRC is consistent with the results of pump–probe experiments using visible few-optical-cycle pulses. The pump–probe experiment suggests a three-stage pathway, whereas the IRC calculation suggests an asynchronous concerted process. It should be clarified in a future study if the difference originates from the different reaction trigger, reaction in solution vs. the gas phase, or for some other reason.

11.1.6.3.2 CLAISEN REARRANGEMENT OF ALLYL PHENYL ETHER

The [3,3]-sigmatropic rearrangement of allyl vinyl ether is known to be a thermally allowed process. That is, it is a photochemically forbidden process following Woodward–Hoffmann rules. Even though it is photochemically forbidden, the rearrangement occurs using visible few-optical-cycle pulses. To confirm that a photochemically forbidden process can be triggered using visible few-optical-cycle pulses, we tried to observe the Claisen rearrangement process of allyl phenyl ether. When allyl phenyl ether is heated, [3,3]-sigmatropic rearrangement (thermally allowed Claisen rearrangement) in the electronic ground state occurs to generate *ortho*-substituted phenol following the mechanism shown in Figure 11.1.6.7a. This rearrangement is thought to proceed through the six-membered TS by a supra-supra facial reaction to generate a keto-intermediate. Instability of the keto-intermediate leads to subsequent keto-enol tautomerization, generating *ortho*-substituted phenol. Therefore, it is expected that the carbonyl stretching mode of a keto-intermediate is observed at the beginning of the reaction. As the keto-intermediate undergoes keto-enol tautomerization, the carbonyl stretching mode disappears.

In contrast, Kharasch et al. [35] reported in 1952 that photoirradiation triggered the rearrangements of allyl phenyl and benzyl phenyl ethers in their electronic excited states. The mechanism of photochemical Claisen rearrangement is shown in Figure 11.1.6.7b [36]. Under photoirradiation, the electronic excited state of allyl phenyl ether forms a pair of radical intermediates (PhO• and

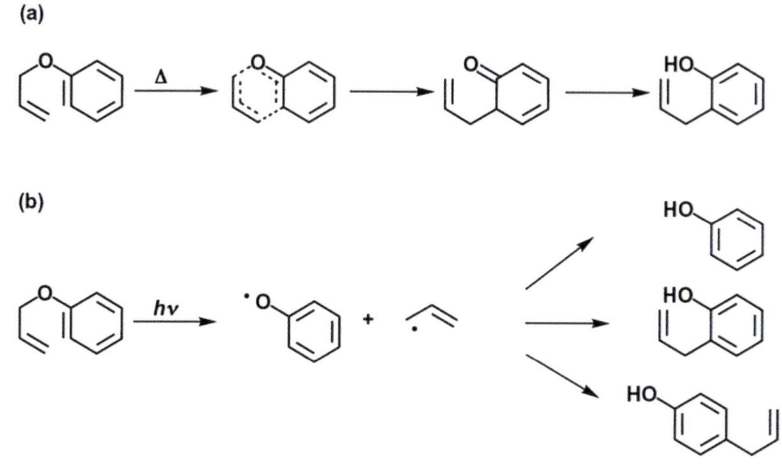

(a)

(b)

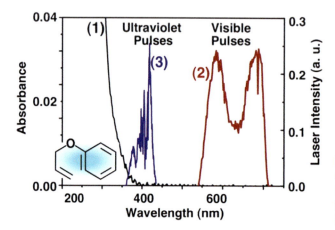

FIGURE 11.1.6.7 (a) Thermally allowed Claisen rearrangement of allyl phenyl ether. (b) Photochemical Claisen rearrangement of allyl phenyl ether.

FIGURE 11.1.6.8 Absorption spectrum of APE (1) and spectrum of visible few-optical-cycle pulses (2) and spectrum of UV few-optical-cycle pulses (3) [37].

CH_2CHCH_2·), and radical reactions generate the parent phenol as well as *ortho*- and *para*-substituted phenols. Therefore, both phenoxy and allyl radicals can be observed.

To compare both cases (photochemically forbidden and allowed), we tried to selectively induce the photochemical and thermally allowed Claisen rearrangement of allyl phenyl ether using UV and visible few-optical-cycle pulses, respectively. Allyl phenyl ether has an absorption band at shorter than 400 nm (Figure 11.1.6.8). Therefore, irradiation of allyl phenyl ether with UV pulses of 360–440 nm triggers electronic excitation, which is followed by photochemical Claisen rearrangement in the electronic excited state. Meanwhile, irradiation of allyl phenyl ether with visible pulses of 525–725 nm does not trigger electronic excitation. Therefore, the visible pulse triggers coherent molecular vibrations in the electronic ground state that should be followed by Claisen rearrangement in the electronic ground state.

First, pump–probe measurement of allyl phenyl ether using UV pulses was performed to observe the photochemical Claisen rearrangement. Time-resolved vibrational spectra were obtained by sliding-window Fourier transformation with a Blackman window function with a FWHM of 400 fs (Figure 11.1.6.9). Following irradiation with UV pulses, we observed only the molecular vibrations of the reactant allyl phenyl ether. Signals were assigned as the C=C stretching vibration ($\nu_{C=C}$) of the allyl group (1670 cm^{-1}), C–H$_2$ deformation vibration (δ_{C-H2}) of the methylene group (1482 cm^{-1}), C–H deformation vibration (δ_{C-H}) of the allyl group (1320 cm^{-1}), C–O–C symmetric stretching vibration ($\nu_{s\,C-O-C}$) of the ether group (1080 cm^{-1}), and C6 symmetric stretching vibration ($\nu_{s\,Ph}$) of the phenyl group (1007 cm^{-1}).

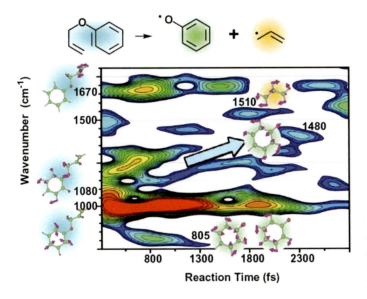

FIGURE 11.1.6.9 Spectrogram of photochemical Claisen rearrangement of allyl phenyl ether induced by UV few-optical cycle pulses [37].

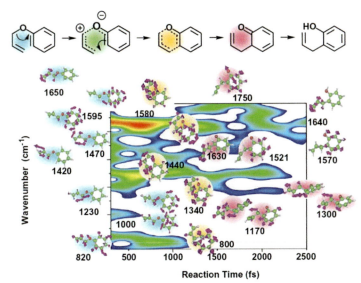

FIGURE 11.1.6.10 Spectrogram of nonthermal Claisen rearrangement in the electronic ground state of allyl phenyl ether induced by visible few-optical-cycle pulses [38].

The highest band at $1000\,cm^{-1}$ ($v_{s\,Ph}$) was separated into two modes. One exhibited a red shift toward $800\,cm^{-1}$, and the other did not move. The $v_{C=C}$ of the allyl group disappeared at a delay of about 1700 fs. At the same time, the allyl radical was generated, and a new band consistent with the C–C stretching vibration (v_{C-C}) of an allyl radical appeared at $1510\,cm^{-1}$. Moreover, $v_{s\,C-O-C}$ observed at $1080\,cm^{-1}$ just after UV-pulse irradiation, exhibited a gradual blue shift as the C–O group formed, and a new band (v_{C-O}) at $1480\,cm^{-1}$ appeared at a delay of about 2000 fs. These results also indicate that the active radical species were generated after about 2000 fs. Thus, irradiation with few-optical cycle pulses induces photoreaction by excitation of the electronic state if the laser spectrum overlaps with the absorption band of the sample [37].

Pump–probe measurement of allyl phenyl ether using visible pulses was then performed to observe thermally allowed Claisen rearrangement. The transient wavenumber changes of molecular vibration modes were analyzed by calculating spectrogram (see Figure 11.1.6.10). Molecular vibrational modes just after visible pulse irradiation are consistent with the reactant allyl phenyl ether. The signals δ_{CH2} ($1420\,cm^{-1}$), δ_{CH2} ($1230\,cm^{-1}$), $v_{s\,C-O-C}$ ($1030\,cm^{-1}$), and $v_{s\,Ph}$ ($1000\,cm^{-1}$) disappeared after a delay of about 700 fs. This implies that the C⁴–O bond is either weakened or broken in the first stage of the reaction. Next, $v_{C=C}$ of the phenyl group, observed at 1595 cm⁻¹ just after visible-pulse irradiation, exhibited a red shift toward 1500 cm⁻¹ in the probe-delay region from 500 to 750 fs. This indicates that a six-membered structure with aromatic C=C bonds formed in the second stage of the reaction. The $v_{C=C}$ of the allyl group at 1650 cm⁻¹ just after visible-pulse irradiation exhibited a red shift to 1580 cm⁻¹, which also supports the formation of a six-membered structure with aromatic C=C bonds. A new band that appeared at about 1750 cm⁻¹ after 1 ps delay can be assigned to $v_{C=O}$, which verifies that a keto-intermediate was generated in the third stage of the reaction. Instability of this keto-intermediate leads to keto-enol tautomerization as the final stage of the reaction. Therefore, after 2000 fs delay, $v_{C=O}$ disappeared and the phenol product was formed. The observed changes in wavenumber indicate that the reaction pathway of allyl phenyl ether [38] after visible-pulse irradiation is equivalent to that of allyl vinyl ether.

Comparison of Figures 11.1.6.9 and 11.1.6.10 reveals the following differences. When allyl phenyl ether was irradiated with visible pulses, $v_{s Ph}$ disappeared and $v_{C=O}$ appeared. In contrast, the $v_{s Ph}$ did not disappear under UV-pulse irradiation, and $v_{C=O}$ did not appear. These differences imply that [3,3]-sigmatropic rearrangement of allyl phenyl ether occurs under visible few-optical-cycle pulses. In addition, photo-allowed Claisen rearrangement occurs using UV few-optical-cycle pulses. These results certainly show that visible few-optical-cycle pulses induced nonthermal Claisen rearrangement in the electronic ground state of allyl phenyl ether. Nonthermal Claisen rearrangement in the electronic ground state without electronic excitation has the possibility of being either a thermal reaction or a third type of reaction triggered by a novel scheme that is different from both photo- or thermal reactions.

11.1.6.3.3 "NONPHOTO NONTHERMAL CLAISEN REARRANGEMENT" AND THERMAL CLAISEN REARRANGEMENT

Nonthermal Claisen rearrangement is different from thermal Claisen rearrangement in the following three ways. (i) Thermal Claisen rearrangement is reported to proceed with an activation energy of 20–40 kcal/mol [14,16,39–42], whereas nonthermal Claisen rearrangement proceeded in the electronic ground state under irradiation with a visible pulse (525–725 nm) with a bandwidth of just 5200 cm⁻¹ = 15 kcal/mol. (ii) Cleavage of the C–O bond took 700–800 fs in the spectrograms of both allyl vinyl ether (Figure 11.1.6.3) and allyl phenyl ether (Figure 11.1.6.10), which is much longer than the time suggested in the trajectory calculation. (iii) The keto-enol tautomerization observed in Figure 11.1.6.10 occurs much faster than that in typical chemical reactions.

To elucidate why (i) the reaction proceeds with such a low activation energy under visible ultrashort pulse irradiation, and (ii) cleavage of the C–O bond takes longer than expected, we studied the pump power dependence of the real-time trace (Figure 11.1.6.11a) and Fourier power spectrum (Figure 11.1.6.11c) of allyl vinyl ether in the reaction triggered by visible few-optical-cycle pulse irradiation.

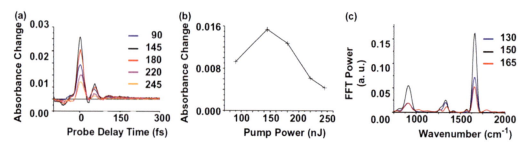

FIGURE 11.1.6.11 Pump power dependence of (a) the real-time trace of allyl vinyl ether in the reaction triggered by visible few-cycle-pulse irradiation and (b) ΔA averaged in the delay region between –20 and 20 fs. (c) Fourier power spectrum of allyl vinyl ether in the reaction triggered by visible few-cycle-pulse irradiation.

The traces of real-time vibrational amplitude show oscillations at a delay of 50–80 fs (see Figure 11.1.6.11a), which has maximum amplitude with a pump pulse of 150 nJ (see Figure 11.1.6.11b). If the excitation is caused by a three-photon absorption process, the Fourier power spectrum should be proportional to the cubic power of the pump power. However, the Fourier power spectrum exhibited maximum intensity when the sample was pumped with a pulse of 150 nJ (see Figure 11.1.6.11c). The observed pump power

dependency can be explained by assuming that when the pump power is higher than 150 nJ, allyl vinyl ether starts to be ionized by a nonlinear process, which suppresses the nonthermal reaction. The reason why the ionization is thought to proceed is as follows. Ionization generally occurs when a molecule is irradiated by light with an intensity of ~10^{14} W/cm^2 [43,44]. When the average laser power is 150 μW, its pulse energy is 150 nJ (=150 μW/kHz). Considering the pulse energy, a pulse duration of 5 fs, and focus area of 100 μm^2, the power density of the laser pulse can be calculated as 0.3×10^{14} W/cm^2 (=150 nJ/100 μm^2/5 fs). Therefore, it is thought that ionization starts to occur when the sample is pumped by a pulse of 150 nJ. We propose the following hypothesis to explain why nonthermal Claisen rearrangement and thermally allowed Claisen rearrangement have the three differences mentioned at the beginning of this section.

i. The reason the nonthermal Claisen rearrangement proceeds with much lower activation energy than that for thermal Claisen rearrangement can be explained by the two possible mechanisms explained below.

   The first possible mechanism is as follows. The impulsive excitation excites only Raman active modes in the sample molecules. Because of low cross-section of the Raman process, a small portion of the molecules are excited by the few-optical-cycle pulse. Raman active modes excited by pulsed excitation cause nonlinear interaction by dynamic mode coupling [45], resulting in the excitation of C–O stretching. In the case of incoherent thermal excitation, even though thermal energy is widely distributed over all of the molecules, the excited molecules have different small vibrational amplitudes with random phases. Under thermal excitation, high wave number vibration modes can be pumped by increasing system temperature. Such excitation generates a small population of high wavenumber modes together with a high population of low wavenumber modes following the Boltzmann distribution. Meanwhile, under coherent pulsed excitation, the Raman excitation populates states equally in the broadband laser bandwidth of 0.7 eV (=5200 cm^{-1}). Therefore, even though the low cross-section of the Raman process generates less total population than the case of thermal excitation, the pulsed excitation can efficiently excite high wavenumber vibration modes. The dynamic mode coupling in the excited molecule results in excitation of other vibration modes. The selection of vibration mode is controlled by the quantum probability in each excited molecule; i.e., the C–O stretching mode will be excited in some molecules, which triggers nonthermal Claisen rearrangement. Furthermore, this nonthermal distribution is formed in a single molecule within a few femtoseconds nonresonantly when a nonlinear process takes place by the combination of the two spectral components with a frequency difference corresponding to the vibration wavenumber. Therefore, by exciting the C–O stretching mode accidentally, C–O bond dissociation can take place, allowing the reaction to proceed with an activation energy much lower than that for thermal Claisen rearrangement.

   The second possible mechanism is as follows. The amplitude of the oscillation at 50–80 fs delay shown in Fig. 11a is proportional to the pump intensity when the ionization of allyl vinyl ether does not proceed (when the pump power is in the range of 90–100 nJ), which indicates that electrons are shaken by the pump pulse. It is thought that electrons of allyl vinyl ether are shaken in the direction of their dipole moments by the electronic field of the visible pulse when allyl vinyl ether is irradiated with a high-intensity visible pulse with an ultrashort pulse width that is much narrower than the period of molecular vibration. Thus, the electrons attached to C and O are driven to oscillate with the nuclei along the direction of the C–O bond axis to break this bond, which causes nonthermal Claisen rearrangement to proceed with a much lower activation energy than that of thermally allowed Claisen rearrangement.

ii. The reason why cleavage of the C–O bond does not occur instantaneously within a few vibrational periods of C–O stretching but at longer delay time of about 700–800 fs is considered to be as follows. Shaking of electrons together with C and O nuclei occurs by dynamic mode coupling between Raman active modes and the C–O stretching mode. Activated Raman vibrations take about several hundred femtoseconds to relax. It is not known how the shaking of electrons by an electric field of the ultrashort visible pulse changes the reaction pathway. Thus, the mechanism of nonthermal Claisen rearrangement induced by the irradiation of coherent ultrashort visible pulse may differ from that of the thermal Claisen rearrangement under incoherent thermal excitation, resulting in delayed cleavage of the C–O bond.

iii. The reason why the keto-enol tautomerization observed in Figure 11.1.6.10 occurs much faster than in common chemical reactions can be explained as follows. In the spectrogram of allyl phenyl ether (Figure 11.1.6.10), almost no vibration modes appear at longer delay than 2 ps, showing that most vibrational coherence is lost. In the spectrogram shown in Figure 11.1.6.10, keto-enol tautomerization proceeds 2.5 ps after excitation, which indicates that the small fraction of molecules in the sample that retains vibrational coherence may have isomerized even though the remaining major fraction of the molecules maintain their original keto structure.

11.1.6.4   CONCLUSION

In this work, we demonstrated that nonthermal Claisen rearrangement replaces thermally allowed Claisen rearrangement in the electronic ground state [46]. Even though it is not clear how similar these Claisen rearrangements are, this work confirms that Claisen rearrangement proceeds in the electronic ground state of a sample when molecular vibrations are excited by pumping with an ultrashort pulse that is much shorter than the molecular vibration period, with photon energy that is much lower than the minimum excitation energy. The collaborative research presented in this subsection was performed by the following people: Izumi Iwakura, Atsushi Yabushita, Jun Liu, Kotaro Okamura, Satoko Kezuka, and Takayoshi Kobayashi [46].

REFERENCES

1. L. Claisen, *Chem. Ber.* **45**, 3157 (1912).
2. F. E. Ziegler, *Chem. Rev.* **88**, 1423 (1988).
3. U. Nubbemeyer, *Synthesis* 961 (2003).
4. A. M. Martín Castro, *Chem. Rev.* **104**, 2939 (2004).
5. C. D. Hurd, and M. A. Pollack, *J. Am. Chem. Soc.* **60**, 1905 (1938).
6. F. W. Schuler, and G. W. Murphy, *J. Am. Chem. Soc.* **72**, 3155 (1950).
7. R. Hoffmann, and R. B. Woodward, *Acc. Chem. Res.* **1**, 17 (1968).
8. K. Fukui. *Acc. Chem. Res.* **4**, 57 (1971).
9. W. v. E. Doering, and W. R. Roth, *Tetrahedron* **18**, 67 (1962).
10. P. Vittorelli, T. Winkler, H.-J. Hansen, and H. Schmid, *Helv. Chim. Acta* **51**, 1457 (1968).
11. H.-J. Hansen, and H. Schmid, *Tetrahedron* **30**, 1959 (1974).
12. J. J. Gajewski, J. Jurayj, D. R. Kimbrough, M. E. Gande, B. Ganem, and B. K. Carpenter, *J. Am. Chem. Soc.* **109**, 1170 (1987).
13. R. L. Vance, N. G. Rondan, K. N. Houk, H. F. Jensen, W. T. Borden, A. Komornicki, and E. Winner, *J. Am. Chem. Soc.* **110**, 2314 (1988).
14. O. Wiest, K. A. Black, and K. N. Houk, *J. Am. Chem. Soc.* **116**, 10336 (1994).
15. H. Hu, M. N. Kobrak, C. Xu, and S. Hammes-Schiffer, *J. Phys. Chem. A* **104**, 8058 (2000).
16. J. G. Hill, P. B. Karadakov, D. L. Cooper, *Theor. Chem. Acc.* **115**, 212 (2006).

17. M. J. S. Dewar, and E. F. Healy, *J. Am. Chem. Soc.* **106**, 7127 (1984).
18. M. J. S. Dewar, and C. Jie, *J. Am. Chem. Soc.* **111**, 511 (1989).
19. J. J. Gajewski, and N. D. Conrad, *J. Am. Chem. Soc.* **101**, 2747 (1979).
20. L. Kupczyk-Subotkowska Jr., W. H. Saunders, H. J. Shine, and W. Subotkowski, *J. Am. Chem. Soc.* **116**, 7088 (1994).
21. M. M. Davidson, and I. H. Hillier, *J. Phys. Chem.* **99**, 6748 (1995).
22. J. J. Gajewski, *Acc. Chem. Res.* **30**, 219 (1997).
23. V. Aviyente, H. Y. Yoo, and K. N. Houk, *J. Org. Chem.* **62**, 6121 (1997).
24. H. Y. Yoo, and K. N. Houk, *J. Am. Chem. Soc.* **119**, 2877 (1997).
25. M. P. Meyer, A. J. DelMonte, and D. A. Singleton, *J. Am. Chem. Soc.* **121**, 10865 (1999).
26. A. Baltuska, T. Fuji, and T. Kobayashi, *Opt. Lett.* **27**, 306 (2002).
27. J. Liu, Y. Kida, T. Teramoto, and T. Kobayashi, *Opt. Expr.* **18**, 4672 (2010).
28. M. J. Frisch, G. W. Trucks, H. B. Schlegel, G. E. Scuseria, M. A. Robb, J. R. Cheeseman, J. A.Montgomery Jr., T. Vreven, K. N. Kudin, J. C. Burant, J. M. Millam, S. S. Iyengar, J. Tomasi, V. Barone, B. Mennucci, M. Cossi, G. Scalmani, N. Rega, G. A. Petersson, H. Nakatsuji, M. Hada, M. Ehara, K. Toyota, R. Fukuda, J. Hasegawa, M. Ishida, T. Nakajima, Y. Honda, O. Kitao, H. Nakai, M. Klene, X. Li, J. E. Knox, H. P. Hratchian, J. B. Cross, V. Bakken, C. Adamo, J. Jaramillo, R. Gomperts, R. E. Stratmann, O. Yazyev, A. J. Austin, R. Cammi, C. Pomelli, J. W. Ochterski, P. Y. Ayala, K. Morokuma, G. A. Voth, P. Salvador, J. J. Dannenberg, V. G. Zakrzewski, S. Dapprich, A. D. Daniels, M. C. Strain, O. Farkas, D. K. Malick, A. D. Rabuck, K. Raghavachari, J. B. Foresman, J. V. Ortiz, Q. Cui, A. G. Baboul, S. Clifford, J. Cioslowski, B. B. Stefanov, G. Liu, A. Liashenko, P. Piskorz, I. Komaromi, R. L. Martin, D. J. Fox, T. Keith, M. A. Al-Laham, C. Y. Peng, A. Nanayakkara, M. Challacombe, P. M. W. Gill, B. Johnson, W. Chen, M. W. Wong, C. Gonzalez, and J. A. Pople. *Gaussian 03, (Revision D.02)*, Gaussian, Inc., Wallingford CT. (2004).
29. I. Iwakura, A. Yabushita, and T. Kobayashi, *J. Am. Chem. Soc.* **131**, 688 (2009).
30. I. Iwakura, A. Yabushita, and T. Kobayashi, *Chem. Lett.* **39**, 374 (2010).
31. M. J. J. Vrakking, D. M. Villeneuve, and A. Stolow, *Phys. Rev. A* **54**, R37 (1996).
32. G. Herzberg, *Infrared and Raman Spectra*, Van Nostrand Reinhold, New York (1945).
33. I. Iwakura, *Phys. Chem. Chem. Phys.* **13**, 5546 (2011).
34. Fig. 11.1.6.5 is not for real bond-order of C_1–C_5. They are only for symbolically depicted to emphasize the differences. They are not linearly scaled either. Figure 5 was just shown as a schematic figure toemphasize that the spectrogram analysis has elucidated that the aromatic TS appears after impulsive excitation even though the bond order of C^4–O^3 starts to decrease earlier than the increase of the bond order of C^1–C^6.
35. M. S. Kharasch, G. Stampa, and W. Nudenberg, *Science* **116**, 309 (1952).
36. F. Galindo, *J. Photochem. Photobiol. C: Photochem. Rev.* **6**, 123 (2005).
37. I. Iwakura, A. Yabushita, J. Liu, K. Okamura, and T. Kobayashi, *Phys. Chem. Chem. Phys.* **14**, 9696 (2012).
38. I. Iwakura, A. Yabushita, and T. Kobayashi, *Chem. Phys. Lett.* **501**, 567 (2011).
39. S. Yamabe, S. Okumoto, and T. Hayashi, *J. Org. Chem.* **61**, 6218 (1996).
40. B. Gómez, P. K. Chattaraj, E. Chamorro, R. Contreras, and P. Fuentealba, *J. Phys. Chem. A* **106**, 11227 (2002).
41. J. J. Gajewski, and N. D. Conrad, *J. Am. Chem. Soc.* **101**, 6693 (1979).
42. C. J. Burrows, and B. K. Carpenter, *J. Am. Chem. Soc.* **103**, 6983 (1981).
43. S. M. Hankin, D. M. Villeneuve, P. B. Corkum, and D. M. Rayner, *Phys. Rev. Lett.* **84**, 5082 (2000).
44. S. M. Hankin, D. M. Villeneuve, P. B. Corkum, and D. M. Rayner, *Phys. Rev. A* **64**, 013405 (2001).
45. T. Kobayashi, A. Shirakawa, H. Matsuzawa, and H. Nakanishi, *Chem. Phys. Lett.* **321**, 85 (2000).
46. I. Iwakura A. Yabushita, J. Liu, K. Okamura, S. Kezuka, and T. Kobayashi, *Pure Appl. Chem.*, ASAP Article **85**, 1991–2004 (2013). DOI: 10.1351/PAC-CON-12-12-01

Section 11.2

Material Processing

11.2.1 Magnetization Dynamics and the Mn³⁺ *d-d* Excitation of Hexagonal HoMnO₃ Single Crystals Using Wavelength-Tunable Time-Resolved Femtosecond Spectroscopy

11.2.1.1 INTRODUCTION

Recently, the multiferroic manganites RMnO$_3$ have attracted great scientific attention due to their manifestations of intriguing and significant coupling between the magnetic and electric order parameters [1–6]. The coexistence of ferroic orders in RMnO$_3$ with a hexagonal smaller ionic radius of R = Sc, Y, and Ho-Lu or orthorhombic larger ionic radius of R = Eu-Dy structure not only gives rise to rich physics of the intimate interactions between charge, orbital, lattice, and spin degrees of freedom but also some fascinating emergent physical properties which might lead to significant potential applications [7,8]. For hexagonal HoMnO$_3$ (*h*-HMO), the paraelectric PE-ferroelectric FE transition occurs at Curie temperature $T_C \sim 875$ K while the long-range antiferromagnetic AFM order appears at a much lower Néel temperature $T_N \sim 76$ K [9].

Here the textbook-like explanation of Curie temperature and Néel temperature is described as follows.

Permanent magnetism is caused by the alignment of magnetic moments and induced magnetism is created when disordered magnetic moments are forced to align in an applied magnetic field. For example, the ordered magnetic moments (ferromagnetic, Figure 11.2.1.1) change and become disordered (paramagnetic, Figure 11.2.1.2) at the Curie temperature. Higher temperatures make magnets weaker, as spontaneous magnetism only occurs below the Curie temperature. Magnetic susceptibility above the Curie temperature can be calculated from the Curie–Weiss law, which is derived from Curie's law.

Antiferromagnetic materials are only antiferromagnetic below their corresponding Néel temperature or magnetic ordering temperature, T_N. This is similar to the Curie temperature as above the Néel Temperature the material undergoes a phase transition and becomes paramagnetic. That is, the thermal energy becomes large enough to destroy the microscopic magnetic ordering within the material.

DOI: 10.1201/9780429196577-82

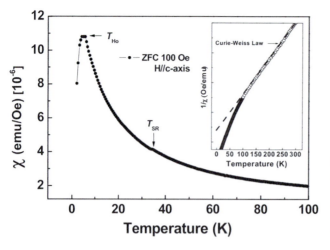

FIGURE 11.2.1.1 The temperature-dependent susceptibility T of h-HMO with a magnetic field of 100 Oe applied along c axis. The inset shows the inverse susceptibility. The dashed line: the Curie Weiss high-temperature extrapolation.

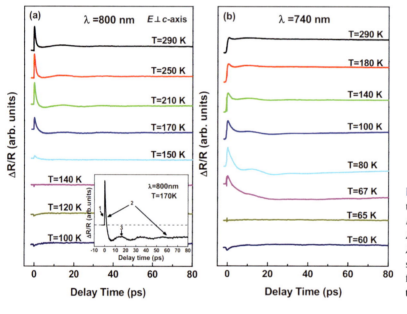

FIGURE 11.2.1.2 The temperature-dependent $\Delta R/R$ measured at (a) $\lambda = 800$ nm and (b) $\lambda = 740$ nm. The inset shows three primary features of dynamics in the $\Delta R/R$ curves.

Although it was revealed by dielectric constant and specific-heat measurements that the low-temperature magnetic phase diagram for h-HMO can be very complicated [10,11] due to the huge paramagnetic signal from Ho^{3+} ions, it is very difficult to unveil the AFM transition directly from the magnetization measurements. Alternatively, the second harmonic generation optical measurements or neutron scattering were used to delineate the magnetic phase transition and various spin arrangements [12,13]. It is conceived that the optical process can be a unique probe for exploring the change of magnetic properties due to its sensitivity to local magnetic ordering [14]. For instance, the interface effect between the multiferroic sample and the electrode which obscures the magneto-electric coupling could be ruled out in optical spectroscopy, when the light probes the sample quite directly without any electrodes [15]. Recently, infrared absorption spectroscopy measurements revealed a temperature-dependent blueshift of the Mn d to d transitions (energy difference E_{dd}) in hexagonal $RMnO_3$ materials [14,16]. In particular, an unexpected extra blueshift behavior at the AFM temperature is observed. The underlying mechanism giving rise to these blueshifts, however, remains to be a matter under debate. Souchkov et al. [17] proposed that the blueshift might originate from the immense magnetic exchange interaction between the Mn moments, and not from the effects of thermal expansion and/or magnetostriction. The magnetic ordering seems to be correlated with the main change of the electronic structure of the Mn ions in hexagonal RMnO$_3$ materials. Consequently, an appropriate method capable of simultaneously unveiling the d-d excitation and the associated magnetic ordering should provide some pivotal insights on these issues.

Fiebig et al. [18] demonstrated that the ultrafast magnetization dynamics in AFM compounds can be probed with the second harmonic generation SHG by nonlinear optical techniques. This kind of characterization was not possible by the usual magneto-optical methods, such as the transmission Faraday effect and the reflective Kerr effect, due to the absence of a macroscopic magnetization. Moreover, a magnetic resonance mode exhibited in multiferroic Ba$_{0.6}$Sr$_{1.4}$Zn$_2$Fe$_{12}$O$_{22}$ has been observed by Talbayev et al. [19] via transient reflectance measurement. Since the magnetic order in the manganites is mainly originated from the d electrons residing in the e_{2g} band, it is possible that by disturbing the configuration of the d electrons with optical excitations some basic insights can be obtained. In this respect, owing to the different characteristic time scales for various degrees of freedom existing in these complex materials, the time-resolved spectroscopy should serve as an ideal tool to resolve the fundamental microscopic dynamics for each order parameter as well as the coupling between them. In particular, the pump-probe studies can yield such important information related to the achievable switching speeds of the multiferroic order parameters.

In this subsection, we report the results of the ultrafast time-resolved evolution of magnetic effects in h-HMO single crystals probed by the wavelength-tunable femtosecond pump-probe technique. By manipulating the Mn^{3+} d-d carrier excitation with the tunable photon energy, the corresponding charge-spin coupling and magnetization dynamics were subsequently identified from various parts of the transient reflectivity change curve $\Delta R/R$, ΔR: the reflectivity change of probe pulses due to pump pulses, R: the reflectivity of probe pulses. Detailed analyses further evidence a temperature-dependent blueshift in the Mn^{3+} d-d transition and an apparent effect on the energy separation of the d-electron levels resulting from the emergence of the long-range and short-range AFM ordering.

11.2.1.2 EXPERIMENTS

The h-HMO single crystals were grown by a traveling solvent optical floating zone method. Figure 11.2.1.1 shows the typical temperature-dependent magnetization $\chi(T)$ of the platelet samples. Although the expected AFM ordering of Mn moments around 76 K is hardly recognized directly from the T curve displayed in Figure 11.2.1.1, a small but distinct kink occurs near 33 K. This has been previously identified as being due to the coupling between the onset of the Ho-AFM order and the Mn^{3+} moments causing the latter to rotate by an angle of 90° at T_{SR} 33 K. The magnetic transition at T_{Ho} 5 K, on the other hand, indicates the complete AFM order of the Ho^{3+} ions [20]. The results evidently indicate the quality and purity of the h-HMO crystal used in this study. For standard pump-probe measurements, a commercial mode-locked Ti:sapphire laser system providing short pulses 30 fs with the repetition rate of 80 MHz and tunable wavelengths from 740 to 815 nm $h = 1.68$–1.52 eV was used. The spectral width of the output pulses was adjusted to 25 nm full width at half maximum for all measurements. To perform the ultrafast spectroscopy, a standard pump-probe setup was employed with the fluences of 50 and 1 nJ/cm^2 for pump beam and probe beam, respectively. The pump beam was focused on the h-HMO single crystals with a diameter of 500 m, and the probe beam with a diameter of 300 m was overlapped with the spot of the pump beam. The polarizations of the pump beam and probe beam which were perpendicular to each other were parallel to the a-b plane of h-HMO single crystals i.e., $E c$ axis. A mechanical delay stage was used to vary the time delay between pump pulses and probe pulses. The reflectivity change of a probe beam was detected by using a photodiode detector and a lock-in amplifier.

11.2.1.3 RESULTS AND DISCUSSION

Figure 11.2.1.2 shows the typical temperature-dependent $\Delta R/R$ for the h-HMO crystals obtained at two different photon energies. Three primary features can be immediately identified in the $\Delta R/R$

curves, namely, the initial rising excitation component, the relaxation component, and the oscillating behavior see the inset of Figure 11.2.1.2a, process 1–3, respectively. For instance, for the case of = 800 nm shown in Figure 11.2.1.2a, the amplitude of the excitation component of R/R process 1 appears to remain constant at high temperatures until it starts to drop noticeably around $T = 170$ K. At $T \sim 150$ K the amplitude of the excitation component has diminished almost completely and with $T < 140$ K it even becomes negative, albeit only barely recognizable. On the contrary, the relaxation component process 2 shows an apparent overshoot to negative $\Delta R/R$ territory which appears to grow gradually with decreasing temperature. Finally, we note that the oscillating component process 3 behaves similarly at all temperatures and disappears altogether with the diminished excitation component. Similar results were observed for other pumping wavelengths [e.g., $\lambda = 740$ nm shown in Figure 11.2.1.2b], except for the temperature at which the amplitude of the excitation component vanishes. The oscillation in $\Delta R/R$ has been previously termed as the coherent acoustic phonon induced by the strain pulse. We note that the period of oscillation grows with the decreasing wavelength, which is consistent with those observed in hexagonal LuMnO$_3$ by Lim et al. [21].

In order to further explore the physical meaning of the excitation component of $\Delta R/R$ process 1, the amplitude of $\Delta R/R$ taken from Figure 11.2.1.2 at zero delay time as a function of temperature was measured with various pumping wavelengths of photon energies. Figure 11.2.1.3a shows that, for each photon energy used, the amplitude of $\Delta R/R$ remains essentially unchanged at higher temperature. It then drops precipitously to across zero amplitude at some characteristic temperature T_0 and becomes negative. For instance, for the case of $\lambda = 815$ nm, the amplitude of $\Delta R/R$ starts to drop steeply below 220 K and across zero at T_0 160 K. Since the photon energy is in the range of d-d transition for the h-RMnO$_3$ absorption peak 1.6 eV at room temperature [22], the absence of the excitation component is thus a clear indication of inadequate photon energy to trigger the d-d transition as depicted schematically in the inset of Figure 11.2.1.3a. The electrons residing on the e_{2g} orbital d_{xy} and $d_{x^2-y^2}$ can transfer to the unoccupied a_{1g} orbital $d_{3z^2-r^2}$ by absorbing pumping photons with energy exceeding E_{dd}. Conversely, this on-site Mn^{3+} E_{dd} transition will be blocked completely

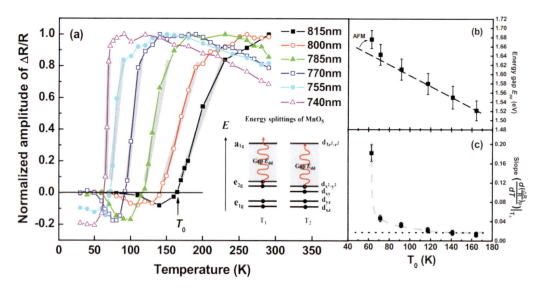

FIGURE 11.2.1.3 (a) The normalized amplitude of $\Delta R/R$ as a function of temperature at various wavelengths taken from Figure 11.2.1.2 at zero delay time. The insets illustrate the splitting of the Mn^{3+} energy levels in the local environment MnO$_5$ for photon energy above and below E_{dd} ($T_1 > T_2$) (b) The energy gap E_{dd} as a function of T_0. E_{dd} is estimated from the wavelengths in a and the error bars are the bandwidth of laser spectrum at various center wavelengths. The dashed line is a guide to the eye emphasizing the linear behavior at high-temperature range. (c) The slope of the temperature-dependent normalized amplitude of $\Delta R/R$ in a the gray thick lines as a function of T_0 at various wavelengths. Dashed line is a guide to the eye emphasizing the behavior of slope.

when the energy gap E_{dd} becomes larger than the energy of pumping photons, leading to the precipitous diminishing in the amplitude of $\Delta R/R$. The fact that T_0 gradually shifts to lower temperatures with increasing photon energy indicates that E_{dd} exhibits a blueshift with decreasing temperature Figure 11.2.1.3b. This is, in fact, consistent with that observed in other h-RMnO$_3$ materials by Fourier transform infrared or optical spectroscopic measurements [14,16,17].

Intuitively, the blueshift of E_{dd} might be attributed to the thermal contraction of the lattice and, hence, the enhanced crystal field effect on splitting the respective d orbitals. Nevertheless, recent investigations have indicated that the blueshift of E_{dd} might be correlated with mechanisms other than simply due to the crystal-field effect associated with the symmetry distortion of the local environment of MnO$_5$ bipyramids [22,23]. Souchkov et al. [17], based on fitting their absorption spectroscopy results and estimation of superexchange energy of h-LuMnO$_3$, argued that the unit cell volume change (only ~0.3%) caused by lowering the temperature from 300 to 2 K is certainly inadequate to fully account for the relatively large change (0.2 eV or about 10% change) observed in E_{dd}. Consequently, they suggested that the blueshift in E_{dd} might be due to the emerging superexchange interaction between neighboring Mn ions, which, in turn, gives rise to a lowering of the e_{2g} levels in the AFM state while the relatively isolated a_{1g} orbital is little affected [17]. They further proposed that even the short-range AFM correlations existing in the frustrated magnetic systems like h-RMnO$_3$ will result in a noticeable shift in the resonance energy. Within the context of this superexchange-induced effect, one expects that the blueshift in E_{dd} might turn on at temperatures much higher than the usually conceived (T_N ~76 K), when the system is in the dynamical short-range ordering state. Indeed, the optical process occurs regionally and it should be sensitive to short-range magnetic order inducing the change of the electronic structure. It is interesting to note that if we regard the photon energy as E_{dd} and plot it as a function of T_0 at which the excitation signal vanishes as shown in Figure 11.2.1.3b, a seemingly linear behavior dashed line in Figure 11.2.1.3b is evident, indicative of a gradual increase in the extent of AFM ordering. Moreover, the behavior starts to deviate from being linear around T_N (~76 K), suggesting an extra enhancement in the blueshift of E_{dd} due to the prevailing of global long-range AFM ordering.

In order to further explore the possible connections between the magnetic ordering state and the excitation component of $\Delta R/R$ the positive component of in $\Delta R/R$ (Figure 11.2.1.2), the slope of $\Delta R/R$ taken at temperatures slightly above $T_0[d(\Delta R/R_p)/dT$ at T_0, indicated by the gray thick lines in Figure 11.2.1.3a] as a function of the pumping photon energy is displayed in Figure 11.2.1.3c. If we attribute the blueshift of energy gap E_{dd} to being due to the short-range AFM ordering emerging at temperatures far above T_N [24], the results shown in Figure 11.2.1.3c can thus be regarded as indications of how the AFM correlation evolves with temperature as it approaches T_N. The fact that $d(\Delta R/R)_p/dT$ increases gradually with reducing temperatures and rises sharply around T_N, thus, follows closely with the emergence of short-range to long-range-ordered AFM with decreasing temperature in this frustrated magnetic system.

Another interesting feature of the $\Delta R/R$ is the relaxation part displayed in Figure 11.2.1.2 process 2 in the inset of Figure 11.2.1.2. In order to extract quantitative information and relaxation dynamics we adopted the widely used three-temperature model [25,26] to analyze the associated relaxation processes involved after excitation. In this three-temperature mode, the electrons, lattice, and spins are artificially separated by defining three independent temperatures T_e, T_l, and T_s that are interconnected by relaxation rates between electrons and lattice (τ_{e-l}). lattice and spins (τ_{l-s}), and electrons and spins (τ_{e-s}). After being excited by laser pulses, the temperature of electron system T_e increases rapidly due to the smaller specific heat ($C_{electron} \ll C_{lattice}$) [25]. Then, the electron-lattice interaction drives the system into thermal equilibrium, where electrons and lattice have the same temperature. As shown in Figure 11.2.1.4, the electron-lattice thermalization within a time scale of 1 ps is depicted schematically by the dashed red, T_e and the dash-dotted green, T_l lines. In the following, the spin system interacts with lattice and reaches thermal equilibrium with characteristic time scale τ_d the disordering time of the spin system through a spin-lattice relaxation process schematic illustration by the short-dashed and blue line T_s in Figure 11.2.1.4 [27]. This raise in the temperature of the spin

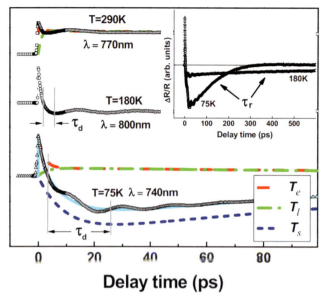

FIGURE 11.2.1.4 (a) The $\Delta R/R$ curves at various temperatures ($T = 290$ K, 180 K, and 75 K). Dashed lines in red and dashdotted lines in green show temperature of electrons (T_e) and lattice (T_l), respectively, due to the energy transfer by electron-lattice relaxation. The short-dashed lines in blue show the temperature of spin system (T_s) due to spin-lattice coupling. The inset shows the reordering time τ_r of the disordered magnetization.

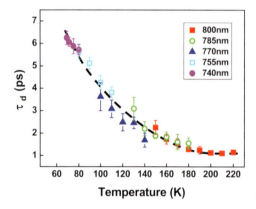

FIGURE 11.2.1.5 The temperature-dependent τ_d for various wavelengths. The dashed line is a guide to emphasizing the behavior of τ_d.

system and the resultant disruption effects in magnetic ordering give rise to a negative $\Delta R/R$ due to the H magnetic field-induced redshift in the optical conductivity spectra which is opposite to the blueshift of optical conductivity spectra at lower temperatures [16]. Additionally, in materials with magnetic ordering, it has been observed that the appearance of a negative component in $\Delta R/R$ due to metastable states around the quasiparticle state formed by the weak electron-phonon interaction is related primarily to the disturbance of magnetic ordering [28]. After the spin-lattice thermalization, the electron, lattice, and spin systems were further cooled with characteristic time scale τ_r through a cold reservoir, e.g., sample holder. It is worth to mention that additional energy relaxation channels such as electron-electron scattering and electron-spin coupling are possible. However, for insulating AFM materials such as h-RMnO$_3$, the energy relaxation channel via the electron spin coupling is largely blocked due to the absence of electron-electron scattering processes [29], which would be otherwise manifested as a sub-ps component in the relaxation process.

Figure 11.2.1.5 shows the temperature dependence of τ_d collected from the studies with various photon energies. It is evident that τ_d increases significantly from 1 ps at 220 K to 6 ps at 70 K, indicating the prominent role played by the state of magnetic ordering. Furthermore, the monotonic and seemingly universal behavior again indicates that it needs longer time to disturb the spin system when the magnetic ordering is more robust at lower temperatures due to the prevailing of the long-range AFM ordering. The magnetic ordering disturbed by spin-lattice coupling increase T_s has to be somehow reordered, i.e., the cooling of spin system. In this case, one expects that, due to the competition between the AFM superexchange interaction and thermal energy, the disturbed magnetic ordering should spend shorter time to reorder at lower temperatures. Indeed, as shown in the inset of Figure 11.2.1.4, we can identify that the characteristic reordering time τ_r is about 120 and 610 ps for $T = 75$ K and $T = 180$ K, respectively. Ju et al. [30] revealed the dynamics of the AFM ordering in the NiO layer by measuring the magnetization rotation in the NiFe/NiO multilayer system. They reported that the reordering time of demagnetization in AFM is on the time scale of 100 ps which is consistent with that obtained in the current pump-probe measurements.

11.2.1.4 SUMMARY

In summary, the ultrafast dynamics associated with the Mn^{3+} *d-d* excitation in *h*-HMO single crystals were studied by measuring *R/R* using the femtosecond spectroscopy [31]. The results clearly display several characteristic excitation and relaxation components with their own distinctive temperature and photon-energy dependences. The amplitude of the initial rising component of $\Delta R/R$ corresponding to the Mn^{3+} *d-d* excitation is caused by strong coupling between the electronic structure and AFM ordering. By combining with the analyses on the subsequent relaxation processes associated with magnetization dynamics, the detected blueshift is believed to originate primarily from the emergence of the AFM superexchange interaction between the neighboring Mn^{3+} ions. Research described in this section was conducted by collaboration among the following people: H. C. Shih, T. H. Lin, C. W. Luo, J.-Y. Lin, T. M. Uen, J. Y. Juang, K. H. Wu, J. M. Lee, J. M. Chen, and T. Kobayashi [31].

REFERENCES

1. T. Lottermoser, T. Lonkai, U. Amann, D. Hohlwein, J. Ihringer, and M. Fiebig, *Nature (London)* **430**, 541 (2004).
2. S.-W. Cheong, and M. Mostovoy, *Nat. Mater.* **6**, 13 (2007).
3. T. Kimura, T. Goto, H. Shintani, K. Ishizaka, T. Arima, and Y. Tokura, *Nature (London)* **426**, 55 (2003).
4. N. A. Spaldin, and M. Fiebig, *Science* **309**, 391 (2005).
5. T. H. Lin, H. C. Shih, C. C. Hsieh, C. W. Luo, J.-Y. Lin, J. L. Her, H. D. Yang, C.-H. Hsu, K. H. Wu, T. M. Uen, and J. Y. Juang, *J. Phys.: Condens. Matter* **21**, 026013 (2009).
6. R. Schmidt, W. Eerenstein, and P. A. Midgley, *Phys. Rev. B* **79**, 214107 (2009).
7. C. W. Nan, M. I. Bichurin, S. Dong, D. Viehland, and G. Srinivasan, *J. Appl. Phys.* **103**, 031101 (2008).
8. N. Hur, S. Park, P. A. Sharma, J. S. Ahn, S. Guha, and S.-W. Cheong, *Nature (London)* **429**, 392 (2004).
9. F. Yen, C. R. dela Cruz, B. Lorenz, Y. Y. Sun, Y. Q. Wang, M. M. Gospodinov, and C. W. Chu, *Phys. Rev. B* **71**, 180407R (2005).
10. B. Lorenz, A. P. Litvinchuk, M. M. Gospodinov, and C. W. Chu, *Phys. Rev. Lett.* **92**, 087204 2004.
11. F. Yen, C. dela Cruz, B. Lorenz, E. Galstyan, Y. Y. Sun, M. Gospodinov, and C. W. Chu, *J. Mater. Res.* **22**, 2163 (2007).
12. M. Fiebig, C. Degenhardt, and R. V. Pisarev, *J. Appl. Phys.* **91**, 8867 (2002).
13. A. Muñoz, J. A. Alonso, M. J. Martínez-Lope, M. T. Casáis, J. L. Martínez, and M. T. Fernández-Díaz, *Chem. Mater.* **13**, 1497 (2001).
14. W. S. Choi, S. J. Moon, S. Seok, A. Seo, D. Lee, J. H. Lee, P. Murugavel, T. W. Noh, and Y. S. Lee, *Phys. Rev. B* **78**, 054440 (2008).
15. R. Schmidt, W. Eerenstein, T. Winiecki, F. D. Morrison, and P. A. Midgley, *Phys. Rev. B* **75**, 245111 (2007).
16. R. C. Rai, J. Cao, J. L. Musfeldt, S. B. Kim, S.-W. Cheong, and X. Wei, *Phys. Rev. B* **75**, 184414 (2007).
17. A. B. Souchkov, J. R. Simpson, M. Quijada, H. Ishibashi, N. Hur, J. S. Ahn, S. W. Cheong, A. J. Millis, and H. D. Drew, *Phys. Rev. Lett.* **91**, 027203 (2003).
18. M. Fiebig, N. P. Duong, T. Satoh, B. B. V. Aken, Ke. Miyano, Y. Tomioka, and Y. Tokura, *J. Phys. D* **41**, 164005 (2008).
19. D. Talbayev, S. A. Trugman, A. V. Balatsky, T. Kimura, A. J. Taylor, and R. D. Averitt, *Phys. Rev. Lett.* **101**, 097603 (2008).

20. B. Lorenz, F. Yen, M. M. Gospodinov, and C. W. Chu, *Phys. Rev. B* **71**, 014438 (2005).
21. D. Lim, R. D. Averitt, J. Demsar, A. J. Taylor, N. Hur, and S. W. Cheong, *Appl. Phys. Lett.* **83**, 4800 (2003).
22. W. S. Choi, D. G. Kim, Sung Seok A. Seo, S. J. Moon, D. Lee, J. H. Lee, H. S. Lee, D.-Y. Cho, Y. S. Lee, P. Murugavel, J. Yu, and T. W. Noh, *Phys. Rev. B* **77**, 045137 (2008).
23. D.-Y. Cho, S.-J. Oh, D. G. Kim, A. Tanaka, and J.-H. Park, *Phys. Rev. B* **79**, 035116 (2009).
24. Th. Lonkai, D. G. Tomuta, and J.-U. Hoffmann, *J. Appl. Phys.* **93**, 8191 (2003).
25. R. D. Averitt and A. J. Taylor, *J. Phys.: Condens. Matter* **14**, R1357 (2002).
26. G. M. Müller, J. Walowski, M. Djordjevic, G.-X. Miao, A. Gupta, A. V. Ramos, K. Gehrke, V. Moshnyaga, K. Samwer, J. Schmalhorst, A. Thomas, A. Hütten, G. Reiss, J. S. Moodera, and M. Münzenberg, *Nat. Mater.* **8**, 56 (2009).
27. A. V. Kimel, A. Kirilyuk, A. Tsvetkov, R. V. Pisarev, and Th. Rasing, *Nature London* **429**, 850 (2004).
28. Y. H. Ren, M. Ebrahim, H. B. Zhao, G. Lüpke, Z. A. Xu, V. Adyam, and Q. Li, *Phys. Rev. B* **78**, 014408 (2008).
29. T. Ogasawara, K. Ohgushi, Y. Tomioka, K. S. Takahashi, H. Okamoto, M. Kawasaki, and Y. Tokura, *Phys. Rev. Lett.* **94**, 087202 (2005).
30. G. Ju, A. V. Nurmikko, R. F. C. Farrow, R. F. Marks, M. J. Carey, and B. A. Gurney, *Phys. Rev. Lett.* **82**, 3705 (1999).
31. H. C. Shih, T. H. Lin, C. W. Luo, J.-Y. Lin, T. M. Uen, J. Y. Juang, K. H. Wu, J. M. Lee, J. M. Chen, and T. Kobayashi, *Phys. Rev. B* **80**, 024427 (2009).

11.2.2 Ultrafast Thermoelastic Dynamics of HoMnO$_3$ Single Crystals Derived from Femtosecond Optical Pump–Probe Spectroscopy

11.2.2.1 INTRODUCTION

Multiferroic oxides have attracted a great deal of attention in recent years because of the strong coupling between the charge, spin, orbital, and lattice degrees of freedom and the accompanying rich and intriguing physics [1–8]. Since the relevant orderings can be manipulated through magnetoelectric coupling (the induction of magnetization by an electric field or of polarization by a magnetic field) [5,6], multiferroic oxides are expected to have great potential for application in the fields of oxide electronics and spintronics, and also in green energy devices designed for reduced power consumption. Therefore, an understanding of the coupling between electric and magnetic ordering is at the heart of realizing these applications of multiferroic oxides. Due to the intrinsic integration and strong coupling between ferroelectricity and magnetism, however, the physics of multiferroicity is extremely complicated. For instance, the linear magnetoelectric effect that was thermodynamically described within the Landau theory framework [9] cannot be observed in BiFeO$_3$ owing to the cycloid antiferromagnetic (AFM) structure that even its crystal symmetry allows. Once the cycloid AFM structure is destroyed by a magnetic field, the linear magnetoelectric effect can be clearly observed in the case of BiFeO$_3$ [10]. In contrast, the linear magnetoelectric effect is forbidden in hexagonal manganites HoMnO$_3$ and thus the detailed mechanism of the electric field control of the magnetic phase remains unclear. Until recently, the origin of their magnetoelectric phenomenon has been further interpreted as due to the so-called magnetoelastic effect [11], which eventually leads to an effective coupling between electric dipole moment and magnetic moment by virtue of combining the elastic coupling between ferroelectric (FE) polarization and strain with magnetic anisotropy. However, all of these anisotropic couplings have never been simultaneously observed in a single measurement.

In the last three decades or so, many studies have demonstrated that ultrafast optical spectroscopy can provide valuable insights into the microscopic dynamics and the underlying functional responses in complex materials. In particular, the ability to simultaneously probe the evolution of multiple degrees of freedom in the time domain, as well as the coupling between them, has made pump–probe spectroscopy a unique and powerful tool for investigating the dynamical properties of multiferroics. For instance, Ogasawara et al. [12] showed that the complex correlation between charge, lattice and spin in various ferromagnetic and ferrimagnetic compounds can be explicitly identified on different characteristic time scales by measuring the transient reflectivity or transmissivity changes.

In this subsection, we derived the ultrafast photo-induced electron and lattice dynamics of hexagonal HoMnO$_3$ (*h*-HMO) single crystals by using the wavelength-tunable femtosecond pump–probe technique. It was found that, around T_N, the anomalous thermoelastic effect in the *ab*-plane and

DOI: 10.1201/9780429196577-83

that along the *c*-axis unambiguously couple with the AFM and FE orderings, as reflected in various components of the temperature-dependent transient reflectivity changes ($\Delta R/R$).

11.2.2.2 EXPERIMENTS

The physical characteristics of the high-quality *h*-HMO single crystals used in this study have been described in detail previously [13]. Briefly, the *h*-HMO single crystals were grown by a traveling solvent optical floating zone method and examined by temperature-dependent magnetization measurements. For pump–probe measurements, we utilized a commercial mode-locked Ti:sapphire laser with 30fs and tunable wavelengths from 740 to 800nm ($h\nu = 1.68$–$1.55\,eV$) with a spectrum width of 25nm as the light source. The fluences of the pump beam and the probe beam were 0.18–0.79 and 0.05 μJ/cm^2, respectively. The pump beam was focused on the *h*-HMO single crystals at a spot with a diameter of 0.5 mm and the probe beam was focused on a spot with a diameter of 0.3 mm that overlapped the spot of the pump beam. The polarizations of the pump beam and the probe beam were perpendicular to each other and parallel to the *ab*-plane of the *h*-HMO single crystals (i.e. $E \perp c$-axis). Moreover, both beams were set in almost normal incidence. The transient reflectivity changes of the probe beam were detected by using a photodiode detector and a lock-in amplifier.

11.2.2.3 RESULTS AND DISCUSSION

11.2.2.3.1 TEMPERATURE- AND WAVELENGTH-DEPENDENT $\Delta R/R$

Figure 11.2.2.1 shows the typical temperature-dependent $\Delta R/R$ obtained at various wavelengths. The rising part of the $\Delta R/R$ curve reflects the d–d excitation of carriers from e_{2g} band to a_{1g} band induced by the optical pump pulses. Thus, the number of excited carriers due to a nonthermal process is related to the amplitude of $\Delta R/R$ at zero delay time (see Figure 11.2.3) [13]. After excitation,

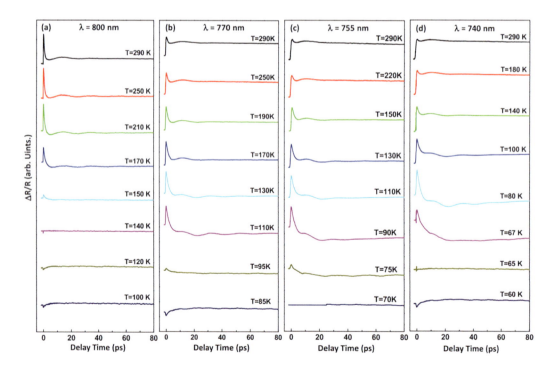

FIGURE 11.2.2.1   The photo-induced $\Delta R/R$ as a function of temperature at various wavelengths: (a) $\lambda = 800$ nm, (b) $\lambda = 770$ nm, (c) $\lambda = 755$ nm and (d) $\lambda = 740$ nm. The temperatures shown in this figure are the steady-state temperatures of samples.

the hot electrons accumulated in the a_{1g} band start to release their energy through thermal processes, such as electron–phonon collisions [14]. During the electron–phonon collision processes, the energy transferred from the hot electrons may generate the thermoelastic effect and may further amplify the lattice vibrations. Here the thermoelastic effect is described briefly like a textbook for the non-specialist as follows.

Thermoelasticity is a combination of elasticity and heat conduction. It is related to the impact of heat on the deformation of an elastic medium and the inverse impact of the deformation on the thermal condition of the considered medium. Thermal stress is produced when the rate (inverse of period) of variation of a heat source in a medium or the rate of variation of thermal boundary conditions on a medium is comparable with that of structural oscillation.

The above-mentioned relaxation processes ($t>0$) as reflected in $\Delta R/R$ are frequently and phenomenologically described by fitting with the following equation [15],

$$\frac{\Delta R}{R}(\lambda,T,t) = A_e(\lambda,T)e^{(-t/\tau_e)} + A_p(\lambda,T)\left(1 - e^{\left(-t/\tau_{p,r}\right)}\right)e^{-(t-\tau_{p,0})/\tau_{p,d}}$$

$$+ A_n(\lambda,T)\left(1 - e^{-t/\tau_{n,r}}\right)e^{-(t-\tau_{n,0})/\tau_{n,d}} + A_o(\lambda,T)e^{-t/\tau_o}\cos(\omega t - \phi)$$

(11.2.2.1)

The first term in the right-hand side of Eq. (11.2.2.1) represents the decay in the number $A_e(\lambda, T)$ of excited electrons with the relaxation time of τ_e. The second term represents the increase and decreases in the number $A_p(\lambda, T)$ of phonons with the rise time $\tau_{p,r}$ and decay time $\tau_{p,d}$, where $\tau_{p,0}$ is the starting time of phonon decay. Due to the electron–phonon coupling, the $\tau_{p,r}$ was set to be equal to τ_e in the data fitting.

The number of excited electrons decreases exponentially (dashed lines in Figure 11.2.2.2) due to the electron–phonon collisions. In the meanwhile, the number of phonons increases initially to respond to the energy gain from the electron–phonon collisions (short dashed lines in Figure 11.2.2.2) and then decays by losing energy to the environment. The third term describes the negative component associated with the magnetic ordering; $A_n(\lambda, T)$ is the strength of magnetic ordering, $\tau_{n,r}$ is the magnetic disordering time, $\tau_{n,d}$ is the magnetic reordering time and $\tau_{n,0}$ is the starting time of magnetic reordering. The final term describes the oscillation component associated with strain pulse propagation; $A_o(\lambda, T)$ is the amplitude of the oscillation component, τ_o is the damping time of the oscillation component, ω is the angular frequency of the oscillation component and φ is the initial phase of the oscillation component. Except for the oscillation term, Eq. (11.2.2.1) is consistent with the widely accepted three-temperature model [16,17], which completely reveals the characteristics of the $\Delta R/R$ curves at $T>T_N$ and $T<T_N$, as shown in Figure 11.2.2.2.

11.2.2.3.2 Attribution of the Negative Component in $\Delta R/R$

As shown in Figure 11.2.2.1, the level of the amplitude of $\Delta R/R$ in the long decay time (>40 ps) regime decreases with decreasing temperature and even overshoots into the 'negative' territory. Thus, in order to describe the lower value of $\Delta R/R$ at $t>5$ ps, an emerging 'negative component' is imposed as $A_n(\lambda, T)$ in Eq. (11.2.2.1) (see Figure 11.2.2.2b).

It is interesting to note that this negative component grows gradually with decreasing temperature and exhibits a dramatic change while the temperature is approaching T_N. This characteristic is clearly demonstrated by the data point (solid squares) displayed in Figure 11.2.2.3. Phenomenologically, a negative $\Delta R/R$ indicates that upon disturbance by the arrival of pump pulses at $t=0$, something happens to make the reflectivity of the probe pulses drop to below its original magnitude. One of the possible reasons for this dramatic reduction in reflectivity is an abrupt increase in absorption, which might be realized if there is a sudden increase of the empty density of states (DOS) within the energy range of the probe beam. As indicated previously, in hexagonal rare-earth manganites, the energy difference between the a_{1g} and e_{2g} bands (E_{dd}) increases with decreasing temperature and is accompanied by a rather significant extra blueshift near T_N [13,18–20]. After pumping, the

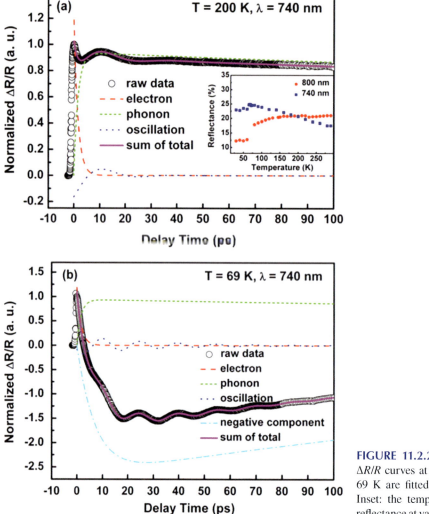

FIGURE 11.2.2.2 The selected $\Delta R/R$ curves at (a) 200 K and (b) 69 K are fitted by Eq. (11.2.2.1). Inset: the temperature-dependent reflectance at various wavelengths.

electron–phonon scattering will raise the lattice temperature and push the Mn atoms back to the position at higher temperatures and, in particular, disrupt the magnetic ordering. Consequently, a redshift of e_{2g} band with increasing temperature [13] will raise the available DOS in the a_{1g} band for absorbing more probe photons ($1A > 0$) with the same wavelength and lead to the emergence of negative $\Delta R/R$ ($1R \propto -1A$ and $A_n(\lambda, T)$ $6 = 0$ in Eq. (11.2.2.1)). Moreover, after a certain pumping fluence, $A_n(\lambda, T)$ increases with increasing pumping fluences and saturates at high fluences, as shown in the inset of Figure 11.2.2.3.

We note that the temperature dependence of the amplitude of negative $\Delta R/R$ [$A_n(\lambda, T)$] is very similar to the displacement of Mn atoms (solid circles in Figure 11.2.2.3) that was revealed by high-resolution neutron diffraction measurements recently [11]. In Ref. [11], it has been pointed out that the drastic position shift of the Mn atoms is directly associated with the magnetic ordering of the Mn moments and is regarded as strong evidence of magnetoelastic coupling. The large position shift of the Mn atoms, in turn, produces a further coupling to electric dipole moments [3] and changes the electronic structure of the system. Although it might appear coincident, the intimate similarity between the anomalous increase in the amplitude of the negative component of $\Delta R/R$ and the large displacement of Mn atoms around T_N (in Figure 11.2.2.3) strongly suggests that they originate from the same magnetoelastic effect. Namely, the dramatic changes in amplitude of the negative

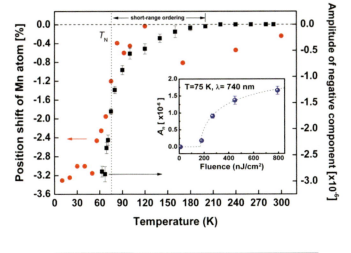

FIGURE 11.2.2.3 The amplitude of the negative component (solid squares) as a function of temperature. The solid circles are taken from Ref. [10]. Note: the T_N of YMnO₃ and HoMnO₃ are 70 and 76 K, respectively. Inset: the pumping fluence-dependent amplitude of the negative component A_n (740 nm, 75 K). The dashed line is a guide to the eyes. The temperatures shown in this figure are the steady-state temperatures of the samples.

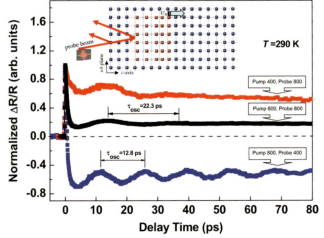

FIGURE 11.2.2.4 The photo-induced $\Delta R/R$ at various pump and probe wavelengths. Inset: schematic illustration of the oscillated $\Delta R/R$ due to a strain pulse.

component around T_N are due to disruption of the established long-range AFM ordering and the accompanying retraction of the magnetoelastic effect. Moreover, it is noted that the onset of the negative $\Delta R/R$ component appearing at temperatures well above T_N is attributed to the emergence of short-range AFM ordering [13].

11.2.2.3.3 ATTRIBUTION OF THE OSCILLATION COMPONENT IN $\Delta R/R$

Another interesting feature appearing in all of the $\Delta R/R$ curves is the damped oscillations (see the dotted lines in Figure 11.2.2.2), which are described by the last term of Eq. (11.2.2.1) and have been ubiquitously observed in various materials. Thomsen et al. [21,22] termed these damped oscillations as the coherent acoustic phonons (CAP) induced by the electronic and lattice stresses. In their model, the stress tensor contributed by the nonthermal and thermal origins can be written as $\sigma_{ij} = \sigma_e + \sigma_p$, with σ_e corresponding to the electronic stress and σ_p corresponding to the thermoelastic stress, respectively. In an isotropic medium, the thermoelastic stress can be expressed as $\sigma_p = -3K\beta 1T_l$, where K is the bulk elastic modulus, $1T_l$ is the lattice temperature rise and β is the linear thermal expansion coefficient. In this model [22], the oscillation of $\Delta R/R$ is caused by the interference between the probe beams reflected from the crystal surface and the rear interface of the propagating strain pulse with the modulated dielectric constant (or FE ordering), as illustrated schematically in the inset of Figure 11.2.2.4. The changes in reflectivity due to the strain can be obtained by solving the Maxwell equations and the closed-form formula for $\Delta R/R$ is as follows [22],

$$\frac{\Delta R(t)}{R} \propto \cos\left(\frac{4\pi n \upsilon_S t}{\lambda} - \phi\right) e^{-\upsilon_s t/\xi} \qquad (11.2.2.2)$$

where λ is the wavelength of the probe pulses, n is the refractive index of samples, υ_s is the speed of sound propagated in the medium and ξ is the penetration depth of probe pulses. Figure 11.2.2.4 shows the typical $\Delta R/R$ curves obtained by using different combinations of pump and probe wavelengths. It is evident that the frequency of the damped oscillations is solely dependent on the wavelength of the probe beam and is rather insensitive to that of the pump beam, which is consistent with the essential features of CAP generation described by the simplified equation,

$$\omega = \frac{4\pi n \upsilon_S}{\lambda} \qquad (11.2.2.3)$$

Figure 11.2.2.5a shows the temperature dependence of the oscillation period τ_{osc} probed by using various wavelengths. Here, τ_{osc} is defined as $\tau_{\mathrm{osc}} = 2\pi/\omega$ with ω obtained by fitting the data with

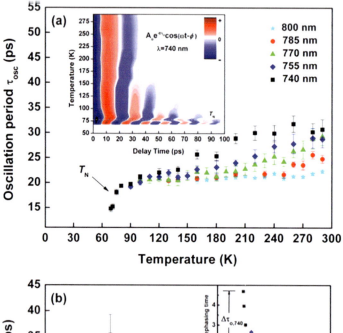

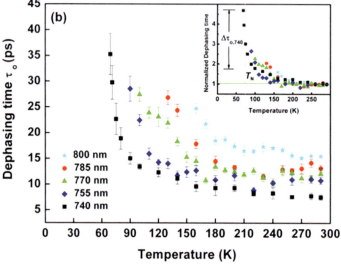

FIGURE 11.2.2.5 (a) The temperature-dependent oscillation period in $\Delta R/R$ at various wavelengths. Inset: the evolution of the oscillation component in $\Delta R/R$ at various temperatures. (b) The temperature-dependent dephasing time of the oscillation component in $\Delta R/R$ at various wavelengths. Inset: the normalized dephasing time as a function of temperature at various wavelengths. The temperatures shown in this figure are the steady-state.

Eq. (11.2.2.1). At room temperature, the τ_{osc} at 740 nm is larger than the τ_{osc} at 800 nm, which is the opposite of the prediction of $\tau_{osc}=\lambda/2nv_s$. Due to the increase in reflectance at 740 nm and the decrease in reflectance at 800 nm with decreasing temperatures shown in the inset of Figure 11.2.2.2a, the wavelength range from 800 nm (1.55 eV) to 740 nm (1.68 eV) is just located at the absorption peak of the optical conductivity spectra [20]. The giant difference between the optical conductivity (σ_1) values at 800 nm and 740 nm and the higher reflectance at 800 nm further leads to $(n_{800nm}/n_{740nm})>(800$ nm$/740$nm$)$.[1] With constant v_s, therefore, $\tau_{osc,800nm}$ would be smaller than $\tau_{osc,740nm}$. This result was also observed for LuMnO$_3$ by Lim et al. [23]. Moreover, the difference between $\tau_{osc,800nm}$ and $\tau_{osc,740nm}$ decreases with decreasing temperature owing to the n_{740nm} getting closer to n_{800nm} (see the inset of Figure 11.2.2.2a) (see footnote 1).

It is evident that, for all the probed wavelengths, τ_{osc} appears to decrease only slightly with decreasing temperature, which is consistent with a similar pump–probe spectroscopy study on hexagonal LuMnO$_3$ reported by Jang et al. [15] and can be interpreted as due to the increase in rigidity of the isostructure lattice as the temperature is lowered. The shortest accessible probe wavelength in our system was 740 nm, which was the only wavelength to delineate such a dramatic change in τ_{osc} for temperatures below T_N. According to Eq. (11.2.2.3), τ_{osc} is a function of the refractive index n of probe pulses and the sound velocity. However, at a wavelength of 740 nm, $\Delta_{n69-100K}/n_{100K}$ is about 2.7% (see footnote 1), which cannot explain the variation in −30% in $\Delta\tau_{osc,69-100K}/\tau_{osc,100K}$. Therefore, such significant shrinkage of τ_{osc} around T_N is most probably dominated by the dramatic increase in v_s along the c-axis owing to the stiffness of the lattice at low temperatures, which is consistent with the Lim et al. results for LuMnO$_3$ [23]. This strongly indicates that the propagation of a strain pulse with modulation of the dielectric constant along the c-axis should be affected by the FE properties with the FE polarization along the c-axis below T_c [3]. Indeed, the temperature-dependent dielectric properties of HoMnO$_3$ are drastically affected by the emergence of AFM ordering around T_N [3]. In any case, the AFM ordering in the ab-plane also influences the temperature-dependent oscillation period of $\Delta R/R$ along the c-axis through the magnetic–elastic coupling and then the elastic–ferroelectric coupling.

According to Eq. (11.2.2.2), the dephasing time τ_0 is proportional to the penetration depth ξ of probe pulses and the inverse sound velocity v_s, which is wavelength independent. At the same temperature, e.g. room temperature, $\tau_{0,800nm}$ is larger than $\tau_{0,740nm}$ because the penetration depth at 800 nm is larger than that at 740 nm in lossy media [24]. For all of the probing wavelengths, τ_0 almost remains constant above 180 K, which is associated with the starting temperature of short-range AFM ordering [13], as shown in Figure 11.2.2.5b and the inset. With decreasing temperature, a significant increase in τ_0 at 740 nm is clearly observed around T_N. To figure out this phenomenon, we consider the attenuation constant as a function of the dielectric constant (ε_1 and ε_2) in lossy media [24]. However, the shrinkage of penetration depth caused by a decrease of ε_1 at 1.68 eV (740 nm) [20] and an increase of ε_2 (estimated from the blueshift of σ_1 spectra in [20] through Kramers–Kronig relation) around T_N cannot explain the large enhancement ($\Delta\tau_{0,69-100K}/\tau_{0,100K}=162\%$) in the dephasing time τ_0. Thus, this dramatic growth of τ_0 around T_N should be dominated by the temperature-dependent v_s. Namely, v_s increases with decreasing temperature owing to the stiffness of the lattice at low temperatures, and it even overshoots around T_N owing to AFM ordering [23], which is consistent with the conclusion on the shrinkage of τ_{osc} around T_N. As mentioned before, Lee et al. [11] in their neutron diffraction experiments also observed extra displacement of the rare-earth and oxygen atoms along the c-axis at T_N. On the basis of the magnetic-ordering-induced atomic displacement, it was argued that the observed magnetoelastic effect is the primary origin of the eventual magneto-electric phenomenon in this intriguing class of materials.

---

[1] Estimated from the temperature-dependent reflectance in the inset of Figure 11.2.2.2a by the Fresnel equation at near normal incidence.

11.2.2.4   CONCLUSION

In the present study presented in this subsection, we have measured temperature-dependent transient reflectivity changes in hexagonal $HoMnO_3$ single crystals to study the photo-induced ultrafast thermoelastic dynamics [25]. The emergence of AFM ordering of Mn^{3+} ions at T_N was clearly delineated in the temperature-dependent evolution of the negative component in $\Delta R/R$ caused by the thermal strain induced by the incident pump beam, which might be the result of a redshift in the d–d transition. The oscillation period and dephasing time in $\Delta R/R$ caused by the strain pulse propagating along the c-axis also exhibit a similar dramatic change around the AFM ordering temperature, indicating that an accompanying variation in FE ordering might have occurred simultaneously.

REFERENCES

1. M. Eerenstein, N. D. Mathur, and J. F. Scott, *Nature* **442**, 759 (2006).
2. N. A. Spaldin, and M. Fiebig, *Science* **309**, 391 (2005).
3. C. dela Cruz, F. Yen, B. Lorenz, Y. Q. Wang, Y. Y. Sun, M. M. Gospodinov, and C. W. Chu, *Phys. Rev. B* **71** 060407 (2005).
4. S. Lee, A. Pirogov, J. H. Han J.-G Park, A. Hoshikawa, and T. Kamiyama, *Phys. Rev. B* **71**, 180413 (2005).
5. T. Lottermoser, T. Lonkai, U. Amann, D. Hohlwein, J. Ihringer, and M. Fiebig, *Nature* **430**, 541 (2004).
6. S.-W. Cheong, and M. Mostovoy, *Nat. Mater.* **6**, 13 (2007).
7. Z.-J. Huang, Y. Cao, Y. Y. Sun, Y. Y. Xue, and C. W. Chu, *Phys. Rev. B* **56**, 2623 (1997).
8. B. Lorenz, A. P. Litvinchuk, M. M. Gospodinov, and C. W. Chu *Phys. Rev. Lett.* **92**, 087204 (2004).
9. K. F. Wang, J.-M. Liu, and Z. F. Ren, *Adv. Phys.* **58**, 321 (2009).
10. Yu F. Popov, A. K. Zvezdin, G. P. Vorob'ev, A. M. Kadomtseva, V. A. Murashev, and D. N. Rakov, *JETP Lett.* **57**, 69 (1993).
11. S. Lee, *Nature* **451**, 805 (2008).
12. T. Ogasawara, K. Ohgushi, Y. Tomioka, K. S. Takahashi, H. Okamoto, M. Kawasaki and Y. Tokura, *Phys. Rev. Lett.* **94**, 087202 (2005).
13. H. C. Shih, T. H. Lin, C. W. Luo, J.-Y. Lin, T. M. Uen, J. Y. Juang K. H. Wu, J. M. Lee, J. M. Chen, and T. Kobayashi, *Phys. Rev. B* **80**, 024427 (2009).
14. P. Ruello, S. Zhang, P. Laffez, B. Perrin, and V. Gusev, *Phys. Rev. B* **79**, 094303 (2009).
15. K.-J. Jang, J. Lim, J. J. Ahn, J.-H. Kim, K.-J. Yee, J. S. Ahn, and S.-W. Cheong, *New J. Phys.* **12**, 023017 (2010).
16. R. D. Averitt, and A. J. Taylor, *J. Phys.: Condens. Matter* **14**, R1357 (2002).
17. G. M. Müller, *Nat. Mater.* **8**, 56 (2009).
18. A. B. Souchkov, J. R. Simpson, M. Quijada, H. Ishibashi, N. Hur, J. S. Ahn, S. W. Cheong, A. J. Millis, and H. D. Drew, *Phys. Rev. Lett.* **91**, 027203 (2003).
19. W. S. Choi, S. J. Moon, S, A. Seo, D. Lee, J. H. Lee, P. Murugavel, T. W. Noh, and Y. S. Lee, *Phys. Rev. B* **78**, 054440 (2008).
20. R. C. Rai J. Cao, J. L. Musfeldt, S.B. Kim, S.-W. Cheong, and X. Wei, *Phys. Rev. B* **75**, 184414 (2007).
21. C. Thomsen, J. Strait, Z. Vardeny, H. J. Maris, J. Tauc, and J. J. Hauseru, *Phys. Rev. Lett.* **53**, 989 (1984).
22. C. Thomsen, H. T. Grahn, H. J. Maris, and J. Tauc, *Phys. Rev. B* **34**, 4129 (1986).
23. D. Lim, R. D. Averitt, J. Demsar, A. J. Taylor, N. Hur, and S. W. Cheong, *Appl. Phys. Lett.* **83**, 4800 (2003).
24. D. K. Cheng, *Field and Wave Electronmagnetics*, Addison-Wesley, Reading, MA (1989).
25. H. C. Shih, T. H. Lin, C. W. Luo, J.-Y. Lin, T. M. Uen, J. Y. Juang, K. H. Wu, J. M. Lee, J. M. Chen, and T. Kobayashi, *New J. Phys*, **13**, 0503003 (2003).

11.2.3 Ultrafast Photoinduced Mechanical Strain in Epitaxial BiFeO$_3$ Thin Films

11.2.3.1 INTRODUCTION

Multiferroic materials possess ferroelastic, ferroelectric, and anti/ferromagnetic orders simultaneously [1] and are promising for the applications of next-generation devices with combined functionalities. Among various multiferroic candidates, BiFeO$_3$ (BFO) stands out because of its strong coupling between structural, ferroelectric, and antiferromagnetic orders at room temperature [2]. Recently, Rovillain et al. [3] demonstrated an important paradigm for magnonics through such strong coupling between magnetic and ferroelectric order parameters, i.e., the spin waves in BFO can be directly controlled by an electric field at room temperature. Moreover, the recent discoveries of photovoltaic effect [4], photo-induced size change [5,6], and photo-assisted THz emission [7] in BFO have received considerable attention because these nontrivial light-BFO interactions may open applications in optoelectronics and optomechanics, e.g., heterostructure diode [8], photovoltaic cells [9], deformable optical cavities [10]. While these discoveries are tantalizing, understanding the physics behind the photon-BFO interactions, including the dynamics of the photo-induced electronic excitation in BFO as well as its coupling with multiferroic orders, e.g., spin, orbit, and electric dipole, remains elusive and is yet to be studied. Femtosecond pump-probe spectroscopy has been established as a protocol to study the interactions between electrons, phonons, and magnons [11–14] and is therefore employed in this work to gain insight into the excitation dynamics in epitaxial BFO thin films.

In this letter, we have investigated the ultrafast photostriction effect in BFO thin films by dual-color pump-probe measurements. We found the anisotropic photostriction in BFO is mainly driven by the optical rectification effect which demonstrates that BFO would be a favorable material for the applications of ultrafast photoelastic, optoelectronic, and optomechanical devices through this non-thermal ultrafast process.

Here brief comments on the rectification and photostriction are described as follows.

Electro-optic rectification (EOR), also referred to as optical rectification, is a NLO process that consists of the generation of a quasi-DC polarization in a NLO medium at the passage of an intense optical beam. For typical intensities, optical rectification is a second-order phenomenon which is the inverse process of the electro-optic effect. This process can be considered to be $\omega - \omega = 0$ where ω and 0 correspond to the laser field and DC polarization. It is the process of reversed phase between the two components in the SHG process of $\omega + \omega = 2\omega$.

The basic definition of photostriction is the generation of strain by irradiation of light. The theory of photostriction can be described in more detail as the combination of the photovoltaic and the piezoelectric effects. The photovoltaic effect is that of light turning into electricity. The piezoelectric effect is that of electricity turning into mechanical motion. In a sense, the photostrictive effect is that of light turning directly into mechanical motion.

There is a reverse process to the photo striction. It is triboluminescence.

Triboluminescence is a phenomenon in which light is generated when a material is mechanically pulled apart, ripped, scratched, crushed, or rubbed (see tribology). The phenomenon is not fully understood, but appears to be caused by the separation and reunification of static electrical

DOI: 10.1201/9780429196577-84

charges. The term comes from the Greek τρίβειν ("to rub"; see tribology) and the Latin lumen (light). Triboluminescence can be observed when breaking sugar crystals and peeling adhesive tapes. Triboluminescence is often used as a synonym for fractoluminescence (a term sometimes used when referring only to light emitted from fractured crystals). Triboluminescence differs from piezoluminescence in that a piezoluminescent material emits light when it is deformed, as opposed to broken. These are examples of mechanoluminescence, which is luminescence resulting from any mechanical action on a solid.

11.2.3.2 EXPERIMENTAL

Samples used in this study are epitaxial BFO (110) thin films grown on $SrTiO_3$ (110) single crystal substrates by pulsed laser deposition. Film thickness was carefully controlled via tuning the deposition time and determined by the x-ray reflectivity technique. Details of the deposition procedure have been reported elsewhere in Ref. [15]. The femtosecond spectroscopy measurement was performed using a commercial Ti:sapphire laser (repetition rate: 5.2 MHz, wavelength: 800 nm, pulse duration: 70 fs) and a homemade dual-color pump-probe system with the standard lock-in technique at 260 K. The fluences of the pump beam and the probe beam are 2 and 0.1 μJ/cm^2, respectively. The pump pulses have corresponding photon energy (3.1 eV) beyond the band gap of BFO (2.67 eV) [16] and hence can generate electronic excitations within 40 nm due to the absorption length at k¼400nm [17]. Excitation dynamics is studied by measuring the photoinduced transient reflectivity changes ($\Delta R/R$) of the probe beam with photon energy of 1.55 eV in the depth of >1μm due to the large absorption length at k¼800nm in BFO [17].

11.2.3.3 RESULTS AND DISCUSSION

Figure 11.2.3.1 shows the photoinduced $\Delta R/R$ in a BFO thin film. Electronic excitations generated by the pump pulses result in a swift rise of $\Delta R/R$ at zero time delay. The observed excitation is triggered by transferring the electrons from 2p valence band of O to p conduction band of Bi [17,18]. At zero time delay, the number of excited electrons generated by this non-thermal process is related to the amplitude of $\Delta R/R$. These high-energy electrons accumulated in the p conduction band of Bi release their energy through the emission of longitudinal-optical (LO) phonons within several picoseconds [19]. The LO phonons further decay into acoustic phonons via anharmonic interactions, i.e., transferring energy to the lattice. This relaxation process can be detected using a probe beam, indicated by

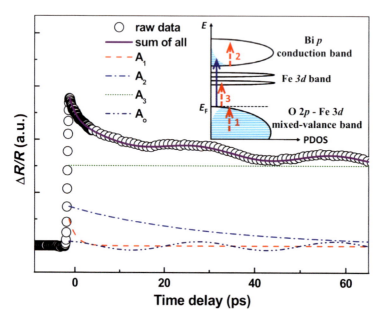

FIGURE 11.2.3.1 The typical $\Delta R/R$ curve is fitted by Eq. (11.2.3.1). The inset schematically shows the electronic band structure of BFO (see Refs. [17] and [18]) and the pump-probe processes for 3.1-eV-pump (solid arrow) and 1.55-eV-probe (dashed arrows).

processes 1 and 2 in the inset of Figure 11.2.3.1. Special care should be taken because the probe beam could be absorbed by additional electronic transitions in BFO. It has been known that the on-site *d-d* transition of Fe^{3+} ions (process 3 in the inset of Figure 11.2.3.1), which should be forbidden due to the total spin of change from S = 5/2 to S = 3/2 [16], can occur in BFO because the parity selection rule is relaxed through the spin-orbit coupling and the octahedral distortion caused by pump pulses [16].

The relaxation processes (*t* > 0) represented by $\Delta R/R$ in BFO display an oscillated feature, which is associated with the strain pulse in oxides [20] and can be phenomenologically described by

$$\frac{\Delta R}{R} = A_1 e^{-t/\tau_1} + A_2 e^{-t/\tau_2} + A_3 + A_o e^{-t/\tau_o} \cos(2\pi / T + \phi) \qquad (11.2.3.1)$$

The 1st term in the right-hand side of Eq. (11.2.3.1) is the decay of the excited electrons with an initial population number, A_1, and a relaxation time, τ_1 (dashed line in Figure 11.2.3.1). The 2nd term is the *d-d* transition of Fe^{3+} ions (process 3) with an absorption probability, A_2, and a corresponding decay time, τ_2 (dash-dotted line in Figure 11.2.3.1). The 3rd term describes energy loss from the hot spot to the ambient environment within the time scale of microsecond, which is far longer than the period (150 ps) of the measurement and hence presented as a constant (dotted line in Figure 11.2.3.1). The last term is the oscillation component associated with strain pulse propagation: A_o is the amplitude of the oscillation (dash-dot-dotted line in Figure 11.2.3.1); τ_o is the damping time; *T* is the period; / is the initial phase of the oscillation.

Figure 11.2.3.2a shows $\Delta R/R$ measurements on BFO thin films with various thicknesses (40–360 nm). A discontinuity is clearly observed in the oscillation feature of $\Delta R/R$ after subtracting the decay background (i.e., the 1st, 2nd, and 3rd terms in Eq. (11.2.3.1)), as shown in the right inset of Figure 11.2.3.2a. The amplitude and period of the first oscillation are both larger than those of the second oscillation. This oscillation is caused by the propagation of strain pulses inside the BFO thin film, namely the interference between the probe beams reflected by the thin film surface and the wavefront of the propagating strain pulse as illustrated by the cartoon in the inset of Figure 11.2.3.2a [21].

When strain pulses propagate through the BFO/STO interface, both the amplitude and period of the $\Delta R/R$ oscillation decrease. Consequently, the time (denoted by $t_{\text{BFO/STO}}$) needed to observe the oscillation

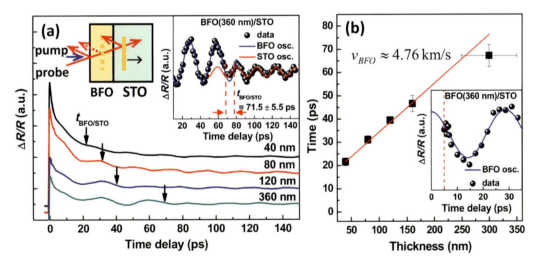

FIGURE 11.2.3.2 (a) The $\Delta R/R$ on (110) BFO thin films with various thicknesses. The arrows indicate the BFO/STO interface. Left inset: schematic illustration of the propagation of strain pulse inside BFO and STO substrate. Right inset: the oscillation signal was obtained by subtracted the decay background (the 1st–3rd terms in Eq. (11.2.3.1)) from $\Delta R/R$ of (a). Solid lines are the sinusoidal fitting. (b) The thickness-dependent strain pulse propagating time ($t_{\text{BFO/STO}}$) through the interface between BFO and STO. Solid line is the linear fitting curve. Inset shows a part of the right inset in (a) on an enlarged scale.

discontinuity can be used to extract the speed of the strain pulse, which is equivalent to the sound velocity, in BFO. Results from this experiment show a linear dependence of $t_{BFO/STO}$ on BFO thickness (Figure 11.2.3.2b). The corresponding strain pulse (or sound) velocity, v_{BFO}, along the [110] direction of BFO is estimated to be 4.76 km/s. The strain pulse velocity can be also calculated by using the strain pulse model [21].

$$v_s = \lambda_{probe} \cos\theta / 2n_{probe} T, \tag{11.2.3.2}$$

where λ_{probe} is the wavelength of probe beam; n_{probe} is the refractive index in probing wavelength; h is the refractive angle of probe beam in samples; T is the period of oscillation signal in $\Delta R/R$. Using $\lambda_{probe} = 800$ nm, $n_{probe} = 2.8$ (Ref. [22]), $\theta = 3.6$ (estimated from the incident angle (10) of the probe beam by Snell's law) and $T = 29.2$ ps, the strain pulse velocity, v_s, is calculated to be 4.88 km/s, which is very close to the result, $v_{BFO} = 4.76$ km/s, obtained from our thickness dependent $t_{BFO/STO}$ measurement. Recently, Smirnova et al. [23] also obtained the strain velocity of 4.31 km/s in BFO ceramics at 300 K by using the pulse-echo technique at a frequency of 10 MHz. A close look at the $\Delta R/R$ signal around zero time delay (shown in the inset of Figure 11.2.3.2b) reveals that the oscillation signal starts within the time scale of ps. This suggests that the time needed to generate strain pulses in BFO is within the time scale of ps. The ultrafast generation of strain pulses in BFO posts interesting questions on the mechanisms that drive the ultrafast photostriction in BFO.

To explore the origin and physics behind the ultrafast photostriction in BFO, further studies of the dependence of $\Delta R/R$ on the azimuth angle (ϕ) as well as the laser polarization angle (θ) have been carried out (Figure 11.2.3.3). The set-up of the $\Delta R/R$ measurements with a variable azimuth angle is illustrated in Figure

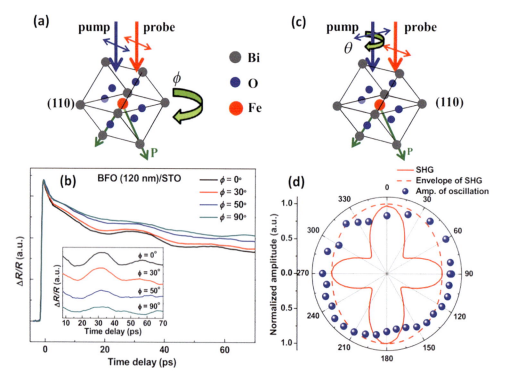

FIGURE 11.2.3.3 (a) The experimental configuration for (b). (b) The measurements of the azimuth angle (ϕ) dependence of $\Delta R/R$ in (110) BFO thin films. Inset: the oscillation signal was obtained by subtracted the decay background (the 1st–3rd terms in Eq. (11.2.3.1)) from $\Delta R/R$ of (b). (c) The experimental configuration for (d). (d) The amplitude of oscillation signal of BFO (in the inset of Figure 11.2.3.2b) in $\Delta R/R$ and the intensity of the second harmonic generation as a function of the polarization angle (θ) of pump beam. 0: the polarizations of both pump and probe beams are parallel to the in-plane component of electric polarization (P) in a (110) BFO thin film. The red-solid line is the θ-dependent SHG intensity. The red-dashed line is the envelope of SHG intensity, and its symmetry is similar to the anisotropic photostriction (solid dots).

11.2.3.3a. Results show that $\Delta R/R$ at $t = 0$ s does not change with the azimuth angle of the substrate (Figure 11.2.3.3b), indicating that the photo-induced excitation is isotropic in BFO thin films. Namely, the total number of photo-excited electron-hole pairs keeps constant with varying the azimuth angle of substrate.

According to the scenarios commonly used as the explanations for strain pulse generation in polar materials, including (i) electron-hole deformation potential mechanism induced by pump pulses, (ii) the electrostrictive effect created by separating the electron-hole pairs, and (iii) the thermal expansion of lattice due to hot carriers transfer energy to lattice [24], the strain pulses generated from photostriction in BFO should be isotropic when the pumping fluence used is the same so that the photoexcited electron-hole pairs are kept constant in number. However, our results show that the oscillation amplitude of $\Delta R/R$ is not isotropic but strongly depends on l, as manifested in the inset of Figure 11.2.3.3b which can be further separated into the isotropic part and anisotropic part with nodes. This finding indicates that the aforementioned models still work for the isotropic strain pulse generation in BFO but are insufficient to explain the observed anisotropic $\Delta R/R$ oscillation in our experiments, and therefore alternative explanations are needed.

Further insight into the anisotropic photostriction in BFO is provided by the study of the correlation between the oscillation amplitude of $\Delta R/R$ and the ferroelectric polarization of the BFO thin films. To avoid the birefringence effect in the (110) BFO thin film, which also contributes to dependent oscillation signals, the polarization of the probe beam and the angle of the sample were both fixed. Only the polarization of the pump beam is rotated (Figure 11.2.3.3c). The oscillation amplitude was found to be maximum when the polarization of the pump beam is rotated 90 and 270 against the in-plane component of the ferroelectric polarization (P), while the minimum of the oscillation amplitude is observed at rotation angles of 0 and 180, as shown in Figure 11.2.3.3d. The observed two-fold symmetry of photostriction in BFO with a minimum-to-maximum ratio of 21% is consistent with the results obtained by Kundys et al. from bulk BFO crystals [5]. The same symmetry and minimum-to-maximum ratio (19%) are also observed in the envelope of second harmonic generation (SHG) pattern [25] in (110) BFO thin films, strongly indicating the anisotropic photostriction effect and the SHG in (110) BFO thin films share a common physical origin.

In nonlinear materials, the second-order polarization (P) can be described by

$$P^{(2)}(t) = \varepsilon_0 \chi^{(2)} E(t) E^*(t) = P_0^{(2)}(t) + P_{2\omega}^{(2)}(t), \qquad (11.2.3.3)$$

where $\chi^{(2)}$ is the nonlinear susceptibility; $E(t)$ and $E^*(t)$ are the optical electric fields. The 1st term at the right side of Eq. (11.2.3.3) is the optically induced polarization to acquire a dc term, i.e., the so-called optical rectification effect [26], and the 2nd term is associated with the SHG. Here, we argue that the optical rectification is responsible for the ultrafast anisotropic photostriction in BFO based on that the optical rectification and the SHG in nonlinear materials are derived from a common nonlinear susceptibility ($\chi^{(2)}$), which is a function of ferroelectric polarization [27]. While an intensity-modulated laser pulse would produce a dc electric field inside materials within the pulse duration of femtosecond time scale, ultrafast strain stress in BFO can be generated via electrostrictive effect which is anisotropic due to the specific direction of ferroelectric polarization (P) as shown in Figures 11.2.3.3a and 11.2.3.3c. Moreover, the nodal feature in the symmetry of optical rectification like the SHG pattern in Figure 11.2.3.3d is smeared by the isotropic excitation of electron-hole pairs, which is revealed in the indifferentiable $\Delta R/R$ at $t = 0$ s for various azimuth angles in Figure 11.2.3.3b, to cause the nodeless photostriction pattern in Figure 11.2.3.3d. Our finding of the optical rectification-driven ultrafast anisotropic photostriction with the time scale of optical coherence (pulse duration) in BFO is far shorter than that reported by Kundys et al. (<0.1 s, obtained under the limited time resolution of measuring systems [5]).

11.2.3.4   CONCLUSION

In summary, we have studied the ultrafast dynamics and photostriction in (110) BFO/STO thin films by dual-color femtosecond spectroscopy. By varying the thin-film thickness, which effectively changes strain pulse propagation time, the sound velocity along [110] direction of BFO thin films is obtained and found to be 4.76 km/s. The ultrafast anisotropic photostriction in BFO is found to be mainly derived from the optical rectification effect. It is expected that our results provide a basic understanding of how photon interacts with multiferroicity in BFO and opens pathways to design ultrafast device with multifunctionality [28].

REFERENCES

1. N. A. Hill, *J. Phys. Chem. B* **104**, 6694 (2000).
2. T. Zhao, A. Scholl, F. ZavaLiche, K. Lee, M. Barry, A. Doran, M. P. Cruz, Y. H. Chu, C. Ederer, N. A. Spaldin, R. R. Das, D. M. Kim, S. H. Baek, C. B. Eom, and R. Ramesh, *Nat. Mater.* **5**, 823 (2006).
3. P. Rovillain, R. de Sousa, Y. Gallais, A. Sacuto, M. A. Masson, D. Colson, A. Forget, M. Bibes, A. Barthlmy, and M. Cazayous, *Nat. Mater.* **9**, 975 (2010).
4. T. Choi, S. Lee, Y. J. Choi, V. Kiryukhin, and S.-W. Cheong, *Science* **324**, 63 (2009).
5. B. Kundys, M. Viret, D. Colson, and D. O. Kundy, *Nat. Mater.* **9**, 803 (2010).
6. B. Kundys, M. Viret, C. Meny, V. Da Costa, D. Colson, and B. Doudin, *Phys. Rev. B* **85**, 092301 (2012). http://dx.doi.org/10.1103/PhysRevB.85.092301
7. D. S. Rana, I. Kawayama, K. Mavani, K. Takahashi, H. Murakami, and M. Tonouchi, *Adv. Mater.* **21**, 2881 (2009). http://dx.doi.org/10.1002/adma.200802094
8. S. R. Basu, L. W. Martin, Y. H. Chu, M. Gajek, R. Ramesh, R. C. Rai, X. Xu, and J. L. Musfeldt, *Appl. Phys. Lett.* **92**, 091905 (2008).
9. S. Y. Yang, L. W. Martin, S. J. Byrnes, T. E. Conry, S. R. Basu, D. Paran, L. Reichertz, J. Ihlefeld, C. Adamo, A. Melville, Y.-H. Chu, C.-H. Yang, J. L. Musfeldt, D. G. Schlom, J. W. Ager III, and R. Ramesh, *Appl. Phys. Lett.* **95**, 062909 (2009).
10. I. Favero, and K. Karrai, *Nat. Photonics* **3**, 201 (2009).
11. C. V. K. Schmising, M. Bargheer, M. Kiel, N. Zhavoronkov, M. Woerner, T. Elsaesser, I. Vrejoiu, D. Hesse, and M. Alexe, *Phys. Rev. Lett.* **98**, 257601 (2007).
12. D. Talbayev, S. A. Trugman, A. V. Balatsky, T. Kimura, A. J. Taylor, and R. D. Averitt, *Phys. Rev. Lett.* **101**, 097603 (2008).
13. C. W. Luo, I. H. Wu, P. C. Cheng, J.-Y. Lin, K. H. Wu, T. M. Uen, J. Y. Juang, T. Kobayashi, D. A. Chareev, O. S. Volkova, and A. N. Vasiliev, *Phys. Rev. Lett.* **108**, 257006 (2012).
14. H. C. Shih, T. H. Lin, C. W. Luo, J.-Y. Lin, T. M. Uen, J. Y. Juang, K. H. Wu, J. M. Lee, J. M. Chen, and T. Kobayashi, *Phys. Rev. B* **80**, 024427 (2009).
15. Y.-H. Chu, M. P. Cruz, C.-H. Yang, L. W. Martin, P.-L. Yang, J.-X. Zhang, K. Lee, P. Yu, L.-Q. Chen, and R. Ramesh, *Adv. Mater.* **19**, 2662 (2007).
16. X. S. Xu, T. V. Brinzari, S. Lee, Y. H. Chu, L. W. Martin, A. Kumar, S. McGill, R. C. Rai, R. Ramesh, V. Gopalan, S. W. Cheong, and J. L. Musfeldt, *Phys. Rev. B* **79**, 134425 (2009).
17. M. O. Ramirez, A. Kumar, S. A. Denev, N. J. Podraza, X. S. Xu, R. C. Rai, Y. H. Chu, J. Seidel, L. W. Martin, S.-Y. Yang, E. Saiz, J. F. Ihlefeld, S. Lee, J. Klug, S. W. Cheong, M. J. Bedzyk, O. Auciello, D. G. Schlom, R. Ramesh, J. Orenstein, J. L. Musfeldt, and V. Gopalan, *Phys. Rev. B* **79**, 224106 (2009).
18. H. Wang, Y. Zheng, M.-Q. Cai, H. Huang, and H. L. W. Chan, *Solid State Commun.* **149**, 641 (2009).
19. F. S. Krasniqi, S. L. Johnson, P. Beaud, M. Kaiser, D. Grolimund, and G. Ingold, *Phys. Rev. B* **78**, 174302 (2008).
20. H. C. Shih, L. Y. Chen, C. W. Luo, K. H. Wu, J.-Y. Lin, J. Y. Juang, T. M. Uen, J. M. Lee, J. M. Chen, and T. Kobayashi, *New J. Phys.* **13**, 053003 (2011).
21. C. Thomsen, H. T. Grahn, H. J. Maris, and J. Tauc, *Phys. Rev. B* **34**, 4129 (1986).
22. A. Kumar, R. C. Rai, N. J. Podraza, S. Denev, M. Ramirez, Y.-H. Chu, L. W. Martin, J. Ihlefeld, T. Heeg, J. Schubert, D. G. Schlom, J. Orenstein, R. Ramesh, R. W. Collins, J. L. Musfeldt, and V. Gopalan, *Appl. Phys. Lett.* **92**, 121915 (2008).
23. E. P. Smirnova, A. Sotnikov, S. Ktitorov, N. Zaitseva, H. Schmidt, and M. Weihnacht, *Eur. Phys. J. B* **83**, 39 (2011).
24. P. Babilotte, P. Ruello, T. Pezeril, G. Vaudel, D. Mounier, J.-M. Breteau, and V. Gusev, *J. Appl. Phys.* **109**, 064909 (2011).
25. The envelope of SHG intensity in Fig. 3(d) with dashed line is plotted by connecting the peaks of the four-fold symmetry pattern in SHG intensity (solid line in Fig. 3(d)) which was also observed in $BiMnO_3$ thin films (Ref. 27).
26. P. N. Butcher, and D. Cotter, *The Elements of Nonlinear Optics, Cambridge Studies in Modern Optics*, Cambridge University Press, Cambridge, MA (1990).
27. A. Sharan, J. Lettieri, Y. Jia, W. Tian, X. Pan, D. G. Schlom, and V. Gopalan, *Phys. Rev. B* **69**, 214109 (2004).
28. L.-Y. Chen, J.- C. Yang, C.- W. Luo, C.- W. Laing, K.- H. Wu, J.-Y. Lin, T. -M. Uen, J. -Y. Juang, Y. -H. Chu, and T. Kobayashi, *Appl. Phys. Lett.*, **102**, 041902 (2012).

11.2.4 Femtosecond Laser-Induced Formation of Wurtzite Phase ZnSe Nanoparticles in Air

11.2.4.1 INTRODUCTION

The nanometer-sized materials have provided an opportunity to study the relation between material properties and size. Due to the uniquely dimension-dependent properties, the nanosized material has the potential for various applications such as nanoelectronics, nano sensors, nano optronics, and chemical catalyst [1–3]. The nanostructures of zinc selenide (ZnSe), in particular, have attracted considerable attention. ZnSe is an important II–VI semiconductor due to its promising optoelectrical and electrical properties of direct wide band gap 2.68 eV at 300 K. The ZnSe nanoparticles can be obtained through chemical reduction [4], vapor synthesis [5], or laser ablation method [6]. Generally, crystalline ZnSe exhibits two structural phases, that is, zinc blende and wurtzite [7]. However, the studies on wurtzite phase ZnSe nanoparticles are few [8–10] owing to that the wurtzite structure is thermodynamically unstable in ambient environments and requires critical growth conditions. Recently, the femtosecond (fs) laser ablation has been extensively used for material processing [11–15]. Because of the unique features, in fs time-scale pulse duration and high power density (~GW/cm^2), the high-quality nanosized materials can be obtained without thermal effects. For instance, nanospike [12], nanowire [14], and nanoparticle [13,15] have been produced through this approach. In this study, we present a handy way to fabricate pure ZnSe nanoparticles in metastable wurtzite structure through femtosecond laser ablation technique. The structural phase transition from zinc blende to wurtzite takes place during the processing and the size of nanoparticles is controllable by adjusting the laser fluences. Additionally, the characteristics and formation mechanism of ZnSe nanoparticles were investigated.

11.2.4.2 EXPERIMENTAL

The fs laser pulses used in this study were provided by a Ti: sapphire regenerative amplifier with 800 nm and 80 fs at a repetition rate of 5 kHz. The linearly polarized laser pulses were focused by a planoconvex fused silica cylindrical lens with a focal length of 100 mm at normal incidence. The spot size on the surface of the samples is 2270 μm×54 μm. The pulse energy was varied from 0 to 260 mJ/cm^2 by using metallic neutral density filters with OD0.1–OD2 (Thorlabs ND series). A double-side polished (100) ZnSe wafer (5 mm×5 mm×0.5 mm) was mounted on a motorized x-y-z translation stage with the scanning speed of 100 μm/s. After laser irradiation, a pale yellow powder, that is, ZnSe nanoparticles, was observed on the surface of a processed ZnSe crystal. Depending on the experimental purpose, the nanoparticles were preserved in two different ways, that is, dissolved in ethanol with ultrasonic waves or picked up with Scotch tape. After removing the ZnSe powder, there were many subwavelength ripples shown on the surface of a laser-ablated ZnSe wafer. The direction of ripples was perpendicular to the polarization of laser beam and laser scanning path which is consistent with the scenario of the interference between the incident beam and scattering laser light [16]. The morphology of ZnSe single crystals, before and after laser ablation, was examined using a scanning electron microscope (SEM). Moreover, the X-ray diffraction, the Raman scattering spectra, transmission electron microscopy (TEM), and the selected area electron diffraction were applied to identify the structure and characteristics of laser-induced ZnSe nanoparticles.

DOI: 10.1201/9780429196577-85

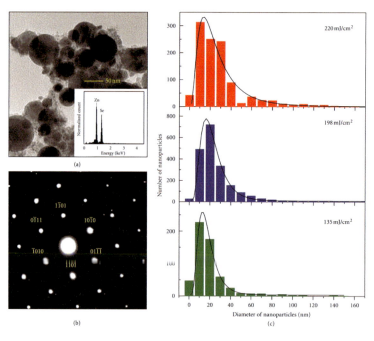

FIGURE 11.2.4.1 (a) TEM images of ZnSe nanoparticles fabricated by the fluence of 220 mJ/cm². Inset: the EDS spectrum shows the composition of ZnSe nanoparticles. (b) TEM diffraction patterns of ZnSe nanoparticles in (a). (c) Size distribution of ZnSe nanoparticles at various laser fluences corresponding to the TEM images in (a) with an area of 3.2 μm × 2.6 μm. The solid lines are the log-normal fitting.

11.2.4.3 RESULTS AND DISCUSSION

Figure 11.2.4.1a shows a typical TEM image of ZnSe nanoparticles with a smooth spherical shape. The diameters of ZnSe nanoparticles are in the range of tens of nanometer. In the selected area electron diffraction (SAED) pattern, a six-fold symmetry was clearly observed in Figure 11.2.4.1b. Through the analysis of distance and angles between the nearest diffraction points and the center biggest point, the crystal structure of ZnSe nanoparticles was identified as a hexagonal (wurtzite) and the orientation of each diffraction point was marked in Figure 11.2.4.1b. Furthermore, the energy dispersive X-ray spectra (EDX) showed that Zn and Se are the only elements detected in the laser-fabricated nanoparticles and the molar ratio of zinc and selenium is close to 1:1 (see the inset of Figure 11.2.4.1a).

During the femtosecond laser irradiation, the dense plasma is formed on the sample surface via multiphoton absorption [17]. The ablated plume was confined near the laser-focused region by surrounding air which prevents other ingredients of air, such as nitrogen, oxygen, and carbon dioxide, to get involved in the formation process of ZnSe nanoparticles under ambient environment. The size distribution of ZnSe nanoparticles fabricated at various fluences was analyzed in Figure 11.2.4.1c. By fitting with the lognormal function, the average diameter of ZnSe nanoparticles was determined to be approximately 16 nm in the case of 135 mJ/cm². With an increase in the laser fluence to 198 and 220 mJ/cm², the average size of the ZnSe nanoparticles slightly increased to 20 and 22 nm, respectively. Although the variation of average size (~6 nm) is smaller than the statistic bar width of 10 nm in Figure 11.2.4.1c (determined by the variation of the average size of ZnSe nanoparticles under the same fluences, which is caused by the fluctuations of laser fluences, pulse durations, spectra, and so on.), the higher fluence indeed generates more ZnSe nanoparticles with the larger size. This indicates that the size distribution of ZnSe nanoparticles can be controlled by adjusting laser fluence and increases as the laser fluence rises. Furthermore, the generation rate of ZnSe nanoparticles using fs laser pulses is approximately $3.63 \times 10^{10} \, s^{-1}$ (or 7.26×10^{6} per pulse) with a fluence of 135 mJ/cm². For the higher fluence of 220 mJ/cm², the generation rate of ZnSe nanoparticles increased by one order of magnitude to $3.63 \times 10^{11} \, s^{-1}$ (or 7.26×10^{7} per pulse).

In Figure 11.2.4.2a, X-ray diffraction patterns revealed the structures of a ZnSe single-crystal wafer and the ZnSe nanoparticles fabricated under various laser fluences are zinc blende and wurtzite, respectively, according to the JCPDS card no. 80-0021 (a=5.618 Å) and no. 80-0008 (a=b=3.974 Å, c=6.506 Å) for ZnSe. It can be clearly seen that the zinc blende phase of a ZnSe single-crystal wafer has been transferred to the wurtzite phase in the ZnSe nanoparticles. Because wurtzite ZnSe is a metastable

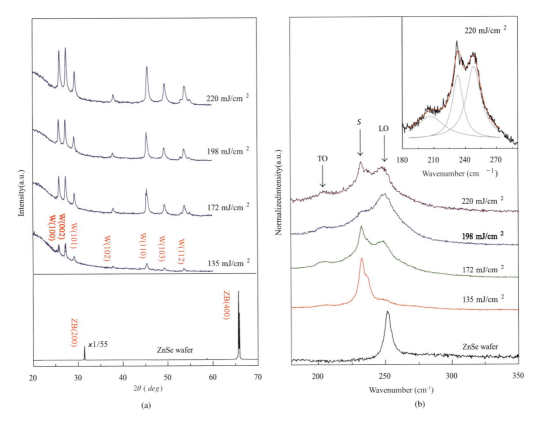

FIGURE 11.2.4.2 (a) X-ray diffraction patterns of ZnSe wafer and ZnSe nanoparticles fabricated at various laser fluences. W: Wurtzite. ZB: Zinc blende. (b) Raman spectra of ZnSe wafer and ZnSe nanoparticles fabricated at various fluences. The 633.0 nm laser was used as the excitation light. Inset: the Raman spectra of ZnSe nanoparticles fabricated at fluence of 220 mJ/cm² on a large scale with Lorentz peak fitting. TO: transverse optical phonon mode. LO: longitudinal optical phonon mode. S: surface phonon mode.

phase under ambient conditions, it can only be observed under high-pressure and high-temperature conditions. However, wurtzite ZnSe nanoparticles can be easily and reliably achieved using femtosecond laser ablation as demonstrated in this study. Additionally, the room-temperature Raman scattering spectra of samples were measured by the micro-Raman system as shown in Figure 11.2.4.2b.

The typical Raman scattering peaks of ZnSe nanoparticles were clearly observed from 200 to 250 cm⁻¹ which are rather different from that of ZnSe single crystal wafer. Moreover, there was no vibration mode due to impurities observed in the Raman spectra which is consistent with the above EDX results. A look at the Raman spectra in the case of 220 mJ/cm² can be fitted well by three Lorentzian lines located at 250, 232, and 203 cm⁻¹. The peaks at 203 and 250 cm⁻¹ are attributed to the transverse optical (TO) phonon mode and longitudinal optical (LO) phonon mode of ZnSe, respectively [18,19]. Another peak located at 232 cm⁻¹ between the LO and TO phonons is assigned to be the surface phonon mode (S) of nanoparticles. This phenomenon, the appearance of surface phonon mode, can be decidedly observed in small-sized material [9,20] due to the high surface-to-volume ratio and the dominant surface properties. To figure out the relation between TO phonon mode and surface phonon mode, the theoretical expression was used as follows [20]:

$$\frac{\omega_S^2}{\omega_T^2} = \frac{\varepsilon_0 + \varepsilon_m((1/L)-1)}{\varepsilon_\infty + \varepsilon_m((1/L)-1)}, \tag{11.2.4.1}$$

where ω_S and ω_T represent the surface phonon and TO phonon frequency, respectively; ε_0 and ε_∞ are static and high-frequency dielectric constant, respectively; ε_m is the dielectric constant of the surrounding medium and L indicates the depolarization factor which is related to the particle shape. In the case of ZnSe, $\omega_T = 203\,\mathrm{cm}^{-1}$, $\varepsilon_0 = 8.6$, $\varepsilon_\infty = 5.7$, and ε_m is equal to 1 for air surrounding condition [18]. Because the laser-fabricated ZnSe nanoparticles are in a spherical shape, according to the TEM image shown in Figure 11.2.4.1a, the depolarization factor (L) is 1/3. Using (11.2.4.1), the calculated surface phonon mode is located at $238\,\mathrm{cm}^{-1}$ which is consistent with the measured value of $232\,\mathrm{cm}^{-1}$. It is worth to note that the LO, TO, and surface phonon modes are strongly fluence dependent. As the laser fluence decreases, the LO and TO peaks gradually shrink; meanwhile, the surface phonon mode increases. In the case of 135 mJ/cm², the surface phonon mode almost dominates the Raman spectra, which is due to the smaller average size and more uniform size distribution as shown in Figure 11.2.4.1c. Additionally, the small-size nanoparticles also result in the red shift of surface phonon mode.

Here brief description of the basics of "surface phonon" is given below.

Surface phonon: A surface phonon is the quantum of a lattice vibration mode associated with a solid surface. Similar to the ordinary lattice vibrations in a bulk solid (whose quanta are simply called "bulk" phonons), the nature of surface vibrations depends on details of periodicity and symmetry of a crystal structure. Surface vibrations are however distinct from bulk vibrations, as they arise from the abrupt termination of a crystal structure at the surface of a solid. Knowledge of surface phonon dispersion gives important information related to the amount of surface relaxation, the existence and distance between an adsorbate and the surface, and information regarding the presence, quantity, and type of defects existing on the surface. In modern semiconductor research, surface vibrations are of interest as they can couple with electrons and thereby affect the electrical and optical properties of semiconductor devices. They are most relevant for devices where the electronic active area is near a surface, as is the case in two-dimensional electron systems and in quantum dots.

According to the early research, ZnSe transforms from a zinc blende structure to the wurtzite structure when the temperature is above the transition temperature (T_{tr}) of 1698K [7]. When ZnSe crystals are irradiated by the femtosecond laser pulses, the temperature of electrons and lattice in ZnSe crystals increases according to the two-temperature model [21]. Because of the much smaller heat capacity in the electron subsystem, the increase of electron temperature is dramatically larger than that of lattice. The electron temperature T_e can be described by [22]

$$T_e = \left\langle \sqrt{T_i^2 + \frac{2(1-R)F}{l_s \gamma} e^{-z/l_s}} \right\rangle, \tag{11.2.4.2}$$

where R is the reflectivity at 800 nm and F is the laser fluences and γ is the linear coefficient of heat capacity due to the electronic subsystem. The mean value is taken for the depth z going from the crystal surface down to the skin depth $l_s \sim 1.87\,\mu\mathrm{m}$ (it was estimated from the nonlinear absorption coefficient β [23]). In this study, taking $T_i = 295$ K, $R = 0$ (which is assumed to be totally absorbed by ZnSe), $F = 220$ mJ/cm², and $\gamma = 29.4$ mJ/(mol K) [24], we obtain $T_e \sim 1200$ K. However, for the structure transition in materials, the key factor is the lattice temperature rather than the electron temperature. Thus, we further consider the increase in the transient temperature ΔT in materials can be estimated according to the relationship of $\Delta T = W/(C \times V)$, where W is the pulse energy, C is the heat capacity, and V is the illuminated volume. For ZnSe at 300 K, C is $\sim 1.89 \times 10^6$ J/m³ K [7], V is $2.29 \times 10^{-13}\,\mathrm{m}^3$ (absorption depth 1.87 μm estimated from the nonlinear absorption coefficient β [23]), and W is in the order of 0.243 mJ (which is assumed to be totally absorbed by ZnSe). Thus, the ΔT is approximately 560 K, which is far below the structural transition temperature of 1698 K. Therefore, a structural transition could not be induced by the increase in temperature. To identify the mechanism underlying the phase transition of ZnSe from zinc blende to wurtzite, we further analyzed the influence of "ablation pressure" [25], which has been studied from various perspectives over the past few decades [26,27]. When solids are irradiated by laser pulses, high-density plasma is formed on the surface of the samples [17]. The compressed plasma in laser-driven implosions has been characterized as the ablating or exploding pusher according to the surface ablation pressure and bulk pressure due to the preheating through electrons.

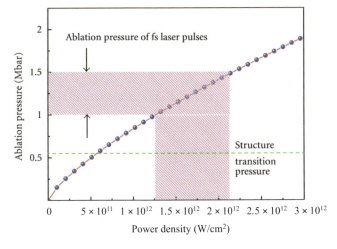

FIGURE 11.2.4.3 Simulated ablation pressure as a function of the laser peak power density according to Eq. (11.2.4.3). The shadow area indicates the range of laser peak power density in this study and corresponding ablation pressure. The dashed line represents the pressure of zinc blende wurtzite phase transition, which was obtained from Ref. [28].

In 2003, Batani et al. [25] derived the shock pressure with the laser and target parameters expressed as

$$P(\text{Mbar}) = 11.6 \left(\frac{I}{10^{14}} \right)^{3/4} \lambda^{-1/4} \left(\frac{A}{2Z} \right)^{7/16} \left(\frac{Z \times t}{3.5} \right)^{-1/8}, \qquad (11.2.4.3)$$

where I is the irradiance on target with the unit of W/cm²; λ is the laser wavelength in μm; A and Z are, respectively, the mass number and the atomic number of the target; t is the time in ns. Figure 11.2.4.3 shows the effective pressure in the irradiated region with the laser peak power density of $0 \sim 3.0 \times 10^{12}$ W/cm². In this study, the maximum pressure induced by the laser reached approximately 1.5 Mbar. According to the studies of Greene et al. in II–VI compounds [28], the solid-solid transition point, that is, the zinc blende-wurtzite phase transition, of ZnSe is approximately 0.55 Mbar. In our experiments, the ablation pressure induced by the femtosecond laser pulses on the ZnSe single crystals was in the range of 1.0–1.5 Mbar as shown in the shadow area of Figure 11.2.4.3. This exceeds the solid-solid transition pressure 0.55 Mbar (the dashed line in Figure 11.2.4.3). Therefore, the wurtzite-phase ZnSe nanoparticles transferred from the zinc blende phase may be caused by high ablation pressure resulting from the femtosecond laser pulses, and the accompanied increase in surface-to-volume ratio in the nanoparticles.

11.2.4.4 CONCLUSION

In this subsection, a simple and reliable approach to obtain the thermally metastable and pure ZnSe nanoparticles was demonstrated. The spherical-shaped wurtzite ZnSe nanoparticles with the average diameter of <25 nm can be well produced through the femtosecond laser ablation technique. While the femtosecond laser pulses are focused on the surface of a ZnSe wafer in air, the ablated plume cannot expand as rapidly as those in a vacuum chamber and then causes an instantaneous high-energy and high-pressure region around the laser-focused point; meanwhile, a large amount of ZnSe nanoparticles was fabricated on the surface of a ZnSe wafer. During the formation of ZnSe nanoparticles, the structural phase further changes from the zinc blende phase to the metastable wurtzite phase due to the ultra-high localized ablation pressure caused by the rapid injection of high laser energy within a femtosecond time scale. The contents of the subsection were reported in the form of a publication in a journal [29]. The research described in this subsection was conducted by collaboration among the following people: H.-I. Wang, W.-T. Tang, P.-S. Tseng, L.-W. Liao, C.-W. Luo, C.-S. Yang and T. Kobayashi [29].

REFERENCES

1. M. Bruchez, M. Moronne, P. Gin, S. Weiss, and A. P. Alivisatos, "Semiconductor nanocrystals as fluorescent biological labels," *Science* **281**(5385), 2013–2016 (1998).

2. R. S. Friedman, M. C. McAlpine, D. S. Ricketts, D. Ham, and C. M. Lieber, "Nanotechnology: high-speed integrated nanowire circuits," *Nature* **434**(7037), 1085 (2005).

3. W. C. W. Chan, and S. Nie, "Quantum dot bioconjugates for ultrasensitive nonisotopic detection," *Science* **281**(5385), 2016–2018 (1998).

4. J. Che, X. Yao, H. Jian, and M. Wang, "Application and preparation of ZnSe nanometer powder by reduction process," *Ceram. Int.* **30**(7), 1935–1938 (2004).

5. D. Sarigiannis, J. D. Peck, G. Kioseoglou, A. Petrou, and T. J. Mountziaris, "Characterization of vapor-phase-grown ZnSe nanoparticles," *Appl. Phys. Lett.* **80**(21), 4024–4026 (2002).

6. K. V. Anikin, N. N. Melnik, A. V. Simakin, G. A. Shafeev, V. V. Voronov, and A. G. Vitukhnovsky, "Formation of ZnSe and CdS quantum dots via laser ablation in liquids," *Chem. Phys. Lett.* **366**(3–4), 357–360 (2002).

7. P. Rudolph, N. Schafer, and T. Fukuda, "Crystal growth of¨ ZnSe from the melt," *Mater. Sci. Eng. R* **15**(3), 85–133 (1995).

8. Y. P. Leung, W. C. H. Choy, I. Markov, G. K. H. Pang, H. C. Ong, and T. I. Yuk, "Synthesis of wurtzite ZnSe nanorings by thermal evaporation," *Appl. Phys. Lett.* **88**(18), (2006).

9. C. X. Shan, Z. Liu, X. T. Zhang, C. C. Wong, and S. K. Hark, "Wurtzite ZnSe nanowires: growth, photoluminescence, and single-wire Raman properties," *Nanotechnology*, **17**(22), 5561–5564 (2006).

10. L. Jin, W. C. H. Choy, Y. P. Leung, T. I. Yuk, H. C. Ong, and J.-B. Wang, "Synthesis and analysis of abnormal wurtzite ZnSe nanowheels," *J. Appl. Phys.* **102**(1), (2007).

11. A. P. Joglekar, H. Liu, G. J. Spooner, E. Meyhofer, G. Mourou, and A. J. Hunt, "A study of the deterministic character of optical damage by femtosecond laser pulses and applications to nanomachining," *Appl. Phys. B* **77**(1), 25–30 (2003).

12. M. Y. Shen, C. H. Crouch, J. E. Carey, and E. Mazur, "Femtosecond laser-induced formation of submicrometer spikes on silicon in water," *Appl. Phys. Lett.* **85**(23), 5694–5696 (2004).

13. C. W. Luo, C. C. Lee, C. H. Li, et al., "Ordered YBCO submicron array structures induced by pulsed femtosecond laser irradiation," *Opt. Express* **16**(25), 20610–20616 (2008).

14. T. Jia, M. Baba, M. Huang, et al., "Femtosecond laser-induced ZnSe nanowires on the surface of a ZnSe wafer in water," *Solid State Commun.* **141**(11), 635–638 (2007).

15. Y. Nakata, T. Okada, and M. Maeda, "Fabrication of dot matrix, comb, and nanowire structures using laser ablation by interfered femtosecond laser beams," *Appl. Phys. Lett.*, **81**(22), 4239–4241 (2002).

16. C. Wang, H.-I. Wang, W.-T. Tang, C.-W. Luo, T. Kobayashi, and J. Leu, "Superior local conductivity in self-organized nano-dots on indium-tin-oxide films induced by femtosecond laser pulses," *Opt. Express* **19**(24), 24286–24297 (2011).

17. A. De Giacomo, M. Dell'Aglio, A. Santagata, and R. Teghil, "Early stage emission spectroscopy study of metallic titanium plasma induced in air by femtosecond- and nanosecond-laser pulses," *Spectrochim. Acta B* **60**(7–8), 935–947 (2005).

18. W. Martienssen and H. Warlimont, *Springer Handbook of Condensed Matter and Materials Data*, Springer, Berlin, Germany (2005).

19. Z. D. Hu, X. F. Duan, M. Gao, Q. Chen, and L.-M. Peng, "ZnSe nanobelts and nanowires synthesized by a closed space vapor transport technique," *J. Phys. Chem. C* **111**(7), 2987–2991 (2007).

20. S. Hayashi and H. Kanamori, "Raman scattering from the surface phonon mode in GaP microcrystals," *Phys. Rev. B* **26**(12), 7079–7082 (1982).

21. S. I. Anisimov, B. L. Kapeliovich, and T. L. Perelman, "Electron emission from metal surfaces exposed to ultrashort laser pulses," *Sov. Phys. JETP* **39**, 375–377 (1974).

22. C. W. Luo, I. H. Wu, P. C. Cheng, et al., "Quasiparticle dynamics and phonon softening in FeSe superconductors," *Phys. Rev. Lett.* **108**(25), (2012).

23. K. Y. Tseng, K. S. Wong, and G. K. L. Wong, "Femtosecond time-resolved Z-scan investigations of optical nonlinearities in ZnSe," *Opt. Lett.* **21**(3), 180–182 (1996).

24. J. A. Birch, "Heat capacities of ZnS, ZnSe and CdTe below 25K," *J. Phys. C* **8**(13), 2043–2047 (1975).

25. D. Batani, H. Stabile, A. Ravasio, et al., "Ablation pressure scaling at short laser wavelength," *Phys. Rev. E* **68**(6), 674031–674034 (2003).

26. M. H. Key, P. T. Rumsby, R. G. Evans, C. L. S. Lewis, J. M. Ward, and R. L. Cooke, "Study of ablatively imploded spherical shells," *Phys. Rev. Lett.* **45**(22), 1801–1804 (1980).

27. J. S. De Groot, K. G. Estabrook, W. L. Kruer, R. P. Drake, K. Mizuno, and S. M. Cameron, "Distributed absorption model for moderate to high laser powers," *Phys. Fluids B* **4**(3), 701–707 (1992).

28. R. G. Greene, H. Luo, and A. L. Ruoff, "High pressure X-ray and Raman study of ZnSe," *J. Phys. Chem. Solids* **56**(3–4), 521–524 (1995).

29. H.- I. Wang, W.-T. Tang, P.-S. Tseng, L.-W. Liao, C.- W. Luo, C.-S. Yang, and T. Kobayashi, "Femtosecond laser-induced formation of wurtzite phase ZnSe," *Appl. Opt.* **51**(26), 6403–6401 (2012).

11.2.5 Controllable Subwavelength-Ripple and -Dot Structures on YBa$_2$Cu$_3$O$_7$ Induced by Ultrashort Laser Pulses

11.2.5.1 INTRODUCTION

Since the introduction of femtosecond (fs) laser amplifiers, nanoscale microstructures on the surfaces of materials have been demonstrated by irradiating the surfaces with femtosecond pulses [1–12]. This has attracted a great deal of attention because periodic structures can be inscribed on almost any material directly and without any masks and chemical photoresists to relieve environmental concerns. For instance, nanoripples [1–7], nanoparticles [8–10], nanocones [11], and nanospikes [12] have been induced by single-beam femtosecond laser pulses in various materials. Nowadays, the interference between the surface electromagnetic wave (surface plasmons) and the incident light has been widely accepted to explain the femtosecond laser-induced subwavelength periodic surface structures (LIPSSs) [2–7]. However, studies on the evolution of ripples under irradiation by circularly polarized light and the formation of dot structures are very few. Therefore, the mechanism of their formation has not been further discussed yet.

In this subsection, controllable subwavelength-ripple and -dot structures on YBa$_2$Cu$_3$O$_7$ in thin films are discussed by irradiation of a single-beam and a dual-beam fs laser, respectively. The dependence of the subwavelength ripples on the incident angle and circularly polarized light is obtained. Finally, a physical model in terms of the interference between the surface plasma wave and the incident light is proposed to explain our results.

11.2.5.2 EXPERIMENTS

The YBCO samples used in this study were prepared by pulsed laser deposition with a KrF excimer laser. The growth and characterization of the (001)-oriented YBCO thin films have been discussed in detail elsewhere [10]. Briefly, the thickness of all films on (001) SrTiO$_3$ (STO) substrates was about 240 nm. The T_c of a YBCO thin film is about 89.5 K, as shown in Figure 11.2.5.1. Figure 11.2.5.2a shows the surface morphology of a typical (001)-oriented YBCO thin film scanned by an ultra-high resolution field emission scanning electron microscope (FEG-SEM, JOEL JSM-7000F). The surface roughness of an as-deposited YBCO thin film was around 4 nm identified by its AFM image.

A commercial regenerative amplified Ti:sapphire laser (Legend USP, Coherent) with 800 nm wavelength, 80 fs pulse duration, ~0.5 mJ pulse energy and 5 kHz repetition rate was used as the irradiation source. After passing through a variable neutral density filter and a quarter-wave plate, the normal incident laser beam was focused onto the sample surface with a spot size of ~200 μm (FWHM) by a convex lens with a 50 mm focal length. The pulse number or irradiation time was precisely controlled by the electric shutter. For the dual-beam scheme, a modified Michelson

DOI: 10.1201/9780429196577-86

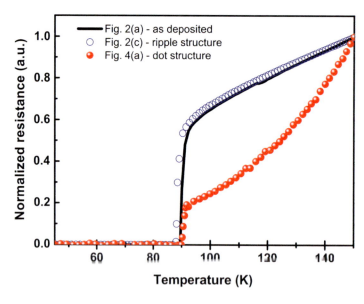

FIGURE 11.2.5.1 The temperature dependence of the normalized resistance for as-deposited YBCO (Figure 11.2.5.2a), YBCO thin film with a ripple structure (Figure 11.2.5.2c) and YBCO thin film with a dot structure (Figure 11.2.5.4a).

interferometer was applied to produce the dot structure on YBCO thin films. The polarizations of both beams could be individually controlled by two quarter-wave plates before the reflection mirrors in both arms of the dual-beam setup. After a beam splitter in the dual-beam setup, both beams were collinearly and simultaneously focused on the sample surfaces by a convex lens with 50 mm focal length. All experiments were performed in the air under atmospheric pressure.

11.2.5.3 RESULTS AND DISCUSSION

The SEM analysis (Figure 11.2.5.2 with a pixel resolution of ~0.04 nm) indicates that the morphology of fs laser-induced surface structures depends on both the number of applied pulses and the laser incident angle (θ is the angle of incidence with respect to the surface normal). Figures 11.2.5.2a–c show the evolution of the ripple structure on YBCO thin films irradiated by a single-beam fs laser with various pulse numbers (N) and a fixed fluence (F) = 75 mJ/cm² at a fixed incident angle of 0°. On increasing the pulse number, the ripple structure becomes clear in the SEM images, which can be evidenced from the appearance of satellite peaks in the 2D Fourier spectra of the insets of Figures 11.2.5.2b and c (there are no satellite peaks in the inset of Figure 11.2.5.2a for an as-deposited YBCO thin film). The spatial period of the ripples, estimated from the position of a satellite peak in the 2D Fourier spectra, is independent of the pulse number and irradiation time as shown in Figure 11.2.5.2g. Once the pulse number was greater than or equal to 5000, i.e. the sample surface was irradiated by the 75 mJ/cm² femtosecond laser pulses for 1 s, ripples with a peak-to-valley height of ~106 nm could be clearly observed on the sample surface. On the other hand, if the fluence and pulse number were, respectively, fixed at ~300 mJ/cm² and 150000, we found that the spatial period decreased with increasing incident angle (θ) (see Figure 11.2.5.2h). However, the observed period of the ripple at θ=0° is significantly smaller than the prediction of $\Lambda = \lambda/(1 + \sin \theta)$ [13]. Besides, the incident angle-dependent period of ripples on YBCO thin films cannot be described by this simplified scattering model (the solid line in Figure 11.2.5.2h). Therefore, the effect of the surface electromagnetic wave, i.e. surface plasmons (SPs), should be taken into account for the formation of subwavelength ripples [2,7]. According to Shimotsuma et al's results [14], the femtosecond incident light easily excites the plasmons on the surface of various materials. As shown in Figure 11.2.5.3c, once the momentum conservation condition for the wavevectors of the linearly polarized laser light (K_i), plasma wave (K_p) and LIPSS (K_L) is satisfied, such a plasmon could couple with the incident light. The interference between the plasmons and the incident light would generate

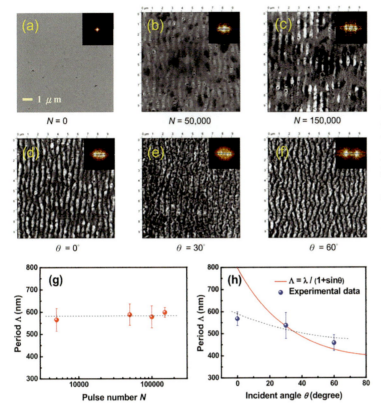

FIGURE 11.2.5.2 (a)–(f) The morphological evolution of structures on YBCO thin films induced by a linear polarized fs laser with various pulse numbers (fixed incident angle $\theta = 0°$, fixed $F = 75$ mJ/cm^2) and with various incident angles (fixed $N = 150000$ and $F = \sim300$ mJ/cm^2). (g) The dependence of the period of the ripples on the pulse number. (h) The dependence of the period of the ripples on the incident angle. Inset: the 2D Fourier spectra which were transferred from their corresponding SEM images (10 μm × 10 μm). The scale bar is applied to all figures. The dashed lines are guides to the eyes.

a periodically modulated electron density to cause the nonuniform melting. After femtosecond laser irradiation, the interference ripple is inscribed on the YBCO surface. Moreover, the superconductivity of the YBCO with the ripple structure in Figure 11.2.5.2c remained nearly unchanged as shown in Figure 11.2.5.1.

Interestingly, when we used a circularly polarized beam, the ripple structure could still be produced, as shown in Figures 11.2.5.3a and b. The orientations of the ripples were observed as $-45°$ and $+45°$ for left- and right-circularly polarized beams, respectively, with respect to the incident plane of the beam. For both cases, the spatial period was 491 nm which was produced by the fs laser pulses with fluences of 185 mJ/cm^2 and a pulse number of 150000.

These results show that the orientation of the ripple structure strongly depends on the polarization state of the incident fs pulses which is consistent with Zhao et al's results on tungsten [12,15]. In principle, the circularly polarized light ($K_{i,c}$) can be decomposed into two perpendicular linearly polarized light beams (E_x and E_y) with retardation of $\lambda/4$ in phase, as shown in Figure 11.2.5.3d. The linearly polarized light beams E_x and E_y can, respectively, induce the LIPSS $K_{L,x}$ and $K_{L,y}$ while the momentum conservation condition in Figure 11.2.5.3c is satisfied. Thus, both $K_{L,x}$ and $K_{L,y}$ with phase retardation of $\lambda/4$ further cause the $K_{L,c}$ according to the momentum conservation condition of $K_{L,c}=K_{L,x}+K_{L,y}$. The wavevector of LIPSS ($K_{L,c}$) at $45°$ is completely consistent with the direction of the satellite peaks in the 2D Fourier spectra (the inset of Figure 11.2.5.3a). Namely, the orientation of ripples is at $-45°$ for the left-circularly polarized beam with respect to the incident plane of the beam. Similarly, the right-circularly polarized beam induces the wavevector of LIPSS ($K_{L,c}$) at $-45°$ according to the momentum conservation condition of $K_{L,c}=-K_{L,x}+K_{L,y}$, which is consistent with the results in Figure 11.2.5.3b.

In order to clarify the above scenario in the formation of $\pm45°$ regular ripples induced by circularly polarized beams, a dual-beam setup with perpendicular linear polarizations was applied to

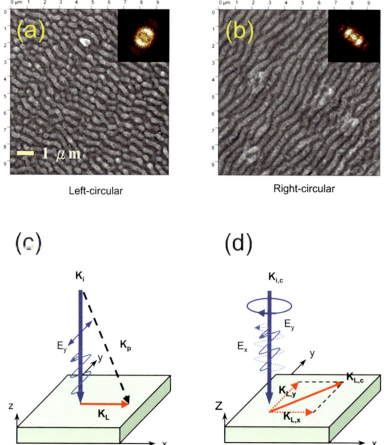

FIGURE 11.2.5.3 The SEM images (10 μm × 10 μm) of fs LIPSS induced by (a) the left- and (b) the right-circularly polarized beams. (c) Schematic of the momentum conservation condition for the wavevectors of linear polarized laser light (K_i), plasma wave (K_p), and LIPSS (K_L). (d) Schematic processes of the LIPSS by the circularly polarized laser light ($K_{i, c}$). The scale bar is applied to all figures.

produce the LIPSS on YBCO thin films. As shown in Figures 11.2.5.4a–d, it is surprising that there are many dots with a height of ~116 nm rather than regular ripples on the surface of the YBCO thin film.

For the case of a dual-beam setup, the $K_{L, x}$ and $K_{L, y}$ without coherence in phase, which are, respectively, induced by two random phase and perpendicular linear-polarization beams (E_x and E_y), would not satisfy the momentum conservation of $K_{L, c} = \pm K_{L, x} + K_{L, y}$ and cannot create the ±45° wavevector of LIPSS ($K_{L, c}$). Therefore, the $K_{L, x}$ and $K_{L, y}$ which are perpendicular to each other would lead to 2D nonuniform melting and then further aggregate to form randomly distributed dots (see the 2D Fourier spectra in the insets of Figures 11.2.5.4e–h) due to the surface tension. For the case of $N = 25000$, the average size of the dots is around 632 nm in diameter estimated by the log-normal fitting as shown in Figure 11.2.5.4e. Additionally, the T_c of YBCO with nanodot structure is almost the same as that of the as-deposited YBCO thin films (see Figure 11.2.5.1). Here the log-normal distribution is discussed in the following way.

The statistical realization of the multiplicative product of many independent random variables, each of which is positive. This is justified by considering the central limit theorem in the log domain sometimes called Gibrat's law. The log-normal distribution is the maximum entropy probability distribution for a random variate X for which the mean and variance of $\ln(X)$ are specified.

If the pulse number is increased, although the average size only slightly increases from 632 to 844 nm, the size distribution markedly broadens (Figures 11.2.5.4f–h). For $N = 300000$, the size of a portion of the dots is of the order of micrometers. However, the larger dots will affect the dot densities on the surfaces of YBCO thin films. For instance, the density of dots increases with an increase

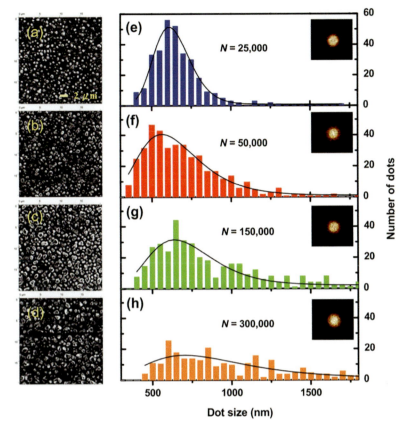

FIGURE 11.2.5.4 (a)–(d) The dot structures on YBCO thin films induced by a dual-beam setup with various pulse numbers (*N*) and a fixed fluence of 87 mJ/cm². (e)–(h) The size distributions correspond to the SEM images (20 μm × 20 μm) (a)–(d), respectively. The solid lines are the log-normal fitting. Insets: the 2D Fourier spectra which were transferred from their corresponding SEM images (a)–(d), respectively. The scale bar is applied to all figures.

in pulse number while the pulse number is greater than or equal to 150000. Once the dots grow large enough to merge with their nearest neighbors and even their next nearest neighbors, the density of dots will significantly shrink, as shown in Figure 11.2.5.4c. Therefore, the size and density of YBCO dots can be controlled by the pulse number of the fs laser.

11.2.5.4 SUMMARY

In summary, the surface morphology of YBCO thin films has been systematically studied under single-beam and dual-beam fs laser irradiation [16]. The generation of ripple and dot periodic structures is decided by the applied laser fluences, pulse numbers and laser polarizations; the period and orientation of ripples and even the size and density of dots can be controlled by these parameters. It was expected that these results may be potentially applied to the manufacture of nanostructure templates for the growth of YBCO thin films and the fabrication of microwave filter devices with array structures or weak-link Josephson junction arrays [15]. The contents of this subsection are the products of the collaborative research by the following people: C.-W. Luo, W.-T. Tang, H.-I. Wang, L.-W. Liao, H.-P. Lo, K.-H. Wu, J.-Y. Lin, J.-Y. Juang, T.-M. Uen and T. Kobayashi [16].

REFERENCES

1. E. M. Hsu, T. H. R. Crawford, H. F. Tiedje and H. K. Haugen, *Appl. Phys. Lett.* **91**, 111102 (2007).
2. S. Sakabe, M. Hashida, S. Tokita, S. Namba, and K. Okamuro, *Phys. Rev. B* **79**, 033409 (2009).
3. X. Jia, T. Q. Jia, Y. Zhang, P. X. Xiong, D. H. Feng, Z. R. Sun, J. R. Qiu, and Z. Z. Xu, *Opt. Lett.* **35**, 1248 (2010).
4. Y. Yang, J. Yang J, Xue, and Y. Guo, *Appl. Phys. Lett.* **97**, 141101.

5. J. Bonse, and J. Kruger, *J. Appl. Phys.* **108**, 034903 (2010).
6. K. Okamuro, M. Hashida, Y. Miyasaka, Y. Ikuta, S. Tokita, and S. Sakabe, *Phys. Rev. B* **82**, 165417 (2010).
7. M. Huang, F. Zhao, Y. Cheng, N. Xu, and Z. Xu, *ACS Nano* **3**, 4062 (2010).
8. Y. Teng, J. Zhou, F. Luo, Z. Ma, G. Lin, and J. Qiu, *Opt. Lett.* **35**, 2299 (2010).
9. T. Q Jia, F. L. Zhao, M. Huang, H. X. Chen, J. R. Qiu, R. X. Li, Z. Z. Xu, and H. Kuroda, *Appl. Phys. Lett.* **88**, 111117 (2006).
10. C. W. Luo, et al., *Opt. Express* **16**, 20610 (2008).
11. B. K. Nayak, M. C. Gupta, and K. W. Kolasinski, *Appl. Phys. A* **90**, 399 (2008).
12. Q. Z. Zhao, S. Malzer, and L. J. Wang, *Opt. Express* **15**, 15741 (2007).
13. G. Zhou, P. M. Fauchet, and A. E. Siegman, *Phys. Rev. B* **26**, 5366 (1982).
14. Y. Shimotsuma, P. G. Kazansky, J. Qiu, and K. Hirao, *Phys. Rev. Lett.* **91**, 247405 (2003).
15. Q. Z. Zhao, S. Malzer, and L. J. Wang, *Opt. Lett.* **32**, 1932 (2007).
16. C.-W. Luo, W.-T. Tang, H.-I. Wang, L.-W. Liao, H.- P. Lo, K. -H. Wu, J.-Y. Lin, J.-Y. Juang, T.- M. Uen, and T. Kobayashi, *Superconductor Sci. Tech.* **25**(11), 115008 (2012).

Section 12

Photobiological Reactions

12.1 Real-Time Vibrational Dynamics in Chlorophyll a Studied with a Few-Cycle Pulse Laser

12.1.1 INTRODUCTION

Chlorophyll a (Chl-a) is the most abundant pigment among the photosynthesis in nature, which plays an essential role in light harvesting and conversion processes in green plants and in various algae. Vibrational properties of the chlorophylls have been investigated extensively by optical spectroscopic techniques, such as resonance Raman (RR) spectroscopy [1–8], spectral hole-burning (HB), and fluorescence line-narrowing (FLN) techniques [9–16]. Specifically, the data achieved using HB and FLN techniques are corresponding to the vibrations in the excited and ground electronic state, respectively; however, these techniques can only be operated at very low temperatures (usually ~4 K). As for the RR spectroscopy, both the Soret-resonant and Q-resonant Raman spectra have been successfully obtained and have reported a wealth of detailed information on the vibrational modes of the electronic ground state of Chl-a molecules. However, nearly all of thus-far reported Raman spectra observed at room temperature are only corresponding to the Soret band in Chl-a.

The vibronic spectra of porphyrin, of which structure is composed of porphyrin are described briefly as follows.

In freebase porphyrins, the Q band is split due to vibrational excitations. Therefore, two bands are produced due to the transition from the ground state to two vibrational states of the excited state [$Q(0,0)$ and $Q(1,0)$]. Further, the presence of the NH protons breaks the symmetry, and as a result, the above-mentioned bands are further split into two bands each. The X and Y components are no longer degenerate and therefore we see four Q bands $Q_x(0,0)$, $Q_y(0,0)$, $Q_x(1,0)$ and $Q_y(1,0)$.

It is difficult to measure the Q-resonant (especially the Q_y-band) Raman spectra of Chl-a at room temperature because of the disturbance by intense fluorescence in the Chl-a solution. In contrast, the fluorescence is significantly quenched and red-shifted in the films. Therefore, the Q_y-excitation RR spectra of Chl-a can only be acquired in a solid film environment by surface-enhanced Raman scattering spectroscopy which also requires the use of very low temperature [5–8].

The above-mentioned methods are powerful for studying the vibrational structure of Chl-a, and plenty of results have been reported; however, none of them can provide the straightforward vibrational information related to the electronic transition of the lowest excited state of the Chl-a Q_y-band at room temperature. Information at room temperature is especially physiologically important because it is the usual temperature of the physiological condition in native complexes.

The featureless Q_y absorption band displays a limited structure even on lowering the sample temperature. To discover the presence of hidden characteristics of the absorption band at physiologically relevant temperatures, the absorption band shape for the Q transition region of Chl-a is usually calculated by using the vibrational frequency modes and Franck-Condon factors obtained from HB and FLN spectroscopies as a function of temperature [13,17] because of the scarcity of vibrational information at room temperature. However, it was reported by Krawczyk [18] using RR spectra of Chl-a at 77 K that the pentacoordinated Chl-a at room temperature would convert into hexacoordinated one at 77 K. Rather recently, Raʾtsep et al. [16] reported that while cooling from 295 to 4.5 K,

DOI: 10.1201/9780429196577-88

the Chl-a in 1-propanol and diethyl ether was converted into a hexacoordinated state, but the one in 2-propanol mainly retained its pentacoordinated status.

Due to the complexity of the temperature-dependent characteristics of Chl-a, there is no current consensus among researchers on how well the data achieved at low temperatures will satisfy the situation at physiologically relevant temperatures. Peterman et al. [13] simulated the temperature dependence of the low-energy part of the Q_y absorption spectrum of light-harvesting complex II very well up to 220 K using the phonon wing from the 4 K emission spectra; however, the simulated and experimental results started to deviate above 220 K. Zucchelli et al. [17] calculated the spectral band shape of Chl-a in the Q-absorption region as a function of temperature to study the Chl-a in penta- and hexacoordinated states, respectively. They concluded that vibrational modes in the range 540–850 cm^{-1} were highly Mg-coordinate-dependent; however, Ra̋tsep et al. [16] denied this conclusion in their recent study and suggested that the vibrational degrees of freedom were not sensitive to the temperature-dependent absorption spectra change as the electronic one was.

To clarify the above-mentioned controversial issues, the most straightforward first step seems to be to obtain the vibrational information coupled to the Q-transition of Chl-a at room temperature directly. Shiu et al. [19] have studied the vibrational frequencies within 98–701 cm^{-1} of the Q_y excited state of Chl-a in ethanol solution. In their experiment, however, they use an excitation pulse with a duration of 25 fs. Only when the duration of the pump pulse is shorter than the vibration period can the coherent molecular vibration of the corresponding mode be excited; consequently, the highest detectable vibrational frequency in their research was limited to ~1300 cm^{-1}. Moreover, the samples were probed at only one wavelength by Shiu et al. [19], and broad-band information was not obtained.

In this study, we used a 6.8 fs laser pulse in the broadband femtosecond pump-probe real-time vibration spectroscopy, which has been proven to be a powerful tool for the observation of dynamical processes in molecules [20–22], to obtain both electronic and vibrational dynamics of Chl-a Q_y-band in solution at 293 K. After the excitation by the ultrashort pump pulse, a wave-packet is prepared in the form of linear superposition of several vibrational levels of different quantum numbers. Then the wave packet moves on the potential energy surface, and the motion modifies the transition spectra and intensities composed of contributions from various vibronic transitions between the vibronic levels in the initial and final electronic states. The contributions of the vibrational levels are proportional to the product of the Franck-Condon factor, which determines the absorption spectrum in the case of Condon approximation is satisfied [23] as well as the spectral intensity of the pump pulse with the vibronic transition energy. Thus, the wavepacket motion of some vibrational modes can be detected by measuring the time-dependent difference absorbance with the weaker probe pulse.

There is another technique to detect the electronic relaxation and vibrational wave-packet dynamics in real time – the time- and frequency-gated spontaneous emission method. However, it has much less sensitivity because of the low efficiency in either the upconversion sum-frequency gate (second-order nonlinear process) or Kerr gate (third-order nonlinear process) in the detection scheme, and it is also difficult to achieve the high temporal resolution which is limited by geometrical path difference in the former and Kerr response time in the latter. This method may also be affected by the elongation of relaxation time due to self-absorption [24]. By using our method, we are not suffering from these effects. In addition, compared with other conventional vibrational spectroscopy, this method has the following advantages:

1. This technique is not limited by the measurement temperatures, so that it can provide both the electronic and vibrational characteristics of Chl-a at physiologically relevant temperatures.
2. RR signals are very frequently overwhelmed by the fluorescence signal, especially in the case of highly fluorescent molecules (e.g., Chl-a). In contrast, the contamination effect of spontaneous fluorescence can be almost totally avoided in real-time vibrational

spectroscopy because of much more intense, highly directional probe light probe beam than spontaneous fluorescence.

3. RR spectroscopy is not capable of reliably detecting the presence of very low-frequency modes due to intense Rayleigh scattering of the excitation beam [25]. However, the low-frequency modes can easily be studied by pump-probe experiment as long as a few quanta of the modes can be covered by the broad laser spectrum with a nearly constant phase.

4. Very small instantaneous frequency change can be detected in real-time spectroscopy; hence, molecular structural change information, for the transition state, can be detected.

5. The dynamics of vibrational modes coupled to the electronic transition can be studied in relation to the decay dynamics of the electronic excited states at the same time, and under exactly the same experimental conditions.

6. The probe-dependent vibrational amplitude can be detected as a function of pump-probe delay time. Therefore, the corresponding vibrational phase information can be obtained.

In this study, using ultrafast excitation and broad-band detection, the dynamics of vibrational modes coupled to the electronic transition of Chl-a Q-band has been studied by real-time vibrational spectroscopy. Both pump and probe pulses possess a broadband of 200 nm, which can cover the whole Q_y absorption band of Chl-a simultaneously. The vibrational modes due to both ground-state and excited-state wave-packet motions have been observed and they are assigned according to the vibrational phase. The data containing both electronic relaxation and vibrational dynamics have been analyzed to obtain a more reliable relaxation mechanism of the excitations than a combination of individual studies of electronic and vibrational relaxations.

To the best of our knowledge, this is the first direct broadband study of the Chl-a vibrational signal coupled with the Q_y-band electronic transition at room temperature.

12.1.2   MATERIALS AND METHODS

In this experiment, both pump and probe pulses are generated from a noncollinear optical parametric amplifier (NOPA) laser system which is seeded by a white-light continuum [26–28], as described in the Supporting Material. The pulse duration of the NOPA output is 6.8 fs and covered the spectral range extending from 539 to 738 nm, as shown in Figure 12.1.1. After the sample, the pump-probe signal is detected by the combination detection system of the polychromator and multichannel lock-in amplifier (see the Supporting Material for further details). The spectral resolution of the total system is ~1.5 nm. The wavelength-dependent difference absorbance of the probe at 128 wavelengths is measured by changing the pump-probe delay times from 200 to 1800 fs with a 0.8-fs step. All the experiments are performed at a constant temperature (293 K).

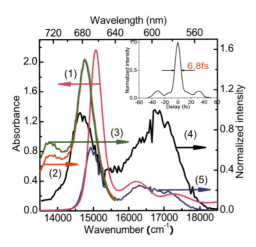

FIGURE 12.1.1   The absorption spectrum (a), fluorescence spectrum (b), and stimulated emission spectrum (c) of Chl-a, the laser spectrum (d), and the absorbed laser photon energy distribution spectrum by Chl-a with 0.5-mm thickness (e). (Inset) Temporal intensity profile of the compressed laser pulse.

Chl-a used in this study is extracted from spinach leaves and subsequently purified using chromatography, according to the method described in Strain and Svec [29]. The solvent used here is a mixture of petroleum ether and 2-propanol with a ratio of 100:5. Optical density is measured to be OD_{1mm} ¼ 2.16 at the $Q_y(0,0)$ transition peak of 664-nm using a 1-mm cell. The stationary absorption and the fluorescence spectra of the Chl-a solution were measured with an absorption spectrometer (UV-3101PC; Shimadzu, Kyoto, Japan) and a fluorophotometer (F-4500; Hitachi, Tokyo, Japan), respectively.

12.1.3 RESULTS AND DISCUSSION

12.1.3.1 STATIONARY ABSORPTION AND FLUORESCENCE SPECTRA AND TIME-RESOLVED DIFFERENCE ABSORPTION SPECTRUM

The stationary absorption, fluorescence, and stimulated emission spectra of Chl-a, together with the NOPA output laser spectrum and the absorbed laser spectrum (0.5-mm flow cell is used in real-time spectroscopy; see the Supporting Material for further details), are shown in Figure 12.1.1. There are three absorption peaks in the laser spectrum range, located at 664, 618, and 580 nm, respectively. There is no doubt that the peak at the longest wavelength belongs to the $Q_y(0–0)$ band of the Chl-a, and the observed absorption spectrum with a $Q_y(0,0)$ transition peak at 664 nm indicates that the sample used in this work consists of the monomer structure.

Figure 12.1.2a shows the two-dimensionally plotted difference absorption DA against the probe delay time and probe photon energy. On top of it, we plot the probe photon energy dependence of time-resolved spectra DA probed from 100 to 1200 fs with an integration time width of 100 fs. Near the two absorption bands at ~617 nm and 580 nm, a weak bleaching signal modulates the positive signal due to excited state absorption. These two bleaching signals are much weaker because of the strong induced absorption in this range. In addition, we note that the most intense negative peak signal at ~664 nm has an asymmetric profile, and the negative signal in the shorter wavelength side is much steeper than the longer side. It can be explained in the following way. The transient signal observed at a probe wavelength longer than 664 nm is due to the mixed contribution of both the ground state bleaching and the stimulated emission, while the DA at a probe wavelength shorter than 664 nm is dominated by the bleaching and the induced absorption.

Figure 12.1.2b shows the real-time DA traces at seven typical wavelengths. The sharp and intense peaks around zero probe-delay time are due to pump-probe coupling induced by the nonlinear process of the pump-probe-pump time ordering interaction of the laser fields and the interference between the scattered pump pulses and the probe pulses. The DA signal observed in the detection time of 200 fs to 1800 fs is nearly constant because of the nanosecond lifetime of the Q_y excited

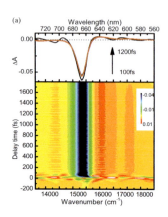

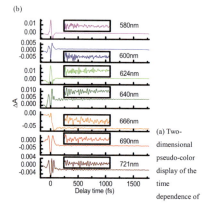

FIGURE 12.1.2 (a) Two-dimensional pseudo-color display of the time dependence of the absorbance changes (probe photon energy versus probe delay time, bottom figure) together with the time-resolved difference absorption spectrum probed from 100 to 1200 fs with an integration time-width of 100 fs (top figure). (b) Real-time traces at seven typical wavelengths. (Inset boxes) Enlarged signals between 100 and 1000 fs.

state [30,31]. It is obvious that the time-dependent DA traces shown in Figure 12.1.2b consist of the slow relaxation component due to electronic decay and the highly oscillating component due to molecular vibrations. The enlarged oscillating signals between 100 and 1000 fs are illustrated in the inset boxes of Figure 12.1.2b. It is noteworthy that the oscillations are especially clearly observed around the Q_y band.

12.1.3.2 ULTRAFAST DYNAMICS OF VIBRATIONAL MODES

As shown in Figure 12.1.2, predominant periodic modulation could be observed in the time-dependent difference absorption, which is caused by the transition probability change induced by molecular vibrations. To gain a better understanding of the vibronic coupling mechanisms, the vibrational mode frequencies and their corresponding amplitudes have been realized by the fast Fourier transform (FFT) analysis of the time-resolved difference absorption spectra.

The FFT analysis is performed in the range from 50 to 1800 fs. (Note that, to avoid possible interference effects between the scattered pump and probe pulses, the difference absorption signals ranging from delay 0 to 50 fs are not included.) The slow decay dynamics of relevant electronic states is removed by subtraction of averaged data over 200-fs time window from the raw data. After that, the FFT power spectra are calculated by using a Hanning window after the zero-padding procedure to increase the frequency resolution. The two-dimensional plot of the Fourier amplitude against the molecular vibration frequency and probe photon energy is presented in Figure 12.1.3a.

The FT amplitude spectra at several typical wavelengths are shown in Figure 12.1.3b. Various strong vibrational modes are highly concentrated at the bleaching band at ~664 nm. Twenty-seven different frequencies of 214, 259, 300, 346, 407, 519, 565, 621, 667, 744, 799, 915, 982, 1043, 1084, 1124, 1175, 1252, 1353, 1415, 1460, 1516, 1582, 1613, 1648, 1699, and 1770 cm^{-1} are obtained. Most of them are similar to those observed by Raman and/or the HB spectroscopy by many groups; however, the modes of 300, 621, 1582, 1648, and 1770 cm^{-1} are observed (to our knowledge) for the first time in this study. Most of the intense frequency signals below 1000 cm^{-1} are assigned to vibrations of the tetrapyrrole skeleton and deformations of the peripheral substituent groups [6,7], while those higher than 1000 cm^{-1} are mainly assigned to the CH_3 bending, CH bending, and CO, CC, and CN stretching vibrations [5,8,32].

In Chl-a, there are only small differences in frequency of the same vibrational modes between the ground and excited states [11,13,14]. Therefore, it is difficult to distinguish whether they are due to the wave-packet motion in the ground or excited state due to our limited frequency resolution. Thanks to the broad probe spectral band detection of real-time traces, we can calculate the probe wavelength dependence of the initial phases of the molecular vibrations. It provides information for identifying the wave-packet motion either in the ground state or in the excited state [33–35]. If the time dependence of the absorbance change is due to the wave packet prepared on the ground-state

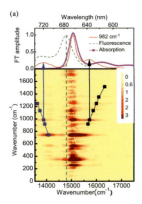

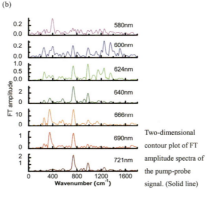

FIGURE 12.1.3 (a) Two-dimensional contour plot of FT amplitude spectra of the pump–probe signal. (Solid line) Mass centers of FT amplitude spectra at side bands. (Top) FT amplitude of the 982 cm^{-1} mode together with the stationary fluorescence and absorption spectra. (b) FT amplitude spectra of the corresponding traces in Figure 12.1.2b.

potential energy surface, the initial vibrational phase will be $5\pi/2$. If the time-dependent amplitude modulation is induced by the excited state wave-packet motion, the phase will be 0 or 5π. Otherwise, it will be due to the mixed contributions from wave-packet motions in the ground and excited states. Using this simple criterion, we can easily assign the vibrational modes of Q_y band. Vibrational modes with frequencies of 214, 259, 300, 346, 407, 565, 1084, 1175, 1460, 1613, and 1770 cm^{-1} result from the excited state wave-packet motion, while those at 519, 621, 799, 1415, 1582, and 1648 cm^{-1} are corresponding to the ground state wave-packet motion; the other modes with 667, 1043, and 1699 cm^{-1} are contributed from wave-packet motions in both ground and excited states. For example, Figure 12.1.4 shows the initial phases of three different values together with their vibrational amplitudes.

The Fourier power spectral line width is inversely proportional to the dephasing time including the effects of temperature in-sensitive inhomogeneity and temperature-sensitive homogeneous dephasing. With increasing temperature, the homogeneous contribution increases because of increased mode-mode coupling. In the Raman spectrum at 15 K [7], the width can be considered to be dominated by inhomogeneous width, and it is calculated to be ~15 cm^{-1}. It is found that the line widths in the real-time spectroscopy in our experiment are ~3 times as broad as those measured in the Raman experiment at low temperature, which is due to the homogeneous origin. Even though the excited-state lifetime of Chl-a is in the nanosecond time range, the electronic relaxation will be affected by the fast vibrational relaxation through vibronic coupling, which will dissipate some of the electronic energy and hence compete with the Chl-a involved energy transfer process. Special attention should be paid to the low-frequency modes (<250 cm^{-1}), which have been classified as intermolecular vibrations [16,36].

Because the rate of the resonance energy transfer through dipole-dipole interaction depends on the distance between the energy donor and acceptor, the energy transfer between Chl-a and its neighboring molecule is expected to be periodically modulated by the low-frequency intermolecular vibrations in their dephasing times of <1 ps.

The high-frequency modes at 744, 915, 982, 1124, 1251, 1353, and 1516 cm^{-1} exhibit more complex spectral distributions than those discussed above, and the simple criterion cannot be performed any more. As shown in Figure 12.1.3a, side bands are clearly observed at both sides of the most intense vibrational band for these modes (only one side for 1353 and 1516 cm^{-1} mode because of probe range). The probe photon energy dependence of FT amplitude for the 982 cm^{-1} mode is shown

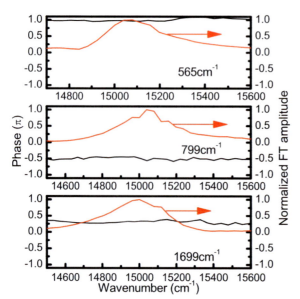

FIGURE 12.1.4 Phases of the vibrational modes with frequencies of 565 (π), 799 ($\pi/2$), and 1699 (neither π nor $\pm\pi/2$) cm^{-1} together with their normalized amplitudes.

at the top of Figure 12.1.3a as an example. In order to investigate the spectral position and the profile of this sideband in detail, the peak position of each vibrational mode has been recorded as a function of the probe photon energy. To minimize the possible error, the side-band mass centers are defined by the first moment given by

$$\bar{\omega}_v = \frac{\displaystyle\int_{\omega_1}^{\omega_2} F_v(\omega) \cdot \omega \, d\omega}{\displaystyle\int_{\omega_1}^{\omega_2} F_v(\omega) d\omega} \tag{12.1.1}$$

where $F_v(\omega)$ is the probe photon frequency (ω)-dependent FT amplitude spectra of the vibrational modes with a vibrational frequency of ω_v.

The calculation results are displayed in Figure 12.1.3a and connected by black and blue lines in the higher and lower probe energy ranges, respectively. These two fitted lines have slopes of -0.96 ± 0.03 and 0.97 ± 0.01, which are very close to ± 1. It means that the side bands of each vibrational mode are spectrally distributed according to a linear function (slope $K = \pm 1$) of their own vibrational frequencies. Therefore, they are symmetric with respect to a center axis.

Because the Huang-Rhys factors in Chl-a are very small (<0.05) [11], the displacement in the coordinate space from the potential minimum of the ground to that of the excited states is small. In the experiment, what we observed is the modulation of the electronic transition probability induced by a wave-packet motion through vibronic coupling.

Therefore, if the vibrational modes appearing at side bands are due to the wave-packet motion in the ground state, the vibrational amplitudes are related to the absorption spectrum from the ground to the excited state transition. The peaks of the vibrational amplitudes are expected to symmetrically appear on both sides of transition energy from the ground to the excited state, displayed as Stokes and anti-Stokes bands. If the wave-packet motion on the excited-state potential surface modulates the stimulated emission intensity, the observed vibrational amplitude at side bands will have peaks symmetric with respect to the stimulated emission peak, also displaying as Stokes and anti-Stokes shifted structures. As shown in Figure 12.1.3a, the measured vibrational signals at side bands are spectrally symmetric with respect to the axis of the stimulated emission peak, so they were assigned to vibrational modes coupling to the stimulated emission from the Q_y electronic excited state. However, the signals at the central band are surprisingly weak; in fact, they nearly vanish in this spectral range.

Why do the vibrational signals disappear in the symmetric center and those at sidebands have a linear spectral distribution? Here, to analyze the complicated results, we use a coupled three-level model to explain the experimental result by the theory of nonlinear ultrafast spectroscopy in the following discussion. In this study, all the vibrational modes in the discussion of the following section showing clear side bands are much larger than $200 \, cm^{-1}$. Because of the experimental temperature of 293 K, the initial population on the vibrational levels with vibrational quantum numbers larger than zero can be neglected for all these involved in the vibrational modes. That means only the lowest vibrational level of the electronic ground state, $|g_0\rangle$, could be regarded as the initial state in the discussion of the following section.

12.1.4 THEORY AND DISCUSSION

It is well known that the wave packet can be generated either via the simultaneous coherent excitation of the vibronic polarization of several vibrational levels in the excited state or via impulsive stimulated Raman scattering in the ground state. The former case is known as a split excited state, which can be represented in terms of V-type. Analogously, the latter case is referred to as a split ground state, which is expressed in terms of Λ-type interaction. In the previous article [37], the

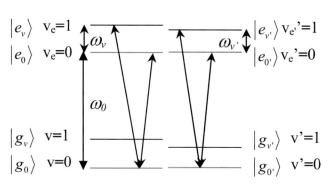

FIGURE 12.1.5 Three-level system describing the excitation processes of the vibrational coherence generation in the excited state. The lowest vibrational state in the ground and excited states is represented by g_0 and e_0; g_v and e_v represent where the vibrational coherence exists, where v and v^0 denote vibrational mode with different frequency. The value u_0 denotes the $Q_y(0–0)$ emission (central band) or the one with emission energy equal to it. The values ω_v and u_v0 represent different vibrational frequencies.

coherent excitation of the vibrations on the excited state was prevented so that only the ground-state vibrations were excited. Hence, the L-type interaction could be used. In our experiment, due to the good overlap of the broad-band laser and the absorption band, the vibrations in the excited states can be coherently excited, as shown in Figure 12.1.5, where u_0 denotes the $Q_y(0–0)$ emission (central band) or the one with emission energy equal to it.

Because there are only small frequency differences of the same vibrational mode between the ground state and that of the excited state [11,13,14], we assume the vibrational energy (ω_v) in the ground and excited states to be identical. In addition, because the Huang-Rhys factors of Chl-a are very significantly small ($S<0.05$), most of the Franck-Condon intensity is concentrated in the 0–0 transition. Therefore, we only considered the 0–0 and 0–1 transition in the following. As mentioned above, only a stimulated emission process should be considered in the following discussions. In total, there are four related NL processes related to our experimental results. As defined in the Supporting Material, they are NL process A (Stokes band, located at the lower probe photon energy range); NL process B (located around the symmetric center); NL process A' (anti-Stokes band, located at the higher probe photon energy range); and NL process B' (located around the symmetric center), respectively.

The observed pump-probe signals, represented by the difference absorption for NL process A and NL process B, can be expressed by [37–44]

$$\Delta A_A(\omega,t) = \frac{2\text{Im}\left[\tilde{E}_{\text{pr}}^*(\omega)\tilde{P}^{(3)}(\omega)\right]}{\left|\tilde{E}_{\text{pr}}(\omega)\right|^2} \propto \exp\left(-\frac{t}{T_{2v}}\right)$$

$$\times \frac{-\left[\omega-(\omega_0-\omega_v)\right]\sin(\omega_v t)+\gamma_2\cos(\omega_v t)}{\left[\omega-(\omega_0-\omega_v)\right]^2+\gamma_2^2}, \tag{12.1.2}$$

$$\Delta A_B(\omega,t) \propto \exp\left(-\frac{t}{T_{2v}}\right)\frac{(\omega-\omega_0)\sin(\omega_v t)+\gamma_2\cos(\omega_v t)}{(\omega-\omega_0)^2+\gamma_2^2}, \tag{12.1.3}$$

respectively, and the dephasing rate $\gamma_2 \equiv 1/T_2 \cdot \tilde{E}_{\text{pr}}^*(\omega)$ and $P^{(3)}(\omega)$ denote the Fourier transforms of $E_{\text{pr}}(t)$ and $P^{(3)}(t)$, respectively. The spectral widths in Eqs. (12.1.2) and (12.1.3) can be determined using the electronic dephasing time T_2, whereas the vibrational dephasing time is given by the bandwidth of the FT amplitude of each mode. Therefore, the electronic dephasing time and the vibrational dephasing time can be independently determined by this principle. Thanks to the measurement of real-time resolved vibrational amplitude, the dependence of the vibrational initial phase upon impulsive excitation on the probe photon energy can also be determined. The total observed difference absorption signal can be expressed by

$$\Delta A(\omega,t) = \Delta A_e(\omega)\exp\left(\frac{-t}{\tau_e}\right) + \sum_i \delta\Delta A_{vi}(\omega,t;\omega_{vi}) \qquad (12.1.4)$$

Here, the first term characterizes the slow dynamic component of the real-time ΔA traces, and τ_e is the electronic decay time depending on the sample. The second term represents the molecular vibration, which is a sum of all the contributions of vibrational modes, v, with frequency ω_{vi} and corresponds to the oscillation component in the ΔA traces. For a certain vibrational mode with a frequency of ω_v, $\delta\Delta A$ is given by

$$\delta\Delta A_v(\omega,t;\omega_v) = \delta A_v(\omega,\omega_v)\exp\left(\frac{-t}{\tau_v(\omega_v)}\right) \times \cos\left[\omega_v t + \varphi(\omega,\omega_v)\right] \qquad (12.1.5)$$

where τ_v is the vibrational dephasing time for the vibrational mode v including both homogeneous and inhomogeneous contributions. The corresponding phase for NL processes A and B of this vibrational mode can be derived from Eqs. (12.1.2), (12.1.3), and (12.1.5) by

$$\phi_A(\omega_v,\omega) = \arctan\left(\frac{\left|\omega-(\omega_0-\omega_v)\right|}{\gamma_2}\right) \qquad (12.1.6)$$

$$\phi_B(\omega_v,\omega) = \arctan\left(\frac{-(\omega-\omega_0)}{\gamma_2}\right) \qquad (12.1.7)$$

Similarly, the corresponding phase for NL processes A^0 and B^0 of vibrational mode ω_v can be derived as

$$\phi_{A'}(\omega_v,\omega) = \arctan\left(\frac{-\left[\omega-(\omega_0+\omega_v)\right]}{\gamma_2}\right) \qquad (12.1.8)$$

$$\phi_{B'}(\omega_v,\omega) = \arctan\left(\frac{(\omega-\omega_0)}{\gamma_2}\right) \qquad (12.1.9)$$

Because it is easy to satisfy the conditions of $\left|\omega-(\omega_0-\omega_v)\right|$ and $\left|\omega-\omega_0\right| \gg \gamma_2$, the difference of the phases should be p between the higher-energy region ($\omega > \omega_0$ for process A and A', $\omega > \omega_0 \pm \omega_v$ for process B and B') and the lower-energy region ($\omega < \omega_0 \pm \omega_v$ for process A and A', $\omega < \omega_0$ for process B and B'). Specifically, the phase change from $-\pi/2$ to $+\pi/2$ is expected at $\sim \omega = \omega_0 - \omega_v$ with the increasing of the laser photon energy because of excited state wave-packet motion in the case of NL process A; and change from $+\pi/2$ to $-\pi/2$ is expected at $\sim\omega = \omega_0 + \omega_v$ for NL process A' with the increasing of the photon energy ω from $\omega_0 < \omega_0 - \omega_v$ to $\omega > \omega_0 + \omega_v$ because the signs of the phases are opposite in both cases. Similarly, the pump-probe signals due to NL processes B and B' also have antiphase relations.

The physical meaning of the phase relation is interpreted as follows. After excitation, a pump could generate molecular vibrational coherence in the Q_y excited state between $v=0$ and $v1$ levels. Then a probe comes to the sample and interacts with the vibrational coherent polarization between $v=0$ and $v1$, resulting in the possibility of laser-Stokes energy exchange as well as laser-anti-Stokes exchange energy. The energy exchange process between the probe and the molecular vibration is then dependent on the molecular vibrational phase, which determines whether the spectra in the range of $\omega = \omega (\pm\omega_v)$ are amplified and/or deamplified. For the antiphase relation between the oscillation signals around the spectral range $\omega \pm \omega_v$, the signal around $\omega-\omega_v$ is amplified if the one around $\omega + \omega_v$ is deamplified at the same time. As for the signals resulting from NL process B and B', because both of them share the same spectral range $\omega = \omega_0$ and are satisfied with antiphase relation, their oscillations are expected to take place concomitantly and then cancel each other out. It is consistent with the experimental result shown in Figure 12.1.3 that, despite the clear vibrational signals on both the lower and higher energy sides due to NL processes A and A', respectively, the vibrations in the center position (straight dashed line) have nearly disappeared.

The phase of the vibrational modes with the frequencies of 982 and 1251 cm^{-1} at $\omega_0 - \omega_v$ range are shown in Figure 12.1.6. We chose these two modes as examples because they show the clearest sidebands around their corresponding $u_0 5 u_v$ range. Both of the vibrational phases have shown a clear phase jump from $-\pi/2$ to $+\pi/2$ (~13800 cm^{-1} for the 1252 cm^{-1} mode and 14020 cm^{-1} for 982 cm^{-1}), as expected. The electronic dephasing time T_2 is determined to be 41 ± 2 fs from the real-time traces of absorbance change in the negative delay time range [45], which is comparable with the one of 54 fs obtained for bacteriochlorophyll [46]. However, the dephasing time is calculated to be ~150 fs from Eq. (12.1.6), which is rather longer than the experimental value. This disagreement may be explained as due to the deviation from the simple three-level model. In a real system, there are other possible contributions to vibration amplitude through modulation of electronic transitions from the excited Q_y to other states, which are not included in the simple three-level model. For example, the induced absorption from the lowest electronic excited state to a higher one may couple with the same vibrational modes and affect the calculation result.

As shown in Figure 12.1.2a, there is another highly oscillating band centered at 14100 cm^{-1} with a full bandwidth of ~300 cm^{-1}, which gives rise to a concentrated vibrational band in addition to those coupled with ground-state bleaching and those symmetrically distributed due to the stimulated emission in Figure 12.1.3. Even though stimulated emission provides the largest contribution in the signal in the $u_0 u_v$ range, induced absorption at ~14100 cm^{-1} is still intense enough to compete with it because the total negative signal is close to zero. In addition, considering the very small spectral overlap between these vibrations and the stimulated emission peak, it is reasonable to assign these vibrations coupled to the transition from the Q_y excited state to a higher electronic state by induced absorption, which is peaked at ~14100 cm^{-1} (~709 nm). Recently, the excited state absorption spectrum of Chl-a has been studied by the Z-scan technique using the white-light continuum, but the excited-state absorption could only be observed in the wavelength region below the ground-state absorption [47]. In this study, the excited-state absorption peak at ~709 nm can be recognized through the evidence from both the electronic and vibrational dynamics simultaneously. Because of the coexistence of stimulated emission and induced absorption, even though the phase jump from $\pi/2$ to $-\pi/2$ still maintains, the jump position and width predicted by the two-electronic level theory have been affected.

The phase characteristic at $\omega_0 + \omega_v$ range has also been studied, but it is too complicated (not shown here) to be compared with the calculation. That might be due to the much stronger influence of the induced absorption in this spectral range. As shown in Figure 12.1.2a, the positive ΔA

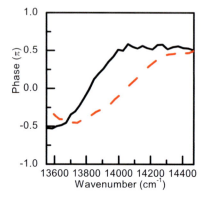

FIGURE 12.1.6 Phase spectra of the mode of 982 cm^{-1} (dashed line) and 1251 cm^{-1} (solid line) observed in the real-time vibration spectra probed at lower energy region.

signal due to induced absorption is most intense in the higher spectral range (15500–$16500\,cm^{-1}$), and all the possible electronic processes, the ground-state bleaching, stimulated emission, and induced absorption, can take place in this photon energy range. Even though the vibrational amplitude position is not much affected, the vibrational phase is much more sensitive, and the influence from other processes is too intense to be neglected for the phase calculation and hence it must be explored with a model far beyond the simple three level (two electronic levels and one vibrational level) system.

12.1.5   CONCLUSIONS

This subsection shows that the usefulness of study in dynamic processes in complex biological molecules like chlorophyll. A femtosecond pump-probe experiment is performed on Chl-a using a 6.8-fs visible pulse. Electronic relaxation and vibrational dynamics are simultaneously observed by real-time vibrational spectroscopy with a 0.8-fs delay step at room temperature. Wave-packet motions in the excited state coupled to the stimulated emission are observed over a broad detection from 539 to 738 nm. The modulation signals are found to nearly vanish around the stimulated emission peak; however, they are symmetrically distributed with respect to the emission peak as a function of their own vibrational frequencies at both sides. The corresponding nonlinear process has been studied using a three-level model, from which the probe wavelength dependence of the phase of the periodic modulation and the spectral distribution of vibrational amplitude has been explained in detail. Because of the scarcity of Q_y-band vibrational information in Chl-a at a physiologically relevant temperature, low-temperature data are generally used in the calculation of the absorption profile at room temperature.

Due to the complexity of the temperature-dependent characteristic of Chl-a, there is no current consensus as to how well the low-temperature data will satisfy the physiologically relevant temperatures situation, and many controversial issues have arisen in these calculations [48]. The Q_y-band vibrational information obtained in this study seems the most straightforward initial step for resolving these and holds promise to reveal hidden features in the absorption band shape of Chl-a with limited structure at physiologically relevant temperature. The research products presented in the present section were conducted by collaboration among the following people: J. Du, T. Teramoto, K. Nakata, E. Tokunaga, and T. Kobayashi [48].

SUPPORTING MATERIAL

Additional information with four figures and detailed experimental procedures are available at http://www.biophysj.org/biophysj/supplemental/S0006-3495(11)00841-1.

REFERENCES

1. M. Lutz. "Resonance Raman spectra of chlorophyll in solution," *J. Raman Spectrosc.* **2**, 497–516 (1974).
2. M. Fujiwara and M. Tasumi. "Meta-sensitive bands in the Raman and infrared spectra of intact and metal-substituted chlorophyll a," *J. Phys. Chem.* **90**, 5646–5650 (1986).
3. A. A. Pascal, L. Caron, and B. Robert. "Resonance Raman spectroscopy of a light-harvesting protein from the brown alga Laminaria saccharina," *Biochemistry.* **37**, 2450–2457 (1998).
4. Y. Koyama, Y. Umemoto, and A. Akamatsu. "Raman spectra of chlorophyll forms," *J. Mol. Struct.* **146**, 273–287 (1986).
5. L. L. Thomas, J. H. Kim, and T. M. Cotton. "Comparative study of resonance Raman and surface-enhanced resonance Raman chlorophyll a spectra using Soret and red excitation," *J. Am. Chem. Soc.* **112**, 9378–9386 (1990).
6. J. R. Diers, Y. Zhu, and D. F. Bocian. "Q_y-excitation resonance Raman spectra of chlorophyll a and bacteriochlorophyll c/d aggregates: Effects of peripheral substituents on the low-frequency vibrational characteristics," *J. Phys. Chem.* **100**, 8573–8579 (1996).

7. C. Zhou, J. R. Diers, and D. F. Bocian. "Q_y-excitation resonance Raman spectra of chlorophyll a and related complexes: Normal mode characteristics of the low-frequency vibrations," *J. Phys. Chem. B.* **101**, 9635–9644 (1997).

8. P. S. Woolley, B. J. Keely, and R. E. Hester. "Surface-enhanced resonance Raman spectroscopic identification of chlorophyll a allomers," *J. Chem. Soc., Perkin Trans.* **2**, 1731–1734 (1997).

9. K. K. Rebane and R. A. Avarmaa. "Sharp line vibronic spectra of chlorophyll and its derivatives in solid solutions," *Chem. Phys.* **68**, 191–200 (1982).

10. R. A. Avarmaa and K. K. Rebane. "High-resolution optical spectra of chlorophyll molecules," *Spectrochim. Acta [A].* **41A**, 1365–1380 (1985).

11. J. K. Gillie, G. J. Small, and J. H. Golbeck. "Nonphotochemical hole burning of the native antenna complex of photosystem I (PSI-200)," *J. Phys. Chem.* **93**, 1620–1627 (1989).

12. A. Pascal, E. Peterman, and B. Robert. "Structure and interactions of the chlorophyll a molecules in the higher plant Lhcb4 antenna protein," *J. Phys. Chem. B.* **104**, 9317–9321 (2000).

13. E. J. G. Peterman, T. Pullerits, and H. van Amerongen. "Electron phonon coupling and vibronic fine structure of light-harvesting complex II of green plants: temperature dependent absorption and high-resolution fluorescence spectroscopy," *J. Phys. Chem.* **101**, 4448–4457 (1997).

14. J. Pieper, J. Voigt, and G. J. Small. "Chlorophyll a Franck-Condon factors and excitation energy transfer," *J. Phys. Chem. B.* **103**, 2319–2322 (1999).

15. I. Renge, K. Mauring, and R. Avarmaa. "Vibrationally resolved optical spectra of Chlorophyll derivatives in different solid media," *J. Phys. Chem.* **90**, 6611–6616 (1986).

16. M. Rätsep, J. Linnanto, and A. Freiberg. "Mirror symmetry and vibrational structure in optical spectra of chlorophyll a," *J. Chem. Phys.* **130**, 194501 (2009).

17. G. R. Zucchelli, R. C. Jennings, and O. Cremonesi. "The calculated in vitro and in vivo chlorophyll a absorption band shape," *Biophys. J.* **82**, 378–390 (2002).

18. S. Krawczyk. "The effects of hydrogen bonding and coordination interaction in visible absorption and vibrational spectra of chlorophyll a," *Biochim. Biophys. Acta.* **976**, 140–149 (1989).

19. Y. J. Shiu, Y. Shi, and S. H. Lin. "Femtosecond spectroscopy study of electronically excited states of chlorophyll a molecules in ethanol," *Chem. Phys. Lett.* **378**, 202–210 (2003).

20. U. Megerle, I. Pugliesi, and E. Riedle. "Sub-50fsbroadbandabsorption spectroscopy with tunable excitation: putting the analysis of ultrafast molecular dynamics on solid ground," *Appl. Phys. B.* **96**, 215–231 (2009).

21. D. Polli, L. Lu̇er, and G. Cerullo. "High-time-resolution pumpprobe system with broadband detection for the study of time-domain vibrational dynamics," *Rev. Sci. Instrum.* **78**, 103108 (2007).

22. A. Yabushita, Y. H. Lee, and T. Kobayashi. "Development of a multiplex fast-scan system for ultrafast time-resolved spectroscopy," *Rev. Sci. Instrum.* **81**, 063110 (2010).

23. T. Kobayashi, J. Zhang, and Z. Wang. "Non-Condon vibronic coupling of coherent molecular vibration in MEH-PPV induced by a visible few-cycle pulse laser," *N. J. Phys.* **11**, 013048 (2009).

24. T. Kobayashi and S. Nagakura. "Reabsorption and high density excitation effects on the time-resolved fluorescence spectra of anthracene crystal," *Molec. Cryst. Liq. Cryst.* **26**, 33–43 (1974).

25. A. Cupane, M. Leone, and R. Schweitzer-Stenner. "Dynamics of various metal-octaethylporphyrins in solution studied by resonance Raman and low-temperature optical absorption spectroscopies: Role of the central metal," *J. Phys. Chem. B.* **102**, 6612–6620 (1998).

26. A. Shirakawa and T. Kobayashi. "Noncollinearly phase-matched femtosecond optical parametric amplification with a 2000 cm^1 bandwidth," *Appl. Phys. Lett.* **72**, 147–149 (1998).

27. A. Shirakawa, I. Sakane, and T. Kobayashi. "Pulse-front-matched optical parametric amplification for sub-10-fs pulse generation tunable in the visible and near infrared," *Opt. Lett.* **23**, 1292–1294 (1998).

28. A. Baltuska, T. Fuji, and T. Kobayashi. "Visible pulse compression to 4fs by optical parametric amplification and programmable dispersion control," *Opt. Lett.* **27**, 306–308 (2002).

29. H. H. Strain and W. A. Svec. "Extraction, separation, estimation and isolation of the chlorophylls," in *The Chlorophylls*, edited by L. P. Vernon and G. R. Seeley, Academic Press, New York, pp. 21–66 (1966).

30. P. Martinsson, J. A. I. Oksanen, and E. Åkesson. "Dynamics of ground and excited state chlorophyll a molecules in pyridine solution probed by femtosecond transient absorption spectroscopy," *Chem. Phys. Lett.* **309**, 386–394 (1999).

31. M. Linke, A. Lauer, and K. Heyne. "Three-dimensionalorientation of the Qy electronic transition dipole moment within the chlorophyll a molecule determined by femtosecond polarization resolved VIS pump-IR probe spectroscopy," *J. Am. Chem. Soc.* **130**, 14904–14905 (2008).

32. Z. L. Cai, H. Zeng, and A. W. Larkum. "Raman spectroscopy of chlorophyll d from Acaryochloris marina," *Biochim. Biophys. Acta.* **1556**, 89–91 (2002).

33. A. T. N. Kumar, F. Rosca, and P. M. Champion. "Investigations of amplitude and phase excitation profiles in femtosecond coherence spectroscopy," *J. Chem. Phys.* **114**, 701–724 (2001).

34. A. T. N. Kumar, F. Rosca, and P. M. Champion. "Investigations of ultrafast nuclear response induced by resonant and nonresonant laser pulses," *J. Chem. Phys.* **114**, 6795–6815 (2001).

35. M. Ikuta, Y. Yuasa, and T. Kobayashi. "Phase analysis of vibrational wave packets in the ground and excited states in polydiacetylene," *Phys. Rev. B.* **70**, 214301 (2004).

36. M. Raˇtsep, J. Pieper, and A. Freiberg. "Excitation wavelength dependent electron-phonon and electron-vibrational coupling in the CP29 antenna complex of green plants," *J. Phys. Chem. B.* **112**, 110–118 (2008).

38. S. Mukamel. *Principles of Nonlinear Optical Spectroscopy*, Oxford University Press, New York (1995).

39. W. W. Parson. *Modern Optical Spectroscopy: With Examples from Biophysics and Biochemistry*, Springer-Verlag, Berlin (2007).

37. N. Ishii, E. Tokunaga, and T. Kobayashi. "Optical frequency- and vibrational time-resolved two-dimensional spectroscopy by real-time impulsive resonant coherent Raman scattering in polydiacetylene," *Phys. Rev. A.* **70**, 023811 (2004).

40. S. Mukamel. "Multidimensional femtosecond correlation spectroscopies of electronic and vibrational excitations," *Annu. Rev. Phys. Chem.* **51**, 691–729 (2000).

41. W. T. Pollard, S.-Y. Lee, and R. A. Mathies. "Wave packet theory of dynamic absorption spectra in femtosecond pump-probe experiments," *J. Chem. Phys.* **92**, 4012–4029 (1990).

42. I. A. Walmsley and C. L. Tang. "The determination of electronic dephasing rates in time-resolved quantum beat spectroscopy," *J. Chem. Phys.* **92**, 1568–1574 (1990).

43. I. A. Walmsley, M. Mitsunaga, and C. L. Tang. "Theory of quantum beats in optical transmission-correlation and pump-probe experiments for a general Raman configuration," *Phys. Rev. A.* **38**, 4681–4689 (1988).

44. M. Mitsunaga and C. L. Tang. "Theory of quantum beats in optical transmission-correlation and pump-probe measurements," *Phys. Rev. A.* **35**, 1720–1728 (1987).

45. T. Kobayashi and A. Yabushita. "Dynamics of vibrational and electronic coherences in the electronic excited state studied in a negative-time range," *Chem. Phys. Lett.* **482**, 143–147 (2009).

46. N. J. Cherepy, A. P. Shreve, and R. A. Mathies. "Electronic and nuclear dynamics of the accessory bacteriochlorophylls in bacterial photosynthetic reaction centers from resonance Raman intensities," *J. Phys. Chem. B.* **101**, 3250–3260 (1997).

47. L. De Boni, D. S. Correa, and C. R. Mendonça. "Excited state absorption spectrum of chlorophyll a obtained with white-light continuum," *J. Chem. Phys.* **126**, 165102–165104 (2007).

48. J. Du, T. Teramoto, K. Nakata, E. Tokunaga, and T. Kobayashi. "Real-time vibrational dynamics in chlorophyll a studied with a few-cycle pulse laser," *Biophys. J.* **101**, 995–1003 (2011).

12.2 Time-Resolved Spectroscopy of Ultrafast Photoisomerization of Octopus Rhodopsin Under Photoexcitation

12.2.1 INTRODUCTION

The visual process in a photoactive cell consists of a series of chemical reactions mediated via several intermediates and culminating in the stimulation of the optic nerve. Rhodopsin (Rh) is a photoreceptive pigment for twilight vision, which consists of the apoprotein opsin and the 11-cis-retinal chromophore. Photoexcitation of Rh leads to a series of intermediates that eventually initiate an enzyme cascade triggering electric excitation [1–5]. The Rh pigment has been studied more than any other visual pigment, including those responsible for color vision, because of its relatively easy preparation. The following section describes the current understanding of the photoisomerization of retinal based on previous studies [6–27].

The first intermediate of Rh is called primeRh [8] (or photoRh [9]) and the second intermediate, identified earlier than the first, is called bathoRh [2]. Comparison of the absorption spectrum of primer [10] and studies of 11-cis-locked Rh [11,12] indicated that the chromophore in primeRh has a distorted all-trans configuration. Therefore, to form batho Rh, the chromophore must undergo cis–trans isomerization. Studies on the cis–trans isomerization yield of the Rh chromophore in solution [13,14] and in protein [15] have shown that the isomerization yield is enhanced by more than an order of magnitude in the protein environment. Other studies have inferred that the first intermediate (primeRh) is the nonthermal state of bathoRh [16,17]. The transitions between the intermediates and their time constants have been reported as follows. After photoexcitation of Rh, a Franck–Condon (FC) state proceeds to a conical intersection (CI) on the potential energy surface of the electronic excited state in ~100 fs as estimated in time-resolved measurements [6,17–20]. Femtosecond transient absorption measurement has elucidated that curve crossing to the ground state takes about 200 fs to form the highly distorted primeRh [21]. Picosecond Raman study clarified that a temporal decay of a few picoseconds corresponds to conversion from primeRh to bathoRh [22].

Relatively few studies [18–21] exist on the ultrafast dynamics of Rh because of the difficulties in preparation due to the fragility of the samples. In place of detailed studies on Rh, bacteriorhodopsin (bR) has been extensively studied as a model system as it is relatively robust in experiments. In the femtosecond time-resolved stimulated Raman study of bR [23], the photoisomerization of bR has been reported to follow the process.

$$
\begin{array}{cccc}
\text{Hv} & <200\ \text{fs} & 500\ \text{fs} & 3\ \text{ps} \\
\end{array}
$$

$$
\text{bR} \rightarrow \text{FC}(H) \longrightarrow - \text{CI}(I_{460}) \longrightarrow J \longrightarrow K \rightarrow
$$

In previous work, we performed ultrafast time-resolved spectroscopy of bR to clarify its electronic dynamics and vibration dynamics in detail [24]. The results elucidated the dynamics of the

DOI: 10.1201/9780429196577-89

intermediates, which have lifetimes consistent with previous reports on dynamic hole-burning [25], time-resolved fluorescence [26], and transient absorption [27]. However, the relaxation dynamics of bR and Rh are expected to be quite different considering the differences in structural transition between the retinal in bR (from all-trans to 13-cis configuration) and that in Rh (from 11-cis to all-trans configuration).

In the present work, we have performed ultrafast time-resolved spectroscopy of octopus Rh using a sub-5-fs laser pulse and a 128-channel detection system. The ultrafast time resolution of sub-5-fs enabled us to study both the electronic and vibration dynamics, and the multichannel detector array allowed simultaneous observation of the signals over all probe wavelengths, avoiding sample degradation. The observed ultrafast dynamics of Rh showed a difference from the results for bR obtained in the previous work, as the present work clarified that the transition from the FC state to CI proceeds faster in octopus Rh than in bR.

12.2.2   EXPERIMENTAL METHODS

12.2.2.1   FEMTOSECOND SPECTROSCOPY APPARATUS

The pulsed light sources and setup for femtosecond time-resolved absorption spectroscopy used in this work are described in previous reports from our group [28–30]. A 4.7-fs, 1-kHz pulse train was generated from a noncollinear optical parametric amplifier (NOPA) with a pulse compressor [28], covering a wavelength range from 528 to 727 nm. The laser spectrum obtained by NOPA is shown in Figure 12.2.1. The energy and peak intensity of the pump pulses at the sample were ~10 nJ and 10 GW/cm^{-2}, respectively, which are approximately 10 times higher than the probe pulses. The pulse energies (intensities) of the pump and probe pulses at the sample position were approximately 20 and 5 nJ (7.5×10^{14} and 8.5×10^{13} photons/cm^{-2}), respectively.

To measure weak pump–probe signals at various wavelengths, we used a multichannel lock-in amplifier, which was specifically designed for the simultaneous detection of low-intensity signals over the entire spectral region. The multichannel lock-in amplifier consists of 128 lock-in amplifiers connected to 128 avalanche photodiodes. In the present experiment, multichannel signals, spectrally resolved by a polychromator (JASCO M25-TP), were detected by the avalanche photodiodes with the lock-in amplifiers in reference to pump pulses modulated at 500 Hz by a mechanical

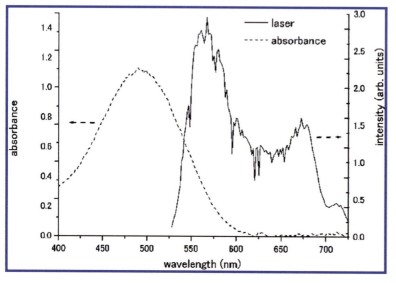

FIGURE 12.2.1 Laser spectrum (solid line) and absorption spectrum (dotted line) of octopus Rh.

chopper. The normalized transmittance changes in the range extending from 528 to 727 nm were measured for −100 to 2000 fs with a 1-fs interval.

12.2.2.2 OCTOPUS RH

The microvillar membranes of the octopus retina (Mizudako, Paroctopus dofleini) were isolated by sucrose flotation (34 wt %, buffer A) (400 mM KLC, 10 mM $MgCl_2$ + 20 µM p-APMSF) repeated twice. The obtained pellet was washed several times with buffer A and with buffer B (10 mM MOPS [pH 7.4], 1 mM DTT, 20 µM p-APMSF). The final products were suspended in buffer B and kept at −80°C in the dark. A part of this frozen sample was thawed and centrifuged. The as-obtained pellet was solubilized in H_2O with 2 mM MOPS [pH 7.4] and 2% sucrose monolaurate (L-1690 or SM1200). The solution was centrifuged, and the supernatant was used as the sample, which was added to a 1 mm-thick stationary optical cell for experiments without circulation. The absorption spectrum of the sample showed no difference between before and after the pump–probe measurement within the margin of the error of about 10%. The peak absorbance at 490 nm of the sample solution was 1.1. The absorption spectrum of the octopus Rh is plotted in Figure 12.2.1 together with the laser spectrum.

12.2.3 RESULTS AND DISCUSSION

Figure 12.2.2a shows the real-time absorption difference spectra, ΔA, in the time region between −100 and 2000 fs over the spectral region from 528 to 727 nm. Real-time traces probed at 577, 610, 641, and 682 nm are shown in Figure 12.2.2b. In all the observed spectral regions, the sign of ΔA was positive in the initial period following photoexcitation, which is thought to reflect induced absorption. Relaxation from the FC state (named the H state in the case of bR) in the first electronic excited state to a geometrically relaxed state such as a twisted state, which is called the I state in the case of bR, is associated with the decrease in the positive ΔA signal. This is induced by the spectral blue shift of the induced absorption. The shift takes place because the induced absorption is affected by the energy stabilization and destabilization of the initial state (I state in the case of bR) and the final state (FC state of the induced absorption from the I state), respectively. The oscillatory features around time zero are considered to be coherent artifacts generated by the interference between probe light and scattered pump light. The photoexcited Rh is reported to decay by the following process [6,7,20]:

$$
\begin{array}{cccccc}
h\nu & < 100\ \text{fs} & 100-300\ \text{fs} & 1-4\ \text{ps} > & 1\ \text{ns} & \\
\text{Rh} \rightarrow \text{FC} & \!\!\!\!\longrightarrow \text{CI} & \longrightarrow & \text{primeRh} & \!\!\!\!\longrightarrow \text{bathoRh} - \rightarrow \!\!\!\!\longrightarrow &
\end{array}
$$

(12.2.1)

where FC indicates the Franck–Condon state. The FC and CI were previously called the H state and J state, respectively, as described above.

12.2.3.1 ELECTRONIC DYNAMICS

To determine the lifetimes and spectra of the decaying components, we fitted the observed time-resolved difference absorption spectra $\Delta A(t, \lambda)$ using a global fitting method [31] over all probe wavelengths. The fitting function included three components as follows, representing two transitions with distinct time constants:

$$
\Delta A(t, \lambda) = \Delta A_0(\lambda) + \Delta A_1(\lambda) \exp(-t/\tau_1) + \Delta A_2(\lambda) \exp(-t/\tau_2)
$$

$$
(\tau_1 < \tau_2)
$$

(12.2.2)

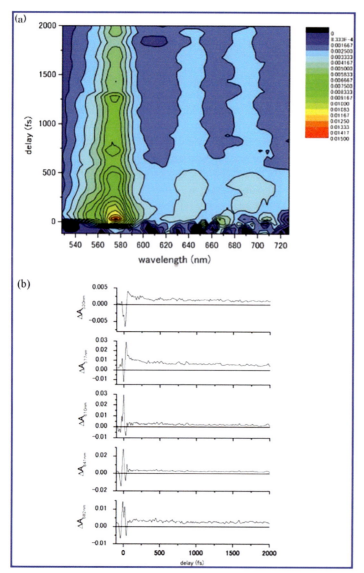

FIGURE 12.2.2 Observed time-resolved difference absorption spectra. (a) Two-dimensional display of measured time-resolved difference absorption spectra $\Delta A(t, \lambda)$. (b) Real-time traces of the absorbance difference for five probe wavelengths.

Finding the condition for least-squares error in the global fit analysis, we have estimated the time constants of τ_1 and τ_2 as 80 fs and 1.1 ps, respectively. The two time constants reflect the sequential relaxation after photoexcitation. The whole process can be expressed by

$$\Delta A(t, \lambda) = \Delta A_a(\lambda)\exp(-t/\tau_1) + \Delta A_b(\lambda) \times [1 - \exp(-t/\tau_1)]\exp(-t/\tau_2)$$
$$+ \Delta A_c(\lambda)[1 - \exp(-t/\tau_1)] \times [1 - \exp(-t/\tau_2)]$$

(12.2.3)

where $\Delta A_i(\lambda)$ ($i = a, b, c$) is the spectrum of each intermediate.

Considering the conditions satisfied in the present case of $\tau_1 \ll \tau_2$, Eq. (12.2.3) can be approximated as

$$\Delta A(t, \lambda) = \Delta A_a(\lambda)\exp(-t/\tau_1) + \Delta A_b(\lambda) \times [\exp(-t/\tau_2) - \exp(-t/\tau_1)]$$
$$+ \Delta A_c(\lambda)[1 - \exp(-t/\tau_2)]$$

(12.2.4)

Using Eqs. (12.2.2) and (12.2.4), we obtain

$$\Delta A_1(\lambda) = \Delta A_a(\lambda) - \Delta A_b(\lambda) \qquad (12.2.5)$$

$$\Delta A_2(\lambda) = \Delta A_b(\bullet)\lambda - \Delta A_c(\lambda) \qquad (12.2.6)$$

$$\Delta A_0(\lambda) = \Delta A_c(\lambda) \qquad (12.2.7)$$

Equations (12.2.5–12.2.7) give

$$\Delta A_a(\lambda) = \Delta A_0(\lambda) + \Delta A_1(\lambda) + \Delta A_2(\lambda) \qquad (12.2.8)$$

$$\Delta A_b(\lambda) = \Delta A_0(\lambda) + \Delta A_2(\lambda) \qquad (12.2.9)$$

Spectra of $\Delta A_0(\lambda)$, $\Delta A_1(\lambda)$, and $\Delta A_2(\lambda)$ were estimated by least-squares fit of $\Delta A(t, \lambda)$ using the two obtained time constants of τ_1 and τ_2. Using Eqs. (12.2.7–12.2.9), $\Delta A_a(\lambda)$, $\Delta A_b(\lambda)$, and $\Delta A_c(\lambda)$ were calculated from the spectra of $\Delta A_0(\lambda)$, $\Delta A_1(\lambda)$, and $\Delta A_2(\lambda)$, and the result is plotted in Figure 12.2.3a. The three spectra were normalized for their comparison in Figure 12.2.3b.

Previous work by Shank and Mathies and others observed the signal rise on a 100 fs time scale in their study of femtosecond transient absorption measured by 35-fs pump pulses and concluded that the wavepacket rapidly leaves the FC region in the 100 fs time scale [32]. Mathies group claimed to have found that the system is carried toward CI in ~50 fs in their study of femtosecond stimulated Raman spectroscopy using 30-fs photochemical pump pulses [7]. It implies that τ_1 (80 fs) obtained in the present work corresponds to the time moving from the FC region to CI. In the present work, we could determine the time constant using a much shorter pulse of sub 5-fs pulses. The spectrum of $\Delta A_a(\lambda)$ in Eq. (12.2.3) has a negative value, which is thought to reflect the induced absorption of the FC state.

Spectra of $\Delta A_b(\lambda)$ and $\Delta A_c(\lambda)$ show similar spectral shape with positive value in the all-probe wavelength region, reflecting induced absorption of reaction intermediates. The spectrum of $\Delta A_c(\lambda)$ around 575 nm is narrower than the spectrum of $\Delta A_b(\lambda)$. The narrowing of the spectrum is considered to be caused by the vibration cooling on the ground state potential surface. Therefore $\Delta A_b(\lambda)$ and $\Delta A_c(\lambda)$ are thought to correspond to the induced absorption of primeRh and bathoRh, respectively. This implies that primeRh becomes thermalized into bathoRh in the time constant of τ_2.

12.2.3.2 Vibration Dynamics

Wavepacket motion in the electronic excited state and that in the electronic ground state modulate the $\Delta A(t)$ traces. Therefore, Fourier analysis of the $\Delta A(t)$ traces reflects the vibrational modes either of the photoactive state or the non-photoactive state. The vibration mode signals of the non-photoactive state are expected to decay without frequency shift or modulation. Therefore, the vibration mode signals showing frequency shift or modulation are thought to be assigned to the dynamics of the vibration mode coupled to the photoactive state. In the following, we discuss the vibration mode signals showing frequency shift or modulation to discuss the dynamics of the vibration mode coupled to the photoactive state.

Fourier power spectra of the time traces of the absorption difference exhibit three prominent modes at around 1550, 1200, and 1000 cm^{-1} assigned to C=C stretching ($v_{C=C}$), C–C stretching (v_{C-C}), and hydrogen out-of-plane (HOOP) modes, respectively. The dynamics of those vibration

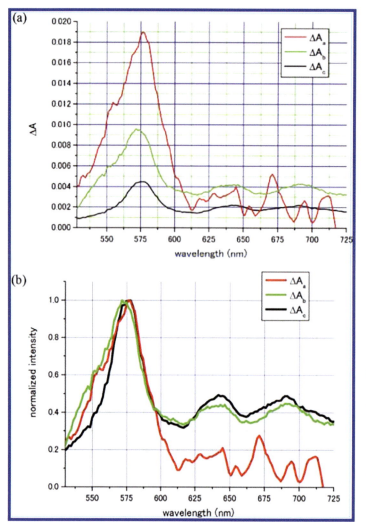

FIGURE 12.2.3 Results of global fitting analysis. (a) Mean square error calculated by scanning the value of τ_1 and τ_2. (b) Spectra of $\Delta A_a(\lambda)$, $\Delta A_b(\lambda)$, and $\Delta A_c(\lambda)$ calculated from the spectra of $\Delta A_i(\lambda)$ ($i=0$, 1, 2). The spectra of $\Delta A_i(\lambda)$ were obtained in least-squares fit of $\Delta A(t, \lambda)$ using τ_1 and τ_2 estimated in panel A.

modes at each probe wavelength can be studied by calculating the instantaneous molecular vibration frequencies using spectrogram analysis [33,34] of the time-resolved difference absorption traces. A Blackman window with a fwhm of 240 fs was used as a gate function in the spectrogram calculation. Figure 12.2.4a shows the calculated spectrograms at a probe wavelength of 530 nm. For delay times shorter than 150 fs, the spectrogram shows the existence of a C=N stretching mode at a frequency of ~1610 cm^{-1}, which quickly decays in a similar fashion to that observed in photoisomerization of bR [24]. This confirms that the primary event after photoexcitation of the octopus Rh is a deformation of the retinal configuration near the C=N bond of the protonated Schiff base.

Just after the photoexcitation, strain in retinal is localized around the Schiff base and propagates to surrounding C–C bonds within a few 100 femtoseconds. Therefore the C=N stretching mode is more distinct than the C=C stretching mode in the early time scale. The reason why the observed frequency of the C=N stretching mode (1610 cm^{-1}) was lower than that of the ground state reported as ~1650 cm^{-1} in Raman study [35] can be explained as follows in the analogy of the explanation for bR [36]. The Schiff base forms a hydrogen bond to its salt bridge partner in the ground state. The primary process after photoexcitation involves a movement of the Schiff base proton away from a counterion, which breaks the hydrogen bond and causes a red shift of the C=N stretching frequency. Figure 12.2.4b shows a spectrogram of the 1410–1560 cm^{-1} range with rescaled intensity to clearly

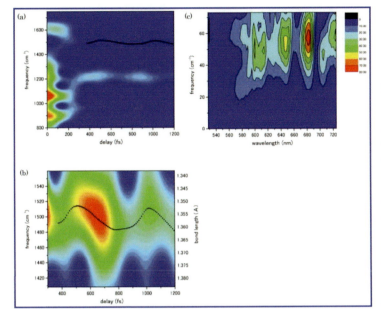

FIGURE 12.2.4 Spectrograms calculated from the ΔA traces at 530 nm. The black dots show the C=C stretching mode. Spectrograms are shown for the (a) 900–1750 cm^{-1} range and (b) the 1410–1560 cm^{-1} range rescaled in intensity to show the frequency modulation of the C=C stretching mode. (c) Two-dimensional Fourier power spectra of the time-resolved difference absorption traces $\Delta A(t)$ over delay times ranging from 800 to 2000 fs.

show the frequency modulation in the C=C stretching mode. The spectrogram is a time-frequency analysis using the sliding Fourier transform window and an artifact can appear in the spectrogram trace as a modulation of frequency and intensity that is caused by a beat between neighboring frequency modes [37]. The modulation period corresponds to the inverse of the frequency difference between the neighboring modes. When the time width of the sliding window is much shorter than the modulation period of the artifact, the spectrogram trace is significantly contaminated by the artifact. The observed frequency modulation in Figure 12.2.4b has a period of 500 fs, which corresponds to 67 cm^{-1}. If the frequency modulation is caused by an artifact, neighboring frequency modes separated by 67 cm^{-1} should be observed in the Fourier power spectrum of the time-resolved difference absorption trace. However, the Fourier power spectrum calculated from the time-resolved trace in the region from 200 to 1200 fs did not show a neighboring mode around 1500 cm^{-1} separated by 67 cm^{-1}. It confirms that the frequency modulation observed in the spectrogram trace is not an artifact caused by the beat between neighboring modes, but reflects a real-time frequency change of the vibration frequency.

After photoisomerization of the retinal is completed, the frequency of the C=C stretching mode was found to be modulated at a period of ~500 fs. The frequency modulation reflects the wavepacket motion on the potential energy surface of the relevant electronic states of all of the ground state, the excited state, prime Rh, and bathoRh, which appear after the photoexcitation [38,39]. This value for octopus Rh is consistent with the period of 550 fs observed in the wavepacket motion results on bovine Rh reported by Shank et al. [40] and is considerably different from the period of ~200 fs in bR [27]. The frequency modulation reflects torsional motion around the $C_{11}=C_{12}$ double bond before thermalization in the all-trans structure. The change in the frequency can be ascribed to modulation of the bond length of the $C_{11}=C_{12}$ double bond using an empirical equation relating the bond length and vibration frequency. The empirical equation is obtained as follows.

The relationship between the bond-stretching force constant k (dyn/cm), bond length d (Å), π-bond order (P), and electronegativity (χ) was found by Gordy as follows [41]:

$$k = a(P+1)\left(\chi/d\right)^{3/2} + b \qquad (12.2.10)$$

Here, a and b are empirical constants equal to 1.67×10^5 and 0.30×10^5, respectively, for stable molecules exhibiting their normal covalencies. Gordy has estimated the average deviation of k calculated from k values observed for 71 cases to be 1.84%. The bond vibration frequency v (cm^{-1}) can then be related to the bond-stretching force constant k (dyn/cm) assuming an isolated oscillator as

$$v = (2\pi^2 c^2 m)^{(-1/2)} k^{1/2}$$ (12.2.11)

where m and c are the mass of the bonded atom and the velocity of light, respectively [42]. Taking Eqs. (12.2.1) and (12.2.2), we can obtain the relation between the bond length d and the vibration frequency v as

$$v = C[a(P + 1)(x/d) + b]^{1/2}$$ (12.2.12)

where $X = 2.5$ and $C = 1.66$. Dewar derived the relationship between the bond length d and the bond order P for single and double bonds as $d = 1.489 - 0.151\ P$ [43]. For a single bond, P is 0, and for a double bond, P is 1. Using this equation, the observed frequency change in the C=C stretching mode leads to a modulation in the bond length of the C_{11}=C_{12} double bond of $\sim 11 \pm 2$ mÅ.

The assignment was certified by the direct observation of the real-time frequency of the HOOP mode and C=C–H in-plane bending mode as discussed below.

How HOOP mode is related in genera to the photo-reception process is briefly described as follows.

The coherent coupling in the excited state is promoted by torsional skeletal and coupled HOOP vibrational modes, in combination with a twisted conformation around the isomerization region. Since such torsion will strongly enhance the infrared intensity of coupled HOOP modes, FTIR difference spectra of rhodopsin, isorhodopsin and several analog pigments in the spectral range of isolated and coupled H–C=C–H wagging were studied and the result was that the coupled HOOP signature in these retinal pigments correlates with the distribution of torsion over counteracting segments in the retinylidene polyene chain.

The in-plane C=C–H bending modes coupled with C–C stretching modes appear at around 1200 cm^{-1}. This is the so-called fingerprint region, which is very sensitive to chromophore conformation. In the observed spectrogram, the HOOP mode and C=C–H in-plane bending mode are merged until ~ 200 fs, followed by the clear separation of the two modes. It elucidates that the rapid torsion along the HOOP coordinate finishes around 200 fs, which is the time when primeRh is reported to appear [7,32].

The intensities of the HOOP mode and C=C–H in-plane bending mode are also modulated at a period of ~ 500 fs, reflecting torsional motion around the C_{11}=C_{12} double bond in thermalization from primeRh to bathoRh.

Therefore, when the system crosses CI at 80 fs after excitation, the 200-fs tortional motion along the HOOP coordinate is nearly half finished, and the system is still in the 11-cis configuration with partial deformation of the quasi planarity of the plane formed by C_{11}=C_{12}–C_{13} bonds, i.e., the torsional motion around the C_{11}=C_{12} double bond with 500-fs period is (partially) finished with the rotation angle of about one-sixth (= 80/500) of 180°.

Figure 12.2.4c shows two-dimensional Fourier power spectra of the time-resolved difference absorption traces $\Delta A(t)$ over delay times ranging from 800 to 2000 fs. The signals centered at around 60 cm^{-1} confirm that these time-resolved traces are also modulated at a period of ~ 500 fs, reflecting wavepacket motion at the same period.

In the case of bovine Rh, the transition time from the FC state to CI is reported to be 50 fs [44], which was estimated as 80 fs for octopus Rh in the present work. For the other transitions with femtosecond and picosecond time constants in bovine Rh, the results are similar to those observed in octopus Rh shown above. Comparing these results with our previous study on transient absorption of bR [27], we can see that the transition from the FC state to CI proceeds faster in octopus Rh (80 fs) than in bR (~ 200 fs).

12.2.4   CONCLUSIONS

In this work, we have measured the time-resolved difference absorption spectra of octopus Rh using a sub-5-fs laser pulse with broadband multichannel detectors [45]. The obtained data matrix has 128 data points in the 528–727-nm region with a 1.56-nm interval and 2100 data points between −100 and 2000 fs with a 1-fs interval. The obtained time-resolved traces were fitted by a double-exponential function using a global fitting method over all probe wavelengths. The time constants of the processes following photoexcitation are comparable between the octopus Rh and bovine Rh results. However, in comparison with the model bR system, the transition time from the FC state to CI in octopus Rh was found to be 80 fs, being about 3 times shorter than in bR, which we measured in transient absorption of bR [27].

The vibration dynamics of photoisomerization were studied by spectrogram analysis. At delay times shorter than 150 fs, the C=N stretching mode appears and quickly disappears within 150 fs. This observation indicates that the primary event after photoexcitation of the octopus Rh is a deformation of the retinal configuration near the C=N bond in the protonated Schiff base. An observed frequency modulation of the C=C stretching mode at a period of ~500 fs was also observed, reflecting torsional motion around the $C_{11}=C_{12}$ double bond of primeRh before thermalization to bathoRh. This value is similar to the period of 550 fs obtained for wavepacket motion observed in transient absorption measurement of bovine Rh [40]. By contrast, the modulation period for bR was observed to be ~200 fs in the transient absorption measurement of bR [27]. On the basis of the frequency change in the C=C stretching mode, the bond length of the $C_{11}=C_{12}$ double bond was found to be modulated on the order of ~10 mÅ. The HOOP and C=C–H modes are merged until 200 fs, therefore the rapid torsion along the HOOP coordinate is thought to be half finished at the delay of 80 fs when the system crosses CI. Their intensities were observed to be modulated at a period of ~500 fs, again affected by the torsional motion around the $C_{11}=C_{12}$ double bond. The research output described in the present section is the product of the collaboration activity among the following people: Atsushi Yabushita, Takayoshi Kobayashi, and Motoyuki Tsuda [45].

REFERENCES

1. T. Yoshizawa and Y. Kito, *Nature* **182**, 1604–1605 (1958).
2. T. Yoshizawa and G. Wald, *Nature* **197**, 1279–1286 (1963).
3. J. Buchert, V. Stefancic, A. G. Doukas, R. R. Alfano, R. H. Callender, J. Pande, H. Akita, V. Balogh-Nair, and K. Nakanishi, *Biophys. J.* **43**, 279–283 (1983).
4. S. Horiuchi, F. Tokunaga, and T. Yoshizawa, *Biochim. Biophys. Acta* **591**, 445–457 (1980).
5. G. Eyring, B. Curry, R. Mathies, R. Fransen, I. Palings, and J. Lugtenburg, *Biochem.* **19**, 2410–2418 (1980).
6. H. Kandori, Y. Shichida, and T. Yoshizawa, *Biochemistry (Moscow)* **66**, 1197–1209 (2001).
7. P. Kukura, D. W. McCamant, S. Yoon, D. B. Wandschneider, and R. A. Mathies, *Science* **310**, 1006–1009 (2005).
8. H. Ohtani, T. Kobayashi, M. Tsuda, and T. G. Ebrey, *Biophys. J.* **53**, 17–24 (1988).
9. Y. Shichida, S. Matsuoka, and T. Yoshizawa, *Photobiochem. Photobiophys.* **7**, 221–228 (1984).
10. W. Sperling, in *Biochemistry and Physiology of Visual Pigments*, edited by H. Langer, Springer-Verlag, Heidelberg, pp. 19–28 (1973).
11. Y. Fukada, Y. Shichida, T. Yoshizawa, M. Ito, A. Kodama, and K. Tsukida, *Biochemistry* **23**, 5826–5832 (1984).
12. B. Mao, M. Tsuda, M. T. G. Ebrey, H. Akita, V. Balogh-Nair, and K. Nakanishi, *Biophys. J.* **35**, 543–546 (1981).
13. R. S. Becker and K. Freedman, *J. Am. Chem. Soc.* **107**, 1477–1485 (1985).
14. Y. Koyama, K. Kubo, M. Komori, H. Yasuda, and Y. Mukai, *Photochem. Photobiol.* **54**, 433–443 (1991).
15. H. J. A. Dartnall, *Vision Res.* **8**, 339–358 (1967).
16. T. Kobayashi, M. Kim, M. Taiji, T. Iwasa, M. Nakagawa, and M. Tsuda, *J. Phys. Chem. B* **102**, 272–280 (1998).
17. L. Zhu, J. Kim, and R. A. Mathies, *J. Raman Spectrosc.* **30**, 777–783 (1999).

18. H. Kandori, H. Sasabe, K. Nakanishi, T. Yoshizawa, T. Mizukami, and Y. Shichida, *J. Am. Chem. Soc.* **118**, 1002–1005 (1996).
19. H. Chosrowjan, N. Mataga, Y. Shibata, S. Tachibanaki, H. Kandori, Y. Shichida, T. Okada, and T. Kouyama, *J. Am. Chem. Soc.* **120**, 9706–9707 (1998).
20. H. Kandori, Y. Furutani, S. Nishimura, Y. Shichida, H. Chosrowjan, Y. Shibata, and N. Mataga, *Chem. Phys. Lett.* **334**, 271–276 (2001).
21. R. W. Schoenlein, L. A. Peteanu, R. A. Mathies, and C. V. Shank, *Science* **254**, 412–415 (1991).
22. J. E. Kim and R. A. Mathies, *J. Phys. Chem. A* **106**, 8508–8515 (2002).
23. S. Shim, J. Dasgupta, and R. A. Mathies, *J. Am. Chem. Soc.* **131**, 7592–7597 (2009).
24. A. Yabushita and T. Kobayashi, *Biophys. J.* **96**, 1447–1461 (2009).
25. W. T. Pollard, C. H. Brito Crus, C. V. Shank, and R. A. Mathies, *J. Chem. Phys.* **90**, 199–208 (1989).
26. M. Du and G. R. Flemming, *Biophys. Chem.* **48**, 101–111 (1993).
27. T. Kobayashi, T. Saito, and H. Ohtani, *Nature* **414**, 531–534 (2001).
28. A. Baltuska and T. Kobayashi, *Appl. Phys. B: Laser Opt.* **75**, 427–443 (2002).
29. T. Kobayashi, A. Shirakawa, H. Matsuzawa, and H. Nakanishi, *Chem. Phys. Lett.* **321**, 385–397 (2000).
30. A. Shirakawa, I. Sakane, M. Takasaka, and T. Kobayashi, *Appl. Phys. Lett.* **74**, 2268–2270 (1999).
31. A. R. Holzwarth, *Adv. Photosynth. Respir.* **3**, 75–92 (2004).
32. L. A. Peteanu, R. W. Schoenlein, Q. Qang, R. A. Mathies, and C. V. Shank, *Proc. Natl. Acad. Sci. U.S.A.* **9**, 11762–11766 (1995).
33. T. Meier and S. Mukamel, *Phys. Rev. Lett.* **77**, 3471–3474 (1996).
34. S. Rutz and E. Schreiber, *Eur. Phys. J. D* **4**, 151–158 (1998).
35. S. O. Smith, M. S. Braiman, A. B. Myers, J. A. Pardoen, J. M. L. Courtin, C. Winkel, J. Lugtenburg, and R. A. Mathies, *J. Am. Chem. Soc.* **109**, 3108–3125 (1987).
36. K. J. Rothschild and H. Marrero, *Proc. Natl. Acad. Sci. U.S.A.* **79**, 4045–4049 (1982).
37. A. Kahan, O. Nahmias, N. Friedman, M. Sheves, and S. Ruhman, *J. Am. Chem. Soc.* **129**, 537–546 (2007).
38. W. T. Pollard, S.-Y. Lee, and R. A. Mathies, *J. Chem. Phys.* **92**, 4012–4029 (1990).
39. J. M. Jean and G. R. Fleming, *J. Chem. Phys.* **103**, 2092–2101 (1995).
40. Q. Wang, R. W. Schoenlein, L. A. Peteanu, R. A. Mathies, and C. V. Shank, *Science* **266**, 422–424 (1994).
41. W. J. Gordy, *Chem. Phys.* **14**, 305–320 (1946).
42. R. H. Baughman, J. D. Witt, and K. C. Yee, *Chem. Phys.* **60**, 4755–4759 (1974).
43. M. J. S. Dewar and S. Hyperconjugation, *Modern Concepts in Chemistry*, Ronald Press, New York, pp. 48–70 (1962); Chapters 3 and 4.
44. G. G. Kochendoerfer and R. A. Mathies, *J. Phys. Chem.* **100**, 14526–14532 (1996).
45. A. Yabushita, T. Kobayashi, and M. Tsuda, *J. Phys. Chem. B* **116**, 1920–1926 (2012).

12.3 Schiff Base Proton Acceptor Assists Photoisomerization of Retinal Chromophores in Bacteriorhodopsin

12.3.1 INTRODUCTION

One of the most crucial photochemical reactions in living matter is the photoisomerization of retinal chromophores in rhodopsin (Rh), which triggers the vision process. Clarifying the primary reaction in the vision process is critical. However, studying the primary mechanism is difficult, because non-reversible photodamage in the Rh sample makes it difficult to acquire the pump-probe signal and thus obtain sufficient signal/noise data to clarify the reaction mechanism from the time-resolved measurement.

A membrane protein, bacteriorhodopsin (BR), is more stable than Rh, and its photochemical reaction is similarly triggered by photoisomerization of retinal. Therefore, as a model of the vision process, the photo-induced reaction in BR has been widely studied both theoretically [1,2] and experimentally [3–9]. The photoisomerization of retinal works as a trigger in the proton pump of BR [10] to produce a chemical potential for ATP synthesis. Being more stable than artificial chemicals, BR is employed for various applications, such as optical memory and switches [7,11].

The photocycle of retinal Rh was previously thought to be initiated by light-induced proton transfer [12], but its initiation is currently associated with the ultrafast isomerization of retinal [13]. Quantum mechanics calculations have demonstrated that the aborted double-bicycle-pedal isomerization in BR occurs together with hydrogen-bond breaking [14]. Thus, it is assumed that retinal isomerization causes steric strain and/or unfavorable electrostatics, which destabilize the protein, resulting in perturbation of the surrounding hydrogen-bond network (HBN). Theoretical studies have revealed that retinal isomerization occurs with a sudden polarization, which triggers positive charge translocation along retinal [15]. This charge translocation is thought to directly activate the opsin photocycle, because such activation occurs even if the chromophores are locked against isomerization in channel rhodopsin [16], Rh [17], and BR [18]. The spatial electrostatic potential (ESP) change during chromophore photoisomerization was recently calculated by Melaccio et al. [19], and the results suggest that an ultrafast (30–40 fs) and substantial change in the ESP triggers HBN destabilization. In this study, we experimentally demonstrated that the sudden polarization dynamics on the chromophores are affected by the surrounding residues controlling the proton transfer channels of BR.

The proton-conducting channel of BR is subdivided into two half-channels by the retinal chromophore [20]. One half-channel is named the extracellular (EC) channel and connects the Schiff base with the EC medium. The other, named the cytoplasmic (CP) channel, connects the Schiff base with the cytoplasm. Proton currents in mutagenized molecules of *Halobacterium salinarum* BR (HsBR) have revealed that aspartic acids (Asps) 85 and 96 (D85 and D96) play a critical role in the proton-translocation mechanism [21–23]. D85 acts as an acceptor of the Schiff base proton in the EC channel, whereas D96 acts as the proton donor of the Schiff base in the CP channel [24–26]. Removal or protonation of D85 prevents deprotonation of the Schiff base, and removal of D96 slows the reprotonation of the Schiff base. For example, the D85E mutant with a pK of >7 prevents

DOI: 10.1201/9780429196577-90

deprotonation of the Schiff base, and mutants lacking D85 do not exhibit proton translocation activity upon illumination of the blue chromophore [24,27,28]. Even though the roles of D85 and D96 as proton acceptor and donor, respectively, are known, it is not known which starts faster after photoexcitation. Differences in the ultrafast dynamics of the wildtype and its mutants can elucidate the effects of the Schiff base proton acceptor and donor, which have never been observed, on the femto- and picosecond timescales.

In this study, transient absorption spectroscopy was performed to study the ultrafast dynamics of BR from Haloquadratum walsbyi (HwBR) and its mutants. HwBR, one of the three microbial Rhs, is considered a traditional BR, like HsBR. The function and crystal structure of HwBR were reported by Hsu et al. [29]. Compared with HsBR, HwBR has higher stability under low-pH conditions. We have studied the D93N and D104N mutants of HwBR, which correspond to the D85N and D96N mutants, respectively, of HsBR. The measured results for HwBR were analyzed using the global fitting method and two-dimensional correlation spectroscopy (2D-CS) [30]. A comparison of the calculated 2D-CS spectra of the wild-type and its mutants clarifies whether the proton donor and/or acceptor in the Schiff base controls the ultrafast dynamics of retinal chromophores.

We have also performed transient absorption spectroscopy on two BRs, HmBRI and HmBRII, from Haloarcula marismortui [31,32]. HmBRII maintains its activity at low pH, similar to HwBR. By contrast, HmBRI is not active at low pH [29]. The ultrafast dynamics of the three BRs (HwBR, HmBRI, and HmBRII) support these findings (HmBRII and HwBR show similar dynamics, but HmBRI shows a difference from HmBRII and HwBR).

Additionally, we previously performed transient absorption spectroscopy on BR, measuring at five probe wavelengths [33] and at 100 probe wavelengths [34]. These studies visualized the primary reaction of BR, identifying the molecular structure change in BR that occurs during ultrafast photoisomerization of its retinal chromophores. However, each scan in those measurements took 1 h per sample. Thus, the reproducibility of the measurements was relatively poor because of the damage accumulated in the sample and/or a spectral change in the light source during the long measurement period.

In the study presented here, we employed the multiplex fast-scan system [35] for the transient absorption spectroscopy of BR, which provides reproducible transient absorption spectroscopy data and avoids damage accumulation. Thus, we could clearly visualize differences in the ultrafast dynamics of the samples.

12.3.2 MATERIALS AND METHODS

12.3.2.1 CHEMICALS USED IN THIS STUDY

The following materials were obtained and used in this study: isopropyl-β-D-thiogalactopyranoside (BioShop, Burlington, Ontario, Canada), ampicillin (BioShop), all-trans retinal (Sigma-Aldrich, St. Louis, MO), phenylmethylsulfonyl fluoride (BioShop), β-mercaptoethanol (BioShop), n-dodecyl-β-d-maltoside (DDM; Anatrace, Maumee, OH), Ni-nitrilotriacetic acid (NTA) resin (70666; Novagen Biosciences, Madison, WI), imidazole (Sigma-Aldrich), morpholineethanesulfonic acid (MES; Sigma), Tris (BioBasic, Markham, Ontario, Canada), and NaCl (Sigma-Aldrich).

12.3.2.2 PLASMID CONSTRUCTIONS

The gene bop of Haloquadratum walsbyi was synthesized by Genomics BioSci & Tech., Ltd., and was cloned into pET-21d (Novagen Biosciences) by introducing a 5' NcoI and 3' XhoI restriction-enzyme cutting site. The gene bop of Haloarcula marismortui was amplified using polymerase chain reaction from the genomic DNA. Subsequently, the gene bop was cloned into pET-21b (Novagen Biosciences) by introducing a 5' NdeI and 3' HindIII restriction-enzyme cutting site.

12.3.2.3 Primers Used for Mutant Constructions

The forward and reverse primer sequences of D93N-HwBR are 5^0-GTG TATTGGGCACGATAT GCTAACTGGCTATTCAC-3^0 and 5^0-GTGAATA GCCAGTTAGCATATCGTGCCCAATAC AC-3^0. The forward and reverse primer sequences of D104N-HwBR are 5^0-AACACCACTGCTTC TGCT TAACATTGGCCTCCTTG-3^0 and 5^0- CAAGGAGGCCAATGTTAAGCA GAAGCAGT GGTGTT-3^0.

12.3.2.4 Protein Purification

Escherichia coli C43(DE3) cells were used for protein expression. The cells were inoculated in Luria-Bertani medium containing 50 mg/mL ampicillin and incubated at 37C overnight. Overexpression of protein was induced using 1 mM isopropyl-b-D-thiogalactopyranoside, and all-trans retinal (final concentration 5–10 mM) was added. After 4–6 h in the dark, cells were collected using centrifugation at 6750 g for 10 min at 4C (R10A3; Hitachi CR-21, Tokyo, Japan). The collected cells were resuspended in buffer (50 mM Tris-HCl, 4 M NaCl, 14.7 mM 2-mercaptoethanol, and 0.2 mM phenylmethylsulfonyl fluoride (pH 7.8)) and broken using an ultrasonic processor (S-3000; Misonix, Farmingdale, NY). For the separation of the membrane fraction, total cell-extract centrifugation was performed at 6750 g for 10 min at 4C. The supernatant was then centrifuged at 169538 g for 1 h at 4C (P70AT; Hitachi CP80WX, Tokyo, Japan). The sediment was dissolved with 2% DDM for at least 12 h at 4C, after which it was centrifuged at 32816 g for 45 min at 4C (R20A2; Hitachi CR-21) to separate the detergent-soluble fraction. BR was purified through affinity purification using the Ni-NTA (Ni^{2b}-nitrilotriacetate) method. The target protein was eluted with 250 mM imidazole and then dialyzed into buffer (50 mM MES, 4 M NaCl, and 0.02% DDM (pH 5.8)) for further assay.

12.3.2.5 Flash-Laser-Induced Photocycle Measurement

Protein was induced using an Nd-YAG laser (532 nm, 6-ns pulse duration, 40 mJ). The purified protein was diluted to reach 0.3 at OD_{lmax}, and the transient absorbance change was recorded at the selected wavelength.

12.3.2.6 Fast-Scan Transient Absorption Spectroscopy

Ultrafast dynamics has been studied in various materials to elucidate the primary reaction mechanisms in photoreactions. For the study of the femtosecond-timescale dynamics, the pump-probe method is one of the most common measurement methods employed. Using a 100-fs pulse, the dynamics of the photoexcited states could be investigated using pump-probe measurement in this study.

When the pump–probe measurements are performed using pulses as short as <10 fs, the measured data reflect not only the femtosecond dynamics of the electronic state but also the molecular vibration frequencies, which indicate changes in the molecular structure during the photoreaction.

The simultaneous measurement of electronic and vibrational dynamics typically takes 1 h per scan, time resolving the molecular vibration period of a few tens of femtoseconds. Thus, if the sample is photofragile, measurements cannot be performed because of the effect of damage accumulation in the sample during the measurement. Moreover, measurement with ultrafast time resolution requires the use of a sub-10-fs pulse, which has ultrahigh peak power and can easily damage the sample.

We have developed a method, fast-scan time-resolved spectroscopy, of performing time-resolved absorption-change measurements with fine time resolution that is >100 times faster than the conventional method. The scanning motion of the optical delay stage and the signal detection timing were synchronized using the repletion of the light source in the ultrashort visible pulse. Thus, during a scan time of 5 s, we can obtain the trace of the transient absorption spectroscopy signal for all probe wavelengths simultaneously. This method enables us to study the ultrafast dynamics of photofragile materials by shortening the measurement time. Measuring the transient absorption spectroscopy signal in the broad probe spectral region also enables the study of the probe photon energy dependence of the ultrafast dynamics to identify the contribution from the electronic ground state and excited states. Thus, we performed transient absorption spectroscopy of the BR samples in the broad visible spectral region by utilizing the fast-scan system. The fast-scan system is described in detail in [35].

12.3.2.7   Visible Broadband Sub-10-Fs Pulse

A broadband visible probe spectrum is required to study the probe-wavelength dependence of transient absorption spectroscopy signals distinguishing contributions from the electronic ground state and excited states. By contrast, the ultrafast dynamics can be observed in transient absorption spectroscopy signals exciting the sample using an ultrashort visible pulse, because the time resolution of the measured signal is limited by the duration of the pump pulse. Thus, we developed a sub-10-fs visible pulse laser using a noncollinear optical parametric amplifier (NOPA). A near-infrared laser pulse from a Ti:sapphire regenerative amplifier generated an ultraviolet pump pulse and a broadband visible seed pulse through the second harmonic generation and self-phase modulation, respectively. The pump pulse and seed pulse were mixed in a nonlinear crystal at a certain angle to amplify the seed pulse in the broadband visible spectral region. The visible broadband laser pulse amplified in the Napa was compressed to be sub-10-fs using a diffraction grating and deformable mirror. Separating the pulse using a beam sampler, we obtained two pulses with equal time profile and different pulse energies of 10 and 1 nJ, which were used as the pump and probe pulses, respectively, in the transient absorption spectroscopy pump-probe measurements. In [36–38], the authors describe in detail the ultrashort pulse laser generated by the NOPA.

12.3.2.8   Software Employed in This Study

To obtain the decay lifetime of the transient absorption spectroscopy signal, we employed single-and double-exponential fitting using a homemade fitting program, which was written using the programming language LabVIEW 2013 (National Instruments, Austin, TX). The fitting procedure solves the nonlinear least-square problem using the Levenberg-Marquardt algorithm.

A program to calculate 2D-CS results was also written using LabVIEW 2013, and it followed the algorithm described in [39]. Calculation of each trace took 2 min on a typical personal computer (Windows 7; central processing unit: Intel Pentium G645, 2.9 GHz; memory: 8 GB).

12.3.3   RESULTS AND DISCUSSION

12.3.3.1   Sequence Alignment of HwBR and Other BRs

We conducted the protein sequence alignment of HwBR with HsBR, HmBRI, and HmBRII. The result identified two highly conserved Asp residues corresponding to D85 and D96 of HsBR (Figure 12.3.1). The amino acid sequence alignment demonstrated a sequence identity of 50.57% with HsBR, where D93 and D104 of HwBR were conserved corresponding to D85 and D96, respectively, of HsBR.

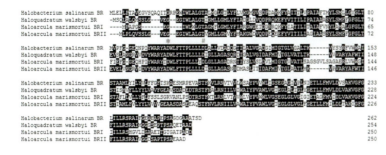

FIGURE 12.3.1 Protein sequence alignment analysis. Protein sequence alignment of HwBR with HsBR, HmBRI, and HmBRII was conducted to identify the two highly conserved Asp residues corresponding to D85 and D96 of HsBR.

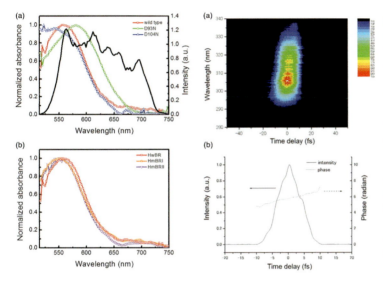

FIGURE 12.3.2 Laser spectrum and absorption spectra of each sample. (a) Laser spectrum (solid black curve) and stationary absorption spectra of the three HwBR samples, wild-type (open red squares), D93N (open green circles), and D104N (open blue triangles). (b) Stationary absorption spectra of the three wild-type samples, HwBR (open red squares), HmBRI (open orange stars), and HmBRII (open purple inverted triangles). To see this figure in color, go online.

12.3.3.2 PROTEIN CONSTRUCTIONS, EXPRESSION, PURIFICATION, AND ULTRAVIOLET-VISIBLE MAXIMUM ABSORBANCE OF WILD-TYPE, D93N, AND D104N

The wild-type, D93N, and D104N of HwBR were constructed into pET-21d plasmids, as described in Section 12.3.2. All three proteins were expressed in *E. coli* C43 strain cells and purified through a Ni-NTA column. The yields of the three proteins were 1 mg/L cultures.

They were first probed using the maximum absorbance of purified proteins in 50 mM MES, 4 M NaCl, and 0.02% DDM at pH 5.8. The ultraviolet-visible-spectrum scanning identified maximum absorbance at 552 nm for the wild-type HwBR, 580 nm for HwBR-D93N, and 552 nm for HwBR-D104N (Figure 12.3.2).

12.3.3.3 GROUND-STATE PHOTOCYCLE OF WILD-TYPE, D93N, AND D104N

Laser-flash photolysis measurements demonstrated that HwBR and D93N have quick photocycle kinetics with recovery τ-values of 0.032 and 0.037 s, respectively, whereas D104N appeared to have slower recovery kinetics with a *t*-value of 0.594 s (Figure 12.3.3).

12.3.3.4 TRANSIENT ABSORPTION SPECTROSCOPY OF WILD-TYPE AND MUTANTS OF HWBR

The effect of mutation on the proton acceptor and donor in the Schiff base during the primary process was studied using transient absorption spectroscopy with an ultrashort visible laser for the three

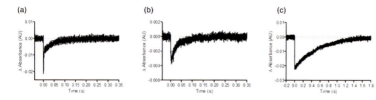

FIGURE 12.3.3 Flash-laser-induced photocycle measurements. The photocycles of (a) wild-type HwBR, (b) D93N-HwBR, and (c) D104N-HwBR were induced using a flash of a 532-nm laser. The laser was induced at 0.00 s. The wavelengths monitored for wild-type HwBR, D93N-HwBR, and D104NHwBR were 552, 580, and 552 nm, respectively. Difference absorbance has arbitrary units.

HwBR samples: wild-type, D93N mutation, and D104N mutation. Each sample was suspended in 50 mM MES buffer stabilized at pH 5.8. The stationary absorption spectra of the sample solutions are presented in Figure 12.3.2, together with the laser spectrum of the ultrashort visible laser pulse. The laser spectrum is sufficiently broad to cover the absorption band of the electronic ground state and excited the sample efficiently through a single photon transition. The bandwidth of the laser extended lower than the absorption band of the electronic ground state, enabling us to observe the dynamics of the absorption band of the electronic excited states, which could have appeared in the red-shifted stationary absorption region. Thus, the dynamics of the electronic excited states could be identified by observing the probe wavelength dependence of the transient absorption spectroscopy signal in the broadband visible spectral region of the probe pulse.

By contrast, the broadband spectral bandwidth of the ultrashort visible pulse lengthens the pulse duration when the pulse is transmitted through a dispersive material such as glass or water. Thus, in the transient absorption spectroscopy of a liquid sample, a glass cell used to store the sample solution should have a short optical pathlength to maintain the ultrahigh time resolution and suppress the lengthening of the pulse duration. In this study, we placed the sample solution in a quartz glass cell with 1-mm optical pathlength (6210–12501; GL Sciences, Tokyo, Japan).

In the optical path of the laser pulse, we inserted a quartz glass plate with a thickness equal to that of the front wall of the quartz glass cell. The pulse, which passed through the quartz glass plate, was estimated using a pulse characterization method, second-harmonic frequency-resolved optical gating (SH-FROG). The SH-FROG trace and pulse envelope retrieved from the trace are displayed in Figure 12.3.4. The pulse was thus estimated to have a sub-10-fs duration in the sample solution.

The ultrafast dynamics of HwBR were measured in the femtosecond (from 300 to 1400 fs) and picosecond (from 1.00 to 14.5 ps) regions. The observed spectral region was 514.5–758.1 nm with a 2.54-nm step, covering the entire broadband probe spectrum.

Figure 12.3.5 presents the 2D transient absorption spectra in the femtosecond (Figure 12.3.5a, c, and e) and picosecond (Figure 12.3.5b, d, and f) regions for the three HwBR samples, wild-type (Figure 12.3.5a and b), D93N mutant (Figure 12.3.5c and d), and D104N mutant (Figure 12.3.5e and f). The transient absorption spectrum is negative in the middle spectral region, which overlaps with the absorption band of the electronic ground state. The negative transient absorption signal is thought to indicate absorbance reduction due to ground-state depletion, and the decay time of the negative transient absorption signal indicates the recovery time of the ground-state population after photoexcitation. Positive transient absorption signals were obtained at both edges of the visible spectral region, indicating induced absorption in the transition from the first excited state to a higher-energy excited state [40,41]. The decay time of the positive transient absorption signal indicates the lifetime of the excited states.

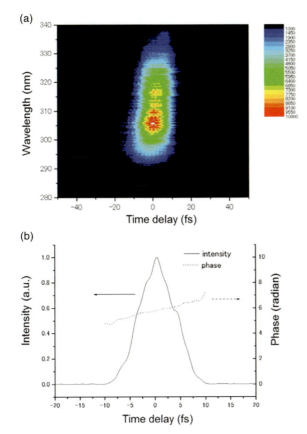

FIGURE 12.3.4 SH-FROG measurement. (a) Measured SH-FROG trace of the ultrashort visible pulse and (b) its time profile. Pulse duration was estimated to be 8.9 fs. To see this figure in color, go online.

In the 2D view of the transient absorption spectra, the negative signal is red-shifted for the D93N mutant compared with the other two samples. The negative signal is thought to indicate ground-state bleaching. The observed red shift in the transient absorption spectrum is consistent with the red shift in the stationary absorption spectrum of D93N.

12.3.3.5 GLOBAL FITTING USING THE TRIPLE-EXPONENTIAL FUNCTION

The transient absorption spectroscopy signal, ΔA (λ, t), in the femtosecond region and that in the picosecond region were connected and fitted using the following triple-exponential function and global fitting method:

$$\Delta A(\lambda,t) = \Delta A_0(\lambda) + \Delta A_1(\lambda)\exp\left(-\frac{t}{\tau_1}\right)$$

$$+ \Delta A_2(\lambda)\exp\left(-\frac{t}{\tau_2}\right) + \Delta A_3(\lambda)\exp\left(-\frac{t}{\tau_3}\right)$$

(12.3.1)

where $\tau_1 < \tau_2 < \tau_3$. The estimated lifetimes τ ($i=1$, 2, 3) are plotted in Figure 12.3.6. Using the obtained time constants, the spectra of the lifetime components ΔA_i ($i=0$, 1, 2, 3) – called the decay-associated spectra (DAS) in the following discussion – were calculated using the least-square method. The calculated DAS are presented in Figure 12.3.7a–c, for the wild-type, D93N mutant, and D104N mutant HwBR samples, respectively.

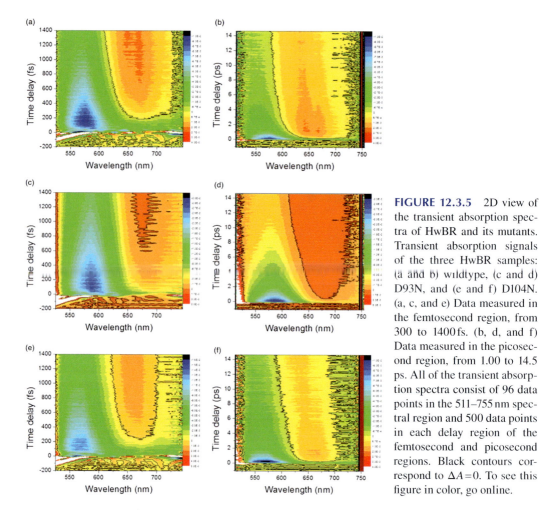

FIGURE 12.3.5 2D view of the transient absorption spectra of HwBR and its mutants. Transient absorption signals of the three HwBR samples: (a and b) wildtype, (c and d) D93N, and (e and f) D104N. (a, c, and e) Data measured in the femtosecond region, from 300 to 1400 fs. (b, d, and f) Data measured in the picosecond region, from 1.00 to 14.5 ps. All of the transient absorption spectra consist of 96 data points in the 511–755 nm spectral region and 500 data points in each delay region of the femtosecond and picosecond regions. Black contours correspond to $\Delta A = 0$. To see this figure in color, go online.

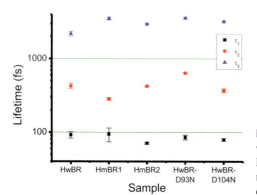

FIGURE 12.3.6 Estimated lifetimes. The lifetimes were estimated for five samples (HwBR wild-type, HmBRI wild-type, HmBRII wild-type, HwBR D93N mutant, and HwBR D104N mutant). To see this figure in color, go online.

The obtained lifetimes – τ_1, τ_2, and τ_3 – reflect the state transitions H/I, I/J, and J/K, respectively, where states H, I, J, and K correspond to the Franck-Condon state, the state at the conical intersection between the electronic ground state and first excited state, the hot vibrational state formed after photoisomerization of retinal chromophores, and a thermalized state, respectively [33,42]. Considering the sequential relaxation process indicated by the transition from state H to state K, we

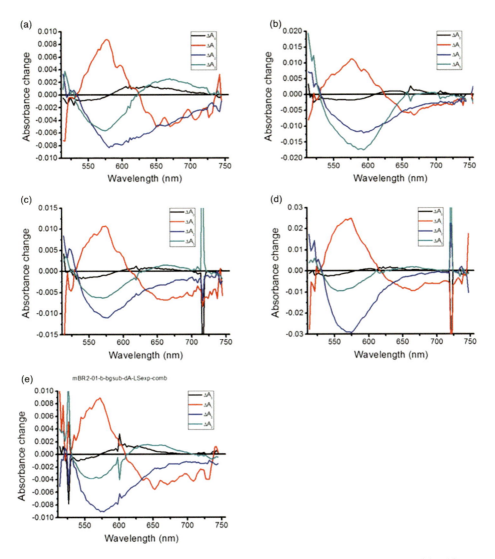

FIGURE 12.3.7 DAS were calculated for five samples: (a) HwBR wild-type, (b) HwBR with D93N mutant, (c) HwBR with D104N mutant, (d) HmBRI wild type, and (e) HmBRII wild-type.

calculated the species-associated spectra (SAS) corresponding to these electronic states using the equations correspond to the SAS of the H, I, J, and K states, respectively. The calculated SAS are presented in Figure 12.3.8.

$$\Delta A(\lambda, t) = \Delta A_H(\lambda)\exp\left(-\frac{t}{\tau_1}\right) + \Delta A_1(\lambda)\left(\exp\left(-\frac{t}{\tau_2}\right) - \exp\left(-\frac{t}{\tau_1}\right)\right)$$

$$+ \Delta A_J(\lambda)\left(\exp\left(-\frac{t}{\tau_3}\right) - \exp\left(-\frac{t}{\tau_2}\right)\right) + \Delta A_K(\lambda)\left(1 - \exp\left(-\frac{t}{\tau_3}\right)\right)$$

$$\Delta A_1(\lambda) = \Delta A_H(\lambda) - \Delta A_I(\lambda)$$

$$\Delta A_2(\lambda) = \Delta A_I(\lambda) - \Delta A_J(\lambda)$$

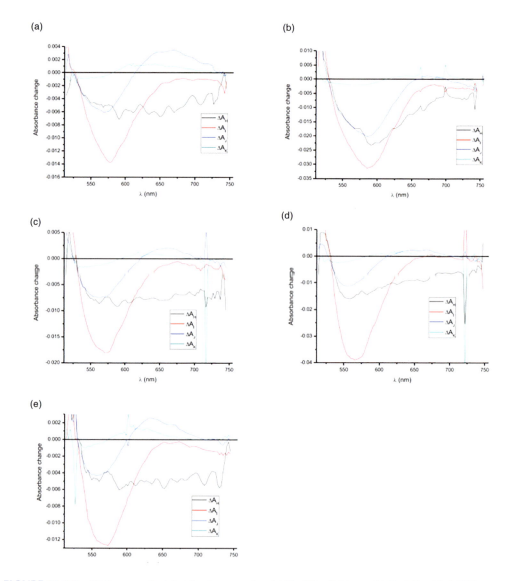

FIGURE 12.3.8 SAS were calculated for five samples: (a) HwBR wild-type, (b) HwBR with D93N mutant, (c) HwBR with D104N mutant, (d) HmBRI wild-type, and (e) HmBRII wild-type. To see this figure in color, go online.

$$\Delta A_3 (\lambda) = \Delta A_J (\lambda) - \Delta A_K (\lambda)$$

$$\Delta A_o (\lambda) = \Delta A_K (\lambda)$$

$$\Delta A_J (\lambda) = \Delta A_o (\lambda) + \Delta A_3 (\lambda)$$

$$\Delta A_I (\lambda) = \Delta A_o (\lambda) + \Delta A_3 (\lambda) + \Delta A_2 (\lambda)$$

$$\Delta A_H (\lambda) = \Delta A_o (\lambda) + \Delta A_3 (\lambda) + \Delta A_2 (\lambda) + \Delta A_1 (\lambda)$$

(12.3.2)

The time constant t_2 is 60% larger for the D93N mutant than for the wild-type and D104N mutant, which indicates that photoisomerization occurring during the *I*/*J* transition is slowed by the inactivation of the Schiff base proton acceptor. By contrast, mutation on the Schiff base proton donor does not affect the femtosecond dynamics with time constants t_1 and t_2. The D104N mutant, in which the original negative charge is eliminated in the residue, did not exhibit a difference during the ultrafast photoisomerization. Comparison with the D93N mutant result leads to the conclusion that the negative charge from the Schiff base proton acceptor residue D93 interacts with the substantial ultrafast change in ESP associated with chromophore isomerization.

The ultrafast ESP change triggers a restructuring of the chromophore cavity HBN involving the protonated Schiff base, which takes place on the picosecond timescale. The time constant τ_3 is 50% larger for the D93N and D104N mutants than for the wild-type. This reveals that the Schiff base proton donor assists in the restructuring of the chromophore-cavity HBN during the thermalization of the vibrational hot state.

The 10-fs-pulse time resolution of the transient absorption enabled us to observe time-dependent changes in the molecular vibration frequency. Spectrogram traces calculated using the transient absorption signal are presented (Figure 12.3.S2) which agree with the conclusion made from the global fitting analysis.

To further support the conclusion, we also analyzed the transient absorption spectra using 2D-CS, as described in the next section.

12.3.3.6 FEMTOSECOND 2D-CS

We analyzed the transient absorption spectra, ΔA (λ, t) using 2D-CS. The calculation procedure can be summarized as follows.

The dynamics spectrum used in 2D-CS is defined as

$$\tilde{\Delta\tilde{A}}\left(\lambda_j,t_k\right)= A\left(\lambda_j,t_k\right)-\bar{A}\left(\lambda_j\right)\tag{12.3.3}$$

where $\bar{A}\left(\lambda_j\right)$ is the spectrum of the reference state of the system. In this study, the spectrum of the reference state was set to the transient absorption spectrum averaged over the corresponding delay region as follows:

$$\bar{A}\left(\lambda_j\right)=\frac{1}{N}\sum_{k=1}^{N}A\left(\lambda_j,t_k\right)\tag{12.3.4}$$

The synchronous and asynchronous spectra, l_1; l_2Þ, are calculated as

$$\Phi\left(\lambda_1,\lambda_2\right)=\frac{1}{N-1}\sum_{k=1}^{N}\bar{A}\left(\lambda_1,t_k\right)\times\bar{A}\left(\lambda_2,t_k\right)\tag{12.3.5}$$

$$\Psi\left(\lambda_1,\lambda_2\right)=\frac{1}{N-1}\sum_{j=1}^{N}\bar{A}\left(\lambda_1,t_j\right)\times\sum_{i=1}^{N}N_{ij}\bar{A}\left(\lambda_2,t_i\right)\tag{12.3.6}$$

The term N_{ij}, the Hilbert–Noda transformation matrix, is given by

$$N_{ij}=\begin{cases}0, & \text{if } i = j \\ \dfrac{1}{\pi\left(j-i\right)}, & \text{if } i \neq j\end{cases}\tag{12.3.7}$$

The synchronous spectrum psi $\phi(\lambda_1, \lambda_2)$ represents simultaneous time-dependent changes in the transient absorption spectroscopy signal at the two probe wavelengths, λ_1 and λ_2. The synchronous spectrum has a positive sign when the transient absorption spectroscopy signal changes in equal directions, either increasing or decreasing, at the two probe wavelengths. By contrast, a negative sign of the synchronous spectrum implies that the transient absorption TA spectroscopy signal changes in opposite directions, with one increasing and the other decreasing.

The asynchronous spectrum $\varphi(\lambda_1, \lambda_2)$ indicates the difference in the time dependence of the transient absorption TA spectroscopy signal probed at λ_1 and that probed at λ_2. If $\phi(\lambda_1, \lambda_2)$ and $\varphi(\lambda_1, \lambda_2)$ have the same sign, the time-dependent change in the transient absorption TA spectroscopy signal at l_1 occurs faster than that at λ_2. If Fðl_1;l_2Þ and Jðl_1;l_2Þ have opposite signs, this order is reversed; that is, the time-dependent change in the transient absorption TA spectroscopy signal at λ_1 occurs more slowly than that at λ_2.

We analyzed the transient absorption spectra in the femtosecond region using 2D-CS. The obtained spectra are called femtosecond 2D-CS spectra. The delay region of the transient absorption spectroscopy signal used for the calculation of the femtosecond 2D-CS spectra was from 57 to 1275 fs.

The delay-time region was set so as to analyze the dynamics of the t_1 and $t_\perp$ lifetime components and avoid the coherent artifacts in the zero-delay region. Figure 12.3.9 presents the synchronous and asynchronous femtosecond 2D-CS spectra calculated using the transient absorption spectra in the femtosecond region.

The positive broadband signal in the synchronous spectrum indicates that the DAS corresponding to the femtosecond region has broad bandwidth with the same sign.

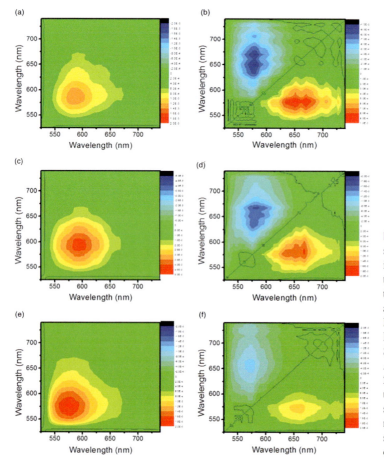

FIGURE   12.3.9 Femtosecond 2D-CS of HwBR and its mutants. The 2D-CS in the femtosecond region of the three HwBR samples: (a and b) wild-type, (c and d) D93N mutant, and (e and f) D104N mutant. (a, c, and e) Synchronous and (b, d, and f) Asynchronous femtosecond 2D-CS patterns calculated using the transient absorption spectra in the femtosecond region from 57 to 1275 fs. To see this figure in color, go online.

This agrees with the DAS of DA$_2$ðlÞ, displayed in Figure 12.3.7a–c. The asynchronous spectrum has a peak at approximately (x, y) ¼ (575, 660 nm). The synchronous and asynchronous spectra have opposite signs at approximately (x, y) ¼ (575, 660 nm), which indicates that the lifetime at 660 nm is shorter than that at 575 nm. The lifetime can also be calculated from the SAS in Figure 12.3.8 as follows. In the femtosecond region, the transient absorption signal at 660 nm is dominated by DA$_H$ with lifetime t_1, whereas the transient absorption (TA) signal at 575 nm in the femtosecond region is dominated by DA$_I$, which decays in t_2. Thus, the transient absorption signal at 660 nm decays faster than that at 575 nm.

The femtosecond 2D-CS spectra demonstrate that the synchronous and asynchronous spectra of the three HwBR samples – wild-type, D93N, and D104N – are similar. This indicates that the relaxation order is common between the three samples in their femtosecond dynamics, with a lifetime of t_2 (400 fs). It also confirms that DA$_H$ and DA$_I$ are similar between the three samples (Figure 12.3.8a–c). The lifetime t_2 is longer in D93N compared with that in the wild-type and D104N, as was discovered from the global fitting analysis in Global fitting using the triple-exponential function.

12.3.3.7 PICOSECOND 2D-CS

The 2D-CS analysis was also performed for the transient absorption spectra in the picosecond region from 1.3 to 13 ps, and the obtained traces are called picosecond 2D-CS spectra. The delay region was set so as to analyze the dynamics of the t_3 (3 ps) lifetime component and avoid the contribution from the dynamics of the femtosecond region. Figure 12.3.10 displays the synchronous

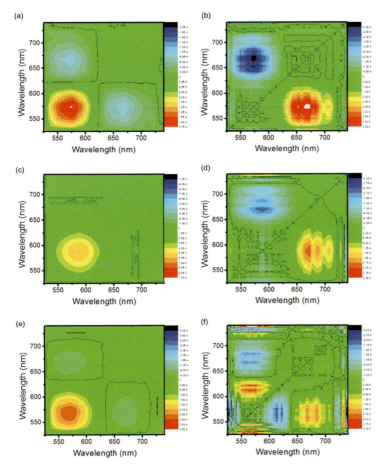

FIGURE 12.3.10 Picosecond 2D-CS of the three HwBR samples: (a and b) wild-type, (c and d) D93N mutant, and (e and f) D104N mutant. (a, c, and e) Synchronous and (b, d, and f) Asynchronous picosecond 2D-CS patterns calculated using the transient absorption spectra in the picosecond region from 1.3 to 13. To see this figure in color go online.

and asynchronous picosecond 2D-CS spectra calculated using the transient absorption spectra in the picosecond region.

Most of the signal peaks in the 2D-CS spectra were found for all three samples. The synchronous spectra have a negative peak at approximately $(x, y) = (575, 670\,\text{nm})$. This demonstrates that the SAS corresponding to picosecond dynamics have different signs at 575 and 670 nm, which agrees with ΔA_J, displayed in Figure 12.3.8a–c. The asynchronous spectra have a peak at approximately $(x, y) = (575, 670\,\text{nm})$. The synchronous and asynchronous spectra have an equal sign at $(x, y) = (575, 670\,\text{nm})$, which indicates that the lifetime is longer at 670 nm than at 575 nm. The contribution of ΔA_K is considerably less than that of ΔA_J at 575 nm, whereas the contributions are comparable at 670 nm (Figure 12.3.8a–c). Therefore, the SAS led to the same conclusion, that the signal at 575 nm decays faster than that at 670 nm.

A peak at $(x, y) = (575, 620\,\text{nm})$ was solely found in the asynchronous spectrum of the HwBR D104N mutant, indicating that the long-life component $(>\tau_3)$ has positive and negative signs at 670 and 620 nm, respectively. This was not found in the global fitting analysis results for the picoseconds-region data, probably because the lifetime is longer than the measured region and thus could not be estimated in the global fitting analysis.

Assuming the existence of an additional 15-ps lifetime in the global fitting analysis, we discovered that the DAS for the 15-ps lifetime has an equal sign between 620 and 670 nm in the HwBR wild-type and D93N, but the opposite sign in the HwBR D104N (data not shown). This feature, observed only in D104N, implies that inactivation of the Schiff base proton donor delays the vibrational cooling, which is consistent with the conclusion obtained from the global fitting analysis in Global fitting using the triple-exponential function.

12.3.3.8   TRANSIENT ABSORPTION SPECTROSCOPY OF WILD-TYPES OF HwBR, HmBRI, AND HmBRII

We performed transient absorption spectroscopy measurements of the wild types of HwBR, HmBRI, and HmBRII. The HwBR and HmBRII are still active at low pH. By contrast, HmBRI loses its activity at low pH. This activity difference at low pH is thought to be caused by the following two differences between the three BRs [43].

1. One difference is the charge distribution on the helices on the CP side. The CP side in HmBRII and HwBR is strongly negatively charged, whereas that in HmBRI is slightly positively charged.
2. The other difference between the three BRs is the existence of two backbone HBNs located in the EC region of the proton pumping path. HBNs were identified in HwBR and HmBRII, but not in HmBRI. The existence of such backbone HBNs is thought to protect the proton acceptor of the Schiff base from external influences, resulting in the pH-independent activity spectra of HwBR and HmBRII but not HmBRI.

Previous studies have not been able to determine whether the ultrafast dynamics are affected by the Schiff base proton donor or acceptor, but this was elucidated by the transient absorption spectroscopy measurements in this study.

The HwBR, HmBRI, and HmBRII samples were individually suspended in 50 mM MES buffer stabilized at pH 5.8. The stationary absorption spectra of the sample solutions are presented in Figure 12.3.2b, together with the laser spectrum of the ultrashort visible laser pulse.

The ultrafast dynamics of the samples was measured in the femtosecond (from 300 to 1400 fs) and picosecond (from 1.00 to 14.5 ps) regions. The observed spectral region was 514.5–758.1 nm with a 2.54-nm step, covering the entire broadband probe spectrum.

Figure 12.3.11 presents the 2D transient absorption spectra in the femtosecond (Figure 12.3.11a, c, and e) and picosecond (Figure 12.3.11b, d, and f) regions for the three samples, HwBR (Figure 12.3.11a and b), HmBRI (Figure 12.3.11c and d), and HmBRII (Figure 12.3.11e and f). Their time

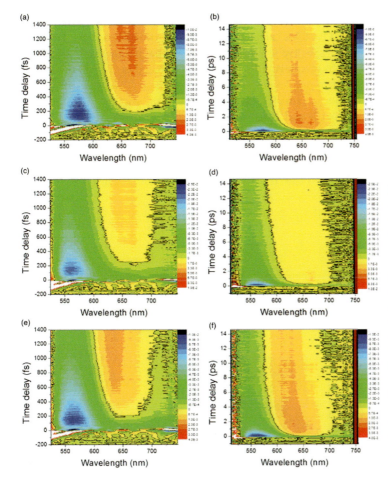

FIGURE 12.3.11 2D transient absorption spectra of the three wild-type samples: (a and b) HwBR, (c and d) HmBRI, and (e and f) HmBRII. (a, c, and e) Data measured in the femtosecond region from 300 to 1400 fs. (b, d, and f) Data measured in the picosecond region from 1.00 to 14.5 ps. All of the transient absorption spectra consist of 96 data points in the 514.5–758.1 nm spectral region and 500 data points in each delay region of the femtosecond and picosecond regions. Black contours correspond to $\Delta A = 0$. To see this figure in color, go online.

traces are presented in Figure 12.3.S3. The transient absorption spectrum is negative in the middle spectral region, which overlaps with the absorption band of the electronic ground state. The negative transient absorption signal indicates the photobleaching of the ground state. The positive transient absorption spectroscopy signal at both ends of the visible laser spectrum reveals the induced absorption in the transition from the first excited state to a higher-energy excited state. The decay time of the positive transient absorption spectroscopy signal indicates the lifetime of the population in the excited states.

The 2D transient absorption spectra of the three samples are similar in the femtosecond region. However, the black curve (contour indicating $\Delta A = 0$) in the picosecond region of the transient absorption spectra reveals difference in the delay dependence in HmBRI compared with HwBR and HmBRII. This reveals that the picosecond decay dynamics in HmBRI are different from those in the other two samples. The difference was more clearly visualized using 2D-CS analysis, as shown in Picosecond 2D-CS.

12.3.3.9 GLOBAL FITTING USING THE TRIPLE-EXPONENTIAL FUNCTION

The transient absorption spectroscopy signal in the femtosecond and picosecond regions was connected and fitted using the triple exponential function and global fitting method, Global fitting using the triple-exponential function.

The estimated time constants τ_i ($i = 0, 1, 2, 3$) are plotted in Figure 12.3.6. Using the obtained time constants, the spectra of the lifetime components $\Delta A_i(\lambda)$ ($i = 0, 1, 2, 3$) were calculated using the least-square method. The calculated spectra are presented in Figure 12.3.7a, d, and e for HwBR, HmBRI, and HmBRII, respectively. The SAS calculated for HwBR, HmBRI, and HmBRII are presented in Figure 12.3.8a, d, and e, respectively. The assignment of the lifetimes in HmBRI and HmBRII is considered to be the same as that in HwBR, which was discussed in Global fitting using the triple-exponential function.

The time constant t_2 is 30% smaller in HmBRI than in the other two samples. This indicates that the HBN in the EC region in HwBR and HmBRII slows the photoisomerization of retinal chromophores.

By contrast, the time constant t_3 in HmBRI is larger than that in HwBR and HmBRII, indicating that the negatively charged helices on the CP side of HwBR and HmBRII accelerate the picosecond relaxation dynamics, which corresponds to the thermalization of the vibrational hot state (the J state) of retinal chromophores.

We also analyzed the transient absorption spectra using 2D-CS as follows.

12.3.3.10 FEMTOSECOND 2D-CS

We analyzed the transient absorption spectra using 2D-CS. The delay region of the transient absorption spectroscopy signals used for the calculation of the femtosecond 2D-CS spectra was 57–1276 fs, which avoided the coherent artifacts in the zero-delay region. Figure 12.3.12 displays the

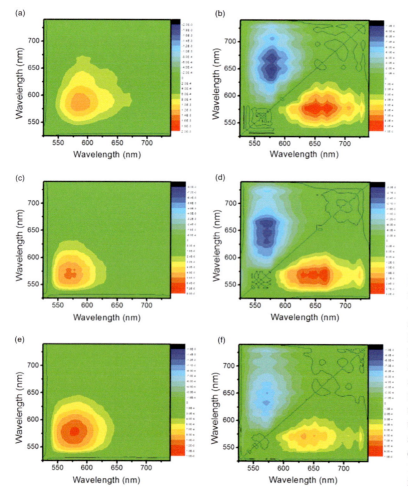

FIGURE 12.3.12 Femtosecond 2D-CS of three wildtype BR samples: (a and b) HwBR, (c and d) HmBRI, and (e and f) HmBRII. (a, c, and e) Synchronous and (b, d, and f) Asynchronous femtosecond 2D-CS patterns calculated using the transient absorption spectra in the femtosecond region from 57 to 1276 fs. To see this figure in color, go online.

synchronous and asynchronous femtosecond 2D-CS spectra calculated using the transient absorption spectra in the femtosecond region.

The synchronous spectra have a single peak at 580 nm, which corresponds to the broadband DAS of $\Delta A_2(\lambda)$ with a negative sign. The synchronous and asynchronous spectra have opposite signs at approximately $(x, y) = (580, 650 \, nm)$, which indicates that the lifetime at 650 nm is shorter than that at 580 nm. The transient absorption signals at 650 and 580 nm are dominated by ΔA_H and ΔA_I, respectively (Figure 12.3.8a, d, and e). Therefore, the SAS also indicated the same conclusion: the transient absorption signal at 650 nm decays faster than that at 580 nm.

The femtosecond 2D-CS spectra demonstrate similar synchronous and asynchronous spectra for HwBR, HmBRI, and HmBRII. This indicates that the relaxation order is common between the three samples in their femtosecond dynamics, with lifetimes of τ_1 (100 fs) and τ_2 (400 fs). It also confirms that ΔA_H and ΔA_I are similar between the three samples (Figure 12.3.8a, d, and e). The lifetime t_2 is shorter in HmBRI compared with that in HwBR and HmBRII, as was discovered from the global fitting analysis in Global fitting using the triple-exponential function.

12.3.3.11 Picosecond 2D-CS

The 2D-CS analysis was also performed for the transient absorption spectra in the picosecond region from 1.3 to 13 ps. The delay region was set so as to analyze the dynamics of the t_3 (3 ps) lifetime component and avoid the contribution from the dynamics of the τ_1 (100 fs) and τ_2 (400 fs) lifetime components. Figure 12.3.13 presents the synchronous and asynchronous picosecond 2D-CS spectra calculated using the transient absorption spectra in the picosecond region.

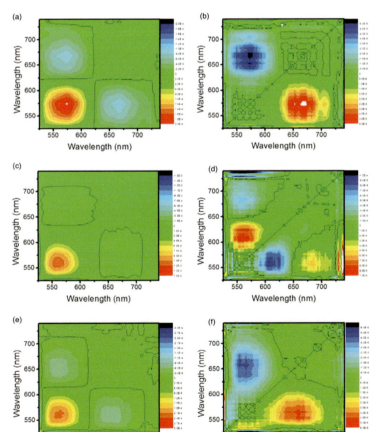

FIGURE 12.3.13 Picosecond 2D-CS of three wildtype samples: (a and b) HwBR, (c and d) HmBRI, and (e and f) HmBRII. (a, c, and e) Synchronous and (b, d, and f) Asynchronous picosecond 2D-CS patterns calculated using the transient absorption spectra in the picosecond region from 1.3 to 13 ps.

Most of the signal peaks in the 2D-CS spectra were observed for all three samples. The synchronous spectra have a positive peak at 575 nm and a negative peak at 670 nm. This reveals that the DAS corresponding to the picosecond dynamics has different signs at 575 and 670 nm, which agrees with the DAS of $\Delta A_3(\lambda)$ displayed in Figure 12.3.7a, d, and e. The asynchronous spectra have a peak at approximately $(x, y) = (575, 670\,\text{nm})$. The synchronous and asynchronous spectra both have a peak at approximately $(x, y) = (575, 670\,\text{nm})$ with the same sign, indicating that the lifetime is longer at 670 nm than at 575 nm. The contribution of ΔA_K was much less than that of ΔA_J at 575 nm, and both of the contributions were comparable at 670 nm (Figure 12.3.8a, d, and e). Therefore, SAS analysis also demonstrates that the transient absorption signal at 575 nm decays faster than that at 670 nm.

A peak at $(x, y) = (575, 620\,\text{nm})$ was solely found in the asynchronous spectrum of HmBRI, indicating that the long life component $(>\tau_3)$ has positive and negative signs at 670 and 620 nm, respectively. This was not found in the global fitting analysis results for the picosecond region data, probably because the lifetime is longer than the measured region and thus could not be estimated in the global fitting analysis.

Assuming the existence of an additional 15-ps lifetime in the global fitting analysis, we discovered that the DAS for the 15-ps lifetime have an equal sign between 620 and 670 nm in the HwBR and HmBRII, but an opposite sign in the HmBRI. Surprisingly, the asynchronous picosecond 2D-CS spectra of the wild-type HmBRI and D104N mutant of HwBR are similar, which can be explained as follows.

The picosecond relaxation dynamics in the wild-type HmBRI are slowed by the positively charged helices on the CP side (Global fitting using the triple-exponential function), whereas those in the D104N mutant of HwBR are slowed by inactivation of the Schiff base proton donor (Global fitting using the triple-exponential function). Therefore, the following two possible mechanisms could explain the observed similarity in the picosecond relaxation dynamics of the wild-type HmBRI and D104N mutant of HwBR.

1. First, positively charged helices on the CP side suppress the activity of residues corresponding to the Schiff base proton donor.
2. Second, the inactivation of the Schiff base proton donor results in a positive charge distribution on the helices on the CP side.

The first proposed mechanism is not probable, because the Schiff base proton donor in HmBRI is active, similar to the case for the other BRs. Thus, we can conclude that the inactivation of the Schiff base proton donor induces a positive charge on the helices of the CP side.

12.3.4 CONCLUSIONS

In this study, we investigated the effect of the Schiff base proton acceptor and donor on the ultrafast dynamics of retinal chromophores. The residues of D93 and D104 correspond to the control by the Schiff base proton acceptor and donor of the proton-translocation subchannels on the EC and CP sides, respectively. A comparison of the ultrafast dynamics of the wild-type, D93N mutant, and D104N mutant clarified the effect of the Schiff base proton acceptor and donor. We analyzed the transient absorption spectra using two methods, global fitting analysis and 2D-CS.

1. First, the ultrafast dynamics were studied and compared for three HwBR samples, wild-type, D93N mutant, and D104N mutant. Global fitting analysis led to the following conclusions. The photoisomerization of retinal chromophores is slowed by the inactivation of the Schiff base proton acceptor but not affected by the Schiff base proton donor, whereas the thermalization of the vibrational hot state is assisted by the Schiff base proton donor. The differences observed in the asynchronous spectra in the picosecond region for D104N

are consistent with this conclusion regarding thermalization. Thus, the negative charge of the Schiff base proton acceptor residue D93 interacts with the ultrafast and substantial change in ESP that is associated with chromophore isomerization. By contrast, the Schiff base proton donor assists in the restructuring of the chromophore cavity HBN during the thermalization of the vibrational hot state.

2. Second, the ultrafast dynamics of the wild types of HwBR, HmBRI, and HmBRII were compared. Global fitting analysis of their transient absorption spectra led to the following conclusions. The HBN in the EC region in HwBR and HmBRII slows the photoisomerization of retinal chromophores, and the negatively charged helices on the CP side of HwBR and HmBRII accelerate the thermalization of the vibrational hot state (J state) of retinal chromophores. The asynchronous spectra in the picosecond region of the three wild-type BRs present clear differences in HmBRI, which supports the stated conclusion regarding thermalization. Moreover, the asynchronous spectrum in the picosecond region of the wild-type HmBRI resembles that of the D104N mutant of HwBR, indicating that inactivation of the Schiff base proton donor induces a positive charge on the helices on the CP side.

The contents of this section is the product of the collaborative research activity of the following people: Chang Hung, Xiao-Ru Chen, Ying-Kuan Ko, Takayoshi Kobayashi, Chii-Shen Yang, and Atsushi Yabushita [44].

REFERENCES

1. R. Gonza´lez-Luque, M. Garavelli, and M. Olivucci. "Computational evidence in favor of a two-state, two-mode model of the retinal chromophore photoisomerization," *Proc. Natl. Acad. Sci. USA*. **97**, 9379–9384 (2000).
2. W. Humphrey, H. Lu, and K. Schulten. "Three electronic state model of the primary phototransformation of bacteriorhodopsin," *Biophys. J.* **75**, 1689–1699 (1998).
3. S. Schenkl, F. van Mourik, and M. Chergui. "Probing the ultrafast charge translocation of photoexcited retinal in bacteriorhodopsin," *Science* **309**, 917–920 (2005).
4. J. Herbst, K. Heyne, and R. Diller. "Femtosecond infrared spectroscopy of bacteriorhodopsin chromophore isomerization," *Science* **297**, 822–825 (2002).
5. G. Haran, K. Wynne, and R. M. Hochstrasser. "Excited state dynamics of bacteriorhodopsin revealed by transient stimulated emission spectra," *Chem. Phys. Lett.* **261**, 389–395 (1996).
6. F. Gai, K. C. Hasson, and P. A. Anfinrud. "Chemical dynamics in proteins: the photoisomerization of retinal in bacteriorhodopsin," *Science* **279**, 1886–1891 (1998).
7. J. A. Stuart, D. L. Marcy, and R. R. Birge. "Volumetric optical memory based on bacteriorhodopsin," *Synth. Met.* **127**, 3–15 (2002).
8. L. Song and M. A. El-Sayed. "Primary step in bacteriorhodopsin photosynthesis: bond stretch rather than angle twist of its retinal excited-state structure," *J. Am. Chem. Soc.* **120**, 8889–8890 (1998).
9. M. Du and G. R. Fleming. "Femtosecond time-resolved fluorescence spectroscopy of bacteriorhodopsin: direct observation of excited state dynamics in the primary step of the proton pump cycle," *Biophys. Chem.* **48**, 101–111 (1993).
10. W. T. Pollard, C. H. B. Cruz, and R. A. Mathies. "Direct observation of the excited-state cis-trans photoisomerization of bacteriorhodopsin: multilevel line shape theory for femtosecond dynamic hole burning and its application," *J. Chem. Phys.* **90**, 199–208 (1989).
11. Y. Huang, S.-T. Wu, and Y. Zhao. "All-optical switching characteristics in bacteriorhodopsin and its applications in integrated optics," *Opt. Express*. **12**, 895–906 (2004).
12. K. Peters, M. L. Applebury, and P. M. Rentzepis. "Primary photochemical event in vision: proton translocation," *Proc. Natl. Acad. Sci. USA*. **74**, 3119–3123 (1977).
13. A. Warshel. "Bicycle-pedal model for the first step in the vision process," *Nature* **260**, 679–683 (1976).
14. P. Altoe`, A. Cembran, and M. Garavelli. "Aborted double bicyclepedal isomerization with hydrogen bond breaking is the primary event of bacteriorhodopsin proton pumping," *Proc. Natl. Acad. Sci. USA*. **107**, 20172–20177 (2010).

15. L. Salem and P. Bruckmann. "Conversion of a photon to an electrical signal by sudden polarisation in the N-retinylidene visual chromophore," *Nature* **258**, 526–528 (1975).

16. K. W. Foster, J. Saranak, and K. Nakanishi. "Activation of Chlamydomonas rhodopsin in vivo does not require isomerization of retinal," *Biochemistry* **28**, 819–824 (1989).

17. V. J. Rao, J. P. Zingoni, and R. S. Liu. "Isomers of 3,7,11-trimethyldodeca-2,4,6,8,10-pentaenal (a linear analogue of retinal) and lower homologues in their interaction with bovine opsin and bacterioopsin," *Photochem. Photobiol.* **41**, 171–174 (1985).

18. A. Aharoni, B. Hou, and Q. Zhong. "Non-isomerizable artificial pigments: implications for the primary light-induced events in bacteriorhodopsin," *Biochemistry (Mosc.).* **66**, 1210–1219 (2001).

19. F. Melaccio, N. Calimet, and M. Olivucci. "Space and time evolution of the electrostatic potential during the activation of a visual pigment," *J. Phys. Chem. Lett.* **7**, 2563–2567 (2016).

20. J. Tittor, U. Schweiger, and E. Bamberg. "Inversion of proton translocation in bacteriorhodopsin mutants D85N, D85T, and D85,96N," *Biophys. J.* **67**, 1682–1690 (1994).

21. T. Mogi, L. J. Stern, and H. G. Khorana. "Aspartic acid substitutions affect proton translocation by bacteriorhodopsin," *Proc. Natl. Acad. Sci. USA.* **85**, 4148–4152 (1988).

22. H. J. Butt, K. Fendler, and D. Oesterhelt. "Aspartic acids 96 and 85 play a central role in the function of bacteriorhodopsin as a proton pump," *EMBO J.* **8**, 1657–1663 (1989).

23. T. S. Marinetti, S. Subramaniam, and H. G. Khorana. "Replacement of aspartic residues 85, 96, 115, or 212 affects the quantum yield and kinetics of proton release and uptake by bacteriorhodopsin," *Proc. Natl. Acad. Sci. USA.* **86**, 529–533 (1989).

24. M. Holz, L. A. Drachev, and H. G. Khorana. "Replacement of aspartic acid-96 by asparagine in bacteriorhodopsin slows both the decay of the M intermediate and the associated proton movement," *Proc. Natl. Acad. Sci. USA.* **86**, 2167–2171 (1989).

25. H. Otto, T. Marti, and M. P. Heyn. "Substitution of amino acids Asp-85, Asp-212, and Arg-82 in bacteriorhodopsin affects the proton release phase of the pump and the pK of the Schiff base," *Proc. Natl. Acad. Sci. USA.* **87**, 1018–1022 (1990).

26. K. Gerwert, B. Hess, and D. Oesterhelt. "Role of aspartate-96 in proton translocation by bacteriorhodopsin," *Proc. Natl. Acad. Sci. USA.* **86**, 4943–4947 (1989).

27. S. Subramaniam, T. Marti, and H. G. Khorana. "Protonation state of Asp (Glu)-85 regulates the purple-to-blue transition in bacteriorhodopsin mutants Arg-82/Ala and Asp-85/Glu: the blue form is inactive in proton translocation," *Proc. Natl. Acad. Sci. USA.* **87**, 1013–1017 (1990).

28. J. K. Lanyi, J. Tittor, and D. Oesterhelt. "Influence of the size and protonation state of acidic residue 85 on the absorption spectrum and photoreaction of the bacteriorhodopsin chromophore," *Biochim Biophys. Acta.* **1099**, 102–110 (1992).

29. M. F. Hsu, H. Y. Fu, and A. H. Wang. "Structural and functional studies of a newly grouped Haloquadratum walsbyi bacteriorhodopsin reveal the acid-resistant light-driven proton pumping activity," *J. Biol. Chem.* **290**, 29567–29577 (2015).

30. I. Noda. "Generalized two-dimensional correlation method applicable to infrared, Raman, and other types of spectroscopy," *Appl. Spectrosc.* **47**, 1329–1336 (1993).

31. F. K. Tsai, H. Y. Fu, and L. K. Chu. "Photochemistry of a dualbacteriorhodopsin system in Haloarcula marismortui: HmbRI and HmbRII," *J. Phys. Chem. B.* **118**, 7290–7301 (2014).

32. V. Shevchenko, I. Gushchin, and V. Gordeliy. "Crystal structure of Escherichia coli-expressed Haloarcula marismortui bacteriorhodopsin I in the trimeric form," *PLoS One.* **9**:e112873 (2014).

33. T. Kobayashi, T. Saito, and H. Ohtani. "Real-time spectroscopy of transition states in bacteriorhodopsin during retinal isomerization," *Nature* **414**, 531–534 (2001).

34. A. Yabushita and T. Kobayashi. "Primary conformation change in bacteriorhodopsin on photoexcitation," *Biophys. J.* **96**, 1447–1461 (2009).

35. A. Yabushita, Y.-H. Lee, and T. Kobayashi. "Development of a multiplex fast-scan system for ultrafast time-resolved spectroscopy," *Rev. Sci. Instrum.* **81**, 063110 (2010).

36. A. Shirakawa, I. Sakane, and T. Kobayashi. "Pulse-front-matched optical parametric amplification for sub-10-fs pulse generation tunable in the visible and near infrared," *Opt. Lett.* **23**, 1292–1294 (1998).

37. G. Cerullo, M. Nisoli, and S. De Silvestri. "Generation of 11 fs pulses tunable across the visible by optical parametric amplification," *Appl. Phys. Lett.* **71**, 3616–3618 (1997).

38. T. Wilhelm, J. Piel, and E. Riedle. "Sub-20-fs pulses tunable across the visible from a blue-pumped single-pass noncollinear parametric converter," *Opt. Lett.* **22**, 1494–1496 (1997).

39. I. Noda. "Two-dimensional codistribution spectroscopy to determine the sequential order of distributed presence of species," *J. Mol. Struct.* **1069**, 50–59 (2014).

40. A. Kahan, O. Nahmias, and S. Ruhman. "Following photoinduced dynamics in bacteriorhodopsin with 7-fs impulsive vibrational spectroscopy," *J. Am. Chem. Soc.* **129**, 537–546 (2007).

41. K. C. Hasson, F. Gai, and P. A. Anfinrud. "The photoisomerization of retinal in bacteriorhodospin: experimental evidence for a threestate model," *Proc. Natl. Acad. Sci. USA.* **93**, 15124–15129 (1996).

42. H. Abramczyk. "Femtosecond primary events in bacteriorhodopsin and its retinal modified analogs: revision of commonly accepted interpretation of electronic spectra of transient intermediates in the bacteriorhodopsin photocycle," *J. Chem. Phys.* **120**, 11120–11132 (2004).

43. H. Y. Fu, H. P. Yi, and C. S. Yang. "Insight into a single halobacterium using a dual-bacteriorhodopsin system with different functionally optimized pH ranges to cope with periplasmic pH changes associated with continuous light illumination," *Mol. Microbiol.* **88**, 551–561 (2013).

44. C.-C. Hung, X.-R. Chen, Y.-K. Ko, T. Kobayashi, C.-S. Yang, and A. Yabushita. "Schiff base proton acceptor assists ultrafast photoisomerization of retinal chromophores in bacteriorhodopsin from Haloquadratum," *Biophys. J.* **112**, 2503–2529 (2017).